T0186049

# A Course in

# Analysis

—————————— Vol. I ——————————

**Introductory Calculus**
**Analysis of Functions of One Real Variable**

# A Course in Analysis

# A Course in Analysis

# Analysis

—— Vol. I ——

## Introductory Calculus
## Analysis of Functions of One Real Variable

**Niels Jacob**
**Kristian P Evans**

Swansea University, UK

**World Scientific**

NEW JERSEY · LONDON · SINGAPORE · BEIJING · SHANGHAI · HONG KONG · TAIPEI · CHENNAI · TOKYO

*Published by*

World Scientific Publishing Co. Pte. Ltd.

5 Toh Tuck Link, Singapore 596224

*USA office:* 27 Warren Street, Suite 401-402, Hackensack, NJ 07601

*UK office:* 57 Shelton Street, Covent Garden, London WC2H 9HE

**British Library Cataloguing-in-Publication Data**

A catalogue record for this book is available from the British Library.

**A COURSE IN ANALYSIS**
**Volume I: Introductory Calculus, Analysis of Functions of One Real Variable**

ISBN 978-981-4689-08-3
ISBN 978-981-4689-09-0 (pbk)

Printed in Singapore

# Preface

We are currently living in times where many undergraduates consider the internet as their main, if not their only source for supporting their academic studies. Furthermore, many publishers prefer short textbooks directly related to modules as the best solution for mathematics textbooks. This project, namely to write and publish a whole course on analysis consisting of up to 6 volumes, therefore, may appear to be going against the grain, perhaps even a Don Quixote's style fight against modernity. However the motivation for developing these volumes has slowly emerged over the last few years by our observations while teaching analysis to undergraduates.

The modular approach to teaching combined with examination pressure has prevented students from seeing crucial connections between topics being taught in different modules, even when prerequisites and dependencies are emphasised. In fact many universities in the U.K. expect their modules to be quite independent. The problem is further amplified by the tendency for lecturers to teach the same module for several years - mainly to reduce the burden of teaching in order to gain more time for research. All this has led to a situation where topics such as analysis of several variables, vector calculus, differential geometry of curves and surfaces are seen by students as rather unrelated topics. They also consider Lebesgue integration, real-variable theory, Fourier analysis as separate topics with no connections, and this list is unfortunately easy to extend. In situations where algebraic concepts (linear algebra, symmetry and groups) are used in higher dimensional analysis is even worse. In short, while in the most exciting recent mathematical research the interplay of many diverse mathematical subject areas becomes more important than ever, in our teaching as it is perceived by the students, we artificially separate closely related mathematical topics and put them into isolated boxes called modules. It is clear that such a common practice prevents even the better students from advancing and getting a deeper insight into mathematics.

Five years ago, after long discussions and preparations we changed the undergraduate mathematics provision at Swansea University. We now think more in terms of courses than modules. Our analysis course runs over five terms as does our algebra course, and both are compulsory for all students. Clearly there are still choices and in particular in the final year students can choose out of quite a few advanced modules. A further, rather important new feature of the new provision is that we leave (whenever possible) each course for each cohort in the hands of one lecturer. The students seem to favour this type of continuity in terms of both the presentation of material and the lecturer, and more importantly they are performing much better than they have done in previous years.

Another problem that needed to be addressed was providing students with problems that fitted to their lecture material. Everyone who has taught mathematics for some time has experienced that many problems eventually do not work out because at some point in the solution a result not yet covered in the lectures is needed. But students still need to have a good number of problems with correct solutions. These should be a mixture of routine exercises, more testing problems going beyond what was so far covered in the lecture and some real challenges. Moreover problems can provide an opportunity to extend the theory or link to other parts of mathematics, but they are only useful when students are confident that they have mastered them correctly. For this reason we have added to every chapter a good number of problems and we have provided complete solutions. In total, for the 32 chapters in volume 1 there are more than 360 problems (often with sub-problems) with complete solutions. This part constitutes more than 25% of the first volume. Note that problems marked with * are more challenging.

Our aim is to provide students and lecturers with a coherent text which can and should serve entire undergraduate studies in Analysis. The Course can also be used as a standard reference work. It might be worth mentioning that for graduate students in analysis such a lack of a modern course was also felt at no other place but Princeton University. E.M. Stein's four-volume course "Princeton Lectures in Analysis" published jointly with R. Shakarchi between 2003 - 2011 is a response to such a real need, i.e. multiple-volume courses are by no means out of date, maybe they are needed more than ever to give students a foundation and a lasting reference for their mathematical education and beyond.

The first named author has taught mathematics, mainly analysis related topics, but also geometry and probability theory, for over 38 years at 7 universities in 2 countries. The material in this course is based on ca. 40 different modules he has taught over the years. For these volumes the material was of course rearranged and amended, but nonetheless to a large extent they reflect still the provision. This first volume covers first year analysis as taught by the first named author with the support of the second named author in Swansea in the academic year 2010/11, an introduction to calculus and analysis of functions of one variable.

Finally we want to thank all who have supported us in writing this volume, in particular the World Scientific Press team.

Niels Jacob
Kristian P. Evans
Swansea, January 2015

# Contents

CONTENTS

# Acknowledgements and Apologies

Calculus and basic analysis of functions of one real variable is a standard topic taught in mathematics across the world. The material is well studied and a lot of textbooks covering the topics exist. The first textbook dealing with "calculus", i.e. analysis of a real-valued function of one variable, was published in 1696 by de l'Hospital. In the last 300 years thousands of such textbooks have been published in all major languages, in addition many collections of problems have been added. This is easy to understand since the topic was and still is rapidly developing, in particular its place within mathematics, and this has of course an impact on its presentation. Thus, there is a need to "rewrite" calculus and analysis textbooks in each generation. However basic results and examples (and hence problems) remain unchanged and still have a place in modern presentations. The tradition in writing textbooks on such a topic is not to give detailed references to original sources, in fact this is almost impossible. In drafting my own lecture notes I made use of many of them, but as all academics know, when drafting lecture notes about standard material we do not usually make a lot of references. Consequently, when now using my notes which are partly three decades old, I do not recall most of the sources I used and combined at that time. There are a number of books that I used as both a student and a lecturer and therefore they have been used here. Thus in the main text there are essentially no references but I do acknowledge the important influence of the following treatises (and I will always refer below to the copy I had used).

Dieudonné, J., Grundzüge der modernen Analysis, 2. Aufl. Logik und Grundlagen der Mathematik Bd. 8. Friedrich Vieweg & Sohn, Braunschweig 1972.

Endl, K., und Luh, W., Analysis I, 3. Aufl. Analysis II, 2. Aufl. Akademische Verlagsgesellschaft, Wiesbaden 1975 und 1974.

Fichtenholz, G.M., Differential- und Integralrechnung I, 8. Aufl. Differential- und Integralrechnung II, 4. Aufl. Differential- und Integralrechnung III, 6. Aufl. Hochschulbücher für Mathematik Bd. 61, 62, 63. VEB Deutscher Verlag der Wissenschaften, Berlin 1973, 1972 und 1973.

Forster, O., Analysis 1, 2. Nachdruck. Analysis 2, 2. Nachdruck. Analysis 3. Friedrich Vieweg & Sohn, Braunschweig 1978, 1979, 1981. (These books will have stronger impact on some passages, in particular in parts dealing with integration theory, since they were much used textbooks when I started my teaching career supporting corresponding modules.)

Heuser, H., Lehrbuch der Analysis. Teil 1 und 2. B.G. Teubner Verlag, Stuttgart 1980 und 1981.

Rudin, W., Principles of Mathematical Analysis, 3rd ed. McGraw-Hill International Editions, Mathematical Series. McGraw-Hill Book Company, Singapore 1976.

Walter, W., Gewöhnliche Differentialgleichungen. Heidelberger Taschenbücher Bd. 110. Springer Verlag, 1972.

Walter, W., Analysis 1, 3. Aufl. Analysis 2, 4. Aufl. Springer Verlag, Berlin, 1992 und 1995.

For compiling the lists of formulae in some of the appendices we used often

Zeidler, E., (ed.), Oxford Users Guide to Mathematics. Oxford University Press, Oxford 2004.

Solved problems are important for students and we used some existing collections of solved problems to supplement our selection. Sometimes these collections served only to get some ideas, on some occasions we picked problems but provided different or modified solutions, but here and there we used complete solutions. The main sources which are very valuable for students are

Kaczor, W.J., and Nowak, M.T., Problems in Mathematical Analysis I, II and III. Students Mathematical Library Vol. 4, 12, and 21. American Mathematical Society, Providence R.I., 2000, 2001, and 2003.

Lipschutz, M.M., Differentialgeometrie. Theorie und Anwendung. (Deutsche Bearbeitung von H.-D. Landschulz.) Schaum's Outline Series. McGraw-Hill Book Company, Duesseldorf, 1980.

Spiegel, M.R. Advanced Calculus. Schaum's Outline Series Theory and Problems. McGraw-Hill Book Company, New York 1963.

Spiegel, M.R., Real Variables. Schaum's Outline Series Theory and Problems. McGraw-Hill Book Company, New York 1969.

Spiegel, M.R., Advanced Mathematics for Engineers and Scientists. Schaum's Outline Series Theory and Problems. McGraw-Hill Book Company, New York 1971.

We would finally like to mention that although we have endeavoured to correct all typos etc via proof-reading, clearly some errors may remain. Please contact us if you find any such mistakes.

Niels Jacob

# List of Symbols

$\mathbb{N}$    natural numbers

$k\mathbb{N} := \{n \in \mathbb{N} \mid n = km \text{ for } m \in \mathbb{N}\}$

$\mathbb{N}_0 := \mathbb{N} \bigcup \{0\}$

$\mathbb{Z}$    integers

$\mathbb{Q}$    rational numbers

$\mathbb{R}$    real numbers

$\mathbb{R}_+$    non-negative real numbers

$\mathbb{R}^n = \mathbb{R} \times \cdots \times \mathbb{R}$    set of ordered $n$-tuples of real numbers

$x^{-1} := \frac{1}{x}$

$x^n := x \cdot x \cdot \ldots \cdot x$ ($n$ factors)

$a^{\frac{1}{n}}$ or $\sqrt[n]{a}$    $n^{\text{th}}$ root of $a$

$x^{\frac{n}{m}} = \sqrt[m]{x^n}$

$x > 0$    $x$ is strictly greater than 0

$x < 0$    $x$ is strictly less than 0

$x \geq 0$    $x$ is non-negative

$x \leq 0$    $x$ is non-positive

$|x|$    absolute value of $x$

$\infty$    infinity

$-\infty$    negative infinity

$n!$    $n$ factorial

$\binom{n}{k}$    binomial coefficient

$\max\{a_1, \ldots, a_n\}$    maximum of $a_1, \ldots, a_n$

$\min\{a_1, \ldots, a_n\}$    minimum of $a_1, \ldots, a_n$

$\displaystyle\sum_{j=1}^{n} a_j$    finite sum of $a_j$

$\displaystyle\sum_{k=1}^{\infty} a_k$    infinite series

$\displaystyle\sum_{j=m}^{k} a_j = a_m + a_{m+1} + \cdots + a_k$

$\displaystyle\prod_{j=1}^{n} a_j$    finite product of $a_j$

$\displaystyle\prod_{j=l}^{n} = a_l \cdot a_{l+1} \cdot \ldots \cdot a_n$

$\prod_{j=1}^{\infty} a_j$    infinite product of $a_j$

$X \times Y$    Cartesian product

$\emptyset$    empty set

$\mathcal{P}(X)$    power set of the set $X$

$\in$    belongs to

$\notin$    does not belong to

$x \bullet y$    binary operation

$\subset$    set subset

$M_1 \setminus M_2$    set subtraction

$\cap$    set intersection

$\cup$    set union

$A^{\complement}$    complement of $A$

$\Longrightarrow$    implies

$xRy$    relation

$\sim$    equivalence relation

$[a]$    equivalence class

$\vee$    or

$\wedge$    and

$\Longleftrightarrow$    equivalence (statements)

$\forall$    for all

$\exists$    there exists

$\neg p$    negation of $p$

$\displaystyle\bigcup_{j=1}^{N} A_j$    finite union of sets $A_j$

$\displaystyle\bigcap_{j=1}^{N} A_j$    finite intersection of sets $A_j$

$B_\epsilon(a) := \{x \in \mathbb{R} \mid |x - a| < \epsilon\}$

$S^1$   circle centred at the origin with radius 1

$\overline{B_\epsilon}(a) := \{x \in \mathbb{R} \mid |x - a| \le \epsilon\}$

$(a, b) := \{x \in \mathbb{R} \mid a < x < b\}$

$[a, b) := \{x \in \mathbb{R} \mid a \le x < b\}$

$(a, b] := \{x \in \mathbb{R} \mid a < x \le b\}$

$[a, b] := \{x \in \mathbb{R} \mid a \le x \le b\}$

$(0, \infty) := \{x \in \mathbb{R} \mid x > 0\}$

$\max D$   maximum of $D$

$\min D$   minimum of $D$

$\sup D$   supremum of $D$

$\inf D$   infimum of $D$

$f : D \to \mathbb{R}$   mappings, see Chapter 4

$D(f)$   domain of $f$

$\Gamma(f)$   graph of $f$

$R(f)$   range of $f$

$f(D)$   image of $D$ under $f$

$f^{-1}(B)$   pre-image of $B$

$Aut(X)$   set of all bijective mappings $f : X \to X$

$f_2 \circ f_1$   composition of $f_1$ with $f_2$

$\chi_A$   characteristic function of a set $A$

$pr_1$   first coordinate projection

$pr_2$   second coordinate projection

$f^{-1}$   inverse mapping

$id_D$   identity mapping

$f|_{D_1}$   restriction of $f$ to $D_1$

$f^+$   positive part of $f$

$f^-$   negative part of $f$

$f \perp g$   $f$ and $g$ orthogonal

$C^k(I)$   $k$-times continuously differentiable functions

$C(I) = C^0(I)$   continuous functions

$C^\infty(I)$   arbitrarily often differentiable functions

$C_b^k(I)$   $k$-times differentiable bounded functions

xix

$M(K; \mathbb{R})$   set of functions from $K$ to $\mathbb{R}$

$M_b(K; \mathbb{R}) := \{f : K \to \mathbb{R} \mid \sup_{x \in K} |f(x)| < \infty\}$

$BV([a, b])$   set of functions of bounded variation on $[a, b]$

$T[a, b]$   step functions on $[a, b]$

$\lim_{y \to x} f(y) = a$   limit of the function $f$

$\lim_{y \to \infty} f(y) = a$   limit of the function $f$ at $\infty$

$f'(x_0)$ or $\frac{df(x_0)}{dx}$   derivative of $f$ with respect to $x$ at $x_0$

$f''(x_0)$ or $f^{(2)}(x_0)$ or $\frac{d^2 f(x_0)}{dx^2}$   second derivative of $f$ at $x_0$

$f^{(k)}(x_0)$ or $\frac{df^k(x_0)}{dx^k}$   $k^{\text{th}}$ derivative of $f$ at $x_0$

$(a_n)_{n \in \mathbb{N}}$   sequence

$(a_{n_l})_{l \in \mathbb{N}}$   subsequence

$\lim_{n \to \infty} a_n = a$   limit of a sequence

$\limsup_{n \to \infty} = \overline{\lim}$   limit superior

$\liminf_{n \to \infty} = \underline{\lim}$   limit inferior

$\lim_{y \searrow x} f(y)$ or $\lim_{\substack{y \to x \\ y > x}} f(y)$   limit from the right

$\lim_{y \nearrow x} f(y)$ or $\lim_{\substack{y \to x \\ y < x}} f(y)$   limit from the left

$Z(t_1, \ldots, t_n)$ or $Z_n$   partition

$m(Z_n)$   mesh size of $Z_n$

$V_Z(f) := \sum_{k=0}^{n-1} |f(x_{k+1}) - f(x_k)|$

$V(f) := \sup_Z V_Z(f)$

$V_a^b(f) := V(f)$

$S_r(g, Z_n)$   Riemann sum of $g$ with respect to $Z_n$

$\int_a^b g(t)dt$   definite integral

$\int g(t)dt$   indefinite integral

$\int_a^{*b}$   upper integral

$\int_{*a}^b$   lower integral

$T^a_{(c_n)}$    power series associated with $c_n$ centred at $a$

$T^{(k)}_{f,c}$    Taylor polynomials

$R^{(n+1)}_{f,c}$    remainder of Taylor's formula

$||x||_1 = |x_1| + \cdots + |x_n|$ for $x = (x_1, \ldots, x_n) \in \mathbb{R}^n$

$||x||_2 = \sqrt{(x_1^2 + \cdots + x_n^2)}$ for $x = (x_1, \ldots, x_n) \in \mathbb{R}^n$

$||x||_\infty = \max\{|x_1|, \ldots, |x_n|\}$ for $x = (x_1, \ldots, x_n) \in \mathbb{R}^n$

$||x||_p := \left(\sum_{\nu=1}^{n} |x_\nu|^p\right)^{\frac{1}{p}}$

$||f||_{K,\infty} := \sup_{x \in K} |f(x)|$

$||f||_p := \left(\int_a^b |f(x)|^p dx\right)^{\frac{1}{p}}$

$\exp x = e^x$    exponential function

$\ln x$    natural logarithm

$a^x := e^{x \ln a}$

$\log_a x$    logarithm of $x$ with respect to the basis $a$

$[x]$    entier-function

sin    sine function

cos    cosine function

tan    tangent function

cot    cotangent function

sec    secant function

csc    co-secant function

arcsin    inverse sine function

arccos    inverse cosine function

arctan    inverse tangent function

arccot    inverse cotangent function

sinh    hyperbolic sine function

cosh    hyperbolic cosine function

tanh    hyperbolic tangent function

coth    hyperbolic cotangent function

sech    hyperbolic secant function

cosech    hyperbolic co-secant function

arsinh   inverse hyperbolic sine function

arcosh   inverse hyperbolic cosine function

artanh   inverse hyperbolic tangent function

$\Gamma(x)$   gamma-function

$J_l(x)$   Bessel function

$B(x, y)$   beta-function

$e$   Euler number

$\gamma$   Euler's constant

# The Greek Alphabet

| | | |
|---|---|---|
| alpha | $\alpha$ | $A$ |
| beta | $\beta$ | $B$ |
| gamma | $\gamma$ | $\Gamma$ |
| delta | $\delta$ | $\Delta$ |
| epsilon | $\epsilon$ | $E$ |
| zeta | $\zeta$ | $Z$ |
| eta | $\eta$ | $H$ |
| theta | $\theta$ | $\Theta$ |
| iota | $\iota$ | $I$ |
| kappa | $\kappa$ | $K$ |
| lambda | $\lambda$ | $\Lambda$ |
| mu | $\mu$ | $M$ |
| nu | $\nu$ | $N$ |
| xi | $\xi$ | $\Xi$ |
| omikron | $O$ | $o$ |
| pi | $\pi$ | $\Pi$ |
| rho | $\rho$ | $P$ |
| sigma | $\sigma$ | $\Sigma$ |
| tau | $\tau$ | $T$ |
| upsilon | $\upsilon$ | $\Upsilon$ |
| phi | $\phi$ | $\Phi$ |
| chi | $\chi$ | $X$ |
| psi | $\psi$ | $\Psi$ |
| omega | $\omega$ | $\Omega$ |

Note that $\varphi$ is also used for $\phi$.

# Part 1: Introductory Calculus

# 1 Numbers - Revision

Before we start with calculus we need to know how to manipulate complicated expressions of real numbers and above all we must become familiar in doing this. We urge students to avoid using calculators in this course. The intention here is to ensure that we understand the basics; much of what is introduced may seem obvious but the concepts will become very useful later in the book. In particular, we will need a lot of familiarity in manipulating expressions where numbers are replaced by functions or later on even by operators. We start to systematically introduce set theory as a common language in modern mathematics. Basic notions from logic on which we rely are taught in other courses, however these are collected in Appendix I.

The **natural numbers** or **positive integers** are the numbers

$$1, 2, 3, 4, 5, \ldots \tag{1.1}$$

We denote the **set of all natural numbers** by $\mathbb{N}$. When we want to indicate that $n$ is a natural number, i.e. an element of the set $\mathbb{N}$, we simply write

$$n \in \mathbb{N}, \tag{1.2}$$

for example

$$12 \in \mathbb{N}. \tag{1.3}$$

The set of all **integers** is denoted by $\mathbb{Z}$ and consists of the numbers

$$\ldots, -5, -4, -3, -2, -1, 0, 1, 2, 3, 4, 5, \ldots \tag{1.4}$$

When $k$ is an integer we write

$$k \in \mathbb{Z}, \tag{1.5}$$

for example

$$-15 \in \mathbb{Z}. \tag{1.6}$$

Note that $-15$ is not a natural number and for this we write

$$-15 \notin \mathbb{N}. \tag{1.7}$$

It is obvious that every natural number is an integer, or more formally

$$n \in \mathbb{N} \text{ implies } n \in \mathbb{Z}. \tag{1.8}$$

We say that $\mathbb{N}$ is a subset of $\mathbb{Z}$ and for this we write

$$\mathbb{N} \subset \mathbb{Z}. \tag{1.9}$$

Clearly, there are other subsets of $\mathbb{Z}$, for example the set of all negative integers, or the set of all even integers, etc. The **rational numbers** are denoted by $\mathbb{Q}$ and this is the set of all fractions

$$q = \frac{k}{n}, \quad \text{where} \quad k \in \mathbb{Z} \quad \text{and} \quad n \in \mathbb{N}. \tag{1.10}$$

Examples of fractions are

$$\frac{3}{7}, \frac{7}{7}, \frac{-1}{8}, \frac{-12}{3}, \text{etc.} \tag{1.11}$$

We also write $-\frac{1}{8}$ for $\frac{-1}{8}$, etc. Note that we face a problem: $-\frac{12}{3}$ and $-\frac{4}{1}$ are different formal expressions which represent the same number, and in addition we want to consider $-\frac{4}{1}$ and $-4$ to be equal. For now we use a naïve approach where we consider two rational numbers $q = \frac{k}{m}$ and $r = \frac{l}{n}$ as equal if $kn = lm$, and further, for $\frac{k}{1}$ we write $k$. The last identification of $\frac{k}{1}$ with $k$, $k \in \mathbb{Z}$, allows us to consider $\mathbb{Z}$ as a subset of $\mathbb{Q}$, i.e.

$$\mathbb{Z} \subset \mathbb{Q}. \tag{1.12}$$

Since $\mathbb{N} \subset \mathbb{Z}$, i.e. every natural number is also an integer, we find further

$$\mathbb{N} \subset \mathbb{Q}. \tag{1.13}$$

Note that the last argument is:

*"a subset of a subset is a subset"*.

It is helpful to introduce at this stage the few notions and notations from set theory that we have used so far in a more systematic way. Unfortunately, there is no simple and unproblematic way to introduce the general notion of a set. For our purposes the original definition of G. Cantor is sufficient:

*We consider a **set** as a collection of well distinguishable objects of our intuition or our thinking as an entity $M$. We call these objects the **elements** of the set $M$.*

If $M$ is a set and $m$ an element of $M$ we write

$$m \in M, \tag{1.14}$$

and if $k$ does not belong to $M$ we write

$$k \notin M. \tag{1.15}$$

Before we can do anything with sets we need to define when two sets $M_1$ and $M_2$ are equal:

*Two sets are* **equal** *if and only if they have the same elements.*

For this we write

$$M_1 = M_2. \tag{1.16}$$

If every element of a set $M_2$ is an element of a set $M_1$ we call $M_2$ a **subset** of $M_1$ and we write

$$M_2 \subset M_1. \tag{1.17}$$

Thus (1.17) means that $m \in M_2$ implies that $m \in M_1$. In the case that $M_3$ is a further set which is a subset of $M_2$, i.e. $M_3 \subset M_2$, then every element of $M_3$ must be an element of $M_1$ too. Hence we have

$$M_3 \subset M_2 \text{ and } M_2 \subset M_1 \text{ implies } M_3 \subset M_1. \tag{1.18}$$

It may happen that $M_2$ is a subset of $M_1$ and $M_1$ is a subset of $M_2$, i.e. $M_2 \subset M_1$ and $M_1 \subset M_2$. In this case every element of $M_1$ is an element of $M_2$ and every element of $M_2$ is an element of $M_1$, hence $M_1 = M_2$, or

$$M_2 \subset M_1 \text{ and } M_1 \subset M_2 \text{ implies } M_1 = M_2. \tag{1.19}$$

So far we have introduced the natural numbers, the integers and the rational numbers. We already know that there are numbers which are not rational, i.e. have no representation as a fraction. Take for example $\pi$ or $\sqrt{2}$. We call these numbers **irrational numbers**. The **real numbers**, denoted by $\mathbb{R}$, is the set consisting of all rational and irrational numbers. Of course, a second thought shows that this is not a proper definition. However, up until now we have had a naïve idea of what the real numbers are, for example the points on a straight line. We will operate with this naïve approach for some time until we can eventually give a proper definition and characterisation of $\mathbb{R}$.

5

This approach is the more justified one since historically an understanding of the nature of $\mathbb{R}$ took mankind a few thousand years of mathematical thinking. Indeed, the understanding of real numbers was one of the most outstanding and important problems in the history of mathematics and it is still a challenge to students.

Therefore, for the moment, the irrational numbers are those real numbers which are not rational. Let $M_2 \subset M_1$, i.e. $M_2$ is a subset of the set $M_1$. The set consisting of elements in $M_1$ not belonging to $M_2$ is denoted by $M_1 \setminus M_2$; we write

$$M_1 \setminus M_2 = \{x \in M_1 \mid x \notin M_2\}. \tag{1.20}$$

In this notation we find that the irrational numbers form the set

$$\mathbb{R} \setminus \mathbb{Q} = \{x \in \mathbb{R} \mid x \notin \mathbb{Q}\}, \tag{1.21}$$

for which we do not introduce an extra symbol. Note that (1.20) suggests a way to characterise sets, for example

$$\{k \in \mathbb{Z} \mid k = 2n \text{ for some } n \in \mathbb{N}\} \tag{1.22}$$

is the set of all even natural numbers. Again, it is easier to slowly get used to this notation than to give a formal definition. The idea is to consider all those elements of a given set $M$ which share a certain property $A$, i.e.

$$\{m \in M \mid m \text{ has the property } A\}. \tag{1.23}$$

Another way to characterise a set is by listing all of its elements, for example

$$\{1, 2, 3\} \quad \text{or} \quad \{x, y\}. \tag{1.24}$$

Before turning to algebraic operations in $\mathbb{R}$, we introduce a further practical notation:

$$x, y \in M \quad \text{means} \quad x \in M \text{ and } y \in M, \tag{1.25}$$

which of course generalises to more than two elements.

We have the following rules for adding real numbers $x, y, z \in \mathbb{R}$

$$(x + y) + z = x + (y + z), \tag{1.26}$$

and

$$x + y = y + x. \tag{1.27}$$

6

From equality (1.26) we deduce that it does not make any difference whether we first add $x$ and $y$ together and then add $z$, or whether we first add $y$ and $z$ together and then add $x$. We say that addition of real numbers is **associative** and (1.26) is called the **associative law of addition**. Equality (1.27) tells us that the order does not matter when adding real numbers, i.e. addition in $\mathbb{R}$ is **commutative**.

There is one (and only one) real number which is very special with respect to addition: we may add this number to any other number $x \in \mathbb{R}$ and the result is again $x$. This number is 0 and we consider 0 as the **neutral element with respect to addition**, i.e.

$$x + 0 = x \quad \text{for all } x \in \mathbb{R}. \tag{1.28}$$

Given a real number $x$, there is always exactly one real number $-x$ such that

$$x + (-x) = 0. \tag{1.29}$$

We call $-x$ the **inverse element to $x$ with respect to addition**. Instead of (1.29) we often write

$$x - x = 0, \tag{1.30}$$

and more generally if $-y$ is the inverse of $y$ and $x$ is a real number we write

$$x - y := x + (-y). \tag{1.31}$$

Note that we have used the symbol ":=" here for the first time. In general $A := B$ means that $A$ is defined by $B$, for example we write

$$2\mathbb{N} := \{ n \in \mathbb{N} \mid n = 2m \text{ for } m \in \mathbb{N} \} \tag{1.32}$$

for the even natural numbers.

We can also multiply real numbers where the rules

$$(x \cdot y) \cdot z = x \cdot (y \cdot z) \tag{1.33}$$

and

$$x \cdot y = y \cdot x \tag{1.34}$$

hold for all $x, y, z, \in \mathbb{R}$. Hence **multiplication is associative and commutative**. For the time being we write $x \cdot y$ for the product of $x$ and $y$ but later on we will adopt the usual notation and will just write $xy$. As in

7

the case of addition, for **multiplication** there exists a **neutral element**, namely the real number 1. Indeed, for all $x \in \mathbb{R}$ we have

$$1 \cdot x = x. \tag{1.35}$$

This leads immediately to the question of the existence of an **inverse element with respect to multiplication** for a real number $x$. We already know the answer; all but one real numbers have an inverse with respect to multiplication. The real number 0 does not. Thus for $x \in \mathbb{R} \setminus \{0\} = \{y \in \mathbb{R} \mid y \neq 0\}$ there exists an element $x^{-1} \in \mathbb{R}$ such that

$$x \cdot x^{-1} = 1. \tag{1.36}$$

Shortly we will investigate why 0 does not have an inverse element with respect to multiplication. A further notation for $x^{-1}$ is

$$\frac{1}{x} := x^{-1}, \; x \neq 0, \tag{1.37}$$

and for $x \cdot y^{-1} = y^{-1} \cdot x$ we write

$$\frac{x}{y} := x \cdot y^{-1}, \; y \neq 0. \tag{1.38}$$

Finally we want to link addition and multiplication. This is done by the **law of distributivity** which states that for all $x, y, z, \in \mathbb{R}$

$$x \cdot (y + z) = (x \cdot y) + (x \cdot z). \tag{1.39}$$

With the standard convention that multiplication precedes addition we write

$$x \cdot (y + z) = x \cdot y + x \cdot z \tag{1.40}$$

or

$$x(y + z) = xy + xz. \tag{1.41}$$

Now we can address the problem why 0 cannot have an inverse with respect to multiplication. Since $1 - 1 = 0$ for any $x \in \mathbb{R}$ it follows that

$$0 \cdot x = (1 - 1)x = x - x = 0.$$

Since $0 \neq 1$ it follows that there is a real number $x$ such that $0 \cdot x = 1$, hence 0 cannot have an inverse with respect to multiplication. This is nothing but

8

the following well-known statement: you cannot divide by 0; the expression $\frac{k}{0}, k \in \mathbb{Z}$ does not make sense.

Let us have a more formal look at addition in $\mathbb{R}$ and multiplication in $\mathbb{R} \setminus \{0\}$. In the set $\mathbb{R}$ we can pick any two elements $x, y \in \mathbb{R}$ and form a new element in $\mathbb{R}$ called $x + y$. This is an example of a binary operation on a given set. In our current naïve approach, given a set $A$ which contains at least one element, i.e. $A$ is non-empty, we call any rule a **binary operation** if it assigns to a pair $(x, y)$ of elements $x$ and $y$ in the set $A$ a new element $z \in A$. For this new element we write for now

$$z = x \bullet y. \tag{1.42}$$

The condition that a set $A$ is non-empty will occur quite often. We formally introduce the **empty set** $\emptyset$ as the set which has no elements and $A$ being non-empty means

$$A \neq \emptyset. \tag{1.43}$$

The set $\mathbb{R} \setminus \{0\}$ is of course non-empty and multiplication on $\mathbb{R} \setminus \{0\}$ gives a further binary operation. We write $(\mathbb{R}, +)$ and $(\mathbb{R} \setminus \{0\}, \cdot)$ to indicate that we want to consider $\mathbb{R}$ with the binary operation "+" and $\mathbb{R} \setminus \{0\}$ with the binary operation "·".

Let us return to $\mathbb{R}$ with the algebraic operation addition satisfying (1.26) - (1.29), the algebraic operation multiplication satisfying (1.33) - (1.36) and the law of distributivity (1.39). We want to derive some simple rules for doing calculations. For $a, b, c, d \in \mathbb{R}$, $b \neq 0$ and $d \neq 0$ we have

$$\frac{a}{b} + \frac{c}{d} = \frac{ad + cb}{bd}. \tag{1.44}$$

Indeed:

$$\frac{a}{b} + \frac{c}{d} = ab^{-1} + cd^{-1}$$

$$= (ab^{-1} + cd^{-1})\frac{bd}{bd}$$

$$= ((ab^{-1} + cd^{-1}) \cdot (bd)) \cdot \frac{1}{bd}$$

$$= (ab^{-1}bd + cd^{-1}bd) \cdot \frac{1}{bd}$$

$$= \frac{ad + cb}{bd},$$

where we used that for every real number $x$, $x \neq 0$, that $\frac{x}{x} = 1$. (Recall that $\frac{1}{1} = 1$ and $\frac{x}{x} = \frac{1}{1}$ if and only if $1 \cdot x = x \cdot 1$ which is of course true.) Let us do an example. For $a = \frac{3}{4}$, $b = \frac{8}{15}$, $c = \frac{9}{2}$ and $d = \frac{2}{3}$ we find

$$\frac{a}{b} + \frac{c}{d} = \frac{\frac{3}{4}}{\frac{8}{15}} + \frac{\frac{9}{2}}{\frac{2}{3}},$$

and it follows that

$$\frac{\frac{3}{4}}{\frac{8}{15}} + \frac{\frac{9}{2}}{\frac{2}{3}} = \frac{3}{4} \cdot \left(\frac{8}{15}\right)^{-1} + \frac{9}{2} \cdot \left(\frac{2}{3}\right)^{-1}$$

$$= \frac{3}{4} \cdot \frac{15}{8} + \frac{9}{2} \cdot \frac{3}{2}$$

$$= \frac{45}{32} + \frac{27}{4}$$

$$= \frac{45 \cdot 4 + 27 \cdot 32}{32 \cdot 4} = \frac{1044}{128} = \frac{261}{32}.$$

Here we have already used the general rule that for $a \neq 0$ and $b \neq 0$ we have that

$$\left(\frac{a}{b}\right)^{-1} = \frac{b}{a}, \tag{1.45}$$

as we know the rule

$$a \cdot \frac{1}{b} = \frac{a}{b} \quad \text{for} \quad a \in \mathbb{R} \quad \text{and} \quad b \in \mathbb{R} \setminus \{0\}. \tag{1.46}$$

In addition we know that

$$\frac{a}{b} \cdot \frac{c}{d} = \frac{ac}{bd}, \ b \neq 0, \ d \neq 0. \tag{1.47}$$

The rule (1.45) is of some more interest, so let us spend some time on it. Recall that $\frac{a}{b} = ab^{-1}$, hence (1.45) claims that $(ab^{-1})^{-1} = ba^{-1}$. We can prove this easily by assuming that the inverse element is uniquely determined:

$$(ab^{-1})(ba^{-1}) = ab^{-1}ba^{-1}$$

$$= a(b^{-1}b)a^{-1} = a \cdot 1 \cdot a^{-1}$$

$$= a \cdot a^{-1} = 1.$$

Next we turn our attention to powers of real numbers. Let $x, y \in \mathbb{R}$ and $n, m \in \mathbb{N}$. We set

$$x^n := x \cdot x \cdot x \cdot \ldots \cdot x \quad (n \text{ factors}). \tag{1.48}$$

Elementary rules are

$$x^n \cdot x^m = x^{n+m} \tag{1.49}$$

and

$$(x \cdot y)^n = x^n \cdot y^n. \tag{1.50}$$

Clearly we have

$$0^n = 0 \quad \text{for all} \ \ n \in \mathbb{N}. \tag{1.51}$$

Suppose that $x \neq 0$, then $x^n \neq 0$ and we may consider the inverse element $(x^n)^{-1}$ of $x^n$. It follows that

$$x^n \cdot (x^n)^{-1} = \frac{x^n}{x^n} = 1,$$

but in addition we have

$$x^n \cdot (x^n)^{-1} = \frac{x \cdot \ldots \cdot x}{x \cdot \ldots \cdot x} = (x \cdot \ldots \cdot x)\frac{1}{x \cdot \ldots \cdot x},$$

thus we find

$$(x^n)^{-1} = \frac{1}{x^n} \tag{1.52}$$

and we write

$$x^{-n} := (x^n)^{-1}. \tag{1.53}$$

The rules (1.49) and (1.50) now extend to all $n, m \in \mathbb{Z}$ provided that $x \neq 0$ and $y \neq 0$. If we agree to define

$$x^0 = 1, \tag{1.54}$$

for all $x \in \mathbb{R}$, then we may summarise our considerations to

$$x^k \cdot x^l = x^{k+l} \tag{1.55}$$

and

$$(x \cdot y)^k = x^k \cdot y^k \tag{1.56}$$

11

for all $x, y \in \mathbb{R} \setminus \{0\}$ and $k, l \in \mathbb{Z}$. For fractions we find that

$$\left(\frac{a}{b}\right)^k = \frac{a^k}{b^k} \tag{1.57}$$

is true for either $a, b \in \mathbb{R} \setminus \{0\}$ and $k \in \mathbb{Z}$, or $a \in \mathbb{R}$, $b \in \mathbb{R} \setminus \{0\}$ and $k \in \mathbb{N}$. Now we may calculate

$$\frac{\left(\frac{3}{2}\right)^2 - \left(\frac{2}{3}\right)^3}{\left(\frac{4}{3}\right)^{-2} + \frac{7}{8}} = \frac{\frac{9}{4} - \frac{8}{27}}{\frac{9}{16} + \frac{7}{8}}$$

$$= \frac{\frac{211}{108}}{\frac{23}{16}} = \frac{844}{621}.$$

Finally we extend our considerations to **fractional powers**. We take it for granted that for $n \in \mathbb{N}$ and $a \geq 0$, $a \in \mathbb{R}$, there exists a unique $b \in \mathbb{R}$, $b \geq 0$, such that $b^n = a$. This number $b$ is denoted by $a^{\frac{1}{n}}$ or $\sqrt[n]{a}$. We call $a^{\frac{1}{n}}$ the $n^{\text{th}}$ **root** of $a$. Note that so far the $n^{\text{th}}$ root is only defined for $a \geq 0$ and it is unique and non-negative. For $a \geq 0$ and $b \geq 0$ and $m, n \in \mathbb{N}$ we have

$$(a \cdot b)^{\frac{1}{n}} = a^{\frac{1}{n}} \cdot b^{\frac{1}{n}}. \tag{1.58}$$

Indeed, we can extend (1.55) and (1.56) to fractional powers. For $x > 0$ and $y > 0$ and $p, q \in \mathbb{Q}$ it follows that

$$x^p \cdot z^q = x^{p+q} \tag{1.59}$$

and

$$(x \cdot y)^p = x^p \cdot y^p. \tag{1.60}$$

Further, for $p = \frac{n}{m}$, $n, m \in \mathbb{N}$, and $x \geq 0$ we write

$$x^{\frac{n}{m}} = \sqrt[m]{x^n}. \tag{1.61}$$

We have already used the notion of "positive" or "negative" real numbers. Let us recollect this **order structure** on $\mathbb{R}$. Given any real number $x \in \mathbb{R}$ then exactly one of the following three statements is true

$$x = 0, \ x > 0, \ x < 0, \tag{1.62}$$

i.e. either $x$ is equal to 0, or it is strictly larger than 0, or it is strictly less than 0. We can represent the real numbers as points on a line, the real line:

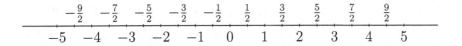

*Figure 1.1*

At the moment we pretend that there is a one-to-one correspondence between the points on the real line and the real numbers. If $x > 0$ we say that $x$ is **positive**, we call $x$ **negative** if $x < 0$. We write $x \geq 0$ if $x > 0$ or $x = 0$ and we write $x \leq 0$ if $x < 0$ or $x = 0$. It is convenient to add the notation $\mathbb{R}_+$ for all non-negative real numbers, i.e. $\mathbb{R} := \{x \in \mathbb{R} \mid x \geq 0\}$. If $x \geq 0$ we call $x$ **non-negative**, if $x \leq 0$ we call $x$ **non-positive**. The following rules hold for $x, y \in \mathbb{R}$:

$$x > 0 \quad \text{then} \quad -x < 0, \; x < 0 \quad \text{then} \quad -x > 0; \tag{1.63}$$

$$x > 0 \quad \text{then} \quad x^{-1} > 0, \; x < 0 \quad \text{then} \quad x^{-1} < 0; \tag{1.64}$$

$$x > 0 \quad \text{and} \quad y > 0 \quad \text{then} \quad x \cdot y > 0; \tag{1.65}$$

$$x > 0 \quad \text{and} \quad y < 0 \quad \text{then} \quad x \cdot y < 0; \tag{1.66}$$

$$x < 0 \quad \text{and} \quad y > 0 \quad \text{then} \quad x \cdot y < 0; \tag{1.67}$$

$$x < 0 \quad \text{and} \quad y < 0 \quad \text{then} \quad x \cdot y > 0; \tag{1.68}$$

$$x > 0 \quad \text{and} \quad y > 0 \quad \text{then} \quad x + y > 0. \tag{1.69}$$

Furthermore we write

$$x < y \quad \text{if} \quad x - y < 0, \tag{1.70}$$

or

$$x > y \quad \text{if} \quad x - y > 0, \tag{1.71}$$

as well as

$$x \leq y \quad \text{if} \quad x < y \quad \text{or} \quad x = y, \tag{1.72}$$

and

$$x \geq y \quad \text{if} \quad x > y \quad \text{or} \quad x = y. \tag{1.73}$$

Clearly we have

$$x > y \quad \text{if and only if} \quad y < x \tag{1.74}$$

and

$$x \geq y \quad \text{if and only if} \quad y \leq x. \tag{1.75}$$

Here are some simple rules for handling inequalities. For $a, b \in \mathbb{R}$ and $x, y \in \mathbb{R}$ we have:

$$x > y \quad \text{implies} \quad x + a > y + a; \tag{1.76}$$

$$x \geq y \quad \text{implies} \quad x + a \geq y + a; \tag{1.77}$$

$$x < y \quad \text{implies} \quad x + a < y + a; \tag{1.78}$$

$$x \leq y \quad \text{implies} \quad x + a \leq y + a; \tag{1.79}$$

$$x > y \quad \text{and} \quad a > b \quad \text{implies} \quad x + a > y + b. \tag{1.80}$$

If $x, y \in \mathbb{R}$ and $a \in \mathbb{R}$, $a > 0$, then we have:

$$x > y \quad \text{implies} \quad a \cdot x > a \cdot y; \tag{1.81}$$

$$x \geq y \quad \text{implies} \quad a \cdot x \geq a \cdot y; \tag{1.82}$$

$$x < y \quad \text{implies} \quad a \cdot x < a \cdot y; \tag{1.83}$$

$$x \leq y \quad \text{implies} \quad a \cdot x \leq a \cdot y. \tag{1.84}$$

We also know that

$$a > b > 0 \quad \text{and} \quad x > y > 0 \quad \text{imply} \quad a \cdot x > b \cdot y. \tag{1.85}$$

However, for $a < 0$ we have

$$x > y \quad \text{implies} \quad a \cdot x < a \cdot y; \tag{1.86}$$

$$x \geq y \quad \text{implies} \quad a \cdot x \leq a \cdot y; \tag{1.87}$$

$$x < y \quad \text{implies} \quad a \cdot x > a \cdot y; \tag{1.88}$$

$$x \leq y \quad \text{implies} \quad a \cdot x \geq a \cdot y. \tag{1.89}$$

In the next section we will often make use of these rules. Here are some simple examples:

i)

$$\frac{3}{4} \leq \frac{7}{8}, \quad \text{hence} \quad 4 \cdot \frac{3}{4} = 3 \leq \frac{7}{2} = 4 \cdot \frac{7}{8},$$

however

$$(-4) \cdot \frac{3}{4} = -3 \geq -\frac{7}{2} = (-4) \cdot \frac{7}{8}.$$

ii)

$$3 + x > 2 + y \quad \text{implies} \quad 1 + x > y \quad \text{or} \quad y - x < 1.$$

iii) Consider $7x-5 > 21x+30$. This inequality is equivalent to $7x > 21x+35$, which is again equivalent to $x > 3x + 5$, or $-5 > 2x$, implying $x < -\frac{5}{2}$. In fact all these manipulations are reversible. Thus the problem: find all $x \in \mathbb{R}$ such that

$$7x - 5 > 2x + 30$$

has the solution $x \in \mathbb{R}$ such that $x < -\frac{5}{2}$. More formally, the set of solutions of the inequality

$$7x - 5 > 2x + 30$$

is given by

$$\left\{ x \in \mathbb{R} \mid x < -\frac{5}{2} \right\}.$$

In this chapter we have summarised what we may have already learned elsewhere about real numbers. We might have even slightly extended these ideas. Some ideas from set theory have been introduced, further, we occasionally pointed out that some of the statements and rules we take for granted need a proper justification, and we indicated some of the more formal aspects, such as relations to binary operations. In Part 1 of our course we will consequently use the following approach: starting from a basic knowledge we will gradually move to more and more precision, indicating any gaps in our work along the way. Eventually we will be prepared for a more mature approach to mathematics in particular in analysis when entering Part 2.

## Problems

1. Is the set $\{\phi\}$ empty?

2. Decide which of the following sets is empty
   a) $\{x \in \mathbb{R} \mid x^2 = 16 \text{ and } 2x+3 = 12\}$, b) $\{x \in \mathbb{Q} \mid x^2 = 9 \text{ and } 3x-6 = 3\}$,
   c) $\{x \in \mathbb{R} \mid x \neq x\}$, d) $\{x \in \mathbb{Z} \mid x^2 = \frac{1}{4}\}$, e) $\{x \in \mathbb{Q} \mid x^2 = \frac{1}{4}\}$.

3. Given the 3 sets

   $A = \{3, 5, 7, 9, 11\}, B = \{z \in \mathbb{Z} \mid z \text{ is odd}\}$ and $C = \{z \in \mathbb{Z} \setminus \{2\} \mid z \text{ is prime}\}$.

   Recall that $z \in \mathbb{Z}$ is an **even number** if $z$ is divisible by 2, otherwise it is an **odd number**. By definition 0 is an even number. State which of the following inclusions are true:
   a) $A \subset B$; b) $A \subset C$; c) $C \subset B$.

4. Let $M = \{n \in \mathbb{N} \,|\, n \geq 5\}$. Find $\mathbb{Z} \setminus M$.

5. With $R = \{k \in \mathbb{N} \,|\, k^2 \leq 10\}$ and $B = \{1, 2, 3, 4, 5, 6\}$ find $B \setminus R$.

6. Determine the following sets:
   a) $\{x \in \mathbb{Z} \,|\, 5x + 7 = 13\}$; b) $\{x \in \mathbb{Q} \,|\, 5x + 7 = 13\}$;
   c) $\{x \in \mathbb{Z} \,|\, 5x + 7 \leq 13\}$.

7. Simplify:

$$a) \; \frac{-7}{3} \left( \frac{27}{8} - \frac{18}{5} \right); \qquad b) \; \frac{\frac{3}{4} + \frac{7}{12}}{\frac{2}{19} - \frac{1}{7}}; \qquad c) \; \frac{4^2 - 3^3}{5^2 + 19}.$$

8.  a) Simplify:

$$\frac{3a + 4(a + b)^2 - 6a(\frac{1}{2} + b) - 2b(a + 2b)}{\frac{1}{2}(a + b)}, \; a + b \neq 0.$$

   b) Show that for $a + b \neq c$

$$\frac{\frac{1}{2}(a^2 - 3b^2 - c^2 - 2ab + 4bc)}{\frac{1}{4}(a + b - c)} = 2a - 6b + 2c$$

   c) Simplify:

$$\frac{a - b}{a + b} + \frac{4ab}{(a + b)^2} - \frac{a + b}{a - b}$$

$$(a \neq b \text{ and } a \neq -b).$$

   d) Simplify:

$$\frac{x^3 - y^3}{y - x} - y^4 x^2 \left( \frac{1}{y^3 x} - \frac{x}{y} + \frac{y}{x} \right)$$

$$(x \neq y, x \neq 0, y \neq 0).$$

9. Simplify:

$$\frac{\frac{1}{9} \left( \frac{8}{11} - \frac{2}{9} \right) \left( \frac{12}{5} - \frac{6}{7} \right)}{\frac{8}{3} \left( \frac{3}{4} - \frac{7}{2} \right)}.$$

10. Simplify:

$$a) \; \left( \frac{2}{3} \right)^3 - \left( \frac{1}{2} \right)^4 + 5 \left( \frac{8}{9} \right); \qquad b) \; \frac{\left( \frac{2}{5} \right)^3 - \left( \frac{3}{8} \right)^2}{\frac{19}{40}}.$$

11. Simplify:

   a)
$$\frac{(a+b)^3 - (b-a)^2(b+a)}{4ab}, \quad ab \neq 0;$$

   b)
$$\frac{\left(\frac{a}{b}\right)^3 - \left(\frac{b}{a}\right)^4}{a^2 b^3}, \quad ab \neq 0.$$

12. Find:

   a) $\sqrt{625}$;   b) $\sqrt{\frac{225}{49}}$;   c) $\sqrt{\frac{a^4 b^6}{(a+b)^2}}$,   $a \geq 0, b \geq 0$ and $a + b \neq 0$.

13. Find every $x \in \mathbb{R}$ such that

   a) $3x - 12 \geq -7$,   b) $\frac{7}{4} + \frac{2}{5}x \leq \frac{3}{8}x$,   c) $(x-3)(x+4) \geq 0$,

   and give a graphical representation of the set of solutions.

14. Let $x > 0, y > 0, z > 0$. Is the term $x^{y^z}$ well defined?

   Hint: try $x = 2, y = 3, z = 2$ and compare $(x^y)^z$ with $x^{(y^z)}$.

15. Prove by using the stated rules for addition and multiplication that

   (a) $\frac{1}{b} + \frac{1}{d} = \frac{d+b}{d \cdot b}$;   $b \neq 0, d \neq 0$.

   (b) $\frac{\frac{a}{b}}{\frac{c}{d}} = \frac{a}{b} \cdot \frac{d}{c}$,   $b \neq 0, c \neq 0, d \neq 0$.

   Hint: first prove that for $x \neq 0$, $(x^{-1})^{-1} = x$.

16. Let $a, b, c \in \mathbb{R}, a > 0$ and $b^2 - 4ac \geq 0$.

   (a) Prove that $ax^2 + bx + c = 0$ for some $x \in \mathbb{R}$ if and only if

$$a\left(x + \frac{b}{2a}\right)^2 - \frac{b^2}{4a} + c = 0.$$

   (b) Use the fact that for $y \geq 0$ there exists exactly one real number $\sqrt{y} \geq 0$ such that $(\sqrt{y})^2 = y$ to find all solutions to the quadratic equation
$$ax^2 + bx + c = 0.$$

# 2 The Absolute Value, Inequalities and Intervals

In order to be able to handle inequalities and to handle terms involving real numbers we need to know whether $x \in \mathbb{R}$ is zero, positive or negative. Let us start with a simple example:

$$x \in \mathbb{R} \quad \text{then} \quad x^2 \geq 0. \tag{2.1}$$

For $x = 0$ there is nothing to prove. If $x > 0$ then by (1.65) we know $x \cdot x = x^2 > 0$, if $x < 0$ then by (1.68) it follows that $x^2 > 0$.

This may look quite trivial but it opens the way to a non-trivial result: let $a, b \in \mathbb{R}$ then we always have

$$ab \leq \frac{a^2 + b^2}{2}. \tag{2.2}$$

Here we say that $\frac{a^2+b^2}{2}$ is an estimate (upper estimate) for $ab$. To show this we firstly see that for $a, b \in \mathbb{R}$ it follows from (2.1) that $(a-b)^2 \geq 0$. However $(a - b)^2 = a^2 - 2ab + b^2$ therefore

$$a^2 - 2ab + b^2 \geq 0,$$

or

$$a^2 + b^2 \geq 2ab,$$

implying

$$ab \leq \frac{a^2 + b^2}{2}. \tag{2.3}$$

For the case $a = 5$ and $b = 6$, we find

$$30 \leq \frac{25 + 36}{2} = 30\frac{1}{2}.$$

This is a reasonably good estimate since intuitively, $30\frac{1}{2}$ is a fairly good estimate of 30; it is not too far way. For $a = -5$ and $b = 6$ we find

$$-30 \leq \frac{25 + 36}{2} = 30\frac{1}{2},$$

which is a rather crude result, i.e. $30\frac{1}{2}$ is a poor estimate for -30. The problem is that on the right hand side we only have positive terms and they cannot give a good estimate of negative terms.

To remedy this situation we introduce one of the most important notation in calculus and analysis.

**Definition 2.1.** *Let $x \in \mathbb{R}$, the **absolute value** of $x \in \mathbb{R}$, denoted by $|x|$, is defined by*

$$|x| := \begin{cases} x, & x > 0; \\ 0, & x = 0; \\ -x, & x < 0. \end{cases} \tag{2.4}$$

Thus for all $x \in \mathbb{R}$ the absolute value $|x|$ is non-negative, i.e.

$$|x| \geq 0 \quad \text{for all} \quad x \in \mathbb{R}. \tag{2.5}$$

Here are some examples: $|\frac{3}{5}| = \frac{3}{5}, |-\frac{7}{8}| = \frac{7}{8}, |0| = 0$.
We claim that we can improve (2.2) by

$$|ab| \leq \frac{a^2 + b^2}{2} \quad \text{for all} \quad a, b \in \mathbb{R}. \tag{2.6}$$

We already know

$$ab \leq \frac{a^2 + b^2}{2}, \tag{2.7}$$

therefore all we need to show is that

$$-ab \leq \frac{a^2 + b^2}{2}.$$

To do this consider $(a + b)^2$. As before we find

$$0 \leq (a + b)^2 = a^2 + 2ab + b^2,$$

and therefore $-2ab \leq a^2 + b^2$, or

$$-ab \leq \frac{a^2 + b^2}{2}. \tag{2.8}$$

Thus, (2.7) and (2.8) imply

$$|ab| \leq \frac{a^2 + b^2}{2}, \tag{2.9}$$

since $|ab|$ can only take the value $ab$ or $-ab$.
Here are some rules for handling the absolute value: For $x, y \in \mathbb{R}$ we find

$$|xy| = |x||y|, \quad \text{in particular} \quad |x| = |-x|. \tag{2.10}$$

We can prove (2.10) by considering 4 cases. First note the table

$$x \geq 0 \qquad\qquad y \geq 0 \qquad\qquad x \leq 0 \qquad\qquad y \leq 0$$

$$|x| = x \qquad\qquad |y| = y \qquad\qquad |x| = -x \qquad\qquad |y| = -y.$$

Now we have

1. $x \geq 0$, $y \geq 0$ then $|x||y| = xy$, and $xy \geq 0$, i.e. $|xy| = xy$;

2. $x \geq 0$, $y \leq 0$ then $|x||y| = x(-y) = -xy$, and $xy \leq 0$, i.e. $|xy| = -xy$;

3. $x \leq 0$, $y \geq 0$ then $|x||y| = (-x)y = -xy$, and $xy \leq 0$, i.e. $|xy| = -xy$;

4. $x \leq 0$, $y \leq 0$ then $|x||y| = (-x)(-y) = xy$, and $xy \geq 0$, i.e. $|xy| = xy$.

For $y \neq 0$ it follows from (2.10) that

$$\left|\frac{x}{y}\right| = \frac{|x|}{|y|}. \tag{2.11}$$

Thus we have for example

$$\left|\frac{3}{7} \cdot \left(-\frac{4}{8}\right)\right| = \frac{3}{7} \cdot \frac{4}{8} \quad \text{or} \quad \left|\frac{-12}{-5}\right| = \frac{|-12|}{|-5|} = \frac{12}{5}.$$

The **triangle inequality** is a very important result: It states that for $x, y \in \mathbb{R}$ we have

$$|x + y| \leq |x| + |y|. \tag{2.12}$$

Again we prove (2.12) by discussing the different cases:

1. $x \geq 0$ and $y \geq 0$ implies $x + y \geq 0$, hence $|x + y| = x + y$, but in this case $|x| = x$ and $|y| = y$, hence $|x| + |y| = x + y$ and we have proved (2.12) with equality.

2. $x \geq 0$ and $y \leq 0$. Two cases may occur : $x + y \geq 0$ or $x + y \leq 0$. In the first case $|x + y| = x + y \leq x - y = |x| + |y|$, in the second case $|x + y| = -(x + y) = -x - y \leq x - y = |x| + |y|$.

3. $x \leq 0$ and $y \geq 0$. This is just the second case with $x$ and $y$ interchanged.

4. $x \leq 0$ and $y \leq 0$. Then $x + y \leq 0$, hence $|x + y| = -x - y$ but $|x| = -x$ and $|y| = -y$, hence $|x + y| = -x - y = |x| + |y|$.

**Basic Properties of the Absolute Value**:

**i)** $|x| \geq 0$ for all $x \in \mathbb{R}$ and $|x| = 0$ if and only if $x = 0$;

**ii)** $|xy| = |x||y|$ for all $x, y \in \mathbb{R}$;

**iii)** $|x + y| \leq |x| + |y|$ for all $x, y \in \mathbb{R}$.

Note that both $x$ and $-x$ have the same absolute value $|x|$. On the real line we find that for any $x \in \mathbb{R}$

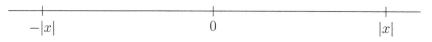

$$-|x| \qquad\qquad 0 \qquad\qquad |x|$$

*Figure 2.1*

Let us change our point of view. Consider on $\mathbb{R}$ the set

$$\{y \in \mathbb{R}|\ |y| = x, \quad x > 0 \text{ fixed}\}. \tag{2.13}$$

This set only consists of two points: $x$ and $-x$. Thus we may use the absolute value to define subsets of $\mathbb{R}$. We may extend this procedure by allowing inequalities:
Let $\varepsilon > 0$ and $a \in \mathbb{R}$ be fixed. Define on $\mathbb{R}$ the subset

$$B_\varepsilon(a) := \{x \in \mathbb{R} \mid |x - a| < \varepsilon\}. \tag{2.14}$$

We want to find all points in $\mathbb{R}$ belonging to the set $B_\varepsilon(a)$. Using the definition of the absolute value we find

$$|x - a| < \varepsilon \quad \text{if and only if} \quad -\varepsilon < x - a < \varepsilon,$$

or

$$|x - a| < \varepsilon \quad \text{if and only if} \quad -\varepsilon + a < x < a + \varepsilon.$$

As the simplest case take $a = 0$. This means that in $B_\varepsilon(0)$ we find all points with absolute value less than $\varepsilon$, or equivalently those whose distance to 0 is less than $\varepsilon$ :

$$B_\varepsilon(0)$$

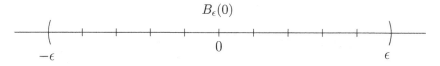

$$-\epsilon \qquad\qquad 0 \qquad\qquad \epsilon$$

*Figure 2.2*

22

But now we see the general interpretation: in $B_\varepsilon(a)$ we find all points which have a distance less than $\varepsilon$ to $a$

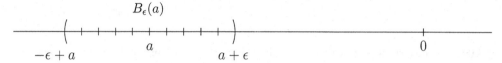

$$B_\varepsilon(a)$$

$$-\epsilon + a \qquad a \qquad a + \epsilon \qquad\qquad 0$$

Figure 2.3

**Example 2.2. A.** Consider the set

$$B_{\frac{1}{2}}(5) = \{x \in \mathbb{R}| \ |x - 5| < \frac{1}{2}\}$$
$$= \{x \in \mathbb{R}| \ -\frac{1}{2} + 5 < x < 5 + \frac{1}{2}\}$$
$$= \{x \in \mathbb{R}| \ \frac{9}{2} < x < \frac{11}{2}\}$$

$$B_{\frac{1}{2}}(5)$$

$$0 \qquad\qquad\qquad\qquad \frac{9}{2} \quad 5 \quad \frac{11}{2}$$

Figure 2.4

**B.** Next we look at

$$B_2\left(-\frac{2}{3}\right) = \left\{x \in \mathbb{R}| \ \left|x - \left(-\frac{2}{3}\right)\right| < 2\right\}$$
$$= \{x \in \mathbb{R}| \ |x + \frac{2}{3}| < 2\}$$
$$= \{x \in \mathbb{R}| \ -2 - \frac{2}{3} < x < 2 - \frac{2}{3}\}$$
$$= \{x \in \mathbb{R}| \ -\frac{8}{3} < x < \frac{4}{3}\}$$

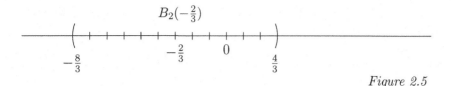

$$B_2\left(-\tfrac{2}{3}\right)$$

*Figure 2.5*

We can now define sets in $\mathbb{R}$ by using inequalities. Let us define for $a, b \in \mathbb{R}$, $a < b$,

$$(a, b) := \{x \in \mathbb{R} \mid a < x < b\} \tag{2.15}$$

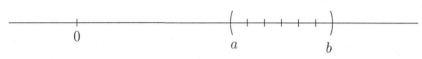

*Figure 2.6*

Thus in $(a, b)$ we find all real numbers $x$ which are larger than $a$ and less than $b$. For example

$$(-3, 8) = \{x \in \mathbb{R} \mid -3 < x < 8\}$$

*Figure 2.7*

With this notation we have

$$(-\varepsilon, \varepsilon) = B_\varepsilon(0)$$

or more generally

$$(-\varepsilon + a, a + \varepsilon) = B_\varepsilon(a)$$

for $\varepsilon > 0$ and $a \in \mathbb{R}$. Note that the numbers $a$ and $b$ do not belong to $(a, b)$. Again we can extend our procedure of defining sets. For $a, b \in \mathbb{R}$, $a < b$ we set

$$[a, b) := \{x \in \mathbb{R} \mid a \leq x < b\}, \tag{2.16}$$

which corresponds to

24

*Figure 2.8*

Also, we may consider

$$(a, b] := \{x \in \mathbb{R} | \ a < x \leq b\}, \tag{2.17}$$

which corresponds to

*Figure 2.9*

Finally we introduce

$$[a, b] := \{x \in \mathbb{R} | \ a \leq x \leq b\}, \tag{2.18}$$

which corresponds to

*Figure 2.10*

**Definition 2.3.** *For $a, b \in \mathbb{R}$, $a < b$, we call*

(a,b) *the **open interval** with end points a and b;*

(a,b] *the (left) **half-open interval** with end points a and b;*

[a,b) *the (right) **half-open interval** with end points a and b;*

[a,b] *the **closed interval** with end points a and b.*

An important remark: in the case of a closed interval the end points belong to the interval (set) whereas in the case of an open interval the end points do not belong to the interval (set).

**Example 2.4.** We find

$$[-3, \frac{2}{3}] = \{x \in \mathbb{R} | \ -3 \leq x \leq \frac{2}{3}\}$$

25

*Figure 2.11*

or

$$\left(\frac{1}{5}, \frac{3}{4}\right] = \{x \in \mathbb{R} \mid \frac{1}{5} < x \leq \frac{3}{4}\}$$

*Figure 2.12*

For the closed interval $[-\varepsilon + a, a + \varepsilon]$ we also write

$$\overline{B_\varepsilon(a)} := [-\varepsilon + a, a + \varepsilon]. \tag{2.19}$$

Often we will encounter the following type of problem: given $\varepsilon_1 > 0$ and $\varepsilon_2 > 0$ as well as $a_1, a_2 \in \mathbb{R}$, find all points $x \in \mathbb{R}$ such that $x \in B_{\varepsilon_1}(a_1)$ and $x \in B_{\varepsilon_2}(a_2)$. We have an easy geometric solution to the problem: it may happen that

*Figure 2.13*

or

*Figure 2.14*

In the first case $B_{\varepsilon_1}(a_1)$ and $B_{\varepsilon_2}(a_2)$ have no points in common i.e. they are **disjoint**. In the second case there are points in the intersection of $B_{\epsilon_2}(a_1)$ and $B_{\epsilon_2}(a_2)$, i.e. these points belong to both sets. In order to find the points in the intersection, we must solve simultaneously the inequalities

$$-\varepsilon_1 + a_1 < x < a_1 + \varepsilon_1 \quad \text{and} \quad -\varepsilon_2 + a_2 < x < a_2 + \varepsilon_2. \tag{2.20}$$

The conditions on $x$ are

$$-\varepsilon_1 + a_1 < x \quad \text{and} \quad -\varepsilon_2 + a_2 < x$$

and

$$x < a_1 + \varepsilon_1 \quad \text{and} \quad x < a_2 + \varepsilon_2$$

therefore

$$\max\{-\varepsilon_1 + a_1, -\varepsilon_2 + a_2\} < x$$

and

$$x < \min\{a_1 + \varepsilon_1, a_2 + \varepsilon_2\}.$$

Thus the solution to (2.20) is

$$x \in (\max\{-\varepsilon_1 + a_1, -\varepsilon_2 + a_2\}, \min\{a_1 + \varepsilon_1, a_2 + \varepsilon_2\}).$$

**Example 2.5. A.** We have $x \in B_2(3)$ and $x \in B_2(4)$ only for

$$x \in (\max\{1, 2\}, \min\{5, 6\}) = (2, 5).$$

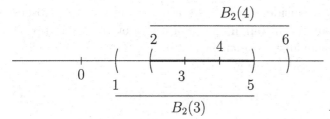

Figure 2.15

**B.** We have $x \in B_2(3)$ and $x \in B_2(8)$ only for

$$x \in (\max\{1, 6\}, \min\{5, 10\}) = (6, 5),$$

but (6,5) is not an interval since $6 > 5$, i.e. there are no points belonging to both sets.

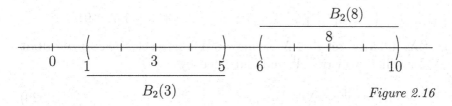

Figure 2.16

27

The set of all points belonging to $B_{\varepsilon_1}(a_1)$ and to $B_{\varepsilon_2}(a_2)$ is denoted by

$$B_{\varepsilon_1}(a_1) \cap B_{\varepsilon_2}(a_2), \tag{2.21}$$

and this set is called the intersection of $B_{\varepsilon_1}(a_1)$ and $B_{\varepsilon_2}(a_2)$. In the case where there are no points in the intersection, i.e. in the case where the intersection is empty, we write

$$B_{\varepsilon_1}(a_1) \cap B_{\varepsilon_2}(a_2) = \emptyset. \tag{2.22}$$

We define the **intersection** of two general sets $A$ and $B$ by

$$A \cap B = \{x \in A \mid x \in B\} = \{x \in B \mid x \in A\} = \{x \mid x \in A \text{ and } x \in B\},$$

i.e. $x \in A \cap B$ if $x \in A$ and $x \in B$. Two sets with an empty intersection are called **disjoint**. Before we continue to discuss intersections of intervals in more detail, we want to introduce a few more ideas from set theory. For two sets $A$ and $B$ we introduce their **union** by

$$A \cup B = \{x \mid x \in A \text{ or } x \in B\}. \tag{2.23}$$

Often it is advantageous to consider the sets we are dealing with as subsets of a given set $X$. For example all our intervals are subsets of $\mathbb{R}$. Suppose $A \subset X$ and $B \subset X$ for which we sometimes write $A, B \subset X$. Then the intersection and union of $A$ and $B$ are given by

$$A \cap B = \{x \in X \mid x \in A \text{ and } x \in B\}, \tag{2.24}$$

$$A \cup B = \{x \in X \mid x \in A \text{ or } x \in B\}. \tag{2.25}$$

For example with $X = \mathbb{N}$, $A = \{1, 2, 3, 5, 7\}$ and $B = \{3, 4, 5, 8, 9\}$ we find

$$A \cap B = \{1, 2, 3, 5, 7\} \cap \{3, 4, 5, 8, 9\} = \{3, 5\}$$

and

$$A \cup B = \{1, 2, 3, 5, 7\} \cup \{3, 4, 5, 8, 9\} = \{1, 2, 3, 4, 5, 7, 8, 9\}.$$

Given a set $X$ and a subset $A \subset X$ we may form a new set, the **complement** of $A$ in $X$ for which we write $A^{\complement}$ and is defined by

$$A^{\complement} := X \setminus A = \{x \in X \mid x \notin A\}. \tag{2.26}$$

Note that $A^\complement$ depends on $X$ therefore we should write $A^{\complement_X}$ or in the more traditional way; $\complement_X A$. For example, $\mathbb{N} \subset \mathbb{Z}$, and

$$\complement_\mathbb{Z} \mathbb{N} = \{z \in \mathbb{Z} \mid z \notin \mathbb{N}\} = \{z \in \mathbb{Z} \mid z \leq 0\} \tag{2.27}$$

whereas for $\mathbb{N} \subset \mathbb{R}$ we find

$$\complement_\mathbb{R} \mathbb{N} = \{x \in \mathbb{R} \mid x \notin \mathbb{N}\} = \mathbb{R} \setminus \mathbb{N} \tag{2.28}$$

and clearly
$$\complement_\mathbb{Z} \mathbb{N} \neq \complement_\mathbb{R} \mathbb{N}.$$

We will use the notation $A^\complement$ when it is clear from the context which set $X$ is meant, i.e. for which $X$ we consider $A$ to be a subset, otherwise we write $X \setminus A$ instead of $A^\complement$.

In Appendix II we have collected many results about operations on sets. Here we summarise some rules and give an outline of some of the proofs. Further proofs are given in Appendix II. The empty set is a special set, basic rules for the empty set which are all discussed in Appendix II are: For any set $X$ the following hold:

$$X \cup \emptyset = X \quad \text{and} \quad X \cap \emptyset = \emptyset. \tag{2.29}$$

Further, $\emptyset \subset X$ for every set $X$ and when considering $\emptyset$ as a subset of $X$ we have $\emptyset^\complement = X$. For every set $X$ we have the obvious relations

$$X \cup X = X \quad \text{and} \quad X \cap X = X, \tag{2.30}$$

and for two sets $X$ and $Y$ we have

$$X \cup Y = Y \cup X \quad \text{and} \quad X \cap Y = Y \cap X. \tag{2.31}$$

Let us have a look at $X \cup Y = Y \cup X$. We prove the equality of the two sets, as mentioned previously, by proving that each is a subset of the other. Thus in the case under consideration we prove

$$X \cup Y \subset Y \cup X \quad \text{and} \quad Y \cup X \subset X \cup Y. \tag{2.32}$$

The next rule for proving such statements is to transform these statements into a formal logical statement: for example $X \cup Y \subset Y \cup X$ corresponds to

$$(x \in X \cup Y) \quad \text{implies} \quad (x \in Y \cup X) \tag{2.33}$$

or equivalently

$$(X \in X \cup Y) \implies (X \in Y \cup X). \tag{2.34}$$

Now let us have a closer look at the statement $x \in X \cup Y$:

$$x \in X \cup Y \quad \text{if and only if} \quad x \in X \quad \text{or} \quad x \in Y, \tag{2.35}$$

or equivalently

$$(x \in X \cup Y) \iff (x \in X) \vee (x \in Y). \tag{2.36}$$

But $x \in X$ or $x \in Y$ is equivalent to $x \in Y$ or $x \in X$, more formally

$$(x \in X) \vee (x \in Y) \iff (x \in Y) \vee (x \in X). \tag{2.37}$$

The latter statement however implies $x \in Y \cup X$. Thus we have proved $x \in X \cup Y$ implies $x \in Y \cup X$, or $X \cup Y \subset Y \cup X$. Analogously, we may prove $Y \cup X \subset X \cup Y$, however this is left as a useful exercise.
We can prove further similar rules for the sets $X, Y, Z$:

$$X \cup (Y \cup Z) = (X \cup Y) \cup Z \tag{2.38}$$

which allows us just to write $X \cup Y \cup Z$, and further

$$X \cap (Y \cap Z) = (X \cap Y) \cap Z \tag{2.39}$$

which similarly allows us just to write $X \cap Y \cap Z$. We can also combine unions and intersections, however more care is needed here:

$$X \cup (Y \cap Z) = (X \cup Y) \cap (X \cap Z) \tag{2.40}$$

and

$$X \cap (Y \cup Z) = (X \cap Y) \cup (X \cap Z). \tag{2.41}$$

Let us prove (2.41): we need to prove

$$X \cap (Y \cup Z) \subset (X \cap Y) \cup (X \cap Z) \tag{2.42}$$

and

$$(X \cap Y) \cup (X \cap Z) \subset X \cap (Y \cup Z). \tag{2.43}$$

Note that

$$x \in X \cap (Y \cup Z) \iff (x \in X) \wedge (x \in Y \cup Z)$$

$$\Longleftrightarrow (x \in X) \wedge ((x \in Y) \vee (x \in Z))$$
$$\Longleftrightarrow ((x \in X) \vee (x \in Y)) \wedge ((x \in X) \vee (x \in Z)),$$

where we used (A.I.10) from Appendix I. However,

$$((x \in X) \vee (x \in Y)) \wedge ((x \in X) \vee (x \in Z)) \Longleftrightarrow x \in (X \cup Y) \cap (X \cup Z). \quad (2.44)$$

Thus we have proved (2.42) as well as (2.43).

Now let us turn to the complement. In the following, $A, B, C$ are all subsets of a fixed set $X$. First we note that

$$(A^{\complement})^{\complement} = A, \quad (2.45)$$

which follows from

$$x \in (A^{\complement})^{\complement} \Longleftrightarrow x \notin A^{\complement} \Longleftrightarrow x \in A.$$

Finally we state **de Morgan's laws**:

$$(A \cap B)^{\complement} = A^{\complement} \cup B^{\complement} \quad (2.46)$$

and

$$(A \cup B)^{\complement} = A^{\complement} \cap B^{\complement}. \quad (2.47)$$

We prove (2.46). The fact that $x \in (A \cap B)^{\complement}$ means

$$x \notin A \cap B \Longleftrightarrow (x \notin A) \vee (x \notin B)$$
$$\Longleftrightarrow (x \in A^{\complement}) \vee (x \in B^{\complement})$$
$$\Longleftrightarrow x \in (A^{\complement} \cup B^{\complement}),$$

therefore we have proved $(A \cap B)^{\complement} \subset (A^{\complement} \cup B^{\complement})$ as well as $(A^{\complement} \cup B^{\complement}) \subset (A \cap B)^{\complement}$.

Let $A_1, \ldots, A_N$ be a finite number of sets. For their union we write

$$\bigcup_{j=1}^{N} A_j = A_1 \cup \cdots \cup A_N, \quad (2.48)$$

and for their intersection we write

$$\bigcap_{j=1}^{N} A_j = A_1 \cap \cdots \cap A_N. \quad (2.49)$$

Thus, $x \in \bigcup_{j=1}^{N} A_j$ if for at least one $j_0 \in \{1, \ldots, N\}$ we have $x \in A_{j_0}$, whereas $x \in \bigcap_{j=1}^{N} A_j$ means that $x \in A_j$ for all $j \in \{1, \ldots, N\}$.

We now return to intervals on the real line. We may determine intersections of intervals:

$$(a, b) \cap (c, d) \quad \text{or} \quad [a, b) \cap [c, d] \quad \text{etc.}$$

In each case we have to solve systems of inequalities

$$x \in (a, b) \cap (c, d) \quad \text{if and only if} \quad a < x < b \quad \text{and} \quad c < x < d,$$

i.e.

$$\max\{a, c\} < x < \min\{b, d\},$$

or

$$x \in [a, b) \cap [c, d] \quad \text{if and only if} \quad a \leq x < b \quad \text{and} \quad c \leq x \leq d,$$

i.e.

$$\max\{a, c\} \leq x < b \quad \text{if} \quad b \leq d$$

or

$$\max\{a, c\} \leq x \leq d \quad \text{if} \quad d < b.$$

Here $\max\{a, c\}$ stands for the larger number, i.e. the maximum of $a$ and $c$, whereas $\min\{b, d\}$ stands for the smaller number, i.e. the minimum of $b$ and $d$.

**Example 2.6.** We have

$$[-2, 5) \cap [3, 6] = [3, 5) \tag{2.50}$$

Figure 2.17

Note that (2.50) is an equality of sets, namely

$$\{x \in \mathbb{R} | -2 \leq x < 5\} \cap \{x \in \mathbb{R} | 3 \leq x \leq 5\} = \{x \in \mathbb{R} | 3 \leq x < 5\}. \tag{2.51}$$

We may also look at unions of intervals which is less problematic since we do not need to solve inequalities however we might have to combine them. For two, say open, intervals $(a, b)$ and $(c, d)$ it may happen that they do not intersect, their union is then just $(a, b) \cup (c, d)$

*Figure 2.18*

If $(a, b) \cap (c, d) \neq \emptyset$, then $(a, b) \cup (c, d)$ is either one of these intervals, namely $(a, b)$ if $(c, d) \subset (a, b)$ or $(c, d)$ if $(a, b) \subset (c, d)$

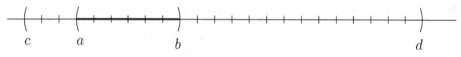

*Figure 2.19*

or $(a, b) \cup (c, d) = (\min(a, c), \max(b, d))$

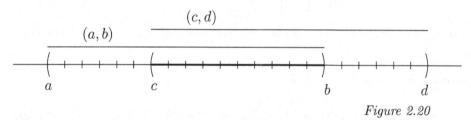

*Figure 2.20*

Note that in the case of closed or half-open intervals we may meet some new possibilities(compared with open intervals). The two intervals $(a, b]$ and $(b, c)$, for example, do not intersect

$$(a, b] \cap (b, c) = \{x \in \mathbb{R} \mid a < x \leq b \text{ and } b < x < c\} = \emptyset,$$

however

$$(a, b] \cup (b, c) = \{x \in \mathbb{R} \mid a < x \leq b \text{ or } b < x < c\}$$

$$= \{x \in \mathbb{R} \mid a < x < c\} = (a, c).$$

Thus the union of two disjoint open intervals is never an interval, while in the case of disjoint half-open intervals the union might be an interval. We will discuss more cases in the exercises.

For convenience let us introduce some further notation

$$(a, \infty) := \{x \in \mathbb{R} | x > a\}, \tag{2.52}$$
$$[a, \infty) := \{x \in \mathbb{R} | x \geq a\}, \tag{2.53}$$
$$(-\infty, b) := \{x \in \mathbb{R} | x < b\}, \tag{2.54}$$
$$(-\infty, b] := \{x \in \mathbb{R} | x \leq b\}, \tag{2.55}$$

and

$$(-\infty, \infty) := \mathbb{R}. \tag{2.56}$$

We call "$\infty$" infinity and "$-\infty$" minus infinity and at the moment it is just a useful name and notation.

We have already used max and min without stating the formal definitions:

$$\max\{a, b\} := \begin{cases} a, & a \geq b; \\ b, & b \geq a, \end{cases} \tag{2.57}$$

and

$$\min\{a, b\} := \begin{cases} a, & a \leq b; \\ b, & b \leq a. \end{cases} \tag{2.58}$$

It is interesting to note that we can express max and min using the absolute value.

**Lemma 2.7.** *For $a, b \in \mathbb{R}$ we have*

$$\max\{a, b\} = \frac{1}{2}(a + b + |a - b|) \tag{2.59}$$

*and*

$$\min\{a, b\} = \frac{1}{2}(a + b - |a - b|). \tag{2.60}$$

*Proof.* We prove (2.59) and leave (2.60) as an exercise. If $a \geq b$ then $\max\{a, b\} = a$. In this case $a - b \geq 0$, hence $|a - b| = a - b$ and

$$\frac{1}{2}(a + b + a - b) = \frac{1}{2}2a = a.$$

If however $b \geq a$ then $\max\{a, b\} = b$. In this case $a - b \leq 0$ hence $|a - b| = b - a$ and we find

$$\frac{1}{2}(a + b + b - a) = \frac{1}{2}2b = b.$$

$\square$

The notations of maximum and minimum easily extend to finite sets of real numbers. If $a_1, \cdots, a_n \in \mathbb{R}$ then

$$\max\{a_1, \cdots, a_n\} := a_k \quad \text{if} \quad a_k \geq a_j \quad \text{for} \quad j = 1, \cdots, n \qquad (2.61)$$

and

$$\min\{a_1, \cdots, a_n\} := a_l \quad \text{if} \quad a_l \leq a_j \quad \text{for} \quad j = 1, \cdots, n. \qquad (2.62)$$

Definition (2.61) tells us that $a_k$ is larger or equal than all other elements $a_1, \cdots, a_n$ in the set $\{a_1, \cdots, a_n\}$ and (2.62) says that $a_l$ is less or equal to all other elements of the set $\{a_1, \cdots, a_n\}$.

**Example 2.8.** The following hold

$$\max\{1, 7, -\frac{3}{5}, 13\} = 13,$$

and

$$\min\{\frac{1}{3}, 2, -5, 13\} = -5.$$

We close this chapter by showing some additional properties of the absolute value. As a rule lower bounds or estimates from below are in general more difficult to obtain. Let us consider the triangle inequality

$$|a + b| \leq |a| + |b|.$$

Since $|a + b| \geq 0$ the estimate

$$-|a| - |b| \leq |a + b|$$

is trivial. The **converse triangle inequality** however is non-trivial:

**Lemma 2.9.** *For all $a, b \in \mathbb{R}$ we have*

$$||a| - |b|| \leq |a - b| \qquad (2.63)$$

*and*

$$||a| - |b|| \leq |a + b|. \qquad (2.64)$$

*Proof.* First note that (2.64) follows from (2.63) and vice versa. In fact we may take the real number $-b$ instead of $b$ in (2.63) to find

$$||a| - |b|| = ||a| - |-b|| \le |a - (-b)| = |a + b|.$$

The proof that (2.64) implies (2.63) follows the same idea.
Now we prove (2.63). By the triangle inequality we know that

$$|a| = |a - b + b| \le |a - b| + |b|$$

implying

$$|a| - |b| \le |a - b|. \tag{2.65}$$

On the other hand we have

$$|b| = |b - a + a| \le |b - a| + |a| = |a - b| + |a|$$

implying

$$-(|a| - |b|) \le |a - b|, \tag{2.66}$$

thus together with (2.65) we have

$$||a| - |b|| \le |a - b|. \tag{2.67}$$

$\square$

# Problems

1. Let $X = \{a, b, c, d, e, f, g, h, i\}$ and consider the subsets
   $A = \{a, b, c, d\}$, $B = \{b, d, f, h\}$ and $C = \{c, d, e, f\}$. Find $A^\complement$,
   $(A \cap C)^\complement$, $B \setminus C$, and $(A \cup B)^\complement$.

2. Find the following subsets of the real line:
   a) $B_4(2) \cap B_3(8)$; b) $(B_2(5) \cap B_7(-2))^\complement$; c) $((-3, \frac{3}{2}) \cup [-\frac{1}{4}, \frac{7}{3}))^\complement$;
   d) $[-2, \frac{7}{3}) \cap [\frac{3}{5}, \frac{15}{4}]$.
   In each case, sketch the solution set.

3. For the sets $A \subset X$ and $B \subset X$, prove the following statements:
   a) $A \cap B \subset A \subset A \cup B$; b) $(A \setminus B) \cap B = \phi$; c) $B \setminus A = B \cap A^\complement$.

4. For $A, B, C \subset X$ prove the following statements:
   a) $(A \cap B) \cup C = (A \cup C) \cap (B \cup C)$; b) $(A \cup B)^{\complement} = A^{\complement} \cap B^{\complement}$.

5. For $A, B \subset X$ (which means that $A \subset X$ and $B \subset X$) prove that the following statements are equivalent
   a) $A \subset B$, b) $A \cap B = A$, c) $B^{\complement} \subset A^{\complement}$, d) $A \cup B = B$, e) $B \cup A^{\complement} = X$, f) $A \cap B^{\complement} = \phi$.

6. Calculate the following values:
   a) $\left| -\frac{5}{8} \right|$; b) $\left| \frac{11}{3} - 3 \right|$; c) $\left| \frac{7}{9} - \frac{12}{5} \right|$; d) $\left| |-3| - |-5| \right|$; e) $\sqrt{a^2}$, $a \in \mathbb{R}$.

7. Prove that for every $\varepsilon > 0$ and all $a, b \in \mathbb{R}$ the following hold

$$|ab| \le \varepsilon a^2 + \frac{1}{4\varepsilon} b^2,$$

and

$$\min\{a, b\} = \frac{1}{2}(a + b - |a - b|).$$

Furthermore, for $a > 0$ prove that

$$a + \frac{1}{a} \ge 2.$$

8. Prove for $a, b, c \in \mathbb{R}$ that

$$|a - c| \le |a - b| + |b - c|$$

and

$$\left| |a - b| - |c| \right| \le \left| |a - b| - c \right| \le |a| + |b| + |c|.$$

9.   a) Find every $x \in \mathbb{R}$ that satisfies

$$8x - 11 > -24x + 89.$$

b) Find every $x \in \mathbb{Z}$ that satisfies

$$-3 \le 7x - 2 < 6x + 5.$$

c) Find every $x \in \mathbb{R}$ that satisfies

$$|x - 3| \le |x + 3|.$$

10.    a) For which values of $x \in \mathbb{R}$ does the inequality

$$2x + 6(2 - x) \geq 8 - 2x$$

hold?

b) Find all values of $x \in \mathbb{R}$ such that

$$x^2 + 2x - 10 < 3x + 2.$$

# 3 Mathematical Induction

Mathematics derives new statements from given ones. The underlying procedure is of course called a **proof**. It is by no means easy to define what a (correct) proof is, and there is no need to do this here. For a working mathematician a proof reduces to the following: you start with some statements either being taken for granted to be true (axioms) or already proven (theorems, propositions, lemmata), and then you apply the usual rules of (mathematical) logic which we have collected in Appendix I in order to arrive at new statements. Very often we have to handle statements $A(n)$ depending on $n \in \mathbb{N}$ or $n \in \{k \in \mathbb{Z} \mid k \geq m$ for some $m \in \mathbb{Z}\}$. For example the statement $A(n)$ could be

$$A(n) : 1 + 2 + \cdots + n = \frac{n(n+1)}{2}, \quad n \in \mathbb{N}. \tag{3.1}$$

To prove that such a statement is true for $\mathbb{N}$ we cannot just check one-by-one that it is true for every natural number however we may use a method called **mathematical induction**. It is possible to show that this method is sufficient for proving statements like $A(n), n \in \mathbb{N}$, however this involves looking at the actual construction of $\mathbb{N}$ and Peano's axioms which goes beyond the scope of this introductory section. For more about Peano's axioms and mathematical induction, see Appendix III. The method of mathematical induction follows from the **axiom of mathematical induction** (one of Peano's axioms): *Suppose that for each $n \geq m$, $m, n \in \mathbb{Z}$, a mathematical statement $A(n)$ is given. If $A(m)$ is true and if for all $n \geq m$ the statement $A(n)$ implies that the statement $A(n + 1)$ is true, then $A(n)$ is true for all $n \geq m$.*

At this stage we will just assume this axiom. An alternative version of the axiom of mathematical induction is:

*Suppose for each $n \geq m$, $m, n \in \mathbb{Z}$, a statement $A(n)$ is given. If $A(m)$ is true and if for all $n \geq m$ the statements $A(m), \ldots, A(n)$ imply the statement $A(n + 1)$, then $A(n)$ holds for all $n \geq m$.*

In simple terms this means that the method of mathematical induction is as follows: we begin by showing that $A(m)$ is true for some $m \in \mathbb{N}$, usually $m = 0$ or $m = 1$ (base case). Next we assume that $A(n)$ is true for arbitrary $n \geq m$ (induction hypothesis) and then prove that $A(n+1)$ is true (induction step).

Let us start with a simple example to see how we can apply the axiom of mathematical induction.

**Example 3.1.** For every $n \geq 0$, $n \in \mathbb{Z}$, the statement

$$A(n) : 11^{n+2} + 12^{2n+1} \text{ is divisible by } 133 \qquad (3.2)$$

holds. Recall that a natural number $l$ is **divisible** by a natural number $m$ if there exists a natural number $k$ such that $l = k \cdot m$. We start by proving that $A(0)$ is true i.e. that

$$11^2 + 12 \text{ is divisible by } 133.$$

Since $11^2 + 12 = 121 + 12 = 133$ this statement is true. Now we assume that for arbitrary but fixed $n \geq 0$, the statement $A(n)$ is true and we want to deduce that $A(n + 1)$ is also true. In other words, we want to prove that if $11^{n+2} + 12^{n+1}$ is divisible by 133 then $11^{(n+1)+2} + 12^{2(n+1)+1}$ is divisible by 133 too. Note that $n$ is not specified, it is arbitrary but fixed. We cannot take a particular $n$, say $n = 5$ or $n = 12543$. So we have to prove that $11^{(n+1)+2} + 12^{2(n+1)+1} = 11^{n+3} + 12^{2n+3}$ is divisible by 133 assuming that $11^{n+2} + 12^{2n+1}$ is divisible by 133. How can we reduce or transform $A(n + 1)$ to a statement to which we can apply $A(n)$?
Here is a suggestion:

$$11^{n+3} + 12^{2n+3} = 11 \cdot 11^{n+2} + 12^2 \cdot 12^{2n+1}$$

$$= 11 \cdot 11^{n+2} + 144 \cdot 12^{2n+1}$$

$$= 11 \cdot 11^{n+2} + (11 + 133) \cdot 12^{2n+1} \qquad (3.3)$$

$$= 11 \cdot 11^{n+2} + 11 \cdot 12^{2n+1} + 133 \cdot 12^{2n+1}$$

$$= 11(11^{n+2} + 12^{2n+1}) + 133 \cdot 12^{2n+1}. \qquad (3.4)$$

The step in (3.3) is crucial, splitting 144 into the sum of 11 and 133 allows us to deduce a statement to which we can apply $A(n)$. Indeed, since by assumption $11^{n+2} + 12^{2n+1}$ is divisible by 133, there exists a $k \in \mathbb{N}$ such that $11^{n+2} + 12^{2n+1} = 133k$, and (3.4) becomes

$$11^{n+3} + 12^{2n+3} = 11^{(n+1)+2} + 12^{2(n+1)+1} \qquad (3.5)$$

$$= 11 \cdot 133 \cdot k + 133 \cdot 12^{2n+1}$$

$$= 133(11k + 12^{2n+1}).$$

Since $11k + 12^{2n+1}$ is a natural number, say $m$, it follows that $11^{(n+1)+2} + 12^{2(n+1)+1} = 133m$, i.e. $A(n+1)$ is correct. Now the principle of mathematical induction yields that $A(n)$ holds for all $n \geq 0$.

This example already gives an insight that mathematical induction as a method of proving a statement $A(n)$ for all $n \geq m$, $n, m \in \mathbb{Z}$, is a way forward. However, depending on the statement $A(n)$ we may need additional knowledge for the proof that $A(n)$ implies $A(n + 1)$. Indeed this is already of course the case when proving $A(m)$.

There is a reason why we have not started with proving (3.1). Although the notation $1 + 2 + \cdots + n$ is intuitively clear, we will introduce a better one. Suppose that $a_1, \ldots a_k \in \mathbb{R}$, which is shorthand for: suppose that for every $j \in \{k \in \mathbb{N} \mid k \leq n\}$ we have $a_j \in \mathbb{R}$. The sum $A$ of these $n$ real numbers is denoted by

$$A := \sum_{j=1}^{n} a_j, \tag{3.6}$$

which is what we mean when writing

$$A = a_1 + \cdots + a_n. \tag{3.7}$$

At the end of this chapter we will discuss an even more formal way to introduce (3.6). Here are some examples on how to use this new notation

**Example 3.2. A.** For $j \in \mathbb{N}$ let $a_j = j$. Then

$$\sum_{j=1}^{n} a_j = \sum_{j=1}^{n} j \tag{3.8}$$

gives of course the expression considered in (3.1).
**B.** Take $a_j = j^2$, $j \in \mathbb{N}$, to find

$$\sum_{j=1}^{n} a_j = \sum_{j=1}^{n} j^2. \tag{3.9}$$

**C.** Now take $b_j = \frac{1}{j}$, $j \in \mathbb{N}$, to form the sum

$$\sum_{j=1}^{n} b_j = \sum_{j=1}^{n} \frac{1}{j}. \tag{3.10}$$

(Of course it does not matter whether we denote the numbers by $a_j$ or $b_j$.)

**D.** Finally, with $c_j = 2^j$, $j \in \mathbb{N}_0 = \mathbb{N} \cup \{0\}$, we can form the sum

$$\sum_{j=0}^{n} c_j = \sum_{j=0}^{n} 2^j. \tag{3.11}$$

The last part of Example 3.2 is interesting. Everyone understands what is meant by

$$\sum_{j=0}^{n} c_j = c_0 + c_1 + \cdots + c_n. \tag{3.12}$$

We can extend this notation: for $m, k \in \mathbb{Z}$, $m \leq k$, let the real numbers $a_m, a_{m+1}, \ldots, a_k$ be given. We set for their sum

$$\sum_{j=m}^{k} a_j = a_m + a_{m+1} + \cdots + a_k. \tag{3.13}$$

For example with $a_j = (j + \frac{1}{2})^{-2}$ we can form

$$
\begin{aligned}
\sum_{j=-3}^{2} a_j &= \sum_{j=-3}^{2} \frac{1}{(j + \frac{1}{2})^2} \\
&= \frac{1}{(\frac{5}{2})^2} + \frac{1}{(\frac{3}{2})^2} + \frac{1}{(\frac{1}{2})^2} + \frac{1}{(\frac{1}{2})^2} + \frac{1}{(\frac{3}{2})^2} + \frac{1}{(\frac{5}{2})^2} \\
&= 4\left(\frac{1}{25} + \frac{1}{9} + 1 + 1 + \frac{1}{9} + \frac{1}{25}\right) = \frac{2072}{225}.
\end{aligned}
$$

It is convenient to include the following convention

$$\sum_{j=m}^{k} a_j = 0 \quad \text{for} \quad k < m. \tag{3.14}$$

Moreover, the associative law for addition implies for $k \leq l \leq m$ that

$$\sum_{j=k}^{m} a_j = \sum_{j=k}^{l} a_j + \sum_{j=l+1}^{m} a_j \tag{3.15}$$

$$(= (a_k + \cdots + a_l) + (a_{l+1} + \cdots + a_m).)$$

Now we return to statement (3.1).

**Example 3.3.** Prove that

$$A(n) : \sum_{j=1}^{n} j = \frac{n(n+1)}{2}, \quad n \geq 1. \tag{3.16}$$

We start by proving $A(1)$, i.e. we note that

$$\sum_{j=1}^{1} j = 1 \quad \text{as well as} \quad \frac{1(1+1)}{2} = 1,$$

i.e. $A(1)$ holds. Now suppose that $A(n)$ holds for arbitrary but fixed $n \in \mathbb{N}$. We want to show that then $A(n+1)$ holds too. Indeed we have

$$\sum_{j=1}^{n+1} j = \sum_{j=1}^{n} j + (n+1),$$

and this is already the crucial step since it allows us to use statement $A(n)$, namely

$$\sum_{j=1}^{n+1} j = \sum_{j=1}^{n} j + (n+1)$$

$$\frac{n(n+1)}{2} + (n+1) = \frac{n(n+1) + 2(n+1)}{2}$$

$$= \frac{(n+1)(n+2)}{2},$$

which is $A(n+1)$.

**Example 3.4.** For $x \neq 1$ the statement

$$A(n) : \sum_{j=0}^{n} x^j = \frac{x^{n+1} - 1}{x - 1}, \quad n \geq 0 \tag{3.17}$$

holds. Recall that $x^0 = 1$, thus we have $\sum_{j=0}^{n} x^j = 1 + x + x^2 + \cdots + x^n$. For $n = 0$ the statement $A(0)$ is correct:

$$\sum_{j=0}^{0} x^j = x^0 = 1 \quad \text{and} \quad \frac{x^1 - 1}{x - 1} = \frac{x - 1}{x - 1} = 1.$$

Now if $\displaystyle\sum_{j=0}^{n} x^j = \frac{x^{n+1} - 1}{x - 1}$ then

$$\sum_{j=0}^{n+1} x^j = \sum_{j=0}^{n} x^j + x^{n+1} = \frac{x^{n+1} - 1}{x - 1} + x^{n+1}$$

$$= \frac{x^{n+1} - 1}{x - 1} + \frac{x^{n+1}(x - 1)}{x - 1}$$

$$= \frac{x^{n+1} + x^{n+2} - x^{n-1} - 1}{x - 1} = \frac{x^{n+2} - 1}{x - 1},$$

i.e. we have proved that $A(n + 1)$ is correct:

$$\sum_{j=0}^{n+1} x^j = \frac{x^{n+2} - 1}{x - 1}.$$

We can also use mathematical induction to prove inequalities or estimates.

**Lemma 3.5.** *Let $a_1, \ldots, a_n \in \mathbb{R}$. Then we have the estimates*

$$\left| \sum_{l=1}^{n} a_l \right| \leq \sum_{l=1}^{n} |a_l| \leq n \max \{|a_1|, \ldots, |a_l|\}. \tag{3.18}$$

*Proof.* For $n = 1$ we find

$$\left| \sum_{l=1}^{1} a_l \right| = |a_1| = 1 \cdot \max \{|a_1|\}.$$

Now suppose that (3.18) holds for arbitrary but fixed $n \in \mathbb{N}$. We find using the triangle inequality (2.12) that

$$\left| \sum_{l=1}^{n+1} a_l \right| = \left| \sum_{l=1}^{n} a_l + a_{n+1} \right| \leq \left| \sum_{l=1}^{n} a_l \right| + |a_{n+1}|$$

$$\leq \sum_{l=1}^{n} |a_l| + |a_{n+1}|,$$

44

where in the last step we used (3.18) for $n$. Now the rest is straightforward since

$$\sum_{l=1}^{n} |a_l| + |a_{n+1}| = \sum_{l=1}^{n+1} |a_l|,$$

and the first estimate is proved for $n + 1$ provided it holds for $n$, hence by mathematical induction the first estimate holds for all $n \in \mathbb{N}$. The second estimate in (3.18) is proved without induction. Let $\max\{|a_1|, \ldots, |a_n|\} = |a_k|$ for some $1 \leq k \leq n$. Replacing each number $|a_l|$ in $\sum_{l=1}^{n} |a_l|$ by $|a_k|$ will increase the sum, i.e.

$$\sum_{l=1}^{n} |a_l| \leq |a_k| + \cdots + |a_k| \leq n \cdot \max\{|a_1|, \ldots, |a_k|\}.$$

$\square$

As in the case for finite sums we can introduce a notation for finite products. Let $a_1, \ldots, a_n \in \mathbb{R}$ be given. We denote their product by

$$\prod_{j=1}^{n} a_j = a_1 \cdot a_2 \cdot \ldots \cdot a_n. \tag{3.19}$$

Clearly, using the associative law for multiplication we have for $m < n$ that

$$\prod_{j=1}^{n} a_j = \left(\prod_{j=1}^{m} a_j\right) \cdot \left(\prod_{j=m+1}^{n} a_j\right). \tag{3.20}$$

Note that the second term on the right hand side of (3.20) is an obvious generalisation of (3.19), compare with the analogous notation for sums, see (3.15).

Hence, for $l < n$, $l, n \in \mathbb{Z}$, and real numbers $a_l, a_{l+1}, \ldots, a_n$ we write for their product

$$\prod_{j=l}^{n} a_j = a_l \cdot a_{l+1} \cdot \ldots \cdot a_n. \tag{3.21}$$

We introduce further for $n \in \mathbb{N}$

$$n! := \prod_{j=1}^{n} j, \tag{3.22}$$

and we call this number $n$ **factorial**. For example we have $6! = 1 \cdot 2 \cdot 3 \cdot 4 \cdot 5 \cdot 6 = 720$. Using (3.20) we find

$$(n+1)! = n!(n+1).  \tag{3.23}$$

Further we define

$$0! = 1.  \tag{3.24}$$

**Definition 3.6.** *For $n, k \in \mathbb{N} \cup \{0\}$, $k \leq n$, we define the* **binomial coefficient** *by*

$$\binom{n}{k} := \frac{n!}{k!(n-k)!},  \tag{3.25}$$

*where we read $\binom{n}{k}$ as* **n over k**. *For $k > n$ we set*

$$\binom{n}{k} = 0.  \tag{3.26}$$

**Example 3.7.** The following hold:

$$
\begin{aligned}
\binom{n}{0} &= \frac{n!}{0!n!} = 1; \\
\binom{n}{1} &= \frac{n!}{1!(n-1)!} = n; \\
\binom{n}{n} &= \frac{n!}{n!(n-n)!} = 1; \\
\binom{2}{1} &= \frac{2!}{1!1!} = 2; \\
\binom{4}{2} &= \frac{4!}{2!(4-2)!} = \frac{1 \cdot 2 \cdot 3 \cdot 4}{2 \cdot 2} = 6.
\end{aligned}
$$

**Lemma 3.8.** *For $1 \leq k \leq n$ the following holds*

$$\binom{n}{k} = \binom{n-1}{k-1} + \binom{n-1}{k}.  \tag{3.27}$$

*Proof.* For $n = k$ it is straightforward:

$$\binom{n}{n} = \binom{n-1}{n-1} + \binom{n-1}{n}$$

46

or

$$1 = 1 + 0.$$

Now for $1 \leq k < n$ we have

$$
\begin{aligned}
\binom{n-1}{k-1} + \binom{n-1}{k} &= \frac{(n-1)!}{(k-1)!(n-k)!} + \frac{(n-1)!}{k!(n-k-1)!} \\
&= \frac{k(n-1)! + (n-k)(n-1)!}{k!(n-k)!} \\
&= \frac{(n-k+k)!(n-1)!}{k!(n+k)!} = \binom{n}{k}.
\end{aligned}
$$

$\square$

We can now prove our first non-trival result. The following formulae should be familiar:

$$(a+b)^2 = a^2 + 2ab + b^2 \quad \text{and} \quad (a-b)^2 = a^2 - 2ab + b^2.$$

These are generalised by:

**Theorem 3.9 (Binomial theorem).** *For $x, y \in \mathbb{R}$ and $n \in \mathbb{N} \cup \{0\}$*

$$(x+y)^n = \sum_{k=0}^{n} \binom{n}{k} x^{n-k} y^k. \tag{3.28}$$

*Proof.* We use mathematical induction. Denote the statement in (3.28) as $A(n)$. For $n = 0$ we have

$$(x+y)^0 = 1 \quad \text{and since} \quad \sum_{k=0}^{0} \binom{0}{k} x^{0-k} y^k = \binom{0}{0} x^0 y^0 = 1$$

the statement $A(0)$ holds. Now we prove that $A(n)$ implies $A(n+1)$:

$$(x+y)^{n+1} = (x+y)^n (x+y) = \left( \sum_{k=0}^{n} \binom{n}{k} x^{n-k} y^k \right)(x+y)$$

$$= \sum_{k=0}^{n} \binom{n}{k} x^{n+1-k} y^k + \sum_{k=0}^{n} \binom{n}{k} x^{n-k} y^{k+1}$$

$$= x^{n+1} + \sum_{k=1}^{n} \binom{n}{k} x^{n+1-k} y^k + \sum_{k=0}^{n-1} \binom{n}{k} x^{n-k} y^{k+1} + y^{n+1}$$

$$= x^{n+1} + \sum_{k=1}^{n} \binom{n}{k} x^{n+1-k} y^k + \sum_{k=1}^{n} \binom{n}{k-1} x^{n-(k-1)} y^k + y^{n+1} \qquad (3.29)$$

$$= x^{n+1} + \sum_{k=1}^{n} \left( \binom{n}{k} + \binom{n}{k-1} \right) x^{n+1-k} y^k + y^{n+1}$$

$$= \binom{n+1}{0} x^{n+1} y^0 + \sum_{k=1}^{n} \left( \binom{n}{k} + \binom{n}{k-1} \right) x^{n+1-k} y^k + \binom{n+1}{n+1} x^0 y^{n+1}$$

$$= \sum_{k=0}^{n+1} \binom{n+1}{k} x^{n+1-k} y^k,$$

proving the result. □

In Remark 3.13 below we clarify the calculation leading to (3.29) in more detail.

**Corollary 3.10.** *The following holds*

$$\sum_{k=0}^{n} \binom{n}{k} = (1+1)^n = 2^n, \qquad (3.30)$$

*and moreover we have*

$$\sum_{k=0}^{n} (-1)^k \binom{n}{k} = (1-1)^n = 0. \qquad (3.31)$$

**Example 3.11.** Using the binomial theorem we get

$$(x+y)^0 = 1,$$

$$(x+y)^1 = x+y,$$

$$(x+y)^2 = x^2 + 2xy + y^2,$$

$$(x - y)^2 = x^2 - 2xy + y^2,$$

$$(x + y)^3 = x^3 + 3x^2y + 3xy^2 + y^3,$$

$$(x + y)^4 = x^4 + 4x^3y + 6x^2y^2 + 4xy^3 + y^4.$$

**Remark 3.12.** The binomial coefficients will play an important part in probability theory and combinatorics.

**Remark 3.13 (Changing the running index in a sum).** In deriving (3.29) we used

$$\sum_{k=0}^{n-1} \binom{n}{k} x^{n-k} y^{k+1} = \sum_{k=1}^{n} \binom{n}{k-1} x^{n-(k-1)} y^k. \tag{3.32}$$

To obtain this result we argue as follows: in the first sum put the running index $k$ equal to $l = k + 1$. Thus, whenever we see $k$ we replace it by $l - 1$ to get

$$\sum_{l-1=0}^{n-1} \binom{n}{l-1} x^{n-(l-1)} y^{l-1+1} = \sum_{l-1=0}^{n-1} \binom{n}{l-1} x^{n-(l-1)} y^l$$

$$= \sum_{l=1}^{n} \binom{n}{l-1} x^{n-(l-1)} y^l,$$

and now put $l = k$.

There is still a need to improve our formal definition of the sum of $n$ real numbers as given in (3.6), the same applies to the definition of their product, see (3.21). We have to introduce the concept of a **recursive definition**. Suppose for $m \le j \le n$, $m, n \in \mathbb{Z}$, mathematical objects $C(j)$ are defined. For example $C(j) = \sum_{l=1}^{j} a_l$ for $1 \le j \le n$ and $a_l \in \mathbb{R}$. It might happen that we can extend the definition to get a new object $C(n + 1)$. In our example we may define

$$C(n + 1) := \sum_{l=1}^{n+1} a_l := a_{n+1} + \sum_{l=1}^{n} a_l = a_{n+1} + C(n). \tag{3.33}$$

Thus we use the already defined objects $C(m), \ldots, C(n)$ to define the new object $C(n+1)$. If we can extend this to all $n \ge m$, i.e. for all $n \ge m$, $m, n \in \mathbb{Z}$, we can define $C(n+1)$ given $C(m), \ldots C(n)$, then we say that the sequence

49

of objects $C(n)$, $n \geq m$, is **recursively defined**, or defined by recursion, or as some authors say defined by mathematical induction.

Here are a few examples in addition to (3.33):

$$\prod_{l=m}^{n+1} a_l := a_{n+1} \cdot \left( \prod_{l=m}^{n} a_l \right) ; \tag{3.34}$$

$$a^{n+1} := a \cdot a^n, \quad n \geq m, \ a \neq 0. \tag{3.35}$$

We can put this into a more formal scheme which indicates that we may prove by mathematical induction that when defining objects $C(j)$ by recursion we indeed have defined all the elements of the sequence $C(j)$, $j \geq m$. The formal proof however we omit. If $C(j)$, $j \geq m$, $j, m \in \mathbb{Z}$, are the objects we want to define we start with

$$A(m) : C(m) \ \text{is defined by some formula,}$$

for example

$$A(1) : \sum_{l=1}^{1} a_l := a_1.$$

In the next step we consider

$$A(n + 1) : C(n + 1) \ \text{which is defined using} \ C(m), \ldots, C(n),$$

for example

$$A(n + 1) : C(n + 1) := \sum_{l=1}^{n+1} a_l := a_{n+1} + \sum_{l=1}^{n} a_l = a_{n+1} + C(n).$$

Note that we will not always need $C(m), \ldots, C(n)$ to define $C(n+1)$; in our example $C(n)$ is sufficient. We can interpret $A(n)$ as the statement: given $A(m), \ldots, A(n-1)$, then it is formally possible to define $A(n)$. The proof that a definition by recursion gives all objects $C(n)$, $n \geq m$, must now show that for all $n \geq m$ the following holds: if we can formally define $A(m), \ldots A(n)$, then we can also formally define $A(n + 1)$.

Next comes an observation which will force us to be a bit cautious. So far mathematical statements are objects which we have not really defined, however we have a naïve but often correct idea of what statements are. Mathematical induction was introduced to prove such (naïve) statements. The situation above i.e. the definition by recursion, is slightly different. The statement we want to prove is that a formal definition is correct, i.e. we need to know what "formally correct" definitions are. Currently, for our course we need not resolve these problems, all we need to know is that sometimes we must be cautious. Those of you who will later study mathematical logic or the foundations of mathematics will read more about this and similar problems.

## Problems

1.  a) Use mathematical induction to prove that for $k \in \mathbb{N} \cup \{0\}$

$$k^3 + (k+1)^3 + (k+2)^3$$

is divisible by 9.

b) Prove by mathematical induction that for every integer $n \geq 0$ the number

$$\frac{n^5}{5} + \frac{n^4}{2} + \frac{n^3}{3} - \frac{n}{30}$$

is an integer.

2. Prove by mathematical induction that

a) for every $x, y \in \mathbb{R}$ and all $n \in \mathbb{N}$ the term $x^n - y^n$ always has $x - y$ as a factor, i.e.

$$x^n - y^n = (x - y)Q_n(x, y)$$

where for $x$ fixed $Q_n(x, y)$ is a polynomial with respect to $y$ and for $y$ fixed $Q_n(x, y)$ is a polynomial with respect to $x$ and $y$.

b) For every $x > 0$ and $y > 0$ and for all $n \in \mathbb{N}$ the following holds

$$(n - 1)x^n + y^n \geq nx^{n-1}y.$$

3. Find the value of each of the following sums:

$$a) \sum_{j=-2}^{2} \frac{1}{2^j}; \quad b) \sum_{k=2}^{5} (a^k - a^{k-2}), \ a \neq 1; \quad c) \sum_{l=1}^{6} (-1)^l \frac{l+1}{l}.$$

4.    a) For $\lambda \in \mathbb{R}$ and $a_1, \ldots, a_N, b_1, \ldots, b_N \in \mathbb{R}$ show that

$$\lambda \sum_{j=1}^{N} a_j = \sum_{j=1}^{N} (\lambda a_j)$$

and

$$\sum_{j=1}^{N} a_j + \sum_{j=1}^{N} b_j = \sum_{j=1}^{N} (a_j + b_j).$$

b) For $x, y \in \mathbb{R}$ simplify

$$(x - y) \sum_{k=0}^{5} x^k y^{5-k}.$$

5. Prove the following identities:

a) $\displaystyle\sum_{k=1}^{n} \frac{1}{(2k-1)(2k+1)} = \frac{n}{2n+1}$;   b) $\displaystyle\sum_{n=1}^{k} n \cdot n! = (k+1)! - 1$;

c) $\displaystyle\sum_{j=1}^{m} (a + (j-1)d) = \frac{1}{2} m(2a + (m-1)d).$

6. Find the value of each of the following products:

a) $\displaystyle\prod_{k=-2}^{2} 2^{-k}$;   b) $\displaystyle\prod_{j=3}^{6} (j-4)$;   c) $\displaystyle\prod_{j=1}^{5} \frac{j+2}{j+4}.$

7. For $\nu, \mu \in \mathbb{R}$ and $a_1, \ldots, a_N \in \mathbb{R}$ show that

$$\prod_{j=1}^{N} (\mu a_j) + \prod_{j=1}^{N} (\nu a_j) = (\mu^N + \nu^N) \prod_{j=1}^{N} a_j.$$

8. Find the value of the following:

a) $7!$  and  $\frac{63!}{60!}$;   b) $\frac{(n+1)! - n!}{n}$;   c) $\frac{(n+1)!}{(n-1)!}.$

9. Prove the following by induction:

$$a) \prod_{k=1}^{n} \frac{2k-1}{2k} = \frac{1}{2^{2n}} \binom{2n}{n}, n \geq 2; \quad b) \prod_{k=1}^{n-1} \left(1 + \frac{1}{k}\right)^k = \frac{n^n}{n!}, n \geq 1.$$

10. Find the binomial expansion of the following:

$$a) (5x^2 + 3y)^4; \quad b) (x-y)^n.$$

11. Prove the following:

a)
$$\binom{n}{k} = \frac{n(n-1) \cdot \ldots \cdot (n-k+1)}{1 \cdot 2 \cdot \ldots \cdot k};$$

b) For $\alpha \in \mathbb{R}$ and $k \in \mathbb{N}$ consider

$$\binom{\alpha}{k} := \frac{\alpha(\alpha-1) \cdot \ldots \cdot (\alpha-k+1)}{1 \cdot 2 \cdot \ldots \cdot k}, \quad \binom{\alpha}{0} := 1,$$

and prove for $k \geq 2$ that

$$\binom{\frac{1}{2}}{k} = (-1)^{k-1} \frac{1 \cdot 3 \cdot \ldots \cdot (2k-3)}{2 \cdot 4 \cdot \ldots \cdot (2k)}.$$

12. Let $p, k \in \mathbb{N}$. Use mathematical induction to prove:

a) $k \geq 1$ and $p \geq 2$ implies $p^k > k$;

b) $k \geq 1$ and $p \geq 3$ implies $p^k > k^2$;

c) for $k \geq 5$ it is true that $2^k > k^2$.

13. Prove the following by induction:

$$a) \sum_{j=1}^{N} \frac{1}{\sqrt{j}} \leq 2\sqrt{N}; \quad b) \prod_{m=1}^{k} (2m)! \geq ((k+1)!)^k.$$

14. Prove the **arithmetic-geometric mean inequality**:
For $k \geq 2$ and $a_1, \ldots, a_k \in \mathbb{R}$, $a_j \geq 0$ where $j = 1, \ldots, k$, the following holds

$$(**) \qquad \sqrt[k]{a_1 \cdot \ldots \cdot a_k} \leq \frac{a_1 + \ldots + a_k}{k}$$

53

or

$$\left(\prod_{j=1}^{k} a_j\right)^{\frac{1}{k}} \le \frac{1}{k}\sum_{j=1}^{k} a_j.$$

Hint: first prove $(**)$ by induction for $n = 2^k$, $k \in \mathbb{N}$. Then for $k \in \mathbb{N}$ choose $k$ such that $n < 2^k$, and with

$$a := \frac{1}{k}\sum_{j=1}^{n} a_j$$

consider $a_1 \cdot \ldots \cdot a_k \cdot a^{2^k - k}$.

15. Define

$$x_n := \frac{1}{2}\left(x_{n-1} + \frac{c}{x_{n-1}}\right), n \in \mathbb{N},$$

with $c > 0$ and $x_0 := 1$. Further set

$$a_n := \frac{c}{x_n}, \ n \in \mathbb{N} \cup \{0\}.$$

Prove $a_n \le a_{n+1} \le x_{n+1} \le x_n$ for $n \ge 1$.

# 4 Functions and Mappings

Let $D \subset \mathbb{R}$, i.e. $D$ is a subset of the real numbers. Often we need to associate with $x \in D$ a new real number which we denote at the moment by $f(x)$.

**Example 4.1. A.** Suppose that a shop offers $n \in \mathbb{N}$ items for sale and we enumerate these items by $1, \cdots, n$, we may then assign a price to each. Thus $D = \{1, \cdots, n\}$ and for $x \in D$ the new number $f(x)$ denotes the price.
**B.** For $x \in D = \mathbb{R}$ we can consider its absolute value $|x|$, i.e. $f(x) = |x|$.
**C.** With $D = \{x \in \mathbb{R} | x \geq 0\}$ we may consider

$$x \longmapsto f(x) = \sqrt{x}$$

i.e. for each $x \in D$ we consider its square root.
**D.** Let $D = \{x \in \mathbb{R} | x \neq 0\} = \mathbb{R} \setminus \{0\}$. With $x \in D$ we may consider its inverse with respect to multiplication, i.e.

$$f(x) = \frac{1}{x} = x^{-1}.$$

Let us agree to the following

**Definition 4.2.** *Let $D \subset \mathbb{R}$. A **function** $f : D \longrightarrow \mathbb{R}$ is a rule which assigns to every $x \in D$ exactly one real value $f(x)$. For this we write $x \mapsto f(x)$ and say that $x$ is mapped onto $f(x)$, or $f(x)$ is the value of $f$ at $x$.*

It is convenient to introduce the notation

$$\begin{aligned} f : D &\longrightarrow \mathbb{R} \\ x &\longmapsto f(x). \end{aligned}$$

Note that when thinking more carefully about the foundations of mathematics this definition causes some problems. However for now it is absolutely sufficient for our purposes. We call $D$ the **domain** of the function $f$, sometimes we write $D(f)$ instead of $D$. Often, if no confusion arises (just as above) we call $f$ a function and omit the domain and the **target set** or **co-domain** $\mathbb{R}$. Two functions $f_j : D_j \to \mathbb{R}, j = 1, 2$, are **equal** if and only if $D_1 = D_2$ and if for all $x \in D_1 = D_2$ we have $f_1(x) = f_2(x)$. Sometimes it is useful to write $f(\cdot)$ instead of $f$.

**Example 4.3. A.** The absolute value is the function

$$|.| : \mathbb{R} \longrightarrow \mathbb{R}$$
$$x \longmapsto |x| \cdot$$

**B.** The square root (function) is given by

$$\sqrt{\cdot} : \mathbb{R}_+ \longrightarrow \mathbb{R}$$
$$x \longmapsto \sqrt{x}$$

where we write $\mathbb{R}_+ = \{x \in \mathbb{R} \mid x \geq 0\}$.
**C.** Consider

$$f_1 : \mathbb{R} \longrightarrow \mathbb{R} \quad \text{and} \quad f_2 : \mathbb{Z} \longrightarrow \mathbb{R}$$
$$x \longmapsto x^2 \qquad \qquad x \longmapsto x^2.$$

Both are functions but they are not equal since $D(f_1) \neq D(f_2)$.
**D.** Let $k \in \mathbb{N}_0 := \mathbb{N} \cup \{0\}$ and $a_0, a_1, \ldots, a_k \in \mathbb{R}$. Then for every $x \in \mathbb{R}$ we can construct a new real number by

$$p(x) := \sum_{j=0}^{k} a_j x^j = a_0 + a_1 x + \ldots + a_k x^k. \tag{4.1}$$

Thus we may define the function

$$p : \mathbb{R} \longrightarrow \mathbb{R}$$
$$x \longmapsto p(x) := \sum_{j=0}^{k} a_j x^j.$$

Functions of this type are called **polynomial functions (on $\mathbb{R}$)** or in short-hand **polynomials**.

We are mainly interested in studying functions defined on some subset of $\mathbb{R}$, often an interval. There is a simple but important way to interpret such a function, namely by considering it as a set of ordered pairs of real numbers:

$$\{(x, f(x)) \mid x \in D\}.$$

Let us first formalise this idea and then we will use it to give a geometric interpretation of a function. For $x \in \mathbb{R}$ and $y \in \mathbb{R}$ we can form the pairs $(x, y)$ and $(y, x)$ where it matters whether $x$ or $y$ is in the first position. The set of all **ordered pairs** of real numbers is called the **Cartesian product** of $\mathbb{R}$ with itself and is denoted by $\mathbb{R} \times \mathbb{R}$ or simply by $\mathbb{R}^2$. Thus $a \in \mathbb{R}^2$ if $a$ is a pair $(x, y)$ of real numbers, $x, y \in \mathbb{R}$. Two pairs $(x_1, y_1)$ and $(x_2, y_2)$ are equal if and only if $x_1 = x_2$ and $y_1 = y_2$. For example $(\sqrt{4}, 1) = (2, \sqrt{1})$ but $(2, 1) \neq (1, 2)$. If $D \subset \mathbb{R}$ we can define a subset of $\mathbb{R}^2$ by

$$D \times \mathbb{R} := \left\{ (x, y) \in \mathbb{R}^2 \, \middle| \, x \in D \text{ and } y \in \mathbb{R} \right\}, \qquad (4.2)$$

and of course this extends easily to $D \subset \mathbb{R}$ and $R \subset \mathbb{R}$:

$$D \times R := \left\{ (x, y) \in \mathbb{R}^2 \, \middle| \, x \in D \text{ and } y \in R \right\}. \qquad (4.3)$$

Now, given a function $f : D \longrightarrow R$ it follows that

$$\left\{ (x, f(x)) \mid x \in D \right\} \subset D \times R \subset \mathbb{R}^2.$$

We call this set the **graph** of $f$ and denote it by $\Gamma(f)$

$$\Gamma(f) := \left\{ (x, f(x)) \mid x \in D \right\}. \qquad (4.4)$$

For a function $f : D \longrightarrow R$ the value at $x$ is the real number $f(x)$ and the graph $\Gamma(f)$ is a subset of the Cartesian product $D \times \mathbb{R}$.

Consider the function $|.| : \mathbb{R} \longrightarrow \mathbb{R}$, $x \longmapsto |x|$. It is defined for all $x \in \mathbb{R}$ but only non-negative real numbers may occur as a value of the function, since $|x| \geq 0$ for $x \in \mathbb{R}$. We introduce the **range** of a function $f : D \longrightarrow \mathbb{R}$ as the set

$$R(f) := \left\{ y \in \mathbb{R} \mid \text{ there exists } x \in D \text{ such that } y = f(x) \right\}. \qquad (4.5)$$

Another way to look at the range of $f$ is to consider it as the image of $D$. In this sense we define the **image** of $D$ under $f$, denoted by $f(D)$, as

$$f(D) = \left\{ y \in \mathbb{R} \mid \text{ exists } x \in D \text{ such that } y = f(x) \right\} = R(f).$$

An important problem is to determine the range of a given function. Let us give a geometrical interpretation of the graph of a function. We have already agreed to interpret a real number $x$ as a point on the real line. Thus it is natural to interpret a pair of real numbers as a point in the plane. The graph $\Gamma(f)$ of a function $f : D \longrightarrow \mathbb{R}$ is the collection of all points $(x, f(x))$ for $x \in D$, thus $\Gamma(f) \subset D \times \mathbb{R}$.

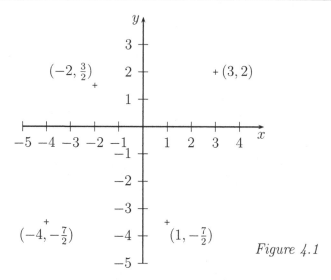

Figure 4.1

Here are some examples with $y = f(x)$ (or $y = g(x), y = h(x)$). In the following figure the function $f$ is the identity on $\mathbb{R}$, $g$ is a parabola, again defined on $\mathbb{R}$, and $h$ is the square root function which is of course only defined on $\mathbb{R}_+ = \{x \in \mathbb{R} \mid x \geq 0\}$.

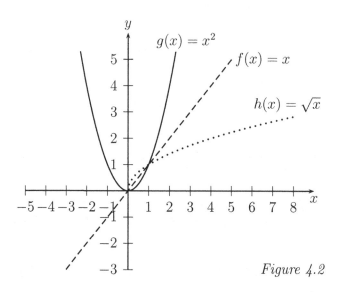

Figure 4.2

In Figure 4.3 the function $f$ is the absolute value with domain $\mathbb{R}$, the function $g$ is a hyperbola defined on $\mathbb{R} \setminus \{0\}$, and $h$ is a cubic polynomial with domain $\mathbb{R}$.

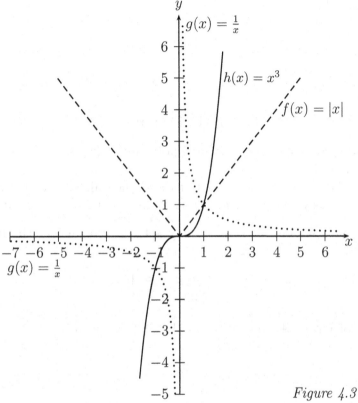

Figure 4.3

It is likely you have already seen these graphs before, but there is a non-trivial question: how do we know that they are correct? A typical domain $D(f)$ contains infinitely many points. We cannot calculate all values $f(x)$. Thus before we can draw the graph we need to understand and discuss the function $f : D \longrightarrow \mathbb{R}$ and its behaviour. The following are natural questions:

- are there lower and upper bounds?

- are there local or global extreme values, i.e. maxima or minima?

- is the function monotone?

- is the graph connected?

  $\vdots$

The last question arises when looking at $f : \mathbb{R} \setminus \{0\} \longrightarrow \mathbb{R}, x \longmapsto \frac{1}{x}$. The graph $\Gamma(f)$ has the two components $\Gamma_+(f) = \left\{ \left(x, \frac{1}{x}\right) \middle| x > 0 \right\}$ and $\Gamma_-(f) = \left\{ \left(x, \frac{1}{x}\right) \middle| x < 0 \right\}$, i.e.

$$\Gamma(f) = \Gamma_+(f) \cup \Gamma_-(f) \tag{4.6}$$

and in addition

$$\Gamma_+(f) \cap \Gamma_-(f) = \emptyset. \tag{4.7}$$

Thus it is not possible to get from a point in $\Gamma_+(f)$ to a point in $\Gamma_-(f)$ while staying in $\Gamma(f)$.

So far we "know" only a few functions and they all look very "nice", i.e. "smooth" and easy to deal with. Here are a few not so nice candidates:

**Example 4.4. A.** Let $A \subset \mathbb{R}$ be any set, its **characteristic function** $\chi_A : \mathbb{R} \longrightarrow \mathbb{R}, x \longmapsto \chi_A(x)$, is defined by

$$\chi_A(x) = \begin{cases} 1, & x \in A \\ 0, & x \notin A \end{cases} \tag{4.8}$$

The graph of $\chi_A$ for $A = [-1, -\frac{1}{2}] \cup \{\frac{1}{2}\} \cup [1, 2]$ is given by

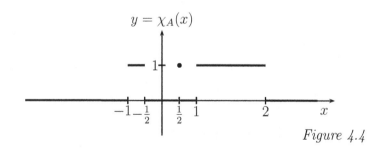

*Figure 4.4*

For $A = \mathbb{Q}$ we get the **Dirichlet function** $\chi_\mathbb{Q} : \mathbb{R} \longrightarrow \mathbb{R}$

$$\chi_\mathbb{Q}(x) = \begin{cases} 1, & x \in \mathbb{Q} \\ 0, & x \in \mathbb{R} \setminus \mathbb{Q} \end{cases} \tag{4.9}$$

however it is not possible to draw this graph.

**B.** The **entier-function** is given by $x \longmapsto [x]$, i.e. $[.] : \mathbb{R} \longrightarrow \mathbb{R}, x \longmapsto [x]$, with $[x]$ being the largest integer less or equal than $x$. Thus $[1] = 1$, in general $[k] = k$ for $k \in \mathbb{Z}$, but $[\frac{1}{2}] = 0$, $[-\frac{1}{2}] = -1$ etc. Note that we always have $x - [x] \in [0, 1)$. Here is the graph of $[x]$ and $x - [x]$:

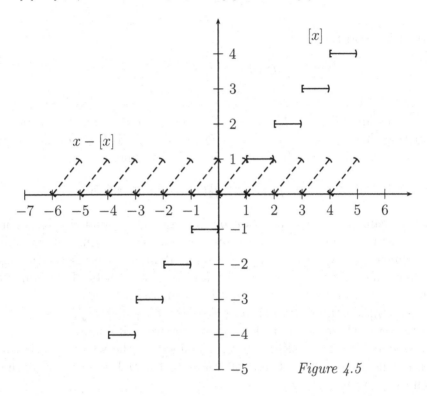

*Figure 4.5*

where [ indicates that the left end point is included and ) indicates that the right point is not included. In addition let us consider the new function $f : \mathbb{R} \longrightarrow \mathbb{R}$, $x \longmapsto x - [x]$. Its graph looks **periodic** with period 1. This means that $f(x + 1) = f(x)$ for all $x \in \mathbb{R}$. Thus for a general function we may ask whether it is periodic.

There are some simple procedures to construct new functions from given ones. Let $f_1, f_2 : D \longrightarrow \mathbb{R}$ be given functions. Note that they have the same domain. We define
i) their **sum** by

$$f_1 + f_2 : D \longrightarrow \mathbb{R}$$
$$x \longmapsto (f_1 + f_2)(x) := f_1(x) + f_2(x) \tag{4.10}$$

61

ii) their **difference** by

$$
\begin{aligned}
f_1 - f_2 : D &\longrightarrow \mathbb{R} \\
x &\longmapsto (f_1 - f_2)(x) := f_1(x) - f_2(x)
\end{aligned}
\tag{4.11}
$$

iii) their **product** by

$$
\begin{aligned}
f_1 \cdot f_2 : D &\longrightarrow \mathbb{R} \\
x &\longmapsto (f_1 \cdot f_2)(x) := f_1(x) \cdot f_2(x).
\end{aligned}
\tag{4.12}
$$

The constant function $f_c : \mathbb{R} \longrightarrow \mathbb{R}$, $x \longmapsto c$, $c \in \mathbb{R}$ fixed, is a polynomial function therefore we have already encountered this function. In particular, when taking $f_c$ for the function $f_1$ in (4.12) we find

$$
(f_c \cdot f_2)(x) = c\, f_2(x) \quad \text{for all } x \in \mathbb{R},
\tag{4.13}
$$

i.e. we can form a new function by multiplying it pointwise by a constant. However this argument is not quite correct: The product of two functions is defined only when they have the same domain. We resolve this problem by introducing the restriction of a function to subsets of its domain. Let $D_1 \subset D \subset \mathbb{R}$ and let $f : D \longrightarrow \mathbb{R}$ be a function. We call $f|_{D_1} : D_1 \longrightarrow \mathbb{R}$, $x \longmapsto f|_{D_1}(x)$, the **restriction** of $f$ to $D_1$ if $f|_{D_1}(x) = f(x)$ for all $x \in D_1$. In the case where no confusion may occur we write simply $f|_{D_1}$ or even $f$. Now, since $f_c : \mathbb{R} \longrightarrow \mathbb{R}$, $x \longmapsto c$, is defined on the whole real line we may restrict it to any subset $D$ and therefore (4.13) makes sense for all functions defined on some $D \subset \mathbb{R}$.

A problem, in fact a more serious one than one may think at the beginning, is to define the quotient of two functions $f_1, f_2 : D \longmapsto \mathbb{R}$. The idea is to define

$$
\left( \frac{f_1}{f_2} \right)(x) = \frac{f_1(x)}{f_2(x)} \quad \text{for } x \in D.
\tag{4.14}
$$

However this does not make sense for $f_2(x) = 0$. We either have to assume $f_2(x) \neq 0$ for all $x \in D$ or we can only define $\frac{f_1}{f_2}$ on $D_q = \{x \in D | f_2(x) \neq 0\}$. In fact the situation is more delicate if we look at the simple case where $f_1, f_2 : \mathbb{R} \longmapsto \mathbb{R}$, $f_1(x) = x$ and $f_2(x) = x$ for all $x \in \mathbb{R}$. Of course $\frac{f_1(x)}{f_2(x)} = 1$ for all $x \neq 0$ but we would like to extend this so that it also holds for $x = 0$. Thus a further problem to study is: when does a given function $f : D \longrightarrow \mathbb{R}$ have an extension to a larger domain $D_1$, $D \subset D_1 \subset \mathbb{R}$? We

call $f_1 : D_1 \longrightarrow \mathbb{R}$ an **extension** of $f : D \longrightarrow \mathbb{R}$ if $D \subset D_1$ and $f_1|_D = f$.
In our example $\frac{f_1}{f_2}$ is only defined on $\mathbb{R} \setminus \{0\}$, but we may extend $\frac{f_1}{f_2}$ to $\mathbb{R}$ by
defining $\left(\frac{f_1}{f_2}\right)(0) := 1$. Again here a new problem arises. There is nothing
to stop us defining $\left(\frac{f_1}{f_2}\right)(0) = 2$ or $\left(\frac{f_1}{f_2}\right)(0) = q$, $q$ being any real number.
In each case we get a function extending $\frac{f_1}{f_2}$ with domain $D\left(\frac{f_1}{f_2}\right) = \mathbb{R} \setminus \{0\}$
to a function with domain $\mathbb{R}$. Extensions are not unique. Thus we may add
conditions to achieve uniqueness. In the above example it is natural to define
$\left(\frac{f_1}{f_2}\right)(0)$ by 1. We long for some criteria providing us with some help to find
natural extensions.

Consider two polynomial functions $p : \mathbb{R} \mapsto \mathbb{R}, x \mapsto \sum_{j=0}^{k} a_j x^j$, and $q : \mathbb{R} \mapsto$
$\mathbb{R}, x \mapsto \sum_{i=0}^{l} b_i x^i$. We can easily define their sum $p + q$, their difference $p - q$
and their product $p \cdot q$. By easily we mean that we can rely on (4.10), (4.11)
and (4.12). In Problem 11 we will show that $p + q$, $p - q$ and $p \cdot q$ are also
polynomials. We will determine their coefficients, i.e. each of the functions
is of the type $x \mapsto \sum_{r=0}^{m} c_r x^r$ where $m$ is determined by $k$ and $l$, whereas the
coefficients $c_r, 0 \leq r \leq m$, are determined by the numbers $a_j, 0 \leq j \leq k$, and
$b_i, 0 \leq i \leq l$, and of course in each case they are different. However since
$q(x)$ might be zero, we have a problem to define the quotient $x \mapsto \frac{p(x)}{q(x)}$.

Thus when discussing functions, the set of their zeroes, or more generally the
set of their $a$−points, i.e. the set $\{x \in D | f(x) = a\}$ is also of importance.
The set of all functions $h : D_h \longrightarrow \mathbb{R}, h(x) = \frac{p(x)}{q(x)}$ for two polynomials $p$ and
$q$, with $D_h := \{x \in \mathbb{R} | q(x) \neq 0\}$ is called the set of all **rational functions**.
Since $q(x) = 1$ for all $x \in \mathbb{R}$ is a polynomial, all polynomials are rational
functions.

Note that we can add polynomials, but in general we cannot add two rational
functions: they might have different domains. However, if $h_1 : D(h_1) \longrightarrow \mathbb{R}$
and $h_2 : D(h_2) \longrightarrow \mathbb{R}$ are two rational functions then

$$h_1|_{D(h_1) \cap D(h_2)} + h_2|_{D(h_1) \cap D(h_2)}$$

is always defined. The same type of argument holds for the difference and
the product of two rational functions, and with the obvious extension in each
case for finitely many ones.

Now look at $p(x) = (x - 1)^2$ and $q(x) = (x - 1)$. Both are polynomials,

$\{x \in \mathbb{R} \mid q(x) = 0\} = \{1\}$. Thus we may define their quotient on $\mathbb{R} \setminus \{1\}$ by

$$\left(\frac{p}{q}\right)(x) = \frac{(x-1)^2}{x-1} = x - 1. \tag{4.15}$$

Obviously we can extend $\frac{p}{q} : \mathbb{R} \setminus \{1\} \mapsto \mathbb{R}$ to $\mathbb{R}$ just by defining $\left(\frac{p}{q}\right)(1) = 0$.
Hence the domain $D(\frac{p}{q}) = \mathbb{R} \setminus \{x \in \mathbb{R} \mid q(x) = 0\}$ does not have to be the natural one for the quotient $\frac{p}{q}$. We will return to this problem later. By being careful with domains we can even define the quotient of two rational functions:

$$\frac{\frac{p_1}{q_1}}{\frac{p_2}{q_2}} = \frac{p_1 q_2}{p_2 q_1}, \tag{4.16}$$

but note that the left hand side requires $q_1(x) \neq 0, p_2(x) \neq 0$ and $q_2(x) \neq 0$, whereas for the right hand side we need at most that $p_2(x)q_1(x) \neq 0$.

We want to extend the idea of a function to arbitrary sets $X$ and $Y$, $X \neq \emptyset$ and $Y \neq \emptyset$. We start by transforming our old definition:
A **mapping** $f : X \mapsto Y, x \mapsto f(x)$, is a rule which associates to every $x \in X$ one and only one $y := f(x) \in Y$. Sometimes we also write

$$f : X \mapsto Y$$
$$x \mapsto f(x)$$

For example we may take $X$ as the set of all bounded open intervals, i.e. $X := \{(a,b) \mid a < b \text{ and } a, b \in \mathbb{R}\}$, and for $Y$ we may take the non-negative real numbers, i.e. $Y = \mathbb{R}_+$. Now we may define $\lambda : X \to \mathbb{R}_+, \lambda((a,b)) = b-a$. Thus the mapping $\lambda$ maps every bounded open interval $(a,b) \subset \mathbb{R}$ onto its length $b - a \in \mathbb{R}$. Another example is the following: Take $X = \mathbb{R}_+$ and $Y$ to be the set of all closed intervals $[0, a], a > 0$, i.e. $Y := \{[0, a] \mid a > 0\}$. Then $f : X \to Y, a \mapsto [0, a]$ is a mapping.

However, as we have already pointed out in the case of functions, the term "rule" is not well defined. Thus we try something else taking into account our experience with functions.
Consider the **Cartesian product** $X \times Y$, i.e.

$$X \times Y := \{(x, y) \mid x \in X \text{ and } y \in Y\}. \tag{4.17}$$

A subset $\mathcal{R} \subset X \times Y$ is called a **relation** of elements in $X$ and $Y$. For example with $X = Y = \mathbb{Z}$ we may look at:

$\mathcal{R}_1 := \{(k, k^2) \mid k \in \mathbb{Z}\} \subset \mathbb{Z} \times \mathbb{Z}$, with graphical representation:

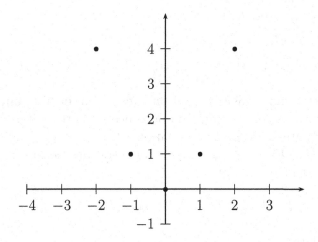

Figure 4.6

or $\mathcal{R}_2 := \{(k, m) \mid k \in \mathbb{Z}, m \in \mathbb{Z}, |m| = |k|\}$, with graphical representation:

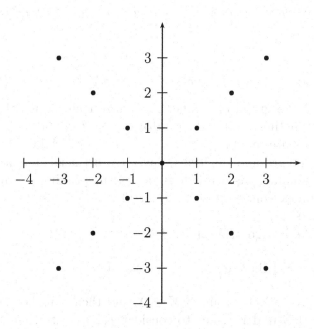

Figure 4.7

65

Given a relation $\mathcal{R} \subset X \times Y$ we sometimes write $x\mathcal{R}y$ instead of $(x, y) \in \mathcal{R}$. A relation $\mathcal{R} \subset X \times X$ is called:

**reflexive** if $x\mathcal{R}x$, i.e. $(x, x) \in \mathcal{R}$, for all $x \in X$;

**symmetric** if $x\mathcal{R}y$ and $y\mathcal{R}x$, i.e. $(x, y) \in \mathcal{R}$ and $(y, x) \in \mathcal{R}$, for all $x, y \in X$;

**transitive** if $x\mathcal{R}y$ and $y\mathcal{R}z$ implies $x\mathcal{R}z$, i.e. if $(x, y) \in \mathcal{R}$ and $(y, z) \in \mathcal{R}$ implies $(x, z) \in \mathcal{R}$.

A reflexive, symmetric and transitive relation is called an **equivalence relation** and these relations are of central importance in mathematics. Often we write "$\sim$" to indicate an equivalence relation.

The relation $R = \{(k, -k) \mid k \in \mathbb{Z}\} \subset \mathbb{Z}$ is symmetric but neither reflexive nor transitive. The identity relation

$$\mathcal{R} = \{(x, x) \mid x \in X\} \subset X \times X$$

is an equivalence relation.

**Definition 4.5.** *A **mapping** $f : X \mapsto Y$ is a relation $\mathcal{R}_f \subset X \times Y$ such that for every $x \in X$ there exists exactly one $y \in Y$ such that $x\mathcal{R}_f y$. We write $y := f(x)$.*

In other words

$$\mathcal{R}_f = \{(x, f(x)) \mid x \in X\} \subset X \times Y$$

is a generalisation of the graph of a function. Once again $X$ is called the **domain** of $f$, $Y$ is sometimes called the **co-domain** of $f$ or the **target set**. In this sense functions are mappings $f : D \to \mathbb{R}$, $D \subset \mathbb{R}$. Making this distinction between mappings and functions may seem artificial, but it might be helpful in the beginning. As a rough guide, when speaking about functions we mean mappings from some set to the real numbers (or the complex numbers in later parts).

The **range** of $f$ or the **image** of $X$ under $f$ is

$$R(f) = \{y \in Y \mid \text{there exists } x \in X \text{ such that } y = f(x)\}.$$

We may restrict $f : X \to Y$ to a subset $Z \subset X$ and then it makes sense to speak of the image of $Z$ under $f$, i.e. to consider $f(Z)$. Note that $f(Z)$ is always a subset of $Y$, not an element of $Y$. Some results for the image are obvious, for example $Z_1 \subset Z_2 \subset X$ implies $f(Z_1) \subset f(Z_2) \subset R(f) = f(X)$.

**Example 4.6. A.** Let $f : \mathbb{R} \to \mathbb{R}$, $x \mapsto x^2$. Then we have

$$f([-2, -1] \cup [1, 2]) = f([-2, -1]) = f([1, 2]) = [1, 4].$$

**B.** Let $A \subset \mathbb{R}$ and $\chi_A : \mathbb{R} \to \mathbb{R}$. We find $\chi_A(B) = \{1\}$ for every $B \subset A$, $\chi_A(C) = \{0\}$ for every $C \subset \mathbb{R}$ such that $C^{\complement} \subset A$.

The **graph** of $f$ is again denoted by $\Gamma(f)$ which is of course $R_f \subset X \times Y$. Clearly every function $f : D \mapsto \mathbb{R}$, $D \subset \mathbb{R}$, is a mapping in this new sense. Let $f : X \to Y$ be a mapping and let $B \subset Y$. The **pre-image** of $B$ is the set

$$f^{-1}(B) := \{x \in X | f(x) \in B\}. \tag{4.18}$$

**Example 4.7. A.** For the parabola $f : \mathbb{R} \to \mathbb{R}, x \mapsto x^2$, we find for $y > 0$ that $f^{-1}(\{y\}) = \{-\sqrt{y}, \sqrt{y}\}$, for $y = 0$ we have $f^{-1}(\{0\}) = 0$, and for $y < 0$ we have that $f^{-1}(\{y\}) = \emptyset$.
**B.** Consider the function $f : \mathbb{R} \to \mathbb{R}, x \mapsto x - [x]$, compare with Example 4.4.B. If $B \subset \mathbb{R} \setminus [0, 1)$ then $f^{-1}(B) = \emptyset$, however for every $y \in [0, 1)$ the pre-image $f^{-1}(\{y\})$ consists of infinitely many points. Indeed for $y \in [0, 1)$ we have

$$f^{-1}(\{y\}) = \{x = y + k | k \in \mathbb{Z}]\}.$$

This is typical behaviour of a periodic function.
**C.** Let $A \subset \mathbb{R}$ be a non-empty set and let $\chi_A$ be the characteristic function of $A$. Then the following hold: $\chi_A^{-1}(\{1\}) = A$; $\chi_A^{-1}(\{0\}) = A^{\complement}$; $\chi_A^{-1}(\{y\}) = \emptyset$ if $y \notin \{0, 1\}$.

Before returning to functions $f : D \mapsto \mathbb{R}$, $D \subset \mathbb{R}$, we want to discuss a further new idea: the power set. Let $X$ be a set. Its **power set** $\mathcal{P}(X)$ is by definition the set of all subsets of $X$ i.e.

$$\mathcal{P}(X) := \{Y \mid Y \subset X\}$$

where we understand that $\emptyset \subset X$ for every set and $X \subset X$. If $X = \{1, 2\}$ then $\mathcal{P}(X) = \{\emptyset, \{1\}, \{2\}, \{1, 2\}\}$.

**Note: elements of $\mathcal{P}(X)$ are sets.**
We may ask the following question: consider $\mathcal{P}(\mathbb{R}^2)$, the power set of $\mathbb{R}^2$, i.e. all subsets of the plane. Can we define for every $A \in \mathcal{P}(\mathbb{R}^2)$, i.e for every subset $A \subset \mathbb{R}^2$, area in a reasonable way? Thus we are looking for a mapping $\mu : \mathcal{P}(\mathbb{R}^2) \mapsto [0, \infty]$, $A \mapsto \mu(A)$, where $\mu(A)$ is the area of $A$. We will see that

this is not possible if we want to maintain basic properties of area. However, this example indicates that mappings might be defined on families of sets but we should not be afraid of working with such mappings.

# Problems

1.    a) Find the product sets $A \times B$ and $B \times A$ for $A = \{3, 4, 5, 6\}$ and $B = \{1, 2, 3\}$ and sketch the set in the plane.

   b) Prove that $\mathbb{N} \times \mathbb{Z} \subset \mathbb{R} \times \mathbb{Q}$.

   c) Let $X = \{1, 2, 3\}$, $Y = \{3, 4, 5\}$, and $Z = \{6, 7\}$.
   Find $(X \cup Y) \times Z$, $X \times (Y \cup Z)$ and $(X \times Z) \cap (Y \times Z)$.

2. For the sets $A, B, C$ and $D$ prove that: a) $(A \cup B) \times C = (A \times C) \cup (B \times C)$ and b) $(A \times B) \cap (C \times D) = (A \cap C) \times (B \cap D)$.

3. For the sets $X, Y$ and $X', Y'$ show that $X' \times Y' \subset X \times Y$ if and only if $X' \subset X$ and $Y' \subset Y$.

4. Sketch the sets

$$\bigcup_{j=1}^{5} (\{j\} \times I_j) \quad \text{and} \quad \bigcup_{j=1}^{5} (I_j \times \{j\})$$

   for $I_j = [j, j+1] \subset \mathbb{R}$, i.e. $I_j = \{x \in \mathbb{R} \mid j \leq x \leq j+1\}$.

5. Let $p \in \mathbb{N}$ be fixed and consider on $\mathbb{Z}$ the following relation: $m\mathcal{R}_p n$ if $m - n$ is divisible by $p$. For this we should use the more commonly used notation

$$m \equiv n \, \mathrm{mod}(p)$$

   (this reads as: $m$ is congruent to $n$ modulo $p$). The reader has probably already seen this relation in an algebra course. Prove that $m \equiv n \, \mathrm{mod}(p)$ is an equivalence relation on $\mathbb{Z}$.

6. Consider the set $\mathbb{Z} \times \mathbb{N}$ and define a relation on $\mathbb{Z} \times \mathbb{N}$ by $(k, m) \sim (l, n)$ if and only if $nk = lm$. Prove that "$\sim$" is an equivalence relation on $\mathbb{Z} \times \mathbb{N}$.

7. Find the power set of: a) the empty set $\phi$; b) the set $\{1, 2, 3\}$.

8. Let $X$ be a set with $N$ elements, Prove that the power set $\mathcal{P}(X)$ of $X$ has $2^N$ elements. Use the fact that the number of subsets of $k$ elements of a set with $N$ elements is $\binom{N}{k}$.

9. Consider the following rule: $x \in \mathbb{R}$ is mapped onto the solution of the quadratic equation $y^2 - 2y + x = 0$. Does this rule define a function on $\mathbb{R}$?

10. Let $p, q : \mathbb{R} \to \mathbb{R}$ be two polynomials. Prove that $p + q$ and $p \cdot q$ are also polynomials.

11. We call a polynomial even if it is of the form

$$p(x) = \sum_{j=0}^{n} a_{2j} x^{2j}, \ a_{2n} \neq 0.$$

   a) Show that $p$ is a polynomial of degree $2n$ and has the unique representation

$$p(x) = \sum_{l=0}^{2n} b_l x^l.$$

   b) Define the function $f : \mathbb{R} \longrightarrow \mathbb{R}$ by

$$x \mapsto f(x) := \sum_{j=0}^{n} a_{2j} |x|^{2j}.$$

Prove that $p = f$ (as functions).

   c) Determine the largest set $D \subset \mathbb{R}$ where $g : \mathbb{R} \longrightarrow \mathbb{R}, \ x \mapsto x^3$, and $h : \mathbb{R} \longrightarrow \mathbb{R}, \ x \mapsto |x|^3$, coincide, i.e. $g|_D = h|_D$.

12. For each of the following rational expressions $q(x)$ find the largest set $D \subset \mathbb{R}$ such that $q : D \longrightarrow \mathbb{R}$ is a well defined function. Where appropriate try to extend $q : D \longrightarrow \mathbb{R}$ to a larger domain in a meaningful way by modifying $q$.

   a) $q_1(x) = \frac{x^3 - 5x^2 - 17}{x^2 + 7}$    b) $q_2(x) = \frac{(x-3)^2 (2x+7)^5}{(x-3)(x+4)(2x+7)^8}$    c) $q_3(x) = \frac{x^2 - x - 12}{(x-4)(x+2)}.$

13. (i) Let $f : X \longrightarrow Y$ be a mapping. For pre-images prove:

 a) $f^{-1}(A \cap B) = f^{-1}(A) \cap f^{-1}(B)$, $A, B \subset Y$;

 b) $f^{-1}(A \cup B) = f^{-1}(A) \cup f^{-1}(B)$, $A, B \subset Y$.

(ii) Let $f : X \longrightarrow Y$ be a mapping and $A, B \subset X$.
For the image prove:

 a) $f(A \cap B) \subset f(A) \cap f(B)$;

 b) $f(A \cup B) = f(A) \cup f(B)$;

 c) $f(\{x\}) = \{f(x)\}$ for $x \in X$.

14. (i) In each of the following cases find the pre-images:

 a) $f : \mathbb{R} \longrightarrow \mathbb{R}$, $x \mapsto x^2 + 1$, find $f^{-1}(\{y\})$ for $y \in \mathbb{R}$;

 b) $g : \mathbb{R} \setminus \{0\} \longrightarrow \mathbb{R}$, $x \mapsto \frac{1}{x}$, find $g^{-1}(\{z\})$ for $z \in \mathbb{R}$;

 c) $h : \mathbb{R} \longrightarrow \mathbb{R}$, $x \mapsto \frac{1}{2}x + 3$, find $h^{-1}((a, b))$ for $(a, b) \subset \mathbb{R}$, $a < b$.

(ii) In each of the following cases find the image of the indicated set:

 a) $f : [0, \infty) \longrightarrow \mathbb{R}$, $x \mapsto \sqrt{x}$, find $f\left(\left[\frac{1}{4}, 9\right]\right)$;

 b) $g : \mathbb{R} \longrightarrow \mathbb{R}$, $x \mapsto \frac{x^2 - 1}{x^2 + 2}$, find $g(\{1, 2, 3, 4\})$;

 c) $h : \mathbb{R} \longrightarrow \mathbb{R}$, $x \mapsto 2^x$, find $h(\mathbb{N})$.

# 5 Functions and Mappings Continued

We continue our considerations on mappings $f : X \mapsto Y$ between two sets $X$ and $Y$. We may consider functions $f : D \mapsto F$, with $D, F \subset \mathbb{R}$ instead of functions $f : D \mapsto \mathbb{R}$. While we can always restrict a given function $f : D \mapsto \mathbb{R}$ or $f : D \mapsto F$ to a subset $D_1 \subset D$, we cannot in general easily restrict the target set or co-domain $F$. For example $f : \mathbb{R} \longrightarrow \mathbb{R}, x \longmapsto x+2$, is a well defined function. However, when shrinking the target set to $[0, 2]$, then $f : \mathbb{R} \longrightarrow [0, 2], x \longmapsto x+2$, does not define a function since for example for $x = 5 \in \mathbb{R}$ the "value" $f(5) = 7$ does not belong to the co-domain $[0, 2]$. However, if we restrict $f$ to the set $[-2, 0]$ then $f|_{[-2,0]} : [-2, 0] \longrightarrow [0, 2]$ is of course once again a function. Nonetheless, for reasons which will become clear later, it makes sense to consider functions with co-domains different to $\mathbb{R}$.

**Definition 5.1.** *Let $f : D \longrightarrow F, D, F \subset \mathbb{R}$, be a function.* **A.** *We call $f$* ***injective*** *or* ***one-to-one*** *if for $x, y \in D$, $x \neq y$, it follows that $f(x) \neq f(y)$.* **B.** *We call $f$* ***surjective*** *or* ***onto*** *if for every $y \in F$ there exists $x \in D$ such that $f(x) = y$.* **C.** *If $f$ is injective and surjective we call $f$* ***bijective****.*

**Remark 5.2. A.** Obviously we can extend the definition of injectivity, surjectivity and bijectivity to general mappings. A mapping $f : X \to Y$ is **injective** if $x_1 \neq x_2$, $x_1, x_2 \in X$, implies $f(x_1) \neq f(x_2)$. The mapping $f$ is **surjective** if for every $y \in Y$ there exists $x \in X$ such that $f(x) = y$. If $f$ is injective and surjective then it is **bijective**.
**B.** A mapping $f : X \to Y$ (or a function $f : D \to F, D, F \subset \mathbb{R}$) is surjective if and only if $R(f) = Y$ (or $R(f) = F$). There is an easy way to make every mapping surjective: shrink the co-domain to the range. This is formally correct but of course in general we do not know $R(f)$ explicitly.

**Example 5.3. A.** Consider the function $f_1 : \mathbb{R} \longrightarrow \mathbb{R}, x \longmapsto x^2$. Since for $x \neq -x$, i.e. $x \neq 0$, it follows that $x^2 = (-x)^2 = x^2$, the function $f_1$ is not injective. Moreover, since $x^2 \geq 0$ for all $x \in \mathbb{R}$, any negative number does not belong to the range of $f_1$, hence $f_1$ is not surjective. However, the function $\tilde{f}_1 : \mathbb{R} \longrightarrow \mathbb{R}_+, x \longmapsto x^2$ is surjective: given $y \geq 0$ there exists a unique $x_y \geq 0$ such that $x_y^2 = y$, namely $x_y = \sqrt{y}$. But $\tilde{f}_1$ is still not injective since $\tilde{f}_1(x_y) = \tilde{f}_1(-x_y)$. Now, we may also reduce the domain of $\tilde{f}_1$ and consider $f_1^* : \mathbb{R}_+ \longrightarrow \mathbb{R}_+, x \longmapsto x^2$. We know that for $y \geq 0$ there exists a unique $x_y = \sqrt{y}$ such that $x_y^2 = y$ and $x_y \geq 0$, hence $f_1^*$ is injective and surjective,

71

i.e. bijective. This example shows the importance of the domain and the range of a function when deciding about its injectivity and its surjectivity, respectively.

**B.** For $a \in \mathbb{R}$ consider $f_a : \mathbb{R} \longrightarrow \mathbb{R}$, $x \longmapsto x + a$. We claim that $f_a$ is always bijective. First, for $x \neq y$ it follows that $f_a(x) = x + a \neq y + a = f_a(y)$, and secondly, given $z \in \mathbb{R}$ the equation $f_a(x) = z$, i.e. $x + a = z$, has the (unique) solution $x = z - a$ and it follows that $f_a(z - a) = (z - a) + a = z$, i.e. $f_a$ is surjective. The last calculation shows what "determine whether $f : D \longrightarrow F$ is surjective" or "find the range $R(f)$" really means: we have to solve the equation $f(x) = y$ for all $y \in F$ such that the solution belongs to $D$.

**C.** For $a \neq 0$ the function $g_a : \mathbb{R} \longrightarrow \mathbb{R}$, $x \longmapsto ax$ is bijective. First note that for $x, z \in \mathbb{R}$, $x \neq z$, it follows that $ax \neq az$, i.e. $g_a$ is injective. To show that $g_a$ is surjective we have to solve for all $y \in \mathbb{R}$ the equation $g_a(x) = y$, i.e. $ax = y$. Clearly the solution is $x = \frac{y}{a}$ provided $a \neq 0$. Thus $g_a$, $a \neq 0$, is bijective.

Note that in the case where $a = 0$ the function $g_0$ is the constant function $g_0 : \mathbb{R} \longrightarrow \mathbb{R}$, $x \longmapsto 0$. This function is neither injective nor surjective. In fact for every $c \in \mathbb{R}$ the constant function $h_c : \mathbb{R} \longrightarrow \mathbb{R}$, $x \longmapsto c$, i.e. $h_c(x) = c$ for all $x \in \mathbb{R}$, is neither injective nor surjective.

**D.** Let $A \subset \mathbb{R}$ be a set and consider $\chi_A : \mathbb{R} \longrightarrow [0, 1]$, the characteristic function of the set $A$. For all $x \in A$ this function is equal to 1, and for all $x \in A^{\complement}$ it has the value 0. Thus it is neither injective nor surjective: it is not injective since either $A$ or $A^{\complement}$ has at least two elements and they are mapped by $\chi_A$ onto the same value. In addition for $\frac{1}{2} \in [0, 1]$ there is no $x \in \mathbb{R}$ such that $\chi_a(x) = \frac{1}{2}$, therefore it is not surjective.

**E.** The absolute value $|.| : \mathbb{R} \longrightarrow \mathbb{R}_+$ is surjective but not injective. Indeed, for $x \neq -x$, i.e. $x \neq 0$, we know that $|x| = |-x|$, i.e. $|\cdot|$ is not injective. On the other hand, for $y \geq 0$ we may take $x = y$ to find $|x| = y$, showing surjectivity.

Next we meet some examples considering general mappings.

**Example 5.4. A.** Let $X = \mathbb{N}$ and $Y = \mathbb{R}$. We consider mappings $f : \mathbb{N} \to \mathbb{R}$, $n \mapsto f(n)$. Such mappings are called **sequences of real numbers** and it is convenient to write $(f(n))_{n \in \mathbb{N}}$ for such a sequence. Later on we will just start with a sequence $(a_n)_{n \in \mathbb{N}}$, $a_n \in \mathbb{N}$, suppressing often that we are working with a mapping, i.e. that $a_n = f(n)$ for some $f : \mathbb{N} \to \mathbb{R}$. Now the question arises whether a mapping $f : \mathbb{N} \to \mathbb{R}$ can be surjective. The answer is no, a proof will be given later, see Theorem 18.35.

**B.** Let $X = \mathbb{R}^2 = \mathbb{R} \times \mathbb{R}$ and $Y = \mathbb{R}$. We define the two **coordinate projections** $pr_1 : \mathbb{R}^2 \mapsto \mathbb{R}, pr_1(x) = x_1$, and $pr_2 : \mathbb{R}^2 \mapsto \mathbb{R}, pr_2 = x_2$, where $x = (x_1, x_2) \in \mathbb{R}^2$. Both projections are surjective.

We give a proof for $pr_1$: given $x_1 \in \mathbb{R}$ we need to find $y = (y_1, y_2) \in \mathbb{R}^2$ such that $pr_1(y) = x_1$. Any pair $y = (x_1, y_2), y_2 \in \mathbb{R}$, will do. However, $pr_1$ and $pr_2$ are not injective. Again, we only deal with $pr_1$ and consider $x = (x_1, x_2)$ and $y = (x_1, y_2)$ with $x_2 \neq y_2$. Then $x \neq y$ but $pr_1(x) = x_1 = pr_1(y)$.

Consider now two functions $f_1 : D_1 \longrightarrow F_1$ and $f_2 : D_2 \longrightarrow F_2$. Suppose in addition that $R(f_1) = D_2$. Given $x \in D_1$ then $f_1(x) \in R(f_1) = D_2$. Hence we may apply $f_2$ to $f_1(x)$, i.e. we may form $f_2(f_1(x))$.

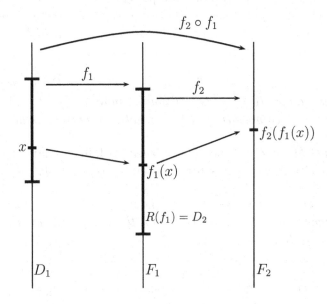

*Figure 5.1*

Thus we have defined a new function from $D_1$ to $F_2$:

**Definition 5.5.** *Let $f_1 : D_1 \longrightarrow F_1$ and $f_2 : D_2 \longrightarrow F_2$ be two functions such that $R(f_1) = D_2$. The function $g : D_1 \longrightarrow F_2$ defined by $g(x) = f_2(f_1(x))$ is called the* **composition** *of $f_1$ with $f_2$ and is denoted by $f_2 \circ f_1$.*

**Remark 5.6.** Once again we can extend our considerations to general mappings $f : X \to Y$ and $g : Y \to Z$. If $R(f) = Y$ then we may define the **composition** $h := g \circ f : X \to Z$ by $h(x) = g(f(x))$.

73

**Remark 5.7. A.** Note that in the case where $f_2 \circ f_1$ is well defined $f_1 \circ f_2$ need **not** be defined. For example take $f_1 : \mathbb{R}_+ \longrightarrow \mathbb{R}$, $x \longmapsto \sqrt{x}$ and $f_2 : \mathbb{R}_+ \longrightarrow \mathbb{R}$, $x \longmapsto -x$. Then $(f_2 \circ f_1)(x) = -\sqrt{x}$. But since $f_2(x) \leq 0$ for all $x \in \mathbb{R}_+$ we cannot apply $f_1$ to $f_2(x)$ for $x > 0$.
**B.** Suppose that $f_1 : \mathbb{R}_+ \longrightarrow \mathbb{R}_+$ and $f_2 : \mathbb{R}_+ \longrightarrow \mathbb{R}_+$ are both surjective. Then $f_2 \circ f_1$ and $f_1 \circ f_2$ are both defined. However they **do not** necessarily coincide. For example, take $f_1(x) = x^2$ and $f_2(x) = 2x$. Then $f_2(f_1(x)) = 2x^2$ whereas $f_1(f_2(x)) = (2x)^2 = 4x^2$. Thus, in general, when both $f_2 \circ f_1$ and $f_1 \circ f_2$ are defined they are different functions.
**C.** We may extend our definition to the situation where $R(f_1) \subset D_2$. Then we can still define $f_2|_{R(f_1)} \circ f_1$. For example consider the two functions $f_1 : \mathbb{R} \longrightarrow \mathbb{R}$, $x \longmapsto x^2$ and $f_2 : \mathbb{R} \longrightarrow \mathbb{R}$, $f_2$ being an arbitrary function. Since $R(f_1) = \mathbb{R}_+$ we have $R(f_1) \subset D(f_2)$. Thus we can form $(f_2|_{\mathbb{R}_+} \circ f_1)(x) = f_2(x^2)$. Soon we will also write $f_2 \circ f_1$ instead of $f_2|_{\mathbb{R}_+} \circ f_1$.

**Lemma 5.8.** *Let $f_1 : D_1 \longrightarrow F_1$ and $f_2 : D_2 \longrightarrow F_2$ be two injective functions. Suppose that $R(f_1) = D_2$. Then the function $f_2 \circ f_1 : D_1 \longrightarrow F_2$ is injective too, i.e. the composition of two injective functions is also injective.*

*Proof.* Let $x, y \in D_1$, $x \neq y$. Since $f_1$ is injective it follows that $f_1(x) \neq f_1(y)$. Now, the injectivity of $f_2$ implies further that $f_2(f_1(x)) \neq f_2(f(y))$. $\qquad \square$

**Lemma 5.9.** *Let $f_1 : D_1 \longrightarrow F_1$ and $f_2 : D_2 \longrightarrow F_2$ be two surjective functions. Suppose that $R(f_1) = D_2$. Then the composed function $f_2 \circ f_1 : D_1 \longrightarrow F_2$ is surjective.*

*Proof.* Let $z \in F_2$. Since $f_2$ is surjective there exists $y \in D_2$ such that $f_2(y) = z$. Now, $D_2 = R(f_1)$ and $f_1$ is surjective. Hence there exists $x \in D_1$ such that $f_1(x) = y \in D_2 = R(f_1)$. Thus we have $f_2(f_1(x)) = z$ implying that $f_2 \circ f_1$ is surjective. $\qquad \square$

**Corollary 5.10.** *Let $f_1 : D_1 \longrightarrow F_1$ and $f_2 : D_2 \longrightarrow F_2$ be two bijective functions such that $R(f_1) = D_2$. Then $f_2 \circ f_1 : D_1 \longrightarrow D_2$ is bijective too.*

*Proof.* We know that in this case $f_2 \circ f_1$ is injective and surjective. $\qquad \square$

**Exercise 5.11.** *Prove that the composition of two injective mappings is injective and that of two surjective mappings is surjective. Deduce that the composition of two bijective mappings is bijective.*

Consider now three functions $f_1 : D_1 \longrightarrow F_1$, $f_2 : D_2 \longrightarrow F_2$, $f_3 : D_3 \longrightarrow F_3$. Suppose that $R(f_1) = F_1 = D_2$ and that $R(f_2) = F_2 = D_3$. Then we may consider the two compositions

$$f_3 \circ (f_2 \circ f_1) : D_1 \longrightarrow F_3 \tag{5.1}$$

$$(f_3 \circ f_2) \circ f_1 : D_1 \longrightarrow F_3. \tag{5.2}$$

From (5.1) we find for all $x \in D_1$

$$(f_3 \circ (f_2 \circ f_1))(x) = f_3 \circ ((f_2 \circ f_1)(x)) = f_3(f_2(f_1(x)))$$

and (5.2) yields for all $x \in D_1$

$$((f_3 \circ f_2) \circ f_1)(x) = (f_3 \circ f_2) \circ (f_1(x)) = f_3(f_2(f_1(x))).$$

Thus we have proved

**Lemma 5.12.** *The composition of functions (mappings) is associative, i.e. for $f_1 : D_1 \longrightarrow F_1$, $f_2 : D_2 \longrightarrow F_2$, $f_3 : D_3 \longrightarrow F_3$ with $R(f_1) = F_1 = D_2$ and $R(f_2) = F_2 = D_3$ we have*

$$f_3 \circ (f_2 \circ f_1) = (f_3 \circ f_2) \circ f_1. \tag{5.3}$$

By Lemma 5.12 we may just write $f_3 \circ f_2 \circ f_1$ for both expressions (5.1) and (5.2). This clearly extends to finitely many functions.

**Example 5.13.** Let $f_1 : \mathbb{R} \to \mathbb{R}_+$, $x \longmapsto 1 + x^2$, and $f_2 : \{x \in \mathbb{R} | x \geq 1\} \to \mathbb{R}_+$, $x \longmapsto \sqrt{x}$. Then $R(f_1) = \{x | x \geq 1\} = D(f_2)$ and we find $(f_2 \circ f_1)(x) = \sqrt{1 + x^2}$. Clearly we may consider $\tilde{f}_2 : \mathbb{R}_+ \to \mathbb{R}_+$, $x \longmapsto \sqrt{x}$, and then we may form $\tilde{f}_2|_{\{x | x \geq 1\}} \circ f_1 = f_2 \circ f_1$. Everyone will agree that the latter approach is simpler and no confusion will arise when we just write $\tilde{f}_2 \circ f_1$, which is however an abuse of notation.

Let $f : D \longrightarrow F$ be a bijective function. Given $y \in F$ we can find a unique $x \in D$ such that $f(x) = y$. This defines a new function mapping $y$ to $x$.

**Definition 5.14.** *Let $f : D \longrightarrow F$ be a bijective function. The function $f^{-1} : F \longrightarrow D$, $x \longmapsto f^{-1}(y)$ where $f^{-1}(y) = x$ if $f(x) = y$ is called the* **inverse function**, *or just the* **inverse**, *of $f$.*

**Remark 5.15.** Once again, this definition extends to arbitrary mappings in the obvious way: let $f : X \mapsto Y$ be bijective. Define $f^{-1} : Y \mapsto X$ by $f^{-1}(y) = x$ if $f(x) = y$.

**Example 5.16. A.** Consider the function $f_a : \mathbb{R} \longrightarrow \mathbb{R}$, $x \longmapsto x + a$. The inverse function $f_a^{-1}$ is determined by finding for $y \in \mathbb{R}$ the value $x \in \mathbb{R}$ such that $y = f_a(x) = x + a$, which gives $x = y - a$. Thus $f_a^{-1} : \mathbb{R} \longrightarrow \mathbb{R}$, $y \longmapsto y - a$, or $f_a^{-1} = f_{-a}$.
**B.** For $a \neq 0$ consider the function $g_a : \mathbb{R} \longrightarrow \mathbb{R}$, $x \longmapsto ax$. The inverse function is determined by solving $y = ax$, i.e. $x = \frac{y}{a}$. Hence $g_a^{-1} : \mathbb{R} \longrightarrow \mathbb{R}$, $y \longmapsto \frac{y}{a}$, i.e. $g_a^{-1} = g_{a^{-1}}$.
**C.** Consider $\sqrt{\cdot} : \mathbb{R}_+ \longrightarrow \mathbb{R}_+$. We want to determine its inverse function. Now we have to solve the equation $y = \sqrt{x}$, i.e. $x = y^2$. Thus the inverse is given by $f : \mathbb{R}_+ \longrightarrow \mathbb{R}_+$, $y \longmapsto y^2$. Note that $\sqrt{\cdot} : \mathbb{R}_+ \longrightarrow \mathbb{R}_+$ is **not** the inverse to $\tilde{f} : \mathbb{R} \longrightarrow \mathbb{R}$, $x \longmapsto x^2$. This function is not bijective. However it is easy to check that $\sqrt{\cdot} : \mathbb{R}_+ \longrightarrow \mathbb{R}_+$ is the inverse of $\tilde{f} : \mathbb{R}_+ \longrightarrow \mathbb{R}_+$, $x \longmapsto x^2$.

Let $f : D \longrightarrow F$ be bijective with inverse $f^{-1} : F \longrightarrow D$. We may consider the two compositions

$$f^{-1} \circ f : D \longrightarrow D \tag{5.4}$$

$$f \circ f^{-1} : F \longrightarrow F. \tag{5.5}$$

In the first case we find $(f^{-1} \circ f)(x) = f^{-1}(f(x))$ and since $f^{-1}(y) = x$ when $f(x) = y$ it follows that $(f^{-1} \circ f)(x) = x$ for all $x \in D$. On the other hand, for $y \in F$ we have $f(f^{-1}(y)) = f(x)$ for $f(x) = y$, hence $(f \circ f^{-1})(y) = y$.

**Definition 5.17.** *Let $D$ be a set. The **identity** (or identity mapping) on $D$ is the function $id_D : D \longrightarrow D$, $x \longmapsto x$.*

Obviously we have for $f : D \longrightarrow F$

$$f \circ id_D = f \text{ and } id_F \circ f = f. \tag{5.6}$$

Therefore, just before giving Definition 5.17 we proved:

$$f^{-1} \circ f = id_D \tag{5.7}$$

and

$$f \circ f^{-1} = id_F. \tag{5.8}$$

76

**Corollary 5.18.** *If $f : D \longrightarrow F$ is bijective then $f^{-1} : F \longrightarrow D$ is also bijective and $(f^{-1})^{-1} = f$. Moreover $f^{-1}$ is uniquely determined.*

*Proof.* Firstly we claim that $f^{-1}$ is injective. For $y_1 \neq y_2$, $y_1, y_2 \in F$, suppose that $f^{-1}(y_1) = f^{-1}(y_2)$. Then by (5.8) we find

$$y_1 = f(f^{-1}(y_1)) = f(f^{-1}(y_2)) = y_2 \tag{5.9}$$

which is a contradiction, hence $f^{-1}$ is injective. Next we claim that $f^{-1}$ is surjective. Given $x \in D$, with $y = f(x)$ we find by (5.7) that

$$f^{-1}(y) = f^{-1}(f(x)) = x, \tag{5.10}$$

i.e. $f^{-1}$ is surjective. A bijective function $g : D \longrightarrow F$ is the inverse of the bijective function $f^{-1} : F \longrightarrow D$ if $g(x) = y$ for $f^{-1}(y) = x$, but $f$ has exactly this property, i.e. $(f^{-1})^{-1} = f$.

Finally we prove that $f^{-1}$ is uniquely determined. Let $g, h : F \longrightarrow D$ be two bijective functions such that $g \circ f = h \circ f = id_D$. We have to prove that $g(y) = h(y)$ for all $y \in F$. Given $y \in F$. Since $f$ is bijective there exists a unique $x \in D$ such that $f(x) = y$. Now it follows that

$$g(y) = g(f(x)) = x = h(f(x)) = h(y)$$

implying that $g = h$. $\qquad\square$

The reader may have noted that Definition 5.17 and its Corollary are now given for $D$ being an arbitrary set, i.e. $f$ being a mapping and not necessarily a function.

**Lemma 5.19.** *Let $f_1 : D_1 \longrightarrow F_1$ and $f_2 : D_2 \longrightarrow F_2$ be bijective mappings such that $R(f_1) = F_1 = D_2$. Then the composition $f_2 \circ f_1 : D_1 \longrightarrow F_2$ has the inverse function*

$$(f_2 \circ f_1)^{-1} = f_1^{-1} \circ f_2^{-1} : F_2 \longrightarrow D_1. \tag{5.11}$$

*Proof.* We know that $f_2 \circ f_1$ is bijective, hence $(f_2 \circ f_1)^{-1}$ exists and is bijective. Since we also know that $(f_2 \circ f_1)^{-1}$ is uniquely determined we may find an expression for $(f_2 \circ f_1)^{-1}$ from the two following calculations:

$$
\begin{aligned}
(f_2 \circ f_1) \circ \left( f_1^{-1} \circ f_2^{-1} \right) &= \left( f_2 \circ \left( f_1 \circ f_1^{-1} \right) \circ f_2^{-1} \right) \\
&= f_2 \circ id_{F_1} \circ f_2^{-1} \\
&= f_2 \circ id_{D_2} \circ f_2^{-1} \\
&= f_2 \circ f_2^{-1} = id_{F_2}
\end{aligned}
$$

77

and

$$\begin{aligned}
\left(f_1^{-1} \circ f_2^{-1}\right) \circ \left(f_2 \circ f_1\right) &= f_1^{-1} \circ \left(f_2^{-1} \circ f_2\right) \circ f_1 \\
&= f_1^{-1} \circ id_{D_2} \circ f_1 = f_1^{-1} \circ id_{F_1} \circ f_1 \\
&= f_1^{-1} \circ f_1 = id_{D_1}
\end{aligned}$$

proving the lemma. $\square$

There are easy ways to understand the concept of injectivity, surjectivity, bijectivity, and inverse functions by looking at the graph of a function.

If $f : D \longrightarrow F$ is surjective then for every value in the target set $F$ considered as a subset of the $y$-axis there must correspond at least one value in the domain $D$ considered as a subset of the $x$-axis:

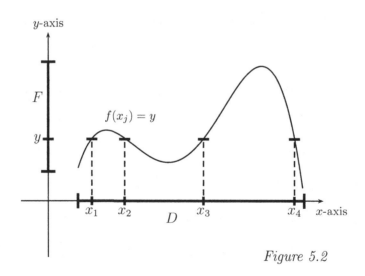

Figure 5.2

If $f : D \longrightarrow F$ is injective then for every value on the $y$-axis belonging also to $F$ there corresponds at most one value in $D$ considered as a subset of the $x$-axis:

78

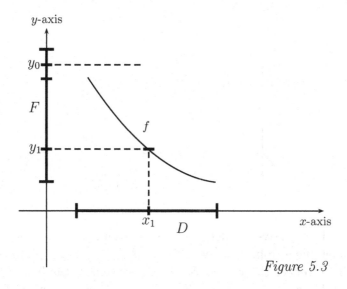

*Figure 5.3*

If $f$ is bijective then for every value $y \in F$, $F$ considered as subset of the $y$-axis, there corresponds one and only one point $x \in D$, $D$ considered as a subset of the $x$-axis:

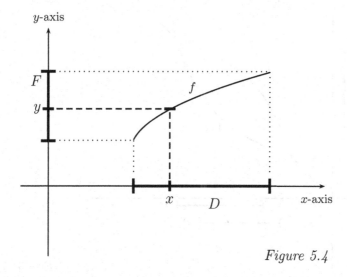

*Figure 5.4*

Before looking further at bijective functions we consider a useful geometric interpretation. Let $(a, b) \in \mathbb{R}^2$. The point $(b, a) \in \mathbb{R}^2$ is obtained by reflecting $(a, b)$ in the line $y = x$:

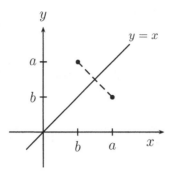

*Figure 5.5*

Let $f : D \longrightarrow F$ be bijective and let $\Gamma(f)$ be its graph. Since $f$ is bijective we may also consider the graph of $f^{-1}$ which is $\Gamma(f^{-1}) = \{(y, f^{-1}(y)) | y \in F\} \subset F \times D$. Now if we reflect the whole coordinate system in the line $y = x$, i.e. in the principal diagonal, we can recover the graph $\Gamma(f^{-1})$ from $\Gamma(f)$.

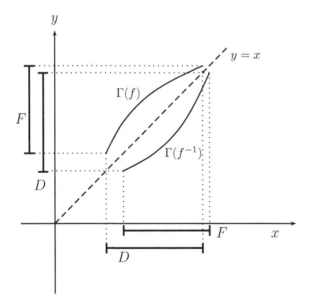

*Figure 5.6*

For example for $\sqrt{\cdot} : \mathbb{R}_+ \longrightarrow \mathbb{R}_+$ we find:

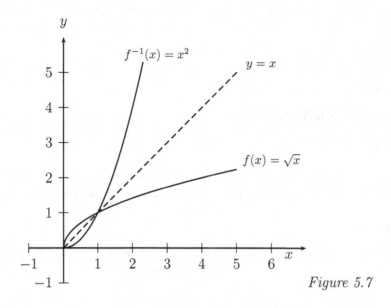

*Figure 5.7*

We end this chapter by looking at some algebraic operations. Let $h : D_1 \longrightarrow F_1$ and $f, g : D_2 \longrightarrow F_2$ be functions such that $R(h) = D_2$. Then we define

$$(f \pm g) \circ h := f \circ h \pm g \circ h, \tag{5.12}$$

$$(f \cdot g) \circ h := (f \circ h) \cdot (g \circ h), \tag{5.13}$$

and if $g(y) \neq 0$ for all $y \in D_2$

$$\left(\frac{f}{g}\right) \circ h := \frac{f \circ h}{g \circ h}. \tag{5.14}$$

For example we may consider $h : \mathbb{R} \longrightarrow \mathbb{R}_+$, $x \longmapsto |x|$, $f : \mathbb{R}_+ \longrightarrow \mathbb{R}$, $x \longmapsto \sqrt{x}$ and $g : \mathbb{R}_+ \longrightarrow \mathbb{R}$, $x \longmapsto \sqrt{1 + x}$ where we get

$$\begin{aligned}
((f \pm g) \circ h)(x) &= \sqrt{|x|} \pm \sqrt{1 + |x|} \\
((f \cdot g) \circ h)(x) &= \sqrt{|x|} \sqrt{1 + |x|} \\
\left(\left(\frac{f}{g}\right) \circ h\right)(x) &= \frac{\sqrt{|x|}}{\sqrt{1 + |x|}}.
\end{aligned}$$

Now let us again consider some more abstract mathematics.

Let us return to general sets and equivalence relations. Let $X \neq \emptyset$ be a set.

We define on its power set $\mathcal{P}(X)$ a relation $\mathcal{R}$ by $(A, B) \in \mathcal{R} \subset \mathcal{P} \times \mathcal{P}$ if and only if there exists $f : A \to B$ which is bijective. (In our other notation we would write $A \sim B$ if there exists a bijective mapping $f : A \to B$.) We claim that $\mathcal{R}$ is an equivalence relation.

$\mathcal{R}$ is reflexive:

$$A\mathcal{R}A: \text{ take } f = id_A : A \mapsto A, x \mapsto id_A(x) = x$$

$\mathcal{R}$ is symmetric:

$A\mathcal{R}B$ and $B\mathcal{R}A$ : if $A\mathcal{R}B$ then there exists a bijective $f : A \mapsto B$, but $f^{-1} : B \mapsto A$ is bijective too.

$\mathcal{R}$ is transitive:

$A\mathcal{R}B$ and $B\mathcal{R}C$ means that there exists $f : A \mapsto B$ and $g : B \mapsto C$ both bijective, then $g \circ f : A \mapsto C$ is bijective too.

This is one of the most important equivalence relations which had an enormous influence on the historical development of set theory. We will return to it later.

To proceed further we need the following considerations. Let $X \neq \emptyset$ be a set and "$\sim$" an equivalence relation on $X$. Let $a \in X$. We denote by $[a]$ the set of all $x \in X$ with $x \sim a$, i.e.

$$[a] := \{x \in X \mid x \sim a\} \tag{5.15}$$

and we call $[a]$ the **equivalence class** of $a$ or generated by $a$. A partition of a set $X$ is a set of subsets of $X$ such that every element of $X$ belongs to only one of these subsets, for example $\{1, 2\}, \{3, 4\}, \{5\}$ would be a partition of $\{1, 2, 3, 4, 5\}$, however neither $\{1\}, \{3, 4\}$ nor $\{1, 2\}, \{2, 3, 4\}, \{5\}$ would be. Formally, we call a family of sets $\mathcal{Z} \subset \mathcal{P}(X)$ a **partition** of $X$ if

1. every $x \in X$ belongs to some $Z \in \mathcal{Z}$, i.e. for $x \in X$ there exists $Z \in \mathcal{Z}$ such that $x \in Z$ or

$$X = \bigcup_{Z \in \mathcal{Z}} Z,$$

with

$$\bigcup_{Z \in \mathcal{Z}} Z := \{x \in X \mid x \in Z \text{ for some } Z \in \mathcal{Z}\}$$

82

2. for $Z_1, Z_2 \in \mathcal{Z}$ we have that either $Z_1 \cap Z_2 = \emptyset$ or $Z_1 = Z_2$, i.e. we have that $x \in Z_1 \cap Z_2$ implies $Z_1 = Z_2$.

Figure 5.8 illustrates a typical partition of a set $X$:

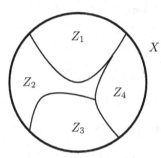

*Figure 5.8*

**Proposition 5.20.** *Let $X \neq \emptyset$ and "$\sim$" be an equivalence relation on $X$. Then $\{[a] \mid a \in X\}$ is a partition of $X$.*

*Proof.* Clearly $a \in [a]$ since $a \sim a$. Hence $X = \bigcup_{a \in X} [a]$. Further, if $c \in [a] \cap [b]$ then $c \sim a$ and $c \sim b$ therefore $a \sim b$ implying $[a] \subset [b]$ as well as $[b] \subset [a]$, i.e. $[a] = [b]$. Thus, if $[a] \cap [b] \neq \emptyset$ then $[a] = [b]$. Thus we have proved that $\{[a] \mid a \in X\}$ is a partition of $X$. $\qquad\square$

It is of interest that given a partition $\mathcal{Z}$ of $X$ then there exists an equivalence relation "$\sim$" on $X$ such that the elements of $\mathcal{Z}$ are exactly the equivalence classes corresponding to "$\sim$". Indeed, we just have to define $x \sim y$ if and only if $x, y \in Z$ for some $Z \in \mathcal{Z}$. Obviously $x \sim x$ since $x \in Z$ for some $Z \in \mathcal{Z}$, and $x \sim y$ if and only if $y \sim x$ since equivalence means to belong to the same set $Z$. Finally, if $x \sim y$ and $y \sim z$ then $x, y \in Z'$ and $y, z \in Z''$, however $y \in Z' \cap Z''$ implying $Z' = Z''$, i.e. $x \sim z$.

**Definition 5.21.** *Let $X \neq \emptyset$ and "$\sim$" be an equivalence relation on $X$ with equivalence classes $[a]$, $a \in X$, inducing the partition $\mathcal{Z}$ of $X$. We call a subset $R \subset X$ a **complete set of representatives** with respect to "$\sim$" if*

*1. $r_1, r_2 \in R$ and $r_1 \sim r_2$ implies $r_1 = r_2$*

*and*

2. $X = \bigcup_{r \in R} [r] = \bigcup_{r \in R} \{x \in X \mid x \sim r\}.$

Now we return to the equivalence relation we considered at the beginning of the chapter.

Dealing with the set of all sets can be quite troublesome and it may lead to some serious problems. Let us fix a set $X \neq \emptyset$ and suppose $X$ is "large". On $\mathcal{P}(X)$, the power set of $X$, i.e.

$$\mathcal{P}(X) = \{A \mid A \subset X\} \tag{5.16}$$

we introduce the equivalence relation

$$A \sim B \text{ if there exists a bijection } f_{AB} : A \to B. \tag{5.17}$$

This is our old example; it induces a partition of $X$. When $X = \mathbb{R}$ and $A \sim B$ we say that $A$ and $B$ have the same **cardinality**. The notion of cardinality of sets can be extended to more general sets than subsets of $\mathbb{R}$, but for our purposes it is sufficient to restrict ourselves to the case of $\mathbb{R}$. Denote by $\mathbb{N}_n := \{1, \ldots, n\}$ the first $n$ natural numbers. Every finite subset $A \subset \mathbb{R}$ is equivalent to one and only one of the sets $\mathbb{N}_n$, $n \in \mathbb{N}$. Indeed, if $A$ has $n$ elements $a_1, \ldots, a_n$ then $j \mapsto a_j$ is a bijection from $\mathbb{N}_n$ to $A$ and $n$ is uniquely determined. Thus $\{\mathbb{N}_n\}_{n \in \mathbb{N}}$ gives a complete set of representatives of the finite subsets of $\mathbb{R}$. In this equivalence relation the representative is just determined by the number of elements of the finite set. Now, $\mathbb{N}$ itself is not finite and determines a further equivalence class, the class of the **countable sets** and of course $\mathbb{N}$ is a representative of this class. A set $Y \subset \mathbb{R}$ is countable if there exists a bijection from $\mathbb{N}$ to $Y$ or equivalently if there exists a bijection from $Y$ to $\mathbb{N}$. The finite sets together with the countable sets, i.e. sets $A \subset \mathbb{R}$, such that $A \sim \mathbb{N}_n$ or $A \sim \mathbb{N}$, are called the **denumerable** subsets of $\mathbb{R}$.

We claim that $\mathbb{Z}$ and $\mathbb{Q}$ are countable. (We identify $\mathbb{Q}$ as a subset of $\mathbb{R}$). How do we prove this surprising statement? There are clearly many "more" integers or fractions than natural numbers, i.e. $\mathbb{N} \subset \mathbb{Z}$, $\mathbb{N} \subset \mathbb{Q}$, $\mathbb{N} \neq \mathbb{Z}$ and $\mathbb{N} \neq \mathbb{Q}$ as well as $\mathbb{Z} \neq \mathbb{Q}$. Note that this is typical for infinite sets: they contain proper subsets which can be mapped bijectively onto them, i.e. they have the same cardinality. Here is a possible bijection $f_{\mathbb{Z}\mathbb{N}}$:

$$f_{\mathbb{Z}\mathbb{N}}(k) := \begin{cases} 2k, & k \in \mathbb{N} \\ 2|k| + 1, & k \in \mathbb{Z} \setminus \mathbb{N}. \end{cases} \tag{5.18}$$

Clearly $f_{ZN}$ is injective, $k \neq k'$ implies $f_{ZN}(k) \neq f_{ZN}(k')$. However $f_{ZN}$ is also surjective: given $n \in \mathbb{N}$, if $n$ is even, i.e. $n = 2k$, $k \in \mathbb{N}$, then $f_{ZN}(k) = n$. If $n$ is odd, i.e. $n = 2k + 1, k \in \mathbb{N}$, then $f_{ZN}(-k) = n$.

The case of $\mathbb{Q}$ is more involved and we only indicate the idea of showing how to prove that all non-negative fractions can be mapped bijectively onto $\mathbb{N}$. Note that there is a lot of multiple counting in the following scheme, i.e. we need to refine the counting process. This enumeration scheme is due to G. Cantor who is together with R. Dedekind the founder of set theory; one of the greatest intellectual achievements of mankind.

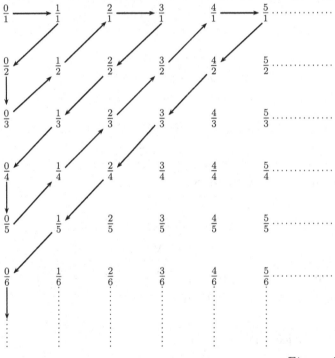

*Figure 5.9*

Now we consider the question: is $\mathbb{R}$ is countable? That is, does a bijective function from $\mathbb{N}$ to $\mathbb{R}$ exist? The answer is no and we will prove this when we discuss decimal fractions in Chapter 18 of Part 2. For this we first need to understand the convergence of series of real numbers. Thus we know that $\mathbb{R}$ has finite and countable subsets but it is not itself countable. We may ask whether there is a subset $C \subset \mathbb{R}$ such that $C$ is not countable and has not the same cardinality as $\mathbb{R}$, i.e. there is no bijection $f_{CR} : C \mapsto \mathbb{R}$. The famous **continuum hypothesis** (CH) states that such a set does not exist. So far a proof does not exist, however K. Gödel proved that in the standard model of set theory which is denoted by ZFC, where Z stands for E. Zermelo, F for A. Fraenkel and C for the Axiom of Choice, CH cannot be disproved. Some thirty years later P. Cohen proved that CH is independent of ZFC.

# Problems

1. Decide whether or not the following functions are injective, surjective or bijective. Sketch the graph in each case.

   a)
   $$f_1 : \mathbb{R} \longrightarrow \mathbb{R}_+$$
   $$x \longmapsto |x - 3| + 2;$$

   b)
   $$f_2 : [1, \infty) \longrightarrow (0, 2]$$
   $$x \longmapsto \frac{2}{x}$$
   where for $a \in \mathbb{R}$ we write $[a, \infty) = \{x \in \mathbb{R} \mid x \geq a\}$;

   c)
   $$f_3 : [-2, 7] \longrightarrow [0, 3]$$
   $$x \longmapsto \sqrt{x + 2}$$

2. a) Consider the mapping $g : \mathbb{Q} \longrightarrow \mathbb{Z}$, $\frac{p}{q} \longmapsto g\left(\frac{p}{q}\right) = p + q$. Is $g$ injective, surjective or bijective?

   b) Let $r : \mathbb{R} \times \mathbb{R} \longrightarrow \mathbb{R} \times \mathbb{R}$, $(x, y) \longmapsto r(x, y) := (y, x)$. Test $r$ for injectivity, surjectivity and bijectivity.

3. a) Given $f : \mathbb{R} \longrightarrow \mathbb{R}$, $x \longmapsto 5x^2 - 2x + 1$, and $g : [-5, \infty) \longrightarrow \mathbb{R}$, $x \longmapsto \sqrt{5 + x}$. Find $f \circ g : [-5, \infty) \longrightarrow \mathbb{R}$.

   b) Consider $f : \mathbb{R} \longrightarrow \mathbb{R}$, $x \longmapsto |x + 3| - 2$ and $h : \mathbb{R} \longrightarrow \mathbb{R}$, $x \longmapsto \sqrt{x^4 + 2}$. Find the largest sets $D_1 \subset \mathbb{R}$ and $D_2 \subset \mathbb{R}$ such that we can form $f \circ h : D_1 \longrightarrow \mathbb{R}$ and $h \circ f : D_2 \longrightarrow \mathbb{R}$. In each case give a formula for the function, i.e. for $f \circ h$ and $h \circ f$.

   c) Find the largest set $D \subset \mathbb{R}$ where we can define $f \circ h$ where $f : [0, \infty) \longrightarrow \mathbb{R}$, $x \longmapsto \sqrt{x}$ and $h : \mathbb{R} \longrightarrow \mathbb{R}$, $x \longmapsto |x + 2| - 1$.

4. (Exercise 5.11) Prove that the composition of two injective mappings is injective and that of two surjective mappings is surjective. Deduce that the composition of two bijective mappings is bijective.

5. Let $X \neq \phi$ be a non-empty set. Denote by $Aut(X)$ the set of all bijective mappings $f : X \longrightarrow X$. The abbreviation $Aut(X)$ comes from **automorphism**, a notion that is dealt with in algebra. Prove that $(Aut(X), \circ)$ is in general a **non-Abelian group**, where "$\circ$" stands for the composition of mappings.
   Note: in order to verify that $Aut(X)$ is a non-Abelian group the following must be proved:

   (i) $f, g, h \in Aut(X)$ implies $(f \circ g) \circ h = f \circ (g \circ h)$;

   (ii) $f, g \in Aut(X)$ implies $f \circ g \in Aut(X)$;

   (iii) there exists $e \in Aut(X)$ such that for all $f \in Aut(X)$ the following holds: $f \circ e = e \circ f = f$;

   (iv) for $f \in Aut(X)$ there exists $k_f \in Aut(X)$ such that $f \circ k_f = k_f \circ f = e$.

6. Let $f : X \longrightarrow Y$ be a mapping.

   a) Prove that $f$ is injective if and only if there exists a mapping $g : Y \longrightarrow X$ such that $g \circ f = id_X$.

   b) Prove that $f$ is surjective if and only if there exists a mapping $h : Y \longrightarrow X$ such that $f \circ h = id_Y$.

7. Consider the mapping $f : \{x \in \mathbb{R} | x > 0\} \longrightarrow \{x \in \mathbb{R} | x > 0\}, x \longmapsto \frac{1}{x}$. Prove that $f = f^{-1}$.

8. For $h : \mathbb{R} \longrightarrow \mathbb{R}, x \longmapsto x^2 + 2$, and $f : \mathbb{R}_+ \longrightarrow \mathbb{R}, x \longmapsto \frac{1}{x}$, $g : \mathbb{R}_+ \longrightarrow \mathbb{R}, x \longmapsto \sqrt{x} + |x - 2|$, find $(f + g) \circ h$, $(f \cdot g) \circ h$ and $\left(\frac{1}{g}\right) \circ h$.

9. Let $X \neq \phi$ and $f : X \longrightarrow \mathbb{R}$. Define two functions

$$f^+ : X \longrightarrow \mathbb{R}, \ x \longmapsto f^+(x) := \frac{|f(x)| + f(x)}{2}$$

and

$$f^- : X \longrightarrow \mathbb{R}, \ x \longmapsto f^-(x) := \frac{|f(x)| - f(x)}{2},$$

which are called the **positive part** of $f$ and the **negative part** of $f$ respectively. Prove that $f^+(x) \geq 0$ and $f^-(x) \geq 0$ for all $x \in X$ and

$$f = f^+ - f^-, \ |f| = f^+ + f^-.$$

10. Find the inverse of each of the following mappings:

   a) $f_1 : \{x \in \mathbb{R} | x \geq 0\} \longrightarrow (0, 1], \ x \longmapsto \frac{1}{1+x^2}$;

   b) $f_2 : \{x \in \mathbb{R} | x \geq 0\} \longrightarrow (0, 2]$ where

$$f_2(x) = \begin{cases} -x + 2, & x \in [0, 1] \\ \frac{1}{x}, & x \in (1, \infty). \end{cases}$$

   Sketch the graph of $f_2$.

   c) $f_3 : \mathbb{N} \longrightarrow \left\{ q \,|\, q = \frac{1}{n^3} \ \text{and} \ n \in \mathbb{N} \right\}, \ n \longmapsto \frac{1}{n^3}$.

11.    a) Let $pr_1 : \mathbb{R}^2 \longrightarrow \mathbb{R}, \ (x, y) \longmapsto pr_1((x, y)) = x$. Denote by $B_1(0) = \{(x, y) \in \mathbb{R}^2 \,|\, x^2 + y^2 \leq 1\}$ the disc with centre $0 = (0, 0) \in \mathbb{R}^2$ and radius 1 and by $S^1 = \{(x, y) \in \mathbb{R}^2 \,|\, x^2 + y^2 = 1\}$ the circle with centre $0 = (0, 0) \in \mathbb{R}^2$ and radius 1. Find $pr_1(B_1(0))$ and $pr_1(S^1)$. (Sketch the situations).

   b) Let $R(g) = \{(x, g(x)) \,|\, x \in [0, 1] \ \text{and} \ g(x) = x^2 + 1\}$. Find $pr_2(R(g))$ where $pr_2 : \mathbb{R}^2 \longrightarrow \mathbb{R}, \ (x, y) \longmapsto pr_2((x, y)) = y$.

12. Let $X$ and $Y$ be two non-empty sets $A \subset X$, $B \subset Y$. For the projections $pr_1 : X \times Y \longrightarrow X, \ (x, y) \longmapsto pr_1((x, y)) = x$, and $pr_2 : X \times Y \longrightarrow Y, \ (x, y) \longmapsto pr_2((x, y)) = y$, prove that

$$pr_1^{-1}(A) = A \times Y \ \text{and} \ pr_2^{-1}(B) = X \times B.$$

13. Let $j : \mathbb{N} \longrightarrow \mathbb{R}$ be a mapping. Prove that the image $j(\mathbb{N}) \subset \mathbb{R}$ is a countable set if $j$ is injective. Does the converse hold, i.e does the countability of $j(\mathbb{N})$ imply that $j$ is injective?

14. Let $D \subset \mathbb{R}$, $D \neq \phi$, and denote by $M(D; \mathbb{R})$ the set of all mappings $f : D \longrightarrow \mathbb{R}$. We define the relation $f \sim g$ for $f, g \in M(D; \mathbb{R})$ as follows: $f \sim g$ if and only if there exists a finite set $A = A_{f,g} = \{x_1, \ldots, x_m\}$ depending on $f$ and $g$, i.e. the points $x_j$ as well as $m$ depend on $f$ and $g$, such that $f|_{D \backslash A} = g|_{D \backslash A}$. Prove that "$\sim$" defines an equivalence relation on $M(D; \mathbb{R})$.

15. Let $X, Y, Z$ be sets. We can define the Cartesian product $(X \times Y) \times Z$. An element of this set has the form $((x, y), z)$ with $(x, y) \in X \times Y$ and $z \in Z$. Define the set $X \times Y \times Z$ as the set of all **ordered triples** $(x, y, z)$ where $x \in X$, $y \in Y$ and $z \in Z$, i.e.

$$X \times Y \times Z = \{(x, y, z) \mid x \in X \wedge y \in Y \wedge z \in Z\}.$$

Prove that
$$J : (X \times Y) \times Z \longrightarrow X \times Y \times Z$$

$$((x, y), z) \longmapsto (x, y, z)$$

is a bijective mapping. (Note that by definition $(x, y, z) = (x', y', z')$ if and only if $x = x'$, $y = y'$ and $z = z'$.)

**Remark**: for finitely many sets $A_1, \ldots, A_N$ we can define their cartesian product by

$$A_1 \times \ldots \times A_N := \{(a_1, \ldots, a_N) \mid a_1 \in A_1, \ldots, a_N \in A_N\},$$

or more formally

$$A_1 \times \ldots \times A_N := \{(a_1, \ldots, a_N) \mid \text{for all } j \in \{1, \ldots, N\} : a_j \in A_j\}.$$

In particular we may work with

$$\mathbb{R}^n := \mathbb{R} \times \ldots \times \mathbb{R} \quad (n \text{ terms});$$

$$\mathbb{Z}^m := \mathbb{Z} \times \ldots \times \mathbb{Z} \quad (m \text{ terms})$$

and more generally

$$A^k := A \times \ldots \times A \quad (k \text{ terms}).$$

# 6 Derivatives

We want to study real-valued functions $f : D \to \mathbb{R}$, $D \subset \mathbb{R}$, more closely. For example we would like to know whether f is monotone increasing or decreasing, attains local extreme values, has zeroes, etc. For all this and for many more problems the concept of the derivative is very helpful. We will spend some time on the construction of the derivative which we will formally define in Definition 6.2. The central idea is to substitute locally, i.e. in a neighbourhood of a point $x_0 \in D$, the graph $\Gamma(f)$ of a function $f$ by a straight line, more precisely by the graph $\Gamma(g)$ of a function $g : \mathbb{R} \to \mathbb{R}$, $x \mapsto ax + b$.

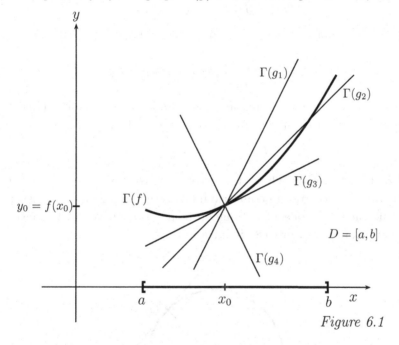

Figure 6.1

As Figure 6.1 shows, many straight lines are possible. We have already indicated one condition that we want to impose: if $x_0 \in D$ is the point of interest i.e. if we want to replace $\Gamma(f)$ in a neighbourhood of $x_0$ by $\Gamma(g)$, then $(x_0, f(x_0))$ should lie on the straight line being selected.

The equation of a straight line passing through $(x_0, f(x_0))$ can be obtained as follows. A straight line should be interpreted as the graph $\Gamma(g)$ of a function $g : \mathbb{R} \to \mathbb{R}$, $x \mapsto g(x) = ax + b$. The condition that $(x_0, f(x_0)) \in \Gamma(g)$ means

$$g(x_0) = ax_0 + b = f(x_0) \tag{6.1}$$

91

which is one equation for the two unknown $a$ and $b$. Thus we need a further condition to determine $g$, i.e. $\Gamma(g)$. Since our aim is to substitute locally, i.e. in a neighbourhood of $x_0 \in D$, $\Gamma(f)$ by $\Gamma(g)$, we may argue as follows. For $|x - x_0|$ small we should have

$$f(x) \approx g(x) = ax + b, \qquad (6.2)$$

where " $\approx$ " stands at the moment for "$f(x)$ being close to $g(x)$". Of course, in addition to (6.2) we assume (6.1).

Thus for $|x - x_0|$ small we should have

$$f(x) - f(x_0) \approx a(x - x_0), \qquad (6.3)$$

which we obtain by subtracting (6.1) from (6.2). For $x \neq x_0$ this yields

$$\frac{f(x) - f(x_0)}{x - x_0} = a + \text{error}. \qquad (6.4)$$

Now if $|x - x_0|$ tends to 0 then the error should also go to 0. This would determine $a$ and from (6.1) we can now calculate $b$ to be

$$b = f(x_0) - ax_0. \qquad (6.5)$$

We need to be precise by what "the error goes to 0 as $|x - x_0|$ goes to 0" means. Before this we give a geometric interpretation for our considerations. We have the intuitive idea of a tangent to a given curve, in the case of a circle we can even give a precise definition:

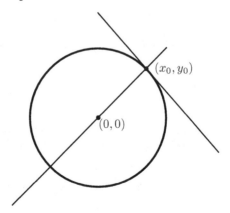

*Figure 6.2*

A straight line is a tangent to a circle at the point $(x_0, y_0)$ if the point $(x_0, y_0)$ belongs to (the graph of) this straight line and this straight line is perpendicular (later we will also say orthogonal) to the straight line through the centre of the circle and the point $(x_0, y_0)$.

For a general curve we cannot use this definition, but we may do the following: consider the graph $\Gamma(f)$ of $f : D \to \mathbb{R}$, $x_0 \in D$.

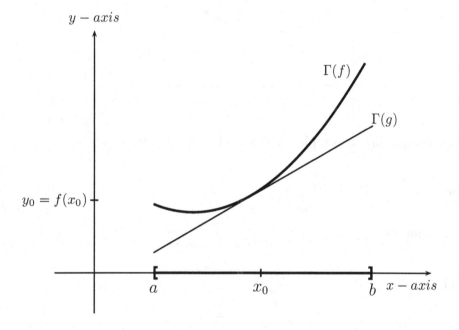

*Figure 6.3*

Instead of $g$, which we do not know, we consider a straight line $\tilde{g}$ nearby given as $\tilde{g}(x) = \tilde{a}x + \tilde{b}$, which has the property that $(x_0, f(x_0)) \in \Gamma(\tilde{g})$, i.e. the point $(x_0, f(x_0))$ lies on the graph of $\tilde{g}$, therefore $\tilde{g}(x_0) = \tilde{a}x_0 + \tilde{b} = f(x_0)$, and for $|x - x_0|$ small $\Gamma(\tilde{g})$ intersects $\Gamma(f)$ (only) in one further point, say $(x_1, f(x_1))$

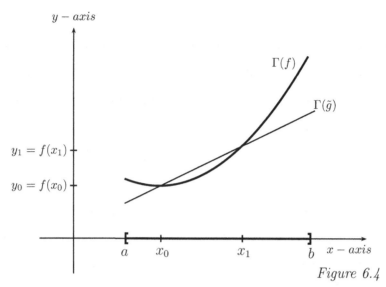

Figure 6.4

This straight line is completely determined by the two conditions:

$$\tilde{g}(x_0) = f(x_0) = \tilde{a}x_0 + \tilde{b}; \qquad (6.6)$$

and

$$\tilde{g}(x_1) = f(x_1) = \tilde{a}x_1 + \tilde{b}. \qquad (6.7)$$

This leads to

$$\tilde{a} = \frac{f(x_0) - f(x_1)}{x_0 - x_1} \qquad (6.8)$$

and

$$\tilde{b} = \frac{f(x_1)x_0 - f(x_0)x_1}{x_0 - x_1}. \qquad (6.9)$$

Thus the error term in (6.4) should be given by $|a - \tilde{a}|$. Intuitively we now take a sequence of points $(x_\nu, f(x_\nu))$, $\nu \in \mathbb{N}$, on $\Gamma(f)$, $x_\nu \neq x_0$ for all $\nu \in \mathbb{N}$, tending to $(x_0, f(x_0))$ and consider the corresponding straight lines $g_\nu(x) = a_\nu x + b$ with

$$a_\nu = \frac{f(x_0) - f(x_\nu)}{x_0 - x_\nu} \qquad (6.10)$$

and

$$b_\nu = \frac{f(x_\nu)x_0 - f(x_0)x_\nu}{x_0 - x_\nu}. \qquad (6.11)$$

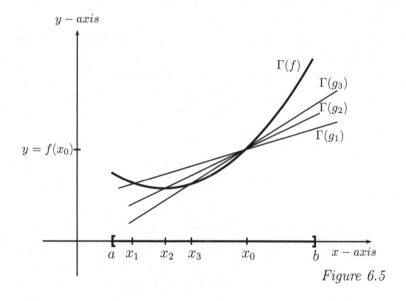

$y - axis$

$\Gamma(f)$

$\Gamma(g_3)$

$\Gamma(g_2)$

$\Gamma(g_1)$

$y = f(x_0)$

$a$ $x_1$ $x_2$ $x_3$ $x_0$ $b$ $x - axis$

*Figure 6.5*

We may think that the tangent is just the "limit line". However, here we encounter one of the main problems in analysis. Do we know that the "limit" exists?

Having these preliminaries in mind we now do the preparations needed for correct and precise statements. We need to understand the concept of a limit of a function:

$$\lim_{y \to x} f(y) = a. \tag{6.12}$$

This should be equivalent to

$$\lim_{y \to x} (f(y) - a) = 0 \tag{6.13}$$

or

$$\lim_{y \to x} |f(y) - a| = 0. \tag{6.14}$$

The latter means: given a small error bound $\varepsilon > 0$, if $y$ is close to $x$ then $|f(y) - a| < \varepsilon$.

So let us give a first definition: we say that the **limit** of $f : D \to \mathbb{R}$ as $y \in D$ approaches $x \in D$ is equal to $a$, i.e.

$$\lim_{y \to x} f(y) = a, \tag{6.15}$$

95

if for every $\varepsilon > 0$ there exists $\delta > 0$ such that $0 < |x - y| < \delta$ implies $|f(y) - a| < \varepsilon$.

We will see later in Part 2 that this definition yields the following simple **rules for limits**:

Let $f, g : D \to \mathbb{R}$ be two functions and assume that

$$\lim_{y \to x} f(y) = a \tag{6.16}$$

and

$$\lim_{y \to x} g(y) = b \tag{6.17}$$

then we have

$$\lim_{y \to x} (f \pm g)(y) = \lim_{y \to x} f(y) \pm \lim_{y \to x} g(y) = a \pm b, \tag{6.18}$$

as well as

$$\lim_{y \to x} (f \cdot g)(y) = \lim_{y \to x} f(y) \cdot \lim_{y \to x} g(y) = a \cdot b. \tag{6.19}$$

If in addition $g(y) \neq 0$ for all $y \in D$ and $b \neq 0$, then

$$\lim_{y \to x} \frac{f(y)}{g(y)} = \frac{\lim_{y \to x} f(y)}{\lim_{y \to x} g(y)} = \frac{a}{b}. \tag{6.20}$$

(Note that we will improve (6.20), we will need only the assumption that $b \neq 0$.)

**Example 6.1. A.** Consider the constant function $h_c : \mathbb{R} \to \mathbb{R}$, $x \mapsto c$, i.e. $h_c(x) = c$ for all $x \in \mathbb{R}$. Then $|h_c(y) - h_c(x)| = |c - c| = 0$ and therefore whatever the value of $|x - y|$ is, $|h_c(y) - h_c(x)| < \varepsilon$ for every $\varepsilon > 0$. Hence

$$\lim_{y \to x} h_c(x) = c. \tag{6.21}$$

**B.** We claim for $f_a : \mathbb{R} \to \mathbb{R}$, $x \mapsto f_a(x) = a + x$ that

$$\lim_{y \to x} f_a(y) = f_a(x). \tag{6.22}$$

Indeed, consider

$$|f_a(y) - f_a(x)| = |a + y - a - x| = |y - x|.$$

Given $\varepsilon > 0$, take $\delta = \varepsilon$ to find that for $|y - x| < \delta$ it follows that

$$|f_a(y) - f_a(x)| = |y - x| < \varepsilon,$$

i.e.

$$|f_a(y) - f_a(x)| < \varepsilon.$$

**C.** Now let $p : \mathbb{R} \to \mathbb{R}$ be a polynomial, i.e.

$$p(x) = \sum_{j=1}^{M} a_j x^j = a_0 + a_1 x + a_2 x^2 + \cdots + a_M x^M.$$

Thus $p$ is the finite sum of finite products of functions for which we know the limits, hence

$$\lim_{y \to x} p(y) = p(x). \tag{6.23}$$

**D.** Consider the characteristic function $\chi_A : \quad \mathbb{R} \to \mathbb{R}, \; x \mapsto \chi_A(x)$, for $A = (0, \infty) := \{x \in \mathbb{R} | x > 0\}$. The graph of $\chi_{(0,\infty)}$ is

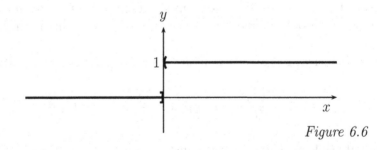

Figure 6.6

Suppose that $\lim_{y \to 0} \chi_{(0,\infty)}(y) = a$ for some $a \in \mathbb{R}$. Then for every $\epsilon > 0$ there exists $\delta > 0$ such that $|y| < \delta$ implies $|\chi_{(0,\infty)}(y) - a| < \epsilon$, i.e. $-\delta < y < \delta$ implies $|\chi_{(0,\infty)}(y) - a| < \epsilon$. Now for $-\delta < y < 0$ we have $\chi_{(0,\infty)}(y) = 0$, implying that $|a| < \epsilon$ for all $\epsilon > 0$, i.e. $a$ must be equal to 0. However, for $0 < y < \delta$, if we have $\chi_{(0,\infty)}(y) = 1$ and with $a = 0$ we would have $|1 - 0| = 1 < \epsilon$ for every $\epsilon > 0$. This is of course a contradiction. Therefore $\lim_{y \to 0} \chi_{(0,\infty)}(y)$ does not exist.

We now return to our original problem and study for a given function $f :$ $D \to \mathbb{R}$ the limit

$$\lim_{y \to x} \frac{f(y) - f(x)}{y - x} \quad \left( = \lim_{y \to x} \frac{f(x) - f(y)}{x - y} \right). \tag{6.24}$$

**Definition 6.2.** *Let* $f : D \to \mathbb{R}$, $D \subset \mathbb{R}$, *be a function and* $x_0 \in D$. *We say that* $f$ *is* **differentiable at** $x_0$ *and has the* **derivative** $f'(x_0)$ **at** $x_0$ *if the following limit*

$$\lim_{\substack{y \to x_0 \\ y \neq x_0}} \frac{f(y) - f(x_0)}{y - x_0} \tag{6.25}$$

*exists. In this case we set*

$$f'(x_0) = \lim_{\substack{y \to x_0 \\ y \neq x_0}} \frac{f(y) - f(x_0)}{y - x_0}. \tag{6.26}$$

**Remark 6.3. A.** It is clear that we have to exclude the value $y = x_0$ in (6.25) otherwise $\frac{f(y) - f(x_0)}{y - x_0}$ may not be defined.
**B.** Note that $\lim_{\substack{y \to x_0 \\ y \neq x_0}} \frac{f(y) - f(x_0)}{y - x_0}$ always means that only points $y \in D$ are considered.
**C.** Note that we have given a pointwise definition, i.e. given $f : D \to \mathbb{R}$, so far we have only defined its derivative at $x_0$ and this is the real number $f'(x_0)$.
**D.** For historical reasons as well as for practical reasons we will often write

$$\frac{df(x_0)}{dx} = f'(x_0) \quad \text{or} \quad \frac{df}{dx}(x_0) = f'(x_0). \tag{6.27}$$

**Example 6.4. A.** Consider the constant function $h_c : \mathbb{R} \to \mathbb{R}$, $x \mapsto h_c(x) = c$. For every $x_0 \in \mathbb{R}$ we find

$$\lim_{\substack{y \to x_0 \\ y \neq x_0}} \frac{h_c(y) - h_c(x_0)}{y - x_0} = \lim_{\substack{y \to x_0 \\ y \neq x_0}} \frac{c - c}{y - x_0} = \lim_{\substack{y \to x_0 \\ y \neq x_0}} h_0(y) = 0,$$

i.e.

$$h'_c(x_0) = 0 \quad \text{for all} \quad x_0 \in \mathbb{R}. \tag{6.28}$$

**B.** For the function $f : \mathbb{R} \to \mathbb{R}$, $x \mapsto ax + b$, $a, b \in \mathbb{R}$, we find for every $x_0 \in \mathbb{R}$

$$\lim_{\substack{y \to x_0 \\ y \neq x_0}} \frac{f(y) - f(x_0)}{y - x_0} = \lim_{\substack{y \to x_0 \\ y \neq x_0}} \frac{ay + b - (ax_0 + b)}{y - x_0}$$

$$= \lim_{\substack{y \to x_0 \\ y \neq x_0}} \frac{a(y - x_0)}{y - x_0} = \lim_{\substack{y \to x_0 \\ y \neq x_0}} h_a(y) = a,$$

i.e.

$$f'(x_0) = a \qquad \text{for all } x_0 \in \mathbb{R}. \tag{6.29}$$

**C.** Let $g : \mathbb{R} \to \mathbb{R}$, $x \mapsto ax^2$, $a \in \mathbb{R}$. Using the formula

$$y^2 - x_0^2 = (y + x_0)(y - x_0)$$

we get for $x_0 \in \mathbb{R}$

$$\lim_{\substack{y \to x_0 \\ y \neq x_0}} \frac{g(y) - g(x_0)}{y - x_0} = \lim_{\substack{y \to x_0 \\ y \neq x_0}} \frac{ay^2 - ax_0^2}{y - x_0} = \lim_{\substack{y \to x_0 \\ y \neq x_0}} \frac{a(y^2 - x_0^2)}{y - x_0}$$

$$= \lim_{\substack{y \to x_0 \\ y \neq x_0}} \frac{a(y + x_0)(y - x_0)}{y - x_0} = \lim_{\substack{y \to x_0 \\ y \neq x_0}} a(y + x_0) = 2ax_0$$

i.e.

$$g'(x_0) = 2ax_0. \tag{6.30}$$

**D.** We want to differentiate the function $h : \mathbb{R} \setminus \{0\} \to \mathbb{R}$, $x \mapsto \frac{1}{x}$. For $x_0 \neq 0$ we find

$$\lim_{\substack{y \to x_0 \\ y \neq x_0}} \frac{\frac{1}{y} - \frac{1}{x_0}}{y - x_0} = \lim_{\substack{y \to x_0 \\ y \neq x_0}} \frac{\frac{x_0 - y}{y \cdot x_0}}{y - x_0}$$

$$= \lim_{\substack{y \to x_0 \\ y \neq x_0}} \frac{-1}{y \cdot x_0} = -\frac{1}{x_0^2},$$

i.e.

$$h'(x_0) = -\frac{1}{x_0^2}, \qquad x_0 \neq 0. \tag{6.31}$$

Recall that by assumption $y \in D(h) = \mathbb{R} \setminus \{0\}$, i.e. $y \neq 0$.

Note that in all these examples we can find the derivative for all points in the domain. Thus in each case we can define a new function. Therefore we give

**Definition 6.5.** *Let* $f : D \to \mathbb{R}$ *be a function. If* $f'(x)$ *exists for all* $x \in D$ *we define the new function* $f'$, *called the* **derivative** *(or first order derivative) of* $f$ *by* $f' : D \to \mathbb{R}, \quad x \mapsto f'(x)$.

By Example 6.4 we may write

$$(c)' = 0 ; \tag{6.32}$$
$$(ax + b)' = a ; \tag{6.33}$$
$$(ax^2)' = 2ax ; \tag{6.34}$$
$$(\frac{1}{x})' = -\frac{1}{x^2}. \tag{6.35}$$

In the next step we want to derive some **rules** for calculating **derivatives**.

**Theorem 6.6.** *Let* $f, g : D \to \mathbb{R}$ *be two functions each differentiable at* $x_0 \in D$ *with derivatives* $f'(x_0)$ *and* $g'(x_0)$, *respectively. Then for all* $a \in \mathbb{R}$ *we have*

$$(af)'(x_0) = af'(x_0) \tag{6.36}$$

*and*

$$(f \pm g)'(x_0) = f'(x_0) \pm g'(x_0). \tag{6.37}$$

*In particular, this means that* $(af)'(x_0)$ *and* $(f \pm g)'(x_0)$ *exist.*

*Proof.* To see (6.36) just note

$$\lim_{\substack{y \to x_0 \\ y \neq x_0}} \frac{(af)(y) - (af)(x_0)}{y - x_0} = \lim_{\substack{y \to x_0 \\ y \neq x_0}} \frac{a(f(y) - f(x_0))}{y - x_0}$$

$$= (\lim_{\substack{y \to x_0 \\ y \neq x_0}} a) \cdot (\lim_{\substack{y \to x_0 \\ y \neq x_0}} \frac{f(y) - f(x_0)}{y - x_0}) = af'(x_0),$$

where we write $\lim_{y \to x_0} a$ for $\lim_{y \to x_0} h_a(y)$ and $h_a(y) = a$ for all $y \in \mathbb{R}$. Now we prove (6.37)

$$\lim_{\substack{y \to x_0 \\ y \neq x_0}} \frac{(f \pm g)(y) - (f \pm g)(x_0)}{y - x_0} = \lim_{\substack{y \to x_0 \\ y \neq x_0}} \left( \frac{f(y) - f(x_0)}{y - x_0} \pm \frac{g(y) - g(x_0)}{y - x_0} \right)$$

$$= \left( \lim_{\substack{y \to x_0 \\ y \neq x_0}} \frac{f(y) - f(x_0)}{y - x_0} \right) \pm \left( \lim_{\substack{y \to x_0 \\ y \neq x_0}} \frac{g(y) - g(x_0)}{y - x_0} \right)$$

$$= f'(x_0) \pm g'(x_0).$$

$\square$

To proceed further, we need the following simple but far reaching observation.

**Lemma 6.7.** *If* $g : D \to \mathbb{R}$ *is differentiable at* $x_0 \in D$ *then*

$$\lim_{\substack{y \to x_0 \\ y \neq x_0}} g(y) = \lim_{y \to x_0} g(y) = g(x_0).$$

*Proof.* Note that for $y \neq x_0$ we have

$$g(y) - g(x_0) = \frac{g(y) - g(x_0)}{y - x_0}(y - x_0).$$

Now $\displaystyle\lim_{\substack{y \to x_0 \\ y \neq x_0}} \frac{g(y) - g(x_0)}{y - x_0} = g'(x_0)$ and $\displaystyle\lim_{\substack{y \to x_0 \\ y \neq x_0}} (y - x_0) = 0$. Consequently we have

$$\lim_{\substack{y \to x_0 \\ y \neq x_0}} (g(y) - g(x_0)) = \lim_{\substack{y \to x_0 \\ y \neq x_0}} \frac{g(y) - g(x_0)}{y - x_0} \lim_{\substack{y \to x_0 \\ y \neq x_0}} (y - x_0) = 0,$$

or

$$\lim_{\substack{y \to x_0 \\ y \neq x_0}} g(y) = g(x_0).$$

$\square$

We want to determine $(fg)'(x_0)$ for two function $f, g : D \to \mathbb{R}$ differentiable at $x_0$. For this firstly consider

$$
\begin{aligned}
\frac{(f \cdot g)(y) - (f \cdot g)(x_0)}{y - x_0} &= \frac{f(y)g(y) - f(x_0)g(x_0)}{y - x_0} \\
&= \frac{(f(y) - f(x_0))g(y) + (g(y) - g(x_0))f(x_0)}{y - x_0} \\
&= \frac{f(y) - f(x_0)}{y - x_0} \cdot g(y) + f(x_0) \cdot \frac{g(y) - g(x_0)}{y - x_0}.
\end{aligned}
$$

Now we can prove **Leibniz's rule**, which is also known as the **product rule**.

**Theorem 6.8.** *Let* $f, g : D \to \mathbb{R}$ *be two functions each differentiable at* $x_0 \in D$. *Then* $(f \cdot g)$ *is differentiable at* $x_0$ *and for* $(f \cdot g)'(x_0)$ *we find*

$$(f \cdot g)'(x_0) = f'(x_0)g(x_0) + f(x_0)g'(x_0). \tag{6.38}$$

*Proof.* Using the calculation made above we have

$$
\lim_{\substack{y \to x_0 \\ y \neq x_0}} \frac{(fg)(y) - (fg)(x_0)}{y - x_0} = \lim_{\substack{y \to x_0 \\ y \neq x_0}} \left( \frac{f(y) - f(x_0)}{y - x_0} \cdot g(y) + f(x_0) \cdot \frac{g(y) - g(x_0)}{y - x_0} \right)
$$

$$
= \lim_{\substack{y \to x_0 \\ y \neq x_0}} \left[ \frac{f(y) - f(x_0)}{y - x_0} \cdot g(y) \right] + \lim_{\substack{y \to x_0 \\ y \neq x_0}} \left[ f(x_0) \cdot \frac{g(y) - g(x_0)}{y - x_0} \right].
$$

Now it follows by Lemma 6.7 that

$$
\lim_{\substack{y \to x_0 \\ y \neq x_0}} \frac{(fg)(y) - (fg)(x_0)}{y - x_0}
$$

$$
= \lim_{\substack{y \to x_0 \\ y \neq x_0}} \left( \frac{f(y) - f(x_0)}{y - x_0} \right) \lim_{\substack{y \to x_0 \\ y \neq x_0}} g(y) + \lim_{\substack{y \to x_0 \\ y \neq x_0}} f(x_0) \lim_{\substack{y \to x_0 \\ y \neq x_0}} \left( \frac{g(y) - g(x_0)}{y - x_0} \right)
$$

$$
= f'(x_0)g(x_0) + f(x_0)g'(x_0).
$$

$\square$

With Lemma 6.7 in mind we add a new, central concept to our considerations.

**Definition 6.9.** *A function* $f : D \to \mathbb{R}$ *is called* **continuous at** $x_0 \in D$ *if* $\lim_{y \to x_0} f(y) = f(x_0)$. *If* $f$ *is continuous for each* $x_0 \in D$ *we call* $f$ *continuous (in* $D$*).*

We can now restate Lemma 6.7 as:

**Corollary 6.10.** *Let* $f : D \to \mathbb{R}$ *be a function. This function is continuous at each point where it is differentiable.*

The class of continuous functions is much larger than the class of differentiable functions and we will discuss these functions in greater detail later on. We will also give an example of a continuous function which is not differentiable.

**Remark 6.11.** A function $f : (a, b) \to \mathbb{R}$ is continuous at $x_0 \in (a, b)$ if and only if

$$
\lim_{y \to x_0} f(y) = f(\lim_{y \to x_0} y) = f(x_0). \tag{6.39}
$$

Next we use Leibniz's rule to calculate further derivatives.

**Example 6.12. A.** The derivative of the function $M_n : \mathbb{R} \to \mathbb{R}$, $x \mapsto x^n$, $n \in \mathbb{N}$, is given by $M_n'(x) = nx^{n-1}$, i.e.

$$(x^n)' = nx^{n-1}. \qquad (6.40)$$

We prove this by mathematical induction. For $n = 1$ we have $M_1(x) = x$, i.e.

$$M_1'(x) = 1 = 1 \cdot x^0.$$

Now assume that $M_m'(x) = mx^{m-1}$. We calculate

$$\begin{aligned}
M_{m+1}'(x) &= (xM_m(x))' \\
&= M_m(x) + xM_m'(x) \\
&= x^m + mx \cdot x^{m-1} \\
&= (m+1)x^m,
\end{aligned}$$

which proves (6.40).

**B.** For a polynomial $p(x) = \sum_{j=0}^N a_j x^j$ we have

$$p'(x) = \sum_{j=0}^N a_j j x^{j-1} = \sum_{j=1}^N a_j j x^{j-1}. \qquad (6.41)$$

The proof consists of the following chain of observations:

$$(a_j x^j)' = a_j j x^{j-1},$$

and for differentiable functions $f_j$ we have

$$\left(\sum_{j=0}^N f_j\right)' = \sum_{j=0}^N f_j',$$

which follows from (6.37). For example we find

$$(5x^2 + 7x^3 - 3x^5)' = 10x + 21x^2 - 15x^4.$$

**C.** Consider the function $x \to \frac{x^2+1}{x}$ for $x \neq 0$, we can write this function as

$$(x^2 + 1) \cdot \left(\frac{1}{x}\right),$$

and to determine its derivative we use also $(\frac{1}{x})' = -\frac{1}{x^2}$ :

$$\left((x^2+1)(\frac{1}{x})\right)' = (x^2+1)'(\frac{1}{x}) + (x^2+1)(\frac{1}{x})'$$
$$= 2x(\frac{1}{x}) + (x^2+1)(-\frac{1}{x^2})$$
$$= 2 - \frac{x^2+1}{x^2}$$
$$= \frac{2x^2 - x^2 - 1}{x^2}$$
$$= \frac{x^2 - 1}{x^2}$$
$$= 1 - \frac{1}{x^2}.$$

**D.** For $n \in \mathbb{N}$ we claim

$$(x^{-n})' = -nx^{-n-1}, \qquad x \neq 0. \tag{6.42}$$

Again we use induction. For $n = 1$ we know

$$(x^{-1})' = -x^{-2} = -x^{-1-1}.$$

Now, if $(x^{-m})' = -mx^{-m-1}$ it follows that

$$(x^{-m-1})' = (\frac{1}{x}(x^{-m}))'$$
$$= -\frac{1}{x^2}(x^{-m}) + \frac{1}{x}(x^{-m})'$$
$$= -x^{-m-2} - \frac{1}{x}mx^{-m-1}$$
$$= -(m+1)x^{-m-2},$$

proving (6.42).

In the next chapter we will discuss more examples after having investigated the derivatives of composed functions.

## Problems

1. Using the rules $(6.18) - (6.20)$ for limits prove:

    a) $\lim\limits_{x \to \frac{3}{4}} \left(\frac{5}{3}x^2 - \frac{7}{12}x\right) = \frac{1}{2}$;   b) $\lim\limits_{x \to 1} \dfrac{1 - x^2}{1 - x} = 2$;

    c) $\lim\limits_{x \to 3} \dfrac{x^3 - 4x^2 + 7x - 13}{-\frac{7}{5}x^2 + \frac{1}{1+x^2}} = \dfrac{2}{25}$.

2. Find the following limits:

    a) $\lim\limits_{x \to 4} \dfrac{x^2 - 2x + 5}{x - 2}$;   b) $\lim\limits_{x \to -3} \dfrac{x^2 - 9}{(x + 5)(x + 3)}$.

3. Consider the function
$$f : \mathbb{R} \longrightarrow \mathbb{R},$$
$$x \mapsto \begin{cases} x^3 - 22, & x \neq 3 \\ 17, & x = 3. \end{cases}$$
Find $\lim\limits_{x \to 3} f(x)$. Is $f$ continuous at $x = 3$?

4.   a) Assume: $f, g : (a, b) \longrightarrow \mathbb{R}$, $a < b$, are two functions such that $|f(x)| \leq g(x)$ for all $x \in (a, b)$, then $\lim\limits_{x \to c} g(x) = 0$ implies $\lim\limits_{x \to c} f(x) = 0$, $a < c < b$.
Prove that for every bounded function $h : (-2, 2) \longrightarrow \mathbb{R}$ it follows that $\lim\limits_{x \to 0}(xh(x)) = 0$. Here, we call $h$ bounded if for some $M \geq 0$ we have $|h(x)| \leq M$ for all $x \in (-2, 2)$.

   b) Use part a) to prove that the function
$$f : \mathbb{R} \longrightarrow \mathbb{R},$$
$$x \mapsto \begin{cases} x \sin \frac{1}{x}, & x \neq 0 \\ 0, & x = 0 \end{cases}$$
is continuous at $x = 0$.

5. By using the definition of the derivative prove that $f : (-1, 1) \longrightarrow \mathbb{R}$, $x \mapsto \frac{3}{4}x^2 - 2$ is differentiable at $x_0 = -\frac{1}{2}$ and find $f'\left(-\frac{1}{2}\right)$.

6.* Consider the characteristic function $\chi_{[0,1]} : \mathbb{R} \longrightarrow \mathbb{R}$. Prove that this function is differentiable for $x \notin \{0, 1\}$ and has derivative 0, while for $x \in \{0, 1\}$ the function is not differentiable.

7.* Consider the function
$$g : \mathbb{R} \longrightarrow \mathbb{R},$$
$$g(x) = \begin{cases} 1, & x \leq 2 \\ x^2 - 3, & x > 2. \end{cases}$$

Prove that $g$ is not differentiable at $x_0 = 2$. Is $g$ continuous at $x_0 = 2$? Hint: you will need to go back to the very definition in order to investigate the continuity of $g$ at $x_0 = 2$.

8. Using rules $(6.36) - (6.38)$ as well as Example 6.12 find the derivatives of the following functions:

   a) $f : (1, 5) \longrightarrow \mathbb{R}$, $f(x) = \frac{7}{5}x^2 - \frac{2}{x^3}$;

   b) $g : (1, 2) \longrightarrow \mathbb{R}$, $g(t) = \frac{t^7 + 12t^3 - 2}{t^5}$;

   c) $h : (2, 7) \longrightarrow \mathbb{R}$, $h(s) = \sum_{j=1}^{M} j s^{-j}$, $M \geq 2$.

9. First prove that $f : \mathbb{R} \longrightarrow \mathbb{R}$, $f(x) = \chi_{\mathbb{R}_+}$ is not differentiable at $x_0 = 0$. Now consider the function $h : \mathbb{R} \longrightarrow \mathbb{R}$, $x \mapsto x^2 f(x) = x^2 \chi_{\mathbb{R}_+}(x)$. Is $h$ differentiable at $x_0 = 0$? If it is, find $h'(0)$.

# 7 Derivatives Continued

In this chapter we want to extend the number of rules for calculating derivatives. Before doing this, let us agree to a slight simplification in our notation. In the following we will often write

$$\lim_{y \to x_0} \frac{f(y) - f(x_0)}{y - x_0} \quad \text{instead of} \quad \lim_{\substack{y \to x_0 \\ y \neq x_0}} \frac{f(y) - f(x_0)}{y - x_0},$$

however we still assume that $y \neq x_0$ when using the simplified notation.

**Example 7.1.** We want to find the derivative of the function $f : \mathbb{R} \longrightarrow \mathbb{R}$, $x \longmapsto (x^2 + 1)^2$. There is an easy way to do this:

$$(x^2 + 1)^2 = x^4 + 2x^2 + 1,$$

hence

$$f'(x) = 4x^3 + 4x.$$

If we instead consider the function $\tilde{f}_k : \mathbb{R} \longrightarrow \mathbb{R}$, $x \longmapsto (x^2 + 1)^k$, $k \in \mathbb{N}$, the calculation becomes more involved, we first calculate $(x^2 + 1)^k = \ldots$ and then take the derivative. Note that $f$ and $\tilde{f}_k$ are composed functions. With $\tilde{g} : \mathbb{R} \longrightarrow \mathbb{R}$, $x \longmapsto x^2 + 1$, we find $f(x) = (\tilde{g}(x))^2$ and $\tilde{f}_k(x) = (\tilde{g}(x))^k$. Thus with $h_k(y) = y^k$ we have $f = h_2 \circ \tilde{g}$ and $\tilde{f}_k = h_k \circ \tilde{g}$. We aim to express for an arbitrary composed function $f = h \circ g$ its derivative by using those of $h$ and $g$. Note that $h_k'(x) = kx^{k-1}$ is simple to calculate as is $\tilde{g}'(x) = 2x$.

**Example 7.2.** Consider $\sqrt{\cdot} : \mathbb{R}_+ \longrightarrow \mathbb{R}$, $x \longmapsto \sqrt{x}$. We want to calculate the derivative of $\sqrt{\cdot}$ at $x_0 \in \mathbb{R}_+$. Thus we have to look at

$$\frac{\sqrt{x} - \sqrt{x_0}}{x - x_0} = \frac{(\sqrt{x} - \sqrt{x_0})(\sqrt{x} + \sqrt{x_0})}{(x - x_0)(\sqrt{x} + \sqrt{x_0})}$$

$$= \frac{x - x_0}{(x - x_0)(\sqrt{x} + \sqrt{x_0})}$$

$$= \frac{1}{\sqrt{x} + \sqrt{x_0}}.$$

Assuming $\lim_{x \to x_0} \sqrt{x} = \sqrt{x_0}$ we find for $x_0 \neq 0$

$$(\sqrt{x})'\big|_{x=x_0} = \frac{1}{2\sqrt{x_0}};$$

or

$$(x^{1/2})' = \frac{1}{2x^{1/2}} = \frac{1}{2}x^{-1/2}, \qquad x > 0. \tag{7.1}$$

Now we want to calculate $\frac{d}{dx}\sqrt{p(x)}$, where $p : \mathbb{R} \longrightarrow \mathbb{R}$ is a differentiable function with $R(p) \subset \{x \in \mathbb{R} | x > 0\}$.

Consider

$$
\begin{aligned}
\frac{\sqrt{p(x)} - \sqrt{p(x_0)}}{x - x_0} &= \frac{(\sqrt{p(x)} - \sqrt{p(x_0)})(\sqrt{p(x)} + \sqrt{p(x_0)})}{(x - x_0)(\sqrt{p(x)} + \sqrt{p(x_0)})} \\
&= \frac{p(x) - p(x_0)}{x - x_0} \cdot \frac{1}{\sqrt{p(x)} + \sqrt{p(x_0)}},
\end{aligned}
$$

and for $x \longrightarrow x_0$ we find assuming $\lim\limits_{x \to x_0} \sqrt{p(x)} = \sqrt{p(x_0)}$, that

$$\lim_{x \to x_0} \frac{\sqrt{p(x)} - \sqrt{p(x_0)}}{x - x_0} = \frac{1}{2\sqrt{p(x_0)}}p'(x_0).$$

If we write for a moment $g(x) = \sqrt{x}$ the above result reads as

$$
\begin{aligned}
(g \circ p)'(x_0) &= \lim_{x \to x_0} \frac{g(p(x)) - g(p(x_0))}{x - x_0} \\
&= \lim_{x \to x_0} \frac{\sqrt{p(x)} - \sqrt{p(x_0)}}{x - x_0} \\
&= \frac{1}{2\sqrt{p(x_0)}}p'(x_0) = g'(p(x_0)) \cdot p'(x_0),
\end{aligned}
$$

where we used that $g'(x) = (\sqrt{x})' = \frac{1}{2\sqrt{x}}$, which we still need to prove.

The previous example suggests the following general result:

$$(f \circ h)'(x) = f'(h(x)) \cdot h'(x)$$

and we are going to prove this now.

**Theorem 7.3 (Chain rule).** *Let $h : D \longrightarrow \mathbb{R}$ be a differentiable function and let $f : G \longrightarrow \mathbb{R}$, $R(h) \subset G$, be a further differentiable function. Then the composed function $f \circ h : D \longrightarrow \mathbb{R}$ is differentiable and*

$$(f \circ h)'(x) = f'(h(x)) \cdot h'(x), \quad x \in D. \tag{7.2}$$

*Proof.* First recall that by Corollary 6.10 both functions $h$ and $f$ are continuous. In particular we have

$$\lim_{x \to x_0} h(x) = h(x_0).$$

Now consider as a first attempt

$$\frac{f(h(x)) - f(h(x_0))}{x - x_0} = \frac{f(h(x)) - f(h(x_0))}{h(x) - h(x_0)} \cdot \frac{h(x) - h(x_0)}{x - x_0},$$

with $y = h(x)$, $y_0 = h(y_0)$ this reads as

$$\frac{f(h(x)) - f(h(x_0))}{x - x_0} = \frac{f(y) - f(y_0)}{y - y_0} \cdot \frac{h(x) - h(x_0)}{x - x_0}.$$

As $x \longrightarrow x_0$ we know that

$$\lim_{x \to x_0} \frac{h(x) - h(x_0)}{x - x_0} = h'(x_0)$$

and since $x \longrightarrow x_0$ implies $y = h(x) \longrightarrow h(x_0) = y_0$ we have

$$
\begin{aligned}
\lim_{x \to x_0} \frac{f(h(x)) - f(h(x_0))}{x - x_0} &= \lim_{y \to y_0} \frac{f(y) - f(y_0)}{y - y_0} \\
&= f'(y_0) \\
&= f'(h(x_0))
\end{aligned}
$$

which yields indeed

$$(f \circ h)'(x_0) = f'(h(x_0)) \cdot h'(x_0).$$

However, there is a problem: $h(x) - h(x_0) \neq 0$ need not be true. Indeed the term $h(x) - h(x_0)$ could be zero for infinitely many values. Thus we have to modify the proof. Define the function

$$f^*(y) := \begin{cases} \frac{f(y) - f(y_0)}{y - y_0} & \text{for } y \neq y_0 \\ f'(y_0) & \text{for } y = y_0 \end{cases}. \tag{7.3}$$

Then we have

$$\lim_{y \to y_0} f^*(y) = f^*(y_0) = f'(y_0) \tag{7.4}$$

and further

$$f(y) - f(y_0) = f^*(y)(y - y_0).\qquad(7.5)$$

Now it follows that

$$
\begin{aligned}
(f \circ h)'(x_0) &= \lim_{x \to x_0} \frac{f(h(x)) - f(h(x_0))}{x - x_0} \\
&= \lim_{x \to x_0} \frac{f^*(h(x))(h(x) - h(x_0))}{x - x_0} \\
&= \lim_{x \to x_0} f^*(h(x)) \lim_{x \to x_0} \frac{h(x) - h(x_0)}{x - x_0} \\
&= f'(h(x_0))h'(x_0),
\end{aligned}
$$

where we used that $\lim_{x \to x_0} h(x) = h(x_0)$ and therefore $\lim_{y \to y_0} f^*(y) = f'(y_0)$ implies $\lim_{x \to x_0} f^*(h(x)) = f'(h(x_0))$. $\qquad\square$

**Example 7.4. A.** In the situation of Example 7.1 we first find

$$\left((x^2 + 1)^2\right)' = 2(x^2 + 1) \cdot 2x = 4x^3 + 4x,$$

and more generally

$$\left((x^2 + 1)^k\right)' = k(x^2 + 1)^{k-1} \cdot 2x = 2xk(x^2 + 1)^{k-1}.$$

**B.** Let $g : \mathbb{R} \longrightarrow \mathbb{R}$, $g(x) \neq 0$ for all $x \in \mathbb{R}$, be differentiable. We want to find $\left(\frac{1}{g(\cdot)}\right)'(x)$. With $f(x) = \frac{1}{x}$ for $x \neq 0$ the function $x \longmapsto \frac{1}{g(x)}$ is the composed function $x \longmapsto (f \circ g)(x)$.
Therefore we find

$$
\begin{aligned}
\left(\frac{1}{g(\cdot)}\right)'(x) &= (f \circ g)'(x) = f'(g(x)) \cdot g'(x) \\
&= -\frac{1}{g^2(x)} \cdot g'(x) = -\frac{g'(x)}{g(x)^2},
\end{aligned}
$$

i.e.

$$\left(\frac{1}{g}\right)' = -\frac{g'}{g^2}, \quad g \neq 0.\qquad(7.6)$$

Thus for $g(x) = x^2 + 1$ we find

$$\left(\frac{1}{x^2 + 1}\right)' = \frac{-2x}{(x^2 + 1)^2}.$$

**C.** Let $g : \mathbb{R} \longrightarrow \mathbb{R}$, $g(x) \neq 0$ for all $x \in \mathbb{R}$, be a differentiable function and let $h : \mathbb{R} \longrightarrow \mathbb{R}$ be a further differentiable function. Then it follows using Leibniz's rule and (7.6)) that

$$
\begin{aligned}
\left(\frac{h}{g}\right)'(x) &= \left(h \cdot \frac{1}{g}\right)'(x) \\
&= h'(x) \cdot \frac{1}{g(x)} + h(x) \left(\frac{1}{g}\right)'(x) \\
&= \frac{h'(x)}{g(x)} - \frac{h(x) \cdot g'(x)}{g^2(x)} \\
&= \frac{g(x) \cdot h'(x) - g'(x) \cdot h(x)}{g(x)^2}.
\end{aligned}
$$

This rule is often called the **quotient rule.**

For example we find

$$
\begin{aligned}
\left(\frac{x^3 - 7x}{x^2 + 3}\right)' &= \frac{(x^2 + 3)(3x^2 - 7) - 2x(x^3 - 7x)}{(x^2 + 3)^2} \\
&= \frac{x^4 + 16x^2 - 21}{x^4 + 6x^2 + 9}.
\end{aligned}
$$

We may use the chain rule to determine the derivative of the inverse function of $f : \mathbb{R} \longrightarrow \mathbb{R}$ provided it exists. Since $f^{-1}(f(x)) = x$ we find by the chain rule

$$\left(f^{-1} \circ f\right)'(x) = (f^{-1})'(f(x)) \cdot f'(x) = (x)' = 1.$$

In the case where $f'(x) \neq 0$ we find

$$\left(f^{-1}\right)(f(x)) = \frac{1}{f'(x)}$$

or with $f(x) = y$, i.e. $x = f^{-1}(y)$ we get

$$\left(f^{-1}\right)'(y) = \frac{1}{f'(f^{-1}(y))}$$

or putting $\phi(y) := f^{-1}(y)$ :

$$\phi'(y) = \frac{1}{f'(\phi(y))}.$$

This calculation has some critical points, but it paves the way to prove:

**Theorem 7.5.** *Let $D \subset \mathbb{R}$ be a closed interval and let $f : D \longrightarrow \mathbb{R}$ be an injective function, i.e. $f : D \longrightarrow R(f)$ is bijective. Suppose that $f$ is differentiable at the point $x_0 \in D$ and that $f'(x_0) \neq 0$. Then the inverse function*

$$\phi := f^{-1} : R(f) \longrightarrow \mathbb{R}$$

*is differentiable at $y_0 := f(x_0)$ and we have*

$$\phi'(y_0) = \frac{1}{f'(x_0)} = \frac{1}{f'(\phi(y_0))}. \tag{7.7}$$

We will provide a complete proof of this theorem later in our course but for the moment we take this result for granted.

**Example 7.6.** Let $f : \mathbb{R}_+ \longrightarrow \mathbb{R}_+$, $x \longmapsto x^2$. For $x \neq 0$ we have $f'(x) = 2x \neq 0$. The inverse function $f^{-1}$ is of course $f^{-1} : \mathbb{R}_+ \longrightarrow \mathbb{R}_+$, $x \longmapsto \sqrt{x}$. From (7.7) we derive with $\sqrt{y_0} = x_0$, i.e. $y_0 = x_0^2$ that

$$(\sqrt{y_0})' = \frac{1}{2x_0} = \frac{1}{2\sqrt{y_0}}$$

confirming our previous result.

We close this chapter by providing an example of a continuous function which is not differentiable.

**Example 7.7.** The function $|\cdot| : \mathbb{R} \to \mathbb{R}$ is not differentiable at $x_0 = 0$. Consider the quotient

$$\frac{|x| - |x_0|}{x - x_0} = \frac{|x|}{x} = \begin{cases} 1, & \text{for } x > 0 \\ -1, & \text{for } x < 0. \end{cases}$$

Suppose that $\lim\limits_{\substack{x \to 0 \\ x \neq 0}} \dfrac{|x|}{x} = a$ for some $a \in \mathbb{R}$. Then for $\epsilon = \frac{1}{2}$ there exists $\delta > 0$ such that for all $x \in \mathbb{R}$ with $|x| < \delta$, i.e. $-\delta < x < \delta$, it follows that

$\left|\frac{|x|}{x} - a\right| < \frac{1}{2}$. In particular, for $-\delta < x < 0$ we have $|-1-a| < \frac{1}{2}$ and for $0 < x < \delta$ we have $|1-a| < \frac{1}{2}$, i.e.

$$-\frac{1}{2} < -1-a < \frac{1}{2} \quad \text{and} \quad -\frac{1}{2} < 1-a < \frac{1}{2}$$

implying that $-\frac{3}{2} < a < \frac{1}{2}$ and $\frac{1}{2} < a < \frac{3}{2}$ which is a contradiction. Therefore $x \mapsto |x|$ is not differentiable at $x_0 = 0$.

The continuity of $x \mapsto |x|$ at $x_0 = 0$ is trivial. We just need to prove that $\lim_{x \to 0} |x| = 0$, i.e. given $\epsilon > 0$ we need to find $\delta > 0$ such that $|x| < \delta$ implies $||x| - |0|| = |x| < \epsilon$. Thus $\delta = \epsilon$ will do.

# Problems

To solve these problems knowledge of derivatives of rational functions and the square root function $x \mapsto \sqrt{x}$, $x > 0$ may be used. Moreover, while solving these problems results from previous questions may be used without proof or justification.

1. Consider the function $h_k : (0, \infty) \longrightarrow \mathbb{R}$, $k \in \mathbb{N}$, $h_k(x) = \sqrt{x^k}$. Find $\frac{d}{dx} h_k(x)$.

2. Find the derivatives of the following functions:

   i) $f : \mathbb{R} \longrightarrow \mathbb{R}$, $f(x) = (1+x^2)^{-\frac{k}{2}}$, $k \in \mathbb{N}$;

   ii) $g : \mathbb{R} \setminus \{0\} \longrightarrow \mathbb{R}$, $g(y) = \sqrt{1 + \frac{1}{y^4}}$;

   iii) $h : \mathbb{R} \longrightarrow \mathbb{R}$, $h(z) = \sqrt{\frac{z^4}{1+z^2}}$.

3. Find the derivatives of the following functions:

   i) $f : \mathbb{R} \longrightarrow \mathbb{R}$, $f(u) = \frac{3u^5 - 7u^9}{1+u^6+u^8}$;

   ii) $g : \mathbb{R} \longrightarrow \mathbb{R}$, $g(v) = \frac{(1+v^2)^{\frac{1}{2}}}{(5+v^2)^{\frac{7}{2}}}$;

   iii) $h : (0, \infty) \longrightarrow \mathbb{R}$, $h(z) = \frac{\sqrt{z^5} - 2z^4}{12 + z^2(1+z^3)}$.

4. The function $f : (0, \infty) \longrightarrow (0, \infty)$, $x \mapsto x^k$, $k \in \mathbb{N}$, is bijective and $f'(x) = kx^{k-1} \neq 0$ for all $x \in (0, \infty)$. Find the derivative of its inverse function $f^{-1}$.

5. In the following denote the inverse function of $x \mapsto x^k$ by $x \mapsto x^{\frac{1}{k}} = \sqrt[k]{x}$, $x > 0$ for $k \in \mathbb{N}$. Find the derivatives of:

i) $f : \mathbb{R} \longrightarrow \mathbb{R}$, $f(s) = (1 + s^2)^{\frac{1}{k}}$;

ii) $g : \mathbb{R} \longrightarrow \mathbb{R}$, $g(t) = \frac{\sqrt{1+t^4}}{\sqrt[5]{1+t^6+t^8}}$;

iii) $h : (0, \infty) \longrightarrow \mathbb{R}$, $h(u) = \frac{u^{\frac{1}{7}}}{\sqrt{\frac{1+u^2}{1+u^4}}}$.

6. For $x > 0$, $k \in \mathbb{N}$ and $l \in \mathbb{N}_0$ we set $x^{\frac{l}{k}} = x^l \cdot x^{\frac{1}{k}} = x^l \sqrt[k]{x}$. Find the derivatives of:

i) $f : (0, \infty) \longrightarrow \mathbb{R}$, $f(x) = x^{\frac{l}{k}}$;  ii) $g : \mathbb{R} \longrightarrow \mathbb{R}$, $g(s) = \frac{(1+s^2)^{-\frac{3}{2}}}{(1+s^4)^5}$.

7. Let $p, q : \mathbb{R} \longrightarrow \mathbb{R}$ be two polynomials such that $q(x) \neq 0$ for all $x \in \mathbb{R}$ and $\frac{p(x)}{q(x)} > 2$ for all $x \in \mathbb{R}$. Find

$$\frac{d}{dx} \sqrt{\frac{p(x)}{q(x)} - 2}.$$

8. Find the derivative of $g : (-1, 1) \longrightarrow \mathbb{R}$, where $g(t) = \sqrt{(t^2 - 1)(2t + 3)^{\frac{1}{2}}}$.

9. Let $f : (0, 1) \longrightarrow (2, 3)$ and $h : (2, 3) \longrightarrow (3, 4)$ be two bijective and differentiable functions such that $f'(x) \neq 0$ for all $x \in (0, 1)$ and $h'(y) \neq 0$ for all $y \in (2, 3)$. For $z \in (3, 4)$ find the derivative of $(h \circ f)^{-1}(z)$.

10. Let $p(x) = \sum_{k=0}^{m} a_k x^k$ be a polynomial and $u : \mathbb{R} \longrightarrow \mathbb{R}$ be a differentiable function. i) Find $\frac{d}{dx}(p(u(x)))$.  ii) Find $\frac{d}{dx} u(p(x))$.  iii) Suppose that $u(x) \neq 0$ for all $x \in \mathbb{R}$. Find $\frac{d}{dx} \frac{1}{u(p(x))}$.

# 8 The Derivative as a Tool to Investigate Functions

In this chapter we discuss how to use the derivative to investigate functions. We will give some motivation for the results and statements, but we postpone most of the proofs until Part 2 of our course. The reason is simple: all proofs will require a deeper understanding of the concept of a limit. However it is helpful to introduce at an early stage certain useful tools. In fact this is the main justification for a calculus course preceding a rigorous analysis course.

**Example 8.1.** Consider the function $f$ corresponding to the given graph $\Gamma(f)$

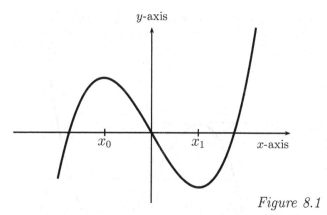

*Figure 8.1*

It looks like the function is unbounded but at $x_0$ the function has a local maximum and at $x_1$ it has a local minimum.

We want to find criteria for these properties to hold. For this we first need some definitions.

**Definition 8.2.** *A function* $f : D \to \mathbb{R}$ *is said to be **bounded** if there exists* $M \geq 0$ *such that* $|f(x)| \leq M$ *for all* $x \in D$.

**Example 8.3. A.** The function $\chi_A : \mathbb{R} \to \mathbb{R}$ is for every set $A \subset \mathbb{R}$ bounded since $|\chi_A(x)| \leq 1$ for all $x \in \mathbb{R}$.
**B.** The function $|\cdot| : \mathbb{R} \to \mathbb{R}, \quad x \mapsto |x|$ is unbounded. Indeed suppose there exists $M \geq 0$ such that $|x| \leq M$ for all $x \in \mathbb{R}$. Then for $x = M + 1$ we would find

$$|M + 1| = M + 1 \leq M$$

115

which is a contradiction.

**Definition 8.4.** *Let $f : (a, b) \to \mathbb{R}$ be a function, $a < b$. We say that $f$ has a **local maximum** at $x_0 \in (a, b)$ (a **local minimum** at $x_1 \in (a, b)$) if there exists $\epsilon > 0$ such that $f(x_0) \geq f(y)$ for all $y \in (a, b)$ satisfying $|x_0 - y| < \epsilon$ ($f(x_1) \leq f(y)$ for all $y \in (a, b)$ such that $|x_1 - y| < \epsilon$).*

In the case that $f$ has either a local maximum or a local minimum at $x_2 \in (a, b)$ we just speak of a **local extreme value** or a **local extremum** at $x_2$. Of central importance is:

**Theorem 8.5.** *Suppose that $f : (a, b) \to \mathbb{R}$ has a local extremum at $x_0 \in (a, b)$. If $f$ is differentiable at $x_0$ then $f'(x_0) = 0$.*

This result fits well to our imagination, look at the graph

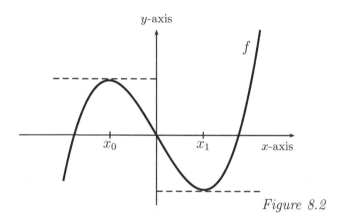

Figure 8.2

The function $f$ has a local maximum at $x_0$ and a local minimum at $x_1$. At these points we expect there to be a horizontal tangent, i.e. a tangent with slope zero.

**Example 8.6. A.** Consider the parabola $f : \mathbb{R} \to \mathbb{R}$, $x \mapsto (x - \alpha)^2 + \beta$. It is differentiable for all $x \in \mathbb{R}$ with derivative $f'(x) = 2(x - \alpha)$, thus $f'(x_0) = 0$ if and only if $x_0 = \alpha$. If we restrict $f$ to any interval (a,b) such that $\alpha \in (a, b)$ then according to Theorem 8.5 the function $f|_{(a,b)}$ might have a local extreme value at $x_0 = \alpha$. In this example the statement is of course easy to prove without using the derivative. Since $(x - \alpha)^2 \geq 0$ for all $x \in \mathbb{R}$ it follows that $f(x) \geq \beta$ for all $x \in \mathbb{R}$ but for $x = \alpha$ we have $f(\alpha) = \beta$ implying that there is a (local) minimum at $x_0 = \alpha$.

**B.** Consider the function $g : \ (-N, M) \to \mathbb{R}, \ M, N \in \mathbb{N}, \ x \mapsto x^3$.
The only zero of $g'(x) = 3x^2$ is $x_0 = 0$. Now $g(0) = 0$, but $g(x) < 0$ for $x < 0$
and $g(x) > 0$ for $x > 0$. Hence the function has no local extreme value at
$x_0 = 0$. This is obvious from its graph:

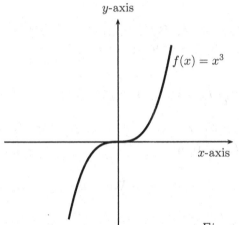

*Figure 8.3*

This example shows that Theorem 8.5 is a necessary but not a sufficient
condition for a local extreme value.
**C.** The function $|\cdot| : \ \mathbb{R} \to \mathbb{R}, \ x \mapsto |x|$ is for all $x > 0$ or $x < 0$ strictly
positive whereas $|0| = 0$. Thus at $x_0 = 0$ it has a local minimum. However
the absolute value is not differentiable at $x_0 = 0$, compare Example 7.7. Thus
we cannot apply Theorem 8.5.

Theorem 8.5 only gives a necessary condition for local extreme values to
exist. We want to find now sufficient criteria for local maxima and minima.
It turns out that for this we need **higher order derivatives**.
Let $f : \ D \to \mathbb{R}$ be a function such that $f'(x)$ exist for all $x \in D$. Then we
can consider $f'$ as a new function $f' : \ D \to \mathbb{R}, \ x \mapsto f'(x)$. Next we may
ask whether $f'$ has at $x_0 \in D$ a derivative, i.e. whether

$$\lim_{\substack{x \to x_0 \\ x \neq x_0}} \frac{f'(x) - f'(x_0)}{x - x_0} = f''(x_0) \tag{8.1}$$

exists. When it does we call $f''(x_0)$ the **second derivative** of $f$ at $x_0$. Instead of $f''(x_0)$ the notation $f^{(2)}(x_0)$ or $\frac{d^2}{dx^2}f(x_0)$ are common. Of course we may iterate the process and define

$$\frac{d^k f}{dx^k}(x_0) = f^{(k)}(x_0) = (f^{k-1})'(x_0)$$

$$= \lim_{\substack{x \to x_0 \\ x \neq x_0}} \frac{f^{k-1}(x) - f^{k-1}(x_0)}{x - x_0}$$

as the $k^{\text{th}}$ **derivative** of $f$ at $x_0$. By definition: $f^{(0)} = \frac{d^0 f}{dx^0} = f$. Note that the definition of higher order derivatives is a definition by recursion:

$$\frac{d^k}{dx^k}f(x) := \frac{d}{dx}\left(\frac{d^{k-1}}{dx^{k-1}}f\right)(x).$$

**Example 8.7. A.** Consider $f : \mathbb{R} \to \mathbb{R}$, $x \mapsto x^2$. Then we find $f'(x) = 2x$, $f''(x) = 2$ and $f^{(3)}(x) = 0$, hence $f^{(k)}(x) = 0$ for $k \geq 3$.

**B.** Consider $g : (0, \infty) \to \mathbb{R}$, $x \mapsto x^{-1}$. We find

$$g'(x) = -1 \cdot x^{-2}, \ g''(x) = (-1)(-2)x^{-3}, \ g^{(3)} = (-1)(-2)(-3)x^{-4}.$$

Clearly we may extend our rules for taking derivatives to higher order derivatives. Here are some of the simple ones:

$$\frac{d^k}{dx^k}(f \pm g) = \frac{d^k f}{dx^k} \pm \frac{d^k g}{dx^k} \tag{8.2}$$

and

$$\frac{d^k}{dx^k}(cf) = c\frac{d^k f}{dx^k}. \tag{8.3}$$

However the following rule is not so simple:

$$\frac{d^k}{dx^k}(f \cdot g) = \sum_{l=0}^{k} \binom{k}{l} \frac{d^{k-l}f}{dx^{k-l}} \frac{d^l g}{dx^l}, \tag{8.4}$$

where $\binom{k}{l}$ denote the binomial coefficients. We will return to this formula in Part 2, see Problem 2 in Chapter 21. Here is the above rule in its simplest form, i.e. when $k = 2$:

118

$$\frac{d^2}{dx^2}(f \cdot g) = \frac{d}{dx}\left[\frac{d}{dx}(f \cdot g)\right]$$
$$= \frac{d}{dx}(f'g + fg')$$
$$= f''g + f'g + f'g' + fg''$$
$$= f''g + 2f'g' + fg''$$
$$= \binom{2}{0}\left(\frac{d^2}{dx^2}f \cdot\right)g + \binom{2}{1}\frac{df}{dx}\frac{dg}{dx} + \binom{2}{2}f\frac{d^2g}{dx^2}.$$

Now let us return to our original problem.

**Theorem 8.8.** *Let* $f : (a,b) \to \mathbb{R}$ *be a differentiable function. Suppose that* $f$ *has a second order derivative at* $x_0 \in (a,b)$*. If* $f'(x_0) = 0$ *and* $f''(x_0) < 0$ *then* $f$ *has a local maximum at* $x_0$*. If* $f'(x_0) = 0$ *and* $f''(x_0) > 0$ *then* $f$ *has a local minimum at* $x_0$*.*

This is sometimes referred to as the second derivative test. We will later, in Part 2, find a geometric interpretation of this result, compare with Remark 23.3.

**Example 8.9.** Again we look at $f : (a,b) \to \mathbb{R}$, $x \mapsto (x-\alpha)^2 + \beta$, $\alpha \in (a,b)$. We know already that $f'(\alpha) = 0$ and $\alpha$ is the only zero of $f'$. In addition we find $f''(x) = \frac{d}{dx}(2(x-\alpha)) = 2$. Hence $f''(\alpha) > 0$ and $f$ has a local minimum at $\alpha$.

The following result, called the **mean value theorem** is useful to study functions in more detail.

**Theorem 8.10.** *Let* $f : [a,b] \to \mathbb{R}$ *be a continuous function differentiable in* $(a,b)$*. Then there exist* $\xi \in (a,b)$ *such that*

$$f(b) - f(a) = f'(\xi)(b - a). \tag{8.5}$$

Writing (8.5) as $\frac{f(b)-f(a)}{b-a} = f'(\xi)$ we get the following intuitive graphical representation (note that both dotted lines are parallel, i.e. they have the same slope):

119

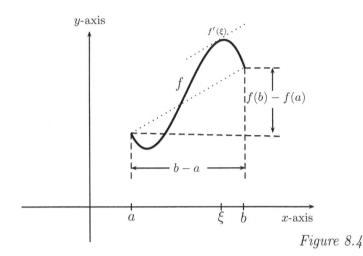

Figure 8.4

**Remark 8.11.** When proving the mean value theorem later in Part 2 of our course, compare with Corollary 22.6, we will carefully discuss the importance of each of the assumptions in the above theorem.

The mean value theorem has important consequences:

**Corollary 8.12.** *Suppose that $f : [a, b] \to \mathbb{R}$ fulfils the assumptions of the mean value theorem. Further suppose that $m \le f'(\eta) \le M$ for all $\eta \in (a, b)$. Then we have the estimates*

$$m(x - y) \le f(x) - f(y) \le M(x - y) \tag{8.6}$$

*for all $x, y \in (a, b), y \le x$.*

*Proof.* We may apply the mean value theorem to $f|_{[y,x]}$ to find first

$$f(x) - f(y) = f'(\xi)(x - y)$$

for some $\xi \in (y, x)$, $\xi = \xi(x, y)$. Since $f'(\xi) \ge m$ and $x - y \ge 0$ this implies

$$m(x - y) \le f(x) - f(y).$$

Further, since $f'(\xi) \le M$ and $x - y \ge 0$ we find in addition

$$f(x) - f(y) \le M(x - y).$$

$\square$

**Corollary 8.13.** *Suppose that* $f : [a, b] \to \mathbb{R}$ *is a function satisfying the assumptions of the mean value theorem. If* $f'(x) = 0$ *for all* $x \in (a, b)$ *then* $f$ *is constant, i.e. there exist* $c \in \mathbb{R}$ *such that* $f(x) = c$ *for all* $x \in [a, b]$.

*Proof.* Using (8.6) with $m = M = 0$ we find $f(x) = f(y)$ for all $x, y \in (a, b)$, i.e. $f(x) = c := f(x_0)$ for $x \in [a, b]$ and some fixed $x_0 \in [a, b]$.  □

Finally we discuss monotone functions.

**Definition 8.14.** *Let* $f : D \to \mathbb{R}$ *be a function,* $D \subset \mathbb{R}$. *We call* $f$

> ***increasing*** *if* $x, y \in D$ *and* $x < y$ *implies* $f(x) \le f(y)$;
> ***strictly increasing*** *if* $x, y \in D$ *and* $x < y$ *implies* $f(x) < f(y)$;
> ***decreasing*** *if* $x, y \in D$ *and* $x < y$ *implies* $f(x) \ge f(y)$;
> ***strictly decreasing*** *if* $x, y \in D$ *and* $x < y$ *implies* $f(x) > f(y)$.

*A function satisfying one of these conditions is called* ***monotone***.

Some authors prefer to call increasing functions non-decreasing and strictly increasing functions just increasing as well as decreasing functions non-increasing and strictly decreasing functions just decreasing.

**Example 8.15. A.** The function $\chi_{(0,\infty)} : \mathbb{R} \to \mathbb{R}$ is increasing but not strictly increasing. This is most easily seen by looking at its graph

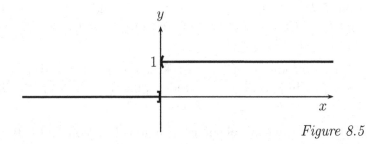

*Figure 8.5*

**B.** The function $f_a : \mathbb{R} \to \mathbb{R}$, $x \mapsto ax$ is strictly increasing for $a > 0$ and strictly decreasing for $a < 0$. Indeed, $a > 0$ implies for $x < y$ that $ax < ay$, whereas $a < 0$ implies for $x < y$ that $ax < ay$.

121

**C.** The function $g : \mathbb{R} \to \mathbb{R}$, $x \mapsto x^2$ is not monotone as is seen from its graph, or verified by an easy calculation. However $g|_{(a,b)}$ is for every $(a,b) \subset \mathbb{R}_+$ strictly increasing, and $g|_{(c,d)}$ is for every $(c,d) \subset \mathbb{R}\backslash\mathbb{R}_+$ strictly decreasing.

**Theorem 8.16.** *Let $f : \mathbb{R} \to \mathbb{R}$ be continuous and differentiable on $(a,b)$. We then have the following statements: if*

$$f'(x) \geq 0 \text{ for all } x \in (a,b) \text{ then } f \text{ is increasing;} \tag{8.7}$$
$$f'(x) > 0 \text{ for all } x \in (a,b) \text{ then } f \text{ is strictly increasing;} \tag{8.8}$$
$$f'(x) \leq 0 \text{ for all } x \in (a,b) \text{ then } f \text{ is decreasing;} \tag{8.9}$$
$$f'(x) < 0 \text{ for all } x \in (a,b) \text{ then } f \text{ is strictly decreasing.} \tag{8.10}$$

For a proof we refer to Part 2, Theorem 22.13.

**Example 8.17.** Consider $f : \mathbb{R} \to \mathbb{R}$, $x \mapsto x^5$. Since $f'(x) = 5x^4$ for all $x \in \mathbb{R}$ it follows that $f'(x) \geq 0$ for all $x \in \mathbb{R}$. In fact $f'(x) > 0$ for all $x \in \mathbb{R}\backslash\{0\}$. Hence $f$ is increasing, in fact strictly increasing. The latter is clear on $(-\infty, 0)$ and $(0, \infty)$, and since $f(0) = 0$ it follows also $f(x) < f(0)$ for $x < 0$ and $f(0) < f(y)$ for $0 < y$.

# Problems

1.   a) Let $f : D_1 \longrightarrow \mathbb{R}$ and $g : D_2 \longrightarrow \mathbb{R}$ be two functions such that $f(D_1) \subset D_2$. Suppose that $g$ is bounded with bound $M \geq 0$, i.e. $|g(x)| \leq M$ for all $x \in D_2$. Prove that $g \circ f : D_1 \longrightarrow \mathbb{R}$ is bounded and find a bound for $g \circ f$.

   b) Consider $f : (1,2) \longrightarrow \mathbb{R}$, $x \mapsto (x-1)^2$ and $g : (0,\infty) \longrightarrow \mathbb{R}$, $y \mapsto \frac{1}{y}$. Show that $f$ is bounded and that $f((1,2)) \subset (0,\infty)$. Is the function $g \circ f : (1,2) \longrightarrow \mathbb{R}$ bounded?

   c) Give an example of a continuous function $f : (a,b) \longrightarrow \mathbb{R}$, $a < b$, with the property that for all $a_1$ and $b_1$ such that $a < a_1 < b_1 < b$ the function $f|_{[a_1,b_1]}$ is bounded but $f$ is unbounded.

2. Let $p : \mathbb{R} \longrightarrow \mathbb{R}$ be a polynomial of degree $k \in \mathbb{N}_0$. Prove that if $n \geq \frac{k}{2}$ then the function $f(x) = \frac{p(x)}{(1+x^2)^n}$ is bounded on $\mathbb{R}$.

3. Find the following derivatives:

a) $\frac{d^2}{dx^2}\left(\frac{x^3+2x-5}{x-1}\right)$, $x \neq 1$; b) $\frac{d^3}{dt^3}(\sqrt{t^4+1})$;

c) $\frac{d}{ds}\left(\frac{|s|^5}{s^2+4}\right)$. Does the function $s \mapsto \frac{|s|^5}{s^2+4}$ have a second derivative for $s = 0$?

4. Let $u, v : \mathbb{R} \longrightarrow \mathbb{R}$ be two twice differentiable functions. Find

$$\frac{d^2}{dx^2}\left((u^2(x) + 1)(v^2(x) + 1)^{-1}\right).$$

5. Let $f : (a, b) \longrightarrow \mathbb{R}$ and $g : (c, d) \longrightarrow \mathbb{R}$ be two twice differentiable functions and suppose that $f((a, b)) \subset (c, d)$. Prove that

$$\left(\frac{d^2}{dx^2}(g \circ f)\right)(x) = g''(f(x))(f'(x))^2 + g'(f(x))f''(x).$$

Now find $\frac{d^2}{dt^2}\left((1 + f^2(t))^{-\frac{1}{2}}\right)$ where $f : \mathbb{R} \longrightarrow \mathbb{R}$ is twice differentiable.

6. Find

$$\frac{d^2}{dx^2}\left(\frac{1}{(u^2(x) + 2)^2}\right) \quad \text{where } u : \mathbb{R} \longrightarrow \mathbb{R}, \ u(x) = \frac{x^2}{\sqrt{1+x^2}}.$$

7. Prove that for $n \in \mathbb{N}_0$ there exists a polynomial $p_n$ of degree $k \leq n$ such that

$$\frac{d^n}{dx^n}\left(\frac{1}{1+x^2}\right) = \frac{p_n(x)}{(1+x^2)^{n+1}}.$$

Now deduce that there exists a constant $c_n \geq 0$ such that

$$\left|\frac{d^n}{dx^n}\left(\frac{1}{1+x^2}\right)\right| \leq \frac{c_n}{(1+x^2)^{\frac{n+2}{2}}}.$$

Hint: a) Use mathematical induction, b) Use Problem 2 of this chapter.

8. Find the local extreme values of:

a) $f : \mathbb{R} \longrightarrow \mathbb{R}$, $f(x) = |x|^3$; b) $g : \mathbb{R} \longrightarrow \mathbb{R}$, $g(s) = \frac{s^2-2s}{2+3s^2}$;

c) $h : (-1, 1) \longrightarrow \mathbb{R}$, $h(u) = (1 + u)\sqrt{1 - u^2}$.

123

9.      a) The function $g : (-1, 1) \longrightarrow \mathbb{R}$, $x \mapsto x^2$, has a minmum at $x_0 = 0$. Find a function $f : (-1, 1) \longrightarrow \mathbb{R}$, $f(t) \geq 0$ for all $t \in (-1, 1)$, such that $f$ is non constant and $f \circ g$ has a maximum at $x_0 = 0$.

     b) Let $f : \mathbb{R} \longrightarrow \mathbb{R}$ and suppose that $f$ has a local maximum at $x_0 \in \mathbb{R}$. Let $c \in \mathbb{R}$. Prove that $h : \mathbb{R} \longrightarrow \mathbb{R}$, $h(x) = f(x - c)$, has a local maximum at $c + x_0$.

10.      a) Suppose that $\sin'(x) := \frac{d}{dx}(\sin x) = \cos x$ for all $x \in \mathbb{R}$ and suppose that $|\cos x| \leq 1$. Use the mean value theorem to deduce that $|\sin x| \leq |x|$ knowing that $\sin 0 = 0$.

     b) Consider $g : [1, 2] \longrightarrow \mathbb{R}$, $g(x) = \sqrt{x}$. Deduce that $\left|\frac{d}{dx} g(x)\right| \leq \frac{1}{2}$ for all $x \in [1, 2]$. Now we use the mean value theorem or its corollaries to estimate $\sqrt{\frac{11}{10}}$ by

$$\frac{19}{20} \leq \sqrt{\frac{11}{10}} \leq \frac{21}{20}.$$

11.      a) For $n \in \mathbb{N}_0$ define $\chi_n := \chi_{[n,\infty)}$ to be the characteristic function of $[n, \infty)$. For $N \in \mathbb{N}$ define

$$X_N(t) := \sum_{n=0}^{N} \chi_n(t).$$

Sketch the graph of $X_5$. Is $X_n$ increasing?

     b) Consider $f_a : \mathbb{R}_+ \longrightarrow \mathbb{R}$, $x \mapsto \frac{x}{1+ax^2}$, where $a > 0$ is a fixed constant. Determine the largest subset of $\mathbb{R}_+$ where $f_a$ is decreasing and the subset where $f_a$ is increasing.

12.      a) Let $f : (a, b) \longrightarrow \mathbb{R}$ and $g : (c, d) \longrightarrow \mathbb{R}$ be two monotone increasing functions such that $g((c, d)) \subset (a, b)$. Prove that $f \circ g : (c, d) \longrightarrow \mathbb{R}$ is monotone increasing.

     b) Let $f, g : \mathbb{R} \longrightarrow \mathbb{R}$ be differentiable functions. Prove that if $f'$ and $g'$ are either both positive or both negative valued functions then $f \circ g$ and $g \circ f$ are monotone increasing.

13. Let $f, g : [a, b] \longrightarrow \mathbb{R}$, $a < b$, be continuous and differentiable on $(a, b)$. Suppose that $f(a) = g(a)$ and $0 \leq f'(x) < g'(x)$ for all $x \in (a, b)$. Prove that $f(x) < g(x)$ for all $x \in (a, b)$.
Hint: use the mean value theorem with $h = g - f$.

# 9 The Exponential and Logarithmic Functions

The functions we will introduce in this and the following chapters i.e. exponential and logarithmic functions, trigonometric functions and hyperbolic functions are the so-called **elementary transcendental functions**. Their definition requires more than just algebraic operations. In fact even the existence of these functions requires a proof. One way to introduce the exponential function is to consider it as the (unique) solution to a simple initial value problem for a first order differential equation. We will later on prove

**Theorem 9.1.** *There exists a function* $f : \mathbb{R} \longrightarrow \mathbb{R}$ *such that*

$$f'(x) = f(x) \text{ for all } x \in \mathbb{R} \text{ and } f(0) = 1. \tag{9.1}$$

**Definition 9.2.** *The function* $f$ *in Theorem 9.1 is called the* **exponential function** *and is denoted by* $\exp$, *i.e.* $\exp : \mathbb{R} \longrightarrow \mathbb{R}, \exp' = \exp$ *and* $\exp(0) = 1$.

**Lemma 9.3.** *For all* $x \in \mathbb{R}$, $\exp(x) \neq 0$ *and*

$$\exp(-x) = \frac{1}{\exp(x)} = (\exp(x))^{-1}.$$

*Proof.* Since for $f = \exp$ we find

$$\begin{aligned}
\frac{d}{dx}(f(x)f(-x)) &= f'(x)f(-x) + f(x)(-f'(-x)) \\
&= f(x)f(-x) - f(x)f(-x) = 0.
\end{aligned}$$

Therefore we know that with some $c \in \mathbb{R}$

$$f(x)f(-x) = c \text{ for all } x \in \mathbb{R}. \tag{9.2}$$

But for $x = 0$ we find $c = (f(0))^2 = 1$. Now it follows from (9.2) that

$$f(x)f(-x) = 1,$$

i.e. $\frac{1}{f(x)} = f(-x)$ or with $f = \exp$

$$\exp(-x) = \frac{1}{\exp(x)}. \tag{9.3}$$

$\square$

125

**Lemma 9.4.** *The function* exp *is unique.*

*Proof.* Suppose that $f$ and $g$ both satisfy (9.1). Then by the previous lemma $\frac{g}{f}$ is defined and we have

$$
\begin{aligned}
\frac{d}{dx}\left(\frac{g}{f}\right)(x) &= \frac{g'(x)f(x) - f'(x)g(x)}{f(x)^2} \\
&= \frac{g(x)f(x) - f(x)g(x)}{f(x)^2} = 0
\end{aligned}
$$

implying $\frac{g}{f}(x) = K$ for all $x \in \mathbb{R}$ and some $K \in \mathbb{R}$, or $g(x) = Kf(x)$. Since $g(0) = f(0) = 1$ we find $g(0) = 1 = Kf(0) = K$, i.e. $K = 1$ and $f = g$. $\square$

Before we can proceed further we state without proof (which we will provide later, in Part 2, Theorem 20.17) the **intermediate value theorem**:

**Theorem 9.5.** *Let $f : [a, b] \longrightarrow \mathbb{R}$ be a continuous function and set $\alpha := f(a)$ and $\beta := f(b)$. Suppose $\alpha < \gamma < \beta$. Then there exists $x_0 \in (a, b)$ such that $f(x_0) = \gamma$. In the case where $\beta < \gamma < \alpha$ we get the same conclusion.*

The intermediate value theorem applied to exp implies that $\exp(x) > 0$ for all $x \in \mathbb{R}$. Indeed, suppose that there is $x_1 \in \mathbb{R}$ such that $\exp(x_1) < 0$. Since $x_0 \neq 0$, we conclude that there must be $x_0 \in (x_1, 0)$ if $x_1 < 0$ or $x_0 \in (0, x_1)$ if $x_1 > 0$, such that $\exp(x_0) = 0$ which is impossible by Lemma 9.3 Hence $\exp(x) > 0$ for all $x \in \mathbb{R}$.

**Lemma 9.6.** *The exponential function is strictly positive and strictly increasing.*

*Proof.* It remains to prove that exp is strictly increasing. But $\exp'(x) = \exp(x) > 0$, implying the result. $\square$

The following result is very important:

**Lemma 9.7 (Functional equation for** exp**).** *For all $x, y \in \mathbb{R}$ we have*

$$\exp(x + y) = \exp(x)\exp(y). \tag{9.4}$$

*Proof.* For $y \in \mathbb{R}$ fixed we consider the function $x \longmapsto g(x) := \exp(y + x)$. It follows that

$$g'(x) = (\exp(y + x))' = \exp(y + x) = g(x),$$

126

hence $g(x) = K \exp(x)$ for some $K \in \mathbb{R}$, compare with the proof of Lemma 9.4. Now, with $x = 0$ we find

$$\exp(y) = g(0) = K \exp(0) = K,$$

or

$$\exp(x + y) = g(x) = \exp(x) \exp(y)$$

proving the lemma. □

Given a function $f : \mathbb{R} \to \mathbb{R}$. We say that $f$ solves **Cauchy's functional equation** if $f(x+y) = f(x)f(y)$ for all $x, y \in \mathbb{R}$. In this sense exp is a solution to Cauchy's functional equation. Note that exp is not the only solution to this functional equation, however it is the only continuous one. We define the **Euler number** $e$ by

$$e := \exp(1). \tag{9.5}$$

Since exp is strictly increasing we have $e > 1$.

**Corollary 9.8.** *For all $n \in \mathbb{N}$ we have*

$$\exp(n) = e^n. \tag{9.6}$$

*Proof.* For $n = 1$ there is nothing to prove. Suppose that $\exp(n) = e^n$ for some $n \in \mathbb{N}$. Then it follows that

$$\exp(n + 1) = \exp(n) \exp(1) = e^n e = e^{n+1}.$$

The principle of mathematical induction now yields the corollary. □

Using (9.3) we deduce from (9.6) that for $m \in \mathbb{N}$

$$\exp(-m) = e^{-m} = \frac{1}{e^m}. \tag{9.7}$$

It is possible to justify for all $x \in \mathbb{R}$

$$\exp(x) = e^x. \tag{9.8}$$

In particular we have

$$e^{x+y} = e^x e^y \text{ and } e^0 = 1. \tag{9.9}$$

We know that exp is strictly increasing and $\exp(x) > 0$ for all $x \in \mathbb{R}$. Assume for a moment that $R(\exp) = \{x \in \mathbb{R} | x > 0\}$. Then we know that $\exp : \mathbb{R} \longrightarrow \{x \in \mathbb{R} | x > 0\}$ is bijective and has a differentiable inverse, i.e. there exists a function $\ln : \{x \in \mathbb{R} | x > 0\} \longrightarrow \mathbb{R}$ with the properties
$$x \longmapsto \ln x$$

$$\ln(\exp x) = x \quad \text{for } x \in \mathbb{R} \tag{9.10}$$

and

$$\exp(\ln y) = y \quad \text{for } y > 0. \tag{9.11}$$

We call ln the **(natural) logarithm**. For its derivative we find using (7.7) that
$$\frac{d}{dy} \ln y = \frac{1}{\exp'(\ln y)} = \frac{1}{\exp(\ln y)} = \frac{1}{y},$$

i.e.

$$(\ln y)' = \frac{1}{y} \quad , y > 0, \tag{9.12}$$

which also implies that ln is strictly increasing on $\{y \in \mathbb{R} | y > 0\}$. Furthermore we have

$$\ln(1) = 0 \tag{9.13}$$

since $1 = \exp(0)$, and we claim for $x, y > 0$ that

$$\ln(x \cdot y) = \ln x + \ln y. \tag{9.14}$$

Fix $y > 0$ and consider $g(x) = \ln(y \cdot x) - \ln x$. Differentiating with respect to $x$ yields
$$g'(x) = y \ln'(y \cdot x) - \ln'(x) = y\frac{1}{yx} - \frac{1}{x} = 0,$$

hence $g'(x) = 0$ for all $x > 0$ implying that $g(x) = c$ for some $c \in \mathbb{R}$, and all $x > 0$. Since $g(1) = \ln y$ we find

$$\ln y = g(1) = \ln(yx) - \ln x$$

or

$$\ln yx = \ln y + \ln x,$$

proving (9.14). Finally we note that for $x > 0$

$$0 = \ln 1 = \ln \frac{x}{x} = \ln x + \ln \frac{1}{x} = \ln x + \ln x^{-1},$$

or

$$\ln x^{-1} = \ln \frac{1}{x} = -\ln x. \tag{9.15}$$

Now let $a > 0$ be given. We define on $\mathbb{R}$ the function $x \longmapsto a^x$ by

$$a^x := e^{x \ln a}. \tag{9.16}$$

It is easy and a good exercise to prove for $x, y \in \mathbb{R}$ that

$$a^{x+y} = a^x a^y \text{ and } a^0 = 1, \tag{9.17}$$

as well as

$$(a^x)^y = a^{xy}. \tag{9.18}$$

Further, $x \longmapsto a^x$ is bijective with range $\{y \in \mathbb{R} | y > 0\}$ and has an inverse function which is denoted by $x \longmapsto \log_a x$. The value of $\log_a x$ is called the logarithm of $x$ with respect to the basis $a$.
(Note that since $a^{-1} = \frac{1}{a}$, it is often convenient to define $x \longmapsto a^x$ and $y \longmapsto \log_a y$ only for $a > 1$.)
For the derivative of $x \longmapsto a^x$ we find

$$\frac{d}{dx}a^x = \frac{d}{dx}e^{x \ln a} = (\ln a)e^{x \ln a} = (\ln a)a^x, \tag{9.19}$$

and this implies for $x \longmapsto \log_a x$

$$\frac{d}{dx}\log_a x = \frac{1}{(a^x)'(\log_a x)} = \frac{1}{(\ln a)a^{\log_a x}} = \frac{1}{(\ln a)x}. \tag{9.20}$$

Here are some examples of derivatives

$$\frac{d}{dx}(x \ln x) = \ln x + x \cdot \frac{1}{x} = 1 + \ln x, \tag{9.21}$$

$$\frac{d}{dx}(x^x) = \frac{d}{dx}\left(e^{x \ln x}\right) = \left(\frac{d}{dx}(x \ln x)\right)e^{x \ln x} = (1 + \ln x)x^x. \tag{9.22}$$

For differentiable functions $u : \mathbb{R} \longrightarrow \mathbb{R}, v : \mathbb{R} \longrightarrow \mathbb{R}_+ \setminus \{0\}$ we find

$$\frac{d}{dx}\left(e^{u(x)}\right) = u'(x)e^{u(x)} \tag{9.23}$$

and

$$\frac{d}{dx}\ln v(x) = v'(x)\frac{1}{v(x)} = \frac{v'(x)}{v(x)}. \tag{9.24}$$

The term $\frac{v'}{v}$ is often called the **logarithmic derivative** of $v$.

Before we can draw the graph of exp and ln, we need to study the asymptotic behaviour of functions.

Let $f : \mathbb{R} \longrightarrow \mathbb{R}$ be a function. We want to study the behaviour of $f(x)$ for $x$ becoming larger and larger, i.e. for $x$ tending to infinity. It may happen that for $x$ tending to infinity $f(x)$ tends to some number $a$ or to infinity, but other cases are possible.

We write

$$\lim_{x \to \infty} f(x) = a \qquad (9.25)$$

if for every $\epsilon > 0$ given, there exists $N = N(\epsilon) \in \mathbb{N}$ such that

$$x > N \text{ implies } |f(x) - a| < \epsilon.$$

**Example 9.9.** We claim for $f(x) = \frac{1}{1+x^2}$ that

$$\lim_{x \to \infty} f(x) = \lim_{x \to \infty} \frac{1}{1 + x^2} = 0.$$

Thus, given $\epsilon > 0$ we need to find $N(\epsilon) \in \mathbb{N}$ such that

$$x > N(\epsilon) \text{ implies } \left| \frac{1}{1 + x^2} - 0 \right| = \frac{1}{1 + x^2} < \epsilon.$$

Since for $x > 0$ it follows that

$$\frac{1}{1 + x^2} < \frac{1}{x}$$

we are done if for $\epsilon > 0$ we can find $N(\epsilon) \in \mathbb{N}$ such that

$$x > N(\epsilon) \text{ implies } \frac{1}{x} < \epsilon.$$

But this is easy: take $N(\epsilon) = \left[ \frac{1}{\epsilon} \right] + 1 > \frac{1}{\epsilon}$. If $x > \left[ \frac{1}{\epsilon} \right] + 1$ then

$$\frac{1}{1 + x^2} < \frac{1}{x} < \frac{1}{\left[ \frac{1}{\epsilon} \right] + 1} < \epsilon.$$

Now, it may happen that $a$ in (9.25) is itself infinity, i.e. we write

$$\lim_{x \to \infty} f(x) = \infty \qquad (9.26)$$

if for every $M > 0$ there exists $N = N_M \in \mathbb{N}$ such that $x > N$ implies $f(x) > M$.

130

**Example 9.10.** We claim for $n \in \mathbb{N}$ that

$$\lim_{x \to \infty} x^n = \infty. \tag{9.27}$$

We have to find for $M > 0$ given a natural number $N = N_M$ such that if $x > N_M$ then $x^n > M$. Take $N = [M] + 1$. Now $x > N_M$ implies $x^n > N_M^n = ([M] + 1)^n > M$ proving (9.27).

In order to study $\lim_{x \to \infty} \exp(x)$ and related limits we need

**Lemma 9.11. A (Bernoulli's inequality).** *Let $a > 0$ and $n \in \mathbb{N}_0$. Then*

$$(1 + a)^n \geq 1 + na. \tag{9.28}$$

**B.** *Let $a > 0$ and $n \in \mathbb{N}_0$. Then*

$$(1 + a)^n \geq 1 + na + \frac{n(n-1)}{2}a^2. \tag{9.29}$$

*Proof.* Since $\frac{n(n-1)}{2}a^2 > 0$ it follows that (9.29) implies (9.28). We now prove (9.29). For $n = 0$ we find

$$1 = (1 + a)^0 = 1 + 0 \cdot a + \frac{0(-1)}{2}a^2 = 1.$$

Now assume that (9.29) holds for some fixed $n \in \mathbb{N}$. For $n + 1$ we find

$$
\begin{aligned}
(1 + a)^{n+1} &= (1 + a)^n(1 + a) \\
&\geq \left(1 + na + \frac{n(n-1)}{2}a^2\right)(1 + a) \\
&= 1 + na + \frac{n(n-1)}{2}a^2 + a + na^2 + \frac{n(n-1)}{2}a^3 \\
&\geq 1 + (n+1)a + \frac{n(n-1) + 2n}{2}a^2 \\
&= 1 + (n+1)a + \frac{(n+1)n}{2}a^2
\end{aligned}
$$

and the result follows by the principle of mathematical induction. $\qquad\square$

**Lemma 9.12.** *We have*

$$\lim_{x \to \infty} e^x = \infty. \tag{9.30}$$

*Proof.* Given $M > 0$. We have to find $N \in \mathbb{N}$ such that $x > N$ implies $e^x > M$. First note that $e = e^1 > e^0 = 1$, i.e. $e = (1 + b)$ for some $b > 0$. The monotonicity of exp implies for $x > N$ using (9.28)

$$e^x > e^N = (1 + b)^N \geq 1 + bN.$$

Thus, given $M > 0$ choose $N \in \mathbb{N}$ such that $1 + bN > M$ to find that

$$x > N \text{ implies } e^x > e^N = (1 + b)^N \geq 1 + bN > M.$$

$\square$

**Remark 9.13.** Note that we have assumed that for $M > 0$ we find $N \in \mathbb{N}$ such that $1 + bN > M$. If $M \leq 1$ then every $N \in \mathbb{N}$ will do, but this case is of course not interesting. If $M > 1$ then $M - 1 > 0$ and we may take $N$ such that $N > \frac{M-1}{b}$, for example $N = 1 + \left[\frac{M-1}{b}\right]$.

**Lemma 9.14.** *We have*

$$\lim_{x \to \infty} \frac{x}{\exp(x)} = \lim_{x \to \infty} xe^{-x} = 0. \tag{9.31}$$

*Proof.* We claim that $\phi(x) := xe^{-x}$ is for $x > 1$ strictly decreasing. This follows from

$$\phi'(x) = e^{-x} - xe^{-x} = e^{-x}(1 - x) < 0,$$

provided $x > 1$. Hence for $x > N > 1$ it follows that

$$0 \leq xe^{-x} < Ne^{-N}.$$

Now, given $\epsilon > 0$ take $N > 1$ such that

$$\frac{2}{b^2} \frac{1}{N-1} < \epsilon, \text{ i.e. } N > \frac{2}{b^2} \frac{1}{\epsilon} + 1,$$

where $b$ is determined by $e = 1 + b$. Now using (9.29)

$$\begin{aligned}
0 &\leq xe^{-x} \leq Ne^{-N} = N(1 + b)^{-N} = \frac{N}{(1 + b)^N} \\
&\leq \frac{N}{1 + Nb + \frac{N(N-1)}{2}b^2} \leq \frac{1}{\frac{N-1}{2}b^2} = \frac{2}{b^2} \frac{1}{N-1} \\
&< \epsilon.
\end{aligned}$$

$\square$

Next we extend our considerations to very small values of $x$. It may happen that $f : \mathbb{R} \longrightarrow \mathbb{R}$ tends to $a \in \mathbb{R}$ when $x$ becomes smaller and smaller. For this we define

$$\lim_{x \longrightarrow -\infty} f(x) = a \tag{9.32}$$

if for every $\epsilon > 0$ there exists $N = N(\epsilon) \in \mathbb{N}$ such that $x < -N$ implies $|f(x) - a| < \epsilon$.

**Lemma 9.15.** *We have*

$$\lim_{x \longrightarrow -\infty} e^x = 0. \tag{9.33}$$

*Proof.* We have to prove that for every $\epsilon > 0$ there exists $N \in \mathbb{N}$ such that $x < -N$ implies $|e^x - 0| = e^x < \epsilon$. With $y := -x > 0$ this is equivalent to $N < y$ implies $e^{-y} < \epsilon$ or $N < y$ implies $\frac{1}{e^y} < \epsilon$, i.e.

$$\lim_{y \longrightarrow \infty} e^{-y} = 0. \tag{9.34}$$

We now prove (9.34). The function $y \longmapsto g(y) = e^{-y}$ is strictly decreasing since $g'(y) = -e^{-y} < 0$. By Bernoulli's inequality we find therefore for $N < y$ and using $e = (1 + b)$ that

$$e^{-y} \leq e^{-N} \leq \frac{1}{1 + Nb} < \frac{1}{Nb}.$$

Hence, given $\epsilon > 0$ choose $N \in \mathbb{N}$ such that $\frac{1}{Nb} < \epsilon$ to find that

$$N < y \text{ implies } e^{-y} < e^{-N} \leq \frac{1}{1 + Nb} < \frac{1}{Nb} < \epsilon.$$

Thus (9.34) and therefore (9.33) is proved. $\qquad\qquad\square$

Note that Lemma 9.15 together with Lemma 9.12 finally proves that the range of exp is equal to $\{x \in \mathbb{R} | x > 0\}$.

Now we can sketch the graph of $x \longmapsto \exp(x)$. It must be strictly positive, strictly increasing, for $x \longrightarrow -\infty$ it tends to 0, at $x = 0$ it has the value 1, and for $x \longmapsto \infty$ it tends to $\infty$:

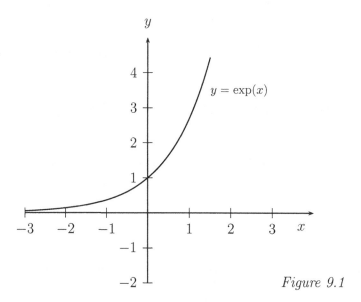

Figure 9.1

By our general considerations we can now also sketch the graph of $x \longmapsto \ln x$. We only have to reflect the graph of exp at the principal diagonal:

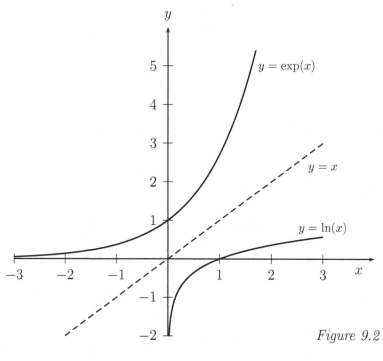

Figure 9.2

Let us calculate some further limits. First we note that

$$\lim_{x \to \infty} \ln x = \infty. \tag{9.35}$$

Given $x > 0$ set $x = \exp y$. Now $x$ tending to infinity implies that $\exp y$ tends to infinity which is only possible when $y$ tends to infinity, but $y = \ln x$. Next consider $\ln x$ for $x$ tending to 0. This is equivalent to considering $\ln \frac{1}{x}$ for $x$ tending to $\infty$. But $\ln \frac{1}{x} = -\ln x$ and $\lim_{x \to \infty} \ln x = \infty$. Thus we find $\ln x \longrightarrow -\infty$ for $x \longrightarrow 0$. For this we write

$$\lim_{x \to 0} \ln x = -\infty. \tag{9.36}$$

**Theorem 9.16.** *We have*

$$\lim_{x \to \infty} \frac{\ln x}{x} = 0. \tag{9.37}$$

*Proof.* Let $x = e^y$, i.e. $y = \ln x$. Then

$$\frac{\ln x}{x} = \frac{y}{e^y} = ye^{-y}.$$

Since $x \longrightarrow \infty$ implies $y \longrightarrow \infty$ we can apply Lemma 9.14 to find

$$\lim_{x \to \infty} \frac{\ln x}{x} = \lim_{y \to \infty} ye^{-y} = 0.$$

$\square$

# Problems

1.    a) Using the definition of $\lim_{x \to \infty} f(x) = \infty$ prove that $\lim_{x \to \infty} (x^2 - 5) = \infty$.

b) Let $p : \mathbb{R} \longrightarrow \mathbb{R}$, $p(x) = \sum_{l=0}^{k} a_l x^l$, be a polynomial of degree $k$ with $a_k > 0$. Prove that $\lim_{x \to \infty} p(x) = \infty$.

c) For $a \in \mathbb{R}$ prove that

$$\lim_{x \to \infty} \frac{1 + a + ax^2}{1 + x^2} = a.$$

2.     a) For $n \in \mathbb{N}$ deduce from Lemma 9.11.B the **Bernoulli inequality**:   $(1+a)^n > 1 + na$, i.e. the strict inequality holds.

    b) Use part a) to prove for $n \geq 2$ that

$$\left(1 + \frac{1}{n^2 - 1}\right)^n > 1 + \frac{1}{n}.$$

3. For $a > 0$ define $a^x := \exp(x \ln a) = e^{x \ln a}$, and prove that $a^{x+y} = a^x a^y$ and $a^0 = 1$.

4. Find the following derivatives:

    a) $\frac{d}{dx} \exp(-\sqrt{x^2 + 1})$;   b) $\frac{d}{du} \exp(-\log_a(1+u^2))$, $a > 0$;   c) $\frac{d^2}{dt^2}\left(\exp\left(-\frac{1}{1+t^2}\right)\right)$.

5. By induction show that for $n \in \mathbb{N}_0$ there exists a polynomial $p_n$ of degree at most $n$ such that

$$\frac{d^n}{dt^n} e^{-x^2} = p_n(x) e^{-x^2}.$$

6. Find the following derivatives:

    a) $\frac{d}{ds} \ln(\sqrt{s^4 + 1} - s^2)$;   b) $\frac{d}{dx}(\ln(a^x))$, $a > 1$;   c) $\frac{d^2}{dy^2} \ln((y^2+1)^{-k})$, $k \in \mathbb{N}$.

7.     a) For $a > 0$ prove that

$$\lim_{x \to \infty} \frac{x}{\exp(ax)} = 0.$$

    b) Use part a) to prove for $a > 0$ and $n \in \mathbb{N}$ that

$$\lim_{x \to \infty} \frac{x^n}{\exp(ax)} = 0.$$

Hint: $\exp x = \exp\left(\frac{x}{n}\right) \cdot \ldots \cdot \exp\left(\frac{x}{n}\right)$.

8. Let $p(x) = \sum_{k=0}^{m} b_k x^k$, $b_m > 0$, be a polynomial. Find

$$\lim_{x \to -\infty} (\exp(p(x))).$$

Hint: distinguish whether $m$ is even or odd.

9. Let $p(x) = \displaystyle\sum_{k=0}^{n} a_k x^k$ be a polynomial and $a_n > 0$. Prove that there exists $R > 0$ such that $p(x) > 0$ for $x \geq R$. Hence for $x \geq R$ the function $x \mapsto \ln(p(x))$ is defined. Now show that $\displaystyle\lim_{x\to\infty} \frac{\ln(p(x))}{x} = 0$.

10.    a) For $x, y > 0$ prove under the assumption that for $a > 0$ it follows that $\ln a^{\frac{1}{2}} = \frac{1}{2}\ln a$ the estimate

$$\frac{\ln x + \ln y}{2} \leq \ln\frac{x+y}{2}.$$

b) For $x > y > 0$ such that $x - y = 1$ prove that

$$\frac{1}{x} \leq \ln x - \ln y \leq \frac{1}{y}.$$

(Use the mean value theorem.)

11. Let $v : \mathbb{R} \longrightarrow \mathbb{R}$ be a differentiable function and suppose that the logarithmic derivative of $v$ is identically 1 and that $v(0) = 1$. Find the function $v$.

# 10 Trigonometric Functions and Their Inverses

Since we have introduced the exponential function as a solution of a differential equation and an initial condition, we may think to introduce sin and cos, as solutions of the differential equations:

$$f' = g, \quad g' = -f \tag{10.1}$$

$$f(0) = 0, \quad g(0) = 1. \tag{10.2}$$

Postponing the existence proof, it is possible to identify $f$ with sin and $g$ with cos, and to prove their basic properties by only using (10.1) and (10.2). We follow however a different method. We introduce both functions by using elementary geometry of the circle and then we will derive some of their properties. It turns out that switching from very classical geometry to calculus leads to some problems, all of which cannot be resolved in this part of the course. However, in Part 2 we will have a more rigorous approach using power series and therefore we may justify our naïve handling of trigonometric functions here.

Consider the circle in $\mathbb{R}^2$ with centre $(0,0)$ and radius 1. The total length of its circumference is $2\pi$. It makes sense to measure the size of an angle $\phi$ by the corresponding arc length. More precisely, let $\phi$ be the angle $\angle CAB$ in Figure 10.1 below and denote by $l(\overset{\frown}{BC})$ the length of the arc $\overset{\frown}{BC}$ connecting $B$ and $C$. For the measure of the size of $\phi$ we take the value $l(\overset{\frown}{BC})$.

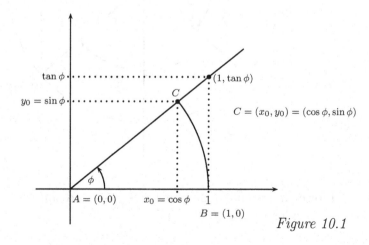

*Figure 10.1*

In this way we find that an angle of $45°$ corresponds to $\frac{\pi}{4}$, an angle of $90°$ corresponds to $\frac{\pi}{2}$ etc. For $0 \leq \phi < 2\pi$ we can now define the following two functions

$$\phi \longmapsto \sin \phi \quad \text{and} \quad \phi \longmapsto \cos \phi$$

where the definitions are easily taken from Figure 10.1: denote by $C = (x_0, y_0)$ the point where the ray starting at $A = (0,0)$ forming the angle $\phi$ with the $x$-axis intersects the circle (as usual angles in the unit circle are measured anticlockwise). Then we define:

$$\sin \phi = y_0, \quad \cos \phi = x_0. \tag{10.3}$$

Figures 10.2 and 10.3 below give a further insight into the values of sin and cos for $0 \leq \phi < 2\pi$. First we look at Figure 10.2:

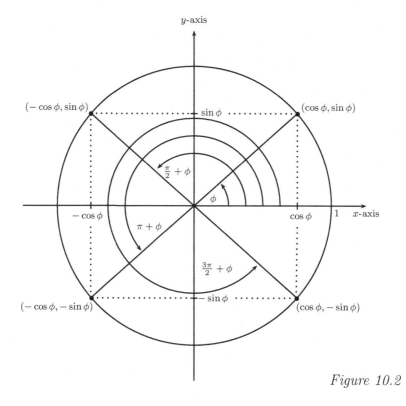

*Figure 10.2*

We find for example that $\cos(\pi - \phi) = -\cos \phi$ and $\sin(\pi + \phi) = -\sin \phi$, etc. Next we consider Figure 10.3:

140

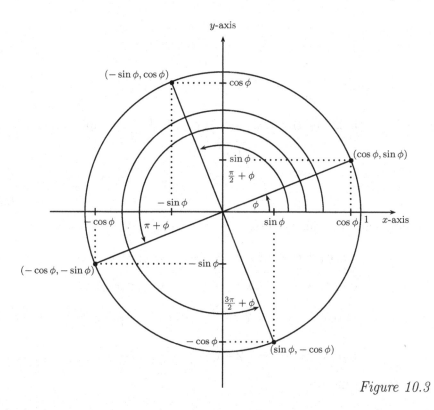

*Figure 10.3*

Here we find for example that $\cos(\frac{\pi}{2}+\phi) = -\sin\phi$ and $\sin(\frac{3\pi}{2}+\phi) = -\cos\phi$. Further similar formulae can be found in Appendix V. Note that in our definition we have excluded $\phi = 2\pi$. We remedy this by extending both functions to all of $\mathbb{R}$ in the following way: let $\phi \in \mathbb{R}$, then there exists a unique $k \in \mathbb{Z}$ such that $\phi \in [2k\pi, 2(k+1)\pi)$, i.e. $2k\pi \leq \phi < 2(k+1)\pi$. We now set

$$\sin\phi := \sin(\phi - 2k\pi), \quad \cos\phi := \cos(\phi - 2k\pi). \qquad (10.4)$$

Note that $\phi - 2k\pi \in [0, 2\pi)$ and therefore $\sin(\phi - 2k\pi)$ and $\cos(\phi - 2k\pi)$ are well defined. From this extension it follows immediately that $\sin : \mathbb{R} \longmapsto \mathbb{R}$ and $\cos : \mathbb{R} \longmapsto \mathbb{R}$ are periodic functions with period $2\pi$, i.e. $\sin(\phi + 2\pi) = \sin\phi$ and $\cos(\phi + 2\pi) = \cos\phi$. Further it follows that

$$|\sin\phi| \leq 1 \text{ and } |\cos\phi| \leq 1 \qquad (10.5)$$

and we have the special values

141

$$\begin{aligned}
\sin 0 &= 0 & \cos 0 &= 1 \\
\sin \tfrac{\pi}{2} &= 1 & \cos \tfrac{\pi}{2} &= 0 \\
\sin \pi &= 0 & \cos \pi &= -1 \\
\sin \tfrac{3\pi}{2} &= -1 & \cos \tfrac{3\pi}{2} &= 0 \\
\sin 2\pi &= 0 & \cos 2\pi &= 1.
\end{aligned}$$

Moreover by Pythagoras' theorem, see Appendix IV, we know

$$x_0^2 + y_0^2 = 1$$

or

$$\cos^2 \phi + \sin^2 \phi = 1. \tag{10.6}$$

We also note the following results:

$$\sin(\phi_1 + \phi_2) = \sin \phi_1 \cos \phi_2 + \cos \phi_1 \sin \phi_2; \tag{10.7}$$

$$\cos(\phi_1 + \phi_2) = \cos \phi_1 \cos \phi_2 - \sin \phi_1 \sin \phi_2; \tag{10.8}$$

$$\sin \phi_1 - \sin \phi_2 = 2 \sin \frac{\phi_1 - \phi_2}{2} \cos \frac{\phi_1 + \phi_2}{2}; \tag{10.9}$$

$$\cos \phi_1 - \cos \phi_2 = -2 \sin \frac{\phi_1 + \phi_2}{2} \sin \frac{\phi_1 - \phi_2}{2}; \tag{10.10}$$

as well as the symmetries

$$\sin(-x) = -\sin x, \quad \cos(-x) = \cos x. \tag{10.11}$$

Again we refer to Appendix V where we have collected more similar formulae. The formulae in (10.11) suggest:

**Definition 10.1.** *Let* $f, g : \mathbb{R} \longmapsto \mathbb{R}$ *be two functions. We call* $f$ *an **even function** if*

$$f(x) = f(-x) \quad \text{for all } x \in \mathbb{R}, \tag{10.12}$$

*and we call* $g$ *and an **odd function** if*

$$g(x) = -g(-x). \tag{10.13}$$

Hence sin is an odd and cos is an even function.

**Lemma 10.2.** *Let* $f_1, f_2 : \mathbb{R} \longmapsto \mathbb{R}$ *be two even functions and let* $g_1, g_2 : \mathbb{R} \longmapsto \mathbb{R}$ *be two odd functions. Then* $f_1 \cdot f_2$ *and* $g_1 \cdot g_2$ *are even, whereas* $f_1 \cdot g_1$ *is odd, i.e. the product of two even or two odd functions is even, the product of an even function with an odd function is odd.*

*Proof.* The following hold

$$(f_1 \cdot f_2)(-x) = f_1(-x)f_2(-x) = f_1(x)f_2(x) = (f_1 \cdot f_2)(x),$$

$$(g_1 \cdot g_2)(-x) = g_1(-x)g_2(-x) = (-g_1(x))(-g_2(x)) = g_1(x)g_2(x),$$

$$(f_1 \cdot g_1)(-x) = f_1(-x)g_1(-x) = f_1(x)(-g_1(x)) = -(f_1 \cdot g_1)(x),$$

proving the lemma. □

Next if we compare in Figure 10.1 $\sin \phi$ with $\phi$, we get

$$|\sin \phi| \le |\phi|. \qquad (10.14)$$

The latter allows us to calculate

$$\lim_{\phi \longmapsto 0} \sin \phi = 0. \qquad (10.15)$$

Indeed, given $\epsilon > 0$ choose $\delta = \epsilon$ to find for $|\phi - 0| = |\phi| < \delta$ that $|\sin \phi - 0| = |\sin \phi| \le |\phi| < \delta = \epsilon$. Thus we have proved that sin is continuous at 0. Since $\cos \phi = \sqrt{1 - \sin^2 \phi}$ we find that

$$\lim_{\phi \longrightarrow 0} \cos \phi = \lim_{\phi \longrightarrow 0} \sqrt{1 - \sin^2 \phi} = 1, \qquad (10.16)$$

i.e. cos is also continuous at 0. This further implies:

**Corollary 10.3.** *The functions* sin *and* cos *are continuous.*

*Proof.* For $\phi_0$ fixed we find with $h = \phi - \phi_0$

$$
\begin{aligned}
\lim_{\phi \longrightarrow \phi_0} \sin \phi &= \lim_{h \longrightarrow 0} \sin(\phi_0 + h) \\
&= \lim_{h \longrightarrow 0} (\sin \phi_0 \cos h + \cos \phi_0 \sin h) \\
&= \sin \phi_0 (\lim_{h \longrightarrow 0} \cos h) + \cos \phi_0 (\lim_{h \longrightarrow 0} \sin h) \\
&= \sin \phi_0
\end{aligned}
$$

proving the continuity of sin. Observing that

$$
\begin{aligned}
\lim_{\phi \longrightarrow \phi_0} \cos \phi &= \lim_{h \longrightarrow 0} \cos(\phi_0 + h) \\
&= \lim_{h \longrightarrow 0} (\cos \phi_0 \cos h + \sin \phi_0 \sin h) = \cos \phi_0
\end{aligned}
$$

we deduce that cos is continuous. □

From elementary geometry we know that a sector $O\widehat{A}B$ with an angle $\phi$, $0 \leq \phi < 2\pi$, of a circle with radius $r$ has area $\frac{1}{2}r^2\phi$, see the following figure for an explanation.

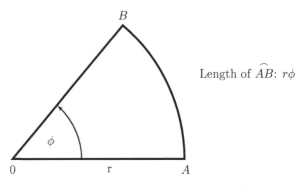

Length of $\widehat{AB}$: $r\phi$

*Figure 10.4*

Now we consider the unit circle and the following figure:

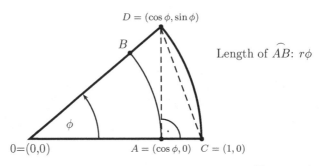

Length of $\widehat{AB}$: $r\phi$

*Figure 10.5*

It is obvious that the area of the sector $O\widehat{A}B$ is less or equal to that of the triangle $OCD$. Since $r = \cos\phi$ and the area of the triangle $OCD$ is given by $\frac{1}{2}\sin\phi$ we find for $0 \leq \phi < \frac{\pi}{2}$

$$\frac{1}{2}\phi\cos^2\phi \leq \frac{1}{2}\sin\phi,$$

144

and for $0 < \phi < \frac{\pi}{2}$

$$\cos^2 \phi \le \frac{\sin \phi}{\phi}. \tag{10.17}$$

Since $\cos^2 \phi$ as well as $\frac{\sin \phi}{\phi}$ are even functions we can extend (10.17) to all $\phi$, $0 < |\phi| < \frac{\pi}{2}$.
We now claim:

**Theorem 10.4.** *We have*

$$\lim_{\phi \longrightarrow 0} \frac{\sin \phi}{\phi} = 1. \tag{10.18}$$

*Proof.* From (10.17) and (10.14) we deduce for $0 < \phi < \frac{\pi}{2}$ that

$$1 - \sin^2 \phi = \cos^2 \phi \le \frac{\sin \phi}{\phi} \le 1.$$

Now it follows that

$$-\sin^2 \phi \le \frac{\sin \phi}{\phi} - 1 \le 0,$$

or

$$0 \le 1 - \frac{\sin \phi}{\phi} = \left| \frac{\sin \phi}{\phi} - 1 \right| \le \sin^2 \phi \le \phi^2.$$

Given $\epsilon > 0$ take $\delta = \sqrt{\epsilon}$ to find for $|\phi| < \delta$, $\phi \ne 0$, that

$$\left| \frac{\sin \phi}{\phi} - 1 \right| \le \sin^2 \phi \le \phi^2 = |\phi|^2 < \delta^2 = \epsilon.$$

$\square$

**Corollary 10.5.** *The function* $\sin : \mathbb{R} \longmapsto \mathbb{R}$ *is differentiable and we have*

$$\sin' = \cos. \tag{10.19}$$

*Moreover,* $\cos : \mathbb{R} \longmapsto \mathbb{R}$ *is differentiable and we have*

$$\cos' = -\sin. \tag{10.20}$$

*Proof.* Note that $\sin 0 = 0$, and therefore Theorem 10.4 states that $\sin$ is differentiable at 0 with derivative 1 which is equal to $\cos 0$. Now, using (10.9) we find

$$\frac{\sin\phi - \sin\phi_0}{\phi - \phi_0} = \frac{2\sin\frac{\phi-\phi_0}{2}\cos\frac{\phi+\phi_0}{2}}{\phi - \phi_0} = \frac{\sin(\phi - \phi_0)/2}{(\phi - \phi_0)/2}\cdot\cos\frac{\phi + \phi_0}{2},$$

which implies

$$\lim_{\phi\longmapsto\phi_0}\frac{\sin\phi - \sin\phi_0}{\phi - \phi_0} = \lim_{\phi\longrightarrow\phi_0}\frac{\sin\frac{\phi-\phi_0}{2}}{\frac{\phi-\phi_0}{2}}\lim_{\phi\longrightarrow\phi_0}\cos\frac{\phi - \phi_0}{2}$$

$$= \lim_{h\longrightarrow 0}\frac{\sin h}{h}\lim_{\phi\longrightarrow\phi_0}\cos\frac{\phi + \phi_0}{2} = \cos\phi_0,$$

where we used the continuity of $\cos$, compare with Corollary 10.3. Knowing that $\sin$ is differentiable and $\sin' = \cos$ allows us to calculate the derivative of $\cos$ by using the chain rule:

$$\frac{d}{dx}\cos x = \frac{d}{dx}(1 - \sin^2 x)^{\frac{1}{2}}$$

$$= (-2\sin x\cos x)\cdot\frac{1}{2}(1 - \sin^2)x)^{-\frac{1}{2}}$$

$$= -\frac{\sin x\cos x}{\cos x} = -\sin x.$$

$\square$

**Corollary 10.6.** *The function $\sin$ has for $\phi = (2k + \frac{1}{2})\pi, k \in \mathbb{Z}$, a (local) maximum and for $\phi = (2k - \frac{1}{2})\pi, k \in \mathbb{Z}$, a (local) minimum. The function $\cos$ has for $2k\pi, k \in \mathbb{Z}$, a (local) maximum and for $(2k+1)\pi, k \in \mathbb{Z}$, a (local) minimum.*

*Proof.* We know $(\sin\phi)' = \cos\phi = 0$ for $\phi = (k + \frac{1}{2})\pi$, $k \in \mathbb{Z}$. Now $(\sin\phi)'' = -\sin\phi$. Hence for $\phi = (2k + \frac{1}{2})\pi$, we find

$$-\sin\frac{\pi}{2} = -\sin(2k + \frac{1}{2})\pi = (\sin)''(2k + \frac{1}{2})\pi = -1 < 0,$$

thus $\sin$ has a local maximum for $\phi = (2k + \frac{1}{2})\pi$. For $\phi = (2k - \frac{1}{2})\pi$ we find

$$\sin\frac{\pi}{2} = -\sin\left(-\frac{\pi}{2}\right) = -\sin(2k - \frac{1}{2})\pi = (\sin'')(2k - \frac{1}{2})\pi = 1 > 0$$

implying that $\sin$ has a local minimum for $\phi = (2k - \frac{1}{2})\pi$. The result for $\cos$ is proved in an analogous way.

$\square$

From our definition of sin and cos it is clear that $\phi = \pi$ is the smallest zero of sin larger than 0, as is $\frac{\pi}{2}$ the smallest zero of cos larger than 0. We also note the formula:

$$\cos\phi = \sin(\phi + \frac{\pi}{2}).\tag{10.21}$$

The graphs $\Gamma(sin)$ and $\Gamma(cos)$ look like:

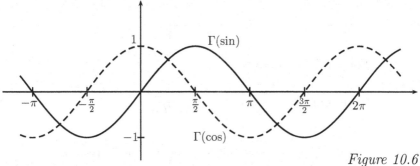

*Figure 10.6*

Consider the function $\sin : \mathbb{R} \longrightarrow \mathbb{R}$. Since it has period $2\pi$ it cannot be injective. Further we know that $\sin\pi = 0$, i.e. $\sin\pi = \sin 0 = -\sin(-\pi) = 0$, implying that sin cannot be injective on $[0, \pi]$. However we claim that $\sin : [-\frac{\pi}{2}, \frac{\pi}{2}] \longrightarrow \mathbb{R}$ is injective, in fact strictly increasing. For this we only need to consider

$$\sin' x = \cos x > 0 \quad \text{for} \ x \in (-\frac{\pi}{2}, \frac{\pi}{2}),$$

implying that $\sin|_{(-\frac{\pi}{2}, \frac{\pi}{2})}$ is strictly increasing. Since $\sin(-\frac{\pi}{2}) = -1$ and $\sin(\frac{\pi}{2}) = 1$ it follows that $\sin : [-\frac{\pi}{2}, \frac{\pi}{2}] \longrightarrow [-1, 1]$ is bijective. Hence it has an inverse function defined on $[-1, 1]$ which we denote by $\sin^{-1}$ or arcsin. In the same way we find that $\cos : [0, \pi] \longrightarrow [-1, 1]$ is strictly decreasing, recall $\cos' x = -\sin x$ and for $x \in (0, \pi)$ we have $\sin x > 0$. Hence there exists the inverse function $\cos^{-1}$ or arccos which is defined on $[-1, 1]$.

**Definition 10.7.** *The function* arcsin *is called the* **arcus-sine function** *and* arccos *is called the* **arcus-cosine function**.

**Theorem 10.8. A.** *The function* $\sin : [-\frac{\pi}{2}, \frac{\pi}{2}] \longrightarrow [-1, 1]$ *is bijective with inverse function* $\arcsin : [-1, 1] \longrightarrow [0, \pi]$ *and for* $-1 < x < 1$ *we have*

$$\frac{d}{dx} \arcsin(x) = \frac{1}{\sqrt{1 - x^2}}.\tag{10.22}$$

**B.** *The function* $\cos : [0, \pi] \longrightarrow [-1, 1]$ *is bijective with inverse function* $\arccos : [-1, 1] \longrightarrow [0, \pi]$ *and for* $-1 < x < 1$ *we have*

$$\frac{d}{dx} \arccos(x) = -\frac{1}{\sqrt{1-x^2}}. \tag{10.23}$$

*Proof.* It remains to prove (10.22) and (10.23). From Theorem 7.5 we know

$$\phi'(x) = \frac{1}{f'(\phi(x))}$$

for $\phi = f^{-1}$. For arcsin we deduce

$$\begin{aligned}
\arcsin'(x) &= \frac{d}{dx} \arcsin x = \frac{1}{\sin'(\arcsin x)} \\
&= \frac{1}{\cos(\arcsin x)} = \frac{1}{\sqrt{1 - \sin^2(\arcsin x)}} \\
&= \frac{1}{\sqrt{1-x^2}}.
\end{aligned}$$

For arccos we find

$$\begin{aligned}
\arccos'(x) &= \frac{d}{dx} \arccos x = \frac{1}{\cos'(\arccos x)} \\
&= \frac{1}{-\sin(\arccos x)} = -\frac{1}{\sqrt{1 - \cos^2(\arccos x)}} \\
&= -\frac{1}{\sqrt{1-x^2}}.
\end{aligned}$$

$\square$

Using sin and cos we may introduce some further functions of importance. Consider first the **tangent function**

$$\tan x := \frac{\sin x}{\cos x}. \tag{10.24}$$

Of course we must assure that $\cos x \neq 0$, thus we define the function tan on the set $\mathbb{R} \setminus \{(k + \frac{1}{2})\pi \,|\, k \in \mathbb{Z}\}$. It is obvious that tan is an odd function since

$$\tan(-x) = \frac{\sin(-x)}{\cos(-x)} = -\frac{\sin x}{\cos x} = -\tan x,$$

and we find on $\mathbb{R} \setminus \{(k + \frac{1}{2})\pi | k \in \mathbb{Z}\}$ that

$$\begin{aligned} \tan'(x) &= \frac{d}{dx}\tan(x) = \frac{d}{dx}\frac{\sin x}{\cos x} \\ &= \frac{\cos x \cos x + \sin x \sin x}{\cos^2 x} = \frac{1}{\cos^2 x}, \end{aligned}$$

i.e.

$$\tan' x = \frac{1}{\cos^2 x}. \tag{10.25}$$

Further we may introduce the **cotangent function**

$$\cot x := \frac{\cos x}{\sin x}, \tag{10.26}$$

which is defined on $\mathbb{R} \setminus \{k\pi | k \in \mathbb{Z}\}$. Once again we find that cot is an odd function and we have

$$\begin{aligned} \cot'(x) &= \frac{d}{dx}\frac{\cos x}{\sin x} = \frac{-\sin x \sin x - \cos x \cos x}{\sin^2 x} \\ &= -\frac{1}{\sin^2 x}, \end{aligned}$$

i.e.

$$\cot'(x) = -\frac{1}{\sin^2 x}. \tag{10.27}$$

From (10.25) it follows that on $(-\frac{\pi}{2}, \frac{\pi}{2})$ the function tan is strictly increasing, hence it has an inverse, the **arcus-tangent function** $\arctan : \mathbb{R} \longrightarrow (-\frac{\pi}{2}, \frac{\pi}{2})$. Note however that we have not yet proved that $R(\tan|_{(-\frac{\pi}{2}, \frac{\pi}{2})}) = \mathbb{R}$. For $\arctan'$ we find by Theorem 7.5 that

$$\arctan'(x) = \frac{1}{\tan'(\arctan x)} = \cos^2(\arctan x).$$

Now, $\cos^2 y = \frac{1}{1+\tan^2 y}$ as follows from

$$\frac{1}{1 + \frac{\sin^2 y}{\cos^2 y}} = \frac{1}{\frac{\cos^2 y}{\cos^2 y} + \frac{\sin^2 y}{\cos^2 y}} = \frac{\cos^2 y}{\sin^2 y + \cos^2 y} = \cos^2 y,$$

which yields

$$\arctan'(x) = \frac{1}{1 + \tan^2(\arctan x)} = \frac{1}{1 + x^2},$$

149

i.e.

$$\arctan'(x) = \frac{1}{1+x^2}. \tag{10.28}$$

From (10.27) we find that for $x \in (0, \pi)$ the function cot is strictly decreasing and hence it has an inverse function arccot, **arcus-cotangent.** For arccot we find

$$\operatorname{arccot}'(x) = \frac{1}{\cot'(\operatorname{arccot} x)} = -\sin^2(\operatorname{arccot} x).$$

Since $\sin^2 y = \frac{1}{1+\cot^2 y}$ we find

$$\operatorname{arccot}' x = -\frac{1}{1+\cot^2(\operatorname{arccot} x)} = -\frac{1}{1+x^2},$$

i.e.

$$\operatorname{arccot}' x = -\frac{1}{1+x^2}. \tag{10.29}$$

We postpone the proof of $R(\tan|_{(-\frac{\pi}{2},\frac{\pi}{2})}) = R(\cot|_{(0,\pi)}) = \mathbb{R}$, until Remark 20.18.B. and we refer to Appendix V where one can find a lot of formulae connecting sin, cos, tan, cot, arcsin, arccos, arctan, arccot. We mention that often a new name is introduced for $x \longmapsto \frac{1}{\sin x}$ and $x \longmapsto \frac{1}{\cos x}$, namely

$$\csc x = \frac{1}{\sin x} \text{ and } \sec x = \frac{1}{\cos x} \tag{10.30}$$

called **co-secant** and **secant function**. We finally consider the following graphs:

arcsin($x$)

arccos($x$)

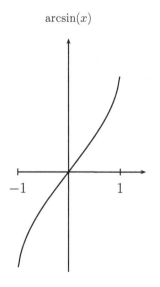

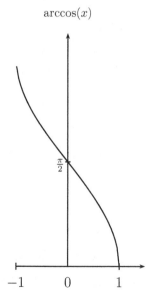

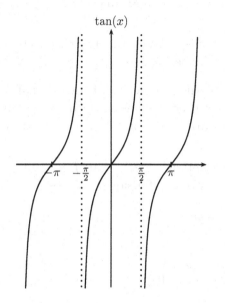

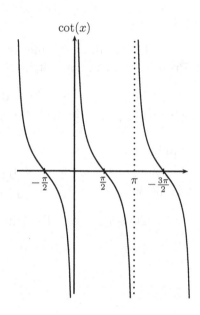

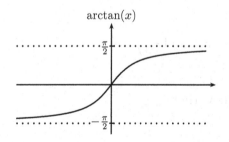

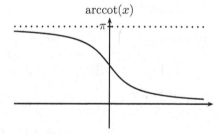

# Problems

1.  a) Let $f : \mathbb{R}_+ \longrightarrow \mathbb{R}$ be any function and let $g : \mathbb{R} \longrightarrow \mathbb{R}_+$ be an even function. Prove that $f \circ g : \mathbb{R} \longrightarrow \mathbb{R}$ is an even function.

    b) Let $f : \mathbb{R} \longrightarrow \mathbb{R}$ and $g : \mathbb{R} \longrightarrow \mathbb{R}$ be odd functions. Is $f \circ g$ an odd function too?

    c) Given an even function $f : \mathbb{R} \longrightarrow \mathbb{R}$ and $(a, b) \subset \mathbb{R}$, $a < 0 < b$. Prove that $f|_{(a,b)}$ cannot have an inverse function.

2.  a) Let $f : \mathbb{R} \longrightarrow \mathbb{R}$ be a differentiable function. Prove that if $f$ is even then $f'$ is odd and if $f$ is odd then $f'$ is even. Deduce that if $f$ is a $k$ times continuously differentiable function and $l \leq k$ is an even number then $f^{(l)}$ is even.

    b) Let $f : \mathbb{R}_+ \longrightarrow \mathbb{R}$ be a function. Show that $f$ has an even extension $g : \mathbb{R} \longrightarrow \mathbb{R}$ and $f|_{(0,\infty)}$ an odd extension $h : \mathbb{R} \longrightarrow \mathbb{R}$.

3.  a) Does the limit $\lim\limits_{x \to \infty} (\sin x)$ exist?

    b) Prove for $k \in \mathbb{N}$ that

    $$\lim_{x \to \infty} \frac{(\sin x)^k}{x} = 0.$$

4.  Using the definitions of sin, cos, tan and cot, and the addition theorems find the values of

    a) $\sin \frac{\pi}{8}$, b) $\cos \frac{\pi}{6}$, c) $\tan \frac{\pi}{3}$, d) $\cot \frac{\pi}{12}$.

5.  Find the values of

    a) $\arcsin \left( \frac{\sqrt{3}}{2} \right)$, b) $\arccos \left( -\frac{1}{2}\sqrt{2} \right)$, c) $\arctan \left( \frac{1}{\sqrt{3}} \right)$, d) $(-\sqrt{3})$.

6.  a) For $x, y \in \mathbb{R}$ prove that $|\sin x - \sin y| \leq |x - y|$.

    b) For $x, y \in [-a, a] \subset \left( -\frac{\pi}{2}, \frac{\pi}{2} \right)$ show that

    $$|\tan x - \tan y| \leq \frac{1}{\cos^2 a} |x - y|.$$

    c) Prove that for all $n \in \mathbb{N}$ and all $x \in \mathbb{R}$ we have $|\sin nx| \leq n |\sin x|$. Does the statement: for all $a > 0$ and all $x \in \mathbb{R}$ $|\sin ax| \leq a |\sin x|$ hold?

7. Let $f : \mathbb{R} \longrightarrow \mathbb{R}$ be a fundtion. Further let $g : \mathbb{R} \longrightarrow \mathbb{R}$ be a periodic function with period $a > 0$. Prove that the function $f \circ g$ is periodic with period $a$. Is the function $g \circ f$ periodic?

8. Find the derivatives (on the natural domains) of the given functions:

a) $\frac{d}{dx} \cos(\ln(1 + x^2))$;   b) $\frac{d}{dt} \frac{\sin(\tan t)}{\sqrt{1 - \cos^4 t}}$;   c) $\frac{d}{ds} \arcsin(\sqrt{1 + \cos s})$;

d) $\frac{d}{du} \arctan(e^{-u^2} \cot u)$.

9. For $n \in \mathbb{N}$ the **Dirichlet kernel** which is of great importance in Fourier analysis is defined on $\left[ -\frac{\pi}{2}, \frac{\pi}{2} \right]$ by

$$D_n(t) := \begin{cases} \frac{\sin(2n+1)t}{\sin t}, & t \in \left[ -\frac{\pi}{2}, \frac{\pi}{2} \right], \, t \neq 0 \\ 2n + 1, & t = 0. \end{cases}$$

Prove that

$$\frac{1}{2} D_n(t) = C_n(2t),$$

where

$$C_n(t) = \frac{1}{2} + \sum_{j=1}^{n} \cos jt,$$

and deduce that $D_n$ is on $\left( -\frac{\pi}{2}, \frac{\pi}{2} \right)$ arbitrarily often differentiable.

Hint: first find $\sum_{j=1}^{n} \cos jt$ and consider $\cos jt \cdot \sin \frac{t}{2}$.

153

# 11 Investigating Functions

In this chapter we want to develop a scheme for investigating a given function in a systematic way. The first problem we have to address is that of the domain. Clearly, if a function is given as $f : D \longrightarrow \mathbb{R}$ we know $D$. However, often we have to handle functions which are obtained from given ones or constructed "indirectly": the exponential function was introduced as a solution of a certain differential equation; the tangent function is the quotient of two functions both having many zeroes; the function $x \longmapsto \frac{x^2}{|x|}$ is not defined for $x = 0$ but easily extended to the function $x \longmapsto |x|$ which is defined for all $x \in \mathbb{R}$.

Thus our starting point should be an expression $f(x)$ defined originally for some subset $\tilde{D} \subset \mathbb{R}$ such that $x \longmapsto f(x)$ is a function on $\tilde{D}$. The first step is to determine the **maximal domain** $D$ **of the expression** $f(x)$, i.e. the largest set $D \subset \mathbb{R}$ such that $\tilde{D} \subset D$ and $f(x)$ is defined on $D$. We distinguish the maximal domain $D$ of the expression $f(x)$ as the domain of the maximal extension of $f : \tilde{D} \longrightarrow \mathbb{R}$.

**Example 11.1.** Consider on $\tilde{D} = \{x \in \mathbb{R} | x \neq 0 \text{ and } x \neq 1\}$ the expression $f(x) = \frac{x^2-1}{x-1}$. This expression is well defined for $x = 0$ and we can extend the domain of this expression easily to $D = \{x \in | x \neq 1\}$, obtaining a function $f : D \longrightarrow \mathbb{R}$. Since $x^2 - 1 = (x-1)(x+1)$ we find that $f(x) = \frac{(x-1)(x+1)}{x-1}$ which is for $x \neq 1$ equal to $x + 1$. However for $x = 1$ the expression $\frac{x^2-1}{x-1}$ is **not** defined and we cannot extend this expression to $\mathbb{R}$ whereas the function $f : D \longrightarrow \mathbb{R}, x \longmapsto \frac{x^2-1}{x-1}$, has an extension to $\mathbb{R}$ by the function $f^* : \mathbb{R} \longrightarrow \mathbb{R}$, $x \longmapsto x + 1$. Indeed, for $x \neq 1$ we have $f^*(x) = x + 1 = \frac{(x+1)(x-1)}{x-1} = \frac{x^2-1}{x-1}$, hence $f^*|_D = f$. This distinction might look a bit artificial, however it is not as we will see later. At the moment we agree to concentrate only on determining the maximal domain $D$ of the expression $f(x)$.

Next we investigate symmetry and monotonicity. So far we know three symmetries: $f$ can be even or odd or periodic (or none of these). Suppose $f : D \longrightarrow \mathbb{R}$ is given. In order for $f$ to be even (odd) we must have that $x \in D$ implies $-x \in D$, and in order for $f$ to have period $a$ we need to have $x \in D$ implies $x + a \in D$. Monotonicity is best checked (if possible) by looking at $f'$.

In general $D$ will be a proper subset of $\mathbb{R}$, i.e. not equal to $\mathbb{R}$. We call a point $x_0 \in D$ an **interior point** or **inner point** of $D$ if there exists $\epsilon > 0$

such that $(-\epsilon + x_0, x_0 + \epsilon) \subset D$. Assume $D = (a, b) = \{x \in \mathbb{R} | a < x < b\}$ is an open interval. We claim that all points of $(a, b)$ are inner points. Indeed, given $x_0 \in (a, b)$.

Figure 11.1

Consider $\epsilon := \frac{1}{2} \min(x_0 - a, b - x_0) > 0$. Then we claim $(-\epsilon + x_0, x_0 + \epsilon) \subset (a, b)$. The proof is simple $x \in (-\epsilon + x_0, x_0 + \epsilon)$ means $-\epsilon + x_0 < x < x_0 + \epsilon$ and with $\epsilon = \frac{1}{2} \min(x_0 - a, b - x_0)$ we find in the case where $\epsilon = \frac{1}{2}(x_0 - a)$ that

$$-\frac{1}{2}(x_0 - a) + x_0 < x < x_0 + \frac{1}{2}(x_0 - a)$$

which yields

$$a < \frac{1}{2}a + \frac{1}{2}x_0 < x < \frac{1}{2}(x_0 - a) + x_0 < \frac{1}{2}b + \frac{1}{2}x_0 < b,$$

hence $x \in (-\epsilon + x_0, x_0 + \epsilon)$, $\epsilon = \frac{1}{2}(x_0 - a)$, implies $x \in (a, b)$. The case where $\epsilon = \frac{1}{2}(b - x_0)$ is proved in the same way and is left as an exercise.

We call $x_0 \in \mathbb{R}$ a **boundary point** of $D \subset \mathbb{R}$ if for every $\epsilon > 0$ the interval $(-\epsilon + x_0, x_0 + \epsilon)$ contains at least a point belonging to $D$ and a point belonging to $D^{\complement}$, recall $D^{\complement} = \{x \in \mathbb{R} | x \notin D\}$. It may happen that a boundary point belongs to $D$ but it need not belong to $D$. Consider the set

$$D = (a, b] = \{x \in \mathbb{R} | a < x \le b\}.$$

By definition $a \notin D$ but $b \in D$. We claim that both $a$ and $b$ are boundary points.

We start with $a$ and choose any $\epsilon > 0$. The set $(-\epsilon + a, a + \epsilon)$ consists of all points $x \in \mathbb{R}$ such that $-\epsilon + a < x < a + \epsilon$, hence all points $-\epsilon + a < x \le a$ belong to $(a, b]^{\complement}$ and all points $a < x < a + \epsilon$ belong to $(a, b]$ provided $\epsilon \le b - a$. Thus $a$ is a boundary point not belonging to $(a, b]$. Now, to see that $b$ is a

156

boundary point take $\epsilon > 0$ and consider $(-\epsilon + b, b + \epsilon)$. These are all points $x$ satisfying $-\epsilon + b < x < b + \epsilon$. Those $x$ satisfying $-\epsilon + b < x \le b$ belong to $(a, b]$, provided $\epsilon < b - a$ and those satisfying $b < x < b + \epsilon$ belong to $(a, b]^{\complement}$. Hence $b$ is also a boundary point and it belongs to $D = (a, b]$.

Note, in both cases we have to modify our argument if $\epsilon$ becomes too large, $\epsilon < \frac{1}{2}(b - a)$ will always be sufficient.

By definition we call $-\infty$ and $+\infty$ the boundary points (at infinity) of the intervals $(-\infty, a)$ or $(-\infty, a]$ and $(b, +\infty)$ or $[b, +\infty)$, respectively, as well as of $\mathbb{R} = (-\infty, \infty)$. This is a slight abuse of the definition but helpful.

Typically the domains $D$ we will have to work with will consist of a finite union of finite or infinite intervals which could be open, closed or half-open. However, countable unions of finite intervals may also occur, think of the tangent function. The set $\partial D$ of all boundary points of $D$ (excluding $-\infty$ and $+\infty$) is called the **boundary** of $D$. The first task is to find all boundary points of $D$.

In the following we will only investigate functions which are continuous on $D$, in fact we will assume the functions to be a few times differentiable. Here is a fact which we will prove in Part 2: if $f : D \longrightarrow \mathbb{R}$ is continuous and $D$ a finite union of bounded and closed intervals then $f$ is bounded, i.e. there exists $M \ge 0$ such that $|f(x)| \le M$ for all $x \in D$.

As the example $f : (0, 1] \longrightarrow \mathbb{R}$, $x \longmapsto \frac{1}{x}$, shows this does not hold for non-closed intervals, and $g : \mathbb{R} \longrightarrow \mathbb{R}$, $x \longmapsto x$, shows that this does not hold for unbounded intervals.

We want to study the continuous function $f : D \longrightarrow \mathbb{R}$ at boundary points of $D$. First consider the case where $D$ is a bounded interval. In the case where $D = [a, b]$ is closed (and bounded) we know that $f$ is bounded and $f(a)$ as well as $f(b)$ are finite values. Suppose that $D$ is not closed, i.e. $D = (a, b]$ or $D = [a, b)$ or $D = (a, b)$. Of course $f$ could still be bounded, but it need not be. If a boundary point does not belong to $D$ everything may happen. However if a boundary point belongs to $D$, $f$ remains "locally" bounded, i.e. bounded at this boundary point (and in a small neighbourhood of it belonging to $D$), but no information is known *a priori* for all of $D$. Indeed, if $a \in D$ (the case $b \in D$ goes analogously) we find that $f|_{[a, \frac{b-a}{2}]}$ is continuous, hence bounded. The simple proof that $f : D \longrightarrow \mathbb{R}$ being continuous implies the continuity of $f|_{\tilde{D}}$, $\tilde{D} \subset D$, is left to the reader.

Let $f : (a, b) \longrightarrow \mathbb{R}$ be a continuous function. Here are some examples of what may happen at the boundary:

**Example 11.2. A.** The function $f : (0,1) \longrightarrow \mathbb{R}, x \longmapsto \frac{1}{x}$, is unbounded as $x \longrightarrow 0$. However, we can control its behaviour as $x \longrightarrow 0$. It is strictly monotone decreasing, i.e. $x < y$ implies $\frac{1}{x} > \frac{1}{y}$. Further it is always non-negative.

**B.** The function $f : (1,\infty) \longrightarrow \mathbb{R}, x \longmapsto \frac{x^2+1}{x-1}$, is unbounded as $x \longrightarrow \infty$. However, we can find its behaviour as $x \longrightarrow \infty$. Since

$$\frac{x^2+1}{x-1} = x\left(\frac{1+\frac{1}{x^2}}{1-\frac{1}{x}}\right)$$

and $\lim\limits_{x \to \infty} \left(\dfrac{1+\frac{1}{x^2}}{1-\frac{1}{x}}\right) = 1$ it follows with $g : (1,\infty) \longrightarrow \mathbb{R}, x \longmapsto x$, that

$$\lim_{x \to \infty} \frac{f(x)}{g(x)} = 1. \tag{11.1}$$

Now, $\lim\limits_{x \to \infty} \dfrac{f(x)}{g(x)} = 1$ means that given $\epsilon > 0$ there exists $N \in \mathbb{N}$ such that for $x > N$ it follows that

$$\left|\frac{f(x)}{g(x)} - 1\right| < \epsilon \text{ or } |f(x) - g(x)| < \epsilon g(x),$$

i.e.

$$-\epsilon g(x) < f(x) - g(x) < \epsilon g(x)$$

or

$$(1 - \epsilon)g(x) < f(x) < (1 + \epsilon)g(x) \text{ for } x > N, \tag{11.2}$$

recall $g(x) = x$ which is positive for $x > N$. This means that for $\epsilon > 0$ given and $x$ sufficiently large, the behaviour of $f$ is controlled by $g$.

**C.** Consider $g : (0,1) \longrightarrow \mathbb{R}, x \longmapsto \sin\frac{1}{x}$. This function is bounded but it does not have a limit or specific asymptotic behaviour as $x \longrightarrow 0$. Indeed, for the sequence $x_n = \frac{1}{n\pi}$ we have $\sin\frac{1}{x_n} = 0$, for the sequence $y_n = \frac{1}{2n+\frac{1}{2}\pi}$ it follows that $\sin\frac{1}{y_n} = 1$, and in fact for every value $z \in [-1,1]$ we can find a sequence $z_n$, $z_n \longrightarrow 0$, such that $\sin\frac{1}{z_n} \longrightarrow z$.

The most interesting case is Example 11.2.B which leads to:

**Definition 11.3.** *Let* $f : (a, b) \longrightarrow \mathbb{R}$ *be a function,* $-\infty \leq a < b \leq \infty$. *We call* $g : (a, b) \longrightarrow \mathbb{R}$, *an* **asymptote** *of* $f$ *at* $a$ *(at $b$) if*

$$\lim_{x \to a} \frac{f(x)}{g(x)} = 1 \tag{11.3}$$

$$\left( \lim_{x \to b} \frac{f(x)}{g(x)} = 1 \right).$$

*If* $g$ *is an asymptote of* $f$ *at* $a$ *we say that as* $x$ *tends to* $a$ *the function* $f$ *behaves* **asymptotically** *as* $g$.

Note that there are more general notions of an asymptote but the one given is sufficient for our purpose.

**Example 11.4.** Consider the polynomial $p : \mathbb{R} \longrightarrow \mathbb{R}$, $x \longmapsto p(x) = \sum_{j=0}^{N} a_j x^j$, with $a_N \neq 0$. We claim that $g(x) = a_N x^N$ is an asymptote of $p$ as $x \longrightarrow +\infty$. We have to prove

$$\lim_{x \to \infty} \frac{p(x)}{a_N x^N} = 1.$$

Since

$$\frac{\sum_{j=0}^{N} a_j x^j}{a_N x^N} = \sum_{j=0}^{N} \frac{a_j}{a_N} x^{j-N}$$

$$= 1 + \sum_{j=0}^{N-1} \frac{a_j}{a_N} x^{j-N}$$

it remains to prove

$$\lim_{x \to \infty} \sum_{j=0}^{N-1} \frac{a_j}{a_N} x^{j-N} = 0.$$

But we know that $\lim_{x \to \infty} x^{j-N} = 0$ for $j < N$. Note that the same argument yields that $g(x) = a_N x^N$ is also an asymptote of $p(x)$ as $x \longrightarrow -\infty$. Further, this example shows that an asymptote is not uniquely determined. Take for simplicity $p(x) = x^2 + 1$, then $x \longmapsto x^2$ is an asymptote, but by a trivial calculation it is easy to see that $x \longmapsto x^2 + c, c \in \mathbb{R}$, is a further one.

Now, given a continuous function $f : D \longrightarrow \mathbb{R}$ where $D$ is maximal and has boundary points $a_1, \ldots, a_N$ ($\pm\infty$ might be included). In order to investigate $f$ we need to determine the behaviour of $f$ at $a_1, \ldots, a_N$. The function might be bounded at some boundary points, it might have asymptotes at other boundary points, but there might also be boundary points where we have quite an irregular behaviour, i.e. we end up with no specific statement. In order to obtain asymptotes we need to calculate limits such as

$$\lim_{x \longrightarrow a} \frac{f(x)}{g(x)}$$

where both $f$ and $g$ may tend to zero as $x \longrightarrow a$, or may tend to infinity as $x \longrightarrow a$. (Note $a = \pm\infty$ is allowed.)

Without proof (see [3, p. 152]) we state

**Theorem 11.5 (de l'Hospital).** *Let $f$ and $g$ be differentiable functions defined on $(a, b)$, $-\infty \leq a < b \leq \infty$, and suppose that $g'(x) \neq 0$ for all $x \in (a, b)$. Suppose that either*

$$\lim_{\substack{x \to a \\ x \neq a}} f(x) = \lim_{\substack{x \to a \\ x \neq a}} g(x) = 0 \tag{11.4}$$

*or*

$$\lim_{\substack{x \to a \\ x \neq a}} g(x) = +\infty \ or \ -\infty. \tag{11.5}$$

*Then*

$$\lim_{\substack{x \to a \\ x \neq a}} \frac{f(x)}{g(x)} = \lim_{\substack{x \to a \\ x \neq a}} \frac{f'(x)}{g'(x)} \tag{11.6}$$

*provided the limit on the right hand side exists. An analogous statement holds for the boundary point $b$.*

**Example 11.6. A.** For $\alpha > 0$ we have

$$\lim_{x \longrightarrow \infty} \frac{e^{\alpha x}}{x} = \lim_{x \longrightarrow \infty} \frac{\alpha e^{\alpha x}}{1} = +\infty. \tag{11.7}$$

**B.** For every polynomial $p(x) = \sum_{j=0}^{N} a_j x^j$, $a_N \neq 0$, and $\alpha > 0$ we have

$$\lim_{x \longrightarrow \infty} \left( \sum_{j=0}^{N} a_j x^j \right) e^{-\alpha x} = 0. \tag{11.8}$$

Indeed

$$\lim_{x \longrightarrow +\infty} \frac{\sum_{j=0}^{N} a_j x^j}{e^{\alpha x}} = \lim_{x \longrightarrow +\infty} \frac{\left(\sum_{j=0}^{N} a_j x^j\right)'}{\alpha e^{\alpha x}}$$

$$= \ldots = \lim_{x \longrightarrow +\infty} \frac{\frac{d^N}{dx^N}\left(\sum_{j=0}^{N} a_j x^j\right)}{\alpha^N e^{\alpha x}}$$

$$= \lim_{x \longrightarrow +\infty} \frac{N! \, a_N}{\alpha^N e^{\alpha x}} = 0.$$

(We are allowed of course to iterate applications of de l'Hospital's rule.)
C. We claim

$$\lim_{\substack{x \to 0 \\ x > 0}} x^x = 1. \tag{11.9}$$

First note that by the continuity of exp we have

$$\lim_{\substack{x \to 0 \\ x > 0}} x^x = \lim_{\substack{x \to 0 \\ x > 0}} \exp(x \ln x)$$

$$= \exp\left(\lim_{\substack{x \to 0 \\ x > 0}} x \ln x\right).$$

Now

$$\lim_{\substack{x \to 0 \\ x > 0}} (x \ln x) = \lim_{\substack{x \to 0 \\ x > 0}} \left(\frac{\ln x}{\frac{1}{x}}\right)$$

$$= \lim_{\substack{x \to 0 \\ x > 0}} \left(\frac{\frac{1}{x}}{-\frac{1}{x^2}}\right) = \lim_{\substack{x \to 0 \\ x > 0}} (-x) = 0,$$

hence

$$\lim_{\substack{x \to 0 \\ x > 0}} x^x = \exp\left(\lim_{x \longrightarrow 0} (x \ln x)\right) = \exp(0) = 1.$$

Now, knowing how to investigate functions at the boundary of their domains we turn to the **interior of the domain**, i.e. all points $x \in D$ which together with a small open interval $(-\epsilon + x, x + \epsilon)$ belong to $D$.
We assume that $f : D \longrightarrow \mathbb{R}$ is twice continuously differentiable. We want to determine local extreme values. For this we know what to do: determine all zeroes $x_1, \ldots, x_K$ of $f'$ in $D$, and then consider $f''(x_j)$. If $f''(x_j) > 0$ then we have a local minimum, if $f''(x_j) < 0$ then we have a local maximum.

Special consideration is needed for points where $f'(x_l) = f''(x_l) = 0$. It is still possible for a function to have a local extreme value at such a point, for example $f : \mathbb{R} \to \mathbb{R}, f(x) = x^4$, has a local (and global) minimum at $x = 0$, however $f'(0) = f''(0) = 0$. On the other hand, for $g : \mathbb{R} \to \mathbb{R}, g(x) = x^3$, we also have $g'(0) = g''(0) = 0$, but at $x = 0$ the function $g$ does not have a local extreme value, in fact it is an example of a **point of inflexion**. If $x < 0$ then $g(x) < 0$ and if $x > 0$ then $g(x) > 0$, while $g(0) = 0$.

Let us summarise our method: given a function $f : \tilde{D} \to \mathbb{R}, \tilde{D} \subset \mathbb{R}$, in order to properly investigate its behaviour we do the following:

- we determine its maximal domain $D$;

- we determine all of its symmetries;

- we investigate whether it is monotone or not;

- we study its behaviour at the boundary points of $D$;

- we look for local extreme values;

- we try to sketch the graph.

We want to investigate the **hyperbolic functions**:

$$\sinh x := \frac{e^x - e^{-x}}{2}; \tag{11.10}$$

$$\cosh x := \frac{e^x + e^{-x}}{2}; \tag{11.11}$$

$$\tanh x := \frac{\sinh x}{\cosh x} = \frac{e^x - e^{-x}}{e^x + e^{-x}}; \tag{11.12}$$

and

$$\coth x := \frac{\cosh x}{\sinh x} = \frac{e^x + e^{-x}}{e^x - e^{-x}}. \tag{11.13}$$

Other hyperbolic functions are:

$$\operatorname{cosech} x := \frac{1}{\sinh x}; \tag{11.14}$$

and

$$\operatorname{sech} x := \frac{1}{\cosh x}. \tag{11.15}$$

We start with sinh. The domain of sinh is obviously $\mathbb{R}$, and since

$$\sinh(-x) := \frac{e^{-x} - e^{-(-x)}}{2} = -\frac{e^x - e^{-x}}{2} = -\sinh(x), \qquad (11.16)$$

sinh is an odd function with $\sinh(0) = 0$. Asymptotes $g_1$ for $x \longrightarrow \infty$ and $g_2$ for $x \longrightarrow -\infty$ are determined by $g_1(x) = \frac{e^x}{2}$ and $g_2(x) = -\frac{e^{-x}}{2}$. Indeed we have

$$\lim_{x \to \infty} \frac{\sinh x}{g_1(x)} = \lim_{x \to \infty} \frac{e^x - e^{-x}}{e^x} = \lim_{x \to \infty} \left(1 - e^{-2x}\right) = 1,$$

and

$$\lim_{x \to -\infty} \frac{\sinh x}{g_2(x)} = \lim_{x \to -\infty} -\frac{e^x - e^{-x}}{e^{-x}} = \lim_{x \to -\infty} \left(1 - e^{2x}\right) = 1.$$

Further we find

$$\sinh'(x) = \frac{e^x + e^{-x}}{2} = \cosh x > 0, \qquad (11.17)$$

implying that sinh is strictly monotone increasing. The graph of sinh looks like:

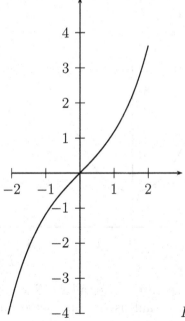

Figure 11.3

For the domain $D$ of cosh we again find that $D = \mathbb{R}$ and from

$$\cosh(-x) = \frac{e^{-x} + e^{-(-x)}}{2} = \frac{e^{-x} + e^{x}}{2} = \cosh(x)$$

we deduce that cosh is an even function which implies that cosh could not be strictly monotone. An asymptote for $x \longrightarrow \infty$ is $g_1(x) = \frac{e^x}{2}$ and for $x \longrightarrow -\infty$ an asymptote is $g_3(x) = \frac{e^{-x}}{2}$. Indeed we find

$$\lim_{x \to \infty} \frac{\cosh x}{g_1(x)} = \lim_{x \to \infty} \frac{e^x + e^{-x}}{e^x} = \lim_{x \to \infty} \left(1 + e^{-2x}\right) = 1$$

and

$$\lim_{x \to \infty} \frac{\cosh x}{g_3(x)} = \lim_{x \to -\infty} \frac{e^x + e^{-x}}{e^{-x}} = \lim_{x \to -\infty} \left(1 + e^{2x}\right) = 1.$$

Since

$$\cosh'(x) = \frac{e^x - e^{-x}}{2} = \sinh(x) \tag{11.18}$$

we find that $x_0 = 0$ is the only zero of $\cosh'$. Further

$$\cosh''(x) = \sinh'(x) = \cosh(x) > 0$$

for all $x$. Hence cosh has a minimum at $x_0 = 0$ with value $\cosh 0 = \frac{e^0 + e^0}{2} = 1$. The graph of cosh is given by

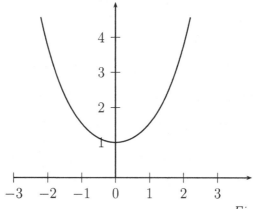

*Figure 11.4*

Next we discuss tanh. Since $\cosh x \neq 0$ for all $x \in \mathbb{R}$ we find that the domain of tanh is again $\mathbb{R}$. Further, tanh is the product of an even and an odd function, hence it is an odd function. From

$$\tanh'(x) = \left(\frac{\sinh x}{\cosh x}\right)' = \frac{\cosh^2 x - \sinh^2 x}{\cosh^2 x} = \frac{1}{\cosh^2 x} \tag{11.19}$$

we deduce that tanh is strictly monotone increasing. Note that we have used

$$\cosh^2 x - \sinh^2 x = 1 \tag{11.20}$$

which is left as an exercise. Since

$$\tanh x = \frac{e^x - e^{-x}}{e^x + e^{-x}}$$

we get for $x \longrightarrow \infty$

$$\lim_{x \to \infty} \tanh(x) = \lim_{x \to \infty} \left(1 \cdot \frac{1 - e^{-2x}}{1 + e^{-2x}}\right) = 1$$

and for $x \longrightarrow -\infty$ we have

$$\lim_{x \to -\infty} \tanh(x) = \lim_{x \to -\infty} \left(-1 \cdot \frac{1 - e^{2x}}{1 + e^{2x}}\right) = -1.$$

Thus $x \longmapsto 1$ is an asymptote for $x \longrightarrow \infty$ and $x \longmapsto -1$ is an asymptote for $x \longrightarrow -\infty$. The graph of tanh looks like

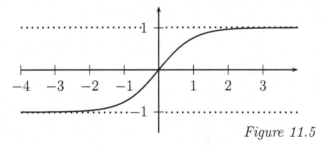

Figure 11.5

Finally we consider $\coth x = \frac{\cosh x}{\sinh x}$. Since $\sinh(0) = 0$, coth is only defined on $\mathbb{R} \setminus \{0\}$. Moreover, as a product of an even and an odd function it is an odd function. Thus we may restrict our discussions to $x > 0$.
The derivative of coth is given by

$$\coth' = \left(\frac{\cosh x}{\sinh x}\right)' = \frac{\sinh^2 x - \cosh^2 x}{\sinh^2 x} = -\frac{1}{\sinh^2 x}, \tag{11.21}$$

which is for $x \neq 0$ always strictly negative, hence $\coth|_{\{x\in\mathbb{R}|x>0\}}$ and $\coth|_{\{x\in\mathbb{R}|x<0\}}$ are strictly decreasing functions.

For $x \longrightarrow \infty$ we find

$$\lim_{x\to\infty} \coth x = \lim_{x\to\infty} \frac{e^x + e^{-x}}{e^x - e^{-x}} = 1$$

implying that $x \longmapsto 1$ is an asymptote for $\coth$ as $x \longrightarrow \infty$.

Further, for $x \longrightarrow 0$, $x > 0$, we find that

$$\frac{\cosh x}{\sinh x} = \frac{e^x + e^{-x}}{e^x - e^{-x}} \longrightarrow +\infty.$$

Thus $\coth$, when it is restricted to $(0, \infty)$ decreases from $+\infty$ to 1 as $x \longrightarrow \infty$. Using $\coth(-x) = -\coth(x)$ we find that $x \longrightarrow -1$ is an asymptote for $x \longrightarrow -\infty$ and that $\coth x \longrightarrow -\infty$ as $x \longrightarrow 0$ for $x < 0$. The graph of $\coth$ is given by

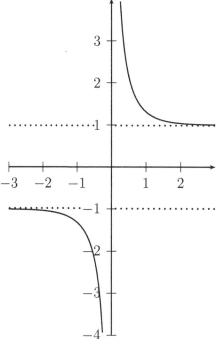

Figure 11.6

Note that Figure 11.6 suggests that $R(\coth)$ has a gap, namely the interval $[-1, 1]$. It is the discontinuity of $\coth$ at $x = 0$ which tolerates such a behaviour.

166

Since sinh is strictly increasing with range $R(\sinh) = \mathbb{R}$ it has an inverse defined on $\mathbb{R}$. By definition

$$\operatorname{arsinh} x := \sinh^{-1} x. \tag{11.22}$$

The notation comes from **area sinus hyperbolicus** and in some books one may see the notation area sinh for arsinh. Using formula (7.7) for the derivative of the inverse function we find

$$\begin{aligned}
\left(\sinh^{-1}\right)'(y) &= \frac{1}{\sinh'\left(\sinh^{-1}(y)\right)} \\
&= \frac{1}{\cosh\left(\sinh^{-1}(y)\right)}.
\end{aligned}$$

Now, $\cosh x = \sqrt{1 + \sinh^2 x}$, recall (11.20), implying that

$$\left(\sinh^{-1}\right)'(y) = \frac{1}{\sqrt{1 + \sinh^2\left(\sinh^{-1}(y)\right)}} = \frac{1}{\sqrt{1 + y^2}}. \tag{11.23}$$

We claim

$$\operatorname{arsinh} x = \ln\left(x + \sqrt{x^2 + 1}\right). \tag{11.24}$$

Note that

$$\ln\left(x + \sqrt{x^2 + 1}\right)' = \frac{\frac{d}{dx}\left(x + \sqrt{x^2 + 1}\right)}{x + \sqrt{x^2 + 1}} = \frac{1}{\sqrt{x^2 + 1}},$$

i.e.

$$\ln\left(x + \sqrt{x^2 + 1}\right)' = \operatorname{arsinh}'(x),$$

and therefore they differ only by a constant:

$$\operatorname{arsinh} x = c + \ln\left(x + \sqrt{x^2 + 1}\right).$$

But $\operatorname{arsinh} 0 = 0$ which gives

$$0 = \operatorname{arsinh} 0 = c + \ln\left(0 + \sqrt{0 + 1}\right) = c,$$

i.e. $c = 0$ and (11.23) holds.

The function tanh is strictly increasing with range $(-1, 1)$, hence it has an inverse

$$\operatorname{artanh} : (-1, 1) \longrightarrow \mathbb{R}, \quad \operatorname{artanh} := \tanh^{-1}. \tag{11.25}$$

We want to find artanh$'$. First note that

$$\tanh'(x) = \frac{1}{\cosh^2 x} = \frac{\cosh^2 x - \sinh^2 x}{\cosh^2 x} = 1 - \tanh^2(x).$$

Now we get by (7.7)

$$\begin{aligned}
\text{artanh}'(y) &= \left(\tanh^{-1}\right)'(y) = \frac{1}{(\tanh')(\tanh^{-1}(y))} \\
&= \frac{1}{(1 - \tanh^2)(\tanh^{-1}(y))} = \frac{1}{1 - y^2},
\end{aligned}$$

i.e. we have

$$\text{artanh}'(x) = \frac{1}{1 - x^2} \qquad \text{for } -1 < x < 1. \tag{11.26}$$

As in the case of arsinh we can prove

$$\text{artanh}(x) = \frac{1}{2} \ln \frac{1+x}{1-x}. \tag{11.27}$$

In the exercises there will be questions related to the inverse function of $\cosh|_{[0,\infty)}$. This function is denoted by arcosh and is defined on $[1, \infty)$. Its derivative is given by

$$\text{arcosh}'(x) = \frac{1}{\sqrt{x^2 - 1}}, \qquad x > 1 \tag{11.28}$$

and we have

$$\text{arcosh}\, x = \ln(x + \sqrt{x^2 - 1}), \qquad x > 1. \tag{11.29}$$

For coth we restrict our attention first to values $x > 1$. Thus $\coth|_{(0,\infty)}$ is considered as a strictly decreasing function with range $(1, \infty)$. This function has an inverse function arcoth and we have

$$\text{arcoth}'(x) = \frac{1}{1 - x^2}, \qquad x > 1 \tag{11.30}$$

and

$$\text{arcoth}(x) = \frac{1}{2} \ln \frac{x+1}{x-1}, \qquad x > 1. \tag{11.31}$$

Using the symmetry of coth we can extend (11.29) and (11.30) to $x < -1$.

# Problems

1. Consider the set $D := [-1, 2) \cup \{3, 4\} \cup [5, 6]$. Find every interior point of $D$ and the boundary $\partial D$ of $D$.

2. For each of the following expressions find the maximal set $D \subset \mathbb{R}$ such that on $D$ the expressions define functions.

   a) $\sqrt{(x^2 - 1)(x^2 + 4x)}$.

   b) $\frac{\cos(\ln(\arctan x))}{x^3 + 4x^2 - 5x}$.

   c) $((\sinh x)(1 - x^4))^{\frac{1}{2}}$.

   d) $\cot(\arcsin x)$.

3. Use l'Hospital's rules to find the following limits. If necessary, iterate an application of these rules.

   a) $\lim\limits_{x \to 1} \dfrac{1 + \cos \pi x}{x^2 - 2x + 1}$;

   b) $\lim\limits_{\substack{t \to 0 \\ t > 0}} \dfrac{\ln(\cos 3t)}{\ln(\cos 2t)}$;

   c) $\lim\limits_{y \to \infty} \dfrac{3y^2 - y + 5}{5y^2 - 6y - 3}$;

   d) $\lim\limits_{u \to 0} \left( \dfrac{1}{\sin^2 u} - \dfrac{1}{u^2} \right)$.

   Hint: rewrite $\frac{1}{\sin^2 u} - \frac{1}{u^2}$ as $\frac{u^2 - \sin 2u}{u^4} \cdot \frac{u^2}{\sin^2 u}$, note that $\lim\limits_{x \to 0} \frac{\sin x}{x} = 1$, and when applying l'Hospital rules to $\frac{u^2 - \sin 2u}{u^4}$ make use of the addition theorems for trigonometric functions.

4.   a) For $g : \mathbb{R} \longrightarrow \mathbb{R}$ find the asymptote as $x \to +\infty$ where

$$g(x) = \ln\left(1 + x^2 + e^{x^2}\right).$$

   b) Find the asymptote as $t \to \pm\infty$ for the function

$$h(t) = e^{-\frac{1}{1+t^2}}.$$

5. Following the method introduced in this chapter investigate the follow-ing functions $f_j : \tilde{D}_j \longrightarrow \mathbb{R}$ and sketch their graphs. (Note that $\tilde{D}_j \subset \mathbb{R}$ is some domain, therefore firstly find the maximal domain $D_j$.)

a) $f_1 : \tilde{D}_1 \longrightarrow \mathbb{R}$, $f_1(x) = \frac{2x^2+12x-2}{15(x^2-1)^{\frac{1}{2}}}$, $\tilde{D}_1 = [2, \infty)$;

b) $f_2 : \tilde{D}_2 \longrightarrow \mathbb{R}$, $f_2(s) = \tan \frac{s^2}{1+s^4}$, $\tilde{D}_2 = \left(\frac{\pi}{6}, \frac{\pi}{4}\right)$;

c) $f_3 : \tilde{D}_3 \longrightarrow \mathbb{R}$, $f_3(t) = \operatorname{arsinh}\left(1 - e^{-t^2}\right)$, $\tilde{D}_3 = \mathbb{R}_+$.

6. Prove the following formulae for hyperbolic functions:

a) $\cosh^2 x - \sinh^2 x = 1$;

b) $\sinh^2 x = \frac{1}{\cosh^2 x - 1}$;

c) $\sinh(x \pm y) = \sinh x \cosh y \pm \cosh x \sinh y$;

d) $\tanh(x - y) = \frac{\tanh x - \tanh y}{1 - \tanh x \tanh y}$.

The following identity may be used:

$$\cosh(x - y) = \cosh x \cosh y - \sinh x \sinh y.$$

# 12 Integrating Functions

Let us start to analyse a natural problem in mathematics. Given a continuous function $g : [a, b] \to \mathbb{R}$, $a, b \in \mathbb{R}$, $a < b$, can we find a function $f : [a, b] \to \mathbb{R}$ such that on $(a, b)$

$$f'(t) = g(t)? \qquad (12.1)$$

Let us assume that we know the value of $f(a)$. A very rough approximation of $f'(t)$, $a < t < b$, is

$$\frac{f(t) - f(a)}{t - a}.$$

Hence (12.1) would give

$$f(t) - f(a) \approx g(t)(t - a) \qquad (12.2)$$

or

$$f(t) \approx f(a) + g(t)(t - a), \qquad (12.3)$$

where $g(t) \approx h(t)$ means that $g$ is close to $h$. There is a simple geometric interpretation of the right hand side of (12.2)

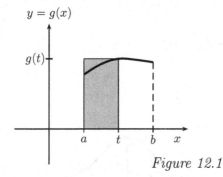

Figure 12.1

The area of the rectangle with vertices $(a, 0), (t, 0), (t, g(t))$ and $(a, g(t))$ is given by $g(t)(t - a)$. Of course, when $t$ varies in $[a, b]$ we obtain a function

$$t \to g(t)(t - a) + f(a). \qquad (12.4)$$

But only for very small values of $t - a$, $t > a$, do we expect the function (12.4) to be a reasonable approximation of a function $f$ satisfying (12.1).

However we may improve the approximation. Given $t \in (a,b)$ as before and take $t_1 \in (a,b)$, $t_1 < t$, and note that

$$f(t) - f(a) = f(t) - f(t_1) + f(t_1) - f(a)$$
$$\approx g(t)(t - t_1) + g(t_1)(t_1 - a).$$

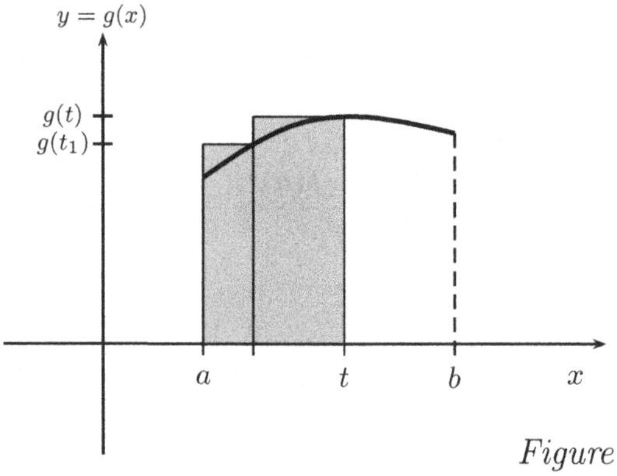

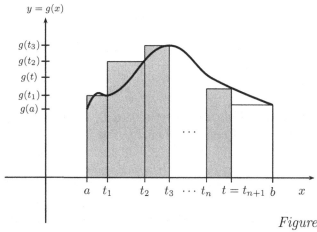

*Figure 12.2*

Iterating this process $n$-times we find with $a < t_1 < t_2 < \cdots < t < b$,

$$f(t) - f(a) = f(t) - f(t_n) + f(t_n) - f(t_{n-1}) + \cdots + f(t_1) - f(a)$$
$$\approx g(t)(t - t_n) + g(t_n)(t_n - t_{n-1}) + \cdots + g(t_1)(t_1 - a)$$
$$= \sum_{j=1}^{n+1} g(t_j)(t_j - t_{j-1}), \tag{12.5}$$

where $t_0 := a$ and $t_{n+1} := t$.

*Figure 12.3*

Now letting $n \to \infty$ such that $\max_{1 \le j \le n+1}(t_j - t_{j-1}) \to 0$, we may conjecture that $f(t) - f(a)$ is given by the area bounded by the sets $\{(x,y) \in \mathbb{R}^2 | x = a\}$, $\{(x,y) \in \mathbb{R}^2 | x = t\}$, $\{(x,y) \in \mathbb{R}^2 | y = 0\}$, and $\{(x,y) \in \mathbb{R}^2 | y = g(t)\}$, or in short the area of the set bounded by the $x$-axis, the function $g$ and the lines $x = a$ and $x = t$.

Although this is the correct conjecture we must overcome some problems to justify this solution. Most of all, we need to define what is meant by "the area bounded by the $x$-axis, the function $g$ and the lines $x = a$ and $x = t$".

Let $g : [a, b] \to \mathbb{R}$ be a continuous function (which must be bounded as every continuous function on a closed and bounded interval is). Let $a = t_0 < t_1 < t_2 < \cdots < t_n < t_{n+1} = b$ be a finite sequence of points in $[a, b]$:

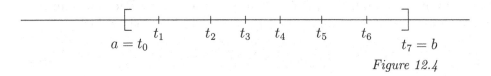

$$a = t_0 \quad t_1 \quad t_2 \quad t_3 \quad t_4 \quad t_5 \quad t_6 \quad t_7 = b$$

*Figure 12.4*

We call such a finite sequence a **partition** of $[a, b]$ into sub-intervals $[t_j, t_{j-1}]$, $j = 1, \dots n + 1$ and we sometimes write $Z(t_1, \cdots t_n)$ or just $Z_n$ for such a partition. The number

$$m(Z_n) := \max\{t_j - t_{j-1} | j = 1, \cdots, n + 1\} \tag{12.6}$$

is called the **mesh size** or **width** of the partition $Z_n$. Given a partition $Z_n$ we can form the (Riemann) sum (of $g$ with respect to $Z_n$)

$$S_r(g, Z_n) := \sum_{j=1}^{n+1} g(t_j)(t_j - t_{j-1}). \tag{12.7}$$

In the case where $g \ge 0$ we already know an interpretation of $S_r(g, Z_n)$ as an approximation of the area bounded by the $x$-axis, the function $g$ and the lines $x = a$ and $x = b$:

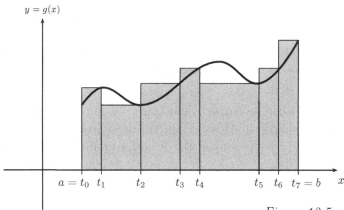

Figure 12.5

In fact we may generalise $S_r(g, Z_n)$ slightly to the general **Riemann sum** of $g$ with respect to $Z_n$ and points $\xi_j \in [t_{j-1}, t_j]$, which is defined by:

$$S(g, Z_n, \xi) := \sum_{j=1}^{n+1} g(\xi_j)(t_j - t_{j-1}) \tag{12.8}$$

where $\xi = (\xi_1, \cdots, \xi_{n+1})$.

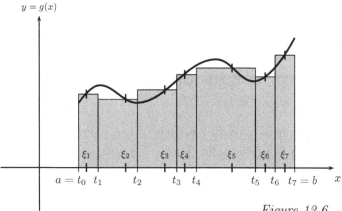

Figure 12.6

**Definition 12.1.** *Let $g : [a, b] \to \mathbb{R}$, $a, b \in \mathbb{R}$, $a < b$, be a continuous function. Suppose there exists a number $I_{a,b}(g) \in \mathbb{R}$ such that for every $\varepsilon > 0$ there exists $\delta > 0$ with the property that if $Z_n$ is any finite partition of $[a,b]$ with mesh size $m(Z_n) < \delta$ then*

$$|I_{a,b}(g) - S_r(g, Z_n)| < \varepsilon.$$

174

*In this case we call $I_{a,b}(g)$ the **(Riemann) integral of** $g$ over the interval $[a,b]$ and denote it by*

$$\int_a^b g(t)dt := I_{a,b}(g).$$ (12.9)

In Chapter 25, in particular Theorem 25.24, we will discuss in detail Riemann sums and their relation to Riemann integrability. Without proof we quote

**Theorem 12.2.** *For every continuous function $g : [a, b] \to \mathbb{R}$, $a, b \in \mathbb{R}$, $a < b$, the integral $\int_a^b g(t)dt$ exists and its value can be calculated by using (12.8) instead of (12.7).*

**Definition 12.3.** *The **area** $A$ of a set in $\mathbb{R}^2$ bounded by the x-axis, a non-negative continuous function $g : [a, b] \to \mathbb{R}$, $a, b \in \mathbb{R}$, $a < b$, and the lines $x=a$ and $x=b$, is by definition*

$$A := \int_a^b g(t)dt.$$ (12.10)

**Remark 12.4. A.** Note that Definition 12.3 is not tautological: it is a non-trivial problem to define the area of an arbitrary subset of $\mathbb{R}^2$.
**B.** Let us agree to define for any function $g : [a, b] \to \mathbb{R}$

$$\int_c^c g(t)dt = 0 \quad \text{for all} \quad c \in [a, b]$$ (12.11)

This definition is justified by the idea that the interval $[c, c]$ has length zero, hence the rectangle with one side of length $g(t)$ and the other of length 0 should have area 0.

Let $g : [a, b] \to \mathbb{R}$ be a continuous function. We define a new function $f : [a, b] \to \mathbb{R}$ by

$$x \to f(x) := \int_a^x g(t)dt.$$ (12.12)

Since $[a, x]$ is a closed and bounded interval and $g|_{[a,x]}$ is continuous $f(x)$ is well defined.
The following theorem is important:

**Theorem 12.5.** *Let $g : [a, b] \to \mathbb{R}$ be a continuous function. Then $f : [a, b] \to \mathbb{R}$ defined by (12.12) is differentiable and we have*

$$f'(x) = g(x),$$ (12.13)

*i.e. we have*

$$\frac{d}{dx} f(x) = \frac{d}{dx} \int_a^x g(t)dt = g(x). \tag{12.14}$$

We will prove this result in Part 2 of our course in a similar way as to how we motivated the introduction of the integral. Note that Theorem 12.5 allows us to calculate integrals. First we give

**Definition 12.6.** *Let $g : [a,b] \to \mathbb{R}$ be a function. We call a differentiable function $f$ a **primitive** of $g$ if $f' = g$.*

Hence by Theorem 12.5, $x \mapsto \int_a^x g(t)dt$ is a primitive of $g$. A primitive of a function $g$ is not unique. If $f$ is a primitive of $g$ then for every constant $c \in \mathbb{R}$ a further primitive of $g$ is given by $f + c$ since $(f + c)' = f'$. It is important that this is the only type of non-uniqueness of a primitive: if $f$ and $h$ are two primitives of $g$ then there exists a constant $c \in \mathbb{R}$ such that $f - h = c$. Indeed, being a primitive implies

$$(f - h)' = g - g = 0,$$

which yields $f - h = c$.

**Theorem 12.7. (Fundamental Theorem of Calculus).** *Let $g : [a,b] \to \mathbb{R}$ be a continuous function and let $h$ be a primitive of $g$. Then we have*

$$\int_a^b g(t)dt = h(b) - h(a). \tag{12.15}$$

*Proof.* We know that $f$ defined by (12.12) is a primitive of $g$. Since $f(a) = 0$ and $f(b) = \int_a^b g(t)dt$ we find in this case that

$$\int_a^b g(t)dt = f(b) - f(a).$$

Now, if $h$ is any further primitive, then $f - h = c$ implying that

$$f(b) - f(a) = h(b) - h(a)$$

and (12.15) follows. $\qquad\square$

Let us now introduce a useful notation. If $f$ is a primitive of $g$ we write

$$\int_a^b g(t)dt = f|_a^b \ . \tag{12.16}$$

Now we use the fundamental theorem to evaluate integrals.

**Example 12.8.** For $k \in \mathbb{N}$ we have

$$\int_a^b x^k dx = \frac{x^{k+1}}{k+1}\Big|_a^b \ , \tag{12.17}$$

i.e. $f(x) = \frac{x^{k+1}}{k+1}$ is a primitive of $g(x) = x^k$ and

$$\int_a^b x^k dx = \frac{b^{k+1}}{k+1} - \frac{a^{k+1}}{k+1} \ . \tag{12.18}$$

We only have to note that

$$f'(x) = \frac{d}{dx}\left(\frac{x^{k+1}}{k+1}\right) = (k+1)\frac{1}{k+1}x^{k+1-1} = x^k.$$

**Example 12.9.** For $0 < a < b$ we have

$$\int_a^b \frac{1}{x}dx = \ln x|_a^b = \ln b - \ln a = \ln \frac{b}{a}. \tag{12.19}$$

Indeed we know that

$$\frac{d}{dx}\ln x = \frac{1}{x} \ \ for \ \ x > 0.$$

Further we have

$$\ln b - \ln a = \ln b + \ln a^{-1} = \ln \frac{b}{a}.$$

**Example 12.10.** Let $k \in \mathbb{Z}$, $k < -1$. Further assume that either $a < b < 0$ or $0 < a < b$ then

$$\int_a^b x^k dx = \frac{x^{k+1}}{k+1}\Big|_a^b \ . \tag{12.20}$$

It is helpful to rewrite (12.20) with $k = -n$, $n \in \mathbb{N}$ and $n > 1$. Then we find

$$\int_a^b x^{-n}dx = \frac{x^{-n+1}}{-n+1}\Big|_a^b \ . \tag{12.21}$$

**Example 12.11.** Let $\alpha > 0$ and define for $x > 0$

$$x^{\alpha} := e^{\alpha \ln x}. \tag{12.22}$$

For $\alpha = n \in \mathbb{N}$ we find

$$e^{n \ln x} = e^{\ln x} \cdots e^{\ln x} = x \cdots x = x^n,$$

thus (12.22) generalises the **power function**. Moreover we have

$$\frac{d}{dx} x^{\alpha} = \frac{d}{dx} e^{\alpha \ln x} = \alpha \frac{1}{x} e^{\alpha \ln x} = \alpha x^{\alpha - 1},$$

which yields for $0 < a < b$ that

$$\int_a^b x^{\alpha} dx = \left. \frac{x^{\alpha+1}}{\alpha + 1} \right|_a^b \tag{12.23}$$

provided $\alpha > 0$. Indeed we find that $\frac{d}{dx}\left(\frac{x^{\alpha+1}}{\alpha+1}\right) = x^{\alpha}$, i.e. $x \mapsto \frac{x^{\alpha+1}}{\alpha+1}$ is a primitive of $x \mapsto x^{\alpha}$. Without proof we note that (12.23) holds for all $\alpha \neq -1$.

We want to return to Example 12.9:

**Example 12.12.** For $a < b < 0$ we have

$$\int_a^b \frac{1}{x} dx = \ln(-x)|_a^b. \tag{12.24}$$

Indeed for $x < 0$ we find $\frac{d}{dx} \ln(-x) = -\frac{1}{-x} = \frac{1}{x}$. We can combine (12.19) with (12.24) to get

$$\int_a^b \frac{1}{x} dx = \left. \ln|x| \right|_a^b, \qquad 0 \notin [a, b]. \tag{12.25}$$

**Example 12.13.** Since $\sin' = \cos$ and $\cos' = -\sin$ we have

$$\int_a^b \sin x dx = -\cos x|_a^b \tag{12.26}$$

and

$$\int_a^b \cos x dx = \sin x|_a^b . \tag{12.27}$$

Taking in (12.27) $a = 0$ and $b = \pi$ we find

$$\int_0^\pi \cos x\, dx = \sin \pi - \sin 0 = 0.$$

Hence there are functions not identical to zero whose integral over a certain interval might be zero.

**Example 12.14.** For exp we find

$$\int_a^b e^x dx = e^x |_a^b = e^b - e^a. \tag{12.28}$$

**Example 12.15.** We find

$$\int_a^b \frac{1}{\sqrt{1-x^2}} dx = \int_a^b \frac{dx}{\sqrt{1-x^2}} = \arcsin x |_a^b \ , \quad [a,b] \subset (-1,1). \tag{12.29}$$

**Example 12.16.** We have

$$\int_a^b \frac{dx}{1+x^2} = \arctan x |_a^b \ . \tag{12.30}$$

**Example 12.17.** We have

$$\int_a^b \frac{dx}{\sqrt{1+x^2}} = \ln(x + \sqrt{1+x^2}) |_a^b = \operatorname{arsinh} x \ |_a^b. \tag{12.31}$$

All these examples are simple to prove: we just use our knowledge about derivatives. Whenever we know of two functions where $f' = g$ we can immediately write

$$\int_a^b g(t) dt = f(x) |_a^b.$$

In the next chapter we will meet rules on how to reduce a given integral to an integral which we can evaluate. Unfortunately this is not always possible. Before doing this, let us introduce a further traditional notation. If $g$ is a continuous function then we denote its generic primitive by

$$\int g(x) dx \quad \text{or} \quad \int g\, dx.$$

Thus $\int$ may have two interpretations: in the form $\int_a^b g(t) dt$ it helps us to define a number, hence $x \mapsto \int_a^x g(t) dt$ defines a unique function; in the form $\int g(x) dx$ it denotes the generic primitive of $g$. Older books tend to call $\int_a^b g(t) dt$ a definite integral and $\int g\, dt$ an indefinite integral.

# Problems

1.      a) Find the Riemann sum of the function $f : [1, 2] \longrightarrow \mathbb{R}$ with $f(t) = 2t^2 - t$ with respect to the partition $t_k = 1 + \frac{k}{n}$, $k = 0, 1, \ldots, n$, and $\xi_k$ being the midpoint of the interval $[t_{k-1}, t_k]$, $k = 1, \ldots, n$.

     b) Let $a < b$ and $h : [a, b] \longrightarrow \mathbb{R}$ be the function $h(t) = \frac{1}{1+t^2}$. For the partition $t_l = \frac{a(m^2 - l^2) + l^2 b}{m^2}$, $l = 0, 1, \ldots, m$, and $\xi_l \in [t_l, t_{l+1}]$, $l = 0, \ldots, m-1$, such that $\xi_l - t_l = \frac{1}{3}(t_{l+1} - t_l)$ and $t_{l+1} - \xi_l = \frac{2}{3}(t_{l+1} - t_l)$ find the corresponding Riemann sum.
(After calculating $t_k - t_{k-1}$, $k = 1, \ldots, m$, and $\xi_k$, $k = 1, \ldots, m$, and forming the sum $\sum_{k=1}^{m} g(\xi_k)(t_k - t_{k-1})$, it will not be possible to simplify much in this expression.)

2. Let $g : [a, b] \longrightarrow \mathbb{R}$ be a function with the Riemann sum

$$S(g, Z_n, \xi) = \sum_{j=1}^{n} g(\xi_j)(t_j - t_{j-1}).$$

Let $a < t_k < b$ be a fixed point in $Z_n$. Prove that

$$S(g|_{[a,t_k]}, Z_n|_{[a,t_k]}, \xi|_{[a,t_k]}) + S(g|_{[t_k,b]}, Z_n|_{[t_k,b]}, \xi|_{[t_k,b]})$$
$$= S(g, Z_n, \xi).$$

Here $Z|_{[a,t_k]}$ is the partition $a = t_0 < t_1 < \cdots < t_k$, $Z|_{[t_k,b]}$ is the partition $t_k < t_{k+1} < \cdots < t_n = b$, and $\xi|_{[a,t_k]}$ as well as $\xi|_{[t_k,b]}$ denote the points $\xi_j$ belonging to $[a, t_k]$ and $[t_k, b]$ respectively.

3.      a) By interpreting integration as the area under a curve (Definition 12.3) find

$$\int_{-2}^{1} |x| \, dx$$

by calculating the area of the triangles $ABC$ and $BDE$ in Figure 12.7 where $A = (-2, 0)$, $B = (0, 0)$, $C = (-2, 2)$, $D = (1, 0)$, $E = (1, 1)$.

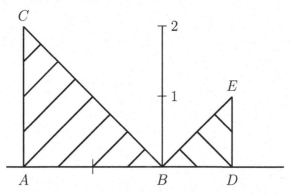

*Figure 12.7*

b) The upper semicircle with radius $R$ in Figure 12.8 is the graph of the function $g : [-R, R] \longrightarrow \mathbb{R}$, $g(r) = \sqrt{R^2 - r^2}$.

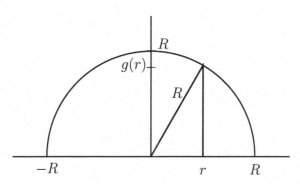

*Figure 12.8*

Again using the interpretation that integration represents area, see Definition 12.3, find $\int_{-R}^{R} g(r)\mathrm{d}r = \int_{-R}^{R} \sqrt{R^2 - r^2}\mathrm{d}r$.

4. By calculating derivatives prove that in each of the following cases $F$ is a primitive of $f$, i.e. $F' = f$.

a) $F(x) = \ln(\cosh x)$, $f(x) = \tanh x$;

b) $F(s) = \frac{a^s}{\ln a}$, $a > 1$, $f(s) = a^s$;

c) $F(u) = \frac{e^u(\sin 5u - 5 \cos 5u)}{26}$, $f(u) = e^u \sin 5u$;

d) $F(r) = -\frac{1}{2} \cos\left(r^2 + 4r - 6\right)$, $f(r) = (r + 2) \sin\left(r^2 + 4r - 6\right)$.

181

# 13   Rules for Integration

There are essentially two sets of rules for integration. The first and easier ones are derived from properties of the summation process. The second set of rules is derived from our rules for taking derivatives and the fundamental theorem.

From Definition 12.1 it follows that we can approximate for a continuous function $g : [a, b] \longrightarrow \mathbb{R}$ the integral $\int_a^b g(t)\, dt$ by (finite) Riemann sums. Since for two continuous functions $g_1 : [a, b] \longrightarrow \mathbb{R}$ and $g_2 : [a, b] \longrightarrow \mathbb{R}$ and for real numbers $\lambda, \mu \in \mathbb{R}$ we have

$$S_r(\lambda g_1 + \mu g_2, Z_n) = \lambda S_r(g_1, Z_n) + \mu S_r(g_2, Z_n)$$

the triangle inequality yields for a given $\epsilon > 0$

$$\left| \int_a^b (\lambda g_1(t) + \mu g_2(t))\, dt - \lambda \int_a^b g_1(t)\, dt - \mu \int_a^b g_2(t)\, dt \right|$$

$$\leq \left| \int_a^b (\lambda g_1(t) + \mu g_2(t))\, dt - S_r(\lambda g_1 + g_2, Z_n) \right|$$

$$+ \left| \lambda \int_a^b g_1(t)\, dt - \lambda S_r(g_1, Z_n) \right| + \left| \mu \int_a^b g_2(t)\, dt - \mu S_r(g_2, Z_n) \right|$$

$$< 3\epsilon$$

provided the mesh size $m(Z_n)$ is small enough. Thus we have proved

$$\int_a^b (\lambda g_1(t) + \mu g_2(t))\, dt = \lambda \int_a^b g_1(t)\, dt + \mu \int_a^b g_2(t)\, dt, \qquad (13.1)$$

i.e. the integral is **linear**. Furthermore, if $g : [a, b] \longrightarrow \mathbb{R}$ is continuous and non-negative, i.e. $g \geq 0$, then it follows that $S_r(g, \pi_n) \geq 0$. Now it follows for a given $\epsilon > 0$ and for a sufficiently small $m(\pi_n)$ that

$$-\epsilon + S_r(g, Z_n) \leq \int_a^b g(t)\, dt$$

implying that $-\epsilon \leq \int_a^b g(t)\, dt$ for all $\epsilon > 0$, i.e.

$$\int_a^b g(t)\, dt \geq 0 \qquad \text{if } g(t) \geq 0 \text{ for all } t \in [a, b]. \qquad (13.2)$$

In all of our considerations we have so far assumed that $a < b$ and in the case where $a = b$ we defined that

$$\int_a^b g(t)\,\mathrm{d}t = 0.$$

We extend our integral definition to the case $a > b$ by defining

$$\int_a^b g(t)\,\mathrm{d}t := -\int_b^a g(t)\,\mathrm{d}t. \tag{13.3}$$

For example we have

$$\int_0^{-1} x\,\mathrm{d}x = -\int_{-1}^0 x\,\mathrm{d}x = -\frac{x^2}{2}\Big|_{-1}^0 = -\frac{0^2}{2} - \left(-\frac{(-1)^2}{2}\right) = \frac{1}{2}.$$

A simple application of (13.1) is

**Example 13.1.** Let $a, b \in \mathbb{R}$ and $p(t) = \sum_{j=0}^N c_j t^j$, $c_j \in \mathbb{R}$, be a polynomial. Then we have

$$\int_a^b p(t)\,\mathrm{d}t = \int_a^b \sum_{j=0}^N c_j t^j\,\mathrm{d}t = \sum_{j=0}^N c_j \int_a^b t^j\,\mathrm{d}t$$

$$= \sum_{j=0}^N \frac{c_j}{j+1} t^{j+1}\Big|_a^b$$

$$= \sum_{j=0}^N \frac{c_j}{j+1} b^{j+1} - \sum_{j=0}^N \frac{c_j}{j+1} a^{j+1} = \sum_{j=0}^N \frac{c_j}{j+1}(b^{j+1} - a^{j+1}).$$

Now we turn to rules following from the fundamental theorem of calculus and rules for taking derivatives. We start with

**Theorem 13.2 (Integration by Parts).** *Let* $f, g : [a, b] \longrightarrow \mathbb{R}$ *be two continuously differentiable functions. Then*

$$\int_a^b f(s)g'(s)\,\mathrm{d}s = f \cdot g\Big|_a^b - \int_a^b g(s)f'(s)\,\mathrm{d}s. \tag{13.4}$$

*Proof.* From Leibniz's rule we know

$$(f\,g)'\,(s) = f'(s)\,g(s) + f(s)\,g'(s).$$

Integrating this equality we get

$$\int_a^b f(s)\,g'(s)\,\mathrm{d}s = \int_a^b (f\,g)'\,\mathrm{d}s - \int_a^b f'(s)\,g(s)\,\mathrm{d}s.$$

Since $f \cdot g$ is a primitive of $(f \cdot g)'$ the fundamental theorem implies

$$\int_a^b (f\,g)'(s)\,\mathrm{d}s = f\,g\,\Big|_a^b$$

which finally yields (13.4). $\qquad\qquad\square$

**Example 13.3.** Let $0 < a < b$. We want to show that

$$\int_a^b \ln x \,\mathrm{d}x = ((x\ln x) - x)\,\Big|_a^b = x((\ln x) - 1)\,\Big|_a^b. \qquad (13.5)$$

For this we take $f(x) = \ln x$ and $g(x) = x$ in (13.4). Since $g'(x) = 1$ and $(\ln x)' = \frac{1}{x}$ we find

$$\begin{aligned}
\int_a^b (\ln x)\,1\,\mathrm{d}x &= (\ln x)x\,\Big|_a^b - \int_a^b \frac{1}{x} x \,\mathrm{d}x \\
&= (\ln x)x\,\Big|_a^b - \int_a^b 1\,\mathrm{d}x = ((x\ln x) - x)\,\Big|_a^b.
\end{aligned}$$

**Example 13.4.** For $a < b$ and with $f(x) = x$ and $g'(x) = \sin x$, i.e. we may take $g(x) = -\cos x$, to find

$$\begin{aligned}
\int_a^b x\sin x\,\mathrm{d}x &= -x\cos x\,\Big|_a^b - \int_a^b 1\,(-\cos x)\,\mathrm{d}x \\
&= -x\cos x\,\Big|_a^b + \int_a^b \cos x\,\mathrm{d}x \\
&= (-x\cos x + \sin x)\,\Big|_a^b.
\end{aligned}$$

**Example 13.5.** For $a < b$ we find

$$
\begin{aligned}
\int_a^b (\cos x)\, \mathrm{e}^x \, \mathrm{d}x &= \cos x \, \mathrm{e}^x \Big|_a^b + \int (\sin x)\, \mathrm{e}^x \, \mathrm{d}x \\
&= \cos x \, \mathrm{e}^x \Big|_a^b + \left( \sin x \, \mathrm{e}^x \Big|_a^b - \int_a^b (\cos x)\, \mathrm{e}^x \, \mathrm{d}x \right) \\
&= (\cos x + \sin x)\, \mathrm{e}^x \Big|_a^b - \int_a^b (\cos x)\, \mathrm{e}^x \, \mathrm{d}x,
\end{aligned}
$$

or

$$
2 \int_a^b (\cos x)\, \mathrm{e}^x \, \mathrm{d}x = (\cos x + \sin x)\, \mathrm{e}^x \Big|_a^b
$$

implying

$$
\int_a^b (\cos x)\, \mathrm{e}^x \, \mathrm{d}x = \frac{(\cos x + \sin x)\, \mathrm{e}^x}{2} \Big|_a^b .
$$

Sometimes integrals "longing" for an integration by parts can be handled easier with a little trick.

**Example 13.6.** For $\alpha, \beta \in \mathbb{R}$ and $a < b$ we have

$$
\sin \alpha x \sin \beta x = \frac{1}{2} \left( \cos(\alpha - \beta)x - \cos(\alpha + \beta)x \right),
$$

compare with (10.10). Therefore we find for $\alpha \neq \beta$ and $\alpha \neq -\beta$

$$
\begin{aligned}
\int_a^b \sin \alpha x \sin \beta x \, \mathrm{d}x &= \frac{1}{2} \int_a^b \left( \cos(\alpha - \beta)x - \cos(\alpha + \beta)x \right) \, \mathrm{d}x \\
&= \frac{1}{2} \left( \frac{\sin(\alpha - \beta)x}{\alpha - \beta} - \frac{\sin(\alpha + \beta)x}{\alpha + \beta} \right) \Big|_a^b .
\end{aligned}
$$

Our next rule for integration is derived from the chain rule.

**Theorem 13.7 (Change of variables, Part 1).** *Let $g : [a, b] \longrightarrow \mathbb{R}$ be a continuous function and let $\phi : [\alpha, \beta] \longrightarrow [a, b]$ be a differentiable function with continuous derivative $\phi'$. Then*

$$
\int_\alpha^\beta g(\phi(t))\phi'(t) \, \mathrm{d}t = \int_{\phi(\alpha)}^{\phi(\beta)} g(x) \, \mathrm{d}x. \tag{13.6}
$$

186

*Proof.* Let $f : [a, b] \longrightarrow \mathbb{R}$ be a primitive of $g$, i.e. $f' = g$. The chain rule yields

$$(f \circ \phi)'(t) = f'(\phi(t))\phi'(t) = g(\phi(t))\phi'(t).$$

Now it follows from the fundamental theorem of calculus that

$$
\begin{aligned}
\int_\alpha^\beta g(\phi(t))\phi'(t)\,\mathrm{d}t &= \int_\alpha^\beta (f \circ \phi)'(t)\,\mathrm{d}t \\
&= (f \circ \phi)\Big|_\alpha^\beta = f(\phi(\beta)) - f(\phi(\alpha)) \\
&= \int_{\phi(\alpha)}^{\phi(\beta)} g(x)\,\mathrm{d}x.
\end{aligned}
$$

$\square$

**Example 13.8.** For a continuous function $g : \mathbb{R} \longrightarrow \mathbb{R}$ we find for $\alpha < \beta$ and $c \in \mathbb{R}$ that

$$\int_\alpha^\beta g(t + c)\,\mathrm{d}t = \int_{\alpha+c}^{\beta+c} g(x)\,\mathrm{d}x. \tag{13.7}$$

Indeed, we just have to take $\phi(t) = t + c$, note $\phi'(t) = 1$, and restrict $g$ to $[\alpha + c, \beta + c]$.

**Example 13.9.** For a continuous function $g : \mathbb{R} \longrightarrow \mathbb{R}$ we find for $\alpha < \beta$ and $c \neq 0$ that

$$\int_\alpha^\beta g(ct)\,\mathrm{d}t = \frac{1}{c}\int_{\alpha c}^{\beta c} g(x)\,\mathrm{d}x. \tag{13.8}$$

This follows from (13.6) with $\phi(t) = ct$, $\phi'(t) = c$ and restricting $g$ to $[\alpha c, \beta c]$ (or $[\beta c, \alpha c]$ if $c < 0$).

**Remark 13.10.** In Examples 13.8 and 13.9 the function $g$ does not have to be defined on all of $\mathbb{R}$. It would be sufficient to consider functions defined on $[\alpha + c, \beta + c]$ and $[\alpha c, \beta c]$, respectively.

**Example 13.11.** Let $\phi : [a, b] \longrightarrow \mathbb{R}$ be a differentiable function with continuous derivative $\phi'$. Assume further that $\phi(t) \neq 0$ for all $t \in [a, b]$. Then

$$\int_a^b \frac{\phi'(t)}{\phi(t)}\,\mathrm{d}t = \ln|\phi(t)|\Big|_a^b. \tag{13.9}$$

187

For this note first that $\frac{\mathrm{d}}{\mathrm{d}t} \ln \phi(t) = \frac{\phi'(t)}{\phi(t)}$ provided $\ln \phi(t)$ is defined, i.e. $\phi(t) > 0$. Now we use in the change of variable formula $g(x) = \frac{1}{x}$ and it follows that

$$\int_a^b \frac{\phi'(t)}{\phi(t)} \, \mathrm{d}t = \int_{\phi(a)}^{\phi(b)} \frac{1}{x} \, \mathrm{d}x = \ln x \Big|_{\phi(a)}^{\phi(b)} = \ln \phi(t) \Big|_a^b.$$

The case where $\phi(t) < 0$ is treated by switching from $\phi(t)$ to $-\phi(t)$. As an immediate consequence of (13.9) we find

$$\int_a^b \frac{x}{1+x^2} \, \mathrm{d}x = \frac{1}{2} \int_a^b \frac{2x}{1+x^2} \, \mathrm{d}x = \frac{1}{2} \ln(1+x^2) \Big|_a^b \tag{13.10}$$

or

$$\int_a^b \cot t \, \mathrm{d}t = \int_a^b \frac{\cos t}{\sin t} \, \mathrm{d}t = \ln|\sin t| \Big|_a^b, \tag{13.11}$$

provided $\sin t$ has no zero in $[a, b]$. Note further that

$$\int_a^b \tan t \, \mathrm{d}t = \int_a^b \frac{\sin t}{\cos t} \, \mathrm{d}t = -\int_a^b \frac{-\sin t}{\cos t} \, \mathrm{d}t = -\ln|\cos t| \Big|_a^b \tag{13.12}$$

provided $\cos t$ has no zero in $[a, b]$.

Before we use the change of variables method in a more sophisticated situations we want to discuss a slightly modified change of variables formula.

**Theorem 13.12 (Change of variables, Part 2).** *Let* $g : [a, b] \longrightarrow \mathbb{R}$ *be a continuous function and let* $\phi : [\alpha, \beta] \longrightarrow \mathbb{R}$ *be a strictly monotone differentiable function with continuous derivative. Suppose that* $\phi(\alpha) = a$ *and* $\phi(\beta) = b$, *i.e.* $\phi^{-1}(a) = \alpha$, $\phi^{-1}(b) = \beta$. *Then*

$$\int_a^b g(x) \, \mathrm{d}x = \int_\alpha^\beta g(\phi(t))\phi'(t) \, \mathrm{d}t = \int_{\phi^{-1}(a)}^{\phi^{-1}(b)} g(\phi(t))\phi'(t) \, \mathrm{d}t. \tag{13.13}$$

*Proof.* Of course (13.13) follows from (13.6) using that $\phi^{-1}$ exists. $\qquad\square$

Let us compare (13.6) with (13.13). In (13.6) we have to identify the function we want to integrate as a term $g(\phi(t))\phi'(t)$, whereas in (13.13) we start with the integral $\int_a^b g(x) \, \mathrm{d}x$ and modify it. But we have to pay a price: we have to find an invertible (bijective) smooth change of variable, i.e. we need to find

$t = t(x) = \phi^{-1}(x)$ to transform $\int_a^b g(x)\,dx$ to the right hand side in (13.13). Note that

$$\frac{dt}{dx} = \frac{d\phi^{-1}}{dx}(x) = \frac{1}{\phi'(\phi^{-1}(x))} = \frac{1}{\phi'(t)}.$$

The transformation of $\int_a^b g(x)\,dx$ could be done in a formal way

$$
\begin{array}{ccc}
g(x) & \rightsquigarrow & g(\phi(t)) \\
dx & \rightsquigarrow & \phi'(t)\,dt \\
a & \rightsquigarrow & \phi^{-1}(a) \\
b & \rightsquigarrow & \phi^{-1}(b).
\end{array}
$$

The second step looks a bit more demanding. In principle we can easily introduce $t = \phi^{-1}(x)$. But now we need $\phi'(t)$, i.e. we have to invert $\phi^{-1}$. In certain examples this is often not needed.

**Example 13.13.** Consider the integral $\int_a^b (x+2)\sin(x^2 + 4x - 6)\,dx$. We choose $t = \phi^{-1}(x) = x^2 + 4x - 6$, i.e. $\frac{dt}{dx} = 2x + 4 = 2(x+2)$.
Now we use

$$\sin(x^2 + 4x - 6) \rightsquigarrow \sin t$$

but instead of

$$dx \rightsquigarrow \phi'(t)\,dt$$

we observe that

$$(x+2)\,dx = \frac{1}{2}\,dt$$

which yields

$$\int_a^b (x+2)\sin(x^2 + 4x - 6)\,dx = \frac{1}{2}\int_{\phi^{-1}(a)}^{\phi^{-1}(b)} \sin t\,dt$$

$$= -\frac{1}{2}\cos t\Big|_{\phi^{-1}(a)}^{\phi^{-1}(b)} = \frac{1}{2}\cos(\phi^{-1}(a)) - \frac{1}{2}\cos(\phi^{-1}(b)).$$

Note that in our example we must ensure that $\phi$ is defined on an interval where it is invertible, since $\phi^{-1}$ is needed. A simple calculation gives

$$
\begin{aligned}
t &= \phi^{-1}(x) = x^2 + 4x - 6 = x^2 + 4x + 4 - 10 \\
&= (x+2)^2 - 10
\end{aligned}
$$

or

$$x = \phi(t) = -2 + \sqrt{t + 10}, \quad t \geq -10 \text{ and } x \geq -2$$

and

$$x = \phi(t) = -2 - \sqrt{t + 10}, \quad t \geq -10 \text{ and } x \leq -2.$$

Hence for $b > a \geq -2$ or $a < b \leq -2$ we may use our calculation. In each case we eventually get

$$\int_a^b (x + 2) \sin(x^2 + 4x - 6) \, dx = \frac{1}{2} \cos(a^2 + 4a - 6) - \frac{1}{2} \cos(b^2 + 4b - 6).$$

We want to optimise our strategy to evaluate integrals further by using the notation

$$\int g(x) \, dx \tag{13.14}$$

for the primitives of $g$, i.e. with this notation we can write for a primitive $f$ of $g$

$$f(x) = \int g(t) \, dt + c \tag{13.15}$$

where $c$ is a constant. (This is not a very well defined notation, but very useful.)

Using in (13.14) a change of variables $t = \phi^{-1}(x)$ we find that

$$\int g(\phi(t)) \phi'(t) \, dt + \tilde{c} \tag{13.16}$$

is a primitive of $g(\phi(t)) \phi'(t)$, and (13.15) and (13.16) differ only by a constant. Thus instead of always transforming the limits of the integral we first work on the level of primitives:

$$\int g(x) \, dx = \int g(\phi(t)) \phi'(t) \, dt.$$

To eventually find $\int_a^b g(x) \, dx$ we observe that

$$\int_a^b g(x) \, dx = h(t) \Big|_{\phi^{-1}(a)}^{\phi^{-1}(b)}$$

where $h$ is any primitive of $g(\phi(t)) \phi'(t)$.

190

**Example 13.14. A.** Consider

$$\int \frac{\cot(\ln x)}{x}\,dx.$$

Using $t = \ln x$, i.e. $dt = \frac{1}{x}\,dx$, and we find by (13.11) that

$$\int \frac{\cot(\ln x)}{x}\,dx = \int \cot t\,dt = \ln|\sin t| + c.$$

**B.** Consider

$$\int_{-1}^{1} \frac{dx}{\sqrt{(x+2)(3-x)}}.$$

Observe first that

$$\int \frac{dx}{\sqrt{(x+2)(3-x)}} = \int \frac{dx}{\sqrt{6-(x^2-x)}} = \int \frac{dx}{\sqrt{\frac{25}{4}-(x-\frac{1}{2})^2}}.$$

Now take $t = x - \frac{1}{2}$, i.e. $dt = dx$ to find

$$\int \frac{dx}{\sqrt{(x+2)(3-x)}} = \int \frac{dt}{\sqrt{\frac{25}{4}-t^2}} = \int \frac{dt}{\frac{5}{2}\sqrt{1-\left(\frac{2t}{5}\right)^2}}$$

$$= \frac{2}{5}\int \frac{dt}{\sqrt{1-\left(\frac{2t}{5}\right)^2}}.$$

By a further change of variables $s = \frac{2t}{5}$, i.e. $ds = \frac{2}{5}dt$ we find

$$\frac{2}{5}\int \frac{dt}{\sqrt{1-\left(\frac{2t}{5}\right)^2}} = \int \frac{ds}{\sqrt{1-s^2}} = \arcsin s + c$$

$$= \arcsin \frac{2t}{5} + c.$$

Therefore we have

$$\int \frac{dx}{\sqrt{(x+2)(3-x)}} = \arcsin \frac{2t}{5} + c = \arcsin \frac{2x-1}{5} + c$$

and finally

$$\int_{-1}^{1} \frac{dx}{\sqrt{(x+2)(3-x)}} = \left. \arcsin \frac{2x-1}{5} \right|_{-1}^{1}$$

$$= \arcsin \frac{1}{5} - \arcsin \left( -\frac{3}{5} \right).$$

$$= \arcsin(\frac{1}{5}) + \arcsin(\frac{3}{5}).$$

**C.** Consider

$$\int 2^{-x} \tanh 2^{1-x} \, dx.$$

Take $t = 2^{1-x}$ which yields $dt = -(\ln 2)2^{1-x} \, dx$, i.e. $2^{-x} \, dx = -\frac{1}{2 \ln 2} dt$ and therefore

$$\int 2^{-x} \tanh 2^{1-x} \, dx = \int (\tanh t) \left( -\frac{1}{2 \ln 2} \right) dt$$

$$= -\frac{1}{2 \ln 2} \ln \cosh t + c = -\frac{1}{2 \ln 2} \ln \cosh \left( 2^{1-x} \right) + c$$

where we used that $(\ln \cosh t)' = \tanh t$.

**D.** Consider

$$\int_{0}^{1/\sqrt{2}} \frac{x \arcsin x^2}{\sqrt{1-x^4}} \, dx.$$

Take $t = \arcsin x^2$ to find $dt = \frac{1}{\sqrt{1-(x^2)^2}} 2x \, dx$, i.e. $dt = \frac{2x \, dx}{\sqrt{1-x^4}}$ and hence

$$\int \frac{x \arcsin x^2}{\sqrt{1-x^4}} \, dx = \frac{1}{2} \int t \, dt = \frac{1}{4} t^2 + c.$$

It follows that

$$\int_{0}^{1/\sqrt{2}} \frac{x \arcsin x^2}{\sqrt{1-x^4}} \, dx = \left. \frac{1}{4} (\arcsin x^2)^2 \right|_{0}^{1/\sqrt{2}}$$

$$\frac{1}{4} \left( \arcsin \frac{1}{2} \right)^2 - \frac{1}{4} (\arcsin 0)^2 = \frac{\pi^2}{144}$$

since $\arcsin 0 = 0$ and $\arcsin \frac{1}{2} = \frac{\pi}{6}$.

**Important Remark.** Using the change of variable formula requires experience and routine which one only gets by doing many examples. There is no general principle on how to find the best change of variables, but of course there are some rules. Nowadays we can use powerful programme packages to evaluate integrals. However, one still needs some experience to handle integrals without using such a package as it will be useful in many theoretical considerations in many fields of mathematics.

A further method we need to learn is related to the decomposition of rational functions into **partial fractions**. Let $P(x)$ and $Q(x)$ be two polynomials and suppose that the degree of $P(x)$ is less than that of $Q(x)$. (Otherwise use polynomial division to decompose $\frac{P(x)}{Q(x)} = g(x) + \frac{R(x)}{Q(x)}$, where $g(x)$ is a polynomial and $R(x)$ is now a polynomial of degree less than $Q(x)$.) From algebra we know that each polynomial in $\mathbb{R}$ with leading coefficient equal to 1 has the unique factorisation

$$Q(x) = (x - z_1)^{p_1} \ldots (x - z_k)^{p_k} (x^2 + \alpha_1 x + \beta_1)^{q_1} \ldots (x^2 + \alpha_l x + \beta_l)^{q_l} \quad (13.17)$$

where the polynomials $x - z_j$, $j = 1, \ldots, k$, and $x^2 + \alpha_j x = \beta_j$, $j = 1, \ldots, l$ have real coefficients and are mutually different, and $p_j, q_l \in \mathbb{N}$. It can be shown that

$$\frac{P(x)}{Q(x)} = \sum_{i=1}^{k} \sum_{j=1}^{p_i} \frac{a_{ij}}{(x - z_i)^j} + \sum_{i=1}^{l} \sum_{j=1}^{q_i} \frac{b_{ij} x + c_{ij}}{(x^2 + \alpha_i x + \beta_i)^j} \quad (13.18)$$

holds with suitable real numbers $a_{ij}, b_{ij}$ and $c_{ij}$. Hence, whenever the integral $\int_a^b \frac{P(x)}{Q(x)} \, dx$ exists we have

$$\int_a^b \frac{P(x)}{Q(x)} \, dx = \sum_{i=1}^{k} \sum_{j=1}^{p_i} \int_a^b \frac{a_{ij}}{(x - z_i)^j} \, dx + \sum_{i=1}^{l} \sum_{j=1}^{q_i} \int_a^b \frac{b_{ij} x + c_{ij}}{(x^2 + \alpha_i x + \beta_i)^j} \, dx. \quad (13.19)$$

In practice we work as in the following example:

$$\frac{3x - 2}{(4x - 3)(2x + 5)^3} = \frac{A}{4x + 3} + \frac{B}{2x + 5} + \frac{C}{(2x + 5)^2} + \frac{D}{(2x + 5)^3}$$
$$= \frac{A(2x + 5)^3 + B(4x - 3)(2x + 5)^2 + C(4x - 3)(2x + 5) + D(4x - 3)}{(4x - 3)(2x + 5)^3}.$$

This leads to the equality

$$3x - 2 = A(2x + 5)^3 + B(4x - 3)(2x + 5)^2 + C(4x - 3)(2x + 5) + D(4x - 3).$$

Expanding the right hand side and comparing coefficients we end up with four linear equations for the four unknowns $A, B, C, D$. Note that $\tilde{Q} = (4x - 3)(2x + 5)^3$ does not have leading coefficient 1 and it is not of type (13.17). However

$$\tilde{Q}(x) = 4\left(x - \frac{3}{4}\right)2^3\left(x + \frac{5}{2}\right)^3 = 32\left(x - \frac{3}{4}\right)\left(x + \frac{5}{2}\right)^3$$
$$= 32Q(x)$$

and $Q(x)$ has leading coefficient 1 and is of type (13.17). In general we can find $\gamma_0 \in \mathbb{R}$ such that for a polynomial $\tilde{Q}(x)$ we get $\tilde{Q}(x) = \gamma_0 Q(x)$ where $Q(x)$ has leading coefficient 1 and is of type (13.17).

For practical purposes switching from $\tilde{Q}$ to $\gamma_0 Q$ is often not needed, but in order to get in (13.17) uniqueness up to the order of factors it is needed. An alternative way is to use in (13.17) for a general polynomial $\tilde{Q}$ the representation $\gamma_0(x - z_1)^{p_1} \cdots (x - z_k)^{p_k}(x^2 + \alpha_1 x + \beta_1)^{q_1} \cdots (x^2 + \alpha_l + \beta_l)^{q_l}$ where $\gamma_0$ is the leading coefficient of $\tilde{Q}$.

Here is a more simple example:

**Example 13.15. A.** Find

$$\int \frac{6 - x}{(x - 3)(2x + 5)}\, \mathrm{d}x.$$

Write

$$\frac{6 - x}{(x - 3)(2x + 5)} = \frac{A}{x - 3} + \frac{B}{2x + 5} = \frac{A(2x + 5) + B(x - 3)}{(x - 3)(2x + 5)}$$

implying

$$6 - x = 5A - 3B + x(2A + B)$$

or

$$5A - 3B = 6 \quad \text{and} \quad 2A + B = -1$$

which yields $A = \frac{3}{11}$ and $B = -\frac{17}{11}$.

Hence

$$\frac{6 - x}{(x - 3)(2x + 5)} = \frac{\frac{3}{11}}{x - 3} - \frac{\frac{17}{11}}{2x + 5},$$

and therefore

$$\int \frac{6-x}{(x-3)(2x+5)}\,dx = \frac{3}{11}\int \frac{1}{x-3}\,dx - \frac{17}{2\cdot 11}\int \frac{2}{2x+5}\,dx$$

$$= \frac{3}{11}\ln|x-3| - \frac{17}{22}\ln|2x+5| + c.$$

**B.** Let $-1, 1 \notin [a,b]$ and consider $\int_a^b \frac{dx}{1-x^2}$. We try

$$\frac{1}{1-x^2} = \frac{1}{(1-x)(1+x)} = \frac{A}{1-x} + \frac{B}{1+x} = \frac{(A+B)+(A-B)x}{1-x^2}$$

which leads to $A + B = 1$ and $A - B = 0$, i.e. $A = B = \frac{1}{2}$.
This implies

$$\int_a^b \frac{1}{1-x^2}\,dx = \frac{1}{2}\int_a^b \frac{1}{1-x}\,dx + \frac{1}{2}\int_a^b \frac{1}{1+x}\,dx$$

$$= \frac{1}{2}\left(\int_a^b \frac{1}{1+x}\,dx - \int_a^b \frac{1}{x-1}\,dx\right)$$

$$= \frac{1}{2}\left(\ln|x+1| - \ln|x-1|\right)\Big|_a^b$$

$$= \frac{1}{2}\ln\left|\frac{x+1}{x-1}\right|\Big|_a^b.$$

## Problems

1. Find

$$\int_0^1 \sum_{k=1}^n (1+k^2)x^{\frac{1}{k^2}}\,dx.$$

2.   a) For $f : [a,b] \longrightarrow \mathbb{R}$ integrable prove that

$$\int_a^b |f(x)|\,dx = \int_a^b f^+(x)\,dx + \int_a^b f^-(x)\,dx.$$

   b) Prove that if $f : [a,b] \longrightarrow \mathbb{R}$ is integrable and satisfies $|f(t)| \leq M$ for all $t \in [a,b]$ then

$$\left|\int_a^b f(t)\,dt\right| \leq M(b-a).$$

195

c) Let $f : [-1, 0] \longrightarrow \mathbb{R}$ be a differentiable function such that $f(-1) = 0$ and $f'(x) \geq 0$ for all $x \in [-1, 0]$. Show that $\int_{-1}^{0} f(x)\mathrm{d}x \geq 0$.

3. By only using symmetry considerations prove that

$$\int_{-1}^{1} \left(1 + \frac{1}{1 + x^2}\right) \sin x^3 \mathrm{d}x = 0.$$

4. Denote the Dirichlet kernel discussed in Problem 3 of Chapter 9 by $D_n$. Use the results of that problem to show that

$$\frac{2}{\pi} \int_{0}^{\frac{\pi}{2}} D_n(t)\mathrm{d}t = 1$$

for all $n \in \mathbb{N}$.

5. For a continuous function $f : \mathbb{R} \longrightarrow \mathbb{R}$ use a straightforward change of variable to find the integrals

$$\int_{a}^{b} f(\alpha t)\mathrm{d}t \quad \text{and} \quad \int_{a}^{b} f(\alpha t + \beta)\mathrm{d}t, \ \alpha \neq 0, \ t \in \mathbb{R},$$

in terms of the integral $\int_{a}^{b} f(t)\mathrm{d}t, \ a < b$.

6. Use integration by parts and where appropriate the results of Problem 5 to evaluate the following integrals:

a) $\displaystyle\int_{0}^{\frac{\pi}{4}} \vartheta \cos \vartheta \mathrm{d}\vartheta$;

b) $\displaystyle\int_{\frac{1}{2}}^{2} x \ln(2x + 1)\mathrm{d}x$;

c) $\displaystyle\int_{0}^{\frac{1}{m}} s \sinh(ms)\mathrm{d}s$;

d) $\displaystyle\int_{1}^{3} \frac{\ln t}{\sqrt{t}}\mathrm{d}t$;

e) $\displaystyle\int_{0}^{\pi} e^{2r} \sin 3r\mathrm{d}r$.

7. For $m, n \in \mathbb{N}$ prove

$$\frac{1}{\pi} \int_{-\pi}^{\pi} (\cos nx)(\cos mx)dx = \begin{cases} 0, & n \neq m \\ 1, & n = m > 0 \\ 2, & n = m = 0 \end{cases}$$

and

$$\frac{1}{\pi} \int_{-\pi}^{\pi} (\sin nx)(\cos mx)dx = 0.$$

8. Find the following primitives:

a) $\int x^2 e^{\lambda x} dx$;

b) $\int \dfrac{dt}{at^2 + bt + c}$, $a, b, c \in \mathbb{R}$,

note that different cases must be considered for different $a, b, c$.

9. Let $g : \mathbb{R} \longrightarrow \mathbb{R}$ be a continuous and periodic function with period $a > 0$, i.e. $g(t + a) = g(t)$ for all $t \in \mathbb{R}$. For all $c \in \mathbb{R}$ show that

$$\int_0^a g(t)dt = \int_c^{c+a} g(t)dt.$$

10. Use a change of variable to evaluate the following integrals:

a) $\displaystyle\int_e^{e^2} \dfrac{dx}{x(\ln x)^3}$;

b) $\displaystyle\int_{\frac{\pi}{3}}^{\frac{\pi}{2}} \dfrac{dt}{5 + 3\cos t}$ (try: $\tan \frac{t}{2} = s$);

c) $\displaystyle\int_0^{\frac{1}{\sqrt{2}}} \dfrac{y \arcsin y^2}{\sqrt{1 - y^4}}dy$ (try: $\arcsin y^2 = v$);

d) $\displaystyle\int_{\frac{1}{2}}^1 \dfrac{ds}{\sqrt{5 - 4s - s^2}}$;

e) $\displaystyle\int_1^4 \dfrac{1}{(1 + x^2)^{\frac{3}{2}}}dx$ (try: $x = \sinh t$).

197

11. Evaluate the following integrals. Note that a change of variables and integration by parts may need to be used.

a) $\int_0^4 3^{\sqrt{2t+1}}dt$;

b) $I := \int_0^\pi \frac{x \sin x}{1 + \cos^2 x}dx.$

Hint: derive the equality $I = \frac{\pi^2}{2} - I.$

12. Use partial fractions to find

$$\int \frac{x+1}{x^4 - x}dx.$$

(The result of Problem 8 b) may eventually become useful.)

13. For $f, g : [a, b] \longrightarrow \mathbb{R}$ being three times continuously differentiable prove

$$\int_a^b f(t)g^{(3)}(t)dt = fg''|_a^b - f'g'|_a^b + f''g|_a^b - \int_a^b f^{(3)}(t)g(t)dt.$$

14. Prove that for $g$ continuously differentiable and $g(s) > 0$ we have

$$\int \frac{g'(s)}{\sqrt{g(s)}}ds = 2\sqrt{g(s)}.$$

Now find

$$\int_{\frac{\pi}{6}}^{\frac{\pi}{2}} \frac{\cos r}{\sqrt{\sin r}}dr.$$

15. Let $f : [-\pi, \pi] \longrightarrow \mathbb{R}$ be a continuously differentiable function such that $|f'(t)| \le M$ for all $t \in [-\pi, \pi]$. Prove that

$$\left| \int_{-\pi}^\pi f(t) \cos nt\, dt \right| \le \frac{2\pi M}{n}.$$

16. For $n \ge 2$, $n \in \mathbb{N}$, find

$$\lim_{x \to \infty} \int_1^x t^{-n}dt.$$

# Part 2: Analysis in One Dimension

# 14 Problems with the Real Line

In Part 1 we omitted several proofs; some were omitted because they are obvious, whereas others were omitted because they depend on tools or results proved in an algebra course and these were therefore perhaps not yet known. However most of the proofs we omitted claim the existence of a real number with certain properties and we could not prove this in Part 1.

We have identified the real numbers with the real line and sometimes we switched from algebraic to geometric arguments, but this is in fact a non-trivial problem. In this chapter we want to analyse this problem in more detail.

Let us summarise, i.e. recollect from Part 1, the basic algebraic properties of the real numbers. On $\mathbb{R}$ we have two operations, **addition** and **multiplication**

$$+ : \mathbb{R} \times \mathbb{R} \longrightarrow \mathbb{R} \qquad \cdot : \mathbb{R} \times \mathbb{R} \longrightarrow \mathbb{R}$$
$$(x, y) \longmapsto x + y \qquad\qquad (x, y) \longmapsto x \cdot y.$$

The rules for addition are for $x, y, z \in \mathbb{R}$

$$(x + y) + z = x + (y + z); \tag{14.1}$$

$$x + 0 = x; \tag{14.2}$$

$$x + (-x) = 0; \tag{14.3}$$

$$x + y = y + x; \tag{14.4}$$

where (14.2) means that in $\mathbb{R}$ there exists an element 0 such that $x + 0 = x$ for all $x \in \mathbb{R}$, i.e. 0 is a **neutral element** with respect to addition. Further we interpret (14.3) as follows: for every $x \in \mathbb{R}$ there exists an **inverse element** $-x$ with respect to addition.

For multiplication we have with $x, y, z \in \mathbb{R}$ the rules

$$(x \cdot y) \cdot z = x \cdot (y \cdot z); \tag{14.5}$$

$$1 \cdot x = x; \tag{14.6}$$

$$x \cdot x^{-1} = 1 \quad \text{for } x \neq 0; \tag{14.7}$$

$$x \cdot y = y \cdot x; \tag{14.8}$$

201

here (14.6) means that there exists $1 \in \mathbb{R}$, $1 \neq 0$, such that 1 is a **neutral element** with respect to multiplication and (14.7) means that each $x \in \mathbb{R}$, $x \neq 0$, has an **inverse element** with respect to multiplication. These two operations are linked by the **law of distribution**

$$x \cdot (y + z) = x \cdot y + x \cdot z. \tag{14.9}$$

It turns out that there are many sets $\mathbb{K}$ with operations $+$ and $\cdot$ satisfying (14.4)-(14.9), we call each such algebraic object $(\mathbb{K}, +, \cdot)$ a (**commutative**) **field**. It can be easily checked that all rational numbers form a field, as do the complex numbers $\mathbb{C}$. In algebra a lot of consequences of these axioms can be learned. These consequences justify our usual calculations in $\mathbb{Q}$, $\mathbb{R}$, or $\mathbb{C}$. Here we take these consequences for granted.

For $\mathbb{R}$ and $\mathbb{Q}$ we also have **axioms of order**:

for every $x \in \mathbb{R}$ ($\in \mathbb{Q}$) one and only one
of the statements $x = 0$, $x > 0$, $x < 0$ holds; $\qquad$ (14.10)

$x > 0$ and $y > 0$ implies $x + y > 0$; $\qquad$ (14.11)

$x > 0$ and $y > 0$ implies $x \cdot y > 0$. $\qquad$ (14.12)

All further properties of the order structure on $\mathbb{R}$ (or $\mathbb{Q}$) can be deduced from (14.10)-(14.12). Recall that we write $x < y$ if $x - y < 0$ and $x > y$ if $x - y > 0$, compare with (1.70) and (1.71). Let us prove some consequences of (14.10)-(14.12) to get some flavour of the arguments involved.
We claim that $x > y$ and $y > z$ implies $x > z$. From each inequality we deduce that $x - y > 0$ and $y - z > 0$ respectively, hence

$$(x - y) + (y - z) = x - z > 0 \text{ or } x > z.$$

Next we show $x > y$ and $a > 0$ implies $ax > ay$. Since $x - y > 0$ and $a > 0$ it follows that $a(x - y) > 0$ or $ax - ay > 0$, i.e. $ax > ay$.
Of course we will continue to use the notation $x \leq y$ and $x \geq y$ as defined in Part 1, Chapter 1.

A (commutative) field $(\mathbb{K}, +, \cdot)$ on which (14.10)-(14.12) hold is called an **ordered field**. Both $\mathbb{R}$ and $\mathbb{Q}$ are ordered fields, but $\mathbb{C}$ is not. This follows for example from the fact that $i^2 = -1$. Indeed, (14.12) implies for $x > 0$ that $x^2 > 0$, for $x = 0$ we have $x^2 = 0$, and for $x < 0$ it follows that $-x > 0$, and therefore $x^2 = (-x)(-x) > 0$. Hence (14.4)-(14.12) imply $x^2 \geq 0$ which does not hold in $\mathbb{C}$. Thus we have a distinction between $\mathbb{C}$ and $\mathbb{R}$ (or $\mathbb{Q}$).

So far we cannot make any distinction between $\mathbb{R}$ and $\mathbb{Q}$. In fact, we do not even know what $\mathbb{R}$ should be. But in $\mathbb{Q}$ we have a problem, in fact we have several quite similar problems:

**Claim:** in $\mathbb{Q}$ there is no element $a$ such that $a^2 = 2$.

Suppose $a \in \mathbb{Q}$ has this property where we may assume that $a > 0$. Then $a = \frac{q}{p}$, $p \neq 0$, $p, q \in \mathbb{N}_0$, and $q$ and $p$ have no common factor. From $a^2 = 2$ we deduce that $q^2/p^2 = 2$ or $q^2 = 2p^2$. This implies that $q^2$ is an even number, hence $q$ is an even number, say $q = 2r$. Now it follows that $4r^2 = 2p^2$ or $p^2 = 2r^2$, i.e. $p^2$, hence $p$ is an even number too, which is a contradiction, therefore the above claim is true.

Now let us turn to our geometric interpretation of $\mathbb{R}$ as the points on the (real) line. Consider the unit square in the plane

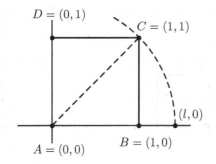

*Figure 14.1*

We know that by Pythagoras' theorem the length $l$ of $\overline{AC}$ is given by $l^2 = 1^2 + 1^2 = 2$, i.e. $l^2 = 2$. Hence this length $l$ is not given by a rational number. Certainly we can consider all rational numbers as points on a line. In doing so, the above consideration shows that on the line containing only rational numbers (points) there are gaps.

On the other hand, given two rational number $q, p \in \mathbb{Q}$, $q < p$, there are infinitely many rational numbers $r \in \mathbb{Q}$ such that $q < r < p$. Indeed, take $r_1 = \frac{q+p}{2}$ and then take instead of $p$ the number $r_1$, now continue this procedure. Doing this $N$-times we find $r_N = q + \frac{p-q}{2^N}$. Thus given $\epsilon > 0$ we can find a rational number $r_N$ such that $|q - r_N| < \epsilon$. Just take $p = q + 1$ and $N$ such that $2^{-N} < \epsilon$.

So we face the following strange situation: not every point on the "line" corresponds to a rational number but we can put into the gap between two rational numbers as many rational numbers as we like.

The number $l$, $l^2 = 2$, lies between two rational numbers. We can argue as follows: the square of the length $d$ of the side $AF$ of the triangle $AEF$ with $A = (0,0)$, $E = (4,0)$ and $F = (4,3)$ is equal $4^2 + 3^2 = 5^2$, i.e. $d = 5$, but $l < d$:

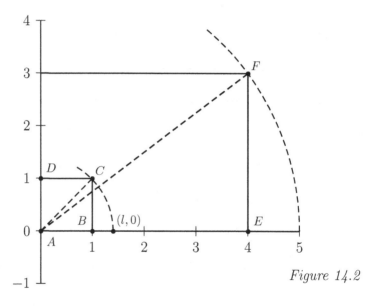

*Figure 14.2*

Thus we have $0 < l < 5$ implying that we can get as close as we wish to $l$ in terms of rational numbers. A word of caution: once again we have mixed geometric arguments with algebraic ones.

We need to resolve these problems, but before we can do this we need more knowledge about properties of the real numbers (if they exist). We continue for a while to pretend as if we already have the real numbers at our disposal and try to deduce new tools so that we are eventually in a position to establish the existence of the real numbers.

First let us add a further axiom

204

**Archimedes' Axiom** Given $x, y \in \mathbb{R}$, $x > 0$ and $y > 0$, there exists a natural number $n \in \mathbb{N}$ such that $nx > y$.

Note that Archimedes' axiom links the order structure of the real numbers with properties of the natural numbers.

**Consequences of Archimedes' Axiom**

1. Given $x \in \mathbb{R}$, $x > 0$, there exists $n \in \mathbb{N}$ such that $n > x$.
2. Given $x \in \mathbb{R}$, there exists a unique $k \in \mathbb{Z}$ such that $k \leq x < k + 1$.
3. For every $\epsilon > 0$ there exists $n \in \mathbb{N}$ such that $\frac{1}{n} < \epsilon$. Indeed: there exists $n \in \mathbb{N}$ such that $n > \frac{1}{\epsilon}$, implying $\frac{1}{n} < \epsilon$.

As before, compare with Example 4.4.B, we denote the unique number $k \in \mathbb{Z}$ in **2.** as $[x]$.

We now extend **Bernoulli's inequality** (Lemma 9.11.A). Let $x \in \mathbb{R}$, $x \geq -1$. Then for all $n \in \mathbb{N}_0$

$$(1 + x)^n \geq 1 + nx. \tag{14.13}$$

*Proof.* For $n = 0$ we have

$$(1 + x)^0 = 1 \geq 1 + 0 \cdot x.$$

Suppose that $(1 + x)^k \geq 1 + kx$. It follows that

$$\begin{aligned}(1 + x)^{k+1} &\geq (1 + kx)(1 + x) = 1 + (k + 1)x + kx^2 \\ &\geq 1 + (k + 1)x\end{aligned}$$

provided $1 + x \geq 0$, i.e. $x \geq -1$. Now (14.13) follows from the principle of mathematical induction. $\qquad\square$

In order to get more used to inequalities let us derive some consequences of Bernoulli's inequality.

**Definition 14.1.** *For $n \in \mathbb{N}$ let $a_1, \ldots, a_n$ be positive real numbers. Their **arithmetic mean** is defined by*

$$A_n := \frac{a_1 + \cdots + a_n}{n} = \frac{1}{n} \sum_{k=1}^{n} a_k, \tag{14.14}$$

*and their **geometric mean** is given by*

$$G_n := (a_1 \cdot \ldots \cdot a_n)^{\frac{1}{n}} = \left( \prod_{j=1}^{n} a_j \right)^{\frac{1}{n}}. \tag{14.15}$$

**Lemma 14.2.** *For positive numbers $a_1, \ldots, a_n \in \mathbb{R}, n \in \mathbb{N}$, the **arithmetic-geometric mean inequality** holds:*

$$\left( \prod_{j=1}^{n} a_j \right)^{\frac{1}{n}} \leq \frac{1}{n} \sum_{k=1}^{n} a_k, \tag{14.16}$$

*or $G_n \leq A_n$.*

*Proof.* The case $n = 1$ is trivial. Let $n \geq 2$ and with $y = x + 1, x > -1$, Bernoulli's inequality reads as

$$y^n \geq 1 + n(y - 1). \tag{14.17}$$

With $y = \frac{A_n}{A_{n-1}} > 0$, $n \geq 2$, we deduce from (14.17) that

$$\left( \frac{A_n}{A_{n-1}} \right)^n \geq 1 + n \left( \frac{A_n}{A_{n-1}} - 1 \right)$$

$$= \frac{A_{n-1} + nA_n - nA_{n-1}}{A_{n-1}} = \frac{nA_n - (n-1)A_{n-1}}{A_{n-1}} = \frac{a_n}{A_{n-1}}$$

implying

$$\begin{aligned} A^n &\geq a_n A_{n-1}^{n-1} \geq a_n a_{n-1} A_{n-2}^{n-2} \\ &\geq a_n a_{n-1} \cdot \ldots \cdot a_1 = G_n^n, \end{aligned}$$

or $G_n \leq A_n$. $\qquad\square$

**Corollary 14.3.** *For $a_1, \ldots, a_n, b_1, \ldots, b_n \in \mathbb{R}, n \in \mathbb{N}$, the **Cauchy-Schwarz inequality** holds:*

$$\left| \sum_{k=1}^{n} a_k b_k \right| \leq \sum_{k=1}^{n} |a_k b_k| \leq \left( \sum_{k=1}^{n} a_k^2 \right)^{\frac{1}{2}} \left( \sum_{k=1}^{n} b_k^2 \right)^{\frac{1}{2}}. \tag{14.18}$$

*Proof.* The first estimate is just the triangle inequality. For $c_1, c_2 \in \mathbb{R}$ we know that

$$|c_1 c_2| \leq \frac{c_1^2 + c_2^2}{2}.$$

206

This implies that for every $j = 1, \ldots, n$

$$|a_j b_j| \left( \left( \sum_{k=1}^{n} a_k^2 \right) \left( \sum_{k=1}^{n} b_k^2 \right) \right)^{\frac{1}{2}} = \left( b_j^2 \sum_{k=1}^{n} a_k^2 \right)^{\frac{1}{2}} \left( a_j^2 \sum_{k=1}^{n} b_k^2 \right)^{\frac{1}{2}}$$

$$\leq \frac{b_j^2 \displaystyle\sum_{k=1}^{n} a_k^2 + a_j^2 \displaystyle\sum_{k=1}^{n} b_k^2}{2},$$

and summing from $j = 1$ to $j = n$ gives

$$\left( \sum_{j=1}^{n} |a_j b_j| \right) \left( \sum_{k=1}^{n} a_k^2 \right)^{\frac{1}{2}} \left( \sum_{k=1}^{n} b_k^2 \right)^{\frac{1}{2}} \leq \frac{\left( \sum_{j=1}^{n} b_j^2 \right) \left( \sum_{k=1}^{n} a_k^2 \right) + \left( \sum_{j=1}^{n} a_j^2 \right) \left( \sum_{k=1}^{n} b_k^2 \right)}{2}$$

$$= \left( \sum_{j=1}^{n} b_j^2 \right) \left( \sum_{n=1}^{n} b_j^2 \right)$$

which implies

$$\left| \sum_{j=1}^{n} a_j b_j \right| \leq \left( \sum_{j=1}^{n} a_j^2 \right)^{\frac{1}{2}} \left( \sum_{j=1}^{n} b_j^2 \right)^{\frac{1}{2}}.$$

$\square$

**Remark 14.4. A.** The Cauchy-Schwarz inequality is often called the Cauchy-Schwarz-Bunyakovsky inequality.
**B.** The proof of Lemma 14.2 is taken from L. Maligranda [9] and that of Corollary 14.3 is taken from M. Lin [8].

**Lemma 14.5 (Minkowski's inequality).** *For real numbers* $a_1, \ldots, a_n, b_1, \ldots, b_n$ *we have*

$$\left( \sum_{k=1}^{n} (a_k + b_k)^2 \right)^{\frac{1}{2}} \leq \left( \sum_{k=1}^{n} a_k^2 \right)^{\frac{1}{2}} + \left( \sum_{k=1}^{n} b_k^2 \right)^{\frac{1}{2}}. \qquad (14.19)$$

*Proof.* If $\sum_{k=1}^{n} (a_k + b_k)^2 = 0$ the statement is trivial.

In the case $\sum_{k=1}^{n}(a_k + b_k)^2 > 0$ we find

$$
\begin{aligned}
\sum_{k=1}^{n}(a_k + b_k)^2 &= \sum_{k=1}^{n}(a_k + b_k)a_k + \sum_{k=1}^{n}(a_k + b_k)b_k \\
&\leq \sum_{k=1}^{n}|a_k + b_k||a_k| + \sum_{k=1}^{n}|a_k + b_k||b_k| \\
&\leq \left(\sum_{k=1}^{n}(a_k + b_k)^2\right)^{\frac{1}{2}}\left(\sum_{k=1}^{n}a_k^2\right)^{\frac{1}{2}} + \left(\sum_{k=1}^{n}(a_k + b_k)^2\right)^{\frac{1}{2}}\left(\sum_{k=1}^{n}b_k^2\right)^{\frac{1}{2}} \\
&= \left(\sum_{k=1}^{n}(a_k + b_k)^2\right)^{\frac{1}{2}}\left(\left(\sum_{k=1}^{n}a_k^2\right)^{\frac{1}{2}} + \left(\sum_{k=1}^{n}b_k^2\right)^{\frac{1}{2}}\right)
\end{aligned}
$$

implying (14.19). $\qquad\square$

**Corollary 14.6.** *For real numbers* $a_1, \ldots, a_k, b_1, \ldots, b_k, c_1 \ldots, c_k$ *we have*

$$
\left(\sum_{k=1}^{n}(a_k - b_k)^2\right)^{\frac{1}{2}} \leq \left(\sum_{k=1}^{n}(a_k - c_k)^2\right)^{\frac{1}{2}} + \left(\sum_{k=1}^{n}(c_k - b_k)^2\right)^{\frac{1}{2}}. \tag{14.20}
$$

*Proof.* We only need to take $a_k - c_k$ for $a_k$ and $c_k - b_k$ for $b_k$ in Minkowski's inequality. $\qquad\square$

We will use the next result quite often, it is a result which depends on Archimedes' axiom.

**Lemma 14.7.** *Let* $a > 1$ *be a real number. For every* $R \in \mathbb{R}, R > 0$, *there exists* $n_0 \in \mathbb{N}$ *such that*

$$
a^{n_0} > R. \tag{14.21}
$$

*Proof.* If we take $x = a - 1 > 0$ in Bernoulli's inequality we find

$$
a^n = (1 + x)^n \geq 1 + nx.
$$

Let $R > 1$ then by Archimedes' axiom we can find an $n_0 \in \mathbb{N}$ such that $n_0 x > R - 1$, i.e.

$$
a^{n_0} \geq 1 + n_0 x > R.
$$

For $R \in (0, 1]$ the statement is trivial. $\qquad\square$

**Corollary 14.8.** *For* $0 < a < 1$ *and* $\epsilon > 0$ *there exists* $n_0 \in \mathbb{N}$ *such that* $a^{n_0} < \epsilon$.

*Proof.* We know that $\frac{1}{a} > 1$ and by Lemma 14.5 there exists $n_0 \in \mathbb{N}$ such that

$$\left(\frac{1}{a}\right)^{n_0} > \frac{1}{\epsilon}, \text{ i.e. } a^{n_0} < \epsilon.$$

$\square$

## Problems

1. Given $x, y \in \mathbb{R}, x < y$. Prove the existence of $z \in \mathbb{R}$ such that $x < z < y$.

2. Using the axioms of an ordered field prove that for $x, y, z \in \mathbb{R}$:
   a) $x < 0$ implies $-x > 0$;
   b) $x^2 > 0$ for all $x \in \mathbb{R}$;
   c) $a < 0$ and $x < y$ implies $ax > ay$.

3. Prove that Archimedes' axiom holds in $\mathbb{Q}$.

4. Show that there is no element $a \in \mathbb{Q}$ such that $a^2 = 3$.

5. Using Bernoulli's inequality prove

$$2n^n \le (n+1)^n \text{ for } n \ge 1,$$

then, by induction show that

$$n! \le 2 \left(\frac{n}{2}\right)^n.$$

6. Use mathematical induction to prove for $x_k \ge 0, k \in \mathbb{N}$, and $n \in \mathbb{N}$ that

$$\prod_{k=1}^{n}(1 + x_k) \ge 1 + \sum_{k=1}^{n} x_k.$$

7.* Prove that the arithmetic-geometric mean inequality implies Bernoulli's inequality and therefore by the proof of Lemma 14.2 it is in fact equivalent to the Bernoulli inequality.
   Hint: first prove the cases $n = 1$ and $n \ge 2$ with $0 < x < 1 - \frac{1}{n}$. Then apply the arithmetic-geometric mean inequality to the $n$ numbers $1 + n(1 + x), 1, \ldots, 1$.

209

8. For $x \in \mathbb{R}$ and $n, m \in \mathbb{N}$ prove that if $-x < n < m$ then

$$\left(1 + \frac{x}{n}\right)^n \leq \left(1 + \frac{x}{m}\right)^m.$$

9. For $a_k \in \mathbb{R}, k = 1, \ldots, n$, prove by using the Cauchy-Schwarz inequality that

$$\left|\sum_{k=1}^{n} a_k\right| \leq \sqrt{n} \left(\sum_{k=1}^{n} a_k^2\right)^{\frac{1}{2}}.$$

Now prove

$$\frac{1}{\sqrt{n}} \sum_{k=1}^{n} |a_k| \leq \left(\sum_{k=1}^{n} a_k^2\right)^{\frac{1}{2}} \leq \sqrt{n} \max\left(|a_1|, \ldots, |a_n|\right).$$

# 15 Sequences and their Limits

By definition a **sequence of real numbers** is a mapping from $\mathbb{N}$ to $\mathbb{R}$, i.e. each $n \in \mathbb{N}$ is mapped on some $a_n \in \mathbb{R}$. Usually we write $(a_n)_{n\in\mathbb{N}}$ for a sequence, but also sometimes $(a_1, a_2, a_3, \dots)$. It is appropriate to consider a little generalisation, namely to consider a mapping from $\{n \in \mathbb{Z} | n \geq k\}$, $k \in \mathbb{Z}$, to $\mathbb{R}$ and we denote the corresponding sequence by $(a_n)_{n\geq k}$.

**Example 15.1. A.** Let $a_n = a$, $a \in \mathbb{R}$ fixed, for all $n \in \mathbb{N}$, then we obtain the constant sequence $(a, a, a, \dots)$.
**B.** Put $a_n = \frac{1}{n}$, $n \in \mathbb{N}$, this gives the sequence $\left(\frac{1}{n}\right)_{n\in\mathbb{N}}$ or $(1, \frac{1}{2}, \frac{1}{3}, \frac{1}{4}, \dots)$.
**C.** The sequence $(-1, 1, -1, 1, \dots)$ could be written as $((-1)^n)_{n\in\mathbb{N}}$. More generally if $(a_n)_{n\in\mathbb{N}}$, $a_n \geq 0$ is a sequence of non-negative numbers we may consider the sequence $((-1)^n a_n)_{n\in\mathbb{N}}$ which has an alternating sign.
**D.** Take $a_n = \frac{n}{n+1}$ for $n \in \mathbb{N}_0$. This leads to the sequence $(0, \frac{1}{2}, \frac{2}{3}, \frac{3}{4}, \frac{4}{5}, \dots)$.
**E.** Let $a \in \mathbb{R}, a \neq 0$. The sequence $(a^n)_{n\in\mathbb{N}_0}$, is called a **geometric sequence**.

Note that we need to know all terms $a_n$ of the sequence $(a_n)_{n\geq k}$; knowing a finite number is not sufficient. In particular, there is no way to find $a_{n+1}$ by only knowing $a_1, \dots, a_n$. For example

$$1, \frac{1}{2}, \frac{1}{3}, \frac{1}{4}, \frac{1}{5}, \frac{1}{6}, \dots$$

does not give us a sequence, by no means can we deduce that the next term is $\frac{1}{7}$. The next term could be any number. For this reason any question in which a finite sequence of real numbers is given and the reader is then asked to find the next number is not valid.

**Example 15.2.** The **Fibonacci numbers** are the sequence defined by $a_0 = 1$, $a_1 = 1$, and $a_n = a_{n-1} + a_{n-2}$ for $n \geq 2$. This sequence is defined by a recursion formula. The first Fibonacci numbers are $1, 1, 2, 3, 5, 8, 13, 21, \dots$.

The Fibonacci numbers form an example of a **recursively** defined sequence. Consider the sequence defined by

$$a_{k+1} = \lambda a_k, \quad k \in \mathbb{N}, \quad a_0 = 1. \tag{15.1}$$

Thus $\frac{a_{k+1}}{a_k} = \lambda$ and the right hand side is independent of $k$. The geometric sequence $(q^k)_{k\in\mathbb{N}_0}, q \in \mathbb{R} \setminus \{0\}$, has the property that $\frac{q^{k+1}}{q^k} = q$. Now $\frac{q^1}{q^0} = q =$

$\lambda$ implies that $a_k = \lambda^k, k \in \mathbb{N}_0$. Next we want to see whether $(q^k)_{n \in \mathbb{N}_0}$ can lead to explicit expressions for a more general recursively defined sequence, for example the Fibonacci numbers:

$$a_{k+2} = a_{k+1} + a_k, \ k \geq 2, \ a_0 = a_1 = 1. \tag{15.2}$$

Taking $a_k = q^k$ in (15.2) we arrive at

$$q^{k+2} = q^{k+1} + q^k, \ k \geq 2,$$

which we may write as

$$q^k(q^2 - q - 1) = 0. \tag{15.3}$$

Since $q^k \neq 0$ we need to find solutions to the quadratic equation $q^2 - q - 1 = 0$ which are $\alpha = \frac{1}{2} + \frac{1}{2}\sqrt{5}$ and $\beta = \frac{1}{2} - \frac{1}{2}\sqrt{5}$. Now $a_k := A\alpha^{k-1} + B\beta^{k-1}$, $A, B \in \mathbb{R}$ satisfies for $k \geq 2$

$$0 = A\alpha^{k-1}(\alpha^2 - \alpha - 1) + B\beta^{k-1}(\beta^2 - \beta - 1),$$

i.e.

$$a_{k+2} = A\alpha^{k+1} + B\beta^{k+1} = A\alpha^k + B\beta^k + A\alpha^{k-1} + B\beta^{k-1} = a_{k+1} + a_k$$

and now we determine $A$ and $B$ such that $a_0 = a_1 = 1$, i.e. we look at the system

$$1 = A + B \text{ and } 1 = A\alpha + B\beta \tag{15.4}$$

which has the solution $A = \frac{\sqrt{5}+1}{2\sqrt{5}}, B = \frac{\sqrt{5}-1}{2\sqrt{5}}$. Hence the Fibonacci numbers are given by

$$a_k = \frac{\left(\frac{1+\sqrt{5}}{2}\right)^k - \left(\frac{1-\sqrt{5}}{2}\right)^k}{\sqrt{5}}, k \geq 0. \tag{15.5}$$

Note that we may extend this approach to tackle more general recursively defined sequences such as

$$\begin{aligned} a_{k+n} &= A_1 a_{k+n-1} + \cdots + A_n a_n \\ a_j &= x_j, j = 0, \ldots, n-1, \end{aligned}$$

by looking at solutions of

$$q^n - A_1 q^{n-1} - A_2 q^{n-2} - \cdots - A_n = 0.$$

An elementary discussion of recursively defined sequences is given in A. I. Markuschewitsch [10].

We now come to one of the fundamental definitions of this course, the **limit of a sequence**.

**Definition 15.3.** *Let $(a_n)_{n \geq k}$ be a sequence of real numbers. The sequence is called **convergent** to $a \in \mathbb{R}$ if for every $\varepsilon > 0$ there exists $N = N(\varepsilon) \in \mathbb{N}$ such that $n \geq N(\epsilon)$ implies*

$$|a_n - a| < \varepsilon. \tag{15.6}$$

*If $(a_n)_{n \in \mathbb{N}}$ converges to $a$ we call $a$ the **limit** of $(a_n)_{n \in \mathbb{N}}$ and we write*

$$\lim_{n \to \infty} a_n = a. \tag{15.7}$$

Before discussing some examples let us give some different formulations of our definition. For $a \in \mathbb{R}$ and $\varepsilon > 0$ we may consider the **open interval** $(a - \varepsilon, a + \varepsilon) := \{x \in \mathbb{R}; \quad |x - a| < \varepsilon\} = \{x \in \mathbb{R}; \quad a - \varepsilon < x < a + \varepsilon\}$.

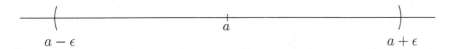

$$a - \epsilon \qquad\qquad\qquad a \qquad\qquad\qquad\qquad a + \epsilon$$

*Figure 15.1*

If $(a_n)_{n \in \mathbb{N}}$ converges to $a$, then given $\varepsilon > 0$, all elements $a_n$, $n \geq N(\varepsilon)$, will lie in the interval $(a - \varepsilon, a + \varepsilon)$. This is equivalent to the statement that for every $\varepsilon > 0$ all but a finite number of the $a_n$'s will lie in $(a - \varepsilon, a + \varepsilon)$.

We call the interval $(a - \varepsilon, a + \varepsilon)$, $\varepsilon > 0$, an $\varepsilon$-**neighbourhood** of $a$. Thus the convergence of $(a_n)_{n \in \mathbb{N}}$ to $a$ means that for every $\varepsilon > 0$ all but finitely many elements of the sequence lie in the corresponding $\varepsilon$-neighbourhood of $a$.

**Definition 15.4.** *A sequence of real numbers is called **divergent** if it has no limit, i.e. it does not converge.*

**Example 15.5. A.** If $a_n = a \in \mathbb{R}$ for all $n \in \mathbb{N}$, then $\lim_{n \to \infty} a_n = a$. Indeed, given $\varepsilon > 0$ then we have

$$|a_n - a| = |a - a| = 0 < \varepsilon \quad \text{for all } n \geq 1.$$

213

**B.** Consider the sequence $\left(\frac{1}{n}\right)_{n\in\mathbb{N}}$. We claim $\lim_{n\to\infty}\frac{1}{n}=0$.

Given $\varepsilon > 0$, let $N(\varepsilon) \in \mathbb{N}$ be such that $N(\varepsilon) > \frac{1}{\varepsilon}$. It follows that

$$\left|\frac{1}{n}-0\right|=\frac{1}{n}<\varepsilon \quad \text{for all } n \geq N(\varepsilon).$$

**C.** The sequence $((-1)^n)_{n\in\mathbb{N}}$ is divergent.

Assume $((-1)^n)_{n\in\mathbb{N}}$ converges to $a \in \mathbb{R}$. Then for $\varepsilon = 1$ there must exist $N \in \mathbb{N}$ such that for all $n \geq N$ it follows that $|(-1)^n - a| < 1$. But for all $n$ we have $|(-1)^{n+1} - (-1)^n| = 2$, and for $n \geq N$

$$2 = |(-1)^{n+1} - (-1)^n| = |((-1)^{n+1} - a) + (a - (-1)^n)|$$
$$\leq |(-1)^{n+1} - a| + |a - ((-1)^n)| < 1 + 1 = 2,$$

which is a contradiction. Hence no $a \in \mathbb{R}$ can be the limit of $((-1)^n)_{n\in\mathbb{N}}$.

**D.** The limit of $\left(\frac{n}{n+1}\right)_{n\in\mathbb{N}}$ is 1, i.e. $\lim_{n\to\infty}\frac{n}{n+1}=1$.

Given $\varepsilon > 0$ we find

$$\left|\frac{n}{n+1}-1\right|=\left|\frac{n-(n+1)}{n+1}\right|=\frac{1}{n+1}<\varepsilon.$$

Hence, if we choose $N(\varepsilon) = \left[\frac{1}{\varepsilon}\right] + 1$, then for each $n \geq N(\varepsilon)$ we have

$$\left|\frac{n}{n+1}-1\right|<\varepsilon.$$

**E.** We have $\lim_{n\to\infty}\frac{n}{2^n}=0$.

For $n > 3$ we know that $n^2 \leq 2^n$. It follows that

$$\frac{n^2}{2^n}\leq 1 \quad \text{or} \quad \frac{n}{2^n}<\frac{1}{n} \text{ for } n > 3.$$

Let $\varepsilon > 0$ be given and take $N(\varepsilon) > \max\{3, \frac{1}{\varepsilon}\}$. Now $n \geq N(\epsilon)$ implies

$$\left|\frac{n}{2^n}-0\right|=\frac{n}{2^n}\leq\frac{1}{n}<\varepsilon.$$

A helpful observation is

**Lemma 15.6.** *For a convergent sequence $(a_n)_{n \geq k}$ of real numbers and any $m \in \mathbb{N}$*

$$\lim_{n \to \infty} a_n = \lim_{n \to \infty} a_{n+m} \qquad (15.8)$$

*holds.*

*Proof.* First define $b_n := a_{n+m}$. Since $\lim_{n \to \infty} a_n = a$ exists, for every $\epsilon > 0$ there exists $N$ such that $n \geq N$ implies $|a - a_n| < \epsilon$. However for these $n$ we have $n + m \geq N$ and therefore

$$|b_n - a| = |a_{n+m} - a| < \epsilon.$$

$\square$

**Definition 15.7.** *A sequence of real numbers $(a_n)_{n \geq k}$ is **bounded above** if there exists $K_1 \in \mathbb{R}$ such that $a_n \leq K_1$ for all $n \in \mathbb{N}$. It is called **bounded below** if there exists $K_2 \in \mathbb{R}$ such that $K_2 \leq a_n$ for all $n \in \mathbb{N}$. We call $(a_n)_{n \geq k}$ **bounded** if it is bounded above and below, i.e. if there is some $K \in \mathbb{R}$ such that $-K \leq a_n \leq K$, or $|a_n| \leq K$.*

**Theorem 15.8.** *Every convergent sequence $(a_n)_{n \geq k}$ is bounded. If $K$ is a bound for $|a_n|$, i.e. $|a_n| \leq K$ for all $n \geq k$, and if $a \in \mathbb{R}$ is the limit of $(a_n)_{n \geq k}$, then $|a| \leq K$.*

*Proof.* Let $(a_n)_{n \geq k}$ be a sequence converging to $a$, i.e. $\lim_{n \to \infty} a_n = a$. By definition, for $\varepsilon = 1$ there exists $N$ such that $|a_n - a| < 1$ for all $n \geq N$. This implies

$$|a_n| = |a_n - a + a| \leq |a| + |a_n - a| \leq |a| + 1$$

for $n \geq N$. Now if we define $M := \max\{|a_1|, \ldots, |a_{N-1}|, |a| + 1\}$, then $|a_n| \leq M$ for all $n \geq k$, i.e. $(a_n)_{n \in \mathbb{N}}$ is bounded.
Further, if $|a_n| \leq K$ for all $n \geq k$ we find

$$|a| \leq |a_n| + |a_n - a| \leq K + |a_n - a|.$$

For $\epsilon > 0$ there exists $N(\epsilon) \in \mathbb{N}$ such that $n \geq N(\epsilon)$ implies $|a_n - a| < \epsilon$ and therefore $n \geq N(\epsilon)$ implies $|a| \leq K + \epsilon$. Since $\epsilon > 0$ is arbitrary we deduce that $|a| \leq K$. $\square$

**Remark 15.9.** Of course, a bounded sequence need not be convergent: $((-1)^n)_{n \in \mathbb{N}}$ is bounded since

$$|(-1)^n| = 1 \quad \text{for all } n \in \mathbb{N},$$

but we already know that this sequence is divergent.

**Example 15.10.** The sequence $(a_n)_{n\geq0}$ of all Fibonacci numbers is divergent since we always have that $a_n \geq n$ for $n \in \mathbb{N}_0$. For $n = 0, 1$ this is trivial. Now suppose $a_n \geq n$ for all $n \leq N$, we find $a_{N+1} = a_N + a_{N-1} \geq N + N - 1 = 2N - 1 \geq N + 1$

**Example 15.11.** We want to study the geometric sequence $(q^n)_{n\in\mathbb{N}}$.
**A.** If $|q| < 1$, then $\lim\limits_{n\to\infty} q^n = 0$.
We know by Corollary 14.8 that for $\varepsilon > 0$ there exists $N \in \mathbb{N}$ such that $|q|^N < \varepsilon$. Now we find

$$|q^n - 0| = |q^n| = |q|^n \leq |q|^N < \varepsilon$$

for all $n \geq N$.
**B.** For $q = 1$ we have $q^n = 1$ and we already know that

$$\lim_{n\to\infty} q^n = \lim_{n\to\infty} 1 = 1.$$

**C.** For $q = -1$, i.e. $q^n = (-1)^n$, we have just shown that $((-1)^n)_{n\in\mathbb{N}}$ is divergent.
**D.** For $|q| > 1$ it follows that $(|q|^n)_{n\in\mathbb{N}}$, hence $(q^n)_{n\in\mathbb{N}}$ is unbounded, compare Lemma 14.7. Therefore $(q^n)_{n\in\mathbb{N}}$ is divergent.

**Example 15.12.** We claim that $\lim\limits_{n\to\infty} \sqrt[n]{n} = \lim\limits_{n\to\infty} n^{\frac{1}{n}} = 1$. For this we set $a_n := \sqrt[n]{n} - 1$. Given $\epsilon > 0$ we need to find $N(\epsilon) \in \mathbb{N}$ such that $n \geq N(\epsilon)$ implies $|a_n| = a_n < \epsilon$ The binomial theorem yields

$$n = (1 + a_n)^n = \sum_{j=0}^{n} \binom{n}{j} a_n^j \geq 1 + \binom{n}{2} a_n^2 = 1 + \frac{n(n-1)}{2} a_n^2.$$

For $n \geq 2$ this implies

$$a_n^2 \leq \frac{2(n-1)}{n(n-1)} = \frac{2}{n},$$

or

$$a_n \leq \frac{\sqrt{2}}{\sqrt{n}}.$$

Thus we need to find $N(\epsilon)$ such that $n \geq N(\epsilon)$ implies $\frac{\sqrt{2}}{\sqrt{n}} < \epsilon$. However with $N_0 \geq \frac{2}{\epsilon^2}$ and $n \geq N_0 \geq 2$ it follows that

$$\frac{\sqrt{2}}{\sqrt{n}} < \frac{\sqrt{2}}{\sqrt{N_0}} \leq \epsilon.$$

So far we have defined the limit of a sequence. But do we know that it is unique?

**Theorem 15.13.** *The limit $a \in \mathbb{R}$ of a sequence $(a_n)_{n \geq k}$ is unique.*

*Proof.* Suppose that $(a_n)_{n \geq k}$ has two limits $a$ and $a'$. Then given $\varepsilon > 0$, since $\lim_{n \to \infty} a_n = a$, there exists $N_1 \in \mathbb{N}$ such that $|a_n - a| < \frac{1}{2}\varepsilon$ for $n \geq N_1$. On the other hand, since we also have $\lim_{n \to \infty} a_n = a'$ there exists $N_2$ such that $|a_n - a'| < \frac{1}{2}\varepsilon$ for $n \geq N_2$. Thus it follows that, if $N \geq \max\{N_1, N_2\}$, then

$$|a - a'| = |(a - a_n) + (a_n - a')| \leq |a - a_n| + |a_n - a'|$$
$$< \frac{1}{2}\varepsilon + \frac{1}{2}\varepsilon = \varepsilon,$$

This is true for all $\varepsilon > 0$ and so $|a - a'| = 0$ or $a = a'$. $\qquad\square$

**Theorem 15.14 (Sum of convergent sequences).** *Let $(a_n)_{n \geq k}$ and $(b_n)_{n \geq k}$ be two convergent sequences with limits $a$ and $b$, respectively, i.e. $\lim_{n \to \infty} a_n = a$ and $\lim_{n \to \infty} b_n = b$. Then the sequence $(c_n)_{n \geq k}$, $c_n := a_n + b_n$, converges to $a + b$, i.e.*

$$\lim_{n \to \infty} c_n = a + b.$$

*Proof.* Given $\varepsilon > 0$. For $\frac{\varepsilon}{2} > 0$ there exist $N_1$ and $N_2$ such that

$$n \geq N_1 \text{ implies } |a - a_n| < \frac{\varepsilon}{2} ,$$

and

$$n \geq N_2 \text{ implies } |b - b_n| < \frac{\varepsilon}{2} .$$

For $N = \max\{N_1, N_2\}$ we find that $n \geq N$ implies

$$|c_n - (a + b)| = |a_n + b_n - (a + b)| = |(a_n - a) + (b_n - b)|$$
$$\leq |a_n - a| + |b_n - b| < \frac{\varepsilon}{2} + \frac{\varepsilon}{2} = \varepsilon,$$

thus $\lim_{n \to \infty} c_n = a + b$. $\qquad\square$

**Example 15.15.** Consider $c_n = \frac{n+1}{n}, n \in \mathbb{N}$. Setting $a_n = 1$ and $b_n = \frac{1}{n}$ we find $c_n = a_n + b_n$. We know $\lim\limits_{n\to\infty} a_n = 1$ and $\lim\limits_{n\to\infty} b_n = \lim\limits_{n\to\infty} \frac{1}{n} = 0$. Thus we find $\lim\limits_{n\to\infty} \frac{n+1}{n} = 0 + 1 = 1$.

**Theorem 15.16 (Product of convergent sequences).** *Let $(a_n)_{n\geq k}$ and $(b_n)_{n\geq k}$ be two convergent sequences. Then the sequence $(a_n \cdot b_n)_{n\geq k}$ converges to $a \cdot b$, i.e.*

$$\lim_{n\to\infty} (a_n \cdot b_n) = (\lim_{n\to\infty} a_n) \cdot (\lim_{n\to\infty} b_n).$$

*Proof.* Put $\lim\limits_{n\to\infty} a_n = a$ and $\lim\limits_{n\to\infty} b_n = b$. We know that $(a_n)_{n\in\mathbb{N}}$ is bounded, hence with some $K > 0$ we have $|a_n| \leq K$ for all $n \in \mathbb{N}$. But $(b_n)_{n\in\mathbb{N}}$ is also a bounded sequence, and without loss of generality we may also assume that $|b_n| \leq K$ for all $n \in \mathbb{N}$. In addition by Theorem 15.8 we also know that $|a| \leq K$ and $|b| \leq K$. The convergence of $(a_n)_{n\in\mathbb{N}}$ and $(b_n)_{n\in\mathbb{N}}$ implies that for $\varepsilon > 0$ there exists $N_1, N_2 \in \mathbb{N}$ such that

$$|a_n - a| < \frac{\varepsilon}{2K} \quad \text{for } n \geq N_1, \text{ and } |b_n - b| < \frac{\varepsilon}{2K} \text{ for } n \geq N_2.$$

Now, for all $n \geq N := \max(N_1, N_2)$ we find

$$\begin{aligned}
|a_n \cdot b_n - a \cdot b| &= |a_n(b_n - b) + (a_n - a) \cdot b| \\
&\leq |a_n||b_n - b| + |a_n - a||b| \\
&\leq K \cdot \frac{\varepsilon}{2K} + K \cdot \frac{\varepsilon}{2K} = \varepsilon.
\end{aligned}$$

$\square$

**Corollary 15.17.** *Let $(a_n)_{n\geq k}$ be a convergent sequence and $\lambda \in \mathbb{R}$. Then the sequence $(\lambda a_n)_{n\geq k}$ converges and the limit is given by*

$$\lim_{n\to\infty} (\lambda a_n) = \lambda \lim_{n\to\infty} a_n.$$

*Proof.* We may apply Theorem 15.16 with $b_n = \lambda$ for all $n \geq k$. $\square$

**Corollary 15.18.** *Let $(a_n)_{n\geq k}$ and $(b_n)_{n\geq k}$ be two convergent sequences. Then the sequence $(a_n - b_n)_{n\geq k}$ is convergent and its limit is*

$$\lim_{n\to\infty} (a_n - b_n) = \lim_{n\to\infty} a_n - \lim_{n\to\infty} b_n.$$

*Proof.* Just combine Corollary 15.17 with Theorem 15.14.  $\square$

**Theorem 15.19.** *Let $(a_n)_{n \geq k}$ and $(b_n)_{n \geq k}$ be convergent sequences and suppose that $\lim\limits_{n \to \infty} b_n \neq 0$. Then there exists $N_0 \in \mathbb{N}$ such that $b_n \neq 0$ for $n \geq N_0$ and the sequence $\left(\frac{a_n}{b_n}\right)_{n \geq N_0}$ is convergent to*

$$\lim_{n \to \infty} \frac{a_n}{b_n} = \frac{\lim\limits_{n \to \infty} a_n}{\lim\limits_{n \to \infty} b_n}.$$

*Proof.* Since $b := \lim\limits_{n \to \infty} b_n$ we find for $\varepsilon := \frac{|b|}{2} > 0$ a number $N_0 \in \mathbb{N}$ such that

$$|b_n - b| < \frac{|b|}{2} \quad \text{for } n \geq N_0.$$

Since $|b| - |b_n| \leq |b_n - b|$ for $n \geq N_0$ we have

$$|b| - |b_n| < \frac{|b|}{2} \quad \text{or} \quad \frac{|b|}{2} < |b_n|,$$

the last statement being equivalent to $\frac{1}{|b_n|} < \frac{2}{|b|}$. In particular for all $n \geq N_0$ we have $b_n \neq 0$ and hence for these $n$ the expression $\frac{a_n}{b_n}$ always makes sense. Next suppose that $a_n = 1$ for all $n$. Given $\varepsilon > 0$ there exists $N_1 \in \mathbb{N}$ such that

$$|b_n - b| < \frac{\varepsilon |b|^2}{2} \quad \text{for } n \geq N_1.$$

Therefore, for $n \geq N := \max\{N_0, N_1\}$ we find

$$\left| \frac{1}{b_n} - \frac{1}{b} \right| = \left| \frac{b - b_n}{b_n b} \right| = \frac{1}{|b_n||b|} |b - b_n| < \frac{2}{|b|^2} |b_n - b|$$

$$< \frac{2}{|b|^2} \cdot \frac{\varepsilon |b|^2}{2} = \varepsilon.$$

Hence we have proved that

$$\lim_{n \to \infty} \frac{1}{b_n} = \frac{1}{b}.$$

Since $\frac{a_n}{b_n} = a_n \cdot \frac{1}{b_n}$, $n \geq N_0$, the general case follows now from Theorem 15.16.  $\square$

**Example 15.20.** Consider the sequence $a_n = \dfrac{7n^2 + 3n}{n^2 - 2}$, $n \in \mathbb{N}$. We may

write $a_n = \dfrac{7 + \frac{3}{n}}{1 - \frac{2}{n^2}}$. Now, $\lim\limits_{n \to \infty} \frac{1}{n} = 0$, implying that $\lim\limits_{n \to \infty} \frac{1}{n^2} = \lim\limits_{n \to \infty} \left( \frac{1}{n} \cdot \frac{1}{n} \right) = 0$

by Theorem 15.16. Further, by Corollary 15.17 we find that $\lim\limits_{n \to \infty} \frac{3}{n} = 0$ and
$\lim\limits_{n \to \infty} \frac{2}{n^2} = 0$. Thus $\lim\limits_{n \to \infty} \left( 7 + \frac{3}{n} \right) = 7$ and $\lim\limits_{n \to \infty} \left( 1 - \frac{2}{n^2} \right) = 1$. According to
Theorem 15.19 we have

$$\lim_{n \to \infty} \frac{7n^2 + 3n}{n^2 - 2} = \frac{\lim\limits_{n \to \infty} \left( 7 + \frac{3}{n} \right)}{\lim\limits_{n \to \infty} \left( 1 - \frac{2}{n^2} \right)} = \frac{7}{1} = 7.$$

**Theorem 15.21.** *Let $(a_n)_{n \geq k}$ and $(b_n)_{n \geq k}$ be two convergent sequences and suppose that $a_n \leq b_n$ for all $n \geq k$. Then we have $\lim\limits_{n \to \infty} a_n \leq \lim\limits_{n \to \infty} b_n$.*

*Proof.* Suppose $b := \lim\limits_{n \to \infty} b_n < a := \lim\limits_{n \to \infty} a_n$. For $\varepsilon := \frac{a-b}{2} > 0$ there exists $N_1, N_2 \in \mathbb{N}$ such that $|a_n - a| < \varepsilon$ for $n \geq N_1$ and $|b_n - b| < \varepsilon$ for $n \geq N_2$. For $n \geq \max\{N_1, N_2\}$ we find that

$$a_n > a - \varepsilon \quad \text{and} \quad b_n < b + \varepsilon.$$

By the definition of $\varepsilon$ we have

$$a - \varepsilon = a - \left( \frac{a-b}{2} \right) = \frac{a+b}{2} = b + \left( \frac{a-b}{2} \right) = b + \varepsilon$$

implying that $b_n < b + \varepsilon = a - \varepsilon < a_n$ which contradicts the assumption $a_n \leq b_n$ and so the theorem is proved. $\square$

**Remark 15.22. A.** In particular $a_n \geq 0$ implies that $\lim_{n \to \infty} a_n \geq 0$.
**B.** Note that in Theorem 15.21 we need not assume that $a_n \leq b_n$ for all $n \geq k$. It is sufficient to assume that $a_n \leq b_n$ for all $N_0 \geq k$, $N_0 \in \mathbb{N}$.
**C.** Note further that $a_n < b_n$ for all $n \geq k$ (or all $n \geq N_0$, $N_0 \geq k$) does not imply that

$$\lim_{n \to \infty} a_n < \lim_{n \to \infty} b_n.$$

To see this take the sequence $a_n = 0$ for all $n \in \mathbb{N}$ and $b_n = \frac{1}{n}$. For all $n \in \mathbb{N}$ we know that $a_n = 0 < \frac{1}{n} = b_n$, but $\lim\limits_{n \to \infty} a_n = 0 = \lim\limits_{n \to \infty} b_n$.

**Corollary 15.23.** *Suppose that $(a_n)_{n \in \mathbb{N}}$ is a convergent sequence and that with two numbers $A$ and $B$ we have $A \leq a_n \leq B$ for all $n \in \mathbb{N}$, $n \geq N_0$. Then*

$$A \leq \lim_{n \to \infty} a_n \leq B.$$

# Problems

1. Let $M$ be a countable set and $f : M \to \mathbb{R}$. Prove that we can arrange $F(M)$ as a sequence, i.e. $f(M) = \{a_k \in \mathbb{R} | k \in \mathbb{N}\}$ with suitable real numbers $a_k$.
   Hint: recall that $M$ is countable if and only if there exists a bijective mapping $g : \mathbb{N} \to M$.

2.    a) Let $(a_n)_{n \geq k}$ be a sequence of real numbers such that $a_n = a$ for all $n \geq M \geq k$. Prove that $(a_n)_{n \geq k}$ converges and find its limit.

   b) Let $(a_n)_{n \geq k}$ be a sequence with limit $a$.
   Consider the sequence

   $$b_n := \begin{cases} c_n, & k \leq n \leq M - 1 \\ a_n, & n \geq M \end{cases}$$

   for any choice of numbers $c_n$, $k \leq n \leq M-1$, and any choice of $M \geq k$. Prove that $\lim_{n \to \infty} b_n = a$.

3. Let $(a_n)_{n \geq k}$ be a sequence converging to 0 and let $(b_n)_{n \geq k}$ be a bounded sequence. Show that $\lim_{n \to \infty} (b_n a_n) = 0$.

4.    a) Suppose that $a = \lim_{n \to \infty} a_n = \lim_{n \to \infty} b_n$. Moreover for $n \geq k$ let $c_n \in \mathbb{R}$ be given satisfying $a_n \leq c_n \leq b_n$. Show that $(c_n)_{n \geq k}$ converges to $a$.

   b) Suppose that $a = \lim_{n \to \infty} a_n$ and $b = \lim_{n \to \infty} b_n$, $a < b$, and suppose that for $n \in \mathbb{N}$ the numbers $c_n \in \mathbb{R}$ satisfy $a_n \leq c_n \leq b_n$. Does this imply the convergence of $(c_n)_{n \in \mathbb{N}}$?

5.    a) Prove that $\lim_{n \to \infty} a_n = a$ implies $\lim_{n \to \infty} |a_n| = |a|$. Now deduce that $\lim_{n \to \infty} a_n = a$ is equivalent to $\lim_{n \to \infty} |a_n - a| = 0$.

   b) Let $(a_n)_{n \in \mathbb{N}}$ be a sequence of real numbers and $a \in \mathbb{R}$. Further let $(\mu_n)_{n \in \mathbb{N}}$ be a sequence of non-negative numbers converging to 0. Suppose that for all $n \in \mathbb{N}$ we have $|a_n - a| \leq \mu_n$. Prove that $\lim_{n \to \infty} a_n = a$.

6. Suppose that $\lim_{n \to \infty} a_n = a$ and $\lim_{n \to \infty} b_n = b$. Prove that $\lim_{n \to \infty} \max\{a_n, b_n\} = \max\{a, b\}$ and $\lim_{n \to \infty} \min\{a_n, b_n\} = \min\{a, b\}$.

Hint: find a representation of the maximum and the minimum of two numbers with the help of the absolute value. Then use the result of Problem 5 a).

7.    a) Prove that $\lim\limits_{n\to\infty} \dfrac{5}{n+6} = 0$, i.e. prove that for every $\epsilon > 0$ there exists $N = N(\epsilon) \in \mathbb{N}$ such that $n \geq N(\epsilon)$ implies $\left|\dfrac{5}{n+6} - 0\right| < \epsilon$.

b) For $\epsilon = \dfrac{1}{1000}$ find $N \in \mathbb{N}$ such that $n \geq N$ implies

$$\left|\frac{4n}{3n+2} - \frac{4}{3}\right| < \frac{1}{1000}.$$

8. For $k \in \mathbb{N}$ prove:

a) $\lim\limits_{n\to\infty} \dfrac{1}{n^k} = 0$;

b) $\lim\limits_{n\to\infty} \dfrac{1}{n^{\frac{1}{k}}} = 0$.

9. Use the theorems about limits and already proved results about limits of sequences to find:

a) $\lim\limits_{n\to\infty} \dfrac{(n+1)^2 - n^2}{n}$;

b) $\lim\limits_{n\to\infty} (\sqrt{n+1} - \sqrt{n})$;

c) $\lim\limits_{n\to\infty} \dfrac{\sum_{j=1}^n j}{n^2}$;

d) $\lim\limits_{n\to\infty} \dfrac{\sum_{j=1}^n j^2}{n^3}$;

e) $\lim\limits_{n\to\infty} \dfrac{1 + 2 \cdot 3^n}{5 + 4 \cdot 3^n}$;

f) $\lim\limits_{n\to\infty} \dfrac{n + 4^n}{5^n}$.

10. Prove that $\lim\limits_{n\to\infty} \sqrt[n]{a} = 1$ for $a \geq 1$.

11. Find the following limit:

$$\lim_{n\to\infty} \prod_{j=1}^n \left(1 - \frac{1}{j+1}\right).$$

222

12. Find

$$\lim_{\nu \to \infty} \frac{\sum_{k=0}^{n} a_k \nu^k}{\sum_{l=0}^{m} b_l \nu^l}, \nu \in \mathbb{N}, a_k, b_l \in \mathbb{R}.$$

Note that the cases $n < m$, $m < n$ and $n = m$ need to be considered separately.

13. Suppose that $\lim_{n \to \infty} a_n = a$. Prove that

$$\lim_{n \to \infty} \frac{\sum_{j=1}^{n} a_n}{n} = a.$$

14. Let $f : (a, b) \to \mathbb{R}$ be a function and $x_0 \in (a, b)$ be fixed. Suppose that there exists $\delta > 0$ such that $|x - x_0| < \delta$ implies $\left| \frac{f(x) - f(x_0)}{x - x_0} - A \right| < \epsilon$. Deduce that then for every $\epsilon > 0$ there exists $N(\epsilon) \in \mathbb{N}$ such that $n \geq N(\epsilon)$ implies

$$\left| n \left( f \left( x_0 + \frac{1}{n} \right) - f(x_0) \right) - A \right| < \epsilon,$$

i.e.

$$\lim_{n \to \infty} \left( n \left( f \left( x_0 + \frac{1}{n} \right) - f(x_0) \right) \right) = A.$$

# 16 A First Encounter with Series

We next want to look at sequences from a different (but equivalent) point of view. Let $(a_n)_{n \in \mathbb{N}}$ be a sequence of real numbers. Starting with $(a_n)_{n \in \mathbb{N}}$ we may introduce a new sequence

$$s_n := \sum_{k=1}^{n} a_k, \quad n \in \mathbb{N},$$

more generally, if $(a_n)_{n \geq l}$, then $s_n := \sum_{k=l}^{n} a_k$. We call $s_n$ the $n^{\text{th}}$ **partial sum** of the (**infinite**) **series** $\sum_{k=1}^{\infty} a_k$. Thus we have a new sequence $(s_n)_{n \in \mathbb{N}}$ and note that at the moment $\sum_{k=1}^{\infty} a_k$ is just a formal expression for this sequence. However, it may happen that the sequence of the partial sums $(s_n)_{n \in \mathbb{N}}$ converges to some limit $s$. In this case we denote the limit also by $\sum_{k=1}^{\infty} a_k$. Thus the symbol $\sum_{k=1}^{\infty} a_k$ will have two meanings: a formal expression for the sequence of partial sums $\left( \sum_{k=1}^{n} a_k \right)_{n \in \mathbb{N}}$ and, if it exists, the limit of the sequence of partial sums.

**Remark 16.1.** Note that every sequence $(a_n)_{n \in \mathbb{N}}$ has a representation as the partial sums of a series, i.e. in a certain sense sequences and series are in a one-to-one correspondence. Indeed, given $(s_n)_{n \in \mathbb{N}}$ define

$$a_n = s_n - s_{n-1}.$$

Then $s_n = \sum_{k=1}^{n} a_k$.

Let us formally state

**Definition 16.2.** *Let $(a_n)_{n \geq k}$ be a sequence of real numbers and denote by $(s_n)_{n \geq k}$ the sequence of partial sums $\sum_{l=k}^{n} a_l$. We call the series $\sum_{l=k}^{\infty} a_l$ **convergent** to $s \in \mathbb{R}$ and denote the limit also by $\sum_{l=k}^{\infty} s_l$ if the sequence $(s_n)_{n \geq k}$ converges to $s$.*

**Example 16.3.** Consider the series $\sum\limits_{k=1}^{\infty} \frac{1}{k(k+1)}$. We then see that

$$
\begin{aligned}
s_n &= \sum_{k=1}^{n} \frac{1}{k(k+1)} = \sum_{k=1}^{n} \left( \frac{1}{k} - \frac{1}{k+1} \right) \\
&= 1 + \sum_{k=2}^{n} \frac{1}{k} - \sum_{k=1}^{n-1} \frac{1}{k+1} - \frac{1}{n+1} \\
&= 1 - \frac{1}{n+1} + \sum_{k=1}^{n-1} \frac{1}{k+1} - \sum_{k=1}^{n-1} \frac{1}{k+1} \\
&= 1 - \frac{1}{n+1} = \frac{n}{n+1},
\end{aligned}
$$

i.e. $(s_n)_{n \in \mathbb{N}} = \left( \frac{n}{n+1} \right)_{n \in \mathbb{N}}$ and therefore the series $\sum\limits_{k=1}^{\infty} \frac{1}{k(k+1)}$ converges and its limit is given by

$$
\sum_{k=1}^{\infty} \frac{1}{k(k+1)} = \lim_{n \to \infty} s_n = \lim_{n \to \infty} \frac{n}{n+1} = 1.
$$

**Theorem 16.4.** *Let $x \in \mathbb{R}$ and $|x| < 1$. Then we have*

$$
\sum_{k=0}^{\infty} x^k = \frac{1}{1-x}.
$$

*Proof.* We first claim

$$
s_n = s_n(x) = \sum_{k=0}^{n} x^k = \frac{1 - x^{n+1}}{1 - x}.
$$

Once this is proved, from Example 15.11.A it follows that

$$
\lim_{n \to \infty} x^{n+1} = x \cdot \lim_{n \to \infty} x^n = 0 \quad \text{for } |x| < 1,
$$

therefore we find that

$$
\lim_{n \to \infty} s_n = \lim_{n \to \infty} \frac{1 - x^{n+1}}{1 - x} = \frac{1}{1 - x}.
$$

The series $\sum_{k=0}^{\infty} x^k$ is called the **geometric series** (with parameter or variable $x \in (-1, 1)$). Now we prove: let $x \in \mathbb{R}$, $x \neq 1$, then for all $n \in \mathbb{N}_0$ we have

$$\sum_{k=0}^{n} x^k = \frac{1 - x^{n+1}}{1 - x}.$$

Indeed, for $n = 0$ we find

$$\sum_{k=0}^{0} x^k = x^0 = 1 = \frac{1 - x^{0+1}}{1 - x} = 1,$$

and further

$$\begin{aligned}
\sum_{k=0}^{n+1} x^k &= \sum_{k=0}^{n} x^k + x^{n+1} = \frac{1 - x^{n+1}}{1 - x} + x^{n+1} \\
&= \frac{1 - x^{n+1}}{1 - x} + \frac{(1 - x)x^{n+1}}{1 - x} \\
&= \frac{1 - x^{n+1} + x^{n+1} - x^{n+2}}{1 - x} \\
&= \frac{1 - x^{n+2}}{1 - x},
\end{aligned}$$

and the result follows by mathematical induction. □

**Remark 16.5.** Let us change our point of view and consider the function $f : \mathbb{R} \setminus \{1\} \to \mathbb{R}$, $f(x) = \frac{1}{1-x}$. If $|x| < 1$ then Theorem 16.4 says that $f$ has a representation by $x \mapsto \sum_{k=0}^{\infty} x^k$ in the sense that $f(x) = \sum_{k=0}^{\infty} x^k$ for $|x| < 1$. We say that for $|x| < 1$ the series $\sum_{k=0}^{\infty} x^k$ converges to the function $f$. Note that $f$ is defined on a much larger set than the series converges, i.e. the series represents $f$ only on a subset of the domain of $f$.

**Example 16.6. A.** The following holds

$$\sum_{k=0}^{\infty} 2^{-k} = \sum_{k=0}^{\infty} \left(\frac{1}{2}\right)^k = 1 + \frac{1}{2} + \frac{1}{4} + \frac{1}{8} + \cdots = \frac{1}{1 - \frac{1}{2}} = 2.$$

**B.** We have

$$\sum_{k=0}^{\infty} (-2)^{-k} = \sum_{k=0}^{\infty} \left(-\frac{1}{2}\right)^k = 1 - \frac{1}{2} + \frac{1}{4} - \frac{1}{8} \pm \cdots = \frac{1}{1 - \left(-\frac{1}{2}\right)} = \frac{2}{3}.$$

**C.** For $\phi \in (0, \pi)$ we know that $|\cos \phi| < 1$ and consequently we find

$$\sum_{k=0}^{\infty} (\cos^2 \phi)^k = \frac{1}{1 - \cos^2 \phi} = \frac{1}{\sin^2 \phi}.$$

Since the convergence of a series is by definition the convergence of the sequence of its partial sums, we may immediately derive some rules for handling convergent series by using known results for sequences:

**Theorem 16.7.** *Let* $\sum_{l=k}^{\infty} a_l$ *and* $\sum_{l=k}^{\infty} b_l$ *be two convergent series and* $\lambda \in \mathbb{R}$ *then the series* $\sum_{l=k}^{\infty}(a_l + b_l)$, $\sum_{l=k}^{\infty}(a_l - b_l)$ *and* $\sum_{l=k}^{\infty}(\lambda a_l)$ *converge. Moreover, for their limits we have*

$$\sum_{l=k}^{\infty}(a_l \pm b_l) = \sum_{l=k}^{\infty} a_l \pm \sum_{l=k}^{\infty} b_l$$

*and*

$$\sum_{l=k}^{\infty} \lambda a_l = \lambda \sum_{l=k}^{\infty} a_l.$$

*Proof.* With $c_n := \sum_{l=k}^{n} a_l$ and $d_n := \sum_{l=k}^{n} b_l$ we have

$$\sum_{l=k}^{n}(a_l \pm b_l) = \sum_{l=k}^{n} a_l \pm \sum_{l=k}^{n} b_l = c_n \pm d_n,$$

implying that

$$\sum_{l=k}^{\infty}(a_l \pm b_l) = \lim_{n \to \infty} \sum_{l=k}^{n}(a_l \pm b_l) = \lim_{n \to \infty}\left(\sum_{l=k}^{n} a_l \pm \sum_{l=k}^{n} b_l\right)$$
$$= \lim_{n \to \infty} \sum_{l=k}^{n} a_l \pm \lim_{n \to \infty} \sum_{l=k}^{n} b_l = \sum_{l=k}^{\infty} a_l \pm \sum_{l=k}^{\infty} b_l.$$

The final assertion is shown in an analogous way. □

**Example 16.8.** Recall the series $\sum_{k=1}^{\infty} \frac{1}{k(k+1)}$. We know that $\sum_{k=1}^{\infty} \frac{1}{k(k+1)} = 1$, and further we have

$$\frac{1}{k(k+1)} = \frac{1}{k} - \frac{1}{k+1}.$$

228

But we will see later that the series $\sum_{k=1}^{\infty} \frac{1}{k}$ is not convergent, hence $\sum_{k=1}^{\infty} \frac{1}{k+1}$ does not converge. Hence

$$\sum_{k=1}^{\infty} \frac{1}{k} - \sum_{k=1}^{\infty} \frac{1}{k+1}$$

does not make sense.

**Definition 16.9.** *A sequence* $(a_n)_{n \geq k}$ *of real numbers is called **divergent to** $+\infty$ ( *to* $-\infty$) *if for any* $K \in \mathbb{R}$ *there exists* $N = N(K) \in \mathbb{N}$ *such that if* $n \geq N$ *then* $a_n > K$ ( $a_n < K$).

For a sequence divergent to $+\infty$ ($-\infty$) we will write

$$\lim_{n \to \infty} a_n = \infty \quad (\lim_{n \to \infty} a_n = -\infty).$$

**Example 16.10. A.** For $m \in \mathbb{N}$ the sequence $(n^m)_{n \in \mathbb{N}}$ diverges to $+\infty$.
**B.** The sequence $(-(2^n))_{n \in \mathbb{N}}$ diverges to $-\infty$.
**C.** The sequence $(s_n)_{n \in \mathbb{N}}$, $s_n := \sum_{k=1}^{n} k = \frac{n(n+1)}{2}$, diverges to $+\infty$.
**D.** The sequence $((-1)^n)_{n \in \mathbb{N}}$ diverges, but it does not diverge to $+\infty$ or $-\infty$.
**E.** The sequence of the Fibonacci numbers diverges to $+\infty$.

**Theorem 16.11.** *Let* $(a_n)_{n \geq k}$ *be a sequence diverging to* $+\infty$ *or* $-\infty$. *Then there exists* $n_0 \in \mathbb{N}$ *such that for all* $n \geq n_0$ *we have* $a_n \neq 0$ *and the sequence* $\left( \frac{1}{a_n} \right)_{n \geq n_0}$ *converges to 0.*

*Proof.* Suppose that $\lim_{n \to \infty} a_n = +\infty$. There exists $n_0 \geq k$ such that $a_n > 0$ for all $n \geq n_0$. In particular we have $a_n \neq 0$ for $n \geq n_0$. Now, given $\varepsilon > 0$, there exists $n_1$ such that $a_n > \frac{1}{\varepsilon}$ for $n \geq n_1$ which implies $\left| \frac{1}{a_n} \right| < \varepsilon$ for $n \geq \max\{n_0, n_1\}$. The other case is shown in an analogous way, or by considering the sequence $(-a_n)_{n \geq k}$. $\square$

**Theorem 16.12.** *Let* $(a_n)_{n \geq k}$ *be a sequence of positive (negative) real numbers such that* $\lim_{n \to \infty} a_n = 0$. *Then the sequence* $\left( \frac{1}{a_n} \right)_{n \geq k}$ *diverges to* $+\infty$ *(or* $-\infty$).

*Proof.* We only handle the case $a_n > 0$ for all $n \geq k$. Let $K > 0$ be given. Since $\lim_{n \to \infty} a_n = 0$ there exists $N \in \mathbb{N}$ such that $|a_n| < \varepsilon := \frac{1}{K}$ for $n \geq N$. Hence

$$\frac{1}{a_n} = \frac{1}{|a_n|} > \frac{1}{\varepsilon} = K \quad \text{for } n \geq N,$$

i.e. $\lim_{n \to \infty} \frac{1}{a_n} = +\infty$. The case $a_n < 0$ follows in a similar way. $\square$

**Example 16.13.** Using Example 15.5.E we find that

$$\lim_{n \to \infty} \frac{2^n}{n} = +\infty.$$

Let us now return to series. Consider a sequence $(a_n)_{n \geq k}$ of non-negative real numbers. The corresponding sequence of partial sums $(s_n)_{n \geq k}$, $s_n = \sum_{l=k}^{n} a_l$, has the property that $m > n$ implies $s_m \geq s_n$ since

$$\sum_{l=k}^{m} a_l = \sum_{l=k}^{n} a_l + \sum_{l=n+1}^{m} a_l \geq \sum_{l=k}^{n} a_l.$$

Suppose that there exists $\kappa > 0$ such that infinitely many $a_l$ satisfy $a_l \geq \kappa$. We claim that in this case $(s_n)_{n \geq k}$ diverges to $+\infty$. Indeed, given $K > 0$ we can find $N_0 \in \mathbb{N}$ such that $\kappa N_0 > K$. Since $a_l \geq \kappa$ for infinitely many $l \geq k$ there exists $N_1 \in \mathbb{N}$ such that in the set $\{a_k, \ldots, a_{N_1}\}$ at least $N_0$ elements satisfy $a_l \geq \kappa$. We introduce the set

$$M(N_0, N_1) := \{l \in \mathbb{N} | l \leq N_1 \text{ and } a_l \geq \kappa\}.$$

For $n \geq N_1$ it follows that

$$\sum_{l=k}^{n} a_l \geq \sum_{l \in M(N_0, N_1)} a_l \geq \sum_{l \in M(N_0, N_1)} \kappa \geq N_0 \kappa > K.$$

Here we used the fact that $M(N_0, N_1)$ has at least $N_0$ elements. The notation $\sum_{l \in M(N_0, N_1)} a_l$ is almost self-explaining: the summation is over all elements of $M(N_0, N_1)$, i.e. we sum up all $a_l$ with $l \in M(N_0, N_1)$.

Therefore for a series $\sum_{l=k}^{\infty} a_l$ of non-negative numbers to converge the following must hold: for every $\epsilon > 0$ there exists $N(\epsilon) \in \mathbb{N}$ such that $n \geq N(\epsilon)$ implies $a_n = |a_n| < \epsilon$, i.e. $\lim_{n \to \infty} a_n = 0$.

Observe that the following two new concepts arose in the considerations above:

- monotonicity of a sequence: $m > n$ implies $a_m \geq a_n$;

- selecting a subsequence: for infinitely many $l$ we have $a_l \geq \kappa$, in other words we can find a sequence of integers $l_j, j \in \mathbb{N}, l_j \geq k$, such that $a_{l_j} \geq \kappa$, i.e. $(a_{l_j})_{j \in \mathbb{N}}$ is a new sequence whose elements are elements of the sequence $(a_l)_{l \in \mathbb{N}}$ and $a_{l_j} \geq \kappa$ for all $j \in \mathbb{N}$.

In the next chapter we will investigate these issues in more detail.

# Problems

1. Let $S_n := \frac{n(n+1)(2n+1)}{6}, n \in \mathbb{N}$, be the $n^{\text{th}}$ partial sum of a sequence $(a_n)_{n \in \mathbb{N}}$. Find $a_n$.

2. Let $a_n \leq b_n$, $n \in \mathbb{N}$, and suppose that $\sum_{n=1}^{\infty} a_n$ and $\sum_{n=1}^{\infty} b_n$ converge. Prove that $\sum_{n=1}^{\infty} a_n \leq \sum_{n=1}^{\infty} b_n$ holds.

3. Use the fact that $\frac{2}{4k^2-1} = \frac{1}{2k-1} - \frac{1}{2k+1}$ to prove that

$$\sum_{k=1}^{\infty} \frac{1}{4k^2 - 1} = \frac{1}{2}.$$

4. Find the limit of the following series:

   a) $\sum_{k=0}^{\infty} \frac{(-1)^k}{5^k}$;

   b) $\sum_{n=0}^{\infty} e^{-nx}, x < 0$;

   c) $\sum_{k=2}^{\infty} \left(\frac{4}{7}\right)^k$.

5. Find all $y \in \mathbb{R}$ for which $\sum_{k=0}^{\infty} \frac{1}{(y-2)^k}$ is a convergent geometric series. When there is convergence find the limit.

6. Series of the type $\sum_{k=1}^{\infty}(a_k - a_{k-1})$ and $\sum_{k=1}^{\infty}(a_k - a_{k+1})$ are called **telescopic series**. Prove that they converge if and only if $\lim_{k \to \infty} a_k$ exists. In this case we have

$$\sum_{k=1}^{\infty}(a_k - a_{k-1}) = \lim_{k \to \infty} a_k - a_0 \text{ and } \sum_{k=1}^{\infty}(a_k - a_{k+1}) = a_1 - \lim_{k \to \infty} a_k.$$

231

7.  a) Find $\sum_{k=0}^{\infty} \left( \frac{1}{2^k} + \frac{(-1)^k}{3^k} \right)$;

b) Under the assumption that $\lim\limits_{n \to \infty} \ln(1 + \frac{1}{n}) = 0$ show that

$$\sum_{k=1}^{\infty} \ln \left( 1 - \frac{1}{k^2} \right) = \ln \frac{1}{2};$$

c) Suppose $\sum_{k=1}^{\infty} \frac{1}{k^2} = A$. Prove that

$$\sum_{k=1}^{\infty} \frac{1}{(2k-1)^2} = \frac{3A}{4}.$$

8.  a) Prove that the sequence $a_n = \frac{n^3 + 2n^2 - 2}{15n^2 + n}$, $n \in \mathbb{N}$, diverges to $+\infty$.

b) Prove that the sequence $\left( \frac{1}{\sin \frac{1}{n}} \right)_{n \in \mathbb{N}}$ diverges to $+\infty$. Hint: for all $x \in \mathbb{R}$, we have $|\sin x| \leq |x|$.

9. Construct sequences $(a_n)_{n \in \mathbb{N}}$ and $(b_n)_{n \in \mathbb{N}}$ of real numbers such that $\lim\limits_{n \to \infty} a_n = \infty$, $\lim\limits_{n \to \infty} b_n = 0$ and

a) $\lim\limits_{n \to \infty} (a_n b_n) = +\infty$;

b) $\lim\limits_{n \to \infty} (a_n b_n) = -\infty$;

c) $\lim\limits_{n \to \infty} (a_n b_n) = c, c \in \mathbb{R}$ is a given number.

# 17   The Completeness of the Real Numbers

We want to discuss the problem of there being "gaps on the real line". Recall that the rational numbers $\mathbb{Q}$ have gaps: there is no rational number $q$ such that $q^2 = 2$. However such a number would represent the length of the diagonal of the unit square, i.e. there is a "need" for such a number to exist. There are other situations where we expect a number with certain properties to exist but we still cannot prove its existence. Consider a sequence $(a_n)_{n\in\mathbb{N}}$, $a_n \in \mathbb{R}$, such that $a_n < a_{n+1}$ for all $n \in \mathbb{N}$ and assume in addition that $a_n \leq M$ for all $n \in \mathbb{N}$.

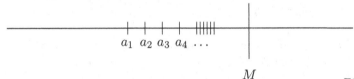

$$M$$

*Figure 17.1*

The distance between $a_n$ and $a_{n+1}$, i.e. $a_{n+1} - a_n > 0$ must become smaller and smaller. Indeed, suppose that for infinitely many $n_k \in \mathbb{N}, k \in \mathbb{N}$, we have $a_{n_k+1} - a_{n_k} \geq \eta > 0$. We claim that there must exist an $N \in \mathbb{N}$ such that $n \geq N$ implies $a_{n+1} \geq M$ which is a contradiction.

By Archimedes' axiom, given $\eta > 0$ there exists $N \in \mathbb{N}$ such that $N\eta \geq M + |a_1|$. Since $a_{n_k+1} - a_{n_k} \geq \eta$ for infinitely many $n_k$ there exists $N_1 \in \mathbb{N}$ such that for at least $N$ elements $l \in \{1, \ldots, N_1\}$, we have $a_{n_l+1} - a_{n_l} \geq \eta$. Now for $n \geq N_1$ we find

$$
\begin{aligned}
a_{n+1} - a_1 &= a_{n+1} - a_n + a_n - a_{n-1} + \cdots + a_2 - a_1 \\
&= \sum_{j=1}^{n} (a_{j+1} - a_j) \\
&\geq \sum_{l=1}^{N} (a_{n_l+1} - a_{n_l}) \geq N\eta \geq M + |a_1|
\end{aligned}
$$

or, since $|x| + x \geq 0$,

$$a_{n+1} \geq M + |a_1| + a_1 \geq M.$$

Thus we know that $a_{n+1} - a_n$ must become smaller and smaller, therefore intuitively we would expect $(a_n)_{n\in\mathbb{N}}$ to have a limit. However does such a

233

limit exist in $\mathbb{R}$?

The following definition is a more formal approach to the statement that the "distance between elements of a sequence becomes smaller and smaller".

**Definition 17.1.** *A sequence $(a_n)_{n \geq k}$ of real numbers is called a **Cauchy sequence** if for every $\epsilon > 0$ there exists $N \in \mathbb{N}$ such that $n, m \geq N$ implies $|a_n - a_m| < \epsilon$.*

**Remark 17.2.** Note that the condition $|a_n - a_m| < \epsilon$ for $n, m \geq N$ is equivalent to $|a_{n+k} - a_n| < \epsilon$ for $n \geq N$ and $k \in \mathbb{N}$.

**Proposition 17.3. A.** *Every convergent sequence is a Cauchy sequence.* **B.** *Every Cauchy sequence is bounded.*

*Proof.* Suppose that $(a_n)_{n \geq k}$ converges to $a$. Given $\epsilon > 0$ we can find $N \in \mathbb{N}$ such that $|a_n - a| < \frac{\epsilon}{2}$ for all $n \geq N$. Thus for $n, m \geq N$ we get

$$|a_n - a_m| + |(a_n - a) - (a_m - a)| \leq |a_n - a| + |a_m - a| < \frac{\epsilon}{2} + \frac{\epsilon}{2} = \epsilon.$$

Now let $(a_n)_{n \geq k}$ be a Cauchy sequence. For $\epsilon = 1$ there exists $N \in \mathbb{N}$ such that $n \geq N$ implies

$$|a_n| - |a_N| \leq |a_n - a_N| < 1$$

which yields for all $l \geq k$

$$|a_l| \leq \max\{1 + |a_N|, |a_k|, \ldots, |a_{N-1}|\}.$$

$\square$

It is quite a difficult problem to decide whether every Cauchy sequence converges. In fact it is impossible to deduce from the axioms for $\mathbb{R}$ in the way we have listed them so far that every Cauchy sequence has a limit in $\mathbb{R}$. Therefore we add the following axiom:

**Axiom of Completeness.** In $\mathbb{R}$ every Cauchy sequence has a limit.

Although we can derive some very important results from this axiom immediately, we of course need to justify the axiom. In Appendix VI we see that the axiom is equivalent to the statement that every set of real numbers bounded from above has a least upper bound - a statement which is more intuitive. In addition we show in Appendix VI how (starting with $\mathbb{Q}$) to construct an ordered field satisfying Archimedes' axiom, the completeness axiom and into which we can embed $\mathbb{Q}$ as a dense subset.

Next we continue by proving some of the important consequences of the completeness axiom. We start with the famous **Theorem of Bolzano and Weierstrass**. We first need to consider the following definition

**Definition 17.4.** *Given a sequence $(a_n)_{n \geq k}$ let $k \leq n_1 < n_2 < n_3 < \dots$ be a strictly increasing sequence of integers $(n_l)_{l \in \mathbb{N}}$. We call the sequence $(a_{n_l})_{l \in \mathbb{N}} = (a_{n_1}, a_{n_2}, \dots)$ a **subsequence** of $(a_n)_{n \geq k}$.*

**Example 17.5.** Consider the sequence $((-1)^n)_{n \in \mathbb{N}}$. The sequence $((-1)^{2k})_{k \in \mathbb{N}}$ and the sequence $((-1)^{2k+1})_{k \in \mathbb{N}}$ are subsequences of $((-1)^n)_{n \in \mathbb{N}}$. Indeed, in the first case we take $n_k = 2k$, in the second case we have $n_k = 2k + 1$. Note that $(-1)^{2k} = 1$ and $(-1)^{2k+1} = -1$, i.e. $((-1)^{2k})_{k \in \mathbb{N}} = (1, 1, \dots)$, and $((-1)^{2k+1})_{k \in \mathbb{N}} = (-1, -1, \dots)$. Thus while $((-1)^n)_{n \in \mathbb{N}}$ is divergent, each of the two subsequences is convergent, but they have different limits.

**Theorem 17.6 (Bolzano-Weierstrass).** *Every bounded sequence $(a_n)_{n \in \mathbb{N}_0}$ in $\mathbb{R}$ has a convergent subsequence.*

*Proof.* We proceed in three steps.
**1.** Since $(a_n)_{n \in \mathbb{N}_0}$ is bounded there exist numbers $A, B \in \mathbb{R}$ such that $A \leq a_n \leq B$ for all $n \in \mathbb{N}_0$. Let us consider the interval $[A, B] := \{x \in \mathbb{R}; A \leq x \leq B\}$. We will use the principle of mathematical induction to construct a sequence $[A_k, B_k]$, $k \in \mathbb{N}_0$, of intervals having the following properties:
**i)** In each interval $[A_k, B_k]$ there are infinitely many elements of $(a_n)_{n \in \mathbb{N}_0}$;
**ii)** $[A_k, B_k] \subset [A_{k-1}, B_{k-1}]$, $k \geq 1$;
**iii)** $B_k - A_k = 2^{-k}(B - A)$.
We start by setting $A_0 = A$ and $B_0 = B$. Now suppose that $[A_k, B_k]$ is already constructed and has the properties i)-iii). Let $M := \frac{A_k + B_k}{2}$ be the centre of the interval. Since in $[A_k, B_k]$ there are infinitely many elements of $(a_n)_{n \in \mathbb{N}_0}$, either $[A_k, M]$ or $[M, B_k]$ (or both) must contain infinitely many elements of $(a_n)_{n \in \mathbb{N}_0}$. Now we define the interval

$$[A_{k+1}, B_{k+1}] := \begin{cases} [A_k, M], & \text{if } [A_k, M] \text{ contains infinitely many} \\ & \text{elements of } (a_n)_{n \in \mathbb{N}_0} \\ [M, B_k], & \text{otherwise.} \end{cases}$$

Obviously $[A_{k+1}, B_{k+1}]$ satisfies i)-iii).
**2.** Now we define inductively a subsequence $(a_{n_k})_{k \in \mathbb{N}_0}$ of $(a_n)_{n \in \mathbb{N}_0}$ such that

$a_{n_k} \in [A_k, B_k]$ for all $k \in \mathbb{N}$. We start with $a_{n_0} := a_0$.

Now suppose that $a_{n_k}$ is already defined. Since in $[A_{k+1}, B_{k+1}]$ there are infinitely many elements of $(a_n)_{n \in \mathbb{N}}$, we may choose some $a_{n_{k+1}} \in [A_{k+1}, B_{k+1}]$ such that $a_{n_{k+1}}$ is an element of the original sequence and $n_{k+1} > n_k$.

**3.** Finally we prove that $(a_{n_k})_{k \in \mathbb{N}_0}$ is a Cauchy sequence. For this let $\varepsilon > 0$ be given and take $N \in \mathbb{N}$ such that $2^{-N}(B - A) < \varepsilon$. For all $k, j \in \mathbb{N}$, $k, j \geq N$, we find

$$a_{n_k} \in [A_k, B_k] \subset [A_N, B_N]$$

$$a_{n_j} \in [A_j, B_j] \subset [A_N, B_N],$$

thus

$$|a_{n_k} - a_{n_j}| \leq |B_N - A_N| = 2^{-N}(B - A) < \varepsilon$$

and we are done. $\qquad\qquad\qquad\qquad\qquad\qquad\qquad\qquad\qquad\qquad \square$

**Remark 17.7.** Clearly the Bolzano-Weierstrass theorem also holds for sequences $(a_n)_{n \geq k}$.

**Definition 17.8. A.** *A number $a \in \mathbb{R}$ is called an **accumulation point** or a **cluster point** or a **limit point** of a sequence $(a_n)_{n \geq k}$ if there exists a subsequence $(a_{n_l})_{l \in \mathbb{N}}$ of $(a_n)_{n \geq k}$, $a_{n_l} \neq a$ converging to $a$, i.e. $\lim_{l \to \infty} a_{n_l} = a$.*
**B.** *A point $a \in \mathbb{R}$ is an **accumulation point** of $B \subset \mathbb{R}$ if there exists a sequence $(b_n)_{n \in \mathbb{N}}$, $b_n \in B$, $b_n \neq a$, converging to $a$.*

**Example 17.9. A.** The sequence $((-1)^n)_{n \in \mathbb{N}}$ has two accumulation points, namely $+1$ and $-1$, compare with Example 17.5. Thus while the limit of a sequence is always unique, a sequence may have a lot of (even infinitely many) accumulation points.
**B.** The sequence $a_n = (-1)^n + \frac{1}{n}$, $n \in \mathbb{N}$, also has the two accumulation points $+1$ and $-1$. Indeed we have

$$\lim_{n \to \infty} a_{2n} = \lim_{n \to \infty} \left( (-1)^{2n} + \frac{1}{2n} \right) = \lim_{n \to \infty} \left( 1 + \frac{1}{2n} \right) = 1,$$

and

$$\lim_{n \to \infty} a_{2n+1} = \lim_{n \to \infty} \left( (-1)^{2n+1} + \frac{1}{2n+1} \right) = \lim_{n \to \infty} \left( -1 + \frac{1}{2n+1} \right) = -1.$$

**C.** The sequence $a_n = n$ has no accumulation point since each of its subsequences is unbounded.

**D.** Consider the sequence

$$a_n = \begin{cases} n^2 & \text{for } n \text{ even} \\ \frac{1}{n} & \text{for } n \text{ odd} \end{cases}.$$

It is unbounded but has one accumulation point, namely 0 since $\lim\limits_{n\to\infty} a_{2n-1} = \lim\limits_{n\to\infty} \frac{1}{2n-1} = 0$.

**Lemma 17.10. A.** *If $(a_n)_{n\geq k}$ converges, then the limit is the only accumulation point of $(a_n)_{n\geq k}$, i.e. every subsequence of a converging sequence converges to the same limit.*
**B.** *If a subsequence of a Cauchy sequence converges, the whole sequence converges (to the same limit).*

*Proof.* Part A is obvious. B. Let $(a_k)_{k\in\mathbb{N}}$ be a Cauchy sequence and suppose that $(a_{k_l})_{l\in\mathbb{N}}$ converges to $a$. For $\epsilon > 0$ there exists $N_1 \in \mathbb{N}$ such that $l \geq N_1$ implies $|a_{k_l} - a| < \frac{\epsilon}{2}$. Since $(a_{k_l})$ is a Cauchy sequence there exists $N_2 \in \mathbb{N}$ such that $n, m \geq N_2$ implies $|a_n - a_m| < \frac{\epsilon}{2}$. Thus for $l \geq N_1$ and $n_l \geq N_2$ we find for all $n \geq N_2$

$$|a_n - a| \leq |a_n - a_{n_l}| + |a_{n_l} - a| < \epsilon.$$

$\square$

**Definition 17.11.** *Let $(a_n)_{n\geq k}$ be a sequence of real numbers. We call $(a_n)_{n\geq k}$*
**monotone increasing** *if $a_n \leq a_{n+1}$ for all $n \geq k$*
**strictly monotone increasing** *if $a_n < a_{n+1}$ for all $n \geq k$*
**monotone decreasing** *if $a_n \geq a_{n+1}$ for all $n \geq k$*
**strictly monotone decreasing** *if $a_n > a_{n+1}$ for all $n \geq k$*

**Remark 17.12.** We call $(a_n)_{n\geq k}$ just **monotone** if one of the four conditions of Definition 17.11 holds.

**Example 17.13. A.** The sequence $a_n = \frac{1}{n}$ is strictly monotone decreasing.
**B.** The Fibonacci sequence is increasing but not strictly increasing.
**C.** The sequence $((-1)^n)_{n\in\mathbb{N}}$ is neither monotone increasing nor decreasing.
**D.** If $(a_n)_{n\in\mathbb{N}}$ is a sequence of positive numbers $a_n > 0$ then the sequence of partial sums $s_n = \sum_{k=1}^{n} a_n$ is strictly monotone increasing.

The next result resolves one of the problems discussed at the beginning of this chapter.

**Theorem 17.14.** *Every monotone and bounded sequence $(a_n)_{n \geq k}$ is convergent.*

*Proof.* We know that $(a_n)_{n \geq k}$ is bounded, hence by the Bolzano-Weierstrass theorem it has a convergent subsequence $(a_{n_k})_{k \in \mathbb{N}}$ and we denote its limit by $a$. We will prove that the whole sequence $(a_n)_{n \geq k}$ converges to $a$. For this let $\varepsilon > 0$ be given and $l_0 \in \mathbb{N}$ such that

$$|a_{n_l} - a| < \varepsilon \quad \text{for all } l \geq l_0.$$

Set $N := n_{l_0}$ then for every $n \geq N$ there exists $l \geq l_0$ such that $n_l \leq n < n_{l+1}$. If $(a_n)_{n \geq k}$ is monotone increasing (decreasing) it follows that

$$a_{n_l} \leq a_n \leq a_{n_{l+1}} \quad (a_{n_l} \geq a_n \geq a_{n_{l+1}}).$$

In either case we find that

$$|a_n - a| \leq \max \left( |a_{n_l} - a|, |a_{n_{l+1}} - a| \right) < \varepsilon$$

which proves the theorem.

$\square$

Next we introduce the principle of nested intervals. Let $I_n := [A_n, B_n]$, $n \in \mathbb{N}_0$, be a family of non-empty (and non-degenerate) intervals $[A_n, B_n] = \{x \in \mathbb{R}; A_n \leq x \leq B_n\}$ with length $l_n = B_n - A_n > 0$.
Suppose that
**i)** $I_{n+1} \subset I_n$, i.e. $A_n \leq A_{n+1} < B_{n+1} \leq B_n$
**ii)** $\lim\limits_{n \to \infty} l_n = 0$.
Such a family $(I_n)_{n \in \mathbb{N}_0}$ is called a family of **nested intervals**. Let us look at the intersection of these intervals

$$I := \bigcap_{n \in \mathbb{N}_0} I_n := \{x \in \mathbb{R} | \quad x \in I_n \text{ for all } n \in \mathbb{N}\}.$$

**Theorem 17.15 (Principle of nested intervals).** *Let $(I_n)_{n \in \mathbb{N}_0}$ be a family of nested intervals. Then there exists exactly one point $x_0 \in I$, i.e. $I = \{x_0\}$.*

*Proof.* Since $A_n \leq A_{n+1} < B_{n+1} \leq B_n$ it follows that the sequence of left end points $(A_n)_{n \in \mathbb{N}_0}$ as well as the sequence of right end points $(B_n)_{n \in \mathbb{N}_0}$ are bounded. Since each of these sequences is monotone it is convergent. Denote their limits by $A$ and $B$, respectively. Clearly we have $A \leq B$. If $A = B$ we are done. Suppose that $A < B$. Then $[A, B] \subset \bigcap_{n \in \mathbb{N}_0} I_n$ and there exists $x_0 \in [A, B]$ such that $A < x_0 < B$. But in this case $A_n < x_0 < B_n$ implying

$$0 < x_0 - A_n < B_n - A_n$$

and

$$A_n - B_n < -B_n + x_0 < 0$$

leading to

$$0 \leq x_0 - \lim_{n \to \infty} A_n \leq \lim_{n \to \infty} (B_n - A_n) = 0$$

and

$$0 = \lim_{n \to \infty} (A_n - B_n) \leq - \lim_{n \to \infty} B_n + x_0 \leq 0,$$

i.e. $x_0 = \lim_{n \to \infty} A_n$ and $x_0 = \lim_{n \to \infty} B_n$, i.e. $A = B$, contradicting the assumption.
$\square$

**Remark 17.16.** It is clear that the principle of nested intervals can be formulated and proved for a sequence $(I_n)_{n \geq k}$.

The proof of the Bolzano-Weierstrass theorem requires the axiom of completeness, hence all other results in this chapter do. The following example shows that we can use the axiom of completeness to find a number $x$ in $\mathbb{R}$ such that $x^2 = 2$.

**Example 17.17.** Let $a > 0$ and $x_0 > 0$ be two real numbers. We define the sequence $(x_n)_{n \in \mathbb{N}_0}$ by

$$x_0 := x_0 \quad \text{and} \quad x_{n+1} := \frac{1}{2}\left(x_n + \frac{a}{x_n}\right).$$

The sequence $(x_n)_{n \in \mathbb{N}_0}$ converges to $a$ and $a$ is the unique positive solution of the equation $x^2 = a$.

We show this result using the following steps.

1. For all $n$ we have $x_n > 0$.
   Indeed, $x_0 > 0$ by assumption and if $x_n > 0$, so is $x_{n+1} = \frac{1}{2}(x_n + \frac{a}{x_n})$.

2. For all $n \geq 1$ we have $x_n^2 \geq a$.

   For this note that

$$x_n^2 - a = \frac{1}{4}(x_{n-1} + \frac{a}{x_{n-1}})^2 - a$$

$$= \frac{1}{4}(x_{n-1}^2 + 2a + \frac{a^2}{x_{n-1}^2}) - a$$

$$= \frac{1}{4}(x_{n-1} - \frac{a}{x_{n-1}})^2 \geq 0.$$

3. For $n \geq 1$ we also have $x_{n+1} \leq x_n$, i.e. the sequence is monotone decreasing, since

$$x_n - x_{n+1} = x_n - \frac{1}{2}(x_n + \frac{a}{x_n}) = \frac{1}{2x_n}(x_n^2 - a) \geq 0$$

   note that $x_n > 0$ and $x_n^2 \geq a$.

4. We conclude that $(x_n)_{n \in \mathbb{N}}$ is a monotone decreasing sequence satisfying $0 \leq x_n \leq x_1$, i.e. it is bounded. Hence it is convergent by Theorem 17.14 and the limit $x$ of $(x_n)_{n \in \mathbb{N}_0}$ satisfies $0 \leq x \leq x_1$.

5. Applying the rules for convergent sequences to the equation

$$x_{n+1} = \frac{1}{2}(x_n + \frac{a}{x_n}),$$

   we obtain

$$x = \frac{1}{2}(x + \frac{a}{x})$$

   i.e. $x^2 = a$. Since $x \geq 0$, $x$ is the positive solution to $x^2 = a$.

**Example 17.18.** Suppose that $a = 2$ in Example 17.17. Starting with $x_0 = 1$, we obtain the sequence: $1, \frac{3}{2}, \frac{17}{12}, \frac{577}{408}, \ldots$ which converges rapidly to $\sqrt{2}$. Note that all the terms of the sequence are rational, but the limit is not.

The following example shows why we cannot just take the limits of the defining equation.

**Example 17.19.** Define a sequence by $x_0 = 2$ and $x_{n+1} = 2x_n - \frac{2}{x_n}$.

If the sequence has a limit $x$ we obtain $x = 2x - \frac{2}{x}$ or $x^2 = 2$. Since all the terms are positive, this would be the positive square root of 2 as before. However, the sequence is $2, 3, 5\frac{1}{3}, 9\frac{7}{15}, \ldots$ which is an increasing sequence and unbounded, as we can prove by induction.

## Problems

1.    a) Consider the sequence $(s_n)_{n \in \mathbb{N}}$, $s_n := \sum_{j=1}^{n} \frac{1}{j}$. Prove that $s_{2n} - s_n > \frac{1}{2}$ and deduce that $(s_n)_{n \in \mathbb{N}}$ diverges to $+\infty$.

   b) Prove that the sequence $(s_n)_{n \in \mathbb{N}}$, $s_n := \sum_{j=1}^{n} \frac{(-1)^{j+1}}{j}$, is a Cauchy sequence.

2. Let $(a_n)_{n \in \mathbb{N}}$ be a sequence of real numbers such that for all $n \geq N$ we have $|a_n - a_{n+1}| < 2^{-n}$. Prove that $(a_n)_{n \in \mathbb{N}}$ is a Cauchy sequence.

3. Let $(a_n)_{n \geq k}$, $(b_n)_{n \geq k}$, and $(c_n)_{n \geq k}$ be sequences of real numbers such that $\lim_{n \to \infty} a_n = a$, $\lim_{n \to \infty} b_n = b$ and for all $n \geq k$ we have $a_n \leq c_n \leq b_n$. Prove that $(c_n)_{n \geq k}$ has a convergent subsequence.

4. Given the sequence $\left( \frac{\sqrt{n}}{n+1} \right)_{n \in \mathbb{N}}$. Show that this is a bounded decreasing sequence and deduce that its limit exists.

5. Consider the sequence $\left( \sum_{k=0}^{n} \frac{1}{k!} \right)_{n \geq 0}$. Prove that this sequence is bounded and deduce that it must have the limit $\sum_{k=0}^{\infty} \frac{1}{k!}$.
   Hint: first show that $k! \geq 2^{k-1}$ for $k \in \mathbb{N}$.

6. Let $(a_n)_{n \in \mathbb{N}}$, $a_n \geq 0$, be a sequence and assume that $(a_n)_{n \in \mathbb{N}}$ has no accumulation points. Prove that $\lim_{n \to \infty} a_n = \infty$.

7. Give an example of a sequence $(a_n)_{n \in \mathbb{N}}$ such that $-2, \frac{1}{3}, 17$ are accumulation points of $(a_n)_{n \in \mathbb{N}}$ and $-3 \leq a_n \leq 19$ for all $n \in \mathbb{N}$.

8.* Let $a > 0$, $k \in \mathbb{N}$ and $x_0^k > a, x_0 > 0$. Define

$$ x_{n+1} := x_n - \frac{x_n^k - a}{k x_n^{k-1}} = \frac{(k-1)x_n^k + a}{k x_n^{k-1}}, n \in \mathbb{N}_0. $$

   Prove that $\lim_{n \to \infty} x_n = a^{\frac{1}{k}} = \sqrt[k]{a}$. Hint: use the following steps:

   i) $x_n > 0$ for all $n \in \mathbb{N}$;

   ii) $- \left( \frac{x_n^k - a}{k x_n^k} \right) \geq 1$;

241

iii) by using Bernoulli's inequality prove that

$$\left(x_n - \frac{x_n^k - a}{k x_n^{k-1}}\right)^k \geq a;$$

iv) $x_n^k \geq a$;

v) $x_{n+1} \leq x_n$.

9.* Prove that the sequence $(a_n)_{n \in \mathbb{N}}$, $a_n = \left(1 + \frac{1}{n}\right)^n$, has the limit $\sum_{j=1}^{\infty} \frac{1}{j!}$. We denote the limit by $e$ where $e$ is called the **Euler number**.

10. Let $a_n = \left(1 + \frac{1}{n}\right)^n$ and $b_n = \left(1 + \frac{1}{n}\right)^{n+1}$. Prove that $([a_n, b_n])_{n \in \mathbb{N}}$ are nested intervals with $\{e\} = \bigcap_{n \in \mathbb{N}} [a_n, b_n]$, and $e$ is the Euler number.

# 18 Convergence Criteria for Series, b-adic Fractions

Our new understanding of the completeness of the real line, in particular the concept of a Cauchy sequence, gives us new tools to handle series. We formulate our first results for sequences $(a_n)_{n \geq k}$. We will soon switch to sequences $(a_n)_{n \in \mathbb{N}}$ or $(a_n)_{n \in \mathbb{N}_0}$, but extending results to the case $(a_n)_{n \geq k}$ is straightforward. We start by formulating the **Cauchy criterion for series**.

**Theorem 18.1.** *Given a sequence $(a_n)_{n \geq k}$ of real numbers. The series $\sum\limits_{n=1}^{\infty} a_n$ converges if and only if for every $\varepsilon > 0$ there exists $N = N(\varepsilon) \in \mathbb{N}$ such that $n \geq m \geq N$ implies*

$$\left| \sum_{k=m}^{n} a_k \right| < \varepsilon. \tag{18.1}$$

*Proof.* Let $s_p := \sum\limits_{l=k}^{p} a_l$ be the $p^{\text{th}}$ partial sum. It follows that

$$s_n - s_{m-1} = \sum_{l=m}^{n} a_l,$$

and the criterion is nothing but the statement that the sequence of partial sums forms a Cauchy sequence. $\square$

**Theorem 18.2.** *If the series $\sum\limits_{l=k}^{\infty} a_l$ converges then $\lim\limits_{l \to \infty} a_l = 0$.*

*Proof.* If $\sum\limits_{l=k}^{\infty} a_l$ converges, then by Theorem 18.1, for every $\varepsilon > 0$ it follows that $\left| \sum\limits_{l=m}^{n} a_l \right| < \varepsilon$ provided $n \geq m \geq N$ for some suitable $N \in \mathbb{N}$. Putting $n = m$ we find that $|a_l| < \varepsilon$ for all $n \geq N$, i.e. $\lim\limits_{l \to \infty} a_l = 0$. $\square$

**Example 18.3. A.** For $|q| < 1$ we know that $\sum_{k=0}^{\infty} q^k = \frac{1}{1-q}$, i.e. the series converges. Moreover, by Example 15.11.A we know that $\lim\limits_{k \to \infty} q^k = 0$.

**B.** The series $\sum\limits_{k=1}^{\infty} (-1)^k$ diverges since the sequence $((-1)^k)_{k \in \mathbb{N}}$ does not converge to 0.

243

**Theorem 18.4.** *Let* $\sum_{l=k}^{\infty} a_l$ *be a series of non-negative numbers* $a_l \geq 0$. *This series converges if and only if it is bounded, i.e. the sequence of its partial sums is bounded.*

*Proof.* Since $a_l \geq 0$ for all $l \geq k$ the sequence of partial sums $s_p = \sum_{l=1}^{p} a_l$ is monotone increasing and bounded, hence by Theorem 17.14 it is convergent. Conversely, if $\sum_{l=1}^{\infty} a_l$ is convergent the corresponding sequence of partial sums must be bounded. $\qquad\square$

**Example 18.5.** The **harmonic series** $\sum_{n=1}^{\infty} \frac{1}{n}$ diverges.

Referring to Problem 1 a) in Chapter 17, we may argue that $(s_n)_{n\geq 1}$ is not a Cauchy sequence, hence it cannot converge. We give here a further proof by showing that the partial sums are unbounded. Consider the special partial sums

$$s_{2^{k+1}} := \sum_{n=1}^{2^{k+1}} \frac{1}{n} = 1 + \frac{1}{2} + \sum_{p=1}^{k} \left( \sum_{n=2^p+1}^{2^{p+1}} \frac{1}{n} \right)$$

$$= 1 + \left( \frac{1}{2} \right) + \left( \frac{1}{3} + \frac{1}{4} \right) + \left( \frac{1}{5} + \frac{1}{6} + \frac{1}{7} + \frac{1}{8} \right) +$$

$$\cdots + \left( \sum_{n=2^k+1}^{2^{k+1}} \frac{1}{n} \right).$$

Each of the terms in brackets is larger than $\frac{1}{2}$. Indeed we have $2^p$ terms to add in the sum $\sum_{n=2^p+1}^{2^{p+1}} \frac{1}{n}$, the smallest of which is $\frac{1}{2^{p+1}}$, hence

$$\sum_{n=2^p+1}^{2^{p+1}} \frac{1}{n} \geq 2^p \cdot \frac{1}{2^{p+1}} = \frac{1}{2}.$$

Therefore we find $s_{2^{k+1}} \geq 1 + \frac{k}{2}$ implying that the partial sums are unbounded and so $\sum_{n=1}^{\infty} \frac{1}{n}$ is divergent.

**Remark 18.6.** Note that $\sum_{n=1}^{\infty} \frac{1}{n}$ is an example of a divergent series $\sum_{l=k}^{\infty} a_l$ with $\lim_{l\to\infty} a_l = 0$. Hence the converse of Theorem 18.2 does not hold.

**Example 18.7.** For all $k \in \mathbb{N}$, $k > 1$, the series $\sum_{n=1}^{\infty} \frac{1}{n^k}$ converges. To see this we apply Theorem 18.4 and prove the boundedness of $\sum_{n=1}^{\infty} \frac{1}{n^k}$ for $k > 1$. For $p \in \mathbb{N}$ such that $N \leq 2^{p+1} - 1$ we find

$$s_N := \sum_{n=1}^{N} \frac{1}{n^k} \leq \sum_{n=1}^{2^{p+1}-1} \frac{1}{n^k}$$

$$= 1 + \left( \frac{1}{2^k} + \frac{1}{3^k} \right) + \cdots + \left( \sum_{n=2^p}^{2^{p+1}-1} \frac{1}{n^k} \right)$$

$$\leq \sum_{q=1}^{p} 2^q \frac{1}{(2^q)^k} = \sum_{q=1}^{p} \left( \frac{1}{2^{k-1}} \right)^q$$

$$\leq \sum_{q=1}^{\infty} (2^{-k+1})^q = \frac{1}{1 - 2^{-k+1}} = \frac{2^{k-1}}{2^{k-1} - 1}.$$

The next result is useful when dealing with **alternating series**, i.e. series in which consecutive terms change sign.

**Theorem 18.8 (Leibniz's criterion for alternating series).** *Let $(a_n)_{n \in \mathbb{N}}$ be a monotone decreasing sequence of non-negative real numbers with $\lim_{n \to \infty} a_n = 0$. Then the series $\sum_{n=1}^{\infty} (-1)^n a_n$ converges.*

*Proof.* Set $s_k := \sum_{n=1}^{k} (-1)^n a_n$. Since

$$s_{2k+2} - s_{2k} = -a_{2k+1} + a_{2k+2} \leq 0$$

it follows that

$$s_0 \geq s_2 \geq s_4 \geq \cdots \geq s_{2k+2} \geq \cdots$$

and analogously, since

$$s_{2k+3} - s_{2k+1} = a_{2k+2} - a_{2k+3} \geq 0$$

we find

$$s_1 \leq s_2 \leq s_3 \leq \cdots \leq s_{2k+3} \leq \cdots$$

In addition we have

$$s_{2k+1} \leq s_{2k}$$

since $s_{2k+1} - s_{2k} = -a_{2k+1} \leq 0$.

The sequence $(s_{2k})_{k \in \mathbb{N}}$ is monotone decreasing and bounded since $s_{2k} \geq s_1$. By Theorem 17.14 it is convergent, hence

$$\lim_{k \to \infty} s_{2k} = S$$

for some $S \in \mathbb{R}$. Analogously we see that $(s_{2k+1})_{k \in \mathbb{N}}$ is monotone increasing and bounded, hence convergent:

$$\lim_{k \to \infty} s_{2k} = S'.$$

Further we find

$$S - S' = \lim_{k \to \infty} (s_{2k} - s_{2k-1}) = \lim_{k \to \infty} a_{2k+1} = 0,$$

i.e. $S = S'$. Now we prove $\sum_{k=1}^{\infty} (-1)^k a_k = S$. For this let $\varepsilon > 0$ be given. Then there exists $N_1(\varepsilon), N_2(\varepsilon) \in \mathbb{N}$ such that $|s_{2k} - S| < \varepsilon$ for $k \geq N_1$, $|s_{2k+1} - S| < \varepsilon$ for $k \geq N_2$. Thus for $k \geq \max(2N_1, 2N_2 + 1)$ we find

$$|s_k - S| < \varepsilon.$$

$\square$

**Example 18.9. A.** The **alternating harmonic** series $\sum_{k=1}^{\infty} \frac{(-1)^k}{k}$ converges. (Also compare with Problem 1 b) in Chapter 17.) **B.** The series $\sum_{k=0}^{\infty} \frac{(-1)^k}{2k+1}$ converges.

**Definition 18.10.** *A series $\sum_{k=1}^{\infty} a_k$ of real numbers is called **absolutely convergent** if the series $\sum_{k=1}^{\infty} |a_k|$ converges.*

**Theorem 18.11.** *Any absolutely convergent series is convergent.*

246

*Proof.* Suppose $\sum\limits_{k=1}^{\infty} a_k$ is absolutely convergent. According to the Cauchy criterion applied to the series $\sum\limits_{k=1}^{\infty} |a_k|$, for $\varepsilon > 0$ there exists a number $N(\varepsilon)$ such that $n \geq m \geq N$ implies

$$\sum_{k=m}^{n} |a_k| < \varepsilon.$$

Now the triangle inequality yields for $n \geq m \geq N$

$$\left| \sum_{k=m}^{n} a_k \right| \leq \sum_{k=m}^{n} |a_k| < \varepsilon,$$

i.e. the Cauchy criterion holds for $\sum\limits_{k=1}^{\infty} a_k$ which implies the convergence of $\sum\limits_{k=1}^{\infty} a_k$ by Theorem 18.1.    □

**Remark 18.12.** The alternating harmonic series shows that the converse of Theorem 18.11 is not true: a convergent series need not be absolutely convergent. Convergent series which are not absolutely convergent are sometimes called conditionally convergent.

**Theorem 18.13 (Comparison test).** *Let $\sum\limits_{k=1}^{\infty} c_k$ be a convergent series of non-negative real numbers $c_k \geq 0$. Further let $(a_k)_{k \in \mathbb{N}}$ be a sequence such that $|a_k| \leq c_k$ for all $k \in \mathbb{N}$. Then the series $\sum\limits_{k=1}^{\infty} a_k$ converges absolutely.*

*Proof.* Given $\varepsilon > 0$ there exists $N(\varepsilon) \in \mathbb{N}$ such that

$$\left| \sum_{k=m}^{n} c_k \right| = \sum_{k=m}^{n} c_k < \varepsilon \quad \text{for } n \geq m \geq N.$$

Therefore we find

$$\sum_{k=m}^{n} |a_k| \leq \sum_{k=m}^{n} c_k < \varepsilon \quad \text{for } n \geq m \geq N,$$

which proves the theorem.    □

The next two tests, the ratio test and the root test are very powerful tools. We will use these tests in this part and later on when dealing with power series, and also in the chapter on complex analysis.

**Theorem 18.14 (Ratio test).** *Let* $\sum\limits_{n=0}^{\infty} a_n$ *be a series such that* $a_n \neq 0$ *for all* $n \geq N_0$. *Suppose that there exists* $\nu$, $0 < \nu < 1$, *such that*

$$\left| \frac{a_{n+1}}{a_n} \right| \leq \nu \quad \text{for all } n \geq N_0.$$

*Then the series* $\sum\limits_{n=0}^{\infty} a_n$ *converges absolutely.*

*Proof.* The convergence of the series $\sum\limits_{n=0}^{\infty} a_n$ does not depend on the first $N_0$ terms. Now

$$\left| \frac{a_{n+1}}{a_n} \right| \leq \nu \quad \text{for all } n > N_0,$$

implies that $|a_{N_0+k}| \leq |a_{N_0}|\nu^k$. Since $0 < \nu < 1$ the series $\sum\limits_{n=N_0}^{\infty} \nu^n$ converges, so by Theorem 18.13 the theorem is proved. $\square$

**Corollary 18.15.** *Let* $(a_n)_{n \in \mathbb{N}_0}$ *be a sequence and suppose that* $\lim\limits_{n \to \infty} \left| \frac{a_{n+1}}{a_n} \right| = a < 1$. *Then the series* $\sum_{n=0}^{\infty} a_n$ *converges absolutely.*

*Proof.* Since $a < 1$ there exists $\epsilon > 0$ such that $0 < a + \epsilon < 1$. For this $\epsilon > 0$ there exists $N \in \mathbb{N}$ such that $n \geq N$ implies

$$\left| \left| \frac{a_{n+1}}{a_n} \right| - a \right| < \epsilon$$

or

$$\left| \frac{a_{n+1}}{a_n} \right| < a + \epsilon < 1$$

and the ratio test then gives the result. $\square$

**Remark 18.16. A.** Note that changing finitely many elements in a sequence or series does not effect its convergence behaviour.

**B.** The series $\sum\limits_{k=1}^{\infty} c_k$ in Theorem 18.13 is called a **majorant** of the series

$$\sum_{k=1}^{\infty} a_k.$$

**C.** Note that the condition $\left|\frac{a_{n+1}}{a_n}\right| < 1$ or $\lim_{n\to\infty}\left|\frac{a_{n+1}}{a_n}\right| = 1$ are not sufficient for the (absolute) convergence of $\sum_{k=1}^{\infty} a_k$ as the harmonic series shows. Here $a_n = \frac{1}{n}$, hence $\left|\frac{a_{n+1}}{a_n}\right| = \frac{a_{n+1}}{a_n} = \frac{n}{n+1} < 1$ as well as $\lim_{n\to\infty}\left|\frac{a_{n+1}}{a_n}\right| = \lim_{n\to\infty}\frac{n}{n+1} = 1$ and $\sum_{n=1}^{\infty}\frac{1}{n}$ diverges.

**Example 18.17.** The series $\sum_{n=0}^{\infty}\frac{n^2}{2^n}$ converges.

If $a_n = \frac{n^2}{2^n}$, then for $n \geq 3$ we have

$$\left|\frac{a_{n+1}}{a_n}\right| = \frac{(n+1)^2 2^n}{2^{n+1} n^2} = \frac{1}{2}\left(1 + \frac{1}{n}\right)^2 \leq \frac{1}{2}\left(1 + \frac{1}{3}\right)^2 = \frac{8}{9} < 1,$$

and so the series is convergent by Theorem 18.14.

**Theorem 18.18 (Root Test).** *Let $(a_n)_{n\in\mathbb{N}_0}$ be a sequence of real numbers and suppose that for all $n \geq N_0$ we have $|a_n|^{\frac{1}{n}} \leq q < 1$. Then $\sum_{n=0}^{\infty} a_n$ converges absolutely.*

*Proof.* For $n \geq N_0$ it follows that $|a_n| \leq q^n$. Therefore $\sum_{n=0}^{N_0-1}|a_n| + \sum_{n=N_0}^{\infty} q^n$ is a convergent majorant for $\sum_{n=0}^{\infty}|a_n|$ and by the comparison test, Theorem 18.13, the result follows. $\square$

**Remark 18.19.** We can replace the condition $|a_n|^{\frac{1}{n}} \leq q < 1$ by $\lim_{n\to\infty}|a_n|^{\frac{1}{n}} < 1$, see Problem 12 b).

**Example 18.20.** For $|r| < 1$ and $a \geq 1$ consider the sequence

$$a_n := \begin{cases} r^n, & n \text{ even} \\ ar^n, & n \text{ odd} \end{cases}$$

It follows that

$$\sqrt[n]{a_n} = \begin{cases} r, & n \text{ even} \\ \sqrt[n]{ar}, & n \text{ odd} \end{cases}$$

therefore using Problem 10 in Chapter 15 we find $\lim_{n\to\infty}\sqrt[n]{|a_n|} = |r| < 1$ and taking Remark 18.19, i.e. Problem 12 b), into account we find that $\sum_{n=1}^{\infty} a_n$ converges absolutely.

The comparison test and its consequences discussed so far cannot help to decide the convergence of $\sum_{n=2}^{\infty} \frac{1}{n \ln n}$ or similar series. However we can establish an integral (comparison) test and this is indeed the most powerful test. The basic idea behind this test is that integrals are limits of sums.

**Theorem 18.21 (Integral Test).** *Let $f : [1, \infty) \to \mathbb{R}$ be a non-negative decreasing function which for every $N \in \mathbb{N}$ is integrable over the interval $[1, N]$. The series $\sum_{n=1}^{\infty} f(n)$ converges if and only if $\lim_{N \to \infty} \int_1^N f(x)dx$ exists and is finite.*

*Proof.* For the interval $[1, N]$ we choose the partition $t_1 = 1 < 2 < 3 < \cdots < N$. Then the sum $f(1) + \cdots + f(N-1)$ is the Riemann sum for $\int_1^N f(x)dx$ with respect to this partition and the points $\xi_j = t_j = j$, whereas the sum $f(2) + \cdots + f(N)$ is the Riemann sum for $\int_1^N f(x)dx$ with respect to the same partition and the points $\xi_j = t_{j+1} = j + 1$. Since $f$ is decreasing it follows that

$$f(2) + \cdots + f(N) < \int_1^N f(x)dx < f(1) + \cdots + f(N-1).$$

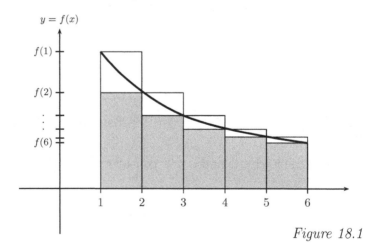

*Figure 18.1*

Since $f \geq 0$ it follows that $\left( \int_1^N f(x)dx \right)_{N \in \mathbb{N}}$ is an increasing sequence. Now if $\sum_{n=1}^{\infty} f(n)$ converges then $\left( \int_1^N f(x)dx \right)_{N \in \mathbb{N}}$ is a bounded increasing sequence and therefore $\lim_{N \to \infty} \int_1^N f(x)dx$ exists. Conversely if $\lim_{N \to \infty} \int_1^N f(x)dx$ exists, i.e. is finite, then the increasing sequence $S_N = \sum_{n=1}^{N} f(n)$ is bounded by $\lim_{N \to \infty} \int_{1}^{N} f(x)dx + f(1)$ and hence $\sum_{n=1}^{\infty} f(n)$ converges.

$\square$

**Example 18.22. A.** The series $\sum_{n=2}^{\infty} \dfrac{1}{n \ln(n)}$ diverges since $x \mapsto \dfrac{1}{x \ln(x)}$ is a decreasing function and $\int_{2}^{N} \dfrac{1}{x \ln(x)} dx = \left[ \ln(\ln(x)) \right]_{1}^{N} = \ln(\ln N) - \ln(\ln 2)$ does not have a finite limit. However the series $\sum_{n=1}^{\infty} \dfrac{1}{n(\ln(n))^2}$ converges since $x \mapsto \dfrac{1}{x(\ln(x))^2}$ is a decreasing function and $\int_{2}^{N} \dfrac{1}{x(\ln(x))^2} dx = \left[ -\dfrac{1}{\ln(x)} \right]_{2}^{N} = -\dfrac{1}{\ln N} + \dfrac{1}{\ln 2}$ converges.

**B.** For $\alpha > 1$ the function $x \mapsto \frac{1}{x^\alpha}$ is on $[1, \infty)$ positive and decreasing. Further

$$\int_{1}^{N} \frac{1}{x^\alpha} dx = \frac{1}{1-\alpha} x^{1-\alpha} \Big|_{1}^{N} = \frac{1}{1-\alpha}(N^{1-\alpha} - 1).$$

Since $\alpha > 1$ it follows that

$$\lim_{N \to \infty} \int_{1}^{N} \frac{1}{x^\alpha} dx = \lim_{N \to \infty} \frac{1}{1-\alpha}(N^{1-\alpha} - 1) = \frac{1}{\alpha - 1}$$

implying the convergence of the series $\sum_{n=1}^{\infty} \frac{1}{n^\alpha}$ for $\alpha > 1$.

**Lemma 18.23.** *For $\alpha > 0$, we have* $\lim_{n \to \infty} \dfrac{\ln(n)}{n^\alpha} = 0$.

*Proof.* Putting $n = e^k$, this is equivalent to $\dfrac{k}{e^{\alpha k}} \to 0$ as $k \to \infty$.

But $\sum_{k=1}^{\infty} \dfrac{k}{e^{\alpha k}}$ is a convergent series by the ratio test:

$$\frac{a_{k+1}}{a_k} = \frac{(k+1)e^{\alpha k}}{e^{\alpha(k+1)} k} = \frac{k+1}{k} \frac{1}{e^\alpha} \to \frac{1}{e^\alpha} < 1.$$

Therefore by Lemma 9.14 $\dfrac{k}{e^{\alpha k}} \to 0$ as $k \to \infty$. $\square$

**Theorem 18.24.** *The sequence $1 + \frac{1}{2} + \frac{1}{3} \cdots + \frac{1}{n} - \ln(n)$ converges to a limit. This limit is denoted by $\gamma$ and is called **Euler's constant**.*

*Proof.* On the interval $[1, n]$ we consider the partition $1 < 2 < \cdots < n$. Then the sum $\frac{1}{2} + \cdots + \frac{1}{n}$ is a Riemann sum for $\int_1^n \frac{dx}{x}$ which is less than the integral and the sum $1 + \frac{1}{2} + \cdots + \frac{1}{n-1}$ is a Riemann sum for $\int_1^n \frac{dx}{x}$ which is greater than the integral, therefore we find

$$\frac{1}{2} + \frac{1}{3} + \cdots + \frac{1}{n} < \ln n < 1 + \frac{1}{2} + \cdots + \frac{1}{n-1}.$$

Now set $a_n := \ln n - \left( \frac{1}{2} + \cdots + \frac{1}{n} \right)$. Note that

$$a_{n+1} = \left( \ln(n+1) - \ln(n) - \frac{1}{n+1} \right) + a_n$$

and since

$$\ln(n+1) - \ln(n) - \frac{1}{n+1} = \int_n^{n+1} \frac{1}{n} dx - \frac{1}{n+1} \geq 0,$$

it follows that $(a_n)_{n \in \mathbb{N}}$ is monotone decreasing. But $a_n > 0$, hence it has a limit. Therefore $1 + \frac{1}{2} + \frac{1}{3} + \frac{1}{n} - \ln n = 1 - a_n$ must also tend to a limit. $\square$

**Remark 18.25.** A numerical approximation for the Euler constant is $\gamma \approx 0.577215664901\ldots$

We know that for a finite sum we can change the order of the summation: addition is commutative. For series this question is a different one, it is summation combined with taking a limit. Thus it is a new, non-trivial question when we ask whether we can rearrange the order of elements "summed up" in a series.

**Definition 18.26.** *Let $\sum_{n=0}^{\infty} a_n$ be a series and $\tau : \mathbb{N}_0 \to \mathbb{N}_0$ be a bijective mapping. The series $\sum_{n=0}^{\infty} a_{\tau(n)}$ is called a **rearrangement** of the series $\sum_{n=0}^{\infty} a_n$.*

**Theorem 18.27.** *Let $\sum_{n=0}^{\infty} a_n$ be an absolutely convergent series with limit $A$, i.e. $\sum_{n=0}^{\infty} a_n = A$. Then every rearrangement of this series also converges to $A$.*

*Proof.* Let $\tau : \mathbb{N}_0 \to \mathbb{N}_0$ be any bijective mapping. We have to prove that

$$\lim_{m \to \infty} \sum_{k=0}^{m} a_{\tau(k)} = A.$$

Let $\varepsilon > 0$. Since $\sum_{k=0}^{\infty} |a_k|$ converges, there exists $N_0 \in \mathbb{N}$ such that $\sum_{k=N_0}^{\infty} |a_k| < \frac{\varepsilon}{2}$. This implies that

$$\left| A - \sum_{k=0}^{N_0-1} a_k \right| = \left| \sum_{k=N_0}^{\infty} a_k \right| \leq \sum_{k=N_0}^{\infty} |a_k| < \frac{\varepsilon}{2}.$$

Now, take $N$ such that $\{\tau(0), \ldots, \tau(N)\} \supset \{0, 1, \ldots, N_0 - 1\}$. For $m \geq N$ we find

$$\left| \sum_{k=0}^{m} a_{\tau(k)} - A \right| \leq \left| \sum_{k=0}^{m} a_{\tau(k)} - \sum_{k=0}^{N_0-1} a_k \right| + \left| \sum_{k=0}^{N_0-1} a_k - A \right|$$

$$\leq \sum_{k=N_0}^{\infty} |a_k| + \frac{\varepsilon}{2} < \varepsilon,$$

and the theorem is proved. $\qquad \qquad \square$

**Remark 18.28.** If $\sum_{n=0}^{\infty} a_n$ is convergent but **not** absolutely convergent, then rearrangements will in general change the limit. In fact, such series can be rearranged to make the limit any value.

**Example 18.29.** Consider $\tau_N : \mathbb{N}_0 \to \mathbb{N}_0$ defined to be

$$\tau(n) := \begin{cases} n + N, & 0 \leq n < N \\ n - N, & N \leq n < 2N \\ n, & n \geq 2N \end{cases}$$

This is a bijective mapping from $\mathbb{N}_0$ to $\mathbb{N}_0$ and in the case of $\sum_{n=0}^{\infty} (-1)^n \frac{1}{n+1}$ the rearranged series for $N = 3$ is

$$\sum_{n=0}^{\infty} (-1)^{\tau(n)} \frac{1}{\tau(n) + 1} = \frac{(-1)^3}{3+1} + \frac{(-1)^4}{4+1} + \frac{(-1)^5}{5+1} + \frac{(-1)^0}{0+1}$$

$$+ \frac{(-1)^1}{1+1} + \frac{(-1)^2}{2+1} + \sum_{n=6}^{\infty} (-1)^n \frac{1}{n+1}$$

$$= -\frac{1}{4} + \frac{1}{5} - \frac{1}{6} + 1 - \frac{1}{2} + \frac{1}{3} + \sum_{n=6}^{\infty} (-1)^n \frac{1}{n+1}.$$

The next result relates rational numbers to real ones from the point of view of approximation.

**Theorem 18.30.** *Every real number can be approximated by a sequence of rational numbers.*

*Proof.* We will show that every positive real number can be approximated by dyadic rationals. Let $r \in \mathbb{R}, r > 0$, be the form $\dfrac{m}{2^n}$ with $n \in \mathbb{Z}$ and $m \in \mathbb{N}$ and let

$$x_n := \sum_{l=-k}^{n} a_l 2^{-l}$$

Here $k$ is the smallest positive integer that

$$x < 2^{k+1}.$$

Then we put $a_{-k} = 1$ and define $a_l$ by

$$a_l = \begin{cases} 1 \text{ if } r - x_{l-1} > 2^{-l} \\ 0 \text{ if } r - x_{l-1} \leq 2^{-l} \end{cases} \tag{18.2}$$

From the construction it follows that

$$x_{n+1} \leq x < x_{n+1} + 2^{-n-1},$$

i.e.

$$|x - x_n| \leq 2^{-n} \quad \text{for } n \geq -k,$$

hence $x = \lim\limits_{n \to \infty} x_n = \sum\limits_{n=-k}^{\infty} a_n 2^{-n}$. $\qquad\qquad\square$

Finally we wish to address a problem about real numbers: their representation as decimal and dyadic numbers or more generally, as b-adic numbers.

**Definition 18.31.** *Let $b \in \mathbb{N}$, $b \geq 2$. A **b-adic fraction** is a series of the type*

$$\pm \sum_{n=-k}^{\infty} a_n b^{-n}$$

with $k \geq 0$ and $a_n \in \mathbb{N}_0$ such that $0 \leq a_n < b$.
If $b$ is fixed then it is sufficient to write

$$\pm a_{-k} a_{-k+1} \cdots a_{-1} a_0 a_1 a_2 a_3 \cdots$$

If $b = 10$ we are dealing with decimal fractions and if $b = 2$ we have dyadic fractions.

**Proposition 18.32.** *Every b-adic fraction converges.*

*Proof.* We show that the sequence of partial sums form a Cauchy sequence. It is sufficient to consider the case of non-negative b-adic fractions. We therefore let $\sum_{n=-k}^{\infty} a_n b^{-n}$ be a b-adic fraction and for $m \geq -k$ we set $s_m = \sum_{n=-k}^{m} a_n b^{-n}$. For $m' \geq m \geq -k$ we find

$$
\begin{aligned}
|s_{m'} - s_m| &= \sum_{n=m+1}^{m'} a_n b^{-n} \\
&\leq \sum_{n=m+1}^{m'} (b-1) b^{-n} \\
&= (b-1) b^{-m-1} \sum_{n=0}^{m'-m-1} b^{-n} \\
&\leq (b-1) b^{-m-1} \frac{1}{1 - b^{-1}} = b^{-m}.
\end{aligned}
$$

For $\epsilon > 0$ we find $N \in \mathbb{N}$ such that $m' \geq m \geq -k$ implies $|s_{m'} - s_m| < \epsilon$, namely if $b^{-m} < \epsilon$ for $m \geq N$, and the result then follows. $\square$

Of central importance is

**Theorem 18.33.** *Let $b \in \mathbb{N}$ where $b \geq 2$ then every real number $x \in \mathbb{R}$ has a representation as a b-adic fraction, i.e.*

$$x = \operatorname{sgn}(x) \sum_{n=-k}^{\infty} a_n b^{-n}$$

*where $k \geq 0$ and $a_n \in \mathbb{N}_0$ such that $0 \leq a_n < b$ and*

$$
\operatorname{sgn}(x) = \begin{cases} 1, & x > 0 \\ 0, & x = 0 \\ -1, & x < 0 \end{cases}.
$$

255

*Proof.* Again we may assume that $x > 0$. By Lemma 14.7 there exists $l \in \mathbb{N}_0$ such that $x < b^{l+1}$. Let $k$ be the smallest non-negative integer such that

$$0 \le x < b^{k+1}.$$

Now we construct a sequence $(a_n)_{n \ge -k}$ of integers $0 \le a_n \le b - 1$ such that for

$$x_m := \sum_{n=-k}^{m} a_n b^{-n}$$

we have

$$x_m \le x < x_m + b^{-m}.$$

Since

$$0 = 0 \cdot b^k < 1 \cdot b^k < \cdots < (b-1)b^k < b \cdot b^k = b^{k+1}$$

is a partition of $[0, b^{k+1}]$ and since $0 \le x < b^{k+1}$, there exists exactly one non-negative integer $0 \le a_{-k} \le b$ such that

$$x_{-k} = a_{-k}b^k < x < (a_{-k} + 1)b^k = x_{-k} + b^k.$$

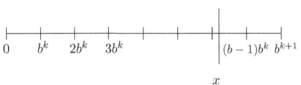

$$x$$

*Figure 18.2*

Thus we have a starting point for induction. Next we suppose that all $a_n$ for $n \le m$ are already constructed such that

$$x_m \le x < x_m + b^{-m}.$$

We now consider the partition

$$x_m < x_m b^{-m-1} < x_m + 2b^{-m-1} < \cdots < x_m + bb^{-m-1} = x_m b^{-m-1}.$$

Then there exists a unique non-negative integer $0 \le a_{m+1} \le b - 1$ such that

$$x_m + a_{m+1}b^{-m-1} \le x < x_m + (a_{m+1} + 1)b^{-m-1}.$$

Since $x_{m+1} = x_m + a_{m+1}b^{-m-1}$ we have

$$x_{m+1} \le x < x_{m+1} + b^{-m-1},$$

256

and the sequence is constructed. By construction we have

$$|x - x_m| < b^{-m} \text{ for all } m \geq -k,$$

which implies $\lim\limits_{m \to \infty} x_m = x$, i.e.

$$x = \sum_{n=-k}^{\infty} a_n b^{-n}.$$

$\square$

**Remark 18.34. A.** For $b = 10$ we can find the **decimal representation** of real numbers and only Theorem 18.33 allows us to work with it as we do. For $b = 2$ we get the **dyadic numbers** or the dyadic representation of real numbers which is important in the representation of numbers in computing. **B.** Theorem 18.33 also implies: given any real number $x$ and $\epsilon > 0$ there exists a rational number $q = q(\epsilon)$ such that $|x - q| < \epsilon$, i.e. we can approximate every real number by rational numbers. In fact we only need to take $\sum_{n=-k}^{N} a_n b^{-n}$ with $N$ such that $\left| x - \sum_{n=-k}^{N} a_n b^{-n} \right| < \epsilon$ since $\sum_{n=-k}^{N} a_n b^{-n} \in \mathbb{Q}$. From this it is evident that every real number in an interval $I \subset \mathbb{R}$ can be approximated by the rational numbers in this interval, i.e. by numbers belonging to $I \cap \mathbb{Q}$. For $b = 2$ this is the content of Theorem 18.30.

Finally we can prove

**Theorem 18.35.** *The real numbers are not countable.*

*Proof.* We prove that $(0,1) \subset \mathbb{R}$ is not countable which of course implies that $\mathbb{R}$ is not countable. Suppose that $(0,1)$ is countable then there exists a sequence $(x_n)_{n \in \mathbb{N}}$ of real numbers $x_n$ such that

$$(0,1) = \{x_n | n \in \mathbb{N}\}.$$

We represent each $x_n$ by its decimal fraction

$$
\begin{aligned}
x_1 &= 0.a_{11}a_{12}a_{13}a_{14}a_{15}\ldots \\
x_2 &= 0.a_{21}a_{22}a_{23}a_{24}a_{25}\ldots \\
x_3 &= 0.a_{31}a_{32}a_{33}a_{34}a_{35}\ldots \\
x_4 &= 0.a_{41}a_{42}a_{43}a_{44}a_{45}\ldots \\
&\vdots \qquad \vdots
\end{aligned}
$$

We define $c \in (0,1)$ by its decimal representation

$$c = 0.c_1 c_2 c_3 c_4 c_5 \ldots$$

with

$$c_k := \begin{cases} 1 \text{ if } a_{kk} \neq 1 \\ 2 \text{ if } a_{kk} = 1. \end{cases}$$

In particular we have $c_k \neq a_{kk}$ for all $k \geq 1$. By assumption there must be some $n \in \mathbb{N}$ such that $x_n = c$ which would imply $a_{nn} = c_n$. This is a contradiction and the theorem is proved. $\qquad\square$

**Remark 18.36.** The procedure used in the proof of Theorem 18.35 is called Cantor's diagonalisation argument (or procedure). In fact it was used 15 years earlier by Paul du Bois-Reymond.

**Corollary 18.37.** *The irrational numbers $\mathbb{R} \setminus \mathbb{Q}$ are not countable.*

This follows from Theorem 18.35 and

**Theorem 18.38.** *For $n \in \mathbb{N}$ let $A_n$ be a countable set then $\cup_{n \in \mathbb{N}} A_n = \{x | x \in A_n \text{ for some } n \in \mathbb{N}\}$ is countable. (I.e. the countable union of a countable set is countable.)*

*Proof.* Each set $A_n$ can be written as a sequence

$$A_n = (a_{nj})_{j \in \mathbb{N}} = (a_{n1}, a_{n2}, a_{n3}, \ldots).$$

Now we can arrange $\cup_{n \in \mathbb{N}} A_n$ in the following way:

$$
\begin{array}{cccccc}
a_{11} & a_{12} & a_{13} & a_{14} & a_{15} & a_{16} \cdots \\
a_{21} & a_{22} & a_{23} & a_{24} & a_{25} & a_{26} \cdots \\
a_{31} & a_{32} & a_{33} & a_{34} & a_{35} & a_{36} \cdots \\
a_{41} & a_{42} & a_{43} & a_{44} & a_{45} & a_{56} \cdots \\
a_{51} & a_{52} & a_{53} & a_{54} & a_{55} & a_{56} \cdots \\
a_{61} & a_{62} & a_{63} & a_{64} & a_{65} & a_{66} \cdots \\
\vdots & \vdots & \vdots & \vdots & \vdots & \vdots
\end{array}
$$

and we construct a bijection to $\mathbb{N}$ as in the case of the rational numbers in $(0,1)$.

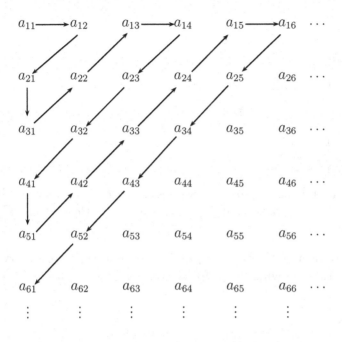

*Figure 18.3*

□

## Problems

1. For $\epsilon > 0$, find $N \in \mathbb{N}$ such that $n \geq N$ implies $\sum_{k=1}^{m} \left(\frac{1}{2}\right)^{n+k} < \epsilon$. Why does this imply the Cauchy criterion holds for the series $\sum_{k=0}^{\infty} 2^{-k}$?

2. Let $(a_n)_{n\in\mathbb{N}}$ be a monotone decreasing sequence of non-negative numbers. Prove that if $\sum_{n=1}^{\infty} a_n$ converges then $\lim_{n\to\infty} (na_n) = 0$.

3. Let $(a_n)_{n\in\mathbb{N}}$ be a sequence of non-negative numbers which is decreasing. Prove that the series $\sum_{n=1}^{\infty} a_n$ converges if and only if the series $\sum_{n=1}^{\infty} 2^n a_{2^n}$ converges.
   Hint: compare $s = \sum_{n=1}^{\infty} a_n$ with the partial sum $s_{2^n}$ and use the monotonicity criterion, i.e. Theorem 18.8.

4. Apply the result of Problem 3 to test the following series for convergence:

a) $\sum_{n=1}^{\infty} \frac{1}{n^\alpha}, \alpha \in \mathbb{R}$;

b) $\sum_{n=2}^{\infty} \frac{1}{n(\ln n)^\alpha}, \alpha \in \mathbb{R}$.

5. Test the following alternating series for convergence:

a) $\sum_{n=1}^{\infty} \frac{(-1)^{n-1}}{n^\alpha}, \alpha \in \mathbb{R}$;

b) $\sum_{n=1}^{\infty} \frac{(-1)^{n+1}}{2n-1}$;

c) $\sum_{n=2}^{\infty} \frac{(-1)^n}{n \ln n}$.

6. Let $(a_n)_{n \geq k}$ and $(b_n)_{n \geq k}$ be two sequences of real numbers such that $0 \leq a_n \leq b_n$. Suppose that $\sum_{n=k}^{\infty} a_n$ diverges. Prove that $\sum_{n=k}^{\infty} b_n$ diverges too.

7. Use a comparison with a convergent or divergent series or otherwise to investigate the following series for convergence:

a) $\sum_{k=1}^{\infty} \frac{(-1)^k k^2}{k^4 + 2k}$;

b) $\sum_{k=1}^{\infty} \frac{k!}{k^k}$;

c) $\sum_{n=1}^{\infty} \frac{\ln(n+1)}{3n^3 + 7}$;

d) $\sum_{n=1}^{\infty} \sin \frac{1}{n^3}$;

e) $\sum_{k=1}^{\infty} \frac{\cos kx}{1+k^2}, x \in \mathbb{R}$;

f) $\sum_{m=1}^{\infty} \frac{e^{mx}}{m^4}, x \in \mathbb{R}$;

g) $\sum_{l=1}^{\infty} \frac{x^2}{l^2 + x^2}, x \in \mathbb{R}$;

h) $\sum_{n=1}^{\infty} \frac{n+5}{(2n+1)\sqrt{n+3}}$.

8. Suppose that $(a_n)_{n \in \mathbb{N}}$ and $(b_n)_{n \in \mathbb{N}}$ are two sequences such that $\sum_{n=1}^{\infty} a_n^2$ and $\sum_{n=1}^{\infty} b_n^2$ converge. Prove the (extended) **Cauchy-Schwarz inequality**

$$\left| \sum_{k=1}^{\infty} a_k b_k \right| \leq \sum_{k=1}^{\infty} |a_k b_k| \leq \left( \sum_{k=1}^{\infty} a_k^2 \right)^{\frac{1}{2}} \left( \sum_{k=1}^{\infty} b_k^2 \right)^{\frac{1}{2}}$$

and the (extended) **Minkowski inequality**

$$\left( \sum_{k=1}^{\infty} |a_k + b_k|^2 \right)^{\frac{1}{2}} \leq \left( \sum_{k=1}^{\infty} a_k^2 \right)^{\frac{1}{2}} + \left( \sum_{k=1}^{\infty} b_k^2 \right)^{\frac{1}{2}}.$$

260

9. Let $(a_k)_{k \in \mathbb{N}}$ be a sequence of real numbers. Prove that the series $\sum_{k=1}^{\infty} \frac{1}{2^k} \frac{|a_k|}{1+|a_k|}$ converges. Furthermore, for two sequences $(a_k)_{k \in \mathbb{N}}$ and $(b_k)_{k \in \mathbb{N}}$ of real numbers prove

$$\sum_{k=1}^{\infty} \frac{1}{2^k} \frac{|a_k + b_k|}{1 + |a_k b_k|} \leq \sum_{k=1}^{\infty} \frac{1}{2^k} \frac{|a_k|}{1 + |a_k|} + \sum_{k=1}^{\infty} \frac{1}{2^k} \frac{|b_k|}{1 + |b_k|}.$$

10. Use the ratio test or otherwise to investigate the convergence of the following series:

   a) $\sum_{n=1}^{\infty} n^6 e^{-n^2}$;

   b) $\sum_{n=1}^{\infty} \frac{4n^2 + 15n - 3}{n^2 (n+1)^{\frac{3}{2}}}$;

   c) $\sum_{k=0}^{\infty} \frac{x^k}{k!}, x \in \mathbb{R}$;

   d) $\sum_{k=0}^{\infty} (-1)^k \frac{x^{2k}}{(2k)!}, x \in \mathbb{R}$.

11. Prove the following: if for a sequence $(a_n)_{n \in \mathbb{N}}$ of real numbers $\left| \frac{a_{n+1}}{a_n} \right| \geq \lambda > 1$ then the series $\sum_{n=1}^{\infty} a_n$ diverges. Use this result to show the divergence of:

   a) $\sum_{n=1}^{\infty} \frac{(-1)^n 3^n}{n^4}$;

   b) $\sum_{n=1}^{\infty} \frac{n^{\frac{3}{2}}}{(n+3)\sqrt{4n+15}}$.

12. Let $(a_n)_{n \in \mathbb{N}}$ be a sequence of real numbers.

   a) Prove that if $|a_n|^{\frac{1}{n}} \geq 1$ then $\sum_{n=1}^{\infty} |a_n|$ diverges.

   b) Prove that if $\lim_{n \to \infty} |a_n|^{\frac{1}{n}} = a < 1$ then $\sum_{n=1}^{\infty} |a_n|$ converges.

13.* Prove **Raabe's test**: suppose that $\left| \frac{a_{n+1}}{a_n} \right| \leq 1 - \frac{a}{n}$ holds for $n \geq N$. If $a > 1$ then $\sum_{n=1}^{\infty} a_n$ converges absolutely.

14. Consider the series

$$\sum_{n=1}^{\infty} \left( \frac{1 \cdot 4 \cdot 7 \cdot \ldots \cdot (3n - 2)}{3 \cdot 6 \cdot 9 \cdot \ldots \cdot 3n} \right)^2.$$

Use Raabe's test to show that it converges. Is it possible to use the ratio test to prove convergence of this series?

261

15. Use the integral test to investigate convergence or divergence of the following series:

   a) $\sum_{k=2}^{\infty} \frac{1}{k(\ln k)^{\alpha}}, \alpha > 1$;

   b) $\sum_{l=1}^{\infty} le^{-l^2}$;

   c) $\sum_{k=2}^{\infty} \frac{\ln k}{k}$;

   d) $\sum_{k=2}^{\infty} \frac{\ln k}{k^2}$.

16.* Let $(a_n)_{n \in \mathbb{N}}$ be a sequence of real numbers for which $\sum_{n=1}^{\infty} a_n$ converges but $\sum_{n=1}^{\infty} |a_n|$ diverges, i.e. the series is not absolutely convergent. Prove that for $c \in \mathbb{R}$ given there exists a rearrangement of $\sum_{n=1}^{\infty} a_n$ the limit of which is $c$.

17. Find the representation of $x = \frac{1}{7}$ as a b-adic fraction when

   a) $b = 2$;

   b) $b = 7$;

   c) $b = 10$.

18. Prove that if $D \subset \mathbb{R}$ is a set which contains an open interval $(a, b)$, i.e. $(a, b) \subset D$, then $D$ is not countable.
   Hint: use the fact that the interval $(0, 1)$ is not countable and construct a bijective mapping $f : (a, b) \to (0, 1)$.

# 19 Point Sets in $\mathbb{R}$

Functions or sequences map subsets of the real line onto subsets of the real line. In order to understand this process better we need to acquire more knowledge of subsets of the real line. This is a task which will accompany us for some time and it is partly more abstract and formal than students are used to at the beginning of their studies. However it is unavoidable in order to gain a deeper understanding of mathematics.

We already know a certain class of subsets of $\mathbb{R}$ and we have seen its importance: intervals.

For $a \leq b$ we define the **closed interval** by

$$[a, b] := \{x \in \mathbb{R} | a \leq x \leq b\}, \tag{19.1}$$

noting that $[a, a] = \{a\}$ is a closed interval. For $a < b$ we have the **open interval**

$$(a, b) := \{x \in \mathbb{R} | a < x < b\}, \tag{19.2}$$

and for $a < b$ we have two kinds of **half-open** intervals, namely

$$[a, b) := \{x \in \mathbb{R} | a \leq x < b\} \tag{19.3}$$

and

$$(a, b] := \{x \in \mathbb{R} | a < x \leq b\}. \tag{19.4}$$

We extend these notions to **infinite** or **unbounded intervals**. For $a \in \mathbb{R}$ we set

$$[a, \infty) := \{x \in \mathbb{R} | x \geq a\}, \tag{19.5}$$
$$(a, \infty) := \{x \in \mathbb{R} | x > a\}, \tag{19.6}$$
$$(-\infty, a] := \{x \in \mathbb{R} | x \leq a\}, \tag{19.7}$$
$$(-\infty, a) := \{x \in \mathbb{R} | x < a\}. \tag{19.8}$$

Moreover we define

$$\mathbb{R}_+ := [0, \infty), \tag{19.9}$$

so that $(0, \infty) = \mathbb{R}_+ \setminus \{0\}$ and we occasionally use

$$(-\infty, \infty) := \mathbb{R}, \tag{19.10}$$

i.e. we consider $\mathbb{R}$ as an interval. The following definition has far reaching consequences.

**Definition 19.1.** *A set $A \subset \mathbb{R}$ is called* **open**, *more precisely an open subset of $\mathbb{R}$, or open in $\mathbb{R}$, if for every $x \in A$ there exists an $\epsilon > 0$ such that the open interval $(x - \epsilon, x + \epsilon)$ belongs entirely to $A$, i.e. $(x - \epsilon, x + \epsilon) \subset A$. By definition the empty set $\emptyset$ is open.*

Clearly $\mathbb{R}$ is an open set. Moreover we find

**Lemma 19.2.** *Every open interval $(a, b) \subset \mathbb{R}$ is an open subset of $\mathbb{R}$.*

*Proof.* First note that there is a need for a proof. At a first glance the notion of an open interval is unrelated to the notion of an open set. But of course we should expect some consistency in our notions. Therefore let $(a, b) \subset \mathbb{R}$ be an open interval. We want to prove that for $x \in (a, b)$ there exists $\epsilon > 0$ such that the open interval $(x - \epsilon, x + \epsilon)$ is a subset of $(a, b)$, i.e. $(x - \epsilon, x + \epsilon) \subset (a, b)$. For this choose $\epsilon := \frac{1}{2} \min(x - a, b - x) > 0$ and it follows that $(x - \epsilon, x + \epsilon) \subset (a, b)$. $\square$

This proof has a clear geometric idea:

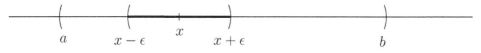

$$a \qquad x - \epsilon \qquad x \qquad x + \epsilon \qquad\qquad b$$

*Figure 19.1*

Note that the proof is also valid for $(-\infty, b)$ or $(a, \infty)$, i.e. both are open sets.

We next want to study some properties of open sets.

**Lemma 19.3. A.** *For a finite collection of open subsets $A_1, \ldots, A_N$ of $\mathbb{R}$ the intersection $\bigcap_{\nu=1}^{N} A_\nu$ is open.*

**B.** *Let $I \neq \emptyset$ be an arbitrary index set and for $j \in I$ let $A_j \subset \mathbb{R}$ be an open set, then the union $\bigcup_{j \in I} A_j$ is an open set in $\mathbb{R}$.*

*Proof.* **A.** Assume that $\bigcap_{\nu=1}^{N} A_\nu \neq \emptyset$, otherwise there is nothing to prove since by definition $\emptyset$ is open. Let $x \in \bigcap_{\nu=1}^{N} A_\nu$, thus $x \in A_\nu$ for all $\nu = 1, \ldots, N$. Since $A_\nu$ is open there exists $\epsilon_\nu > 0$ such that $(x - \epsilon_\nu, x + \epsilon_\nu) \subset A_\nu$. For $\epsilon := \min_{1 \leq \nu \leq N} \epsilon_\nu > 0$ we find

$$x \in (x - \epsilon, x + \epsilon) \subset \bigcap_{\nu=1}^{N} (x - \epsilon_\nu, x + \epsilon_\nu) \subset \bigcap_{\nu=1}^{N} A_\nu,$$

264

implying the openess of $\bigcap_{\nu=1}^{N} A_\nu$.

**B.** Now let $I \neq \emptyset$ be any index set and for $j \in I$ let $A_j \subset \mathbb{R}$ be open. Consider

$$A := \bigcup_{j \in I} A_j := \{x \in \mathbb{R} | x \in A_{j_0} \text{ for some } j_0 \in I\}, \qquad (19.11)$$

and assume that at least one set $A_{j_1}$ is non-empty, otherwise $A = \emptyset$ and nothing remains to prove. Take $x \in A$, then for some $j_0 \in I$ we have $x \in A_{j_0}$ and since $A_{j_0}$ is open there exists an open interval $(x - \epsilon, x + \epsilon) \subset A_{j_0}$ which yields $(x - \epsilon, x + \epsilon) \subset \bigcup_{j \in I} A_j = A$ and the lemma is proved. $\square$

**Example 19.4. A.** If $a_1 < b_1 < a_2 < b_2$ then the two intervals $(a_1, b_1)$ and $(a_2, b_2)$ are open and disjoint. Their union $(a_1, b_1) \cup (a_2, b_2)$ is open too but it is not an interval anymore.

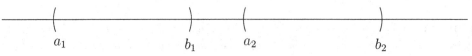

*Figure 19.2*

Moreover, the set $\bigcup_{n=1}^{\infty} (n - \frac{1}{n}, n + \frac{1}{n})$ is open.

**B.** Consider the open intervals $(1 - \frac{1}{n+1}, 1 + \frac{1}{n+1})$. Their intersection is given by

$$\{1\} = \bigcap_{n=1}^{\infty} (1 - \frac{1}{n+1}, 1 + \frac{1}{n+1})$$

(compare also with Problem 4). The set $\{1\}$ does not contain an open interval, hence we cannot expect that an infinite intersection of open sets is open.

**C.** The following type of construction will be used (in a modified form) quite often. Let $a < b$ and $f : [a, b] \to \mathbb{R}$ be a function. Let $\epsilon > 0$ and for $x \in [a, b]$ consider the open interval $(f(x) - \epsilon, f(x) + \epsilon) \subset \mathbb{R}$. It follows that $\bigcup_{x \in [a,b]} (f(x) - \epsilon, f(x) + \epsilon) \subset \mathbb{R}$ is an open set. The image of $f$, i.e. $f([a, b])$ is a subset in $\mathbb{R}$ and clearly $f([a, b]) \subset \bigcup_{x \in [a,b]} (f(x) - \epsilon, f(x) + \epsilon)$. Thus we can consider $f([a, b])$ as a subset of an open set and every $y = f(x) \in f([a, b])$ is the centre of an open interval of length $2\epsilon$ entirely belonging to this open set. Clearly $f([a, b])$ does not have to be open, just consider $f : [a, b] \to \mathbb{R}$, $f(x) = c \in \mathbb{R}$ for all $x \in [a, b]$. Then $f([a, b]) = \{c\}$ which is not open.

Recall that by Definition 17.8.B a point $a \in \mathbb{R}$ is an accumulation point of $B \subset \mathbb{R}$ if there exists a sequence $(b_n)_{n \in \mathbb{N}}$, $b_n \in B, b_n \neq a$, converging to $a$.

265

**Definition 19.5.** *A set $B \subset \mathbb{R}$ is called **closed**, more precisely a closed subset of $\mathbb{R}$, or closed in $\mathbb{R}$, if it contains all its accumulation points.*

**Theorem 19.6.** *A set $B \subset \mathbb{R}$ is closed if and only if its complement $B^{\complement}$ is open. Consequently $A \subset \mathbb{R}$ is open if $A^{\complement}$ is closed.*

*Proof.* Suppose $B$ is closed and $x \in B^{\complement}$, then $x$ is not an accumulation point of $B$, i.e. there is no sequence $(b_n)_{n\in\mathbb{N}}$, $b_n \in B$, converging to $x$, and so there exists an interval $(x - \epsilon, x + \epsilon)$ which contains no point of $B$, i.e $(x - \epsilon, x + \epsilon) \subset B^{\complement}$, and so $B^{\complement}$ is open. Conversely, suppose $B^{\complement}$ is open and $a$ is an accumulation point of $B$. Then, if $a \in B^{\complement}$, there exists an open interval $(a - \epsilon, a + \epsilon)$ contained in $B^{\complement}$, which contradicts the fact that $a$ is an accumulation point of $B$, i.e. the existence of a sequence $(b_n)_{n\in\mathbb{N}}, b_n \in B, b_n \neq a$, converging to $a$. Hence $a \in B$ and $B$ is closed. The final statement follows from $(A^{\complement})^{\complement} = A$. $\qquad\square$

**Lemma 19.7.** *The sets $\emptyset$ and $\mathbb{R}$ are closed and any closed interval is closed. Moreover, the union of finitely many closed sets is closed and the intersection of an arbitrary collection of closed sets is closed.*

*Proof.* We have $\emptyset^{\complement} = \mathbb{R}$ and $\mathbb{R}^{\complement} = \emptyset$ implying that $\emptyset$ and $\mathbb{R}$ are closed. For the interval $[a, b]$ we can write $[a, b] = ((-\infty, a) \cup (b, \infty))^{\complement}$ implying that $[a, b]$ is closed. Also $(-\infty, b] = (b, \infty)^{\complement}$, so that $(-\infty, b]$ is closed. Similarly $[a, \infty)$ is closed.

Now let $B_\nu \subset \mathbb{R}$, $\nu = 1, \ldots, N$, be a family of closed sets. Then

$$\left(\bigcup_{\nu=1}^{N} B_\nu\right)^{\complement} = \bigcap_{\nu=1}^{N} B_\nu^{\complement},$$

and since $B_\nu^{\complement}$ is open, $\left(\bigcup_{\nu=1}^{N} B_\nu\right)^{\complement}$ is open, and hence $\left(\bigcup_{\nu=1}^{N} B_\nu\right)$ is closed. For an arbitrary collection $B_j \subset \mathbb{R}$, $j \in I$, of closed sets we have

$$\bigcap_{j\in I} B_j = \{x \in \mathbb{R} \mid x \in B_j \text{ for all } j \in I.\}$$

and therefore

$$\left(\bigcap_{j\in I} B_j\right)^{\complement} = \bigcup_{j\in I} B_j^{\complement},$$

and since each $B_j^c$ is open it follows from Lemma 19.3 that $\left( \bigcap\limits_{j \in I} B_j \right)^{\complement}$ is open,

hence $\bigcap\limits_{j \in I} B_j$ is closed. $\qquad\qquad\qquad\qquad\qquad\qquad\qquad\qquad\qquad$ □

**Remark 19.8.** In Problem 1 we will prove that $[a, b)$ and $(a, b]$ are neither open nor closed.

**Example 19.9. A.** A single point $a \in \mathbb{R}$ forms a closed set $\{a\}$ since $\{a\} = ((-\infty, a) \cup (a, \infty))^{\complement}$. This implies that any finite union of points $a_1, \ldots, a_N$ is closed:

$$\{a_\nu | \nu = 1, \ldots, N\} = \bigcup_{\nu=1}^{N} \{a_\nu\}.$$

**B.** Let $a_\nu \in \mathbb{R}$, $\nu \in \mathbb{N}$ and assume for some $\delta > 0$ that $|a_\nu - a_{\nu+1}| \geq \delta$. Then $\bigcup_{\nu=1}^{\infty} \{a_\nu\}$ is a closed set. (Compare with Problem 3).

**Definition 19.10.** *A set $U \in \mathbb{R}$ is called a **neighbourhood** of $x \in \mathbb{R}$ if there exists an open set $A \subset U$ containing $x$, i.e. $x \in A \subset U$.*

Obviously every open set is a neighbourhood of all its points. However the closed interval $[a, b]$ is only a neighbourhood of the points belonging to $(a, b) \subset [a, b]$. It is not a neighbourhood in $\mathbb{R}$ of $\{a\}, \{b\}$ or any subset containing $a$ or $b$ (or both). From our considerations above we have

**Theorem 19.11.** *Let $U \subset \mathbb{R}$ be a neighbourhood of $x \in \mathbb{R}$ then there exists an open interval $(x - \delta, x + \delta) \subset U, \delta > 0$. Further, by Theorem 18.33 we know that there exists a dyadic fraction*

$$y = \mathrm{sgn}(x) \sum_{l=-k}^{N} a_l 2^{-l}, \quad a_l \in \mathbb{N}_0, \qquad (19.12)$$

*such that $|x - y| < \delta$, i.e. $y \in U$, implying that in every neighbourhood of a real number we can find a rational number.*

Next we want to understand the idea of boundedness for subsets of the real line.

**Definition 19.12.** *A set $D \subset \mathbb{R}$ is called **bounded from above (bounded from below)** if there exists $K \in \mathbb{R}$ such that*

$$x \leq K (x \geq K) \text{ for all } x \in D. \qquad (19.13)$$

*We call $K$ an **upper (lower) bound** for $D$. If $D$ is bounded from above and from below we call $D$ **bounded**.*

**Remark 19.13. A.** Upper and lower bounds are not uniquely determined. In fact if $D$ is bounded from above by $K$ then $K' > K$ is a further upper bound and if $D$ is bounded from below by $M$ then $M' < M$ is a further lower bound.

**B.** A set $D \subset \mathbb{R}$ is bounded if and only if for some $K$ we have $|x| \leq K$ for all $x \in D$. Indeed, since $A \leq x \leq B$ for some $A \leq B$, we may also take $K := \max(|A|, |B|)$ to find $-K \leq x \leq K$ for $x \in D$.

**C.** Note further that a sequence $(a_n)_{n \in \mathbb{N}}$ is bounded if and only if the set $\{a_\nu | \nu \in \mathbb{N}\} \subset \mathbb{R}$ is bounded in $\mathbb{R}$.

**D.** Let $a < b$ be real numbers then the corresponding open, closed and half-open intervals $(a, b), [a, b], [a, b)$ and $(a, b]$ are all bounded with lower bound $a$ and upper bound $b$. However in some cases the bound belongs to the interval, in other cases it does not. The intervals $(-\infty, a)$ and $(-\infty, a]$ are not bounded sets, but they are bounded from above, while $(b, \infty)$ and $[b, \infty)$ are not bounded but bounded from below.

The last remark raises the following interesting question: Suppose that $D \subset \mathbb{R}$ is bounded above. We would like to know whether there exists a smallest upper bound, i.e. $K \in \mathbb{R}$ being an upper bound of $D$ with the property that if $K' < K$ then $K'$ cannot be an upper bound of $D$.

Of fundamental importance is the following theorem which once again needs the completeness of $\mathbb{R}$.

**Theorem 19.14.** *Every non-empty set $D \subset \mathbb{R}$ which is bounded from above has a least upper bound. Every non-empty set $D \subset \mathbb{R}$ which is bounded from below has a greatest lower bound.*

**Definition 19.15.** *Let $D \subset \mathbb{R}$ be a subset. The least upper bound of $D$ is called its **supremum**, its greatest lower bound is called its **infimum**. The supremum of a set $D$ is denoted by $\sup D$, the infimum is denoted by $\inf D$.*

*Proof of Theorem 19.14.* We show the case where $D$ is bounded from above. Since $D \neq \emptyset$ and bounded from above there exists $x_0 \in D$ and $K_0 \in \mathbb{R}$, an upper bound of $D$, such that $x_0 \leq K_0$, hence $r := K_0 - x_0 \geq 0$. We now take the arithmetic mean $\frac{K_0 + x_0}{2}$ which may or may not be an upper bound for $D$. If it is, we call it $K_1$. If it is not an upper bound for $D$, there exists $x_1 \in D$, $x_1 > x_0$, larger than $\frac{K_0 + x_0}{2}$. In this case we set $K_1 := K_0$, i.e. we do not change the upper bound. We repeat this process to obtain a decreasing sequence of upper bounds and an increasing sequence of elements belonging to $D$, and we will prove that they converge to the same limit. Our demonstration uses mathematical induction:

We construct

   i) a sequence $x_0 \leq x_1 \leq x_2 \leq \cdots$ of elements in $D$, and

   ii) a sequence $K_0 \geq K_1 \geq K_2 \geq \cdots$ of upper bounds of $D$ such that
$$K_n - x_n \leq 2^{-n}r \text{ for all } n \in \mathbb{N}, r = K_0 - x_0. \tag{19.14}$$

Starting with $x_0$ and $K_0$ let us assume that $x_0, \ldots x_n, \in D$ and $K_0, \ldots, K_n$, upper bounds of $D$, are already constructed such that (19.14) holds. Define
$$M := \frac{K_n + x_n}{2}.$$

There are two possibilities: if $M$ is an upper bound of $D$, we put $x_{n+1} := x_n$ and $K_{n+1} := M$; if $M$ is not an upper bound of $D$, we put $K_{n+1} := K_n$ and choose $x_{n+1} \in D$ with $x_{n+1} > M$. In each case we have
$$x_n \leq x_{n+1}, \quad K_n \geq K_{n+1} \text{ and } K_{n+1} - x_{n+1} \leq 2^{-n-1}r.$$

The sequence $(K_n)_{n \in \mathbb{N}}$ is monotone decreasing and bounded since $x_0 \leq K_n \leq K_0$. Hence $(K_n)_{n \in \mathbb{N}_0}$ converges to some $K \in \mathbb{R}$. Since for $x \in D$ we always have $x \leq K_n$, it follows that $x \leq \lim_{n \to \infty} K_n = K$, i.e. $K$ is an upper bound for $D$. To show that it is the least upper bound, suppose $K' < K$. Then there exists $n_0 \in \mathbb{N}$ such that $2^{-n_0}r < K - K'$, which yields
$$x_n \geq K_n - 2^{-n}r \geq K - 2^{-n}r > K',$$

so that $K'$ is not an upper bound. Hence $K = \sup D$. Note that (19.14) implies $\lim_{n \to \infty} K_n = \lim_{n \to \infty} x_n$. $\square$

**Example 19.16. A.** For a closed interval $[a, b]$, $a \leq b$, we have $\sup[a, b] = b$ and $\inf[a, b] = a$.

**B.** For an open interval $(a, b)$, $a < b$, we find $\sup(a, b) = b$ and $\inf(a, b) = a$. We show that $b = \sup(a, b)$. Clearly, $b$ is an upper bound for $(a, b)$. Suppose that $b' < b$. It follows that
$$x := \max \left( \frac{a+b}{2}, \frac{b'+b}{2} \right) \in (a, b)$$

and $b' < x$, hence $b'$ could not be an upper bound.

**Example 19.17.** The following holds

$$\sup\left\{\frac{n^2}{n^2+1}\Big|n \in \mathbb{N}\right\} = 1.$$

Suppose $0 < \epsilon < 1$ is given. Since $\lim_{n\to\infty}\frac{n^2}{n^2+1} = 1$ and since $\left(\frac{n^2}{n^2+1}\right)_{n\in\mathbb{N}}$ is an increasing sequence it follows that there exists $N(\epsilon)$ such that $n \geq N(\epsilon)$ implies $\epsilon < \frac{n^2}{n^2+1}$, hence $\epsilon < 1$ cannot be an upper bound, while 1 is clearly an upper bound. This example easily extends. Let $(a_n)_{n\in\mathbb{N}}$ be a sequence of real numbers converging to $a$, i.e. $\lim_{n\to\infty} a_n = a$. Suppose that $a_n \leq a$ for all $n \in \mathbb{N}$ then $\sup\{a_n | n \in \mathbb{N}\} = a$ (compare with Problem 10).

The examples show that sometimes $\inf D$ or $\sup D$ belong to $D$, sometimes not.

**Definition 19.18. A.** *If* $D \subset \mathbb{R}$ *and* $x = \sup D \in D$, *then we call* $x$ *the* **maximum** *of* $D$ *and write* $x = \max D$. *In this case we have* $\sup D = \max D$. *If* $D \subset \mathbb{R}$ *and* $y = \inf D \in D$, *then we call* $y$ *the* **minimum** *of* $D$ *and write* $y = \min D$. *In this case we have* $\inf D = \min D$.
**B.** *If a set* $D$ *is not bounded from above we write* $\sup D = \infty$, *if it is not bounded from below, we write* $\inf D = -\infty$.

If $D$ is bounded from above, it need not have a maximum. However there is always a sequence in $D$ converging to $\sup D$ as shown in the proof of Theorem 19.14. A similar statement holds for the minimum and infimum.

We now turn to sequences. A sequence may have or may not have a limit, or it may have several converging subsequences. The following notions of limit superior and limit inferior will help to clarify the situation.

**Definition 19.19.** *Let* $(a_n)_{n\in\mathbb{N}}$ *be a sequence of real numbers. We define its* **limit superior** *by*

$$\limsup_{n\to\infty} a_n := \lim_{n\to\infty}\left(\sup\{a_k | k \geq n\}\right) \tag{19.15}$$

*and its* **limit inferior** *by*

$$\liminf_{n\to\infty} a_n := \lim_{n\to\infty}\left(\inf\{a_k | k \geq n\}\right). \tag{19.16}$$

**Remark 19.20. A.** An alternative notation is

$$\overline{\lim} = \limsup \quad \text{and} \quad \underline{\lim} = \liminf.$$

**B.** The sequence $(\sup\{a_k | k \geq n\})_{n \in \mathbb{N}}$ is monotone decreasing whereas the sequence $(\inf\{a_k | k \geq n\})_{n \in \mathbb{N}}$ is monotone increasing. Therefore

$$\limsup_{n \to \infty} a_n \quad \text{and} \quad \liminf_{n \to \infty} a_n$$

exist either as limits in $\mathbb{R}$ or as "improper limits" $+\infty$ or $-\infty$, i.e. the sequence $(\sup\{a_k | k \geq n\})_{n \in \mathbb{N}}$ diverges to $\pm\infty$, and/or the sequence $(\inf\{a_k | k \geq n\})_{n \in \mathbb{N}}$ diverges to $\pm\infty$.

**Example 19.21. A.** Consider the sequence $(a_n)_{n \in \mathbb{N}}$, where $a_n = (-1)^n \left(1 + \frac{1}{n^2}\right)$. We find

$$\sup\{a_k | k \geq n\} = \begin{cases} 1 + \frac{1}{n^2}, & \text{if } n \text{ is even} \\ 1 + \frac{1}{(n+1)^2}, & \text{if } n \text{ is odd,} \end{cases}$$

hence $\limsup_{n \to \infty} a_n = 1$. Further we find

$$\inf\{a_k | k \geq n\} = \begin{cases} -\left(1 + \frac{1}{n^2}\right), & \text{if } n \text{ is odd} \\ -\left(1 + \frac{1}{(n+1)^2}\right), & \text{if } n \text{ is even,} \end{cases}$$

hence $\lim_{n \to \infty} \inf a_n = -1$.

**B.** For the sequence $(a_n)_{n \in \mathbb{N}}$, $a_n = n$, we find

$$\sup\{a_k | k \geq n\} = \infty \quad \text{and} \quad \inf\{a_k | k \geq n\} = n,$$

which yields

$$\limsup_{n \to \infty} a_n = \liminf_{n \to \infty} a_n = +\infty.$$

**Theorem 19.22. A.** *Let $(a_n)_{n \in \mathbb{N}}$, $a_n \in \mathbb{R}$, be a bounded sequence and denote by $A$ the set of all its accumulation points. It holds that*

$$\limsup_{n \to \infty} a_n = \sup A \tag{19.17}$$

*and*

$$\liminf_{n \to \infty} a_n = \inf A. \tag{19.18}$$

**B.** *A sequence $(a_n)_{n \in \mathbb{N}}$ of real numbers $a_n \in \mathbb{R}$ converges to a limit $a \in \mathbb{R}$ if and only if*

$$\limsup_{n \to \infty} a_n = \liminf_{n \to \infty} a_n = a. \tag{19.19}$$

*Proof.* **A.** We prove (19.17), the proof of (19.18) is similar. With $A_n :=$ $\sup\{a_k | k \geq n\}$ we have by the definition of lim sup that

$$a' := \limsup_{n \to \infty} a_n = \lim_{n \to \infty} A_n.$$

Since $(a_n)_{n \in \mathbb{N}}$ is bounded it follows that $A_n \in \mathbb{R}$ as well as $a' \in \mathbb{R}$. We claim that $a'$ is an accumulation point, i.e. $a' \in A$, and that $a \leq a'$ for all $a \in A$. By definition $A$ is the set of all limits of converging subsequences of $(a_n)_{n \in \mathbb{N}}$. Therefore, to prove $a' \in A$ it is sufficient to show that for every $N \in \mathbb{N}$ and every $\epsilon > 0$ there exists $n' \geq N, n' = n_{N,\epsilon}$, such that $|a_{n'} - a'| < \epsilon$. Indeed by this we get a subsequence of $(a_n)_{n \in \mathbb{N}}$ converging to $a'$. Since $\lim_{n \to \infty} A_n = a'$ we find $m \geq N$ such that $|A_m - a'| < \frac{\epsilon}{2}$ and the definition of $A_m$ implies the existence of $n, n \geq m$, such that $|a_n - A_m| < \frac{\epsilon}{2}$ which yields for $n \geq N$ that $|a_n - a'| < \epsilon$. Thus we have proved $a' \in A$. Let $a \in A$ be an accumulation point of $(a_n)_{n \in \mathbb{N}}$. Then there exists a subsequence $(a_{n_k})_{k \in \mathbb{N}}$ of $(a_n)_{n \in \mathbb{N}}$ such that $\lim_{k \to \infty} a_{n_k} = a$. By definition of $A_{n_k}$ we have $A_{n_k} \geq a_{n_k}$. This implies

$$a' = \lim_{n \to \infty} A_n = \lim_{k \to \infty} A_{n_k} \geq \lim_{k \to \infty} a_{n_k} = a,$$

but $a' \in A$ and $a \leq a'$ for all $a \in A$ implies $a' = \sup A$.
**B.** In the case where $(a_n)_{n \in \mathbb{N}}$ converges to $a \in \mathbb{R}$, we know by Theorem 15.8 that $(a_n)_{n \in \mathbb{N}}$ is bounded and further $A = \{a\}$. Thus applying part A we get

$$\limsup_{n \to \infty} a_n = \sup A = a = \inf A = \liminf a_n.$$

Now suppose that (19.19) holds. We set as before $A_n := \sup\{a_k | k \geq n\}$ and further $B_n := \inf\{a_k | k \geq n\}$. In other words $\lim_{n \to \infty} A_n = \lim_{n \to \infty} B_n$. Given $\epsilon > 0$ there exists $N \in \mathbb{N}$ such that $|a - A_N| < \epsilon$ and $|a - B_N| < \epsilon$. Since $B_N \leq a_n \leq A_N$ for all $n \geq N$, it follows that $-(a - B_N) \leq a_n - a \leq A_N - a$ or $|a_n - a| < \epsilon$ for all $n \geq N$, i.e. $(a_n)_{n \in \mathbb{N}}$ converges to $a$. $\square$

The proof of Theorem 19.22 gives an alternative characterisation of lim sup and lim inf.

**Corollary 19.23.** *Let $(a_n)_{n \in \mathbb{N}}$ be a bounded sequence. Its greatest accumulation point is $\limsup_{n \to \infty} a_n$ and its smallest accumulation point is $\liminf_{n \to \infty} a_n$. Moreover we have*

$$\limsup_{n \to \infty} a_n = -\liminf_{n \to \infty}(-a_n). \tag{19.20}$$

*In order to see (19.20) note that passing from $(a_n)_{n \in \mathbb{N}}$ to $(-a_n)_{n \in \mathbb{N}}$ is a reflection about 0 which reverses all order relations.*

Finally we want to provide some results which are useful to know, but we provide the proofs only in Appendix VIII. We start with

**Definition 19.24. A.** *Let $A \subset \mathbb{R}$ be a non-empty set. We call a pair $\{O_1, O_2\}$ of non-empty open and disjoint subsets of $\mathbb{R}$ a **splitting** of $A$ if $A \subset O_1 \cup O_2$ and $A \cap O_1$ as well as $A \cap O_2$ is non-empty.*
**B.** *A non-empty subset $A \subset \mathbb{R}$ is called **connected** if $A$ does not have a splitting.*

**Theorem 19.25.** *A non-empty subset of $\mathbb{R}$ is connected if and only if it is an interval.*

**Corollary 19.26.** *A subset $A \subset \mathbb{R}$ is both open and closed if and only if $A$ is either empty, i.e. $A = \emptyset$, or $A$ is all of $\mathbb{R}$, i.e. $A = \mathbb{R}$.*

*Proof.* Both $\emptyset$ and $\mathbb{R}$ are open and closed. Indeed $\emptyset$ is open by definition, hence $\emptyset^{\complement} = \mathbb{R}$ is closed. However $\mathbb{R} = \bigcup_{n \in \mathbb{N}}(-n, n)$ is the union of open sets, hence open, implying that $\mathbb{R}^{\complement}$ is closed. Suppose that $A$ is open and closed, hence $A^{\complement}$ is open and closed and the connected set $\mathbb{R}$ has the splitting $\mathbb{R} = A \cup A^{\complement}$. Hence either $A$ or $A^{\complement}$ is empty, hence $A$ is either $\mathbb{R}$ or $\emptyset$. $\square$

We finally have

**Theorem 19.27.** *Every open set $A \subset \mathbb{R}$ is a denumerable union of disjoint open intervals.*

In Appendix VIII we will provide a proof of Theorem 19.25 and Theorem 19.27.

# Problems

1. Prove that for $a < b$ the half-open interval $[a, b)$ is neither open nor closed.

2. Is $\mathbb{Q} \subset \mathbb{R}$, i.e. the set of all rational numbers, a closed or an open subset of $\mathbb{R}$?

3. Let $a_\nu \in \mathbb{R}, \nu \in \mathbb{N}$, assume $a_\nu < a_{\nu+1}$ for $\nu \in \mathbb{N}$ and $\lim_{\nu \to \infty} = \infty$. Prove that $\bigcup_{\nu=1}^{\infty}\{a_\nu\}$ is closed.

4. Give an example of a sequence $(B_\nu)_{\nu \in \mathbb{N}}$ of closed sets in $\mathbb{R}$ such that $\cup_{\nu \in \mathbb{N}} B_\nu$ is not closed.

5. Let $(a_\nu)_{\nu \in \mathbb{N}}$, $a_\nu \in \mathbb{R}$, be a sequence converging to $a \in \mathbb{R}$. Is $\{a_\nu | \nu \in \mathbb{N}\}$ closed in general? Prove that $\{a_\nu | \nu \in \mathbb{N}\} \cup \{a\}$ is closed.

6. Let $A$ and $B$ be two non-empty sets of real numbers and define $A + B := \{c = a + b | a \in A, b \in B\}$. Prove that if $A$ and $B$ are both bounded then $A + B$ is bounded too.

7.  a) Given the set $M := (-3, 2) \cup [4, 6] \cup \{10\} \subset \mathbb{R}$. Prove that $(-3, 2) \cup (4, 6)$ is the largest open set contained in $M$ and that $[-3, 2] \cup [4, 6] \cup \{10\}$ is the smallest closed set which contains $M$.

   b) Prove that $\bigcap_{n \in \mathbb{N}} (-\frac{1}{n}, \frac{1}{n}) = \{0\}$.

8.  a) Consider the set

$$G := \left\{ y \in \mathbb{R} \mid y = \frac{1}{x}, x \geq \frac{1}{2} \right\}.$$

Find $\inf G$ and $\sup G$. Does $G$ have a maximum or minimum?

   b) Find a sequence $(a_n)_{n \in \mathbb{N}}$, $a_n \in \mathbb{R}$, with 3 accumulation points such that $\sup\{a_n | n \in \mathbb{N}\} = 3$, $\inf\{a_n | n \in \mathbb{N}\} = 0$, $\limsup\limits_{n \to \infty} a_n = \frac{5}{2}$ and $\liminf\limits_{n \to \infty} a_n = \frac{1}{2}$.

9. For each of the following sequences $(a_n)_{n \in \mathbb{N}}$, $a_n \in \mathbb{R}$, determine $\sup\{a_n | n \in \mathbb{N}\}$, $\inf\{a_n | n \in \mathbb{N}\}$, $\limsup\limits_{n \to \infty} a_n$ and $\liminf\limits_{n \to \infty} a_n$:

   a) $a_n = 2 - \frac{n-1}{10}$;

   b) $a_n = \frac{(-1)^{n-1}}{n+1}$;

   c) $a_n = \frac{2}{3}(1 - \frac{1}{10^n})$.

10. Let $(a_n)_{n \in \mathbb{N}}$ be a sequence of real numbers converging to $a$, i.e. $\lim_{n \to \infty} a_n = a$. Suppose that $a_n \leq a$ for all $n \in \mathbb{N}$. Prove that $\sup\{a_n | n \in \mathbb{N}\} = a$.

11. Let $(a_n)_{n \in \mathbb{N}}$ be a sequence. Prove that $a = \limsup_{n \to \infty} a_n$ if and only of for every $\epsilon > 0$ the estimate $a_n < a + \epsilon$ holds for all but finitely many $n \in \mathbb{N}$.

12. Let $(a_n)_{n\in\mathbb{N}}$ and $(b_n)_{n\in\mathbb{N}}$ be two sequences and let $\lambda > 0$. Prove

   a) $\limsup_{n\to\infty}(\lambda a_n) = \lambda \limsup_{n\to\infty} a_n$;

   b) $\limsup_{n\to\infty}(a_n + b_n) \leq \limsup_{n\to\infty} a_n + \limsup_{n\to\infty} b_n$;

   c) $\limsup_{n\to\infty}(a_n + b_n) \geq \limsup_{n\to\infty} a_n + \liminf_{n\to\infty} b_n$;

   d) if $\lim_{n\to\infty} b_n = b$, i.e. the limit exists, then

$$\limsup_{n\to\infty}(a_n + b_n) = \limsup_{n\to\infty} a_n + \lim_{n\to\infty} b_n.$$

   Hint: use Problem 11.

13. The set $A := [0, 1] \cup \{2\} \cup (3, 4) \subset \mathbb{R}$ is not an interval, hence not connected. Give a splitting $\{O_1, O_2\}$ of $A$.

# 20 Continuous Functions

In Chapter 6 we encountered the concept of a continuous function, see Definition 6.9. This notion depends on the idea of a limit of a function (at some point of its domain) which was introduced in Chapter 6. Recall: a function $f : D \to \mathbb{R}, D \subset \mathbb{R}$, has the limit $a$ as $y \in D$ approaches $x$ if for every $\epsilon > 0$ there exists $\delta > 0$ such that $0 < |x - y| < \delta$ implies $|f(y) - a| < \epsilon$. First we want to relate this definition to limits of sequences.

**Theorem 20.1.** *Let $D \subset \mathbb{R}$ and $f : D \to \mathbb{R}$ be a function and suppose that for $x \in \mathbb{R}$ there exists a sequence $(x_k)_{k \in \mathbb{N}}$, $x_k \in D$, $x_k \neq x$, converging to $x$. The function has the limit $a \in \mathbb{R}$ as $y \in D$ approaches $x$, i.e. $\lim_{y \to x} f(y) = a$, if and only if for every sequence $(x_n)_{n \in \mathbb{N}}$, $x_n \in D \setminus \{x\}$, converging to $x$, i.e. $\lim_{n \to \infty} x_n = x$, it follows that $\lim_{n \to \infty} f(x_n) = a$.*

*Proof.* Suppose that for $\varepsilon > 0$ there exists $\delta > 0$ such that $0 < |y - x| < \delta$, $y \in D$, implies $|f(y) - a| < \varepsilon$. Let $\lim_{n \to \infty} x_n = x$, $x_n \in D$. Then there exists $N = N(\delta)$ such that for $n \geq N(\delta)$ it follows that $|x_n - x| < \delta$. By assumption it follows that $|f(x_n) - a| < \varepsilon$ for $n \geq N(\delta) = N(\delta(\varepsilon))$, i.e. $\lim_{n \to \infty} f(x_n) = a$.

Suppose now that for every sequence $(x_n)_{n \in \mathbb{N}}$, $x_n \in D$, with $\lim_{n \to \infty} x_n = x$ it follows that $\lim_{n \to \infty} f(x_n) = a$. We have to prove that for every $\varepsilon > 0$ there exists $\delta > 0$ such that $0 < |y - x| < \delta$ implies $|f(x) - a| < \varepsilon$. Suppose this does not hold. Then there exists $\varepsilon > 0$ such that for no value of $\delta > 0$ do we have $|f(y) - a| < \varepsilon$ for all $y \in D$ with $0 < |y - x| < \delta$. Thus for every $n \in \mathbb{N}$ there exists $x_n \in D$ such that

$$|x_n - x| < \frac{1}{n} \quad \text{and} \quad |f(x_n) - a| \geq \varepsilon.$$

This implies that $\lim_{n \to \infty} x_n = x$ and therefore $\lim_{n \to \infty} f(x_n) = a$, but $|f(x_n) - a| \geq \varepsilon$ for some $\varepsilon > 0$ which is a contradiction. $\square$

We now have the following characterisations of continuity of $f$ at a point $x$:

**Theorem 20.2.** *A function $f : D \to \mathbb{R}$, $D \subset \mathbb{R}$, is continuous at $x \in D$ if either of the following equivalent conditions holds:*

    *i) for every $\varepsilon > 0$ there exists $\delta = \delta(\varepsilon) > 0$ such that for $y \in D$ the condition $0 < |y - x| < \delta$ implies $|f(y) - f(x)| < \varepsilon$;*

*ii) for every sequence $(x_n)_{n\in\mathbb{N}}$, $x_n \in D$, converging to $x \in D$ it follows that $(f(x_n))_{n\in\mathbb{N}}$ converges to $f(x)$, i.e. $\lim_{n\to\infty} x_n = x$ implies $\lim_{n\to\infty} f(x_n) = f(x)$.*

Note that statement i) is just Definition 6.9.

**Definition 20.3.** *We call $f : D \to \mathbb{R}$, $D \subset \mathbb{R}$, **continuous on** $D$ if $f$ is continuous for each $x \in D$. The set of all continuous functions on $D$ is denoted by $C(D)$.*

From Example 6.1.C we can deduce that every polynomial $p : \mathbb{R} \to \mathbb{R}$ is continuous. In particular, this applies to the constant function $x \mapsto c$, $c \in \mathbb{R}$, the identity $x \mapsto x$ and $x \mapsto x^2$. Furthermore, it is easy to see that $x \mapsto |x|$ is continuous on $\mathbb{R}$. Indeed the converse triangle inequality yields $||x| - |y|| \le |x - y|$, thus given $\epsilon > 0$ choose $\delta = \epsilon$ to find for $0 < |x - y| < \delta$ that $||x| - |y|| \le |x - y| < \epsilon$.

**Corollary 20.4.** *Let $f : D \to \mathbb{R}$ be continuous at $x \in D$ and $f(x) \neq 0$. Then $f(y) \neq 0$ for all $y$ in a neighbourhood of $x$, i.e. there exists $\delta > 0$ such that $f(y) \neq 0$ for all $y \in D$, $|x - y| < \delta$.*

*Proof.* For $\varepsilon := |f(x)| > 0$ there exists $\delta > 0$ such that $y \in D$ and $0 < |y - x| < \delta$ implies $|f(y) - f(x)| < \varepsilon$. It follows that

$$|f(y)| \ge |f(x)| - |f(y) - f(x)| > 0 \quad \text{for } y \in D, \, 0 < |y - x| < \delta.$$

$\square$

Before we prove deeper results on continuous functions we want to investigate more the concept of the limit of a function.

Let $f : D \to \mathbb{R}$, $D \subset \mathbb{R}$, be a function and let $x \in \mathbb{R}$ be an accumulation point of $D$ in the sense that there exists a sequence $(x_k)_{k\in\mathbb{N}}$, $x_k \in D \setminus \{x\}$, such that $\lim_{k\to\infty} x_k = x$. Let $D_1, D_2 \subset D$ be such that $x$ is an accumulation point of both $D_1$ and $D_2$ and suppose that $D_1 \cap D_2 = \emptyset$. If $\lim_{y\to x} f(y) = a$ then $\lim_{y\to x} f|_{D_1}(y) = a$ and $\lim_{x\to y} F|_{D_2}(y) = a$. Of special interest is the case where $D_1$ and $D_2$ are subsets of open intervals with $x$ being the right end point of the interval containing $D_1$ and the left end point of the interval containing $D_2$, still $x$ is supposed to be an accumulation point of $D_1$ and $D_2$.

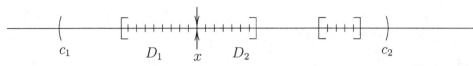

*Figure 20.1*

If $\lim_{y \to x} f(y) = a$ then in the case $\lim_{y \to x} f|_{D_1}(y)$ we are approaching $x$ from the left, i.e. $y < x$, and in the case of $\lim_{y \to x} f|_{D_2}(y)$ we are approaching $x$ from the right, i.e. $x < y$. This leads to

**Definition 20.5. A.** *We say that $f : D \to \mathbb{R}$ has a **limit from the right** if for every sequence $(x_n)_{n \in \mathbb{N}}$, $x_n \in D$ and $x_n > x$, with $\lim_{n \to \infty} x_n = x$ it follows that $\lim_{n \to \infty} f(x_n) = a$. We write*

$$\lim_{y \searrow x} f(y) = a \quad or \quad \lim_{\substack{y \to x \\ y > x}} f(y) = a, . \tag{20.1}$$

*In the case where $a = f(x)$ we call $f$ **right continuous** or **continuous from the right** at $x$.*
**B.** *We say that $f : D \to \mathbb{R}$ has a **limit from the left** if for every sequence $(x_n)_{n \in \mathbb{N}}$, $x_n \in D$ and $x_n < x$, with $\lim_{n \to \infty} x_n = x$ it follows that $\lim_{n \to \infty} f(x_n) = a$. We write*

$$\lim_{y \nearrow x} f(y) = a \quad or \quad \lim_{\substack{y \to x \\ y < x}} f(y) = a. \tag{20.2}$$

*In the case where $a = f(x)$ we call $f$ **left continuous** or **continuous from the left** at $x$.*

**Lemma 20.6.** *Let $x \in \mathbb{R}$ be an accumulation point of $D \subset \mathbb{R}$ and $f : D \to \mathbb{R}$ a function. The function $f$ has a limit $a$ at $x$ if and only if it has a limit from the right and a limit from the left at $x$ and both coincide and are equal to $a$.*

*Proof.* We already know that if $f$ has a limit $a$ at $x$, then it also has a limit from the right and from the left at $x$ and these limits are equal to $a$. Now suppose that $\lim_{\substack{y \to x \\ y > x}} f(y) = \lim_{\substack{y \to x \\ y < x}} f(y) = a$. Let $(x_k)_{n \in \mathbb{N}}, x_k \in D$, be any sequence converging to $x$. For $\epsilon > 0$ there exists $\delta_1 > 0$ and $\delta_2 > 0$ such that for $x_k \in D$, $x_k > x$ and $|x_k - x| < \delta_1$ it follows that $|f(x_k) - a| < \epsilon$, and for $x_k \in D$, $x_k < x$ and $|x_k - x| < \delta_2$ it follows that $|f(x_k) - a| < \epsilon$. Thus for $\delta = \min(\delta_1, \delta_2)$ it follows that $x_k \in D$ and $|x_k - x| < \delta$ implies $|f(x_k) - a| < \epsilon$. Since $(x_k)_{k \in \mathbb{N}}$ converges to $x$, given $\epsilon > 0$ we find $N \in \mathbb{N}$ such that $k \geq N$

implies $|x_k - x| < \delta$, thus given $\epsilon > 0$ we find $N \in \mathbb{N}$ such that $k \geq N$ implies $|f(x_k) - a| < \epsilon$ proving the lemma. $\qquad\square$

**Definition 20.7.** *We say that* $f : D \to \mathbb{R}$ *has a* ***limit*** $a$ ***at*** $\infty$ *if for each sequence* $(x_n)_{n \in \mathbb{N}}$, $x_n \in D$ *and* $\lim_{n \to \infty} x_n = \infty$, *it follows that* $\lim_{n \to \infty} f(x_n) = a$. *We write*

$$\lim_{y \to \infty} f(y) = a. \qquad (20.3)$$

*Analogously we define* $\lim_{y \to -\infty} f(y) = a$.

**Example 20.8. A.** Consider $x \mapsto [x]$. Then $\lim_{x \searrow 1} [x] = 1$ and $\lim_{x \nearrow 1} [x] = 0$.

Indeed, for any sequence $(x_n)_{n \geq 0}$, $x_n > 1$ and $\lim_{n \to \infty} x_n = 1$ it follows for $n$ sufficiently large that $[x_n] = 1$, if however $x_n \to 1$ and $x_n < 1$ then $[x_n] = 0$ for $n$ large.

**B.** Let $P(x) = x^k + a_1 x^{k-1} + \ldots + a_{k-1} x + a$, $k \geq 1$, be a polynomial. It follows that

$$\lim_{x \to \infty} P(x) = \infty$$

and

$$\lim_{x \to -\infty} P(x) = \begin{cases} +\infty & \text{for } k \text{ even} \\ -\infty & \text{for } k \text{ odd.} \end{cases}$$

*Proof.* For $x \neq 0$ we write

$$P(x) = x^k g(x) = x^k \left( 1 + \frac{a_1}{x} + \frac{a_2}{x^2} + \ldots + \frac{a_k}{x^k} \right).$$

If $x \geq c := \max\left(1, 2k|a_1|, \ldots, 2k|a_k|\right)$ it follows that

$$g(x) \geq \frac{1}{2},$$

hence for these $x$ we have

$$P(x) \geq \frac{1}{2} x^k \geq \frac{x}{2}.$$

Thus, if $x_n \to \infty$ then $P(x_n) \geq \frac{x_n}{2} \to \infty$, or $\lim_{n \to \infty} P(x_n) = \infty$. Since $P(-x) = (-1)^k Q(x) = (-1)^k (x^k - a_1 x^{k-1} + \ldots + (-1)^{k-1} a_{k-1} + (-1)^k a_k$ the second statement follows from the first. $\qquad\square$

**Theorem 20.9.** *Let $f, g : D \to \mathbb{R}$ be two functions continuous at $x \in D$, and let $\lambda \in \mathbb{R}$. The following functions are continuous at $x$:*

$$f + g, \; \lambda f, \; f \cdot g.$$

*In addition, if $g(x) \neq 0$, then $\frac{f}{g}$ is also continuous at $x$.*

*Proof.* Let $(x_n)_{n \in \mathbb{N}}$, $x_n \in D$, be a sequence converging to $x$. It follows from the limit theorems for sequences that

$$\lim_{n \to \infty} (f + g)(x_n) = \lim_{n \to \infty} f(x_n) + \lim_{n \to \infty} g(x_n) = f(x) + g(x) = (f + g)(x),$$

$$\lim_{n \to \infty} (\lambda f)(x_n) = \lambda \lim_{n \to \infty} f(x_n) = \lambda f(x),$$

$$\lim_{n \to \infty} (f \cdot g)(x_n) = \left( \lim_{n \to \infty} f(x_n) \right) \cdot \left( \lim_{n \to \infty} g(x_n) \right) = f(x) \cdot g(x) = (f \cdot g)(x),$$

$$\lim_{n \to \infty} \left( \frac{f}{g} \right)(x_n) = \frac{\lim_{n \to \infty} f(x_n)}{\lim_{n \to \infty} g(x_n)} = \frac{f(x)}{g(x)} = \frac{f}{g}(x),$$

note that by assumption $\left( \dfrac{f}{g} \right)(x_n)$ is well defined for $n$ large enough. $\qquad \square$

**Remark 20.10.** In case that $f, g \in C(D)$, i.e. $f$ and $g$ are continuous on $D$, then Theorem 20.9 implies that $f + g, \lambda f, f \cdot g \in C(D)$. Thus $C(D)$ forms an algebra with the natural operations. In particular $C(D)$ is a vector space.

**Corollary 20.11.** *All rational functions $x \mapsto \dfrac{P(x)}{Q(x)}$ where $P$ and $Q$ are polynomials are continuous on the set $\mathbb{R} \setminus \{x_0 \in \mathbb{R} \mid Q(x_0) = 0\}$.*

**Theorem 20.12.** *Let $f : D \to \mathbb{R}$ and $g : E \to \mathbb{R}$ be two functions such that $f(D) \subset E$. Suppose that $f$ is continuous at $x \in D$ and that $g$ is continuous at $y := f(x) \in E$. Then the function $g \circ f : D \to \mathbb{R}$ is continuous at $x$.*

*Proof.* Let $(x_n)_{n \in \mathbb{N}}$, $x_n \in D$, be a sequence with $\lim_{n \to \infty} x_n = x$. Since $f$ is continuous at $x$ it follows that $\lim_{n \to \infty} f(x_n) = f(x)$. Setting $y_n := f(x_n)$ it follows that $\lim_{n \to \infty} y_n = y$ and the continuity of $g$ at $y$ implies that $\lim_{n \to \infty} g(y_n) = g(y)$, hence

$$\lim_{n \to \infty} g(f(x_n)) = \lim_{n \to \infty} (g \circ f)(x_n) = (g \circ f)(x).$$

$$\square$$

**Example 20.13. A.** If $f : D \to \mathbb{R}$ is continuous, then so is $|f|$.
**B.** The continuity of $|f| : D \to \mathbb{R}$, $x \mapsto |f(x)|$ however does not imply the continuity of $f$.

**Theorem 20.14.** *For $a < b$, let $f : [a,b] \to \mathbb{R}$ be a continuous function with $f(a) < 0$ and $f(b) > 0$ (or $f(a) > 0$ and $f(b) < 0$). Then there exists $\xi \in [a,b]$ such that $f(\xi) = 0$.*

*Proof.* Suppose that $f(a) < 0$ and $f(b) > 0$. We will construct a sequence of closed intervals $([a_n, b_n])_{n \in \mathbb{N}}$ with the properties

(i) $[a_n, b_n] \subset [a_{n-1}, b_{n-1}]$ for $n \geq 1$;

(ii) $b_n - a_n = 2^{-n}(b - a)$;

(iii) $f(a_n) \leq 0$ and $f(b_n) \geq 0$.

We start with $[a_0, b_0] = [a, b]$. Suppose that $[a_n, b_n]$ has already been constructed and set $m := \frac{a_n + b_n}{2}$. If $f(m) \geq 0$, then take $[a_{n+1}, b_{n+1}] = [a_n, m]$, if $f(m) < 0$, then take $[a_{n+1}, b_{n+1}] = [m, b_n]$. Obviously (i)–(iii) are fulfilled. The sequence $(b_n)_{n \in \mathbb{N}}$ is monotone decreasing and bounded, The sequence $(a_n)_{n \in \mathbb{N}}$ is monotone increasing and bounded, hence both sequences are convergent and because of (ii) they have the same limit. Let

$$\xi := \lim_{n \to \infty} a_n = \lim_{n \to \infty} b_n.$$

Since $f$ is continuous it follows that

$$\lim_{n \to \infty} f(a_n) = \lim_{n \to \infty} f(b_n) = f(\xi).$$

In addition

$$f(\xi) = \lim_{n \to \infty} f(a_n) \leq 0 \leq \lim_{n \to \infty} f(b_n) = f(\xi),$$

so that $f(\xi) = 0$. $\qquad\qquad\square$

**Remark 20.15.** This result allows us to decide whether the equation $f(x) = 0$ has a solution in the domain $[a, b]$ of $f$. Suppose that for some $c_1 \in [a, b]$ we have $f(c_1) > 0$ ($f(c_1) < 0$) and for some $c_2 \in [a, b], c_2 > c_1$ we have $f(c_2) < 0$ ($f(c_2) > 0$), then $f|_{[c_1,c_2]}$ satisfies the conditions of Theorem 20.14 and hence $f(x) = 0$ must have a solution $\xi \in [c_1, c_2] \subset [a, b]$.

**Example 20.16.** If $f : \mathbb{R} \to \mathbb{R}$, $x \mapsto x^n + c_1 x^{n-1} + \ldots + c_n$ is a polynomial and $n$ is odd, then $f$ has a zero, i.e. there exists some $z \in \mathbb{R}$ such that $f(z) = 0$. Indeed, since $\lim_{x\to\infty} f(x) = +\infty$ and $\lim_{x\to-\infty} f(x) = -\infty$, there exists a closed interval $[a, b]$, $a < b$, such that $f(a) < 0$ and $f(b) > 0$, which implies the result by Theorem 20.14.

We can now provide a proof of the **intermediate value theorem**, see Theorem 9.5.

**Theorem 20.17.** *Let $f : [a, b] \to \mathbb{R}$, $a < b$, be a continuous function and let $\eta$ be any real number between $f(a)$ and $f(b)$. Then there exists $\xi \in [a, b]$ such that $f(\xi) = \eta$.*

*Proof.* Suppose that $f(a) < \eta < f(b)$ and define $g : [a, b] \to \mathbb{R}$ by $g(x) = f(x) - \eta$. Then it follows that $g(a) < 0 < g(b)$ and Theorem 20.14 gives the result, since $g(\xi) = 0$ if and only if $f(\xi) = \eta$. $\qquad\square$

**Remark 20.18. A.** The content of Theorem 20.17 allows the following reformulation: the image of an interval under a continuous function is an interval. In light of Theorem 19.25 we may further rephrase the result as: a continuous function maps connected sets onto connected sets. In this formulation the result has a generalisation far beyond the situation discussed so far.
**B.** We can use the intermediate value theorem to determine the range of a function. Suppose that $f : (a, b) \to \mathbb{R}$ is continuous and $\lim_{x\to a} f(x) = -\infty$ as well as $\lim_{x\to b} f(x) = \infty$. Then the range of $f$ must be $\mathbb{R}$. Indeed, given any $\xi \in \mathbb{R}$ we can find $a_1$ and $b_1$, $a < a_1 < b_1 < b$, such that $f(a_1) \leq \xi \leq f(b_1)$. Hence, by Theorem 20.17 there exists $x_0 \in [a_1, b_1] \subset (a, b)$ such that $f(x_0) = \xi$, i.e. $\xi$ is in the range of $f$. In Chapter 10 we have used this already to determine the range of tan and cot.

We recall the definition of a bounded function, see Definition 8.2.

**Definition 20.19.** *A function $f : D \to \mathbb{R}$ is **bounded** if $f(D) \subset \mathbb{R}$ is bounded, i.e. if there is $M \geq 0$ such that*

$$|f(x)| \leq M \quad \text{for all } x \in D.$$

**Theorem 20.20.** *Every continuous function defined on a closed and bounded interval is bounded and there are $p, q \in [a, b]$ such that*

$$f(p) = \sup\{f(x) \mid x \in [a, b]\} = \max\{f(x) \mid x \in [a, b]\}$$

*and*

$$f(q) = \inf\{f(x) \mid x \in [a, b]\} = \min\{f(x) \mid x \in [a, b]\}.$$

*Proof.* We prove the result for the maximum. For the minimum we only have to consider $-f$ instead of $f$. Set

$$A := \sup\{f(x) \mid x \in [a, b]\} \in \mathbb{R} \cup \{\infty\}.$$

Take a sequence $(x_n)_{n \in \mathbb{N}}$, $x_n \in [a, b]$, such that

$$\lim_{n \to \infty} f(x_n) = A.$$

The sequence $(x_n)_{n \in \mathbb{N}}$ is bounded, hence by the Bolzano-Weierstrass theorem there is a subsequence $(x_{n_k})_{k \in \mathbb{N}}$ converging to some $p \in [a, b]$, i.e.

$$\lim_{k \to \infty} x_{n_k} = p \in [a, b].$$

The continuity of $f$ implies now

$$A = \lim_{k \to \infty} f(x_{n_k}) = f(p),$$

i.e. $f(p) = \sup f([a, b]) = \max f([a, b])$. $\qquad\square$

Continuous functions on bounded closed intervals have the "best" properties you may imagine. The reason behind this is compactness, a notion we will investigate now.

**Definition 20.21.** *Let $D \subset \mathbb{R}$ be any set. We call a collection of open sets $A_\nu \subset \mathbb{R}$, $\nu \in I$, an **open covering** of $D$ if*

$$D \subset \bigcup_{\nu \in I} A_\nu.$$

**Definition 20.22.** *A set $K \subset \mathbb{R}$ is **compact** if for every open covering $(A_\nu)_{\nu \in I}$ of $K$ we may select a **finite subcovering** of $K$, i.e. there exists $\nu_1, \ldots, \nu_N \in I$ such that*

$$K \subset \bigcup_{k=1}^{N} A_{\nu_k}.$$

**Remark 20.23.** The important point in the definition of compactness is that for **every** open covering we may select a finite subcovering of $K$.

**Proposition 20.24.** *Every compact set $K \subset \mathbb{R}$ is bounded and closed.*

*Proof.* Since $(-n, n)_{n \in \mathbb{N}}$ is an open covering of $K$, we may select a finite subcovering $(-n_1, n_1), \ldots, (-n_N, n_N)$ such that $K \subset \bigcup_{k=1}^{N}(-n_k, n_k)$ $= (-n_{N_0}, n_{N_0})$ where $n_{N_0} = \max_{1 \leq k \leq N} n_k$. Thus $K \subset (-n_{N_0}, n_{N_0})$, and so $|x| \leq n_{N_0}$ for all $x \in K$, i.e. $K$ is bounded. Next we prove that $K^c$ is open. Take $x \in K^c$. For every $y \in K$ it follows that $|x - y| > \varepsilon_y > 0$ (for some $\varepsilon_y > 0$) and the open intervals $\left(x - \frac{\varepsilon_y}{2}, x + \frac{\varepsilon_y}{2}\right)$ and $\left(y - \frac{\varepsilon_y}{2}, y + \frac{\varepsilon_y}{2}\right)$ are disjoint.

Figure 20.2

Clearly $\left(y - \frac{\varepsilon_y}{2}, y + \frac{\varepsilon_y}{2}\right)_{y \in K}$ is an open covering of $K$. By the compactness of $K$ we may take a finite subcovering $\left(y_1 - \frac{\varepsilon_{y_1}}{2}, y_1 + \frac{\varepsilon_{y_1}}{2}\right), \ldots, \left(y_N - \frac{\varepsilon_{y_N}}{2}, y_N + \frac{\varepsilon_{y_N}}{2}\right)$ of $K$. It follows that

$$B_x := \bigcap_{j=1}^{N} \left(x - \frac{\varepsilon_{y_j}}{2}, x + \frac{\varepsilon_{y_j}}{2}\right)$$

is open and $x \in B_x$. In addition

$$B_x \cap \bigcup_{j=1}^{N} \left(y_j - \frac{\varepsilon_{y_j}}{2}, y_j + \frac{\varepsilon_{y_j}}{2}\right) = \emptyset$$

implying that $B_x \cap K = \emptyset$, or $B_x \subset K^c$. Thus we have proved that the complement of $K$ is open, i.e. $K$ is closed. $\qquad\square$

In preparing the converse to Proposition 20.24 we show

**Proposition 20.25.** *Every bounded closed interval $[a, b] \subset \mathbb{R}$ is compact.*

*Proof.* We prove the proposition by contradiction. Suppose that there is an open covering $(A_\nu)_{\nu \in I}$ of $[a, b]$ which has no finite subcovering. For $m = \frac{a+b}{2}$ it follows that at least one of the intervals $[a, m]$ and $[m, b]$ cannot be covered by a finite subcovering of $(A_\nu)_{\nu \in I}$. Call this interval $I_1$. By induction we get a sequence of closed intervals $(I_j)_{j \in \mathbb{N}}$ with the following properties:

(i) $[a, b] \supset I_1 \supset I_2 \supset \ldots$

285

(ii) $I_j$ is not covered by a finite subcovering of $(A_\nu)_{\nu \in I}$

(iii) for $x, y \in I_j$ it follows that $|x - y| < 2^{-j}(b - a)$.

By the principle of nested intervals, Theorem 17.15, there is one point $x_0$ which lies in $\bigcap_{j \in \mathbb{N}} I_j$. Therefore, for some $j_0$ we have $x_0 \in A_{j_0}$. Since $A_{j_0}$ is open there is some $\varepsilon > 0$ such that $|y - x_0| < \varepsilon$ implies $y \in A_{j_0}$. Taking $n$ such that $2^{-n}(b - a) < \varepsilon$, then it follows from (iii) that $I_n \subset A_{j_0}$ which contradicts (ii). $\qquad \square$

Now we may prove the famous **Heine-Borel Theorem.**

**Theorem 20.26.** *A set $K \subset \mathbb{R}$ is compact if and only if it is bounded and closed.*

*Proof.* We know already that compact sets are bounded and closed, so it remains to prove that a closed and bounded set is compact. Let $(A_\nu)_{\nu \in I}$ be an open covering of the closed and bounded set $K$. Since $K$ is bounded, there exists a closed interval $[a, b] \subset \mathbb{R}$ such that $K \subset [a, b]$. The family of open sets $(A_\nu)_{\nu \in I}$, together with $A_p := \mathbb{R} \setminus K$ form an open covering of $\mathbb{R}$, since $\bigcup_{j \in I} A_j \cup A_p \supset K \cup K^c = \mathbb{R}$. Therefore, $(A_\nu)_{\nu \in I \cup \{p\}}$ is also an open covering of $[a, b]$ and by Proposition 20.25 it contains a finite subcovering $(A_{\nu_j})_{\nu_j \in I_N}$ where $I_N$ is a finite subset of $I \cup \{p\}$. If $p \in I_N$, then, since $K \cap A_p = \emptyset$, we can remove $A_p$ and we still have a finite covering of $K$. $\qquad \square$

Our first application of compactness is related to uniform continuity.

**Definition 20.27.** *A function $f : D \to \mathbb{R}$ is called **uniformly continuous** on $D$ if for every $\varepsilon > 0$ there exists $\delta > 0$ such that for $x, y \in D$ the inequality $|x - y| < \delta$ implies $|f(x) - f(y)| < \varepsilon$.*

**Remark 20.28. A.** The important difference of continuity on $D$ and uniform continuity lies in the fact that in the latter case $\delta$ is independent of $x \in D$. **B.** If $f : D \to \mathbb{R}$ is uniformly continuous on $D$, then it is obviously continuous on $D$. However the converse is false.

**Example 20.29.** The function $f : (0, 1] \to \mathbb{R}$, $x \to \frac{1}{x}$, is continuous on $(0, 1]$. Indeed, for $p \in (0, 1]$ and $\varepsilon > 0$ it follows with $\delta := \min \left( \frac{p}{2}, \frac{p^2}{2} \varepsilon \right)$ that

$$|f(x) - f(p)| = \left| \frac{1}{x} - \frac{1}{p} \right| = \left| \frac{x - p}{xp} \right| \le \frac{2|x - p|}{p^2} < \frac{2\delta}{p^2} \le \varepsilon,$$

where we use that $|x - p| < \delta \leq \frac{p}{2}$ implies $-\frac{p}{2} < x - p$ or $\frac{p}{2} < x$, i.e. $\frac{1}{x} < \frac{2}{p}$. Thus, $f$ is continuous on $(0, 1]$.

Now, suppose that $f$ is uniformly continuous on $(0, 1]$. Then there would be some $\delta > 0$ such that for all $x, y \in (0, 1]$ and $|x - y| < \delta$ it would follow that

$$|f(x) - f(y)| = \left| \frac{1}{x} - \frac{1}{y} \right| < 1.$$

For $n \in \mathbb{N}$ we have

$$\left| \frac{1}{n} - \frac{1}{2n} \right| = \frac{1}{2n} \quad \text{and} \quad \left| \frac{1}{\frac{1}{n}} - \frac{1}{\frac{1}{2n}} \right| = n,$$

thus for $\frac{1}{2n} < \delta$ it follows that

$$\left| f \left( \frac{1}{n} \right) - f \left( \frac{1}{2n} \right) \right| = n \geq 1,$$

which contradicts $|f(x) - f(y)| < 1$.

**Theorem 20.30.** *Every continuous function $f : K \to \mathbb{R}$ on a compact set $K \subset \mathbb{R}$ is uniformly continuous and bounded.*

*Proof.* Let $\varepsilon > 0$. Since $f$ is continuous for each $x \in K$ there is $\delta_{x,\varepsilon}$ such that $y \in K$ and $|x - y| < \delta_{x,\varepsilon}$ implies $|f(x) - f(y)| < \frac{\varepsilon}{2}$. Denote by $I(x)$ the interval $(x - \frac{\delta_{x,\varepsilon}}{2}, x + \frac{\delta_{x,\varepsilon}}{2})$. Clearly $(I(x))_{x \in K}$ is an open covering of $K$. By compactness there is a finite subcovering

$$\left( x_l - \frac{\delta_{x_l,\varepsilon}}{2}, x_l + \frac{\delta_{x_l,\varepsilon}}{2} \right)_{l \in \{1,\ldots,N\}}.$$

Take $\delta := \frac{1}{2} \min (\delta_{x_1,\varepsilon}, \ldots, \delta_{x_N,\varepsilon})$. For $|x - y| < \delta$ it follows that for some $1 \leq j \leq N$ we have

$$x \in \left( x_j - \frac{\delta_{x_j,\varepsilon}}{2}, x_j + \frac{\delta_{x_j,\varepsilon}}{2} \right)$$

and further

$$|x_j - y| \leq |x - y| + |x - x_j| < \delta + \frac{\delta_{x_j,\varepsilon}}{2} < \delta_{x_j,\varepsilon},$$

and therefore

$$|f(y) - f(x)| \leq |f(y) - f(x_j)| + |f(x) - f(x_j)| < \frac{\varepsilon}{2} + \frac{\varepsilon}{2} = \varepsilon$$

proving that $f$ is uniformly continuous. Next we prove that $f$ is bounded. For $\epsilon = 1$ and $x \in K$ there exists $\delta_x > 0$ such that $y \in K$ and $|x - y| < \delta_x$ implies $|f(x) - f(y)| < 1$. The intervals $J(x) := (x - \delta_x, x + \delta_x), x \in K$, form an open covering of $K$. Hence, since $K$ is compact, we can cover $K$ by finitely many of these intervals, say $J(x_1), \ldots, J(x_N))$. On $J(X_j)$ we have $|f(y) - f(x_j)| < 1$ or $|f(y)| \leq 1 + |f(x_j)|$, implying $|f(y)| \leq 1 + \max_{1 \leq j \leq N} |f(x_j)|$ for all $y \in K$. $\square$

Finally in this chapter we prove

**Theorem 20.31.** *Let* $f : [a,b] \to \mathbb{R}, f([a,b]) = [A,B]$, *have an inverse function* $f^{-1}$, *i.e.* $f^{-1} : [A,B] \to \mathbb{R}$ *and* $f \circ f^{-1} = id_{[A,B]}$ *and* $f^{-1} \circ f = id_{[a,b]}$. *If* $f$ *is continuous, so is* $f^{-1}$.

*Proof.* Suppose that $f^{-1}$ is not continuous. Then there is $y \in [A,B]$ and a sequence $(y_n)_{n \in \mathbb{N}}, y_n \in [A,B]$, such that $\lim_{n \to \infty} y_n = y$ and for some $\varepsilon > 0$

$$|f^{-1}(y_n) - f^{-1}(y)| > \varepsilon.$$

Since $f^{-1}(y_n) \in [a,b]$, a subsequence $(f^{-1}(y_{n_k}))_{k \in \mathbb{N}}$ converges by the Bolzano-Weierstrass theorem:

$$\lim_{k \to \infty} f^{-1}(y_{n_k}) = c,$$

and $|c - f^{-1}(y_{n_k})| \geq \varepsilon$. Further $f(f^{-1}(y_{n_k})) = y_{n_k}$ and the continuity of $f$ implies

$$y = \lim_{k \to \infty} y_{n_k} = \lim_{k \to \infty} f(f^{-1}(y_{n_k})) = f(c),$$

i.e. $f^{-1}(y) = f^{-1}(f(c)) = c$ contradicting $|c - f^{-1}(y)| \geq \varepsilon$ and the theorem is proved. $\square$

# Problems

1.  Let $f : [a,b] \to \mathbb{R}, a < b$, be a function. Prove that $f$ is continuous at $x \in [a,b]$ if and only if for every sequence $(x_n)_{n \in \mathbb{N}}, x_n \in (a,b)$, converging to $x$ the following holds

    $$\lim_{n \to \infty} f(x_n) = f(\lim_{n \to \infty} x_n).$$

2.* Let $D \subset \mathbb{R}$ be an open set. Prove that $f : D \to \mathbb{R}$ is continuous if and only if the pre-image of every open set in $\mathbb{R}$ is again open, i.e. $f^{-1}(U)$ is open whenever $U \subset \mathbb{R}$ is open.

3.   Give an $\epsilon - \delta$ definition for $f : D \to \mathbb{R}$ having a right (left) limit at $x \in D$.

4.   a) Consider the function $\chi_{[0,1]\cap\mathbb{Q}} : [0, 1] \to \mathbb{R}$ i.e.

$$\chi_{[0,1]\cap\mathbb{Q}} = \begin{cases} 1, & x \in [0, 1] \cap \mathbb{Q} \\ 0, & x \in [0, 1] \text{ and } x \notin \mathbb{Q}. \end{cases}$$

Prove that $\chi_{[0,1]\cap\mathbb{Q}}$ is not continuous at any point $x \in [0, 1]$.

b) Define $f : \mathbb{R} \to \mathbb{R}$ by

$$f(x) := \begin{cases} x, & x \in \mathbb{Q} \\ 0, & x \in \mathbb{R} \setminus \mathbb{Q}. \end{cases}$$

Prove that $f$ is only continuous at $x = 0$.

5.   Let $g : [0, 1] \to \mathbb{R}$ be an arbitrary bounded function. Prove that $f : [0, 1] \to \mathbb{R}$, $f(x) = xg(x)$, is continuous at $x = 0$.

6.*   a) Let $f : (a, b) \to \mathbb{R}$ be a monotone function and $x_0 \in (a, b)$. Prove that

$$\lim_{\substack{x \to x_0 \\ x > x_0}} f(x) \text{ exists.}$$

b) Let $f : D \to \mathbb{R}$ be a function and $x \in D$. We call $x$ a **point of discontinuity** of $f$ if $f$ is not continuous at $x$. Now let $I \subset \mathbb{R}$ be an interval (bounded or unbounded) and let $g : I \to \mathbb{R}$ be a monotone function. Prove that $g$ has at most countable points of discontinuity.

7.*   Let $I \subset \mathbb{R}$ be an interval (bounded or unbounded). We call $f : I \to \mathbb{R}$ a **càdlàg function** (*continu à droite, limites à gauche*) if for all $x \in I$ the function $f$ is continuous from the right and has a limit from the left, i.e.

$$f_r := \lim_{\substack{y \to x \\ y > x}} f(y) = f(x) \quad \text{and} \quad f_l := \lim_{\substack{y \to x \\ y < x}} f(y) \text{ exists.}$$

Prove that if $f : I \to \mathbb{R}$ is a monotone function then there exists a monotone function $h : I \to \mathbb{R}$ which is càdlàg and coincides with $f$ apart from at a countable number of points of discontinuity.
Hint: use the result of Problem 6.

8.    a) Let $f, g : D \to \mathbb{R}$ be two continuous functions. Prove that $\varphi, \psi : D \to \mathbb{R}$ defined by $\varphi(x) = \max(f(x), g(x))$ and $\psi(x) = \min(f(x), g(x))$ are also continuous on $D$.

b) Let $f : D \to \mathbb{R}$ be a function and define

$$f_+(x) := \begin{cases} f(x), & \text{if } f(x) \geq 0 \\ 0, & \text{if } f(x) < 0 \end{cases}$$

and

$$f_-(x) := \begin{cases} -f(x), & \text{if } f(x) \leq 0 \\ 0, & \text{if } f(x) > 0. \end{cases}$$

Show that $f = f_+ - f_-$ and $|f| = f_+ + f_-$. Moreover, show that $f$ is continuous if and only if $f_+$ and $f_-$ are continuous.

9.   Let $f, g : [a, b] \to \mathbb{R}$ be two continuous functions and suppose that $f|_{[a,b] \cap \mathbb{Q}} = g|_{[a,b] \cap \mathbb{Q}}$. Prove that $f = g$, i.e. if $f$ and $g$ coincide on rational points of their domains, then they coincide everywhere.

10.  Let $f : [0, a] \to \mathbb{R}$ be a continuous function. Prove that $f$ has a unique continuous extension to $[-a, a]$ as an even function and that $f - f(0)$ has a unique continuous extension to $[-a, a]$ as an odd function.

11.  Let $D \subset \mathbb{R}$ be a non-empty set and $C(D)$ the set of all continuous functions $f : D \to \mathbb{R}$.

a) Prove that $C(D)$ with its natural operations forms an $\mathbb{R}$-algebra.

b) Let $a : D \to \mathbb{R}$ be a fixed continuous function and define $A_{op} : C(D) \to C(D)$ by $A_{op}u = au$, i.e. $A_{op}u(x) = a(x)u(x)$. Prove that $A_{op}$ is a linear operator on $C(D)$.

12.  Let $f : D \to \mathbb{R}, D \subset \mathbb{R}$, be a function. We call $x_0 \in D$ a **fixed point** of $f$ if $f(x_0) = x_0$.

a) Give a geometric or graphical interpretations for a fixed point.

b) Prove that $h : D \to \mathbb{R}$ has an $a$-point, i.e. there exists $x_0 \in D$ such that $h(x_0) = a$, if and only if $g : D \to \mathbb{R}, g(x) = h(x) + x - a$ has a fixed point.

13.    a) Let $f, g : [a, b] \to \mathbb{R}$ be two continuous functions such that $f(a) < g(a)$ and $f(b) > g(b)$. Prove that there exists $x_0 \in [a, b]$ such that $f(x_0) = g(x_0)$.

b) Prove that there exists at least one $x_0 \in [\frac{\pi}{2}, \frac{3\pi}{2}]$ solving the equation $\sin x = \frac{1}{2 + \cos^4 x}$.

14.    a) Consider the two sets $A := \{\frac{1}{n} | n \in \mathbb{N}\} \subset \mathbb{R}$ and $B := A \cup \{0\}$. Using the basic definition of compactness prove that $A$ is not compact but $B$ is.

b) Let $(a_n)_{n \in \mathbb{N}}$, $a_n \in \mathbb{R}$, be a sequence of real numbers converging to $a_0 \in \mathbb{R}$. Prove that $\{a_k | k \in \mathbb{N}_0\}$ is compact.

15.    For every $N \in \mathbb{N}$ an open covering of $(0, 1)$ is given by $(U_x)_{x \in [0,1]}, U_x = (x - \frac{3}{4N}, x + \frac{3}{4N})$. Prove that $(U_{\frac{k}{n}})_{k=0,\ldots,N}$ is an open subcovering of $(0, 1)$ but $(0, 1)$ is not compact.

16.    Let $(K_\nu)_{\nu \in I}$ be a family of compact sets $K_\nu \subset \mathbb{R}$. Prove that $\bigcap_{\nu \in I} K_\nu$ is compact, but in general $\bigcup_{\nu \in I} K_\nu$ is not compact.

17.    a) Let $f : K \to \mathbb{R}$ be a continuous function defined on a compact set $K$. If $f(x) > 0$ for all $x \in K$ then there exists $\alpha > 0$ such that $f(x) \geq \alpha > 0$ for all $x \in K$, i.e. if a continuous function is strictly positive on a compact set, it is bounded away from 0.

b) Prove that if $f : D \to \mathbb{R}$ is uniformly continuous and $D$ is bounded, then $f$ is bounded.

18.    For $a \in \mathbb{R}$ consider $f : [-a, \infty) \to \mathbb{R}, f(x) = \sqrt{x + a}$. Prove that $f$ is uniformly continuous.

19.    Let $f : [a, b] \to \mathbb{R}$ be a continuous function. We call $f$ **piecewise linear** if there exists a partition $a = x_0 < x_1 < \cdots < x_N = b$ of $[a, b]$ and real numbers $\alpha_k$ and $\beta_k$ such that $f|_{[x_{k-1}, x_k]} = \alpha_k x + \beta_k, k = 1, \ldots, N$. Let $g : [a, b] \to \mathbb{R}$ be a continuous function. Prove that for every $\epsilon > 0$ there exists a piecewise linear function $\varphi : [a, b] \to \mathbb{R}$ such that for all $x \in [a, b]$

$$|g(x) - \varphi(x)| \leq \epsilon.$$

20. We call a function $f : D \to \mathbb{R}$ **Lipschitz continuous** if for some $\kappa > 0$ we have $|f(x) - f(y)| \leq \kappa |x - y|$ for all $x, y \in D$. Prove that a Lipschitz continuous function is uniformly continuous.

21.     a) Let $f : D \to \mathbb{R}$ be a uniformly continuous function. Prove that for every $D' \subset D$ the function $f|_{D'} : D' \to \mathbb{R}$ is uniformly continuous too.

b) Let $g : (a, b] \to \mathbb{R}$ be a continuous function. Suppose that $\lim_{\substack{x \to a \\ x > a}} g(x)$ exists. Prove that $g$ is uniformly continuous.
Hint: show that $g$ has a continuous extension to $[a, b]$.

# 21 Differentiation

Let $D \subset \mathbb{R}$ and $f : D \to \mathbb{R}$ be a function. We know by examples that even continuous functions may look rather complicated. Thus we may ask the question whether it is possible to approximate locally a given function by a simpler function. Obviously straight lines (considered as graphs of functions) are the simplest functions on $\mathbb{R}$. They are given by

$$g_{a,b} : \quad \mathbb{R} \to \mathbb{R}$$
$$x \mapsto ax + b$$

with $a, b \in \mathbb{R}$. We want to make our considerations for a moment more complicated and relate our point of view to linear algebra. Given an $n$-dimensional vector space $(V, \mathbb{R})$ over the reals. A mapping $A : V \to V$ is called **linear** if $A(\lambda x + \mu y) = \lambda Ax + \mu Ay$ holds for all $\lambda, \mu \in \mathbb{R}$ and $x, y \in V$. Choosing a fixed basis in $V$ we know that with respect to this basis $A$ has a representation as an $n \times n$-matrix. Now, $\mathbb{R}$ is a real vector space of dimension 1 and taking $1 \in \mathbb{R}$ as basis any matrix is just a real number. Thus all linear mappings $A_a : \mathbb{R} \to \mathbb{R}$ have the matrix representation $x \mapsto A_a x = ax$ where $a \in \mathbb{R}$ represents $A_a$.

Therefore we may interpret a straight line as the graph of the composition of two mappings: A linear mapping $x \mapsto ax$ and a translation $T_b : \mathbb{R} \to \mathbb{R}$, $x \mapsto x + b$, i.e. we consider

$$T_b \circ A_a : \quad \mathbb{R} \to \mathbb{R}$$
$$x \mapsto T_b(A_a x) = T_b(ax) = ax + b.$$

We call these mappings the affine mappings $h_{a,b} := T_b \circ A_a, a, b \in \mathbb{R}$, on $\mathbb{R}$. Thus straight lines are the graphs of affine maps. More generally:

**Definition 21.1.** *Let $(V, \mathbb{R})$ be a vector space over $\mathbb{R}$. We call $F : V \to V$ an **affine mapping** if $Fx = Ax + b$ holds for a linear mapping $A : V \to V$ and a vector $b \in V$.*

Let us return to our original problem. Given $f : D \to \mathbb{R}$ and $x_0 \in D$. We are looking for an affine mapping $h_{a,b} : \mathbb{R} \to \mathbb{R}$ such that in a neighbourhood of $x_0$ the function $x \mapsto h_{a,b}(x)$ is a good approximation of $x \mapsto f(x)$. In particular we require $f(x_0) = h_{a,b}(x_0)$. Thus in a neighbourhood of $x_0$ we want to have that

$$|f(x) - h_{a,b}(x)| = |f(x) - (ax + b)| \quad \text{is small, and}$$

293

$$f(x_0) = ax_0 + b.$$

Thus

$$|f(x) - f(x_0) - a(x - x_0)| \quad \text{should be small,}$$

which leads to

$$\left| \frac{f(x) - f(x_0)}{x - x_0} - a \right| |x - x_0|$$

should be small. Now suppose that $\phi_{x_0} : D \to \mathbb{R}$ is a function such that

$$\lim_{x \to x_0} \frac{\phi_{x_0}(x)}{x - x_0} = 0. \tag{21.1}$$

Consider for some $c$

$$f(x) = f(x_0) + c(x - x_0) + \phi_{x_0}(x).$$

It follows that

$$\frac{f(x) - f(x_0)}{x - x_0} - c = \frac{\phi_{x_0}(x)}{x - x_0}. \tag{21.2}$$

Since by our assumption $\lim_{x \to x_0} \frac{\phi_{x_0}(x)}{x - x_0} = 0$, we find that in a neighbourhood of $x_0$ the expression

$$\frac{f(x) - f(x_0)}{x - x_0} - c$$

will be small and $x \mapsto h_{c, f(x_0) - c x_0}(x)$ would be locally an affine linear approximation of $f$ at $x_0$. However, in order that we may argue as before, it is clear from (21.2) and (21.1) that

$$\lim_{x \to x_0} \frac{f(x) - f(x_0)}{x - x_0} = c \tag{21.3}$$

must hold.

The existence of the limit (21.3) is by no means clear. Take the function $x \mapsto |x|$ and $x_0 = 0$. For $x > 0$ we find

$$\lim_{\substack{x \to x_0 \\ x > 0}} \frac{f(x) - f(x_0)}{x - x_0} = \lim_{\substack{x \to x_0 \\ x > 0}} \frac{|x| - |0|}{x - 0} = \lim_{x \to x_0} \frac{x}{x} = 1$$

but for $x < 0$ we have

$$\lim_{\substack{x \to x_0 \\ x < 0}} \frac{f(x) - f(x_0)}{x - x_0} = \lim_{\substack{x \to x_0 \\ x < 0}} \frac{-x}{x} = -1$$

thus $\lim_{x \to 0} \frac{|x|}{x}$ does not exist. The following definition is crucial.

294

**Definition 21.2.** *Let $D \subset \mathbb{R}$ and $f : D \to \mathbb{R}$ be a function. We call $f$* ***differentiable*** *at $x_0$ if the limit*

$$f'(x_0) := \lim_{x \to x_0} \frac{f(x) - f(x_0)}{x - x_0} \tag{21.4}$$

*exists. The number $f'(x_0)$ is called the* ***derivative of $f$ at*** *$x_0$. If $f$ is at all points of $D$ differentiable we call the function $f' : D \to \mathbb{R}, x \mapsto f'(x)$, the* ***derivative of $f$***.

Instead of $f'(x_0)$ or $f'$ we also will write $\frac{df}{dx}(x_0)$ or $\frac{df}{dx}$, respectively.

From the previous considerations the following theorem is almost clear.

**Theorem 21.3.** *Let $D \subset \mathbb{R}, x_0 \in D$, and suppose that there is at least one sequence $(x_n)_{n \in \mathbb{N}}, x_n \in D \setminus \{x_0\}$ converging to $x_0$. A function $f : D \to \mathbb{R}$ is differentiable at $x_0$ if and only if there is a constant $c \in \mathbb{R}$ and a function $\phi_{x_0} : D \to \mathbb{R}$ satisfying*

$$\lim_{x \to x_0} \frac{\phi_{x_0}(x)}{x - x_0} = 0, \tag{21.5}$$

*such that*

$$f(x) = f(x_0) + c(x - x_0) + \phi_{x_0}(x). \tag{21.6}$$

*In this case we have $c = f'(x_0)$.*

*Proof.* Suppose that $f$ is differentiable at $x_0 \in D$ and set $c = f'(x_0)$. Defining

$$\phi_{x_0}(x) := f(x) - f(x_0) - f'(x_0)(x - x_0)$$

we find

$$\lim_{x \to x_0} \frac{\phi_{x_0}(x)}{x - x_0} = \lim_{x \to x_0} \left( \frac{f(x) - f(x_0)}{x - x_0} - f'(x_0) \right) = 0,$$

and obviously

$$f(x) = f(x_0) + f'(x_0)(x - x_0) + \phi_{x_0}(x).$$

Now suppose that (21.6) holds with $\phi_{x_0}$ satisfying (21.5). Then we find immediately that

$$\lim_{x \to x_0} \left( \frac{f(x) - f(x_0)}{x - x_0} - c \right) = \lim_{x \to x_0} \frac{\phi_{x_0}(x)}{x - x_0} = 0$$

or

$$\lim_{x \to x_0} \frac{f(x) - f(x_0)}{x - x_0} = c, \quad \text{i.e.} \quad c = f'(x_0).$$

$\square$

Thus the concept of differentiation is that of an affine approximation. It is more convenient to speak about **linear approximations** with the interpretation that $f(x) - f(x_0)$ is approximated by $f'(x_0)(x - x_0)$ and $f'(x_0) : \mathbb{R} \to \mathbb{R}$ is considered to be a linear mapping.

We may also provide a geometric interpretation. Suppose that $f'(x_0)$ exists. The term

$$\frac{f(x) - f(x_0)}{x - x_0},$$

which is called the **difference quotient** of $f$ at $x_0$, gives the slope of the straight line through the points $(x_0, f(x_0))$ and $(x, f(x))$:

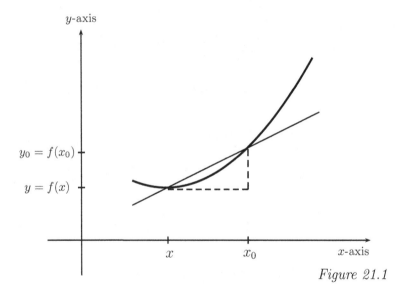

*Figure 21.1*

In the limiting case, if $f$ is differentiable at $x_0$, for $x \to x_0$ these straight lines will "converge" to a straight line with slope $f'(x_0)$ passing through $(x_0, f(x_0))$. This straight line is called the tangent line of (the graph of) $f$ at the point $(x_0, f(x_0))$ :

**Definition 21.4.** *Let $f : D \to \mathbb{R}$ be a function differentiable at $x_0 \in D$. The graph of the function $g_{x_0} : \mathbb{R} \to \mathbb{R}$, $g_{x_0}(t) = f'(x_0)(t - x_0) + f(x_0)$ is called the **tangent** or **tangent line** of $f$ at $x_0$.*

Note that the tangent is defined *after* we have introduced the notion of differentiability. An important consequence of Theorem 21.3 is

**Corollary 21.5.** *A function differentiable at $x_0$ is continuous at $x_0$.*

*Proof.* Since $f$ is differentiable at $x_0$ we have

$$f(x) = f(x_0) + f'(x_0)(x - x_0) + \phi_{x_0}(x)$$

where $\lim_{x \to x_0} \frac{\phi_{x_0}(x)}{x - x_0} = 0$. In particular we have $\lim_{x \to x_0} \phi_{x_0}(x) = 0$ which leads to

$$\lim_{x \to x_0} f(x) = f(x_0) + f'(x_0) \lim_{x \to x_0} (x - x_0) + \lim_{x \to x_0} \phi_{x_0}(x) = f(x_0).$$

$\square$

Let us recollect some concrete derivatives already discussed in Chapter 6, Example 6.1.

**Example 21.6. A.** The function $f_1, x \mapsto c \in \mathbb{R}$, is differentiable on $\mathbb{R}$ and we have $f_1'(x) = 0$ for all $x$:

$$f_1'(x_0) = \lim_{x \to x_0} \frac{f_1(x) - f_1(x_0)}{x - x_0} = \lim_{x \to x_0} \frac{c - c}{x - x_0} = 0.$$

In this case $\phi_{x_0}(x) = 0$ for all $x \in \mathbb{R}$.
**B.** The function $f_2 : \mathbb{R} \to \mathbb{R}, x \mapsto cx$, is differentiable on $\mathbb{R}$ and we have $f_2'(x) = c$ for all $x$:

$$f_2'(x_0) = \lim_{x \to x_0} \frac{f_2(x) - f_2(x_0)}{x - x_0} = \lim_{x \to x_0} \frac{c(x - x_0)}{x - x_0} = c.$$

Again we find $\phi_{x_0}(x) = 0$ for all $x \in \mathbb{R}$. This should not be a surprise: since $f$ is linear, i.e. its graph is a straight line, we just take this straight line as an approximation.
**C.** For $f_3 : \mathbb{R} \to \mathbb{R}, x \mapsto x^2$, we find

$$\frac{f_3(x) - f_3(x_0)}{x - x_0} = \frac{x^2 - x_0^2}{x - x_0} = \frac{(x - x_0)(x + x_0)}{x - x_0} = x + x_0$$

which yields

$$f_3'(x_0) = \lim_{x \to x_0} \frac{x^2 - x_0^2}{x - x_0} = \lim_{x \to x_0} (x + x_0) = 2x_0.$$

Here we have $\phi_{x_0}(x) = (x - x_0)^2$.

**D.** For $f_4, x \mapsto \frac{1}{x}$, defined on $\mathbb{R} \setminus \{0\}$ we find

$$\frac{\frac{1}{x} - \frac{1}{x_0}}{x - x_0} = \frac{\frac{x - x_0}{x x_0}}{x - x_0} = -\frac{1}{x x_0},$$

and it follows for $x_0 \neq 0$ that

$$f_4'(x_0) = \lim_{x \to x_0} \frac{\frac{1}{x} - \frac{1}{x_0}}{x - x_0} = \lim_{x \to x_0} \left( -\frac{1}{x x_0} \right) = -\frac{1}{x_0^2}.$$

In this case we find $\phi_{x_0}(x) = \left( \frac{1}{x_0^2} - \frac{1}{x x_0} \right) (x - x_0)$.

Note that $x \mapsto |x|$ provides us with a function which is not differentiable at $x_0 = 0$.

Elementary rules for differentiation are collected in the following theorem, all results were proved in Chapters 6 and 7.

**Theorem 21.7.** *Let $f, g : D \to \mathbb{R}$ be differentiable at $x_0 \in D$ and $\lambda \in \mathbb{R}$. Then the functions*

$$f + g, \ \lambda f, \ f \cdot g : D \to \mathbb{R}$$

*are differentiable at $x_0$ and we have*

$$(f + g)'(x_0) = f'(x_0) + g'(x_0); \tag{21.7}$$

$$(\lambda f)'(x_0) = \lambda f'(x_0); \tag{21.8}$$

$$(f \cdot g)'(x_0) = f'(x_0)g(x_0) + f(x_0)g'(x_0) \quad \textbf{(Leibniz's rule)}. \tag{21.9}$$

*If in addition $g(x_0) \neq 0$, then $x \mapsto \frac{f}{g}(x)$ is also differentiable at $x_0$ and we have*

$$\left( \frac{f}{g} \right)'(x_0) = \frac{f'(x_0)g(x_0) - f(x_0)g'(x_0)}{g^2(x_0)}. \tag{21.10}$$

**Corollary 21.8.** *All polynomial functions $x \mapsto \sum_{\nu=0}^{m} a_\nu x^\nu$ are differentiable and so are all rational functions $x \mapsto \frac{\sum_{\nu=0}^{m} a_\nu x^\nu}{\sum_{\mu=0}^{k} b_\mu x^\mu}$ on their domain, i.e. the set $\{x \in \mathbb{R} | \sum_{\nu=0}^{k} b_\mu x^\mu \neq 0\}$. Moreover, we have for $n \in \mathbb{N}$*

$$(x^n)' = n x^{n-1}, \quad x \in \mathbb{R},$$

*and*

$$\left( \frac{1}{x^n} \right)' = -n x^{-n-1}, \quad x \in \mathbb{R} \setminus \{0\}.$$

Next we recollect the **chain rule**, see Theorem 7.3.

**Theorem 21.9.** *Let* $f : D \to \mathbb{R}$ *and* $g : E \to \mathbb{R}$ *be two functions such that* $f(D) \subset E$. *Suppose that* $f$ *is differentiable at* $x_0 \in D$ *and that* $g$ *is differentiable at* $y_0 := f(x_0) \in E$. *Then the function*

$$g \circ f : D \to \mathbb{R}$$

*is differentiable at* $x_0$ *and we have*

$$(g \circ f)'(x_0) = g'(f(x_0))f'(x_0). \tag{21.11}$$

**Example 21.10.** For $\alpha \in \mathbb{R}$ consider $f : [0, \infty] \to \mathbb{R}$, $x \mapsto x^\alpha$. Since $x^\alpha = e^{\alpha \ln x} = \exp(\alpha \ln x)$ we find

$$\begin{aligned}
\frac{d}{dx} x^\alpha &= \exp(\alpha \ln x)' \\
&= \exp'(\alpha \ln x) \frac{d}{dx}(\alpha \ln x) \\
&= \frac{\alpha}{x} \exp(\alpha \ln x) = \frac{\alpha}{x} x^\alpha = \alpha x^{\alpha-1}.
\end{aligned}$$

Furthermore we restate Theorem 7.5:

**Theorem 21.11.** *For a strictly monotone and continuous function* $f : I \to \mathbb{R}$ *which is differentiable at* $x_0$ *with* $f'(x_0) \neq 0$ *the inverse function is differentiable at* $y_0 = f(x_0)$ *and*

$$(f^{-1})'(y_0) = \frac{1}{f'(f^{-1}(y_0))} \tag{21.12}$$

*holds.*

Let $f : D \to \mathbb{R}$ be a differentiable function. Then $f' : D \to \mathbb{R}$ is again a function and we may ask whether $f'$ is differentiable. If yes, we call the derivative $(f')'$ of $f'$ the second derivative of $f$ and write simply $f''$. By induction we define **higher order derivatives**(if they exist)

$$\frac{d^k}{dx^k} f(x) := f^{(k)}(x) := \frac{d}{dx}\left(\frac{d^{k-1}}{dx^{k-1}} f(x)\right) = \frac{d}{dx} f^{(k-1)}(x).$$

If $f$ has a $k^{\text{th}}$ derivative as defined above we call $f$ **$k$-times differentiable**

299

**Corollary 21.12.** *Let $f, g : D \to \mathbb{R}$ be two k-times differentiable functions. Then $f \cdot g : D \to \mathbb{R}$ is k-times differentiable and we have*

$$\frac{d^k}{dx^k}(f \cdot g)(x) = \sum_{l=0}^{k} \binom{k}{l} f^{(k-l)}(x) g^{(l)}(x).$$

**Exercise 21.13.** *Prove Corollary 21.12.*

Let us introduce some useful notations. For any interval $I \subset \mathbb{R}$ and $k \in \mathbb{N}$ we denote the set of functions $f$ that are $k$-times differentiable with the $k^{\text{th}}$ derivative continuous on $I$ by $C^k(I)$. Clearly, $C^k(I)$ is a vector space over $\mathbb{R}$, and Leibniz's rule (Corollary 21.12), shows that $C^k(I)$ is an algebra. Further $C(I) = C^0(I)$ denotes the space of all continuous functions $f : I \to \mathbb{R}$, and we set

$$C^\infty(I) := \bigcap_{k \in \mathbb{N}} C^k(I).$$

Moreover we set

$$C_b^k(I) := \{u \in C^k(I) | u^{(l)} \text{ is bounded for } l = 0, \dots, k\},$$

recall $u^{(0)} := u$.

**Exercise 21.14.** *Show that for $0 \leq k \leq \infty$ the set $C^k(I)$ with its natural operations, i.e. pointwise addition and multiplication is an $\mathbb{R}$ algebra.*

Note that functions that are differentiable need not have continuous derivatives.

**Example 21.15.** Let

$$f(x) := \begin{cases} x^2 \sin(\frac{1}{x}) & , x \neq 0 \\ 0 & , x = 0 \end{cases}.$$

Then, by the rules $f$ is differentiable for all $x \neq 0$. In fact

$$f'(x) = 2x \sin\left(\frac{1}{x}\right) - \cos\left(\frac{1}{x}\right)$$

and we have

$$f'(0) = \lim_{x \to 0} \frac{x^2 \sin\left(\frac{1}{x}\right) - 0}{x} = \lim_{x \to 0} x \sin\left(\frac{1}{x}\right) = 0.$$

But $f'(x)$ does not have a limit as $x \to 0$.

Finally we want to state without proof a formula which allows us to calculate higher order derivatives of composed functions. Let $I_1 \subset \mathbb{R}$ and $I_2 \subset \mathbb{R}$ be two intervals and $f : I_2 \to \mathbb{R}$ and $g : I_1 \to \mathbb{R}$ be two arbitrarily often differentiable functions such that $g(I_1) \subset I_2$. The $n^{\text{th}}$-derivative of $h := f \circ g : I_1 \to \mathbb{R}$ is given by

$$h^{(n)}(x) = \sum \frac{n!}{k_1! \cdots k_n!} f^{(k)}(g(x)) \left(\frac{g^{(1)}(x)}{1!}\right)^{k_1} \cdot \left(\frac{g^{(2)}(x)}{2!}\right)^{k_2} \cdots \left(\frac{g^{(n)}(x)}{n!}\right)^{k_n} \quad (21.13)$$

where $k = k_1 + \cdots + k_n$ and the summation is over all $k_1, \ldots, k_n$ such that $k_1 + 2k_2 + \cdots + nk_n = n$. The formula (21.13) is called the **Faà di Bruno formula**, a proof of which is given in W. J. Kaczor and M. T. Nowak [6, p. 227-231].

# Problems

1. In the situation of Figure 21.1 find the straight line $S_{x_0,x} : \mathbb{R} \to \mathbb{R}$ passing through the points $(x_0, f(x_0))$ and $(x, f(x))$. Further prove that if $g_{x_0}$ is the tangent line of $f$ at $x_0$ then $\lim_{x \to x_0} S_{x_0,x}(t) = g_{x_0}(t)$.

2. Prove Leibniz's rule for higher order derivatives: for $f, g \in C^m(I)$, $m \in \mathbb{N} \cup \{\infty\}$, and $k \in \mathbb{N}$, $k \leq m$

$$\frac{d^k}{dx^k}(f \cdot g)(x) = (f \cdot g)^{(k)}(x) = \sum_{l=0}^{k} \binom{k}{l} f^{(k-l)}(x) g^{(l)}(x).$$

3. Show that $C^k(I)$, $0 \leq k \leq \infty$, is an $\mathbb{R}$-algebra.

4. Find $a, b, c, d \in \mathbb{R}$ such that $f : \mathbb{R} \to \mathbb{R}$ given by

$$f(x) := \begin{cases} ax + b, & x \leq 0 \\ cx^2 + dx, & 0 < x \leq 1 \\ 1 - \frac{1}{x}, & x > 1 \end{cases}$$

is differentiable.

5. For $1 \leq j \leq n$, let $f_j : \mathbb{R} \to \mathbb{R}$, $f_j(x) > 0$, be differentiable. Prove

$$\left( \frac{\left( \prod\limits_{k=1}^{n} f_k \right)'}{\prod\limits_{k=1}^{n} f_k} \right)(x) = \sum_{k=1}^{n} \frac{f_k'(x)}{f_k(x)}.$$

6.    a) Let $f : (a, b) \to \mathbb{R}$ be differentiable at $x_0 \in (a, b)$. Show that

$$\lim_{h \to 0} \frac{f(x_0 + h) - f(x_0 - h)}{2h} = f'(x_0).$$

   b) Give an example of a function $g : (a, b) \to \mathbb{R}$ such that for some $x_0 \in (a, b)$

$$\lim_{h \to 0} \frac{g(x_0 + h) - g(x_0 - h)}{2h} = A$$

exists but $g'(x)$ does not, i.e. $g$ is not differentiable at $x_0$.

7. Let $h : [-a, a] \to \mathbb{R}$ be a bounded function. Prove that $f : [-a, a] \to \mathbb{R}$, $f(x) = x^2 h(x)$, is differentiable at $x_0 = 0$ and $f'(0) = 0$.

8.    a) Prove that the derivative of an even function $f : \mathbb{R} \to \mathbb{R}$ is odd and that the derivative of an odd function $g : \mathbb{R} \to \mathbb{R}$ is even.

   b) Let $f : \mathbb{R} \to \mathbb{R}$ be an $a$-periodic function, i.e. $f(x + a) = f(x)$ for all $x \in \mathbb{R}$. Suppose that $f(x) \neq 0$ for all $x \in \mathbb{R}$ and that $f$ is differentiable. Prove that $f'$ is also $a$-periodic.

9. For $|x| < 1$ prove that $\sum_{k=1}^{\infty} k x^k = \frac{x}{(1-x)^2}$.
   Hint: recall $\sum_{k=0}^{\infty} x^k = \frac{1}{1-x}$ for $|x| < 1$.

10.    a) Prove that for $k \in \mathbb{N}$ there exists a polynomial $P_k$ of degree at most $k$ such that

$$\frac{d^k}{dx^k}(1 + x^2)^{-\frac{1}{2}} = \frac{P_k(x)}{(1 + x^2)^{\frac{2k+1}{2}}}$$

and derive that

$$\left| \frac{d^k}{dx^k}(1 + x^2)^{-\frac{1}{2}} \right| \leq c_k \frac{1}{(1 + x^2)^{\frac{k+1}{2}}}.$$

b) Let $f \in C_b^m(\mathbb{R})$, $m \in \mathbb{N}$. Use the Faà di Bruno formula to prove

$$|f^{(m)}((1+x^2)^{-\frac{1}{2}})| \leq c_m \frac{1}{(1+x^2)^{\frac{m+1}{2}}}.$$

# 22 Applications of the Derivative

In this chapter we first recollect results from Part 1 and we provide some of the missing proofs. Moreover, we will add some further applications of the derivative to problems in geometry. We start with

**Definition 22.1.** *Let* $f : (a, b) \to \mathbb{R}$ *be a function. The function* $f$ *has a* **local maximum (local minimum)** *at* $x \in (a, b)$ *if there is an* $\varepsilon > 0$ *such that*

$$f(x) \geq f(y) \quad (f(x) \leq f(y)) \text{ whenever } |y - x| < \varepsilon. \qquad (22.1)$$

*If in* (22.1) *equality holds only for* $x = y$, *then we will speak of an* **isolated maximum (minimum)**. *By a* **local extreme value** *we mean either a local maximum or a local minimum.*

Of fundamental importance is now

**Theorem 22.2.** *Suppose* $f : (a, b) \to \mathbb{R}$ *has a local extreme value at* $x \in (a, b)$ *and that* $f$ *is differentiable at* $x$. *Then we have* $f'(x) = 0$.

*Proof.* If $f$ is constant on $(a, b)$, then any $c$ will do. If not, suppose first that $f$ has a local maximum at $x \in (a, b)$. Take $\varepsilon > 0$ such that $(x - \varepsilon, x + \varepsilon) \subset (a, b)$ and

$$f(y) \leq f(x) \text{ for all } y \in (x - \varepsilon, x + \varepsilon).$$

It follows that

$$f'_+(x) = \lim_{\substack{y \to x \\ y > 0}} \frac{f(y) - f(x)}{y - x} \leq 0 \qquad (22.2)$$

and

$$f'_-(x) = \lim_{\substack{y \to x \\ y < 0}} \frac{f(y) - f(x)}{y - x} \geq 0. \qquad (22.3)$$

The differentiability of $f$ implies now

$$f'_+(x) = f'_-(x) = f'(x),$$

hence $f'(x) = 0$. If $f$ has a local minimum apply the proof to $-f$. $\qquad \square$

**Remark 22.3. A.** Note that $f'(x) = 0$ does not imply that $f$ has a local extreme value at $x$: take $f(x) = x^3$ and $x = 0$.
**B.** The geometric interpretation of Theorem 22.2 is that at a local extreme

value $f$ has a horizontal tangent.

C. Suppose that $f : [a, b] \to \mathbb{R}$ is continuous and differentiable on $(a, b)$. We know that $f$ has a global maximum and minimum on $[a, b]$. But if one of these global extreme values lies at a boundary point $a$ or $b$, then it is not necessary that $f'(a)$ or $f'(b)$ is zero. Here we must understand $f'(a)$ and $f'(b)$ as one-sided limits of the difference quotient. Consider $f : [0, 1] \to \mathbb{R}$, $x \mapsto x$, which has the global minimum at $x_1 = 0$ and the global maximum at $x_2 = 1$. But $f'(y) = 1$ for all $y \in [0, 1]$.

D. We say that $f$ is **differentiable from the right** at $x$ (**differentiable from the left**) if the limit (22.2) ((22.3)) exists.

**Theorem 22.4 (Rolle's Theorem).** *Let $f : [a, b] \to \mathbb{R}$ be a continuous function, differentiable on $(a, b)$, with $f(a) = f(b)$. Then there is a point $x \in (a, b)$ such that $f'(x) = 0$.*

*Proof.* Since $[a, b]$ is compact and $f$ is continuous, $f$ has a maximum and a minimum. Since $f(a) = f(b)$ and if $f$ is not constant, either the maximum or the minimum is attained at some point $x_0 \in (a, b)$. Hence by Theorem 22.2 we must have $f'(x_0) = 0$. $\qquad\square$

**Theorem 22.5.** *Let $f, g : [a, b] \to \mathbb{R}$ be two continuous functions differentiable on $(a, b)$. Then there is a point $x \in (a, b)$ such that*

$$(f(b) - f(a)) \, g'(x) = (g(b) - g(a)) \, f'(x). \tag{22.4}$$

*Proof.* Consider the continuous function $h : [a, b] \to \mathbb{R}$,

$$h(t) = [f(b) - f(a)] \, g(t) - [g(b) - g(a)] \, f(t)$$

which is differentiable on $(a, b)$. Since

$$h(a) = f(b)g(a) - f(a)g(b) = h(b),$$

we can apply Theorem 22.4, and hence there is $x \in (a, b)$ such that $h'(x) = 0$ i.e.

$$0 = h'(x) = (f(b) - f(a))g'(x) - (g(b) - g(a))f'(x),$$

implying the theorem. $\qquad\square$

Theorem 22.5 is sometimes called the **second or the generalised mean value theorem**. Setting $g(x) = x$ in this theorem gives

**Corollary 22.6 (Mean value theorem).** *If $f$ is a real valued continuous function on $[a,b]$ which is differentiable in $(a,b)$, then there is a point $x \in (a,b)$ such that*

$$f(b) - f(a) = (b-a)f'(x). \tag{22.5}$$

**Corollary 22.7.** *If $f$ is as in Corollary 22.6 and furthermore $m \leq f'(z) \leq M$ holds for all $z \in (a,b)$, then we have the estimates*

$$m(y-x) \leq f(y) - f(x) \leq M(y-x) \tag{22.6}$$

*for all $x, y \in [a,b], x < y$.*

*Proof.* Apply the mean value theorem to $f|_{[x,y]}$ to find

$$f(y) - f(x) = (y-x)f'(z), \text{ some } z \in (x,y)$$

and use $m \leq f'(z) \leq M$. $\qquad\square$

**Corollary 22.8.** *If $f'(x) = 0$ for all $x \in (a,b)$, then $f$ is constant on $(a,b)$.*

*Proof.* Apply Corollary 22.7 with $m = M = 0$. $\qquad\square$

This corollary has an interesting consequence. Suppose that both $f_1, f_2 : (a,b) \to \mathbb{R}$ satisfy the equation $f'(x) = h(x)$ for all $x \in (a,b)$ where $h : (a,b) \to \mathbb{R}$ is a given function. It follows that $f'_1(x) - f'_2(x) = 0$ for all $x \in (a,b)$, hence $f_1 - f_2 = c$ for some $c \in \mathbb{R}$. Thus two solutions of the differential equation $f'(x) = h(x), x \in (a,b)$, differ only by a constant.

We always have a problem evaluating the quotient $\dfrac{0}{0}$. The usual example is when we want to evaluate the limit as $x \to x_0$ of a quotient $\dfrac{f(x)}{g(x)}$ when $f(x_0) = g(x_0) = 0$. If the functions are differentiable then there is a useful corollary to Theorem 22.5:

**Theorem 22.9 (L'Hospital's rule).** *If $f$ and $g$ are differentiable in a neighbourhood of $x_0$ and $f(x_0) = g(x_0) = 0$, then*

$$\lim_{x \to x_0} \frac{f(x)}{g(x)} = \lim_{x \to x_0} \frac{f'(x)}{g'(x)} \tag{22.7}$$

*provided the limit on the right hand side exists.*

*Proof.* Suppose first that $x > x_0$. By Theorem 22.5 applied to the interval $[x_0, x]$, there exists $y$, $x_0 < y < x$, such that

$$\frac{f'(y)}{g'(y)} = \frac{f(x) - f(x_0)}{g(x) - g(x_0)} = \frac{f(x)}{g(x)} .$$

As $x \to x_0, x > x_0$, it follows that $y \to x_0$, thus if $\lim_{y \to x_0} \frac{f'(y)}{g'(y)}$ exists, it is equal to $\lim_{x \to x_0} \frac{f(x)}{g(x)}$. A similar argument works when $x < x_0$. $\square$

Already in Part 1 we made use of these rules, see Theorem 11.5 and Example 11.6.

Sometimes we may have to use L'Hospital's rule more than once.

**Example 22.10.** The following holds

$$\lim_{x \to 0} \frac{\sin(x) - x}{x^3} = \lim_{x \to 0} \frac{\cos(x) - 1}{3x^2} = \lim_{x \to 0} \frac{-\sin(x)}{6x} = -\frac{1}{6} ,$$

where we used $\lim_{x \to \infty} \frac{\sin(x)}{x} = 1$, compare with Theorem 10.4.

**Remark 22.11.** We cannot use L'Hospital's rule to establish non-convergence, as it is possible that $\lim_{x \to x_0} \frac{f(x)}{g(x)}$ exists while $\lim_{x \to x_0} \frac{f'(x)}{g'(x)}$ does not.

**Example 22.12.** Let

$$f(x) := \begin{cases} x^2 \sin\left(\frac{1}{x}\right) & , x \neq 0 \\ 0 & , x = 0 \end{cases}$$

and $g(x) = x$. Then $f(0) = g(0) = 0$ and both functions are differentiable at 0. Now $\frac{f(x)}{g(x)} = x \sin(x) \to 0$ as $x \to 0$, but

$$\frac{f'(x)}{g'(x)} = f'(x) = 2x \sin\left(\frac{1}{x}\right) - \cos\left(\frac{1}{x}\right)$$

which does not have a limit as $x \to 0$.

Other forms of l'Hospital's rule are:

1. If $f$ and $g$ are differentiable and $\lim_{x \to \infty} f(x) = \lim_{x \to \infty} g(x) = 0$, then

$$\lim_{x \to \infty} \frac{f(x)}{g(x)} = \lim_{x \to \infty} \frac{f'(x)}{g'(x)} \qquad (22.8)$$

when the limit on the right hand side exists.

2. If $f$ and $g$ are differentiable and $\lim_{x \to \infty} f(x) = \lim_{x \to \infty} g(x) = \infty$, then

$$\lim_{x \to \infty} \frac{f(x)}{g(x)} = \lim_{x \to \infty} \frac{f'(x)}{g'(x)} \qquad (22.9)$$

when the limit on the right hand side exists.

Next we will characterise monotone functions using the derivative.

**Theorem 22.13.** *Let $f : [a, b] \to \mathbb{R}$ be continuous and differentiable on $(a, b)$. If $f'(x) > 0$ for all $x \in (a, b)$ (or $f'(x) \geq 0, f'(x) \leq 0, f'(x) < 0$), then $f$ is on $[a, b]$ strictly monotone increasing (monotone increasing, monotone decreasing, strictly monotone decreasing).*

*Proof.* We discuss only the case $f'(x) > 0$ for all $x \in (a, b)$, the other cases are analogous. Suppose that $f$ is not strictly monotone increasing. Then there are $x_1, x_2 \in (a, b), x_1 < x_2$ such that $f(x_1) \geq f(x_2)$. By the mean value theorem we find some $y \in (x_1, x_2)$ such that

$$f'(y) = \frac{f(x_2) - f(x_1)}{x_2 - x_1} \leq 0,$$

which is a contradiction. $\qquad\qquad\qquad\qquad\qquad\qquad\qquad \square$

**Exercise 22.14.** *Show that, if $f$ is monotone decreasing on $[a, b]$ and is differentiable on $(a, b)$ then $f'(x) \leq 0$ for all $x$.*

**Remark 22.15.** If $f$ is strictly increasing and differentiable, we need not have $f'(x) > 0$ for all $x$. The function $f(x) = x^3$ is strictly monotone increasing and differentiable, but $f'(0) = 0$.

**Theorem 22.16.** *Suppose that $f : (a, b) \to \mathbb{R}$ is twice differentiable at $x \in (a, b)$. In addition assume that*

$$f'(x) = 0 \text{ and } f''(x) > 0 \quad (f''(x) < 0).$$

*Then $f$ has an isolated local minimum (maximum) at $x$.*

*Proof.* We consider the case $f''(x) > 0$, the second case goes analogously. By assumption we have

$$f''(x) = \lim_{y \to x} \frac{f'(y) - f'(x)}{y - x} > 0.$$

Hence there is an $\varepsilon > 0$ such that

$$\frac{f'(y) - f'(x)}{y - x} > 0 \text{ for all } y, 0 < |y - x| < \varepsilon.$$

Since $f'(x) = 0$ it follows that

$$f'(y) < 0 \text{ for } x - \varepsilon < y < x$$

and

$$f'(y) > 0 \text{ for } x < y < x + \varepsilon.$$

Therefore $f$ is strictly monotone decreasing in $[x-\varepsilon, x]$ and strictly monotone increasing in $[x, x + \varepsilon]$ :

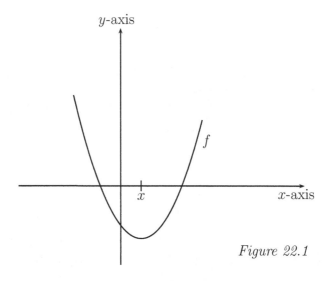

Figure 22.1

Thus, $f$ has an isolated local minimum at $x$. $\quad\square$

**Remark 22.17.** As the function $x \mapsto x^4$ shows we may have a minimum, here at $x_0 = 0$, and $f'(x_0) = f''(x_0) = 0$. Thus if $f''(x_0) = 0$ for a twice differentiable function with $f'(x_0) = 0$ we cannot in general make a statement about whether $f$ has an extreme value at $x_0$ or not.

Let $f : [a, b] \to \mathbb{R}$ be a twice continuously differentiable function. We want to study its graph $\Gamma(f) \subset \mathbb{R}^2$ as a geometrical object.

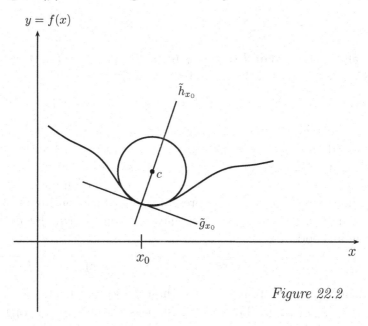

*Figure 22.2*

Locally, i.e. in a neighbourhood of $x_0$ we can replace $\Gamma(f)$ by the tangent line $\tilde{g}_{x_0}$ to give an approximation of $\Gamma(f)$. Recall that $\tilde{g}_{x_0}$ is the straight line

$$\tilde{g}_{x_0} = \{(t, g_{x_0}(t)) | g_{x_0}(t) = f'(x_0)t + f(x_0) - x_0 f'(x_0), t \in \mathbb{R}\} = \Gamma(g_{x_0}) \quad (22.10)$$

which we also interpret as the graph $\Gamma(g_{x_0})$ of the function $t \mapsto g_{x_0}(t)$, $g_{x_0}(t) = f'(x_0)t + f(x_0) - x_0 f'(x_0)$. In the case where $f'(x_0) = 0$ then $\tilde{g}_{x_0}$ is a line parallel to the $x$-axis and the line $\tilde{n}_{x_0} := \{(x_0, t) | t \in \mathbb{R}\}$ is parallel to the $y$-axis and passes through $(x_0, f(x_0))$ and they are perpendicular to each other. However, $\tilde{n}_{x_0}$ is not the graph of a function. For $f'(x_0) \neq 0$ we can consider the straight line

$$\tilde{n}_{x_0} = \left\{ (t, n_{x_0}(t)) | n_{x_0}(t) = -\frac{1}{f'(x_0)} t + f(x_0) + \frac{x_0}{f'(x_0)} \right\} = \Gamma(n_{x_0}),$$

which is the graph of $t \mapsto n_{x_0}(t)$, $n_{x_0}(t) = -\frac{1}{f'(x_0)}t + f(x_0) + \frac{x_0}{f'(x_0)}$. We find that $\tilde{g}_{x_0}$ and $\tilde{n}_{x_0}$ intersect at the point $(x_0, f(x_0))$ and they are perpendicular. The latter follows from the fact that $\tilde{g}_{x_0}$ has direction vector $\begin{pmatrix} 1 \\ f'(x_0) \end{pmatrix}$ and $\tilde{n}_{x_0}$ has direction vector $\begin{pmatrix} 1 \\ -\frac{1}{f'(x_0)} \end{pmatrix}$ implying that their scalar product in $\mathbb{R}^2$ is 0:

$$\left< \begin{pmatrix} 1 \\ f'(x_0) \end{pmatrix}, \begin{pmatrix} 1 \\ -\frac{1}{f'(x_0)} \end{pmatrix} \right> = 1 - \frac{f'(x_0)}{f'(x_0)} = 0.$$

We call $\tilde{n}_{x_0}$ the **normal line** of $f$ (or $\Gamma(f)$) at $x_0$ (or $(x_0, f(x_0))$). In Volume II, Chapter 39, we will understand why it is of advantage to replace the direction vector of $\tilde{n}_{x_0}$ by $\begin{pmatrix} -f'(x_0) \\ 1 \end{pmatrix} = -f(x_0)\begin{pmatrix} 1 \\ -\frac{1}{f(x_0)} \end{pmatrix}$.

If in a neighbourhood of $x_0$ the graph $\Gamma(f)$ is not a straight line, it may be argued that we can approximate the graph $\Gamma(f)$ even better by a circle $\kappa_{x_0}$ passing through $(x_0, f(x_0))$. Suppose that the circle is given by the set $\{(x, y) \in \mathbb{R}^2 | (x - c_1)^2 + (y - c_2)^2 = r^2\}$ and suppose further that in a neighbourhood of $x_0$ we can represent $y$ as a twice continuously differentiable function of $x$, $y = h(x)$. Thus we have $(x - c_1)^2 + (h(x) - c_2)^2 = r^2$, or in a neighbourhood of $x_0$ we have

$$|y - c_2| = |h(x) - c_2| = \sqrt{r^2 - (x - c_1)^2} \quad \text{or} \quad h(x) = \pm\sqrt{r^2 - (x_1 - c_1)^2} + c_2.$$

For being a better approximation than $\tilde{g}_{x_0}$ we must have $h(x_0) = f(x_0)$ and $h'(x_0) = f'(x_0)$, i.e. the circle must pass through $(x_0, f(x_0))$ and have the same tangent line at $(x_0, f(x_0))$ as $f$ has. To improve the approximation we add the condition $h''(x_0) = f''(x_0)$. Now we want to determine $c_1, c_2$ and $r$. Differentiating $(x - c_1)^2 + (h(x) - c_2)^2 = r^2$ twice we find

$$(x - c_1) + h'(x)(h(x) - c_2) = 0 \tag{22.11}$$

and

$$1 + h'^2(x) + h''(x)(h(x) - c_2) = 0. \tag{22.12}$$

For $x_0$ this implies

$$1 + f'^2(x_0) + f''(x_0)(f(x_0) - c_2) = 0,$$

or if $f''(x_0) \neq 0$

$$c_2 = f(x_0) + \frac{1 + f'^2(x_0)}{f''(x_0)}, \tag{22.13}$$

and then
$$x_0 - c_1 + f'(x_0)(f(x_0) - c_2) = 0,$$
or again if $f''(x_0) \neq 0$

$$c_1 = x_0 - f'(x_0)\frac{1 + f'^2(x_0)}{f''(x_0)}, \qquad (22.14)$$

and finally, if $f''(x_0) \neq 0$,

$$r = \frac{(1 + f'^2(x_0))^{\frac{3}{2}}}{|f''(x_0)|}. \qquad (22.15)$$

The condition $f''(x_0) \neq 0$ is of course natural when assuming that locally we can improve the approximation by a straight line. The circle

$$\kappa_{x_0} := \{(x,y) \in \mathbb{R}^2 | (x - c_1)^2 + (y - c_2)^2 = r^2\}$$
$$= \left\{ (x,y) \in \mathbb{R}^2 \left| \left( x - x_0 + f'(x_0)\frac{1 + f'^2(x_0)}{f''(x_0)} \right)^2 \right.\right.$$
$$\left. + \left( y - f(x_0) - \frac{1 + f'^2(x_0)}{f''(x_0)} \right)^2 = \frac{(1 + f'^2(x_0))^3}{|f''(x_0)|^2} \right\}$$

is called the **circle of curvature** or **osculating circle**. Further we call $(c_1, c_2)$ the **centre of curvature**, $r$ is called the **radius of curvature** and $\frac{1}{r}$ is called the **curvature** of $f$ at $x_0$ (or of $\Gamma(f)$ at $(x_0, f(x_0))$).
If we also assume that $f'(x_0) \neq 0$, then we find

$$n_{x_0}(c_1) = -\frac{1}{f'(x_0)}c_1 + \frac{x_0}{f'(x_0)} + f(x_0)$$
$$= -\frac{1}{f'(x_0)} \left( x_0 - f'(x_0)\frac{(1 + f'^2(x_0))}{f''(x_0)} \right) + \frac{x_0}{f'(x_0)} + f(x_0)$$
$$= \frac{1 + f'^2(x_0)}{f''(x_0)} + f(x_0) = c_2,$$

thus the centre of curvature lies on the normal line.

# Problems

1. Use the generalised mean value theorem to prove that for $f \in C^2([a,b])$ satisfying $f(a) = f(b)$ and $f'(a) = f'(b) = 0$ there exists $x_1, x_2 \in (a,b)$, $x_1 \neq x_2$, such that $f''(x_1) = f''(x_2)$.

2. Let $f : [a, b] \to \mathbb{R}$ be a function satisfying the estimate $|f(x) - f(y)| \le c|x - y|^{1+\alpha}$ for all $x, y \in [a, b]$. Prove that $f$ is constant, i.e. $f(x) = c_0$ for some $c_0 \in \mathbb{R}$ and all $x \in [a, b]$.

3. Let $f : (a, b) \to \mathbb{R}$ be differentiable with bounded derivative $f'$, i.e. $|f'(x)| \le M$ for all $x \in (a, b)$. Prove that $f$ is Lipschitz continuous and hence uniformly continuous, see Problem 20 in Chapter 20 for the definition of Lipschitz continuity.

4. For $0 < p < q$ and $x > 0$ use the mean value theorem to show

$$\left(1 + \frac{x}{p}\right)^p < \left(1 + \frac{x}{q}\right)^q .$$

Hint: apply the mean value theorem to $y \mapsto \ln(1 + y)$ on $[0, \frac{x}{q}]$ and on $[\frac{x}{q}, \frac{x}{p}], x > 0$.

5. For $\alpha, \beta > 0$ prove:

   a)
   $$\lim_{x \to \infty} \frac{e^{\alpha x}}{x^\beta} = +\infty;$$

   b)
   $$\lim_{x \to \infty} \frac{(\ln x)^\beta}{x^\alpha} = 0;$$

   c)
   $$\lim_{\substack{x \to 0 \\ x > 0}} x^x = 1.$$

6. Find the following limit:

$$\lim_{x \to 7} (8 - x)^{\frac{1}{x-7}} .$$

7. Let $f \in C^2(\mathbb{R})$ such that $f(0) = 1$, $f'(0) = 0$, and $f''(0) = -1$. Prove that for any $a \in \mathbb{R}$

$$\lim_{\substack{x \to \infty \\ x > 0}} \left( f\left(\frac{a}{\sqrt{x}}\right) \right)^x = e^{-\frac{a^2}{2}} .$$

(This problem is taken from [6].)

314

8. Show that if $f$ is monotone decreasing on $[a, b]$ and is differentiable on $[a, b]$ then $f'(x) \leq 0$ for all $x \in (a, b)$.

9. A function $f \in C^\infty((0, \infty))$ is said to be **completely monotone** if for all $k \in \mathbb{N}_0$ the following holds

$$(-1)^k \frac{d^k f(t)}{dt^k} \geq 0. \tag{22.16}$$

A function $f \in C^\infty((0, \infty))$ is called a **Bernstein function** if $f \geq 0$ and for all $k \in \mathbb{N}$

$$(-1)^k \frac{d^k f(t)}{dt^k} \leq 0. \tag{22.17}$$

Prove that for $a > 0$ the function $t \mapsto e^{-at}$ is completely monotone and the function $t \mapsto 1 - e^{-at}$ is a Bernstein function. Furthermore show that $t \mapsto t^\alpha$, $0 < \alpha \leq 1$, is another Bernstein function.

10.   a) Determine all local extreme values of $f(x) = x^{\frac{1}{3}}(1 - x)^{\frac{2}{3}}$.

   b) Find the maximum of $f : \mathbb{R} \to \mathbb{R}$ given by

$$f(x) = \frac{1}{1 + |x|} + \frac{1}{1 + |x - 1|}.$$

(This problem is taken from [6].)

11. Let $g : [-1, 1] \to \mathbb{R}$, $g(x) = \sqrt{1 - x^2}$. For $x_0 \in (-1, 1)$ find the tangent line, the normal line and the circle of curvature of $g$ at $x_0$.

12. Consider the hyberbola $f : (0, \infty) \to \mathbb{R}$, $f(x) = \frac{1}{x}$. For $x_0 \in (0, \infty)$ find the normal line and the curvature of $f$ at $x_0$.

# 23  Convex Functions and some Norms on $\mathbb{R}^n$

Let us begin with

**Definition 23.1.** *Let $I \subset \mathbb{R}$ be an interval and $f : I \to \mathbb{R}$ be a function. We call $f : I \to \mathbb{R}$ **convex** if for all $x_1, x_2 \in I$ and all $\lambda \in (0, 1)$ the inequality*

$$f(\lambda x_1 + (1 - \lambda)x_2) \leq \lambda f(x_1) + (1 - \lambda)f(x_2) \tag{23.1}$$

*holds. If $-f$ is convex, we call $f$ **concave**.*
*Obviously $(23.1)$ is also correct for $\lambda = 1$ and $\lambda = 0$.*

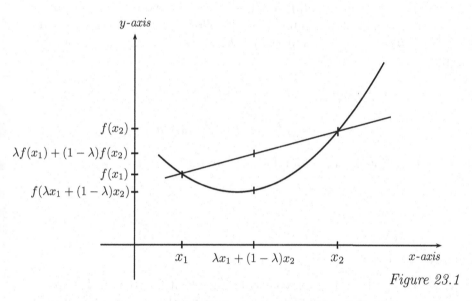

Figure 23.1

**Theorem 23.2.** *Let $I \subset \mathbb{R}$ be an open interval and $f : I \to \mathbb{R}$ a twice differentiable function. The function $f$ is convex if and only if $f''(x) \geq 0$ for all $x \in I$.*

*Proof.* Suppose that $f''(x) \geq 0$ for all $x \in I$. It follows that $f'$ is monotone increasing on $I$. For $x_1, x_2 \in I, x_1 < x_2$, and $0 < \lambda < 1$ we put $x := \lambda x_1 + (1 - \lambda)x_2$ and so $x_1 < x < x_2$. By the mean value theorem there exists $y_1 \in (x_1, x)$ and $y_2 \in (x, x_2)$ such that

$$\frac{f(x) - f(x_1)}{x - x_1} = f'(y_1) \leq f'(y_2) = \frac{f(x_2) - f(x)}{x_2 - x}.$$

But $x - x_1 = (1 - \lambda)(x_2 - x_1)$ and $x_2 - x = \lambda(x_2 - x_1)$, which leads to

$$\frac{f(x) - f(x_1)}{1 - \lambda} \leq \frac{f(x_2) - f(x)}{\lambda},$$

i.e.

$$\lambda f(x) - \lambda f(x_1) \leq (1 - \lambda)f(x_2) - (1 - \lambda)f(x),$$

or

$$\lambda f(x) + (1 - \lambda)f(x) = f(x) \leq \lambda f(x_1) + (1 - \lambda)f(x_2),$$

hence $f$ is convex.

Now suppose that $f : I \to \mathbb{R}$ is convex. Further assume that for some $x_0 \in I$ we have $f''(x_0) < 0$. For $c := f'(x_0)$ and $\phi(x) := f(x) - c(x - x_0), x \in I$, it follows that $\phi'(x_0) = 0$ and $\phi''(x_0) = f''(x_0) < 0$. Therefore the function $\phi$ must have an isolated local maximum at $x_0$. It follows that there exists $h > 0$ such that $[x_0 - h, x_0 + h] \subset I$ and $\phi(x_0 - h) < \phi(x_0), \phi(x_0 + h) < \phi(x_0)$, which implies

$$f(x_0) = \phi(x_0) > \frac{1}{2}(\phi(x_0 - h) + \phi(x_0 + h))$$

$$= \frac{1}{2}(f(x_0 - h) + f(x_0 + h)).$$

Taking $x_1 := x_0 - h, x_2 := x_0 + h$ and $\lambda = \frac{1}{2}$, we find

$$x_0 = \lambda x_1 + (1 - \lambda)x_2$$

and

$$f(\lambda x_1 + (1 - \lambda)x_2) > \lambda f(x_1) + (1 - \lambda)f(x_2)$$

which is a contradiction. □

**Remark 23.3.** The criterion for convexity (concavity) given in Theorem 23.2 we may combine with our sufficient criterion for the existence of a local minimum (maximum). If $f : (a, b) \to \mathbb{R}$ is twice continuously differentiable and has a critical point at $c \in (a, b)$, i.e. $f'(c) = 0$, then the graph $\Gamma(f)$ of $f$ has at $c$ a horizontal tangent. If $f$ is convex (concave) in a neighbourhood of $c$ the graph of $f$ must lie above (below) this horizontal tangent, hence at $c$ the function $f$ has a local minimum (maximum). Thus our sufficient criterion for the existence of a local minimum (maximum) at $c$, i.e. Theorem 8.8, has a natural geometric interpretation: if $f$ has at $c$ a horizontal tangent and if $f''(c) > 0$ ($f''(c) < 0$) then $f$ is locally, i.e. in a neighbourhood of $c$, convex (concave) and therefore $f$ has at $c$ a minimum (maximum).

The basic definition of convexity does not require differentiability and not even continuity, it is a geometric statement expressed by an inequality. If we consider Figure 23.1 then inequality (23.1) says that for all $x \in [x_1, x_2]$ the graph of $f$ lies below the line segment connecting $(x_1, f(x_1))$ to $(x_2, f(x_2))$. This line segment is the graph of the function

$$g(t) = f(x_1) + \frac{f(x_1) - f(x_2)}{x_1 - x_2}(t - x_1), t \in [x_1, x_2]. \qquad (23.2)$$

Hence convexity means

$$f(t) \leq f(x_1) + \frac{f(x_1) - f(x_2)}{x_1 - x_2}(t - x_1) \qquad (23.3)$$

for all $t \in [x_1, x_2]$.

**Lemma 23.4.** *A function $f : I \to \mathbb{R}$ is convex if and only if for any three points $x < z < y$, $x, y, z \in I$, the inequalities*

$$\frac{f(x) - f(z)}{x - z} \leq \frac{f(x) - f(y)}{x - y} \leq \frac{f(z) - f(y)}{z - y} \qquad (23.4)$$

*hold.*

*Proof.* From (23.3) we deduce with $x = x_1, y = x_2, z = t$ that

$$\frac{f(x) - f(z)}{x - z} = \frac{f(z) - f(x)}{z - x} \leq \frac{f(x) - f(y)}{x - y},$$

which is the first inequality in (23.4). Since

$$f(x) - f(y) = \frac{f(x) - f(y)}{x - y}(x - y) = \frac{f(x) - f(y)}{x - y}(z - y + x - z)$$

we find

$$f(x) + \frac{f(x) - f(y)}{x - y}(z - x) = f(y) + \frac{f(x) - f(y)}{x - y}(z - y),$$

and with (23.3) it follows that

$$f(z) \leq f(y) + \frac{f(x) - f(y)}{x - y}(z - y),$$

319

or

$$f(z) - f(y) \leq \frac{f(x) - f(y)}{x - y}(z - y).$$

Taking into account that $z - y < 0$, we eventually arrive at

$$\frac{f(x) - f(y)}{x - y} \leq \frac{f(z) - f(y)}{z - y},$$

proving the second inequality in (23.4). Now suppose that (23.4) holds and take $z = \alpha x + (1 - \alpha)y$ to find

$$\frac{f(x) - f(\alpha x + (1 - \alpha)y)}{x - \alpha x - (1 - \alpha)y} \leq \frac{f(x) - f(y)}{x - y} \leq \frac{f(\alpha x + (1 - \alpha)y) - f(y)}{\alpha x + (1 - \alpha)y - y},$$

which yields

$$\frac{f(x) - f(\alpha x + (1 - \alpha)y)}{(1 - \alpha)(x - y)} \leq \frac{f(\alpha x + (1 - \alpha)y) - f(y)}{\alpha(x - y)},$$

and since $x - y < 0$ we arrive at

$$\alpha(f(x) - f(\alpha x + (1 - \alpha)y)) \geq (1 - \alpha)(f(\alpha x + (1 - \alpha)y) - f(y)),$$

or

$$f(\alpha x + (1 - \alpha)y) \leq \alpha f(x) + (1 - \alpha)f(y),$$

proving the convexity of $f$. $\qquad\square$

**Theorem 23.5.** *Let $I$ be an interval with end points $a < b$ and let $f : I \to \mathbb{R}$ be a convex function. For every $x \in (a, b)$ the function is differentiable from the right and from the left.*

*Proof.* Take $x \in (a, b)$ and $t_1, t_2 \in I$ such that $x < t_1 < t_2$. From (23.4) we deduce

$$\frac{f(x) - f(t_1)}{x - t_1} \leq \frac{f(x) - f(t_2)}{x - t_2},$$

in other words, the function $F : [x, b] \cap I \to \mathbb{R}$, $F(t) = \frac{f(x) - f(t)}{x - t}$, is monotone increasing. Further, for $x_1 \in I$ with $x_1 < x$ it follows again by (23.4) that

$$\frac{f(x_1) - f(x)}{x_1 - x} \leq \frac{f(x) - f(t)}{x - t} = F(t),$$

implying that $F$ is bounded from below. Hence

$$\lim_{\substack{t \to x \\ t>0}} F(t) = \lim_{\substack{t \to x \\ t>0}} \frac{f(x) - f(t)}{x - t} = f'_+(x)$$

exists. Analogously we see that $f$ is differentiable from the left, i.e. $f'_-(x)$ exists for $x \in (a, b)$. $\qquad \square$

**Corollary 23.6.** *Let $f : I \to \mathbb{R}$ be convex with $I$ being an interval with end points $a < b$, then $f|_{(a,b)}$ is continuous.*

*Proof.* With the same argument as in the proof of Corollary 21.5 we deduce that if $f$ is differentiable from the right (left) at $x \in (a, b)$ then $f$ is continuous from the right (left) at $x$. Hence being continuous from the right and from the left, $f$ must be continuous at $x$. (A more detailed proof is given in Problem 3.) $\qquad \square$

**Remark 23.7.** Using Problem 6 in Chapter 20 and some further considerations it is possible to prove that a convex function $f$ as in Theorem 23.5 is at most a countable set non-differentiable. (Compare with D. J. H. Garling, [4, Corollary 7.2.4, p. 184]).

**Proposition 23.8.** *Let $I \subset \mathbb{R}$ be an interval and $f, g, f_n : I \to \mathbb{R}$, $n \in \mathbb{N}$ be convex functions. Then $f + g$ and $\alpha f, \alpha \geq 0$, are convex functions and if $F(x) := \lim_{n \to \infty} f_n(x)$ exists and is finite for every $x \in I$, then $F : I \to \mathbb{R}$ is convex too.*

*Proof.* The convexity of $f + g$ and $\alpha f$ follows from the defining inequalities

$$f(\lambda x_1 + (1 - \lambda)x_2) \leq \lambda f(x_1) + (1 - \lambda)f(x_2)$$

and

$$g(\lambda x_1 + (1 - \lambda)x_2) \leq \lambda g(x_1) + (1 - \lambda)g(x_2)$$

by adding and multiplying by $\alpha \geq 0$, respectively. Moreover, if $\lim_{n \to \infty} f_n(x) = F(x) < \infty$ exists for all $x \in I$ we can pass to the limit in

$$f_n(\lambda x_1 + (1 - \lambda)x_2) \leq \lambda f_n(x_1) + (1 - \lambda)f_n(x_2)$$

and the inequality is preserved for the limit function.

$\qquad \square$

**Remark 23.9.** Note that by Corollary 23.6 convex functions provide us with a class of functions for which the pointwise limit of sequences belonging to this class is always continuous.

**Proposition 23.10.** *Let $I \subset \mathbb{R}$ be an interval and $J \neq \emptyset$ an index set. Suppose that for each $j \in J$ a convex function $f_j : I \to \mathbb{R}$ is given. Then if*

$$g(x) := \sup\{f_j(x) | j \in J\} < \infty \qquad (23.5)$$

*is finite for each $x \in I$, then $g : I \to \mathbb{R}$ is convex.*

*Proof.* Let $\epsilon > 0$. There exists $f_j, j \in J$, such that for all $x_1, x_2 \in I$ and $\lambda \in (0, 1)$ the following holds

$$f_j(\lambda x_1 + (1 - \lambda)x_2) \geq g(\lambda x_1 + (1 - \lambda)x_2) - \epsilon,$$

which implies by the convexity of $f$

$$
\begin{aligned}
g(\lambda x_1 + (1 - \lambda)x_2) - \epsilon &\leq f_j(\lambda x_1 + (1 - \lambda)x_2) \\
&\leq \lambda f_j(x_1) + (1 - \lambda)f_j(x_2) \\
&\leq \lambda g(x_1) + (1 - \lambda)g(x_2).
\end{aligned}
$$

Since $\epsilon > 0$ is arbitrary we eventually get

$$g(\lambda x_1 + (1 - \lambda)x_2) \leq \lambda g(x_1) + (1 - \lambda)g(x_2).$$

$\square$

The following simple inequality turns out to be quite useful:

**Lemma 23.11.** *Let $p, q \in (1, \infty)$ such that $\frac{1}{p} + \frac{1}{q} = 1$, then we have for all $x, y \geq 0$*

$$x^{\frac{1}{p}} y^{\frac{1}{q}} \leq \frac{x}{p} + \frac{y}{q}. \qquad (23.6)$$

*Proof.* We may assume $x, y > 0$. For the function $\ln : (0, \infty) \to \mathbb{R}$ we find $\frac{d^2}{dx^2}(\ln(x)) = -\frac{1}{x^2} < 0$, thus the function $\ln$ is concave, implying that

$$\ln(\frac{1}{p}x + \frac{1}{q}y) \geq \frac{1}{p}\ln x + \frac{1}{q}\ln y,$$

or

$$\exp(\ln(\frac{1}{p}x + \frac{1}{q}y)) \geq \exp(\frac{1}{p}\ln x + \frac{1}{q}\ln y),$$

leading to (23.6) $\square$

322

The following considerations are just the beginning of a better understanding of the concept of a limit and convergence.

On $\mathbb{R}$, the natural distance between two numbers is the absolute value of their difference. Using this we are able to define convergence. In other spaces, we need a notion of 'distance' or **metric**. Even in $\mathbb{R}^2$ we have a choice:

the Euclidean distance: $d(x, y) = \sqrt{(x_1 - y_1)^2 + (x_2 - y_2)^2}$;

the distance: $d(x, y) = |x_1 - y_1| + |x_2 - y_2|$;

and the sup metric: $d(x, y) = \max\{|x_1 - y_1|, |x_2 - y_2|\}$.

In these cases we actually only need to define the distance from a point to the origin.

**Definition 23.12.** *A mapping* $\| \cdot \| : \mathbb{R}^n \to \mathbb{R}$ *is called a* **norm** *on* $\mathbb{R}^n$ *if*

1. $\|x\| \geq 0$ *for all* $x \in \mathbb{R}^n$ *and* $\|x\| = 0$ *if and only if* $x = 0$;

2. $\|\lambda x\| = |\lambda|\|x\|$ *for all* $x \in \mathbb{R}^n, \lambda \in \mathbb{R}$;

3. $\|x + y\| \leq \|x\| + \|y\|, x, y \in \mathbb{R}^n$ *(**triangle inequality**).*

Given a norm we define the metric $d(x, y) = \|x - y\|$. Corresponding to the distances above, we write $\|x\|_2 = \sqrt{(x_1^2 + x_2^2)}$, $\|x\|_1 = |x_1| + |x_2|$, and $\|x\|_\infty = \max\{|x_1| + |x_2|\}$.

The **unit sphere** in $\mathbb{R}^n$ with respect to a given norm is the locus of points at distance 1 from 0, i.e. $\{x \in \mathbb{R}^n | d(x, 0) = \|x\| = 1\}$. The unit spheres for the three norms $\|x\|_2, \|x\|_1$, and $\|x\|_\infty$ are respectively:

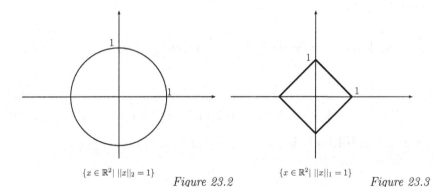

$\{x \in \mathbb{R}^2 | \|x\|_2 = 1\}$      *Figure 23.2*          $\{x \in \mathbb{R}^2 | \|x\|_1 = 1\}$      *Figure 23.3*

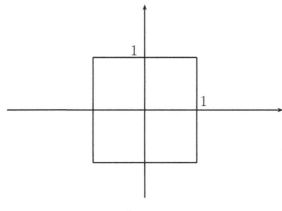

$$\{x \in \mathbb{R}^2 | \ ||x||_\infty = 1\}$$

*Figure 23.4*

**Definition 23.13.** *Let $p \geq 1$ be a real number and $x = (x_1, \ldots, x_n) \in \mathbb{R}^n$. We define*

$$||x||_p := \left( \sum_{\nu=1}^{n} |x_\nu|^p \right)^{\frac{1}{p}}. \tag{23.7}$$

**Remark 23.14. A.** For $p = 2$ we find

$$||x||_2 = \left( \sum_{\nu=1}^{n} |x_\nu|^2 \right)^{\frac{1}{2}},$$

and therefore $||x - y||_2$ is the Euclidean distance of $x$ and $y$ in $\mathbb{R}^n$.

**B.** Obviously we have

$$||\lambda x||_p = \left( \sum_{\nu=1}^{n} |\lambda x_\nu|^p \right)^{\frac{1}{p}} = |\lambda| \, ||x||_p, \lambda \in \mathbb{R}, \tag{23.8}$$

and

$$||x||_p \geq 0 \text{ for all } x \in \mathbb{R}^n \text{ and } ||x||_p = 0 \text{ if and only if } x = 0 \in \mathbb{R}^n. \tag{23.9}$$

**Theorem 23.15 (Hölder's inequality).** *Let $p, q \in (1, \infty)$, $\frac{1}{p} + \frac{1}{q} = 1$. For $x, y \in \mathbb{R}^n$ it follows that the inequality*

$$\sum_{\nu=1}^{n} |x_\nu y_\nu| \leq ||x||_p ||y||_q \tag{23.10}$$

*holds.*

324

*Proof.* Suppose that $\|x\|_p \neq 0$ and $\|y\|_q \neq 0$, otherwise (23.10) is trivial. Consider

$$\xi_\nu := \frac{|x_\nu|^p}{\|x\|_p^p}, \eta_\nu := \frac{|y_\nu|^q}{\|y\|_q^q}.$$

It follows that $\sum_{\nu=1}^{n} \xi_\nu = \sum_{\nu=1}^{n} \eta_\nu = 1$. Applying (23.6) to $\xi_\nu$ and $\eta_\nu$ we obtain

$$\frac{|x_\nu \cdot y_\nu|}{\|x\|_p \|y\|_q} = \xi_\nu^{\frac{1}{p}} \eta_\nu^{\frac{1}{q}} \leq \frac{\xi_\nu}{p} + \frac{\eta_\nu}{q}$$

and summing over all $\nu$ we have

$$\frac{1}{\|x\|_p \|y\|_q} \sum_{\nu=1}^{n} |x_\nu \cdot y_\nu| \leq \frac{1}{p} + \frac{1}{q} = 1$$

which implies Hölder's inequality. $\qquad\qquad\square$

**Remark 23.16.** For $p = 2$ Hölder's inequality reduces to the **Cauchy-Schwarz inequality** (compare with Corollary 14.3)

$$\left| \sum_{\nu=1}^{n} x_\nu y_\nu \right| \leq \sum_{\nu=1}^{n} |x_\nu y_\nu| \leq \|x\|_2 \|y\|_2.$$

Next we extend Minkowski's inequality from the Euclidean norm $\|\cdot\|_2$ (Lemma 14.5) to the norm $\|\cdot\|_p, 1 \leq p < \infty$.

**Theorem 23.17 (Minkowski's inequality).** *Let $p \in [1, \infty)$. Then we have for all $x, y \in \mathbb{R}^n$ the inequality*

$$\|x + y\|_p \leq \|x\|_p + \|y\|_p. \tag{23.11}$$

*Proof.* For $p = 1$ we apply the triangle inequality

$$\sum_{\nu=1}^{n} |x_\nu + y_\nu| \leq \sum_{\nu=1}^{n} |x_\nu| + \sum_{\nu=1}^{n} |y_\nu|.$$

Now, for $p > 1$ and $q = \frac{p}{p-1}$, i.e. $\frac{1}{p} + \frac{1}{q} = 1$, we consider $z \in \mathbb{R}^n$, $z_\nu = |x_\nu + y_\nu|^{p-1}$, $\nu = 1, \ldots, n$. It follows that

$$z_\nu^q = |x_\nu + y_\nu|^{q(p-1)} = |x_\nu + y_\nu|^p,$$

or

$$\|z\|_q = \|x + y\|_p^{\frac{p}{q}}.$$

Next we first apply the triangle inequality and then Hölder's inequality to obtain

$$\sum_{\nu=1}^{n} |x_\nu + y_\nu||z_\nu| \le \sum_{\nu=1}^{n} |x_\nu z_\nu| + \sum_{\nu=1}^{n} |y_\nu z_\nu|$$
$$\le (\|x\|_p + \|y\|_p)\|z\|_q.$$

Using the definition of $z$, we find

$$\|x + y\|_p^p \le (\|x\|_p + \|y\|_p)\|x + y\|_p^{\frac{p}{q}},$$

and since $p - \frac{p}{q} = 1$ the theorem is proved. □

**Corollary 23.18.** *For $1 \le p < \infty$ a norm is given on $\mathbb{R}^n$ by $\|\cdot\|_p$.*

**Definition 23.19.** *Let $\|\cdot\|$ be any norm on $\mathbb{R}^n$.*

***A.*** *A sequence $(x_k)_{k\in\mathbb{N}}, x_k \in \mathbb{R}^n$, **converges** in $\mathbb{R}^n$ with respect to the norm $\|\cdot\|$ to $x \in \mathbb{R}^n$ if for every $\varepsilon > 0$ there exists $N(\varepsilon) \in \mathbb{N}$ such that for $k \ge N(\varepsilon)$*

$$\|x_k - x\| < \varepsilon.$$

***B.*** *Let $D \subset \mathbb{R}^n$ be a set and $x_0 \in D$. We call $f : D \to \mathbb{R}$ **continuous in** $x_0$ with respect to the norm $\|\cdot\|$ if for every $\varepsilon > 0$ there exists $\delta = \delta(\varepsilon, x_0)$ such that $x \in D$ and $0 < \|x - x_0\| < \delta$ implies*

$$|f(x) - f(x_0)| < \varepsilon.$$

*If $f$ is continuous at all points we just call it **continuous on** $D$.*

**Example 23.20.** Let $L : \mathbb{R}^n \to \mathbb{R}$ be a linear mapping, i.e. $L \in (\mathbb{R}^n)^*$. Then $L$ is continuous with respect to any of the norms $\|\cdot\|_p, 1 \le p < \infty$.

*Proof.* Choose a basis $\{b_1, \ldots, b_n\} \subset \mathbb{R}^n$. Then

$$L(x) = L(\sum_{\nu=1}^{n} x_\nu b_\nu) = \sum_{\nu=1}^{n} x_\nu L(b_\nu).$$

Now, for $p > 1$ we use Hölder's inequality and obtain

$$|L(x) - L(y)| = |L(x - y)| \leq \sum_{\nu=1}^{n} |x_\nu - y_\nu||L(b_\nu)|$$

$$\leq \left(\sum_{\nu=1}^{n} |L(b_\nu)|^q\right)^{\frac{1}{q}} \|x - y\|_p,$$

where $\frac{1}{p} + \frac{1}{q} = 1$ or

$$|L(x) - L(y)| \leq M\|x - y\|_p.$$

Hence, given $\varepsilon > 0$, take $\delta = \frac{\varepsilon}{M}$ to find for $\|x - y\|_p < \delta$

$$|L(x) - L(y)| \leq M\|x - y\|_p < \varepsilon.$$

For $p = 1$ we just find

$$|L(x) - L(y)| = |\sum_{\nu=1}^{n}(x_\nu - y_\nu)L(b_\nu)|$$

$$\leq \max_{\nu=1,\ldots,n} |L(b_\nu)|\|x - y\|_1$$

implying the continuity of $L$ in $(\mathbb{R}^n, \|\cdot\|_1)$. □

**Example 23.21.** (Compare with Problem 9 b))Every norm $\|\cdot\|$ on $\mathbb{R}^n$ is continuous, i.e. the mapping $\|\cdot\| : \mathbb{R}^n \to \mathbb{R}$, $x \mapsto \|x\|$, is continuous. Indeed, the triangle inequality gives

$$|\|x\| - \|y\|| \leq \|x - y\|$$

which implies the continuity.

**Exercise 23.22.** *Prove that $(x_k)_{k\in\mathbb{N}}, x_k \in \mathbb{R}^n, x_k = (x_k^{(1)}, \ldots, x_k^{(n)})$ converges with respect to $\|\cdot\|_p$, $1 \leq p < \infty$, to $x = (x^{(1)}, \ldots x^{(n)}) \in \mathbb{R}^n$, if and only if for all $\nu$, $1 \leq \nu \leq n$, the sequences $(x_k^{(\nu)})_{k\in\mathbb{N}}$, $x_k^{(\nu)} \in \mathbb{R}$, converges in $\mathbb{R}$ to $x_\nu$.*

## Problems

1. Prove that the convexity of $f : I \to \mathbb{R}$ implies **Jensen's inequality**: for every $m \in \mathbb{N}, m \geq 2$, and any choice of points $x_1, \ldots, x_m \in I$ and all $0 \leq \lambda_j \leq 1, j = 1, \ldots, m$, such that $\lambda_1 + \cdots + \lambda_m = 1$ it follows that

   $$f(\lambda_1 x_1 + \cdots + \lambda_m x_m) \leq \lambda_1 f(x_1) + \cdots + \lambda_m f(x_m). \qquad (23.12)$$

   Hint: use mathematical induction with respect to $m$.

2. Give a direct proof that a convex function $f : I \to \mathbb{R}, I \subset \mathbb{R}$ being an interval with end points $a < b$, is continuous on $(a, b)$. Hint: use (23.4) to estimate $\left| \frac{f(x) - f(y)}{x - y} \right|$ against a constant.

3. Let $f : \mathbb{R} \to \mathbb{R}$ be a convex function and suppose that at $x_0 \in \mathbb{R}$ the function $f$ attains a local minimum. Show that $x_0$ is in fact a global minimum.

4.    a) Using the fact that $x \mapsto \ln x$ is on $(0, \infty)$ a concave function, give a simple proof of the **arithmetic-geometric mean inequality**, see Lemma 14.2, i.e. prove for $x_1, \ldots, x_n > 0$ that

$$\left( \prod_{k=1}^{n} x_k \right)^{\frac{1}{n}} \leq \frac{1}{n} \sum_{k=1}^{n} x_k.$$

b) Prove that $f : (0, \infty) \to \mathbb{R}, \; f(x) = x \ln x$ is convex and derive

$$(x + y) \ln \left( \frac{x + y}{2} \right) \leq x \ln x + y \ln y. \tag{23.13}$$

5. For $a \in [1, \frac{3}{2}]$ consider $f_a : [-1, 1] \to \mathbb{R}, f_a(x) = e^{ax}$. Prove that $f_a$ is convex and that

$$\sup_{a \in [1, \frac{3}{2}]} f_a(x) = \begin{cases} e^{\frac{3}{2}x}, & x \in [0, 1] \\ e^x, & x \in [-1, 0] \end{cases}.$$

6. Let $f, h : \mathbb{R} \to \mathbb{R}$ be convex and assume in addition that $f$ is increasing. Prove that $h \circ f$ is convex.

7. For $k \in \mathbb{N}$ let $|| \cdot ||_k$ be a norm on $\mathbb{R}^n$. Prove that

$$d(x, y) := \sum_{k=1}^{\infty} \frac{1}{2^k} \frac{||x - y||_k}{1 + ||x - y||_k}$$

is a metric on $\mathbb{R}^n$, i.e. $d(x, y) \geq 0$ and $d(x, y) = 0$ if and only if $x = y$, $d(x, y) = d(y, x)$, and the triangle inequality $d(x, z) \leq d(x, y) + d(y, z)$ holds. Hint: to prove the triangle inequality use the fact that $f \mapsto f(t) = \frac{t}{1+t}$ is increasing on $[0, \infty)$.

8. For the Euclidean norm $||\cdot||$ on $\mathbb{R}^n$ prove **Peetre's inequality**

$$\frac{1+||x||^2}{1+||y||^2} \leq 2(1+||x-y||^2). \qquad (23.14)$$

9.    a) Let $||\cdot||_{(1)}$ and $||\cdot||_{(2)}$ be two norms on $\mathbb{R}^n$. Prove that by

$$||x|| := ||x||_{(1)} + ||x||_{(2)}$$

and

$$|||x||| := \max(||x||_{(1)}, ||x||_{(2)})$$

two further norms are given on $\mathbb{R}^n$.

   b) Prove the converse triangle inequality

$$||x|| - ||y|| \leq |\,||x|| - ||y||\,| < ||x-y||.$$

10. Let $(x_k)_{k\in\mathbb{N}}, x_k \in \mathbb{R}^n$, $x_k = (x_k^{(1)}, \ldots, x_k^{(n)})$ be a sequence in $\mathbb{R}^n$. Prove that $(x_k)_{k\in\mathbb{N}}$ converges to $x \in \mathbb{R}^n$, $x = (x^{(1)}, \ldots, x^{(n)})$, in the norm $||\cdot||_p, 1 \leq p < \infty$, if and only if

$$\lim_{k\to\infty} |x_k^{(j)} - x^{(j)}| = 0$$

for $1 \leq j \leq n$.

11. Let $(x_k)_{k\in\mathbb{N}}, x_k = (x_k^{(1)}, \ldots, x_k^{(n)}) \in \mathbb{R}^n$, be a sequence converging in the norm $||\cdot||_p, p \in [1, \infty)$, to some $x = (x^{(1)}, \ldots, x^{(n)}) \in \mathbb{R}^n$. Suppose that $||\cdot||$ is a further norm on $\mathbb{R}^n$ satisfying the inequality $||y|| \leq c||y||_p$ for all $y \in \mathbb{R}^n$ with some $c > 0$. Prove that $(x_k)_{k\in\mathbb{N}}$ converges to $x$ with respect to $||\cdot||$.

# 24 Uniform Convergence and Interchanging Limits

A lot of the material in this chapter can be skipped during a first reading. Of importance are the definitions of pointwise and uniform convergence, the fact that uniform convergence can be described as convergence with respect to the supremum norm and the result that the uniform limit of continuous functions is continuous, Theorem 24.6. However here is the correct place to add some further material to be considered later.

In the following let $K \neq \emptyset$ be a set. We may consider functions $f, g : K \to \mathbb{R}$ and for $\alpha \in \mathbb{R}$ it follows that the functions $f \pm g$, $f \cdot g$ and $\alpha f$ can be defined on $K$ by

$$(f \pm g)(x) \quad := \quad f(x) \pm g(x), \tag{24.1}$$

$$(f \cdot g)(x) \quad := \quad f(x)g(x), \tag{24.2}$$

$$(\alpha f)(x) \quad := \quad \alpha f(x). \tag{24.3}$$

Note that we use the algebraic operation for real numbers (the target set of our functions) to implement an algebraic structure on the set of functions $f : K \to \mathbb{R}$. Of course this is not new to us, see Chapter 4. If we denote the set of functions from $K$ to $\mathbb{R}$ by $M(K; \mathbb{R}) := \{f | f : K \to \mathbb{R}\}$, it is easy to see that with the natural or pointwise operations (24.1)-(24.3) $M(K; \mathbb{R})$ is an $\mathbb{R}$-algebra, in particular it is an $\mathbb{R}$-vector space. The elements of this vector space are functions. For example, if $K = I \subset \mathbb{R}$ is an interval we find that $C(I) \subset M(I; \mathbb{R})$ is a subspace, in fact a sub-algebra. Recall that $C(I)$ stands for the vector space of all continuous functions from $I$ to $\mathbb{R}$ (also see Problem 11, Chapter 20). The idea of considering functions as elements of a vector space (or an algebra) is new to us - our next step is to consider sequences of functions as sequences of elements in a vector space. For $n \in \mathbb{N}$ let $f_n : K \to \mathbb{R}$ be a function. We may ask whether such a sequence $(f_n)_{n \in \mathbb{N}}$ or $(f_n)_{n \geq k}, k \in \mathbb{Z}$, of functions has a limit, however what does this mean? So far we only know limits of sequences of real numbers or of vectors in $\mathbb{R}^n$ with respect to a norm $|| \cdot ||_p$, see Definition 23.19. Thus instead of looking at $(f_n)_{n \in \mathbb{N}}$ we may look at $(f_n(x))_{n \in \mathbb{N}}$, $x \in K$, which is a sequence of real numbers. More precisely for every $x \in K$ we have a sequence of real numbers, i.e. we are dealing with a family (indexed by $K$) of sequences of real numbers. We can define (at least) two types of convergence, and in each case the limit is again a function $f : U \to \mathbb{R}$.

**Definition 24.1. A.** *We say that* $(f_n)_{n \in \mathbb{N}}$ *converges pointwise* *on $K$ to $f$ if for all $x \in K$ the sequences $(f_n(x))_{n \in \mathbb{N}}$ converge to $f(x)$, i.e. for every $x \in K$ and every $\varepsilon > 0$ there exists $N = N(x, \varepsilon) \in \mathbb{N}$ such that $n \geq N$ implies*

$$|f_n(x) - f(x)| < \varepsilon.$$

**B.** *The sequence* $(f_n)_{n \in \mathbb{N}}$ *is said to* *converge* *uniformly* *to $f$ if for every $\varepsilon > 0$ there is $N(\varepsilon) \in \mathbb{N}$ such that $n \geq N$ implies for all $x \in K$*

$$|f_n(x) - f(x)| < \varepsilon.$$

The important difference is that in the case of uniform convergence $N$ is independent of $x$. Clearly, uniform convergence implies pointwise convergence.

**Example 24.2.** For $n \geq 2$ define $f_n : [0, 1] \to \mathbb{R}$ by
$f_n(x) = \max(n - n^2|x - \frac{1}{n}|, 0)$, see Figure 24.1

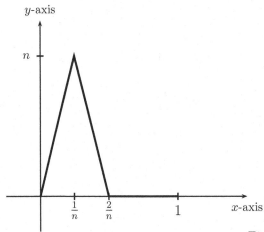

*Figure 24.1*

The sequence $(f_n)_{n \in \mathbb{N} \setminus \{1\}}$ converges pointwise on $[0, 1]$ to $f = 0$. Indeed, for $x = 0$ we have $f_n(x) = 0$ for all $n$. Further, for every $x \in (0, 1]$ there exists $N = N(x) \geq 2$ such that

$$\frac{2}{n} \leq x \quad \text{for} \quad n \geq N(x)$$

implying that for $n \geq N(x)$

$$n - n^2|x - \frac{1}{n}| \leq n - n^2(\frac{2}{n} - \frac{1}{n}) = n - n = 0,$$

hence $f_n(x) = 0$ for $n \geq N(x)$, which yields $\lim\limits_{n \to \infty} f_n(x) = 0$.

However $(f_n)_{n \in \mathbb{N}}$ does not converge uniformly to $f = 0$ since for **no** $n \geq 2$ we have $|f_n(x) - 0| < 1$ for all $x \in [0, 1]$, note that $\max\limits_{x \in [0,1]} f_n(x) = n$.

The last remark leads to a different description of uniform convergence.

**Lemma 24.3.** *The sequence $(f_n)_{n \in \mathbb{N}}$ converges uniformly to $f$ if and only if for every $\varepsilon > 0$ there is $N = N(\varepsilon) \in \mathbb{N}$ such that for all $n \geq N$*

$$\sup_{x \in K} |f_n(x) - f(x)| < \varepsilon.$$

*Proof.* Suppose that $(f_n)_{n \in \mathbb{N}}$ converges uniformly to $f$. Then for $\varepsilon > \varepsilon' > 0$ there exists $N = N(\varepsilon')$ such that

$$|f_n(x) - f(x)| < \varepsilon' \text{ for all } x \in K \text{ and } n \geq N,$$

hence

$$\sup_{x \in K} |f_n(x) - f(x)| \leq \varepsilon' < \varepsilon \text{ for all } n \geq N.$$

Conversely, since

$$|f_n(x) - f(x)| \leq \sup_{y \in K} |f_n(y) - f(y)|$$

for all $x \in K$ it follows that if $\sup\limits_{y \in K} |f_n(y) - f(y)| < \varepsilon$, then $|f_n(x) - f(x)| < \varepsilon$ for all $x \in K$. $\qquad\square$

It turns out that uniform convergence can be considered as convergence with respect to a suitable norm.

**Definition 24.4.** *Let $K \neq \emptyset$ be a set and $f : K \to \mathbb{R}$ be a function. We set*

$$\|f\|_{K,\infty} := \sup_{x \in K} |f(x)|. \qquad (24.4)$$

*If the set $K$ is fixed we just write $\|f\|_\infty$ instead of $\|f\|_{K,\infty}$. We call $\|f\|_\infty$ the **supremum norm** or just the **sup norm** of $f$.*

**Lemma 24.5.** *On the set* $M_b(K;\mathbb{R}) := \{f : K \to \mathbb{R} | \sup_{x \in K} |f(x)| < \infty\}$ *a norm is given by* $\|\cdot\|_{K,\infty}$, *i.e. the following hold:*

$$\|f\|_{K,\infty} \geq 0 \text{ and } \|f\|_{K,\infty} = 0 \text{ if and only if } f(x) = 0 \text{ for all } x \in K;$$
$$\text{i.e. } f \text{ is the 0-element in } M_b(K;\mathbb{R}); \tag{24.5}$$
$$\|\lambda f\|_{K,\infty} = |\lambda| \|f\|_{K,\infty} \text{ for } \lambda \in \mathbb{R} \text{ and } f \in M_b(K;\mathbb{R}); \tag{24.6}$$
$$\|f + g\|_{K,\infty} \leq \|f\|_{K,\infty} + \|g\|_{K,\infty} \text{ for all } f, g \in M_b(K;\mathbb{R}). \tag{24.7}$$

*Proof.* Clearly $\|f\|_\infty \geq 0$ and $\|f\|_\infty = 0$ means that $|f(x)| = 0$ for all $x \in K$, implying $f(x) = 0$ for all $x \in K$. Further, for $\lambda \in \mathbb{R}$ we find

$$\|\lambda f\|_\infty = \sup_{x \in K} |\lambda f(x)| = |\lambda| \sup_{x \in K} |f(x)| = |\lambda| \|f\|_\infty.$$

Finally, for $f, g \in M_b(K;\mathbb{R})$ it follows that

$$\|f + g\|_\infty = \sup_{x \in K} |f(x) + g(x)| \leq \sup_{x \in K} |f(x)| + \sup_{x \in K} |g(x)| = \|f\|_\infty + \|g\|_\infty.$$

$\square$

Note that the triangle inequality implies the **converse triangle inequality**, i.e.

$$\|f\|_{\infty,K} - \|g\|_{\infty,K} \leq |\, \|f\|_{\infty,K} - \|g\|_{\infty,K}| \leq \|f - g\|_{\infty,K},$$

compare with Lemma 2.9 or Problem 9 b) in Chapter 23. The next theorem shows the importance of uniform convergence.

**Theorem 24.6.** *Let* $(f_n)_{n \in \mathbb{N}}$ *be a sequence in* $C(I)$, *where* $I \subset \mathbb{R}$ *is an interval and suppose that* $(f_n)_{n \in \mathbb{N}}$ *converges uniformly to* $f : I \to \mathbb{R}$. *Then* $f$ *is continuous, i.e. the uniform limit of continuous functions is continuous.*

*Proof.* Let $x \in I$. We have to prove: given $\varepsilon > 0$ then there exists $\delta = \delta(x, \varepsilon) > 0$ such that

$$|f(x) - f(x')| < \varepsilon \text{ for all } x' \in I, |x - x'| < \delta.$$

Since $(f_n)_{n \in \mathbb{N}}$ converges uniformly to $f$, there exists $N \in \mathbb{N}$ such that

$$|f_N(y) - f(y)| < \frac{\varepsilon}{3} \text{ for all } y \in I.$$

Since $f_N$ is continuous at $x$, there exists $\delta > 0$ such that

$$|f_N(x) - f(x')| < \frac{\varepsilon}{3} \text{ for all } y \in I, |x - x'| < \delta.$$

Therefore, for all $x' \in I$ such that $|x - x'| < \delta$ it follows that

$$|f(x) - f(x')| \leq |f(x) - f_N(x)| + |f_N(x) - f_N(x')| + |f_N(x') - f(x')|$$
$$< \frac{\varepsilon}{3} + \frac{\varepsilon}{3} + \frac{\varepsilon}{3} = \varepsilon$$

and the theorem is proved. □

**Example 24.7.** Consider on $[0, 1]$ the sequence of functions $f_n(x) = x^n$. This sequence of continuous functions converges pointwise, namely for $x \in [0, 1)$ we find $\lim_{n \to \infty} x^n = 0$ where as for $x = 1$ we have $\lim_{n \to \infty} x^n = 1$.

The limit function is $f(x) = \begin{cases} 0 & , x \in [0, 1) \\ 1 & , x = 1 \end{cases}$ and it is discontinuous.

We want to study uniform convergence more closely, and as already mentioned, the following could be skipped in a first reading. As we will see there is a small problem when dealing with uniform convergence and boundedness of sequences.

**Example 24.8.** Consider the sequence $(f_n)_{n \in \mathbb{N}_0}$ where $f_n : \mathbb{R} \to \mathbb{R}$ and $f_0(x) = e^x$ and $f_n(x) = \frac{1}{n} \sin nx$. Given $\epsilon > 0$ we take $N(\epsilon) = \left[\frac{1}{\epsilon}\right] + 1$ to find that for $n > N(\epsilon)$ it follows that $\frac{1}{n} < \epsilon$, and consequently, for $n > N(\epsilon)$ we have

$$|f_n(x) - 0| = \frac{1}{n}|\sin nx| \leq \frac{1}{n} < \epsilon.$$

Hence $(f_n)_{n \in \mathbb{N}_0}$ converges uniformly to the function $x \mapsto 0$ for all $x \in \mathbb{R}$. However $f_0$ is unbounded. Thus for $n \geq N(\epsilon)$ we have $\sup_{x \in \mathbb{R}} |f_n(x)| < \epsilon$, i.e. $f_n, n \geq N(\epsilon)$, is bounded and clearly the limit function is bounded, but not all functions of the sequence $(f_n)_{n \in \mathbb{N}_0}$ must be bounded.

In general we have

**Lemma 24.9.** *Let $(f_n)_{n \in \mathbb{N}}, f_n \in M(K; \mathbb{R})$, be a sequence converging uniformly to $f \in M(K; \mathbb{R})$. If for all $n \geq N_0$ the functions $f_n$ are bounded, i.e. $n \geq N_0$ implies $f_n \in M_b(K; \mathbb{R})$, then the limit $f$ must be a bounded function too, i.e.*

$$|f(x)| \leq \sup_{x \in K} |f(x)| = ||f||_\infty < \infty. \tag{24.8}$$

*Proof.* By uniform convergence we know that for $\epsilon = 1$ there exists $N_1 \in \mathbb{N}$ such that $||f - f_{n_0}||_\infty \leq 1$ for $n \geq N_1$. For $n_0 \geq \max(N_0, N_1)$ we find

$$||f||_\infty \leq ||f - f_{n_0}||_\infty + ||f_{n_0}||_\infty \leq 1 + ||f_{n_0}||_\infty.$$

$\square$

**Corollary 24.10.** *If a sequence $f_n \in M_b(K; \mathbb{R})$ converges uniformly to $f \in M(K; \mathbb{R})$ then $f \in M_b(K; \mathbb{R})$ and the sequence is bounded in the sense that $||f_n||_\infty \leq C < \infty$ with $C$ independent of $n$. Moreover we have $||f||_\infty \leq C$.*

*Proof.* The first part follows from

$$||f_n||_\infty \leq ||f_n - f||_\infty + ||f||_\infty$$

and Lemma 24.9. To prove $||f||_\infty \leq C$ note that for $\epsilon > 0$ there exists $N(\epsilon)$ such that $n \geq N(\epsilon)$ implies by the converse triangle inequality that

$$||f||_\infty - ||f_n||_\infty \leq ||f - f_n||_\infty < \epsilon$$

or

$$||f||_\infty \leq \epsilon + ||f_n||_\infty \leq \epsilon + C,$$

however $\epsilon > 0$ was arbitrary which implies $||f||_\infty \leq C$. $\square$

In order to simplify matters, in the following we will only investigate uniform convergence in $M_b(K; \mathbb{R})$. As a first result we prove that the Cauchy criterion holds for uniform convergence in $M_b(K; \mathbb{R})$.

**Theorem 24.11.** *A sequence $(f_n)_{n \in \mathbb{N}}$, $f_n \in M_b(K; \mathbb{R})$, converges uniformly with limit $f \in M_b(K; \mathbb{R})$ if and only if for every $\epsilon > 0$ there exists $N(\epsilon)$ such that $n, m \geq N(\epsilon)$ implies $||f_n - f_m||_\infty < \epsilon$.*

*Proof.* Suppose that $(f_n)_{n \in \mathbb{N}}$ converges uniformly to $f$. For $\epsilon > 0$ there exists $N(\epsilon)$ such that $n \geq N(\epsilon)$ implies $||f - f_n||_\infty < \frac{\epsilon}{2}$ which yields for $n, m \geq N(\epsilon)$ that

$$||f_n - f_m||_\infty = ||f_n - f + f - f_m||_\infty \leq ||f_n - f||_\infty + ||f_m - f||_\infty < \frac{\epsilon}{2} + \frac{\epsilon}{2} = \epsilon.$$

Conversely suppose that for $\epsilon > 0$ there exists $N(\epsilon)$ such that $n, m \geq N(\epsilon)$ implies $||f_n - f_m||_\infty < \epsilon$. This gives for every $x \in K$ and $n, m \geq N(\epsilon)$,

$$|f_n(x) - f_m(x)| \leq ||f_n - f_m||_\infty < \epsilon, \tag{24.9}$$

i.e. for every $x \in K$ the sequence $(f_n(x))_{n \in \mathbb{N}}$ is a Cauchy sequence in $\mathbb{R}$, hence has a limit $f(x)$. We define the function $f : K \to \mathbb{R}$ by $x \mapsto f(x)$, and we want to prove that $(f_n)_{n \in \mathbb{N}}$ converges uniformly to $f$. In (24.9) we may pass to the limit as $m \to \infty$ to find

$$|f_n(x) - f(x)| \leq \epsilon, \tag{24.10}$$

which yields

$$\|f_n - f\|_\infty = \sup_{x \in K} |f_n(x) - f(x)| \leq \epsilon, \tag{24.11}$$

i.e. $(f_n)_{n \in \mathbb{N}}$ converges uniformly to $f$. $\qquad\square$

**Definition 24.12.** *A sequence* $(f_n)_{n \in \mathbb{N}}$, $f_n \in M_b(K; \mathbb{R})$, *is called a **Cauchy sequence** with respect to the norm* $\| \cdot \|_\infty$ *if for every* $\epsilon > 0$ *there exists* $N(\epsilon) \in \mathbb{N}$ *such that* $n, m \geq N(\epsilon)$ *implies* $\|f_n - f_m\|_\infty < \epsilon$.

We proved in Theorem 24.11 that on the vector space $M_b(K; \mathbb{R})$ equipped with the sup norm $\| \cdot \|_\infty$ every Cauchy sequence with respect to the sup norm has a limit in $M_b(K; \mathbb{R})$ with respect to the sup norm. In this sense we call $(M_b(K; \mathbb{R}), \| \cdot \|_\infty)$ a **complete normed space** or **Banach space**.

**Lemma 24.13.** *Let* $(f_n)_{n \in \mathbb{N}}, (g_n)_{n \in \mathbb{N}}$ *be two sequences in* $M_b(K; \mathbb{R})$ *which converge uniformly to* $f$ *and* $g$, *respectively. Then* $(f_n + g_n)_{n \in \mathbb{N}}$ *converge uniformly to* $f + g$ *and* $(f_n \cdot g_n)_{n \in \mathbb{N}}$ *converge uniformly to* $f \cdot g$. *In particular, for* $\lambda \in \mathbb{R}$ *the sequence* $(\lambda f_n)_{n \in \mathbb{N}}$ *converges uniformly to* $\lambda f$.

*Proof.* In light of Theorem 24.6 we need to prove the convergence of $(f_n + g_n)_{n \in \mathbb{N}}$ to $f + g$ and the convergence of $(f_n \cdot g_n)_{n \in \mathbb{N}}$ to $f \cdot g$ with respect to the norm $\| \cdot \|_\infty$. We proceed as in the proofs of the analogous results for sequences of real numbers by replacing the absolute value by the norm $\| \cdot \|_\infty$. For $\epsilon > 0$ there exists $N(\epsilon)$ such that $n \geq N(\epsilon)$ implies $\|f - f_n\|_\infty < \epsilon$ and $\|g - g_n\|_\infty < \epsilon$ which implies by the triangle inequality

$$\|(f_n + g_n) - (f + g)\|_\infty \leq \|f_n - f\|_\infty + \|g_n - g\|_\infty < \epsilon + \epsilon = 2\epsilon,$$

i.e. $(f_n + g_n)_{n \in \mathbb{N}}$ converges uniformly to $f + g$. Moreover, since $(g_n)_{n \in \mathbb{N}}$ is bounded with respect to $\| \cdot \|_\infty$, i.e. $\|g_n\|_\infty \leq c_0$, and with $\|f\|_\infty \leq c_1$ it follows that

$$\|f_n g_n - fg\|_\infty = \|f_n g_n - fg_n + fg_n - fg\|_\infty$$
$$\leq \|(f_n - f)g_n\|_\infty + \|f(g_n - g)\|_\infty.$$

Since for $h_1, h_2 \in M_b(K, \mathbb{R})$ we have

$$||h_1 h_2||_\infty = \sup_{x \in K} |h_1(x) h_2(x)| \leq \left( \sup_{x \in K} |h_1(x)| \right) \left( \sup_{x \in K} |h_2(x)| \right) = ||h_1||_\infty ||h_2||_\infty$$

it follows

$$||f_n g_n - fg||_\infty \leq ||g_n||_\infty ||f_n - f||_\infty + ||f||_\infty ||g_n - g||_\infty$$
$$\leq c_0 ||f_n - f||_\infty + c_1 ||g_n - g||_\infty < (c_0 + c_1)\epsilon,$$

implying the uniform convergence of $(f_n \cdot g_n)_{n \in \mathbb{N}}$ to $f \cdot g$. $\square$

We have seen in Example 24.7 that there are pointwise convergent sequences of continuous functions which are not uniformly convergent and whose limit is not continuous. If we combine Proposition 23.8 and Corollary 23.6 we see that the pointwise limit of convex functions is continuous, i.e. uniform convergence is not needed to get continuity. The argument is that convex functions are continuous and that pointwise limits of convex functions are convex. The next result gives a further example that using additional information we sometimes get that pointwise convergence implies uniform convergence.

**Proposition 24.14.** Let $f_n : [a, b] \to \mathbb{R}, a < b$ be a sequence of increasing functions converging pointwise to a continuous function $f : [a, b] \to \mathbb{R}$ then the convergence is uniform.

**Remark 24.15. A.** This result also holds for sequences of decreasing functions.
**B.** Note that we do not require $f_n$ to be continuous, i.e. we may have a sequence of non-continuous functions converging uniformly to a continuous function.

*Proof of Proposition 24.14.* As a continuous function on a compact interval, $f$ is uniformly continuous. Thus for $\epsilon > 0$ there exists $\delta > 0$ such that $|x - y| < \delta, x, y \in [a, b]$, implies $|f(x) - f(y)| < \frac{\epsilon}{2}$. Now we choose a partition of $[a, b]$ with points $a = x_0 < x_1 < \cdots < x_k = b$ such that $|x_j - x_{j-1}| < \delta$ for $j = 1, \ldots k$. Using the pointwise convergence of the sequence $(f_n)_{n \in \mathbb{N}}$ we deduce

$$\lim_{n \to \infty} f_n(x_j) = f(x_j), \; j = 1, \ldots, k.$$

338

Thus there exists $N_0$ such that $n \geq N_0$ implies

$$|f_n(x_j) - f(x_j)| < \frac{\epsilon}{2}, \; j = 1, \ldots, k. \tag{24.12}$$

For $x \in [a, b]$ we find $j$ such that $x_{j-1} \leq x < x_j$ and the monotonicity of $f_n$ implies by (24.12) that

$$f(x_{j-1}) - \frac{\epsilon}{2} < f_n(x_{j-1}) \leq f_n(x) \leq f(x_j) < f(x_j) + \frac{\epsilon}{2}.$$

As a pointwise limit of increasing functions $f$ must be increasing, compare with Problem 4, i.e. $f(x_{j-1}) \leq f(x) \leq f(x_j)$ which now yields using the uniform continuity of $f$

$$-\epsilon < f(x_{j-1}) - f(x_j) - \frac{\epsilon}{2} \leq f_n(x) - f(x) \leq f(x_j) - f(x_{j-1}) + \frac{\epsilon}{2} < \epsilon$$

or $|f_n(x) - f(x)| < \epsilon$ for $x \in [a, b]$ and $n \geq N_0$, i.e. for $n \geq N_0$ we have $||f_n - f||_\infty < \epsilon$. $\qquad\square$

**Exercise 24.16.** *Let $f_n : [a, b] \to \mathbb{R}$ be a sequence of monotone increasing functions converging pointwise to $f : [a, b] \to \mathbb{R}$. Show that $f$ is increasing.*

The Bolzano-Weierstrass theorem, Theorem 17.6, states that every bounded sequence in $\mathbb{R}$ has a convergent subsequence. We may ask whether such a result holds for uniformly convergent sequences of functions too. In fact this is not the case, but with certain additional conditions the result can be rescued. This is the famous Arzela-Ascoli theorem which will be discussed later in our course.

Let us return to Theorem 24.6. We can interpret this result differently as follows:

$$\lim_{n \to \infty} \lim_{x \to x_0} f_n(x) = \lim_{x \to x_0} \lim_{n \to \infty} f_n(x), \tag{24.13}$$

i.e. under uniform convergence we are allowed to interchange the order of the limits.

We have seen that pointwise convergence is not sufficient to justify (24.13). However, what about differentiability? When does the following hold?

$$\lim_{n \to \infty} f_n'(x) = \left( \lim_{n \to \infty} f_n \right)'(x). \tag{24.14}$$

It turns out that uniform convergence of $(f_n)_{n \in \mathbb{N}}$ or of $(f_n')_{n \in \mathbb{N}}$ is not sufficient:

**Example 24.17.** For $n \in \mathbb{N}$ consider $f_n : \mathbb{R} \to \mathbb{R}$, $f_n(x) = \frac{1}{n} \sin nx$. Since $\|f_n\|_\infty = \sup_{x \in \mathbb{R}} \left| \frac{1}{n} \sin nx \right| = \frac{1}{n}$ it follows that $(f_n)_{n \in \mathbb{N}}$ converges uniformly on $\mathbb{R}$ to $f_0(x) = 0$ for all $x \in \mathbb{R}$ and $f_0$ is differentiable with derivative $f_0'(x) = 0$ for all $x \in \mathbb{R}$. However $f_n'(x) = \frac{1}{n} n \cos nx = \cos nx$ and $(f_n')_{n \in \mathbb{N}}$ is not even pointwise convergent, note that for $x = \pi$ it follows for even $n$ that $\cos n\pi = 1$, while for odd $n$ we have $\cos n\pi = -1$.

**Example 24.18.** Now consider $g_n : \mathbb{R} \to \mathbb{R}, g_n(x) = n \cos \frac{1}{n^2} x$. Then $\lim_{n \to \infty} g_n(x)$ does not in general exist, i.e. we cannot define a limit function $g$. However, $g_n'(x) = -\frac{1}{n} \sin \frac{1}{n^2} x$, and since

$$\|g_n'(x)\|_\infty = \sup_{x \in \mathbb{R}} \frac{1}{n} \left| \sin \frac{1}{n^2} x \right| = \frac{1}{n},$$

it follows that $(g_n')_{n \in \mathbb{N}}$ converges uniformly to 0, i.e. the function $x \mapsto 0$ for all $x \in \mathbb{R}$.

It turns out that the pointwise convergence of $(f_n)_{n \in \mathbb{N}}$ and the uniform convergence of $(f_n')_{n \in \mathbb{N}}$ will be sufficient to imply (24.14), but to prove this we will need more tools.

# Problems

1. Show that the sequence $(g_n)_{n \in \mathbb{N}}, g_n : \mathbb{R} \to \mathbb{R}$, where

$$g_n(x) = \begin{cases} \frac{x}{n}, & \text{if } n \text{ is even} \\ \frac{1}{n}, & \text{if } n \text{ is odd} \end{cases}$$

   is pointwise convergent but not uniform.

2. Prove that the pointwise limit of the sequence $(f_n)_{n \in \mathbb{N}}, f_n : [0, 1] \to \mathbb{R}$, $f_n(x) = \frac{1}{1 + (nx - 1)^2}$, is $f(x) = \begin{cases} \frac{1}{2}, & x = 0 \\ 0, & x \in (0, 1], \end{cases}$ and deduce that the convergence cannot be uniform.

3. Test the following for uniform convergence:

   a) $f_n(x) = x^n(1 - x)$, on $[0, 1]$;

   b) $g_n(x) = \frac{nx^2}{1 + nx}$ on $[0, 1]$;

c) $h_n(x) = \arctan \frac{4x}{x^2+n^4}$ on $\mathbb{R}$;

d) $k_{\alpha,n}(x) = \frac{1}{n^\alpha} \cos(a_n x)$ on $\mathbb{R}$ for any sequence $(a_n)_{n\in\mathbb{N}}, a_n \in \mathbb{R}, \alpha > 0$.

4. Prove that if $(f_n)_{n\in\mathbb{N}}, f_n : I \to \mathbb{R}$, where $I \subset \mathbb{R}$ is an interval, is increasing and converges pointwise on $I$ to $f : I \to \mathbb{R}$, then $f$ is increasing too.

5. Consider the polynomial $p(x) = \sum_{k=0}^{N} c_k x^k$ of degree $N$. Show that there exists a sequence of polynomials $p_n(x) = \sum_{k=0}^{N} c_{k,n} x^k$, with rational coefficients $c_{k,n} \in \mathbb{Q}$ converging uniformly on $[0, 1]$ to $p$.

6. Let $I \subset \mathbb{R}$ be an interval and suppose that the sequence $f_n : I \to \mathbb{R}$ of continuous functions converges uniformly to the continuous function $f : I \to \mathbb{R}$. Let $(x_n)_{n\in\mathbb{N}}, x_n \in I$, be a sequence converging to $x \in I$. Prove that

$$\lim_{n\to\infty} f_n(x_n) = f(x).$$

7. Let $f_n \in C((a,b)), a < b$ and $n \in \mathbb{N}$, be a sequence of continuous functions with the property that for every compact interval $[\alpha, \beta] \subset (a, b)$ the sequence $\left(f_n|_{[\alpha,\beta]}\right)_{n\in\mathbb{N}}$ converges uniformly to a function $g_{\alpha,\beta}$. Prove that then $(f_n)_{n\in\mathbb{N}}$ converges pointwise on $(a, b)$ to a continuous function $f \in C((a, b))$.

8. Let $f : \mathbb{R} \to \mathbb{R}$ be a continuously differentiable function such that $f'$ is uniformly continuous on $\mathbb{R}$. Prove that $g_n : \mathbb{R} \to \mathbb{R}, g_n(x) := n\left(f\left(x + \frac{1}{n}\right) - f(x)\right)$ converges uniformly on $\mathbb{R}$ to $f'$.

9. Consider $f_n : [-1, 1] \to \mathbb{R}$ where $f_n(x) = \frac{x}{1+nx}$. Prove that $(f_n)_{n\in\mathbb{N}}$ converges uniformly to the zero function, i.e. to the function $x \mapsto h(x) = 0$ for all $x \in [-1, 1]$, while $(f'_n)_{n\in\mathbb{N}}$ converges pointwise to
$$g(x) := \begin{cases} 1, & x = 0 \\ 0, & x \in [-1, 1] \setminus \{0\}. \end{cases}$$

# 25 The Riemann Integral

In this and the following chapter we want to rigorously derive the results already discussed and used in Chapters 12 and 13. Our starting point is to determine the area $A$ bounded by the graph $\Gamma(f)$ of a function $f : [a, b] \to \mathbb{R}$, $a < b$, the interval $[a, b]$, the line segment joining $(a, 0)$ and $(a, f(a))$ and the line segment joining $(b, 0)$ and $(b, f(b))$, see Figure 25.1.

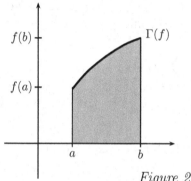

*Figure 25.1*

We take for granted that the area of a rectangle with vertices $(a, 0), (b, 0)$, $(b, c), (a, c), a < b, b < c$, is given by

$$A = (b - a)(c - b).$$

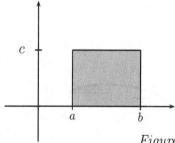

*Figure 25.2*

Interpreting the line segment connecting $(a, c)$ with $(b, c)$ as the graph of the function $f_c : [a, b] \to \mathbb{R}, f_c(x) = c$, we find for the area of this rectangle $A = f_c(a)(b - a)$, in fact $A = f_c(\xi)(b - a)$ for every $\xi \in [a, b]$. Furthermore, when looking at Figure 25.3

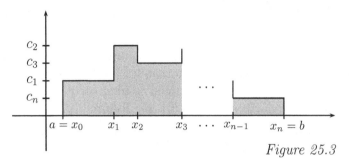

*Figure 25.3*

we of course agree that the area $A$ is given by

$$A = \sum_{k=1}^{n} c_k(x_k - x_{k-1}) \tag{25.1}$$

or with $f : [a, b] \to \mathbb{R}, f|_{(x_{k-1}, x_k)}(x) = c_k,$

$$A = A(f) = \sum_{k=1}^{n} f(\xi_k)(x_k - x_{k-1}), \xi \in (x_{k-1}, x_k). \tag{25.2}$$

Now we have the obvious idea: in order to find the area bounded by $\Gamma(f), f : [a, b] \to \mathbb{R}, f(x) \geq 0$, and the interval $[a, b]$ as well as the line segment connecting $(a, 0)$ and $(a, f(a))$ and the line segment connecting $(b, 0)$ and $(b, f(b))$, see Figure 25.1, we approximate $\Gamma(f)$ by the graphs of piecewise constant functions, see Figure 25.4, and try to pass to the limit.

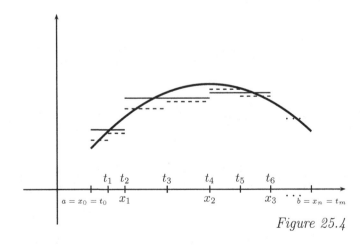

*Figure 25.4*

344

This idea eventually leads to a solution of the original problem, but we must first overcome a few difficulties. One problem is that the area $A(f)$ we are looking for is not yet defined. In fact only after we define a proper approximation process for a (large) class of functions can we define the "area under the graph" of a function. We therefore need to find for a class of functions a way of approximating them with piecewise constant functions $\varphi$ such that the area $A(\varphi)$ associated with $\varphi$ by (25.2) converges to a quantity $A(f)$ which we can interpret as the "area under $\Gamma(f)$".

First let us consider piecewise constant functions.

**Definition 25.1.** *Let $[a, b]$, $a < b$, be a closed and bounded, hence compact interval. We call a finite set of numbers or points $a = x_0 < x_1 < \cdots < x_{n-1} < x_n = b$ a **partition** $Z$ of $[a, b]$.*

We denote partitions by $Z = Z(x_0, \ldots, x_n)$ with the understanding that $x_0 = a$ and $x_n = b$. When we want to emphasise the corresponding interval $[a, b]$ we write $Z(a, x_1, \ldots, x_{n-1}, b)$. On $[a, b]$ we can consider several partitions $Z_1 = Z(x_0, \ldots, x_n)$, $Z' = Z'(t_0, \ldots, t_m)$ (clearly $a = x_0 = t_0, b = x_n = t_m$). Given two partitions $Z_1(x_0, \ldots, x_n)$ and $Z_2(t_0, \ldots, t_m)$ we can construct the **joint partition** $Z(y_0, \ldots, y_k)$ by

$$Z(y_0, \ldots, y_k) = Z_1(x_0, \ldots, x_n) \cup Z_2(t_0, \ldots, t_m) = \{x_0, \ldots, x_n\} \cup \{t_0, \ldots, t_m\},$$

and clearly $y_0 = x_0 = t_0 = a, y_k = x_n = t_m = b$. Given a partition $Z = Z(x_0, \ldots, x_n)$ of $[a, b]$. We call

$$\max\{x_k - x_{k-1} | k = 1, \ldots n\} \tag{25.3}$$

the **mesh size** or **width** of $Z$. If $x_k - x_{k-1} = \eta$ is independent of $k$ we call $Z$ an **equidistant partition** of $[a, b], a < b$.

**Definition 25.2.** *Let $\varphi : [a, b] \to \mathbb{R}$ be a function. We call $\varphi$ a **step function** on $[a, b]$ if there exists a partition $Z = Z(x_0, \ldots, x_n)$ of $[a, b]$ such that $f|_{(x_{k-1}, x_k)}$, is constant, $k = 1, \ldots n$, i.e. $f(x) = c_k$ for all $x \in (x_{k-1}, x_k)$ and some $c_k \in \mathbb{R}$.*

The set of all step functions on $[a, b]$ is denoted by $T[a, b]$. Note that in Definition 25.2 no statement about the values $f(x_k), x_k \in Z$, is made, except that they are real numbers.

**Remark 25.3. A.** It is worth mentioning that $\varphi \in T[a, b]$ may have different step function representations. Take the constant function $\varphi(x) = c$ for all $x \in [a, b]$. It is a step function with respect to the partition $Z = Z(x_0, x_1) = \{a, b\}$, but for any finite number of points $x_0 < x_1 < \cdots < x_{n-1} < x_n$ we can consider $\varphi$ as a step function with respect to that partition, i.e. $\varphi|_{(x_{k-1}, x_k)} = c_k = c$. In general if $\varphi$ is a step function with respect to $Z_1$ and $Z_2$ is a partition such that $Z_1 \subset Z_2$ then we can also represent $\varphi$ as a step function with respect to $Z_2$.

**B.** Given a step function $\varphi : [a, b] \to \mathbb{R}$ with respect to the partition $Z = Z(x_0, \ldots, x_n)$ with $\varphi|_{(x_{k-1}, x_k)} = c_k$. We can write $\varphi$ as

$$\varphi(x) = \sum_{k=1}^{n} c_k \chi_{(x_{k-1}, x_k)}(x) + \sum_{k=0}^{n} \varphi(x_k) \chi_{\{x_k\}}(x), \tag{25.4}$$

where as usual $\chi_A(x) = \begin{cases} 1, & x \in A \\ 0, & x \notin A \end{cases}$ denotes the characteristic function of the set $A$.

We have seen that the set of all functions $f : [a, b] \to \mathbb{R}, a < b$, form an $\mathbb{R}$-vector space, in fact even an algebra, with respect to pointwise operations, i.e. $(f + g)(x) = f(x) + g(x), (\lambda f)(x) = \lambda f(x), (f \cdot g)(x) = f(x)g(x)$.

**Lemma 25.4.** *The step functions $T[a, b]$ are a subspace of the vector space of all real-valued functions defined on $[a, b]$.*

*Proof.* We have to prove that $\varphi, \psi \in T[a, b], \lambda \in \mathbb{R}$ imply that $\varphi + \psi \in T[a, b]$ and $\lambda \varphi \in T[a, b]$. Let $\varphi$ be given with respect to the partition $Z_1$ and $\psi$ with respect to the partition $Z_2$. We now consider $\varphi$ and $\psi$ as step functions with respect to the joint partition $Z = Z_1 \cup Z_2 = \{t_0, \ldots, t_k\}$. For $1 \leq l \leq n$ we have $\varphi|_{(t_{l-1}, t_l)} = c_l$ and $\psi|_{(t_{l-1}, t_l)} = d_l$ for some $c_l, d_l \in \mathbb{R}$, and therefore $(\varphi + \psi)|_{(t_{l-1}, t_l)} = c_l + d_l$, i.e. with respect to $Z$ the function $\varphi + \psi$ is also a step function. Obviously, with $\varphi \in T[a, b]$ it follows that $\lambda \varphi \in T[a, b]$, since $\varphi|_{(x_{j-1}, x_j)} = c_j$ implies $(\lambda \varphi)|_{(x_{j-1}, x_j)} = \lambda c_j$. $\square$

**Exercise 25.5.** *Prove that $T[a, b]$ is an algebra.*

The next result is crucial for the following reason. It tells us that a continuous function can always be "sandwiched" between two step functions such that these two step functions differ only by a prescribed magnitude.

**Theorem 25.6.** *Let* $f : [a, b] \to \mathbb{R}, a < b$, *be a continuous function. For* $\epsilon > 0$ *there exists* $\varphi, \psi \in T[a, b]$ *such that*

$$\varphi(x) - f(x) \le \psi(x) \quad \text{for all } x \in [a, b] \tag{25.5}$$

*and*

$$\psi(x) - \varphi(x) - |\psi(x) - \varphi(x)| \le \epsilon \quad \text{for all } x \in [a, b]. \tag{25.6}$$

*Proof.* As a continuous function on a compact set, $f$ is uniformly continuous. Hence for $\epsilon > 0$ there exists $\delta > 0$ such that $x, y \in [a, b]$ and $|x - y| < \delta$ imply $|f(x) - f(y)| < \frac{\epsilon}{2}$. We divide $[a, b]$ into $n$ equally long intervals with length less than $\delta$:

$$t_k := a + k\frac{b - a}{n}, k = 0, 1, \ldots, n$$

where $n$ is chosen such that $\frac{b-a}{n} < \delta$. This gives an equidistant partition $Z = Z(t_0, \ldots, t_n)$ of $[a, b]$ and we will define $\varphi$ and $\psi$ with respect to $Z$. For $1 \le k \le n$ we set

$$c_k := f(t_k) + \frac{\epsilon}{2}, c'_k := f(t_k) - \frac{\epsilon}{2},$$

and

$$\varphi(a) = \psi(a) := f(a), \tag{25.7}$$

as well as for $x \in (t_{k-1}, t_k], k = 1, \ldots, n$

$$\varphi(x) = c'_k, \quad \psi(x) = c_k. \tag{25.8}$$

The definition of $c_k, c'_k$ yields

$$|\varphi(x) - \psi(x)| \le \epsilon \quad \text{for all } x \in [a, b].$$

For $x = a = t_0$ we have $\varphi(x) = f(x) = \psi(x)$, hence $\varphi(x) \le f(x) \le \psi(x)$. For $x \in (t_{k-1}, t_k]$ it follows that $|x - t_k| < \delta$ and therefore

$$-\frac{\epsilon}{2} < f(x) - f(t_k) < \frac{\epsilon}{2},$$

or

$$\varphi(x) = c'_k = f(t_k) - \frac{\epsilon}{2} < f(x) < f(t_k) + \frac{\epsilon}{2} = c_k = \psi(x),$$

i.e. $\varphi(x) \le f(x) \le \psi(x)$ for all $x \in [a, b]$. $\qquad \square$

Now we define an integral for step functions $\varphi \in T[a,b]$ with the aim to extend it at least to all continuous functions on $[a,b]$.

**Definition 25.7.** Let $\varphi \in T[a,b]$ be given with respect to the partition $Z = Z(x_0, \ldots, x_n)$ by $\varphi|_{(x_{k-1},x_k)} = c_k$. The **integral** of $\varphi$ is defined by

$$\int_a^b \varphi(x)dx := \sum_{k=1}^n c_k(x_k - x_{k-1}). \tag{25.9}$$

Note that the integral does not depend on the values $f(t_k), k = 0, \ldots, n$. However, the integral as defined by (25.9) seems to depend on $Z$, but $\varphi$ can be represented with respect to other partitions. So we need to prove that the integral only depends on $\varphi$ and not on the chosen partition to represent $\varphi$.

**Lemma 25.8.** The definition of $\int_a^b f(x)dx$ is independent of the choice of partition representing $\varphi$.

*Proof.* Let $Z_1(x_0, \ldots, x_n)$ and $Z_2(t_0, \ldots, t_m)$ be two partitions of $[a,b]$ such that $\varphi|_{(x_{k-1},x_k)} = c_k$ and $\varphi|_{(t_{l-1},t_l)} = c_l'$. We have to prove

$$\sum_{i=1}^n c_i(x_i - x_{i-1}) = \sum_{j=1}^m c_j'(t_j - t_{j-1}).$$

Suppose first that $Z_1 \subset Z_2, x_i = t_{k_i}$. It follows that

$$x_{i-1} = t_{k_{i-1}} < t_{k_{i-1}+1} < \cdots < t_{k_i} = x_i, 1 \leq i \leq n,$$

and

$$c_j' = c_i \quad \text{for } k_{i-1} < j \leq k_i,$$

implying

$$\sum_{j=1}^m c_j'(t_j - t_{j-1}) = \sum_{i=1}^n \sum_{j=k_{i-1}+1}^{k_i} c_j'(t_j - t_{j-1}) = \sum_{i=1}^n c_i(x_i - x_{i-1}).$$

The general case follows by using $Z = Z_1 \cup Z_2$ as a third partition and apply the case just proven to $Z$ and $Z_1$ as well as $Z$ and $Z_2$. $\square$

On $T[a,b]$ the integral is linear and positivity preserving, i.e. the integral of non-negative functions is non-negative.

**Theorem 25.9.** *For $\varphi, \psi \in T[a, b]$ and $\lambda \in \mathbb{R}$ the following hold:*

$$\int_a^b (\varphi + \psi)(x)dx = \int_a^b \varphi(x)dx + \int_a^b \psi(x)dx \qquad (25.10)$$

*and*

$$\int_a^b (\lambda\varphi)(x)dx = \lambda \int_a^b \varphi(x)dx. \qquad (25.11)$$

*Further, if $\varphi \geq 0$, i.e. $\varphi(x) \geq 0$ for all $x \in [a, b]$, then we have*

$$\int_a^b \varphi(x)dx \geq 0. \qquad (25.12)$$

*Proof.* For (25.10) we need to represent $\varphi$ and $\psi$ with respect to the same partition and then we can use as for the proofs of (25.11) and (25.12) the fact that the summation process is additive, homogeneous and positivity preserving, i.e. the sum of non-negative numbers is non-negative.  $\square$

**Corollary 25.10.** *Let $\varphi, \psi \in T[a, b]$ and $\varphi \leq \psi$, i.e. $\varphi(x) \leq \psi(x)$ for all $x \in [a, b]$ then we have*

$$\int_a^b \varphi(x)dx \leq \int_a^b \psi(x)dx. \qquad (25.13)$$

*Proof.* Since $\psi - \varphi \geq 0$, using (25.10)-(25.12) we find

$$0 \leq \int_a^b (\psi(x) - \varphi(x))dx = \int_a^b \psi(x)dx - \int_a^b \varphi(x)dx.$$

$\square$

Now we want to extend the integral to a larger class of functions. We try to use the following idea: given $f : [a, b] \to \mathbb{R}$ and a step function $\psi : [a, b] \to \mathbb{R}$ such that $f \leq \psi$. The infimum of the integrals of all $\psi$ with this property should approximate "the area under $\Gamma(f)$" from above. On the other hand the supremum of the integrals of all $\varphi \in T[a, b], \varphi \leq f$, should approximate "the area under $\Gamma(f)$" from below, see Figures 25.5 and 25.6.

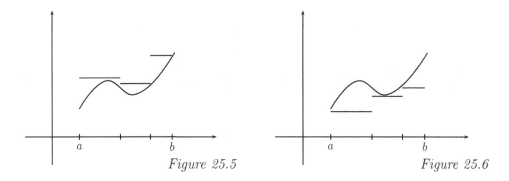

Figure 25.5          Figure 25.6

This leads us to introduce the upper and lower integral as well as the Riemann integral.

**Definition 25.11. A.** *Let* $f : [a, b] \to \mathbb{R}$ *be a bounded function. The* ***upper integral*** *of* $f$ *is defined as*

$$\int\limits_a^{b*} f(x)dx := \inf \left\{ \int_a^b \varphi(x)dx \,|\, \varphi \in T[a, b] \text{ and } \varphi \geq f \right\} \qquad (25.14)$$

*and the* ***lower integral*** *of* $f$ *is defined by*

$$\int\limits_{*a}^{b} f(x)dx := \sup \left\{ \int_a^b \varphi(x)dx \,|\, \varphi \in T[a, b] \text{ and } \varphi \leq f \right\}. \qquad (25.15)$$

***B.*** *We call a bounded function* $f : [a, b] \to \mathbb{R}$ ***Riemann integrable*** *if*

$$\int\limits_a^{b*} f(x)dx = \int\limits_{*a}^{b} f(x)dx. \qquad (25.16)$$

350

*In this case we write*

$$\int_a^b f(x)dx := \int_{*a}^b f(x)dx = \int_a^{*b} f(x)dx \qquad (25.17)$$

*and call the left hand side of (25.17) the **Riemann integral** of $f$ (over $[a, b]$).*
*C. Let $f : [a, b] \to \mathbb{R}$ be a non-negative Riemann integrable function. The area $A(f)$ bounded by $\Gamma(f)$, the interval $[a, b]$, the line segment connecting $(a, 0)$ and $(a, f(a))$ and the line segment connecting $(b, 0)$ and $(b, f(b))$ is defined by*

$$A(f) = \int_a^b f(x)dx.$$

*(Compare with Definition 12.3.)*

Note that from Definition 25.11.A we always have

$$\int_{*a}^b f(x)dx \le \int_a^{*b} f(x)dx. \qquad (25.18)$$

Thus the aim is to determine the class of bounded functions where equality holds in (25.18) and to discuss properties of the Riemann integral. First however we give two examples.

**Example 25.12. A.** For $f \in T[a, b]$ clearly we have

$$\int_{*a}^b f(x)dx = \int_a^{*b} f(x)dx = \int_a^b f(x)dx,$$

thus step functions are Riemann integrable.
**B.** For $\chi_{\mathbb{Q}\cap[0,1]} : [0, 1] \to \mathbb{R}$, i.e.

$$\chi_{\mathbb{Q}\cap[0,1]} = \begin{cases} 1, & x \in \mathbb{Q} \cap [0, 1] \\ 0, & x \in [0, 1], x \notin \mathbb{Q} \end{cases}$$

it follows that

$$\int_0^{*1} \chi_{\mathbb{Q}\cap[0,1]}(x)dx = 1 \text{ and } \int_{*0}^1 \chi_{\mathbb{Q}\cap[0,1]}(x)dx = 0,$$

and therefore $\chi_{\mathbb{Q}\cap[0,1]}$ is not Riemann integrable.

**Theorem 25.13.** *Let $f, g : [a, b] \to \mathbb{R}$ be bounded functions. Then we have*

$$\int_{a}^{b*} (f + g)dx \leq \int_{a}^{b*} f dx + \int_{a}^{b*} g dx, \quad (subadditivity) \qquad (25.19)$$

*and for $\lambda \geq 0$*

$$\int_{a}^{b*} (\lambda f)dx = \lambda \int_{a}^{b*} f dx, \quad (positive\ homogeneity). \qquad (25.20)$$

*Proof.* Let us write $\int^{*} f$ for $\int_{a}^{b*} f dx$ if it is clear what is meant, analogously

we will write $\int_{*} f$. Now, to prove part (25.19) it is sufficient to show that

$$\int^{*} (f + g)dx \leq \int^{*} f dx + \int^{*} g dx + \varepsilon \text{ for all } \varepsilon > 0.$$

We know that there are $\varphi, \psi \in T[a, b]$, $\varphi \geq f$, $\psi \geq g$ such that

$$\int \varphi \leq \int^{*} f + \frac{\varepsilon}{2}, \text{ and } \int \psi \leq \int^{*} g + \frac{\varepsilon}{2}.$$

Since $\varphi + \psi \geq f + g$ it follows that

$$\int^{*} (f + g)dx \leq \int^{*} f dx + \int^{*} g\, dx + \varepsilon.$$

To prove (25.20) it is sufficient to show that

$$\lambda \int^{*} f - \varepsilon \leq \int^{*} (\lambda f) \leq \lambda \int^{*} f + \varepsilon \text{ for all } \varepsilon > 0.$$

Further we may assume that $\lambda > 0$. By definition there is $\varphi \in T[a, b]$ such that

$$\int \varphi \leq \int^{*} f + \frac{\varepsilon}{\lambda}.$$

Since $\lambda\varphi \geq \lambda f$ it follows that

$$\int^* (\lambda f) \leq \int (\lambda\varphi) = \lambda \int \varphi \leq \lambda \left( \int^* f + \frac{\varepsilon}{\lambda} \right) = \lambda \int^* f + \varepsilon,$$

and analogously we may prove

$$\lambda \int^* f \leq \int^* (\lambda f) + \varepsilon.$$

$\square$

**Corollary 25.14.** *Let $f, g : [a, b] \to \mathbb{R}$ be bounded functions. Then it follows that*

$$\int_* (f + g) dx \geq \int_* f \, dx + \int_* g \, dx, \tag{25.21}$$

$$\int_* (\lambda f) dx = \lambda \int_* f \, dx \text{ for all } \lambda \geq 0, \tag{25.22}$$

*and for $\lambda < 0$ we have*

$$\int^* \lambda f = \lambda \int_* f \text{ and } \int_* \lambda f = \lambda \int^* f. \tag{25.23}$$

*Proof.* We only need to note the equality

$$\int_* f = - \int^* (-f)$$

which follows from the definition. $\square$

Suppose that $f : [a, b] \to \mathbb{R}$ is Riemann integrable, i.e.

$$\int_*^b f(x) dx = \int_a^b f(x) dx = \int_a^{*b} f(x) dx.$$

Using the definition of inf and sup, given $\epsilon > 0$, we can find $\psi, \varphi \in T[a, b]$, $\varphi \leq f \leq \psi$, such that

$$\int_a^b f(x) \leq \int_a^b \varphi(x) dx + \frac{\epsilon}{2} \text{ and } \int_a^b \psi(x) dx - \frac{\epsilon}{2} \leq \int_a^b f(x) dx,$$

hence we have:

**Theorem 25.15.** *The function $f : [a, b] \to \mathbb{R}$ is Riemann integrable if and only if for every $\epsilon > 0$ there exists step functions $\varphi, \psi \in T[a, b]$ such that $\varphi \le f \le \psi$ and*

$$\int_a^b (\psi(x) - \varphi(x))dx = \int_a^b \psi(x)dx - \int_a^b \varphi(x)dx \le \epsilon. \tag{25.24}$$

This together with Theorem 25.6 gives

**Theorem 25.16.** *A continuous function $f : [a, b] \to \mathbb{R}$ is Riemann integrable.*

*Proof.* By Theorem 25.6, given $\varepsilon > 0$, there are step functions $\varphi, \psi \in T[a, b]$ such that $\varphi \le f \le \psi$ and $\psi(x) - \varphi(x) \le \frac{\varepsilon}{b-a}$ for all $x \in [a, b]$. It follows that

$$\int_a^b \psi(x)dx - \int_a^b \varphi(x)dx = \int_a^b (\psi(x) - \phi(x))dx \le \int_a^b \frac{\varepsilon}{b - a}dx = \varepsilon.$$

$\square$

Furthermore we have

**Theorem 25.17.** *Every monotone function $f : [a, b] \to \mathbb{R}$ is Riemann integrable.*

*Proof.* Suppose that $f$ is increasing (for decreasing functions the proof goes analogously). By $x_k := a + \frac{(b-a)}{n}, k = 0, 1, \ldots, n$, an equidistant partition of $[a, b]$ is given. We now define the two step functions $\varphi, \psi \in T[a, b]$ by

$$\varphi(x) := f(x_{k-1}), \quad x_{k-1} \le x < x_k,$$
$$\psi(x) := f(x_k), \quad x_{k-1} \le x < x_k,$$

as well as $\varphi(b) = \psi(b) = f(b)$. Since $f$ is monotone increasing we find $\varphi \le f \le \psi$. Furthermore we have

$$\int_a^b \psi(x)dx - \int_a^b \varphi(x)dx = \sum_{k=1}^n f(x_k)(x_k - x_{k-1}) - \sum_{k=1}^n f(x_{k-1})(x_k - x_{k-1})$$

$$= \frac{b - a}{n} \left( \sum_{k=1}^n f(x_k) - \sum_{k=1}^n f(x_{k-1}) \right)$$

$$= \frac{b - a}{n}(f(b) - f(a)).$$

Now given $\epsilon > 0$ we can find $N \in \mathbb{N}$ such that for $n \geq N$ it follows that

$$\frac{b-a}{n}(f(x_n) - f(x_0)) = \frac{b-a}{n}(f(b) - f(a)) < \epsilon,$$

i.e. we have

$$\int_a^b \psi(x)dx - \int_a^b \varphi(x)dx < \epsilon$$

and by Theorem 25.15 the result follows. $\qquad\square$

Note that now we have two classes of integrable functions which are not necessarily continuous: step functions and monotone functions.

**Theorem 25.18.** *The set of all Riemann integrable functions $f : [a, b] \to \mathbb{R}$ form a real vector space. In addition we have for two Riemann integrable functions $f, g : [a, b] \to \mathbb{R}$*

$$f \leq g \text{ implies } \int_a^b f(x)dx \leq \int_a^b g(x)dx. \qquad (25.25)$$

*Proof.* From Theorem 25.13 and Corollary 25.14 we deduce immediately

$$\int_* f + \int_* g \leq \int_* (f + g) \leq \int^* (f + g) \leq \int^* f + \int^* g$$

and

$$\int_* f = \int^* f, \int_* g = \int^* g \text{ implies that } \int_* (f + g) = \int^* (f + g)$$

as well as

$$\int_a^b (f(x) + g(x))dx = \int_a^b f(x)dx + \int_a^b g(x)dx.$$

By Theorem 25.13 and Corollary 25.14 we find for $\lambda > 0$

$$\lambda \int^* f dx = \int_* (\lambda f)dx \leq \int^* (\lambda f)dx = \lambda \int^* f dx,$$

so $\lambda f$ is integrable and $\int \lambda f dx = \lambda \int^* f dx = \lambda \int f dx$. For $\lambda < 0$ we use Corollary 25.14, in particular (25.23). Finally, (25.25) follows once we

355

know that $f \geq 0$ implies $\int_a^b f(x)dx \geq 0$. But for $f \geq 0$ there always exists $\varphi \in T[a,b]$ such that $0 \leq \varphi \leq f$, hence

$$0 \leq \int_{*a}^b f(x)dx = \int_a^b f(x)dx.$$

$\square$

Recall that the **positive part** of $f : D \to \mathbb{R}$ is defined by

$$f_+(x) := \begin{cases} f(x), & f(x) > 0 \\ 0, & f(x) \leq 0 \end{cases} \qquad (25.26)$$

and the **negative part** is defined by

$$f_-(x) := \begin{cases} -f(x), & f(x) < 0 \\ 0, & f(x) \geq 0. \end{cases} \qquad (25.27)$$

Clearly $f^+ \geq 0, f_- \geq 0$ and $f = f_+ - f_-$ as well as $|f| = f_+ + f_-$.

**Theorem 25.19.** *If $f, g : [a, b] \to \mathbb{R}$ are Riemann integrable functions then $f_+, f_-$ and $|f|^p, 1 \leq p < \infty$, as well as $f \cdot g$ are Riemann integrable.*

*Proof.* By our assumptions, given $\varepsilon > 0$ there are step functions $\varphi, \psi \in T[a,b]$ such that $\varphi \leq f \leq \psi$ and

$$\int_a^b (\psi - \varphi)(x)dx \leq \varepsilon.$$

The functions $\phi_+, \psi_+$ are also step functions and we have $\varphi_+ \leq f_+ \leq \psi_+$. In addition it follows that

$$\int_a^b (\psi_+ - \varphi_+)(x)dx \leq \int_a^b (\psi - \varphi)(x)dx \leq \varepsilon.$$

Analogously we find that $f_-$ is integrable. Now, it follows that $|f|$ is integrable, recall $|f| = f_+ + f_-$. We want to prove that for $p \geq 1$ the function

356

$|f|^p$ is integrable. Suppose first that $0 \le f \le 1$. Then for $\varepsilon > 0$ there are step functions $\phi, \psi \in T[a,b]$ such that

$$0 \le \varphi \le f \le \psi \le 1$$

and

$$\int_a^b (\psi - \varphi)(x)dx \le \frac{\varepsilon}{p}.$$

It follows that $\varphi^p$ and $\psi^p$ are step functions and $\varphi^p \le f^p \le \psi^p$. Since

$$\frac{d}{dx}x^p = px^{p-1}$$

the mean value theorem yields

$$\psi^p - \varphi^p \le p(\psi - \varphi),$$

note that $x^{p-1} \le 1$ for $0 \le x \le 1$. Hence we find

$$\int_a^b (\psi^p - \varphi^p)(x)dx \le p\int_a^b (\psi - \varphi)(x)dx < \varepsilon,$$

thus $|f|^p = f^p$ is integrable. Now, for arbitrary $f$ we find that $|f|^p = f_+^p + f_-^p$, hence we may reduce the general case to non-negative functions. Further, if $f \ge 0$ but $\sup\limits_{x\in[a,b]} f(x) > 1$, we may consider $g(x) := \frac{f(x)}{\sup\limits_{x\in[a,b]} f(x)}$, i.e. $0 \le g(x) \le 1$. It follows that $\int_a^b |g|^p(x)dx$ exists, but

$$\int_a^b |g|^p(x)dx = \int_a^b |f(x)|^p \cdot \frac{1}{\sup\limits_t f(t)}dx = \frac{1}{\sup\limits_t f(t)} \int_a^b |f(x)|^p dx$$

and the integrability of $|f|^p$ is proved. Since $f \cdot g = \frac{1}{2}((f+g)^2 - (f-g)^2)$ the integrability of $f \cdot g$ follows from the integrability of $|f|^2$.  $\square$

**Corollary 25.20.** *For a Riemann integrable function $f : [a,b] \to \mathbb{R}$ the* **triangle inequality for integrals** *holds, i.e.*

$$\left| \int_a^b f(x)dx \right| \le \int_a^b |f(x)|dx. \tag{25.28}$$

357

*Proof.* From Theorem 25.19 we deduce

$$\left| \int_a^b f(x)dx \right| = \left| \int_a^b (f^+ - f^-)(x)dx \right| = \left| \int_a^b f^+(x)dx - \int_a^b f^-(x)dx \right|$$

$$\leq \int_a^b f^+(x)dx + \int_a^b f^-(x)dx = \int_a^b |f(x)|dx.$$

$\square$

We now prove the **mean value theorem for integrals**:

**Theorem 25.21.** *Let* $f, \phi : [a, b] \to \mathbb{R}$ *be continuous functions and suppose that* $\phi \geq 0$. *Then there exists* $\xi \in [a, b]$ *such that*

$$\int_a^b f(x)\phi(x)dx = f(\xi) \int_a^b \phi(x)dx. \qquad (25.29)$$

*In particular, for* $\phi = 1$ *it follows that*

$$\int_a^b f(x)dx = f(\xi)(b - a) \text{ for some } \xi \in [a, b]. \qquad (25.30)$$

*Proof.* Define

$$m := \inf\{f(x); x \in [a, b]\},$$

and

$$M := \sup\{f(x); x \in [a, b]\}.$$

It follows that

$$m\phi \leq f\phi \leq M\phi,$$

hence

$$m \int_a^b \phi(x)dx \leq \int_a^b f(x)\phi(x)dx \leq M \int_a^b \phi(x)dx.$$

Thus there is $\mu \in [m, M]$ such that

$$\int_a^b f(x)\phi(x)dx = \mu \int_a^b \phi(x)dx.$$

Now, the intermediate value theorem for continuous functions, Theorem 20.17 or Theorem 9.5, gives the existence of $\xi \in [a, b]$ such that $f(\xi) = \mu$ which proves the theorem. □

When combining the mean value theorem and the fundamental theorem of calculus we get a powerful tool to derive estimates. For this reason we postpone applications of the mean value theorem until the next chapter. However we state a very useful and often applied consequence of (25.30).

**Corollary 25.22.** *Let $f \in C([a, b])$ and $h > 0$ such that $x, x+h \in [a, b]$ then*

$$\lim_{h \to 0} \frac{1}{h} \int_x^{x+h} f(t)dt = f(x). \tag{25.31}$$

*Proof.* By (25.30) we find

$$\frac{1}{h} \int_x^{x+h} f(t)dt = f(\xi) \text{ for some } \xi \in [x, x + h],$$

and the continuity of $f$ implies the result. □

Given an integrable function $f : [a, b] \to \mathbb{R}$. It is obviously not easy to find step functions being close to $f$. Therefore we introduce a further way of approximating the integral $\int_a^b f(x)dx$ by using certain values of $f$.

**Definition 25.23.** *Let $f : [a, b] \to \mathbb{R}$ be a function and $Z = Z(x_0, \ldots, x_n)$ be a partition of $[a, b]$. Further, for $1 \leq k \leq n$ let $\xi_k \in [x_{k-1}, x_k]$. Then*

$$\sum_{k=1}^n f(\xi_k)(x_k - x_{k-1}) \tag{25.32}$$

*is called the **Riemann sum** of $f$ with respect to the partition $Z$ and points $\xi_k, k = 1, \ldots, n$.*
*As before we denote the mesh size of the partition $Z$ by*

$$\eta := \eta(Z) := \max_{1 \leq k \leq n} (x_k - x_{k-1}). \tag{25.33}$$

**Theorem 25.24.** *Let $f : [a, b] \to \mathbb{R}$ be a Riemann integrable function. Then for every $\varepsilon > 0$ there is $\delta > 0$ such that for every partition $Z = Z(x_0, \ldots, x_n)$*

*with mesh size less than or equal to $\delta$, i.e. $\eta(Z) \leq \delta$, and any choice of general points $\xi_k \in [x_{k-1}, x_k]$*

$$\left| \int_a^b f(x)dx - \sum_{k=1}^n f(\xi_k)(x_k - x_{k-1}) \right| \leq \varepsilon$$

*holds.*

*Proof.* Given $\varepsilon > 0$ there are step functions $\phi, \psi \in T[a, b]$ such that

$$\phi \leq f \leq \psi \text{ and } \int_a^b (\psi - \phi)(x)dx \leq \frac{\varepsilon}{2}.$$

Without loss of generality we may assume that $\phi$ and $\psi$ are given with respect to the same partition

$$a = t_0 < t_1 < \ldots < t_m = b.$$

Since $f$ is bounded

$$M := \sup\{|f(x)||x \in [a, b]\} \geq 0$$

is finite and we may assume $M \neq 0$. We claim that for

$$\delta := \frac{\varepsilon}{8Mm}$$

the assertion of the theorem holds.

For this let $Z(x_0, \ldots, x_n)$ be any partition of $[a, b]$ such that $\eta(Z) \leq \delta$ and take general points $\xi_k \in [x_{k-1}, x_k]$. We define the step function $F \in T[a, b]$ by

$$F(x_k) = 0 \text{ and } F(x) = f(\xi_k) \text{ for } x_{k-1} < x < x_k.$$

It follows that

$$\int_a^b F(x)dx = \sum_{k=1}^n f(\xi_k)(x_k - x_{k-1})$$

is the Riemann sum of $f$ with respect to the partition $Z(x_0, \ldots, x_n)$ and points $\xi_k \in [x_{k-1}, x_k]$. The step function $F$ has the properties:

360

1. $\varphi(x) - 2M \leq F(x) \leq \psi(x) + 2M, x \in [a, b]$,

2. if $[x_{k-1}, x_k] \subset (t_{j-1}, t_j)$ for some $j$, then

$$\varphi(x) \leq F(x) \leq \psi(x) \text{ for all } x \in [x_{k-1}, x_k].$$

Denote by $A \subset [a, b]$ the set

$$A := \bigcup \{(x_{k-1}, x_k) | \text{ there is } j \text{ such that } [x_{k-1}, x_k] \subset (t_{j-1}, t_j)\}$$

and define $s \in T[a, b]$ by

$$s(x) := \begin{cases} 0 & , x \in A \\ 2M & , x \notin A \end{cases}.$$

It follows by $1)$ and $2)$ that

$$\varphi(x) - s(x) \leq F(x) \leq \psi(x) + s(x) \text{ for all } x \in [a, b].$$

There are at most $2m$ intervals $[x_{k-1}, x_k]$ where $s$ is not $0$, thus

$$\int_a^b s(x)dx \leq 2M(2m\delta) \leq \frac{\varepsilon}{2},$$

which implies

$$\int_a^b \varphi(x)dx - \frac{\varepsilon}{2} \leq \int_a^b F(x)dx \leq \int_a^b \psi(x)dx + \frac{\varepsilon}{2}.$$

The choice of $\varphi$ and $\psi$ yields further

$$\int_a^b f(x)dx \leq \int_a^b \varphi(x)dx + \frac{\varepsilon}{2} \quad \text{and} \quad \int_a^b \psi(x)dx \leq \int_a^b f(x)dx + \frac{\varepsilon}{2}$$

or

$$| \int_a^b f(x)dx - \int_a^b F(x)dx | \leq \varepsilon$$

proving the theorem. $\qquad\qquad\qquad\qquad\qquad\qquad\qquad\qquad$ $\square$

We want to use Theorem 25.24 to generalise **Minkowski's** and **Hölder's in-equality** to integrable functions. Let $f : [a, b] \to \mathbb{R}$ be a Riemann integrable function and $p \geq 1$. We define

$$\|f\|_p := \left( \int_a^b |f(x)|^p dx \right)^{\frac{1}{p}}. \tag{25.34}$$

**Proposition 25.25.** *For $f, g : [a, b] \to \mathbb{R}$ Riemann integrable and $p \geq 1$ we have*

$$\|f + g\|_p \leq \|f\|_p + \|g\|_p, \tag{25.35}$$

*and for $1 < p < \infty, q := \frac{p}{p-1}$, it follows that*

$$\left| \int_a^b f(x)g(x)dx \right| \leq \int_a^b |f(x)g(x)|dx \leq \|f\|_p \|g\|_q. \tag{25.36}$$

*Proof.* We just have to approximate the integrals by a Riemann sum and then we have to pass to the limit. We prove Minkowski's inequality in detail:

$$\|f + g\|_p = \left( \int_a^b |f(x) + g(x)|^p dx \right)^{\frac{1}{p}}$$

$$\leq \left( \sum_{k=1}^n |f(\xi_k) + g(\xi_k)|^p (x_k - x_{k-1}) \right)^{\frac{1}{p}} + \varepsilon \tag{25.37}$$

$$= \left( \sum_{k=1}^n (|f(\xi_k)(x_k - x_{k-1})^{\frac{1}{p}} + g(\xi_k)(x_k - x_{k-1})^{\frac{1}{p}}|^p \right)^{\frac{1}{p}} + \varepsilon$$

$$\leq \left( \sum_{k=1}^n |f(\xi_k)|^p (x_k - x_{k-1}) \right)^{\frac{1}{p}} + \left( \sum_{k=1}^n |g(\xi_k)|^p (x_k - x_{k-1}) \right)^{\frac{1}{p}} + \varepsilon$$

$$\leq \left( \int_a^b |f(x)|^p dx \right)^{\frac{1}{p}} + \left( \int_a^b |g(x)|^p dx \right)^{\frac{1}{p}} + 2\varepsilon$$

$\square$

**Exercise 25.26.** *For $f, g : [a, b] \to \mathbb{R}$ Riemann integrable prove Hölder's inequality, i.e. (25.36).*

Interchanging limits and integrals is an important topic and we will return to this on many occasions. Here we state a first result.

**Theorem 25.27.** *Let $f_n : [a, b] \to \mathbb{R}, n \in \mathbb{N}$, be a sequence of continuous functions converging uniformly on $[a, b]$ to $f : [a, b] \to \mathbb{R}$, then we have*

$$\int_a^b f(x)dx = \int_a^b \left( \lim_{n \to \infty} f_n(x) \right) dx = \lim_{n \to \infty} \int_a^b f_n(x)dx.$$

*Proof.* By Theorem 24.6 we know that $f$ is continuous, hence integrable and it follows

$$\left| \int_a^b f(x)dx - \int_a^b f_n(x)dx \right| \leq \int_a^b |f(x) - f_n(x)|dx \leq (b - a)\|f - f_n\|_\infty$$

and since $\|f - f_n\|_\infty \to 0$ as $n \to \infty$ the theorem is proved. $\square$

Finally we want to consider the integral as a set function. First we note the trivial fact that if $T[a, b]$ and $a < c < b$ then $\varphi|_{[a,c]} \in T[a, c]$ and

$$\int_a^c \varphi|_{[a,c]}(x)dx = \int_a^b (\varphi \chi_{[a,c]})(x)dx.$$

Therefore if $f : [a, b] \to \mathbb{R}$ is Riemann integrable and for $\epsilon > 0$ given $\varphi, \psi \in T[a, b]$ are such that $\varphi \leq f \leq \psi$ and $\int_a^b (\psi - \varphi)(x)dx < \epsilon$, then $\varphi_{[a,c]} \leq f|_{[a,c]} \leq \psi|_{[a,c]}$ and

$$\int_a^c (\psi|_{[a,c]} - \varphi|_{[a,c]})(x)dx = \int_a^b ((\psi - \varphi)\chi_{[a,c]})(x)dx$$
$$\leq \int_a^b (\psi - \varphi)(x)dx < \epsilon.$$

Hence for $c \in (a, b)$ the function $f|_{[a,c]}$ is integrable and moreover

$$\int_a^b f(x)dx = \int_a^c f(x)dx + \int_c^b f(x)dx. \tag{25.38}$$

This easily extends to

**Proposition 25.28.** *Let* $f : [a, b] \to \mathbb{R}$ *be integrable and* $(I_j)_{j=1,\ldots,N}$ *be a finite partition of* $[a, b]$ *into intervals* $I_j = [a_j, b_j]$ *such that* $[a, b] = \bigcup_{j=1}^{N} I_j$ *and* $(a_j, b_j) \cap (a_l, b_l) = \emptyset$ *for* $j \neq l$, *then we have*

$$\int_a^b f(x)dx = \sum_{j=1}^{N} \int_{I_j} f(x)dx, \qquad (25.39)$$

*where with* $I_j = [a_j, b_j]$ *we write* $\int_{I_j} f(x)dx$ *for* $\int_{a_j}^{b_j} f(x)dx$.

Rewriting (25.39) as

$$\int_{\bigcup_{j=1}^{N} I_j} f(x)dx = \sum_{j=1}^{N} \int_{I_j} f(x)dx \qquad (25.40)$$

we may interpret (25.39) as (25.40) as set-additivity of the integral. Later on we will extend (25.40) to countable many sub-intervals $I_j, j \in \mathbb{N}$. Furthermore, we will try to replace intervals by more general sets.

Two helpful definitions are

$$\int_a^b f(x)dx := -\int_b^a f(x)dx \quad \text{if} \quad b < a \qquad (25.41)$$

and

$$\int_a^a f(x)dx := 0. \qquad (25.42)$$

Note that Proposition 25.28 implies for $f : [a, b] \to \mathbb{R}$, $f \geq 0$, that for $a \leq c \leq d \leq b$ we have

$$\int_c^d f(x)dx \leq \int_a^b f(x)dx. \qquad (25.43)$$

# Problems

1. Prove that the product of two step functions is a step function, i.e. $\varphi, \psi \in T[a, b]$ implies $\varphi \cdot \psi \in T[a, b]$, and deduce that $T[a, b]$ is an algebra.

2. Let $f : [a, b] \to \mathbb{R}$ be Riemann integrable and $y_1, \ldots, y_n \in [a, b]$. Define $\tilde{f} : [a, b] \to \mathbb{R}$ by

$$\tilde{f} := \begin{cases} f(x), & x \in [a, b] \setminus \{y_1, \ldots, y_N\} \\ c_j, & x = y_j \end{cases}$$

for some $c_j \in \mathbb{R}$. Prove that $\tilde{f}$ is Riemann integrable and $\int_a^b \tilde{f}(x)dx = \int_a^b f(x)dx$.

3. We call $f : [a, b] \to \mathbb{R}$ **piecewise continuous** if there exists a partition $Z = Z(x_0, \ldots, x_n)$ of $[a, b]$ such that $f|_{(x_{k-1}, x_k)}$ is continuous and $\lim_{\substack{x \to x_{k-1} \\ x > x_{k-1}}} f(x)$ and $\lim_{\substack{x \to x_{k-1} \\ x < x_k}} f(x)$ exist.

   a) Give an example of a piecewise continuous function which is not continuous.

   b) Let $f : [a, b] \to \mathbb{R}$ be a bounded piecewise continuous function. Prove that $f$ is Riemann integrable

4. If $f : [a, b] \to \mathbb{R}$ is Riemann integrable and $f(x) \geq \gamma > 0$ for all $x \in [a, b]$, then $\frac{1}{f}$ is Riemann integrable.

5. Does $\int_a^b |f(x)|dx = 0$ imply for a Riemann integrable function $f : [a, b] \to \mathbb{R}$ that $f(x) = 0$ for all $x \in [a, b]$?

6. Prove for $f \in C([a, b])$ that $\int_a^b |f(x)|dx = 0$ implies that $f(x) = 0$ for all $x \in [a, b]$. Deduce that

$$\|f\|_{L^1} := \int_a^b |f(x)|dx$$

is a norm on the vector space $C([a, b])$.

7.    a) Let $f : [a, b] \to \mathbb{R}$ be a Riemann integrable function and $(Z_n)_{n \in \mathbb{N}}$ be the sequence of the partition $Z_n = Z(x_0^{(n)}, \ldots, x_n^{(n)})$ where $x_j^{(n)} := a + \frac{j}{n}(b - a), 0 \leq j \leq n$. Further let $S_n(f) := \sum_{j=1}^n f(x_j^{(n)})(x_j^{(n)} - x_{j-1}^{(n)})$. Prove that

$$\lim_{n \to \infty} S_n(f) = \int_a^b f(x)dx.$$

b) Denote by $\fint_a^b f(x)dx := \frac{1}{b-a}\int_a^b f(x)dx$ the mean value of $f :$ $[a, b] \to \mathbb{R}$ which we assume to be Riemann integrable. With $x_j^{(n)} :=$ $a + \frac{j}{n}(b - a), j = 0, \ldots, n$, prove that

$$\fint_a^b f(x)dx = \lim_{n \to \infty} \frac{1}{n} \sum_{j=1}^{n} f(x_j^{(n)}).$$

8. For two Riemann integrable functions $f, g : [a, b] \to \mathbb{R}$ prove Hölder's inequality:

$$\left| \int_a^b f(x)g(x)dx \right| \leq \int_a^b |f(x)g(x)|dx \leq \left( \int_a^b |f(x)|^p dx \right)^{\frac{1}{p}} \left( \int_a^b |g(x)|^q dx \right)^{\frac{1}{q}}$$

with $\frac{1}{p} + \frac{1}{q} = 1, 1 < p$.

9.     a) Use Hölder's inequality to prove for a Riemann integrable function $f : [a, b] \to \mathbb{R}$ the estimate

$$\int_a^b |f(x)|^p dx \leq (b - a)^{\frac{q-p}{q}} \left( \int_a^b |f(x)|^q dx \right)^{\frac{p}{q}}$$

where $1 \leq p < q$. Hint: note that $|f(x)|^p = 1 \cdot |f(x)|^p$ and $x \mapsto 1$ is integrable.

b) Prove that for two Riemann integrable functions $f, g : [a, b] \to \mathbb{R}$ and every $\epsilon > 0$ we have

$$\int_a^b |f(x)g(x)|dx \leq \epsilon \int_a^b |f(x)|^2 dx + \frac{1}{4\epsilon} \int_a^b |g(x)|^2 dx.$$

10. Let $f : [a, b] \to \mathbb{R}$ be Riemann integrable. For $k \in \mathbb{N}$ prove

$$\left( \int_a^b f(x) \sin kx \, dx \right)^2 + \left( \int_a^b f(x) \cos kx \, dx \right)^2 \leq (b - a) \int_a^b f^2(x)dx.$$

11. Let $h : [a, b] \to \mathbb{R}$ be Riemann integrable and $f : [c, d] \to \mathbb{R}$ be a convex function such that $h([a, b]) \subset [c, d]$. Show **Jensen's inequality for integrals**

$$f\left( \frac{1}{b-a} \int_a^b h(t)dt \right) \leq \frac{1}{b-a} \int_a^b f(h(t))dt. \tag{25.44}$$

Hint: use Jensen's inequality for sums, see Problem 1 in Chapter 23.

12. On $[0, 1]$ consider the sequence $(f_n)_{n \in \mathbb{N}}$ of functions

$$f_n(x) := \begin{cases} 4n^2x, & 0 \le x \le \frac{1}{2n} \\ -4n^2x + 4n, & \frac{1}{2n} \le x \le \frac{1}{n} \\ 0, & \frac{1}{n} \le x \le 1. \end{cases}$$

Sketch the graph of $f_n$ and prove that $f_n$ is continuous. Furthermore show that $\lim_{n \to \infty} f_n(x) = 0$ for every $x \in [0, 1]$, i.e. $f_n$ converges on $[0, 1]$ pointwise to the zero function. Verify by using a simple geometric interpretation (calculating the area of a triangle) that $\int_0^1 f_n(x)dx = 1$. Hence we have an example of a sequence converging pointwise, the integrals converge, but the integral of the limit is not equal to the limit of the integrals.

13. Prove that if a sequence of Riemann integrable functions $f_n : [a, b] \to \mathbb{R}$ converges uniformly to $f : [a, b] \to \mathbb{R}$ then $f$ is Riemann integrable and

$$\lim_{n \to \infty} \int_a^b f_n(x)dx = \int_a^b f(x)dx \left( = \int_a^b \left( \lim_{n \to \infty} f_n(x) \right) dx \right).$$

# 26 The Fundamental Theorem of Calculus

We want to investigate the relation of integration and differentiation. In the following $I \subset \mathbb{R}$ will denote any interval (open, closed or half-open) with distinct end points. Note that $I$ need not be bounded.

**Theorem 26.1.** *Let $f : I \to \mathbb{R}$ be a continuous function and $a \in I$. If $F : I \to \mathbb{R}$ is defined by*

$$F(x) := \int_a^x f(t)dt$$

*then $F$ is differentiable with $F' = f$. In particular $F$ is continuous.*

*Proof.* For $h \neq 0$ we find

$$\frac{F(x+h) - F(x)}{h} = \frac{1}{h} \left( \int_a^{x+h} f(t)dt - \int_a^x f(t)dt \right) = \frac{1}{h} \int_x^{x+h} f(t)dt.$$

By the mean value theorem, Theorem 25.21, there is $\xi_h \in [x, x+h]$ (or $\xi_h \in [x+h, x]$ if $h < 0$) such that

$$\int_x^{x+h} f(t)dt = hf(\xi_h).$$

Since $\lim_{h \to 0} \xi_h = x$ and since $f$ is continuous it follows that

$$F'(x) = \lim_{h \to 0} \frac{1}{h} \int_x^{x+h} f(t)dt = \lim_{h \to 0} \frac{1}{h} hf(\xi_h) = f(x).$$

$\square$

**Definition 26.2.** *Let $F : I \to \mathbb{R}$ be a differentiable function. If $F' = f, f : I \to \mathbb{R}$, then we call $F$ a **primitive** of $f$.*

**Proposition 26.3.** *Two primitives of $f$ differ only by a constant.*

*Proof.* Let $c \in \mathbb{R}$ be a constant and $F' = f$, then $(F + c)' = f$ i.e. $F + c$ is a primitive of $f$. Conversely, if $F$ and $G$ are two primitives of $f$, then $F' - G' = f - f = 0$, hence $(F - G) = c$ or $F = G + c$. $\square$

Now we may prove the **fundamental theorem of calculus**, compare with Theorem 12.7.

**Theorem 26.4.** *Let $f : I \to \mathbb{R}$ be a continuous function with primitive $F$. For all $a, b \in I$, $a < b$, we have*

$$\int_a^b f(x)dx = F(b) - F(a).$$

*Proof.* For $x \in I$ set

$$F_0(x) = \int_a^x f(t)dt, (a \in I \text{ fixed}).$$

Then $F_0$ is a primitive of $f$ such that

$$F_0(a) = 0 \quad \text{and} \quad F_0(b) = \int_a^b f(t)dt.$$

If $F$ is any primitive of $f$, then there is $c \in \mathbb{R}$ such that $F - F_0 = c$ which yields

$$F(b) - F(a) = F_0(b) - F_0(a) = F_0(b) = \int_a^b f(t)dt.$$

$\square$

A useful notation is

$$\int_a^b f(x)dx = F|_a^b, \tag{26.1}$$

and more generally

$$h|_a^b := h(b) - h(a). \tag{26.2}$$

Let us restate (with full proofs) some rules for integration that have already been proved in Chapter 13.

**Proposition 26.5 (Integration by Parts).** *For two continuously differentiable functions* $f, g : [a, b] \to \mathbb{R}$ *we have*

$$\int_a^b f(x)g'(x)dx = f \cdot g\big|_a^b - \int_a^b f'(x)g(x)dx$$

$$= f(b)g(b) - f(a)g(a) - \int_a^b f'(x)g(x)dx. \qquad (26.3)$$

*Proof.* With $F := f \cdot g$ we find

$$F'(x) = f'(x)g(x) + f(x)g'(x).$$

This yields by the fundamental theorem

$$\int_a^b f'(x)g(x)dx + \int_a^b f(x)g'(x)dx = F\big|_a^b = f \cdot g\big|_a^b.$$

$\square$

**Proposition 26.6 (Integration by Substitution).** *For* $f : I \to \mathbb{R}$ *a continuous function and* $\phi : [a, b] \to \mathbb{R}$ *a continuously differentiable function, i.e.* $\phi \in C^1[a, b]$, *such that* $\phi([a, b]) \subset I$ *we have*

$$\int_a^b f(\phi(t))\phi'(t)dt = \int_{\phi(a)}^{\phi(b)} f(x)dx. \qquad (26.4)$$

*Proof.* Let $F$ be a primitive of $f$. Using the chain rule we find for $F \circ \phi : [a, b] \to \mathbb{R}$

$$(F \circ \phi)'(t) = F'(\phi(t))\phi'(t) = f(\phi(t))\phi'(t)$$

which implies by the fundamental theorem

$$\int_a^b f(\phi(t))\phi'(t)dt = F \circ \phi\big|_a^b = F(\phi(b)) - F(\phi(a)) = \int_{\phi(a)}^{\phi(b)} f(x)dx.$$

$\square$

In Chapter 13 we have already given a lot of applications of these rules. Here we are interested in more theoretical applications.

**Proposition 26.7. A.** *Let $h : [-a, a] \to \mathbb{R}$ be an even continuous function. Then we find*

$$\int_{-a}^{a} h(t)dt = 2 \int_{0}^{a} h(t)dt. \tag{26.5}$$

***B.*** *For an odd continuous function $g : [-a, a] \to \mathbb{R}$ it follows that*

$$\int_{-a}^{a} g(t)dt = 0. \tag{26.6}$$

*Proof.* **A.** We know that

$$\int_{-a}^{a} h(t)dt = \int_{-a}^{0} h(t)dt + \int_{0}^{a} h(t)dt.$$

Now the change of variable $t \mapsto -s$ gives

$$\int_{-a}^{0} h(t)dt = - \int_{a}^{0} h(-s)ds = - \int_{a}^{0} h(s)ds = \int_{0}^{a} h(s)ds,$$

i.e.

$$\int_{-a}^{a} h(t)dt = \int_{0}^{a} h(t)dt + \int_{0}^{a} h(s)ds = 2 \int_{0}^{a} h(t)dt.$$

**B.** Since

$$\int_{-a}^{a} g(t)dt = \int_{-a}^{0} g(t)dt + \int_{0}^{a} g(t)dt$$

we are done if we can show that $\int_{-a}^{0} g(t)dt = - \int_{0}^{a} g(t)dt$. The change of variable $t \mapsto -s$ yields however

$$\int_{-a}^{0} g(t)dt = - \int_{a}^{0} g(-s)ds = \int_{a}^{0} g(s)ds = - \int_{0}^{a} g(t)dt.$$

$\square$

A further symmetry we have encounter was periodicity. We claim first

**Proposition 26.8.** *Let* $f : \mathbb{R} \to \mathbb{R}$ *be a continuous function with period* $c > 0$. *If for some* $a \in \mathbb{R}$ *we have*

$$\int_{a-c}^{a} f(t)dt = 0 \tag{26.7}$$

*then every primitive* $F$ *and* $f$ *has period* $c$ *too.*

*Proof.* Let $F$ be any primitive of $f$. If follows that

$$F(x+c) - F(a) = \int_{a}^{x+c} f(t)dt = \int_{a}^{x+c} f(t-c)dt = \int_{a-c}^{x} f(t)dt$$
$$= F(x) - F(a-c),$$

or

$$F(x+c) - F(x) = F(a) - F(a-c) = \int_{a-c}^{a} f(t)dt = 0. \tag{26.8}$$

$\square$

**Remark 26.9. A.** The function $g(x) = 1 + \cos x$ is $2\pi$-periodic since

$$g(x + 2\pi) = 1 + \cos(x + 2\pi) = 1 + \cos x = g(x).$$

A primitive of $g$ is $G(x) = x + \sin x$ since $G'(x) = 1 + \cos x$. However,

$$G(x + 2\pi) = x + 2\pi + \sin(x + 2\pi) = x + 2\pi + \sin x \neq G(x).$$

Moreover we have

$$\int_{0}^{2\pi} (1 + \cos x)dx = \int_{a}^{a+2\pi} (1 + \cos x)dx = 2\pi \neq 0$$

for all $a \in \mathbb{R}$. Hence we cannot expect Proposition 26.8 to hold for all periodic functions.

**B.** From (26.8) it follows that $\int_{b}^{b+c} f(t)dt = 0$ for all $b \in \mathbb{R}$ if $\int_{a-c}^{a} f(t)dt = 0$ for an $a \in \mathbb{R}$.

Let $f : \mathbb{R} \to \mathbb{R}$ be a continuous function with period $c > 0$ and for some $a \in \mathbb{R}$ set $A := \int_{a-c}^{a} f(t)dt$, $A$ need not be zero. The function $f_A : \mathbb{R} \to \mathbb{R}$, $x \mapsto f_A(x) = f(x) - \frac{A}{c}$ is once again periodic with period $c$:

$$f_A(x + c) = f(x + c) - \frac{A}{c} = f(x) - \frac{A}{c} = f_A(x).$$

Moreover it holds

$$\int_{a-c}^{a} f_A(x)dx = \int_{a-c}^{a} (f(x) - \frac{A}{c})dx = A - A = 0.$$

Hence we may apply Proposition 26.8 to $f_A$ to find that

$$0 = \int_{b}^{b+c} f_A(t)dt = \int_{b}^{b+c} (f(x) - \frac{A}{c})dx = \int_{b}^{b+c} f(x)dx - A, \qquad (26.9)$$

and we have proved

**Corollary 26.10.** *Let $f : \mathbb{R} \to \mathbb{R}$ be a continuous function with period $c > 0$. For every $b \in \mathbb{R}$ it holds*

$$\int_{b}^{b+c} f(x)dx = A, \qquad (26.10)$$

*i.e. the integrals of $f$ over any interval of length $c$ have all the same value.*

From the last few results we may pick up an important message: symmetry may be used to simplify the evaluation of integrals.

We have seen that the Riemann integral is positivity preserving, with the consequence that for two integrable functions $f, g : I \to \mathbb{R}$, the inequality $f \leq g$ implies

$$\int_{I} f(x)dx \leq \int_{I} g(x)dx. \qquad (26.11)$$

**Corollary 26.11.** *A.* *Let $f : [a, b] \to \mathbb{R}$ be a continuous function such that $m \leq f(x) \leq M$ with $m, M \in \mathbb{R}$. Then it holds*

$$m(b - a) \leq \int_{a}^{b} f(t)dt \leq M(b - a). \qquad (26.12)$$

*B.* *Let $f : [a, b] \to \mathbb{R}$ be a continuous function such that $|f(x)| \leq M$ for all $t \in [a, b]$, $M \in \mathbb{R}$. Then we have*

$$\left| \int_{a}^{b} f(t)dt \right| \leq M(b - a). \qquad (26.13)$$

*Proof.* **A.** Integrating the inequality $m \leq f(t) \leq M$ we find

$$m(b - a) = \int_{a}^{b} mdt \leq \int_{a}^{b} f(t)dt \leq \int_{a}^{b} Mdt = M(b - a).$$

**B.** Using the triangle inequality for integrals we find with (26.11)

$$\left| \int_a^b f(t)dt \right| \leq \int_a^b |f(t)|dt \leq \int_a^b M dt = M(b - a).$$

□

This corollary has many nice applications.

**Example 26.12.** We claim that for all $x, y \in \mathbb{R}$

$$| \sin x - \sin y | \leq |x - y|. \tag{26.14}$$

Indeed for $x \geq y$ we find

$$| \sin x - \sin y | = \left| \int_y^x \cos t \, dt \right| \leq \int_y^x 1 dt = x - y = |x - y|,$$

and for $x \leq y$ it follows that

$$| \sin x - \sin y | = | \sin y - \sin x | \leq |y - x| = |x - y|.$$

Analogously we find for all $x, y \in \mathbb{R}$

$$| \cos x - \cos y | \leq |x - y|. \tag{26.15}$$

**Example 26.13.** Let $1 < a < b$. We claim

$$1 - \frac{a}{b} = \frac{1}{b}(b - a) \leq \ln \frac{b}{a} \leq \frac{1}{a}(b - a) = \frac{b}{a} - 1. \tag{26.16}$$

*Proof.* First note that

$$\ln \frac{b}{a} = \ln b - \ln a = \int_a^b \frac{1}{x} dx.$$

Now we estimate the integral. Since $\frac{1}{b} < \frac{1}{x} < \frac{1}{a}$ for $a < x < b$ it follows that

$$\frac{1}{b}(b - a) = \frac{1}{b} \int_a^b 1 dx \leq \int_a^b \frac{1}{x} dx \leq \frac{1}{a} \int_a^b 1 dx = \frac{1}{a}(b - a),$$

implying (26.16). □

**Example 26.14.** For all $a \geq 0$ and $t \geq 0$ we have

$$\frac{at}{1+at} \leq 1 - e^{-at} \leq at. \tag{26.17}$$

Since

$$1 - e^{-at} = \int_0^{at} e^{-x} dx \leq \int_0^{at} 1 dx = at$$

we get the right inequality. To get the left inequality note that $t \mapsto e^{at} - 1 - at$ is on $[0, \infty)$ for all $a \geq 0$ monotone increasing since $\frac{d}{dt}(e^{at} - 1 - at) = a(e^{at} - 1) \geq 0$. For $t = 0$ we have $e^{at} - 1 - at|_{t=0} = 0$, hence we find

$$0 \leq e^{at} - 1 - at$$

or

$$(1 + at) \leq e^{at},$$

i.e. $(1 + at)e^{-at} \leq 1$, which is equivalent to

$$0 \leq 1 - (1 + at)e^{-at}$$

or

$$at \leq 1 + at - (1 + at)e^{-at},$$

leading eventually to

$$\frac{at}{1+at} \leq 1 - e^{-at}.$$

**Example 26.15.** For $a \geq 0$ and $t \geq 0$ it holds

$$\left| \frac{e^{at} - 1 + at}{t} \right| \leq \frac{1}{2} a^2 t. \tag{26.18}$$

*Proof.* Observe that

$$0 \leq \int_0^{at} (1 - e^x) dx = at + e^{-at} - 1,$$

and therefore using (26.17)

$$|e^{at} - 1 + at| = \int_0^{at} (1 - e^x) dx \leq \int_0^{at} x dx = \frac{(at)^2}{2}$$

which implies (26.18). $\qquad \square$

Note that all these examples work along the same idea: we want to estimate the difference $f(x) - f(y)$ for a given function $f$. If we can identify $f$ as a primitive, i.e. for some function $g$ we have

$$f(x) - f(y) = \int_x^y g(t)dt,$$

then we can use estimates for $g$ to control this difference.

The following inequality is a first, rather crude version of the **Poincaré inequality** in one dimension.

**Proposition 26.16.** *Let $f \in C^1([a,b])$ and suppose that $f(a) = f(b) = 0$. Then there exists a constant $\gamma_0 > 0$ such that*

$$\left(\int_a^b |f(x)|^2 dx\right)^{\frac{1}{2}} \leq \gamma_0 \left(\int_a^b |f'(x)|^2 dx\right)^{\frac{1}{2}}. \tag{26.19}$$

*Proof.* We note that

$$\int_a^b |f(x)|^2 dx = \int_a^b f^2(x)dx = \int_a^b \left(\frac{d}{dx}x\right) f^2(x)dx$$

$$= -\int_a^b x\frac{d}{dx}f^2(x)dx = -2\int_a^b xf(x)f'(x)dx,$$

or by the Cauchy-Schwarz inequality

$$\int_a^b |f(x)|^2 dx \leq 2\max(|a|,|b|)\int_a^b |f(x)||f'(x)|dx$$

$$\leq 2\max(|a|,|b|)\left(\int_a^b |f(x)|^2 dx\right)^{\frac{1}{2}}\left(\int_a^b |f'(x)|^2 dx\right)^{\frac{1}{2}}$$

implying (26.19) with $\gamma_0 = 2\max(|a|,|b|)$. $\qquad\square$

**Remark 26.17.** Using the notation $||f||_p = \left(\int_a^b |f(x)|^2 dx\right)^{\frac{1}{p}}$ the Poincaré inequality reads as

$$||f||_{L^2} \leq \gamma_0 ||f'||_{L^2}, \tag{26.20}$$

and as we will see in Problem 10 this implies that on $C_0^1([a,b]) := \{f \in C^1([a,b])|f(a) = f(b) = 0\}$ a norm is given by $||f'||_{L^2}$. Clearly $||f'||_{L^2}$ is not a norm on $C^1([a,b])$ since every constant function $f_c$, $f_c(x) = c$, (restricted to $[a,b]$) belongs to $C^1([a,b])$ and has derivative $f'_c = 0$, i.e. $||f'_c||_{L^2} = 0$ but $f_c \neq 0$ for $c \neq 0$.

Consider $f : [a, b] \to \mathbb{R}$ a continuous function and let $\alpha, \beta : [c, d] \to [a, b]$ be two continuously differentiable functions such that $\alpha(x) \leq \beta(x)$ for all $x \in [c, d]$. We define the function $G : [c, d] \to \mathbb{R}$ by

$$G(x) := \int_{\alpha(x)}^{\beta(x)} f(t)dt. \tag{26.21}$$

**Proposition 26.18.** *The function $G$ in* (26.21) *is differentiable and*

$$G'(x) = \beta'(x)f(\beta(x)) - \alpha'(x)f(\alpha(x)). \tag{26.22}$$

*Proof.* For a primitive $F$ of $f$, i.e. $F' = f$, it follows that

$$G(x) = F(\beta(x)) - F(\alpha(x))$$

and the chain rule yields

$$G'(x) = \beta'(x)F'(\beta(x)) - \alpha'(x)F'(\alpha(x))$$
$$= \beta'(x)f(\beta(x)) - \alpha'(x)f(\alpha(x)).$$

$\square$

We now consider limits of sequences of differentiable functions.

**Theorem 26.19.** *Let $f_n : [a, b] \to \mathbb{R}$ be a sequence of continuously differentiable functions converging pointwise to $f : [a, b] \to \mathbb{R}$. Suppose that the sequence $(f_n')_{n \in \mathbb{N}}$ converges uniformly. Then $f$ is differentiable and we have*

$$f'(x) = \lim_{n \to \infty} f_n'(x) \quad \text{for all } x \in [a, b]. \tag{26.23}$$

*Proof.* Define $f^*(x) := \lim_{n \to \infty} f_n'(x)$. If follows that $f^* : [a, b] \to \mathbb{R}$ is a continuous function. Further, for $x \in [a, b]$ we have

$$f_n(x) = f_n(a) + \int_a^x f_n'(t)dt.$$

By Theorem 25.27 we know that $\int_a^x f'(t)dt \to \int_a^x f^*(t)dt$, implying

$$f(x) = f(a) + \int_a^x f^*(t)dt.$$

Now the fundamental theorem yields $f'(x) = f^*(x)$ and the theorem is proved. $\square$

We refer to Problem 9 in Chapter 24 for an example showing that uniform convergence of $(f_n)_{n \in \mathbb{N}}$ and pointwise convergence of $(f_n')_{n \in \mathbb{N}}$ is not sufficient for (26.23) to hold. Theorem 26.19 will become particularly powerful when applied to series of functions.

# Problems

1. Prove

    a) If $f \in C^k([a,b])$ then all primitives of $f$ belong to $C^{k+1}([a,b])$, $k \geq 0$.

    b) Every $f \in C([a,b])$ determines by its primitive a one-dimensional affine subspace of $C^1([a,b])$.

2. Define on $C([a,b])$ the mapping $T : C([a,b]) \to C([a,b])$ by $(Tf)(x) := f(a) + \int_a^x e^{-t}f(t)dt$, $x \in [a,b]$. We call $g \in C([a,b])$ a fixed point of $T$ if $Tg = g$, i.e. $(Tg)(x) = g(x)$ for all $x \in [a,b]$. Prove that if $g \in C^k([a,b])$ is a fixed point it must belong to $C^{k+1}([a,b])$, hence $g \in C^\infty([a,b]) = \bigcap_{k=1}^\infty C^k([a,b])$.

3. For $f \in C(\mathbb{R})$, $f \geq 0$, define for the right half-open interval $I = [a,b)$, $a,b \in \mathbb{R}$

$$\mu(I) := \int_a^b f(t)dt \quad \text{and} \quad \mu(\emptyset) := 0.$$

Moreover, for $I_1 = [a_1,b_1)$, $I_2 = [a_2,b_2)$, $b_1 < a_2$, we define

$$\mu(I_1 \cup I_2) = \mu(I_1) + \mu(I_2).$$

    a) Prove that the union of two bounded right half-open intervals is either disjoint or a right half-open interval and deduce that the union of finitely many right half-open intervals is the union of finitely mutually disjoint right half-open intervals or consists of one right half-open interval. Moreover the intersection of two bounded right half-open intervals is either empty or right half-open.

    b) Let $I_1$ and $I_2$ be two bounded right half-open intervals. Prove that

$$\mu(I_1 \cup I_2) + \mu(I_1 \cap I_2) = \mu(I_1) + \mu(I_2).$$

    c) For $a_0 \in \mathbb{R}$ fixed define the function $\mu_{a_0}(x) := \mu([a_0,x))$. Show that $\mu_{a_0} \in C^1(\mathbb{R})$ and $\mu_{a_0}'(x) = f(x)$.

4. Let $f : [-a,a] \to \mathbb{R}$ be a continuous function and suppose that for all $x \in (0,a]$ we have $\int_{-x}^x f(t)dt = 0$. Prove that $f$ is an odd function. Hint: first prove that if $g : [a,b] \to \mathbb{R}$ is continuous and if for all

379

$\alpha < \beta, \alpha, \beta \in [a, b]$, it follows that $\int_\alpha^\beta g(t)dt = 0$ then $g(t) = 0$ for all $t \in [a, b]$.

5.  a) For $\rho > 1$ and $0 \le x < y \le 1$ prove that

$$y^\rho - x^\rho \le \rho(y - x).$$

b) For $-\frac{\pi}{4} \le x < y \le \frac{\pi}{4}$ show that

$$y - x \le \frac{2}{\sqrt{2}}(\sin y - \sin x).$$

6.  Let $f : [a, b] \to \mathbb{R}$ be a continuous function. Prove that a primitive $F$ of $f$ is Lipschitz continuous and

$$|F(x) - F(y)| \le ||f||_\infty |x - y| \text{ for all } x, y \in [a, b].$$

7.  Let $f, g : [a, b] \to \mathbb{R}$ be two non-identical zero Riemann integrable functions. We call $f$ and $g$ **orthogonal** if $\int_a^b f(x)g(x)dx = 0$. If $f$ and $g$ are orthogonal we write $f \perp g$. We agree that $0$ is orthogonal to every $f : [a, b] \to \mathbb{R}$.

a) Prove that

$$f(x) = \begin{cases} c, & x \in [a, \frac{b-a}{2}] \\ 0, & x \in [\frac{b-a}{2}, b] \end{cases}$$

and

$$g(x) = \begin{cases} 0, & x \in [a, \frac{b-a}{2}] \\ -c, & x \in (\frac{b-a}{2}, a] \end{cases}$$

are orthogonal.

b) Suppose that $f : [-a, a] \to \mathbb{R}$ is even and $g : [-a, a] \to \mathbb{R}$ is odd. Show that they are orthogonal.

c) For $f \in C([a, b])$ define $\{f\}^\perp := \{g \in C([a, b]) | f \perp g\}$. Prove that $\{f\}^\perp$ is a subspace of $C([a, b])$.

8.  Let $f \in C^1([a, b])$ and suppose that $f \perp f'$. Prove that this implies $|f(a)| = |f(b)|$.

9.  Prove that on $C_0^1([a,b]) := \{f \in C^1([a,b])|f(a) = f(b) = 0\}$ a norm is given by $||f'||_{L^2}$. Hint: use Corollary 26.10 and Problem 6 in Chapter 25.

10. Let $\alpha, \beta : [a,b] \to \mathbb{R}$ be two differentiable functions such that $\alpha < \beta$. In addition suppose that $\alpha$ is decreasing and $\beta$ is increasing. Let $f : [\alpha(a), \beta(b)] \to \mathbb{R}$ be a non-negative function. Prove that $G : [\alpha(b), \beta(b)] \to \mathbb{R}$,

$$G(x) := \int_{\alpha(x)}^{\beta(x)} f(t)dt$$

is increasing.

11.* The following result sharpens Theorem 26.19 considerably: let $f_n : [a,b] \to \mathbb{R}$ be a sequence of differentiable functions. Suppose that for an $x_0 \in [a,b]$ the sequence $(f_n(x_0))_{n \in \mathbb{N}}$ converges and that $(f_n')_{n \in \mathbb{N}}$ converges uniformly on $[a,b]$ to some function. Then there exists a differentiable function $f : [a,b] \to \mathbb{R}$ such that $(f_n)_{n \in \mathbb{N}}$ converges uniformly to $f$ and we have

$$f'(x) = \lim_{n \to \infty} f_n'(x) \text{ for all } x \in [a,b].$$

Hint: prove that $(f_n)_{n \in \mathbb{N}}$ forms a Cauchy sequence with respect to the norm $|| \cdot ||_\infty$ and apply Theorem 24.11.

12. Consider the sequence $S_N(x) := \sum_{k=0}^{N} x^k$ defined on $(-1,1)$. Let $[a,b] \subset (-1,1)$ be a compact interval. Prove that $S_N(x)$ as well as $S_N'(x)$ converge uniformly on $[a,b]$. Deduce that for $m \in \mathbb{N}, m \geq 2$ it follows that

$$\sum_{k=1}^{\infty} \frac{k}{m^k} = \frac{m}{(m-1)^2}.$$

# 27 A First Encounter with Differential Equations

The fundamental theorem paves the way to solve some differential equations, more precisely some **ordinary differential equations**. First let us re-interpret the fundamental theorem. So far we have used this theorem (see Part 1) to evaluate integrals. We may ask the following question:

Given a function $h : [a,b] \to \mathbb{R}$, can we find a function $u : [a,b] \to \mathbb{R}$ such that

$$u'(x) = h(x) \tag{27.1}$$

holds for all $x \in (a,b)$?

Obviously $u$ will not be unique since for $v := u + c$, $c \in \mathbb{R}$, we find

$$v'(x) = u'(x) + c' = u'(x) = h(x).$$

However, two functions satisfying (27.1) can only differ by a constant. Thus, if we prescribe for example $u(a) = u_0$, $u_0 \in \mathbb{R}$, in the class of all differentiable functions on $(a,b)$ with continuous extension to $[a,b]$ there will be at most one function solving the **initial value problem**

$$u' = h, \quad u(a) = u_0. \tag{27.2}$$

In the case that $h$ is continuous we can find a solution to (27.2) by integration:

$$u(x) = u_0 + \int_a^x h(t)dt, \tag{27.3}$$

which follows from differentiating (27.3). Indeed by the fundamental theorem we find that $x \mapsto \int_a^x h(t)dt$ is differentiable and further

$$\frac{d}{dx}u(x) = \frac{d}{dx}\left(u_0 + \int_a^x h(t)dt\right) = h(x)$$

as well as

$$u(a) = u_0 + \int_a^a h(t)dt = u_0.$$

Note that we can "derive" the solution (27.3) to (27.2) by "integrating" (27.2):

$$\int_a^x u'(t)dt = \int_a^x h(t)dt$$

383

implying

$$u(x) - u(a) = \int_a^x h(t)dt,$$

or

$$u(x) = u(a) + \int_a^x h(t)dt = u_0 + \int_a^x h(t)dt.$$

Of course (27.2) is a rather simple initial value problem. We may want to solve a more general initial value problem, namely

$$g(u(x))u'(x) = h(x) \quad \text{and} \quad u(a) = u_0, \tag{27.4}$$

where $u : [a, b] \to \mathbb{R}$ is sought to be continuous on $[a, b]$ and differentiable on $(a, b)$. We assume again $h : [a, b] \to \mathbb{R}$ to be a continuous function and we assume $g : \mathbb{R} \to \mathbb{R}$ to be continuous too. We may integrate (27.4) to obtain

$$\int_a^x g(u(t))u'(t)dt = \int_a^x h(t)dt. \tag{27.5}$$

Let us have a closer look at the integral on the left hand side. If we consider $z = u(t)$ as a new variable then the rule for integration by substitution, compare with Proposition 26.6 or Theorem 13.12, yields

$$\int_{u(a)}^{u(x)} g(z)dz = \int_a^x g(u(t))u'(t)dt, \tag{27.6}$$

which implies

$$\int_{u(a)}^{u(x)} g(z)dz = \int_a^x h(t)dt. \tag{27.7}$$

Now let $G$ be a primitive of the continuous function $g$ and let $H$ be a primitive of the continuous function $h$. Then (27.7) becomes

$$G(u(x)) - G(u(a)) = H(x) - H(a), \tag{27.8}$$

i.e.

$$G(u(x)) = H(x) + G(u_0) - H(a). \tag{27.9}$$

If we add the assumption that $G$ has an inverse, we find

$$u(x) = G^{-1}(H(x) + G(u_0) - H(a)). \tag{27.10}$$

384

First we note that in (27.10) for $x = a$ we get

$$u(a) = G^{-1}(H(a) + G(u_0) - H(a)) = G^{-1}(G(u_0)) = u_0.$$

Next we suppose that $G^{-1}$ is differentiable, for example the condition $G'(y) = g(y) \neq 0$ for all $y \in \mathbb{R}$ will be sufficient. It follows from (27.10) that

$$
\begin{aligned}
\frac{d}{dx}u(x) &= \frac{d}{dx}G^{-1}(H(x) + G(u_0) - H(a)) \\
&= \left((G^{-1})'(H(x) + G(u_0) - H(a))\right)\frac{d}{dx}(H(x) + G(u_0) - H(a)) \\
&= \frac{1}{G'\left(G^{-1}(H(x) + G(u_0) - H(a))\right)}h(x) \\
&= \frac{1}{g\left(G^{-1}(H(x) + G(u_0) - H(a))\right)}h(x) \\
&= \frac{1}{g(u(x))}h(x),
\end{aligned}
$$

or

$$g(u(x))\frac{d}{dx}u(x) = h(x),$$

i.e. by (27.10) we have indeed a solution to (27.4). We have added two new conditions: $G$ has an inverse and $G^{-1}$ is differentiable. However, these two conditions are not independent: if $g(y) \neq 0$ for all $y \in \mathbb{R}$, it must be either strictly positive or strictly negative. Since $G$ is the primitive of $g$, i.e. $G' = g$, in the first case $G$ is strictly monotone increasing and in the second case it is strictly monotone decreasing. In each case however $G$ has an inverse which is differentiable. Thus we have proved the following existence and uniqueness result:

**Theorem 27.1.** *Let* $h : [a, b] \to \mathbb{R}, a < b$, *be a continuous function and* $u_0 \in \mathbb{R}$. *Suppose that* $g : \mathbb{R} \to \mathbb{R}$ *is a continuous function and* $g(y) \neq 0$ *for all* $y \in \mathbb{R}$. *In this case the initial value problem*

$$g(u(x))u'(x) = h(x), \quad u(a) = u_0 \tag{27.11}$$

*has the unique solution*

$$u(x) = G^{-1}(H(x) + G(u_0) - H(a)), \tag{27.12}$$

*where* $H$ *is a primitive of* $h$ *and* $G$ *is a primitive of* $g$.

**Remark 27.2.** Note that after we have derived a candidate for $u$ as a solution to (27.4) we have to verify that this function is indeed a solution. This is typical for solving differential equations: to derive a formula for $u$ we need to do some calculations, but since we do not know what $u$ is, we may not be able to justify these calculations. Thus we pretend as if all steps in the calculation are allowed, and once we have derived a formula we try (and we have) to verify that this formula gives a solution.

**Remark 27.3.** Let us return to formula (27.8) (or (27.9)). In the case that $h$ and $g$ are continuous, hence have a primitive, this formula makes sense for any $u : [a,b] \to \mathbb{R}, u(a) = u_0$. Hence we can call a function, not necessarily differentiable, a generalised solution to (27.4) if (27.8) holds. Non-differentiable solutions of (partial) differential equations are of importance, however we first need to understand more about differentiable solutions.

For solving differential equations, say the initial value problem (27.4), sometimes a rather formal approach is helpful. With $y = u(x)$ we write (27.4) as

$$g(y)\frac{dy}{dx} = h(x), \tag{27.13}$$

and

$$y_0 = u(a). \tag{27.14}$$

We now write (27.13) formally as

$$g(y)dy = h(x)dx \tag{27.15}$$

and take primitives on both sides, i.e. look at

$$\int g(y)dy = \int h(x)dx + C \tag{27.16}$$

(of course we only need to include one constant). Thus we have a formal algorithm:

To solve $g(u(x))u'(x) = h'(x)$, look at $g(y)\frac{dy}{dx} = h(x)$, and integrate $g(y)dy = h(x)dx$, i.e. for the integration process the variables are separated and consequently this method is called **separation of variables**. Once we have separated the variables we try to evaluate the two integrals in (27.16) and we then return to (27.8).

386

Before discussing some examples we want to give a simple generalisation of the method. Consider the differential equation

$$h_2(x)g_2(u(x))u'(x) - h_1(x)g_1(u(x)) = 0 \tag{27.17}$$

where $h_1, h_2 : [a, b] \to \mathbb{R}$ and $g_1, g_2 : \mathbb{R} \to \mathbb{R}$ are continuous functions. We formally transform this equation to

$$\frac{g_2(u(x))}{g_1(u(x))}u'(x) = \frac{h_1(x)}{h_2(x)}, \tag{27.18}$$

and with $g = \frac{g_2}{g_1}$ and $h = \frac{h_1}{h_2}$ we are back to the first case provided $g_1(y) \neq 0$ for all $y \in \mathbb{R}$ and $h_2(x) \neq 0$ for all $x \in [a, b]$. Thus we derive formally the condition

$$\int \frac{g_2(y)}{g_1(y)}dy = \int \frac{h_1(x)}{h_2(x)}dx + C, \quad y = u(x),$$

i.e. we are looking for a primitive $G$ of $\frac{g_2}{g_1}$ as well as for a primitive $H$ of $\frac{h_1}{h_2}$, and then we try to solve the equation

$$G(y) = H(x) + C$$

for $y$. If this is possible we obtain a function $y = u(x)$ and eventually we can try to adjust the initial value $u_0$ by choosing $C$ such that $u(a) = u_0$. Thus the strategy is to find $G^{-1}$ and then to justify or verify that

$$u(x) = G^{-1}(H(x) + C)$$

solves (27.17).

**Example 27.4.** We want to solve

$$3u^2(x)u'(x) = \cos x, \quad u(0) = 1 \tag{27.19}$$

for $x \in \mathbb{R}$. With $g(z) = 3z^2$ and $h(x) = \cos x$ we find the primitives $G(z) = z^3$ and $H(x) = \sin x$, respectively. The inverse of $G$ is of course $G^{-1}(s) = s^{\frac{1}{3}}$ and since $G(u_0) = G(1) = 1$ and $H(0) = 0$ it follows that

$$u(x) = (1 + \sin x)^{\frac{1}{3}} \tag{27.20}$$

is a candidate for a solution to (27.19). An easy calculation shows that $u(0) = 1$ and

$$\frac{du}{dx} = \frac{d}{dx}(1 + \sin x)^{\frac{1}{3}} = \frac{1}{3}(1 + \sin x)^{-\frac{2}{3}}\cos x$$

or

$$3u^2(x)\frac{d}{dx}u(x) = \cos x.$$

Note that we cannot apply Theorem 27.1 since $g(z) = 3z^2$ has a zero at $z_0 = 0$. The short calculation above is still formal, we still need to specify for which values of $x$ it holds for. The problem are points where $\sin x = -1$, i.e. $x = \frac{3\pi}{2} + 2k\pi, k \in \mathbb{Z}$. At these points $u$ given by (27.20) is not differentiable. However for $x \neq \frac{3\pi}{2} + 2k\pi$ we have

$$3u^2(x)\frac{du(x)}{dx} = 3(1 + \sin x)^{\frac{2}{3}} \cdot \frac{1}{3}(1 + \sin x)^{-\frac{2}{3}} \cos x = \cos x$$

which implies that although $\frac{du}{dx}$ does not exist for $x = \frac{3\pi}{2} + 2k\pi$ we still have

$$\lim_{x \to \frac{3\pi}{2} + 2k\pi} \left(3u^2(x)\frac{du(x)}{dx}\right) = \lim_{x \to \frac{3\pi}{2} + 2k\pi} \cos x = \cos \frac{3\pi}{2} \, (= 0).$$

With this interpretation we can claim that $u(x) = (1 + \sin x)^{\frac{1}{3}}$ satisfies $3u^2 u' = \cos$ on the entire real axis even if it is not differentiable at certain points.

**Example 27.5.** Our aim is to find a solution to the initial value problem

$$\frac{du(x)}{dx} + 3u(x) = 8, \quad u(0) = 2. \tag{27.21}$$

We may write this equation as

$$1\frac{dy}{dx} + (3y - 8)1 = 0, \quad y = u(x), y_0 = u(0),$$

and formally we find

$$\frac{dy}{8 - 3y} = dx,$$

or

$$\int \frac{dy}{8 - 3y} = \int dx + C,$$

i.e.

$$-\frac{1}{3}\ln(8 - 3y) = x + C.$$

388

With $y_0 = 2$ we find $-\frac{1}{3}\ln(8-6) = -\frac{1}{3}\ln 2 = C$, and we arrive at

$$\ln(8-3y) - \ln 2 = \ln\left(\frac{8-3y}{2}\right) = -3x,$$

i.e.

$$\frac{8-3y}{2} = e^{-3x},$$

leading to

$$y = u(x) = \frac{2}{3}(4 - e^{-3x}).$$

Thus we conjecture that $u(x) = \frac{2}{3}(4 - e^{-3x})$ solves (27.21). The verification is straightforward:

$$u(0) = \frac{2}{3}(4 - 1) = 2$$

and

$$\frac{du(x)}{dx} = \frac{d}{dx}\left(\frac{2}{3}\left(4 - e^{-3x}\right)\right) = -\frac{2}{3}(-3)e^{-3x} = 2e^{-3x},$$

which yields

$$\frac{du(x)}{dx} + 3u(x) = 2e^{-3x} + 3\left(\frac{2}{3}(4 - e^{-3x})\right)$$
$$= 3e^{-3x} + 8 - 2e^{-3x} = 8.$$

**Remark 27.6.** A word of caution: in our course the exponential function was introduced as the solution to the initial value problem $u' = u, u(0) = 1$. Thus we still owe a proof of its existence and consequently of the existence of ln and the corresponding integrals. This will be done shortly in Chapter 29.

We close this chapter with a very useful formula. The calculation leading to the justification of (27.10) needs an evaluation of

$$\frac{d}{dx}\int_a^{u(x)} g(z)\,dz,$$

which is of course done with the help of the fundamental theorem. We want to consider a more general expression, namely

$$x \mapsto \int_{v(x)}^{u(x)} g(z)\,dz$$

with differentiable functions $v, u$ such that $v \leq u$. Let $G$ be a primitive of the continuous function $g$. It follows that

$$\int_{v(x)}^{u(x)} g(z)dz = G(u(x)) - G(v(x)).$$

Consequently we find

$$\frac{d}{dx} \int_{v(x)}^{u(x)} g(z)dz = \frac{d}{dx}\left(G(u(x)) - G(v(x))\right)$$
$$= G'(u(x))u'(x) - G'(v(x))v'(x)$$
$$= g(u(x))u'(x) - g(v(x))v'(x).$$

Hence we have

**Proposition 27.7.** *Let $g : [a, b] \to \mathbb{R}, a < b$, be a continuous function and let $u, v : [a, b] \to \mathbb{R}$ be two differentiable functions such that $v(x) \leq u(x)$ holds for all $x \in [a, b]$. The function $F : [a, b] \to \mathbb{R}$ given by*

$$F(x) := \int_{v(x)}^{u(x)} g(z)dz$$

*is differentiable in $(a, b)$ and we have for $a < x < b$*

$$F'(x) = \frac{d}{dx}F(x) = \frac{d}{dx} \int_{v(x)}^{u(x)} g(z)dz = g(u(x))u'(x) - g(v(x))v'(x). \quad (27.22)$$

Even in the case of Theorem 27.1 we will in general have no explicit formula for the solution of (27.11) and of course this applies to (27.17) too. Neither should we expect to find $G$ or $H$ explicitly, nor will we have an explicit formula for $G^{-1}$. Nonetheless, for a function satisfying, say (27.17), we can derive some properties. Here are some first observations. Suppose $u : [a, b] \to \mathbb{R}$ is differentiable and solves

$$u'(x) = h(x), \quad x \in (a, b). \quad (27.23)$$

If $h \geq 0 (> 0, \leq 0, < 0)$ then $u$ must be monotone increasing (strictly increasing, decreasing, strictly decreasing). This is trivial, for us the observation that we can find properties of a solution to (27.23) without having an explicit formula is the important one. A similar, more surprising result is the following: if $u : [a, b] \to \mathbb{R}$ satisfies

$$u'(x) + f(x)u(x) = 0, \quad x \in (a, b) \tag{27.24}$$

and if $f : (a, b) \to \mathbb{R}$ is $k$-times continuously differentiable then $u$ is on $(a, b)$ $(k + 1)$-times continuously differentiable. The proof is as follows: since $u$ is on $(a, b)$ continuous and differentiable, from (27.24) we derive

$$u'(x) = -f(x)u(x)$$

and the right hand side $-fu$ is continuously differentiable. Hence $u'$ is continuously differentiable and we find

$$u''(x) = -f'(x)u(x) - f(x)u'(x)$$
$$= -f'(x)u(x) + f^2(x)u(x).$$

Now we observe that $-f'u + f^2u$ is continuously differentiable implying that $u''$ is continuously differentiable and the following holds

$$u'''(x) = \left(-f''(x) + 3f'(x)f(x) - f^3(x)\right)u(x).$$

We can iterate this process until on the right hand side the $k^{\text{th}}$ derivative of $f$ appears, which is of course the case when forming $u^{(k+1)}$

$$u^{(k+1)}(x) = \frac{d^k}{dx^k}u'(x) = \frac{d^k}{dx^k}(-f(x)u(x))$$
$$= -\sum_{l=0}^{k} \binom{k}{l} f^{(l)}(x)u^{(k-l)}(x),$$

and we see that we can replace $u^{(k-l)}$ by an expression involving derivatives of $f$ up to order $k - l$ and $u$. In particular, if $f$ is arbitrarily often differentiable we find that $u$ is too. Thus, even for satisfying equation (27.24) we only need the first derivative of $u$, depending on the smoothness, i.e. the order of differentiability of $f$, $u$ must have higher order derivatives too. From now on we will encounter a few problems involving ordinary differentiable equations, and step by step we will establish a theory of ordinary differential equations.

# Problems

1. For $c < 0$ consider the function $u_c : \mathbb{R} \to \mathbb{R}$ defined by

$$u_c(x) := \begin{cases} \frac{x^2}{4}, & x > 0 \\ 0, & c \leq x \leq 0 \\ -\frac{(x-c)^2}{4}, & x < c. \end{cases}$$

Prove that $u_c$ is differentiable and satisfies

$$u'_c(x) = \sqrt{|u_c(x)|}.$$

Prove further that every $u_c$ satisfies $u_c(2) = 1$. Now deduce that the initial value problem

$$v'(x) = \sqrt{|v(x)|}, x \in (2,3), v(2) = 1,$$

is solvable but the solution is not unique.

2. Let $f, h : [a,b] \to \mathbb{R}$ be continuous functions and consider the differential equation

$$f(x)u'(x) + h(x)u(x) = 0, \quad x \in (a,b). \tag{27.25}$$

Prove that if $u_1, u_2 : (a,b) \to \mathbb{R}$ are two solutions to (27.25) then for every $\lambda, \mu \in \mathbb{R}$ the function $\lambda u_1 + \mu u_2$ is a further solution.

3. Let $p_0, p_1 : \mathbb{R} \to \mathbb{R}$ be continuous functions and $p_0(x) \neq 0$ for all $x \in \mathbb{R}$. Prove that a solution to

$$p_0(x)u'(x) + p_1(x)u(x) = 0, \quad x \in \mathbb{R}, \quad u(a) = u_a \in \mathbb{R},$$

is given by

$$u(x) := u_a e^{-\int_a^x \frac{p_1(t)}{p_0(t)} dt}. \tag{27.26}$$

4. By using the separation of variables method, if possible, find a solution to the following initial value problems. In each case give a (reasonable) domain for the solution.

   a) $xu'(x) = 2u(x), \quad u(1) = 3$;

b) $y'(t) = 2y^2(t)$,   $y(0) = -1$;

c) $\varphi'(s) = \frac{\varphi(s)}{\tan s}$,   $\varphi(\frac{\pi}{4}) = \frac{\pi}{4}$;

d) $5x^4(r)x'(r) = r\cos r$,   $x(\frac{\pi}{2}) = 1$.

5.   a) If $g$ is a continuous function defined on $\mathbb{R}$ find

$$\frac{d}{dx} \int_{\cos x}^{\sqrt{x^2+1}} g(z)dz.$$

b) For the differentiable functions $u, v : \mathbb{R} \to \mathbb{R}$, $v(x) \le u(x)$ for all $x \in \mathbb{R}$, find

$$\frac{d}{dx} \int_{v(x)}^{u(x)} \frac{1}{1+t^2}dt.$$

6. Let $h : \mathbb{R} \to \mathbb{R}$ be an odd, continuous function and let $u : \mathbb{R} \to \mathbb{R}$ be a non-negative, continuously differentiable function. Prove by a direct calculation that

$$\frac{d}{dx} \int_{-u(x)}^{u(x)} h(t)dt = 0,$$

hence $x \mapsto \int_{-u(x)}^{u(x)} h(t)dt$ is a constant function. Now give reasons without doing any calculation that we have in fact $\int_{-u(x)}^{u(x)} h(t)dt = 0$ for all $x \in \mathbb{R}$.

7. Suppose that $u : [0, \infty) \to \mathbb{R}$ is a continuous function which is continuously differentiable in $(0, \infty)$. If $u$ is a solution of the initial value problem

$$u' = \frac{1}{1+u^{2k}}, \quad u(0) = 1, \quad k \in \mathbb{N},$$

prove that then $u$ is a strictly monotone increasing, arbitrarily often differentiable function which is convex. (You are not expected to find an explicit expression for $u$.)

# 28 Improper Integrals and the Γ-Function

So far we have only integrated certain classes of bounded functions which are defined on a compact interval. We want to extend our notion of integrals to unbounded functions as well as to non-compact intervals.

**Definition 28.1.** *Let $I$ be a bounded interval with end points $a < b$ and $f : I \to \mathbb{R}$ be a fundtion. We assume:*

*a) for every $c, d \in I, c < d$, the function $f|_{[c,d]}$ is continuous;*

*b) for some $\alpha \in I$ the following two limits exist:*

$$\lim_{c \to a} \int_c^\alpha f(t)dt \quad and \quad \lim_{d \to b} \int_\alpha^d f(t)dt. \tag{28.1}$$

*Then we define the integral of $f$ over the interval $I$ by*

$$\int_a^b f(t)dt := \int_I f(t)dt := \lim_{c \to a} \int_c^\alpha f(t)dt + \lim_{d \to b} \int_\alpha^d f(t)dt, \tag{28.2}$$

*where $\alpha \in I$ is any point.*

**Remark 28.2. A.** Since for $\alpha, \beta \in I$ the following identity holds

$$\int_\beta^d f(t)dt = \int_\beta^\alpha f(t)dt + \int_\alpha^d f(t)dt \tag{28.3}$$

we can replace (28.1) by the condition that

$$\lim_{c \to a} \int_c^\alpha f(t)dt \quad and \quad \lim_{d \to b} \int_\beta^d f(t)dt$$

exist for $\alpha, \beta \in I$.

**B.** The definition of $\int_a^b f(t)dt$ is independent of the choice of $\alpha$ since for $\alpha, \gamma \in I$ it follows that

$$\int_c^d f(t)dt = \int_c^\alpha f(t)dt + \int_\alpha^d f(t)dt = \int_c^\gamma f(t)dt + \int_\gamma^d f(t)dt.$$

**C.** In the case that $f|_{[a,\alpha]}$ or $f|_{[\alpha,b]}$ is already integrable, i.e. $\int_a^\alpha f(t)dt$ or $\int_\alpha^b f(t)dt$ exist, then we need only require that $\lim_{d \to b} \int_\alpha^d f(t)dt$ or $\lim_{c \to a} \int_c^\alpha f(t)dt$ exist and we can define

$$\int_a^b f(t)dt := \int_a^\alpha f(t)dt + \lim_{d \to b} \int_\alpha^d f(t)dt \tag{28.4}$$

and

$$\int_a^b f(t)dt := \lim_{c \to a} \int_c^\alpha f(t)dt + \int_\alpha^b f(t)dt, \qquad (28.5)$$

respectively.

**D.** Of course it is possible to reduce all limits under consideration to limits of the type

$$\lim_{\epsilon \to 0} \int_{a+\epsilon}^\alpha f(t)dt \quad \text{or} \quad \lim_{\epsilon \to 0} \int_\alpha^{b-\epsilon} f(t)dt, \qquad (28.6)$$

with $\epsilon > 0$ such that $a + \epsilon, b - \epsilon \in I$.

This definition allows us already to integrate certain continuous functions defined on open or half-open intervals, and even some unbounded functions are included.

**Example 28.3.** Let $R > 0$ and $0 < \alpha < 1$. Consider the unbounded, continuous function $f_\alpha : (\alpha, R] \to \mathbb{R}, x \mapsto \frac{1}{x^\alpha}$. For $c \in (0, R]$ it follows that

$$\int_c^R f_\alpha(x)dx = \int_c^R \frac{1}{x^\alpha}dx = \frac{1}{1-\alpha} \cdot \frac{1}{x^{\alpha-1}}\Big|_c^R = \frac{1}{1-\alpha}\left(R^{1-\alpha} - c^{1-\alpha}\right).$$

Since $1 - \alpha > 0$ we find

$$\lim_{c \to 0} \int_c^R f_\alpha(x)dx = \frac{1}{1-\alpha}R^{1-\alpha},$$

hence

$$\int_0^R f_\alpha(x)dx = \int_0^R \frac{1}{x^\alpha}dx = \frac{1}{1-\alpha}R^{1-\alpha}. \qquad (28.7)$$

**Example 28.4.** The following holds

$$\int_{-1}^1 \frac{dx}{\sqrt{1-x^2}} = \lim_{\epsilon \to 0} \int_{-1+\epsilon}^0 \frac{dx}{\sqrt{1-x^2}} + \lim_{\epsilon \to 0} \int_0^{1-\epsilon} \frac{dx}{\sqrt{1-x^2}}$$

$$= -\lim_{\epsilon \to 0} \arcsin(-1+\epsilon) + \lim_{\epsilon \to 0} \arcsin(1-\epsilon)$$

$$= -\left(-\frac{\pi}{2}\right) + \frac{\pi}{2} = \pi,$$

i.e.

$$\int_{-1}^1 \frac{dx}{\sqrt{1-x^2}} = \pi. \qquad (28.8)$$

**Example 28.5.** We want to investigate the integral of $x \mapsto \ln(\sin x)$ over the interval $(0, \frac{\pi}{2})$. For $x \in (0, \frac{\pi}{2})$ the range of $\sin x$ is $(0, 1)$, but $\lim\limits_{\substack{y \to 0 \\ y > 0}} (\ln y) = -\infty$.

We first note for $\epsilon > 0$ that the substitution $x = \frac{\pi}{2} - y$ yields

$$I_\epsilon := \int_\epsilon^{\frac{\pi}{2} - \epsilon} \ln(\sin x)\,dx = -\int_{\frac{\pi}{2} - \epsilon}^\epsilon \ln(\sin(\frac{\pi}{2} - y))\,dy$$

$$= \int_\epsilon^{\frac{\pi}{2} - \epsilon} \ln(\sin(\frac{\pi}{2} - x))\,dx = \int_\epsilon^{\frac{\pi}{2} - \epsilon} \ln(\cos x)\,dx,$$

or

$$2I_\epsilon = \int_\epsilon^{\frac{\pi}{2} - \epsilon} (\ln(\sin x) + \ln(\cos x))\,dx$$

$$= \int_\epsilon^{\frac{\pi}{2} - \epsilon} \ln(\sin x \cos x)\,dx = \int_\epsilon^{\frac{\pi}{2} - \epsilon} \ln\left(\frac{\sin 2x}{2}\right)\,dx$$

$$= \int_\epsilon^{\frac{\pi}{2} - \epsilon} \ln(\sin 2x)\,dx - \int_\epsilon^{\frac{\pi}{2} - \epsilon} \ln 2\,dx$$

$$= \int_\epsilon^{\frac{\pi}{2} - \epsilon} \ln(\sin 2x)\,dx - \ln 2(\frac{\pi}{2} - \epsilon + \epsilon)$$

$$= \int_\epsilon^{\frac{\pi}{2} - \epsilon} \ln(\sin 2x)\,dx - \frac{\pi}{2} \ln 2.$$

We now study the remaining integral: the substitution $2x = t$ gives

$$\int_\epsilon^{\frac{\pi}{2} - \epsilon} \ln(\sin 2x)\,dx = \frac{1}{2} \int_{2\epsilon}^{\pi - 2\epsilon} \ln(\sin t)\,dt$$

$$= \frac{1}{2} \left( \int_\epsilon^{\frac{\pi}{2} - \epsilon} \ln(\sin t)\,dt + \int_{\frac{\pi}{2} + \epsilon}^{\pi - \epsilon} \ln(\sin t)\,dt \right)$$

$$- \frac{1}{2} \left( \int_\epsilon^{2\epsilon} \ln(\sin t)\,dt + \int_{\pi - 2\epsilon}^{\pi - \epsilon} \ln(\sin t)\,dt \right),$$

and using the first part (or the substitution $x = \pi - t$) we find

$$\int_\epsilon^{\frac{\pi}{2} - \epsilon} \ln(\sin 2x)\,dx = \int_\epsilon^{\frac{\pi}{2} - \epsilon} \ln(\sin x)\,dx - \frac{1}{2} \left( \int_\epsilon^{2\epsilon} \ln(\sin t)\,dt + \int_{\pi - 2\epsilon}^{\pi - \epsilon} \ln(\sin t)\,dt \right),$$

397

implying

$$2I_\epsilon = I_\epsilon - \frac{\pi}{2}\ln 2 - \frac{1}{2}\left(\int_\epsilon^{2\epsilon}\ln(\sin t)dt + \int_{\pi-2\epsilon}^{\pi-\epsilon}\ln(\sin t)dt\right),$$

or

$$I_\epsilon = -\frac{\pi}{2}\ln 2 - \frac{1}{2}\left(\int_\epsilon^{2\epsilon}\ln(\sin t)dt + \int_{\pi-2\epsilon}^{\pi-\epsilon}\ln(\sin t)dt\right).$$

We now claim that

$$\lim_{\epsilon\to 0}\int_\epsilon^{2\epsilon}\ln(\sin t)dt = \lim_{\epsilon\to 0}\int_{\pi-2\epsilon}^{\pi-\epsilon}\ln(\sin t)dt = 0. \tag{28.9}$$

Since $\lim_{y\to 0}(y^\alpha \ln y) = 0$ for any $\alpha > 0$ we first note that $\lim_{\epsilon\to 0}((\sin 2\epsilon)\ln(\sin 2\epsilon)) = 0$, implying of course that $\lim_{\epsilon\to 0}\epsilon\ln(\sin 2\epsilon) = 0$. Next we note that

$$\max_{t\in[\epsilon,2\epsilon]}|\ln(\sin t)| = -\ln(\sin 2\epsilon),$$

and

$$\max_{t\in[\pi-2\epsilon,\pi-\epsilon]}|\ln(\sin t)| = -\ln(\sin 2\epsilon),$$

which implies (28.9) and consequently we find

$$\int_0^{\frac{\pi}{2}}\ln(\sin x)dx = -\frac{\pi}{2}\ln 2.$$

In a further step we want to extend the integral for certain functions defined on unbounded intervals.

**Definition 28.6. A.** *Let $f : [a,\infty) \to \mathbb{R}$ ($g : (-\infty,b] \to \mathbb{R}$) be a continuous function. If the limit*

$$\lim_{R\to\infty}\int_a^R f(x)dx \quad \left(\lim_{R\to\infty}\int_{-R}^b g(x)dx\right) \tag{28.10}$$

*exists we denote it by*

$$\int_a^\infty f(x)dx := \lim_{R\to\infty}\int_a^R f(x)dx \quad \left(\int_{-\infty}^b g(x)dx = \lim_{R\to\infty}\int_{-R}^b g(x)dx\right).$$
$$\tag{28.11}$$

**B.** Let $f : (a, b) \to \mathbb{R}$ be a function where $a \in \mathbb{R} \cup \{-\infty\}$ and $b \in \mathbb{R} \cup \{\infty\}$. Suppose that for every $c, d \in \mathbb{R}, c < d$, such that $[c, d] \subset (a, b)$, the function $f|_{[c,d]}$ is continuous. If for some $\alpha \in (a, b)$ the limits

$$\int_a^b f(x)dx := \lim_{c \to a} \int_c^\alpha f(x)dx \qquad (28.12)$$

and

$$\int_\alpha^b f(x)dx := \lim_{d \to b} \int_\alpha^d f(x)dx \qquad (28.13)$$

exist, then we define

$$\int_a^b f(x)dx := \int_a^\alpha f(x)dx + \int_\alpha^b f(x)dx \qquad (28.14)$$

$$:= \lim_{c \to a} \int_c^\alpha f(x)dx + \lim_{d \to b} \int_\alpha^d f(x)dx.$$

**Remark 28.7.** As before, if one of the integrals $\int_a^\alpha f(x)dx$ or $\int_\alpha^b f(x)dx$ exist as the Riemann integral of a continuous function defined on a compact interval, then in (28.14) we need to consider only one limit.

**Definition 28.8.** *Any of the integrals defined in Definition 28.1 or Definition 28.6 we call the **improper (Riemann) integral** of $f$.*

**Example 28.9.** For $\alpha > 1$ we have

$$\int_1^\infty \frac{dx}{x^\alpha} = \frac{1}{\alpha - 1}. \qquad (28.15)$$

Indeed, for $R > 0$ the integral $\int_1^R \frac{dx}{x^\alpha}$ exists and we find

$$\int_1^R \frac{dx}{x^\alpha} = \frac{1}{1 - \alpha} \cdot \frac{1}{x^{\alpha-1}} \Big|_1^R = \frac{1}{\alpha - 1} \left(1 - \frac{1}{R^{\alpha-1}}\right).$$

Since $\lim\limits_{R \to \infty} \frac{1}{R^{\alpha-1}} = 0$, note that $\alpha - 1 > 0$, it follows that

$$\int_1^\infty \frac{dx}{x^\alpha} = \lim_{R \to \infty} \int_1^R \frac{dx}{x^\alpha} = \lim_{R \to \infty} \frac{1}{\alpha - 1} \left(1 - \frac{1}{R^{\alpha-1}}\right) = \frac{1}{\alpha - 1}.$$

**Example 28.10.** The following holds

$$\int_{-\infty}^{\infty} \frac{dx}{1+x^2} = \pi. \qquad (28.16)$$

We have for $R > 0$

$$\int_{-\infty}^{\infty} \frac{dx}{1+x^2} = \lim_{R\to\infty} \int_{-R}^{0} \frac{dx}{1+x^2} + \lim_{R\to\infty} \int_{0}^{R} \frac{dx}{1+x^2}$$
$$= -\lim_{R\to\infty} \arctan(-R) + \lim_{R\to\infty} \arctan R$$
$$= -\left(-\frac{\pi}{2}\right) + \frac{\pi}{2} = \pi.$$

**Example 28.11.** For $\alpha > 0$ we have

$$\int_{0}^{\infty} e^{-\alpha t} dt = \frac{1}{\alpha}. \qquad (28.17)$$

Indeed, for $R > 0$ it follows that

$$\int_{0}^{R} e^{-\alpha t} dt = -\frac{1}{\alpha} e^{-\alpha t}\Big|_{0}^{R} = \frac{1}{\alpha}(1 - e^{-\alpha R})$$

and passing to the limit $R \to \infty$ we find (28.17).

In most cases we cannot do explicit calculations to check whether or not an improper integral exists. Thus we need criteria for the convergence or divergence of improper integrals.

In the following $I$ is an interval with end points $a \in \mathbb{R} \cup \{-\infty\}$ and $b \in \mathbb{R} \cup \{\infty\}$, $a < b$, and $f : I \to \mathbb{R}$ is a function which is continuous on any compact interval $[c, d] \subset I$. For the improper integral of $f$ over $I$ (if it exists) we will write $\int_I f(x)dx$. Our first criterion is the **Cauchy criterion for improper integrals**.

**Theorem 28.12.** *The improper integral $\int_I f(x)dx$ exists (converges) if for every $\alpha \in (a, b)$ we have:*

*For every $\epsilon > 0$ there exists $s_0, y_0 \in (a, b)$ such that $b > t > s > s_0 > a$ and $a < z < y < y_0 < b$ imply*

$$\left| \int_{s}^{t} f(x)dx \right| < \epsilon \quad and \quad \left| \int_{z}^{y} f(x)dx \right| < \epsilon.$$

*Proof.* The first condition is equivalent to the existence of the limit

$$\lim_{t_0 \to b} \int_{s_0}^{t_0} f(x)dx,$$

while the second condition is equivalent to the existence of the limit

$$\lim_{z_0 \to a} \int_{z_0}^{y_0} f(x)dx.$$

$\square$

**Theorem 28.13.** *Suppose that $f \geq 0$. Then for every $\alpha \in (a,b)$ the integral $\int_\alpha^b f(x)dx$ converges if there exists a constant $M > 0$ such that for all $\beta \in (\alpha, b)$*

$$\int_\alpha^\beta f(x)dx \leq M \tag{28.18}$$

*holds.*

*Proof.* Since $f \geq 0$ the function $\beta \mapsto \int_\alpha^\beta f(x)dx$ is monotone increasing and it is bounded, hence the limit

$$\lim_{\beta \to b} \int_\alpha^\beta f(x)dx$$

exists, see Problem 6 in Chapter 20. $\square$

**Definition 28.14.** *We call $\int_I f(x)dx$ **absolutely convergent** if $\int_I |f(x)|dx$ converges.*

**Lemma 28.15.** *If $\int_I f(x)dx$ converges absolutely, then it converges.*

*Proof.* This follows from the Cauchy criterion with the help of the triangle inequality. Since $\int_I |f(x)|dx$ converges the Cauchy criterion holds for $|f|$, i.e. for $\alpha \in (a,b)$ we have: for every $\epsilon > 0$ there exists $s_0, y_0 \in (a,b)$ such that $b > t > s > s_0 > a$ and $a < z < y < y_0 < b$ imply

$$\int_s^t |f(x)|dx < \epsilon \quad \text{and} \quad \int_z^y |f(x)|dx < \epsilon, \tag{28.19}$$

and consequently

$$\left| \int_s^t f(x)dx \right| < \epsilon \quad \text{and} \quad \left| \int_z^y f(x)dx \right| < \epsilon. \tag{28.20}$$

$\square$

**Remark 28.16.** In Theorem 25.19 we have proved that if $f$ is Riemann integrable on the compact interval $[a, b]$, then $|f|$ is Riemann integrable on $[a, b]$ too. We will see in Problem 9 that this does not hold for improper integrals. Moreover, while the product of two Riemann integrable functions on a compact interval is also Riemann integrable, the product of two improper integrable functions need not be improper integrable. Indeed, by Example 28.3 the function $f_{\frac{1}{2}} := \frac{1}{\sqrt{x}}$ is improper integrable on $(0, 1]$, however $(f_{\frac{1}{2}} \cdot f_{\frac{1}{2}})(x) = \frac{1}{x}$ is not improper integrable on $(0, 1]$:

$$\int_{\epsilon}^{1} \frac{1}{x} dx = \ln 1 - \ln \epsilon = \ln \frac{1}{\epsilon}$$

and the limit $\epsilon \to 0$ does not exist.

The following criterion is useful:

**Theorem 28.17.** *Let $g : I \to \mathbb{R}$, $g(x) \geq 0$, and suppose that $\int_I g(x)dx$ converges. If $|f(x)| \leq g(x)$ for all $x \in I$ then $\int_I f(x)dx$ converges absolutely.*

*Proof.* We use once again the Cauchy criterion. Since the integral $\int_I g(x)dx$ exists, the Cauchy criterion holds, so we can replace in (28.19) the function $|f|$ by $g$. Now we need to observe that

$$\left| \int_s^t f(x)dx \right| \leq \int_s^t |f(x)|dx \leq \int_s^t g(x)dx$$

as well as

$$\left| \int_z^y f(x)dx \right| \leq \int_z^y |f(x)|dx \leq \int_z^y g(x)dx.$$

$\square$

**Corollary 28.18.** *Suppose that $h : I \to \mathbb{R}$, $h(x) \geq 0$, is continuous (integrable would be sufficient) and that $\int_I h(x)dx$ diverges. If $h(x) \leq f(x)$ for all $x \in I$, then $\int_I f(x)dx$ diverges too.*

*Proof.* If we assume the contrary, Theorem 28.17 would imply the convergence of $\int_I h(x)dx$. $\square$

**Example 28.19. A.** The integrals

$$\int_1^{\infty} \frac{\cos x}{x^2} dx \quad \text{and} \quad \int_1^{\infty} \frac{\sin x}{x^2} dx$$

402

exist. Indeed, if $P(\sin x, \cos x) = \sum_{k,l=0}^{N} A_{k,l} \sin^k(a_k x) \cos^l(b_l x)$ and $\alpha > 1$ then

$$\int_r^\infty \frac{P(\sin x, \cos x)}{x^\alpha} dx, \quad r > 0,$$

exists. We only need to observe that

$$|P(\sin x, \cos x)| \leq \sum_{k,l=0}^{N} |A_{k,l}|$$

and that for $\alpha > 1$ the integral $\int_r^\infty \frac{1}{x^\alpha} dx$ converges.

**B.** The integral $\int_0^\infty \frac{\sin x}{x} dx$ converges. We use the Cauchy criterion to show this. Let $0 < s < t$, then integration by parts gives

$$\int_s^t \frac{\sin x}{x} dx = \frac{-\cos x}{x}\Big|_s^t - \int_s^t \frac{\cos x}{x^2} dx$$

which implies

$$\left| \int_s^t \frac{\sin x}{x} dx \right| \leq \frac{1}{s} + \frac{1}{t} + \int_s^t \frac{dx}{x^2}$$

$$= \frac{1}{s} + \frac{1}{t} + \left( -\frac{1}{x}\Big|_s^t \right) = \frac{2}{s}.$$

Thus given $\epsilon > 0$, choose $s_0 \geq \frac{2}{\epsilon}$ to find that for $t > s > s_0$ it follows

$$\left| \int_s^t \frac{\sin x}{x} dx \right| \leq \frac{2}{s} < \epsilon.$$

**Remark 28.20.** Of course we can also use the integral test, Theorem 18.21, to test improper integrals for convergence.

We now want to introduce one of the most important functions in mathematics and for this we need some preparation. For $x \in \mathbb{R}$ consider $y \mapsto \cos xy$. It follows that

$$\int_1^2 \cos(xy) dy = \frac{1}{x} \sin xy \Big|_1^2 = \frac{\sin 2x - \sin x}{x}.$$

Hence we have defined a new function (at least) on $\mathbb{R} \setminus \{0\}$ by

$$x \mapsto \int_1^2 \cos(xy) dy \left( = \frac{\sin 2x - \sin x}{x} \right).$$

More generally, for each $x \in I$, $I \subset \mathbb{R}$ an interval, let a continuous function $g_x : (a, b) \to \mathbb{R}, y \mapsto g_x(y)$, be given, $-\infty \leq a < b \leq \infty$. In addition assume for each $x \in I$ that the integral $\int_a^b g_x(y)dy$ exists. Now we may consider the new function $H : I \to \mathbb{R}$ defined by $H(x) = \int_a^b g_x(y)dy$.

**Lemma 28.21.** *For $x > 0$ the improper integral*

$$\Gamma(x) := \int_0^\infty t^{x-1}e^{-t}dt \tag{28.21}$$

*exists.*

*Proof.* For $t > 0$ we have the estimate

$$t^{x-1}e^{-t} \leq t^{x-1}, \tag{28.22}$$

implying for $\epsilon > 0$ that

$$\int_\epsilon^1 t^{x-1}e^{-t}dt \leq \int_\epsilon^1 t^{x-1}dt = \frac{1}{x}(1 - \epsilon^x) \leq \frac{1}{x}.$$

Since the function $\epsilon \mapsto \int_\epsilon^1 t^{x-1}e^{-t}dt$ is monotone, bounded and continuous it follows that

$$\int_0^1 t^{x-1}e^{-t}dt = \lim_{\epsilon \to 0} \int_\epsilon^1 t^{x-1}e^{-t}dt$$

exists and is finite. Further, using that $\lim_{t\to\infty} t^{x+1}e^{-t} = 0$ implies that for some $N \in \mathbb{N}$ the condition $t \geq N$ yields

$$t^{x-1}e^{-t} \leq \frac{1}{t^2}, \tag{28.23}$$

we find for $R > N$ that

$$\int_1^R t^{x-1}e^{-t}dt = \int_1^N t^{x-1}e^{-t}dt + \int_N^R t^{x-1}e^{-t}dt$$

$$\leq \int_1^N t^{x-1}e^{-t}dt + \int_N^R \frac{1}{t^2}dt$$

$$= \int_1^N t^{x-1}e^{-t}dt + \frac{1}{N} - \frac{1}{R} \leq C(N) < \infty.$$

As before we observe that $R \mapsto \int_1^R t^{x-1} e^{-t} dt$ is monotone, bounded and continuous, hence

$$\lim_{R \to \infty} \int_1^R t^{x-1} e^{-t} dt = \int_1^\infty t^{x-1} e^{-t} dt$$

exists. Thus

$$\Gamma(x) := \lim_{\epsilon \to 0} \int_\epsilon^1 t^{x-1} e^{-t} dt + \lim_{R \to \infty} \int_1^R t^{x-1} e^{-t} dt$$

is well defined. □

**Definition 28.22.** *The function* $\Gamma : (0, \infty) \to \mathbb{R}$ *defined by* (28.21), *i.e.*

$$\Gamma(x) := \int_0^\infty t^{x-1} e^{-t} dt, \qquad (28.24)$$

*is called the* Γ-*function*.

**Theorem 28.23.** *For $x > 0$ we have*

$$\Gamma(x + 1) = x\Gamma(x). \qquad (28.25)$$

*Proof.* Integration by parts yields

$$\int_\epsilon^R t^x e^{-t} dt = -t^x e^{-t} \Big|_\epsilon^R + x \int_\epsilon^R t^{x-1} e^{-t} dt.$$

For $\epsilon \to 0$ and $R \to \infty$ we find

$$\Gamma(x + 1) = \lim_{\epsilon \to 0} \lim_{R \to \infty} \int_\epsilon^R t^x e^{-t} dt$$

$$= \lim_{\epsilon \to 0} \lim_{R \to \infty} \left( t^x e^{-t} \Big|_\epsilon^R \right) + x \lim_{\epsilon \to 0} \lim_{R \to \infty} \int_\epsilon^R t^{x-1} e^{-t} dt$$

$$= \lim_{\epsilon \to 0} \lim_{R \to \infty} \left( -R^x e^{-R} + \epsilon^x e^{-\epsilon} \right) + x\Gamma(x)$$

$$= x\Gamma(x),$$

which proves (28.25). □

Since

$$\Gamma(1) = \lim_{R \to \infty} \int_0^R e^{-t}dt = \lim_{R \to \infty} (1 - e^{-R}) = 1, \tag{28.26}$$

we deduce from (28.25) for $n \in \mathbb{N}$ that

$$\Gamma(n+1) = n\Gamma(n) = n(n-1)\Gamma(n-1) = \cdots = n(n-1)(n-2)\cdots 1 \cdot \Gamma(1) = n!$$

**Corollary 28.24.** *If $n \in \mathbb{N}$ then*

$$\Gamma(n+1) = n!. \tag{28.27}$$

**Lemma 28.25.** *We have the following*

$$\int_{-\infty}^{\infty} e^{-x^2} dx = \Gamma\left(\frac{1}{2}\right). \tag{28.28}$$

*Proof.* The substitution $x = t^{\frac{1}{2}}$ yields

$$\int_{\epsilon}^R e^{-x^2} dx = \frac{1}{2} \int_{\epsilon^2}^{R^2} t^{-\frac{1}{2}} e^{-t} dt,$$

or

$$\int_0^{\infty} e^{-x^2} dx = \frac{1}{2} \int_0^{\infty} t^{-\frac{1}{2}} e^{-t} dt = \frac{1}{2}\Gamma(\frac{1}{2}).$$

Since

$$\int_{-\infty}^{\infty} e^{-x^2} dx = 2 \int_0^{\infty} e^{-x^2} dx$$

the result follows. $\qquad \square$

**Remark 28.26.** We will prove in Theorem 30.14 that $\Gamma\left(\frac{1}{2}\right) = \sqrt{\pi}$, implying

$$\int_{-\infty}^{\infty} e^{-x^2} dx = \sqrt{\pi}. \tag{28.29}$$

**Definition 28.27.** *Let $I \subset \mathbb{R}$ be an interval and $F : I \to (0, \infty)$ be a function. We call $F$ **logarithmic convex** if $\ln F : I \to \mathbb{R}$ is convex.*

**Remark 28.28.** If $F : I \to (0, \infty)$ is logarithmic convex we have for $0 < \lambda < 1$ and $x, y \in I$ that

$$\ln F(\lambda x + (1 - \lambda)y) \leq \lambda \ln F(x) + (1 - \lambda) \ln F(y)$$

or

$$\ln F(\lambda x + (1 - \lambda)y) \le \ln(F(x)^\lambda F(y)^{1-\lambda}),$$

implying

$$F(\lambda x + (1 - \lambda)y) \le F(x)^\lambda f(y)^{1-\lambda}. \tag{28.30}$$

**Theorem 28.29.** *The $\Gamma$-function is logarithmic convex.*

*Proof.* First we note that $\Gamma(x) = \int_0^\infty t^{x-1}e^{-t}dt > 0$ for $x > 0$. Next, for $x, y \in (0, \infty)$ and $0 < \lambda < 1$ we set $p := \frac{1}{\lambda}$ and $q := \frac{1}{1-\lambda}$, i.e. $\frac{1}{p} + \frac{1}{q} = 1$. Define $f(t) = t^{\frac{x-1}{p}}e^{-\frac{t}{p}}$ and $g(t) = t^{\frac{y-1}{q}}e^{-\frac{t}{q}}$. For $\epsilon > 0$ and $R > \epsilon$ Hölder's inequality yields

$$\int_\epsilon^R f(t)g(t)dt \le \left( \int_\epsilon^R f(t)^p dt \right)^{\frac{1}{p}} \left( \int_\epsilon^R g(t)^q dt \right)^{\frac{1}{q}},$$

but

$$f(t)g(t) = t^{\frac{x}{p}+\frac{y}{q}-1}e^{-t} = t^{\lambda x + (1-\lambda)y - 1}e^{-t},$$
$$f(t)^p = t^{x-1}e^{-t},$$
$$g(t)^q = t^{y-1}e^{-t},$$

i.e. we find

$$\int_\epsilon^R t^{\lambda x + (1-\lambda)y - 1}e^{-t}dt \le \left( \int_\epsilon^R t^{x-1}e^{-t}dt \right)^\lambda \left( \int_\epsilon^R t^{y-1}e^{-t}dt \right)^{1-\lambda}.$$

For $\epsilon \to 0$ and $R \to \infty$ we eventually arrive at

$$\Gamma(\lambda x + (1 - \lambda)y) \le \Gamma(x)^\lambda \Gamma(y)^{1-\lambda},$$

i.e. $\Gamma$ is logarithmic convex. $\qquad\square$

In Chapters 30 and 31 we will return to the $\Gamma$-function.

# Problems

1.    a) Let $a < b$. Prove that the improper integral

$$\int_a^b \frac{dx}{(x - a)^\alpha}$$

converges for $\alpha < 1$ and diverges for $\alpha \geq 1$.

b) Prove the existence of the improper integral

$$\int_0^2 \frac{dx}{\sqrt{x(2-x)}}.$$

c) Prove that for every $\alpha \in \mathbb{R}$ the integral

$$\int_0^\infty x^\alpha dx$$

diverges.

d) Show that

$$\int_0^\infty e^{-ax} \cos(wx)dx = \frac{a}{a^2 + w^2}.$$

2. Let $f : [0, \infty) \to \mathbb{R}$ be a continuous function satisfying with some $\beta \in \mathbb{R}$ the estimate $|f(r)| \leq c_0(1+r^2)^{\frac{\beta}{2}}$. Prove that for $\beta + 1 < \alpha$ the integral

$$(*) \quad \int_0^\infty \frac{f(r)}{(1+r^2)^{\frac{\alpha}{2}}} dr$$

converges absolutely. Now suppose that $f$ is a polynomial of degree $m \in \mathbb{N}$. For which $\alpha \in \mathbb{R}$ does $(*)$ converge?

3. Use mathematical induction to prove for $\alpha > -1$ and $k \in \mathbb{N}_0$

$$\int_0^1 x^k(1-x)^\alpha dx = \frac{k!}{(\alpha+1)(\alpha+2)\cdot\ldots\cdot(\alpha+k+1)}.$$

4. Prove that for $a > 0$

$$\int_0^\infty \frac{\ln x}{x^2 + a^2} dx \quad \text{and} \quad \int_0^\infty \frac{\sin^2 t}{t^2 + a^2} dt$$

converge.

5. Let $g : [-1, 1] \to \mathbb{R}$ be an even, continuous function, $g(0) \neq 0$. Prove that

$$\int_{-1}^{0} \frac{g(x)}{x} dx \quad \text{and} \quad \int_{0}^{1} \frac{g(x)}{x} dt$$

diverge and consequently we cannot define

$$\int_{-1}^{1} \frac{g(x)}{x} dx.$$

Find now

$$\lim_{\epsilon \to 0} \left( \int_{-1}^{-\epsilon} \frac{g(x)}{x} dx + \int_{\epsilon}^{1} \frac{g(x)}{x} dx \right).$$

6. Let $f : [0, \infty) \to \mathbb{R}$ be a continuous function. Prove that if $\lim\limits_{x \to \infty} x^\alpha f(x) = c_0 \in \mathbb{R}$ and $\alpha > 1$, then

$$(**) \quad \int_{0}^{\infty} f(x) dx$$

converges. However, if $\lim\limits_{x \to \infty} x^\alpha f(x) = c_0 \neq 0$ and $\alpha \leq 1$, then $(**)$ diverges.

7. Use the result of Problem 6 to test for the convergence or divergence of:

a) $\int_{1}^{\infty} \frac{\ln x}{1+x} dx$;

b) $\int_{0}^{\infty} \frac{1 - \cos y}{y^2} dy$;

c) $\int_{-\infty}^{-1} \frac{e^t}{t} dt$.

8. Prove that the integral

$$\int_{0}^{\infty} \frac{\sin x}{x} dx$$

does not converge absolutely.
Hint: note that

$$\int_{0}^{\infty} \left| \frac{\sin x}{x} \right| dx = \sum_{n=0}^{\infty} \int_{n\pi}^{(n+1)\pi} \left| \frac{\sin x}{x} \right| dx.$$

409

Now prove that

$$\int_{n\pi}^{(n+1)\pi} \left| \frac{\sin x}{x} \right| dx = \int_0^\pi \frac{\sin t}{t + n\pi} dt$$

and test the resulting series for divergence.

9. Prove the following **quotient test for improper integrals**: Let $f, g : (a, b] \to \mathbb{R}$ be two non-negative continuous functions. If $\lim\limits_{x \to a} \dfrac{f(x)}{g(x)} = c_0 > 0$, then $\int_a^b f(x) dx$ exists if and only if $\int_a^b g(x) dx$ exists. If $\lim\limits_{x \to a} \frac{f(x)}{g(x)} = 0$ and $\int_a^b g(x) dx$ converges, then $\int_a^b f(x) dx$ converges. If $\lim\limits_{x \to a} \dfrac{f(x)}{g(x)} = \infty$ and $\int_a^b g(x) dx$ diverges then $\int_a^b f(x) dx$ diverges.

10. Show that

$$\int_0^1 \frac{ds}{\sqrt{-\ln s}} = \Gamma\left(\frac{1}{2}\right).$$

11. For $x > 0$ and $y > 0$ prove the existence of the improper integral

$$B(x, y) := \int_0^1 t^{x-1}(1 - t)^{y-1} dt$$

and deduce $B(x, y) = B(y, x)$. Further, by using the substitution $t = \sin^2 \vartheta$ show that

$$\int_0^{\frac{\pi}{2}} \sin^{2m-1} \vartheta \cos^{2n-1} \nu d\vartheta = \frac{1}{2} B(m, n).$$

12.    a) Prove that the product of two logarithmic convex functions is logarithmic convex.

    b) Show that a twice continuously differentiable function $f : I \to \mathbb{R}, I \subset \mathbb{R}$ an interval, is logarithmic convex if $f > 0$ and $f f'' - (f')^2 \geq 0$.

    c) Prove that the limit of a sequence of logarithmic convex functions is logarithmic convex.

# 29 Power Series and Taylor Series

We have already discussed at several occasions sequences of functions and we know that sequences are closely related to series. We now want to start to discuss series of functions. Let $(f_n)_{n \in \mathbb{N}_0}$, $f_n : K \to \mathbb{R}, K \subset \mathbb{R}$, be a sequence of functions. We may study the partial sums

$$S_f^N(x) := \sum_{n=0}^{N} f_n(x). \tag{29.1}$$

**Theorem 29.1 (Weierstrass' convergence criterion or Weierstrass' M-test).** *Let $(f_n)_{n \in \mathbb{N}_0}$, $f_n : K \to \mathbb{R}$, be a sequence of functions and suppose that*

$$\sum_{n=0}^{\infty} \|f_n\|_\infty < \infty. \tag{29.2}$$

*Then the series, i.e. the sequence $(S_f^N)_{N \in \mathbb{N}_0}$ of partial sums, converges absolutely and uniformly on $K$ to a function $F : K \to \mathbb{R}$.*

*Proof.* First we prove that $\sum_{n=0}^{\infty} f_n(x)$ converges pointwise, i.e. for every $x \in K$, to some function $F : K \to \mathbb{R}$. Since $|f_n(x)| \le \|f_n\|_\infty$ the series $\sum_{n=0}^{\infty} f_n(x)$ converges absolutely by the comparison test, see Theorem 18.13. Therefore we can define for $x \in K$

$$F(x) := \sum_{n=0}^{\infty} f_n(x),$$

which is a function $F : K \to \mathbb{R}$. Next we prove that the convergence is uniform, i.e. the sequence of partial sums $(S_f^N)_{N \in \mathbb{N}_0}$ converges uniformly. Since $\sum_{n=0}^{\infty} \|f_n\|_\infty < \infty$ there exists $\tilde{N}(\varepsilon)$ such that

$$\sum_{n=N+1}^{\infty} \|f_n\|_\infty < \varepsilon \text{ for } N \ge \tilde{N}(\varepsilon).$$

411

Therefore, for $N \geq \tilde{N}(\varepsilon)$ it follows that

$$\|S_f^N - F\|_\infty = \sup_{x \in K} |S_f^N(x) - F(x)|$$

$$= \sup_{x \in K} \left| \sum_{n=N+1}^{\infty} f_n(x) \right| \leq \sum_{n=N+1}^{\infty} \sup_{x \in K} |f_n(x)|$$

$$= \sum_{n=N+1}^{\infty} \|f_n\|_\infty < \varepsilon.$$

$\square$

**Example 29.2.** The series $\sum_{n=1}^{\infty} \frac{\cos nx}{n^2}$ converges uniformly on $\mathbb{R}$ since with $f_n(x) = \frac{\cos nx}{n^2}$ we have

$$\|f_n\|_\infty = \sup_{x \in \mathbb{R}} \left| \frac{\cos nx}{n^2} \right| = \frac{1}{n^2}$$

and

$$\sum_{n=1}^{\infty} \frac{1}{n^2} < \infty.$$

Now we return to power and Taylor series.

**Definition 29.3.** *Let $(c_n)_{n \in \mathbb{N}}$ be a sequence of real numbers and $a \in \mathbb{R}$. We call*

$$T_{(c_n)}^a(x) := \sum_{n=0}^{\infty} c_n(x - a)^n, x \in \mathbb{R}, \tag{29.3}$$

*the **(formal) power series** associated with $(c_n)_{n \in \mathbb{N}}$ and centre $a$.*

Most important of course is the question for which $x \neq a$ the formal power series $T_{(c_n)}^a(x)$ converges.

**Theorem 29.4.** *Let $(c_n)_{n \in \mathbb{N}_0}$ be a sequence of real numbers and $a \in \mathbb{R}$. If $T_{(c_n)}^a(x)$ converges for some $x_1 \neq a$, then it converges for all $x \in \mathbb{R}$ such that $|x-a| \leq \varrho < |x_1-a|$, i.e. it converges for all $x \in [a-\varrho, a+\varrho], 0 < \varrho < |x_1-a|$. Moreover, the convergence is absolute and uniform on $[a - \varrho, a + \varrho]$ and the same holds for the series*

$$T_{(nc_n)}^a(x) = \sum_{n=1}^{\infty} nc_n(x - a)^{n-1}. \tag{29.4}$$

412

*In particular* $x \mapsto T^a_{(c_n)}(x)$ *and* $x \mapsto T^a_{(nc_n)}(x)$ *are for every* $0 < \rho < |x_1 - a|$ *on* $[a - \rho, a + \rho]$ *continuous functions.*

*Proof.* We set $f_n(x) := c_n(x - a)^n$, hence formally we find with $f(x) := T^a_{(c_n)}(x)$ that $f = \sum\limits_{n=0}^{\infty} f_n$. Since $\sum\limits_{n=0}^{\infty} f_n(x_1)$ converges by our assumption, there exists $M \geq 0$ such that $|f_n(x_1)| \leq M$ for all $n \in \mathbb{N}_0$. For $0 < \rho < |x_1 - a|$ and $x \in [a - \rho, a + \rho]$ it follows that

$$|f_n(x)| = |c_n(x-a)^n| = |c_n(x_1-a)^n| \left| \frac{x-a}{x_1-a} \right|^n \leq M\vartheta^n$$

where $\vartheta := \frac{\varrho}{|x_1-a|} < 1$. Thus we have

$$\|f_n\|_{[a-\varrho,a+\varrho],\infty} = \sup_{x \in [a-\rho,a+\rho]} |f_n(x)| \leq M\vartheta^n$$

implying that $\sum_{n=0}^{\infty} \|f_n\|_{[a-\varrho,a+\varrho],\infty} \leq M \frac{1}{1-\vartheta}$, and hence by Theorem 29.1 the series $\sum\limits_{n=0}^{\infty} f_n$ converges absolutely and uniformly on $[a - \varrho, a + \varrho]$, and since $f_n$ is continuous on $[a - \rho, a + \rho]$ the function $f$ is continuous too, see Theorem 24.6.

Now define $g_n(x) := nc_n(x - a)^{n-1}$, and $g = \sum\limits_{n=0}^{\infty} g_n$. As before we may prove that

$$\|g_n\|_{[a-\varrho,a+\varrho],\infty} \leq nM\vartheta^{n-1}$$

and the ratio test implies the convergence of $\sum\limits_{n=0}^{\infty} nM\vartheta^{n-1}$. Note that $\frac{(n+1)M\vartheta^n}{nM\vartheta^{n-1}} = \frac{n+1}{n}\vartheta < 1$ for $n$ large since $\vartheta < 1$ and $\lim\limits_{n\to\infty} \frac{n+1}{n} = 1$. Now, Theorem 29.1 together with Theorem 24.6 implies the result. $\square$

**Definition 29.5.** *Let* $T^a_{(c_n)}$ *be a formal power series. We call the set of all* $x \in \mathbb{R}$ *for which* $T^a_{(c_n)}$ *converges the **domain of convergence** of* $T^a_{(c_n)}$.

**Corollary 29.6.** *Let* $f(x) = \sum\limits_{n=0}^{\infty} c_n(x - a)^n$ *be a power series converging in* $[a - \varrho, a + \varrho], \varrho > 0$, *uniformly. Then we have for* $a - \varrho \leq b < c \leq a + \varrho$

$$\int_b^c f(x)dx = \sum_{n=0}^{\infty} c_n \int_b^c (x-a)^n dx = \sum_{n=0}^{\infty} \frac{c_n}{n+1} \left( (c-a)^{n+1} - (b-a)^{n+1} \right).$$

$$(29.5)$$

413

This corollary follows from Theorem 29.4 and Theorem 25.27.

**Corollary 29.7.** *Let* $f(x) = \sum\limits_{n=0}^{\infty} c_n(x-a)^n$ *be a power series converging uniformly in* $[a-\varrho, a+\varrho]$. *Then we find for* $x \in (a-\varrho, a+\varrho)$ *that*

$$f'(x) = \sum_{n=1}^{\infty} n c_n (x-a)^{n-1} \qquad (29.6)$$

*and the series* $\sum\limits_{n=1}^{\infty} n c_n(x-a)^{n-1}$ *converges uniformly in* $[a-\varrho, a+\varrho]$.

This corollary follows from Theorem 29.4 and Theorem 26.19

**Corollary 29.8.** *Let* $f(x) = \sum\limits_{n=0}^{\infty} c_n(x-a)^n$ *be as in Corollary 29.7. Then* $f : (a-\varrho, a+\varrho) \to \mathbb{R}$ *is arbitrarily often differentiable and we have for* $k \in \mathbb{N}_0$

$$c_k = \frac{1}{k!} f^{(k)}(a). \qquad (29.7)$$

*Proof.* A repeated application of Corollary 29.7 yields first the existence of all derivatives and then

$$f^{(k)}(x) = \sum_{n=k}^{\infty} n(n-1) \cdot \ldots \cdot (n-k+1) c_n (x-a)^{n-k}$$

which gives for $x = a$

$$c_k = \frac{1}{k!} f^{(k)}(a).$$

$\square$

**Example 29.9.** For $|x| < 1$ we find

$$\sum_{n=1}^{\infty} n x^n = x \sum_{n=1}^{\infty} n x^{n-1} = x \frac{d}{dx} \sum_{n=0}^{\infty} x^n$$

$$= x \frac{d}{dx} \left( \frac{1}{1-x} \right) = \frac{x}{(1-x)^2}$$

which for example yields $\sum_{n=1}^{\infty} \frac{n}{2^n} = 2$.

Having Corollary 29.8 in mind, we note that the power series allow us to define arbitrarily often differentiable functions. This opens the road to eventually prove the existence of the exponential function $\exp : \mathbb{R} \to \mathbb{R}$.

**Theorem 29.10.** *There exists a unique function* $\exp : \mathbb{R} \to \mathbb{R}$ *with* $\exp' = \exp$ *and* $\exp(0) = 1$.

*Proof.* We set

$$\exp(x) := \sum_{k=0}^{\infty} \frac{x^k}{k!}. \tag{29.8}$$

First we claim that this power series converges for all $x \in \mathbb{R}$. Indeed for $x \in \mathbb{R}$ fixed we find

$$\left| \frac{\frac{x^{k+1}}{(k+1)!}}{\frac{x^k}{k!}} \right| = |x| \frac{1}{k+1},$$

and therefore, if $k \geq 2|x|$ it follows that $\frac{|x|}{k+1} \leq \frac{k}{2(k+1)} < \frac{1}{2}$, and the ratio test implies the convergence of $\sum_{k=0}^{\infty} \frac{x^k}{k!}$. Now we deduce from Theorem 29.4 that this convergence is uniform on every compact interval. Consequently, by Corollary 29.7 we find

$$\exp'(x) = \sum_{k=1}^{\infty} k \frac{x^{k-1}}{k!} = \sum_{k=1}^{\infty} \frac{x^{k-1}}{(k-1)!} = \sum_{k=0}^{\infty} \frac{x^k}{k!} = \exp(x).$$

and $\exp(0) = 1$. $\qquad\qquad\qquad\qquad\qquad\qquad\qquad\qquad\qquad\qquad \square$

We will see later how we can use power series to solve certain differential equations (some examples are given in the Problems).

Our aim is to discuss Taylor's formula and the Taylor series. The starting point is the fundamental theorem of calculus.

Let $I = [a, b]$ be an interval and $f : (a, b) \to \mathbb{R}$ be of the class $C^2$, i.e. $f \in C^2((a, b))$, and suppose that $f$, $f'$ and $f''$ have continuous extensions on $[a, b]$. Since $f'$ is a primitive of the continuous function $f''$ and since by the fundamental theorem

$$f(x) = f(c) + \int_c^x f'(t)dt, \quad c, x \in (a, b)$$

a further application of the fundamental theorem yields

$$f(x) = f(c) + \int_c^x \left( f'(c) + \int_c^t f''(s)ds \right) dt$$

$$= f(c) + f'(c)(x - c) + \int_c^x \left( \int_c^t f''(s)ds \right) dt.$$

Let $M_{f''} = ||f''||_{[a,b],\infty}$. For $x, t \geq c$ we get

$$\left| R_{f,c}^{(2)}(x) \right| = \left| \int_c^x \left( \int_c^t f''(s)ds \right) dt \right| \leq M_{f''} \int_c^x \left( \int_c^t 1 ds \right) dt$$

$$= M_{f''} \int_c^x (t - c)dt = M_{f''} \frac{(x - c)^2}{2},$$

which yields

$$f(x) = f(c) + f'(c)(x - c) + R_{f,c}^{(2)} \tag{29.9}$$

where

$$|R_{f,c}^{(2)}(x)| \leq M_{f''} \frac{(x - c)^2}{2} = M_{f''} \frac{|x - c|^2}{2}. \tag{29.10}$$

Note that it is easy to see that (29.9) and (29.10) hold for all $x \in [a, b]$.

Moreover with

$$M_{f'} := ||f'||_{[a,b],\infty} \quad \text{and} \quad R_{f,c}^{(1)}(x) = \int_c^x f'(t)dt$$

we have

$$|R_{f,c}^{(1)}(x)| \leq M_{f'}|x - c|. \tag{29.11}$$

Here is the interpretation of these results: If $|x - c|$ is small we can approximate $f(x)$ by $f(c)$, and we might get a better approximation by $f(c) + f'(c)(x - c)$ :

$$|f(x) - f(c)| \leq M_{f'}|x - c|,$$

and

$$|f(x) - (f(c) + f'(c)(x - c))| \leq M_{f''} \frac{(x - c)^2}{2}.$$

Recall that for $|x - c| < \epsilon$ and $\epsilon < 1$ it follows that $|x - c|^2 < |x - c|$. The main question is whether we can get an even better approximation when increasing the order of derivatives and iterating the above process.

416

**Theorem 29.11 (Taylor's formula).** *Let $f : [a, b] \to \mathbb{R}$ be a function such that $f|_{(a,b)} \in C^{n+1}((a, b))$ and $f, f', \ldots, f^{(n+1)}$ have continuous extensions to $[a, b]$. Then for every $c, x \in (a, b)$ the following holds*

$$f(x) = f(c) + \frac{f'(c)}{1!}(x - c) + \frac{f''(c)}{2!}(x - c)^2 + \cdots + \frac{f^{(n)}}{n!}(x - c)^n + R_{f,c}^{(n+1)}(x)$$
(29.12)

*where the **remainder term** is given by*

$$R_{f,c}^{(n+1)}(x) = \frac{1}{n!}\int_c^x (x - t)^n f^{(n+1)}(t)dt.$$
(29.13)

*Proof.* We use mathematical induction. The fundamental theorem yields

$$f(x) = f(c) + \int_c^x f'(t)dt$$

which is (29.12), (29.13) for $n = 0$. Now suppose that (29.12), (29.13) hold for $n - 1 \in \mathbb{N}$. Consider

$$R_{f,c}^{(n)}(x) = \frac{1}{(n-1)!}\int_c^x (x - t)^{n-1} f^{(n)}(t)dt$$

$$= -\int_c^x \frac{d}{dt}\left(\frac{(x - t)^n}{n!}\right) \cdot f^{(n)}(t)dt.$$

Integration by parts gives

$$R_{f,c}^{(n)}(x) = -\int_c^x \frac{d}{dt}\left(\frac{(x - t)^n}{n!}\right) \cdot f^{(n)}(t)dt$$

$$= -f^{(n)}(t) \cdot \frac{(x - t)^n}{n!}\Big|_{t=c}^{t=x} + \int_c^x \frac{(x - t)^n}{n!} \cdot f^{(n+1)}(t)dt$$

$$= \frac{f^{(n)}(c)}{n!}(x - c)^n + R_{f,c}^{(n+1)}(x).$$

Thus

$$f(x) = \sum_{j=0}^{n-1} \frac{f^{(j)}(c)}{j!}(x - c)^j + R_{f,c}^{(n)}(x)$$

$$= \sum_{j=0}^{n-1} \frac{f^{(j)}(c)}{j!}(x - c)^j + f^{(n)}(c)\frac{(x - c)^n}{n!} + R_{f,c}^{(n+1)}(x)$$

$$= \sum_{j=0}^{n} \frac{f^{(j)}(c)}{j!}(x - c)^j + R_{f,c}^{(n+1)}(x),$$

417

and the theorem is proven. □

**Definition 29.12.** *Let* $f : [a, b] \to \mathbb{R}$ *be a (n+1)-times continuously differentiable function on* $(a, b)$ *and let* $f, f', \ldots, f^{(n+1)}$ *have continuous extensions to* $[a, b]$. *The first* $n$ **Taylor polynomials** *of* $f$ *around* $c \in [a, b]$ *are given by*

$$T_{f,c}^{(k)}(x) := \sum_{j=0}^{k} \frac{f^{(j)}(c)}{j}(x - c)^j, k = 1, \ldots, n. \tag{29.14}$$

Thus we have

$$f(x) = T_{f,c}^{(n)}(x) + R_{f,c}^{(n+1)}(x) \tag{29.15}$$

with

$$|R_{f,c}^{(n+1)}(x)| \leq M_{f^{(n+1)}} \frac{|x - c|^{n+1}}{(n+1)!} \tag{29.16}$$

where $M_{f^{(n+1)}} = \|f^{(n+1)}\|_{[a,b],\infty}$.

**Corollary 29.13.** *Let* $f$ *be as in Definition 29.12. If* $f^{(n+1)}(x) = 0$ *for all* $x \in [a, b]$ *then* $f$ *is a polynomial of degree less than or equal to* $n$.

*Proof.* In this case $R_{f,x_0}^{(n+1)}(x) = 0$ for all $x \in [a, b]$. □

Here are some examples of Taylor polynomials:

$$T_{\exp,0}^{(k)}(x) = \sum_{j=0}^{k} \frac{1}{j!} x^j; \tag{29.17}$$

$$T_{g,0}^{(k)}(x) = \sum_{l=0}^{k} \frac{\prod_{j=0}^{l-1}(\frac{1}{2} - j)}{l!} x^l, \ g(x) = \sqrt{1 + x}; \tag{29.18}$$

$$T_{h,0}^{(k)}(x) = \sum_{j=1}^{k} \frac{(-1)^{j-1} x^j}{j}, \ h(x) = \ln(1 + x); \tag{29.19}$$

$$T_{\sin,0}^{(2k+1)}(x) = \sum_{j=0}^{k} (-1)^j \frac{x^{2j+1}}{(2j+1)!}; \tag{29.20}$$

$$T_{\cos,0}^{(2k)}(x) = \sum_{j=0}^{k} (-1)^j \frac{x^{2j}}{(2j)!}. \tag{29.21}$$

We want to understand how good the Taylor polynomial approximates the function and for this we need to estimate the remainder term. Sometimes, instead of using $R_{f,c}^{(n+1)}$ it is more helpful to use the **Lagrange form of the remainder term**.

**Theorem 29.14.** *Let $f$ be as in Definition 29.12 and $x, x_0 \in [a, b]$. Then there is $\xi \in [x, x_0]$ or $\xi \in [x_0, x]$ such that*

$$f(x) = \sum_{k=1}^{n} \frac{f^{(k)}(x_0)}{k!}(x - x_0)^k + \frac{f^{(n+1)}(\xi)}{(n + 1)!}(x - x_0)^{n+1}. \qquad (29.22)$$

*Proof.* Let us suppose that $x_0 < x$, the other case goes analogously. By the mean value theorem for integrals we find

$$R_{f,x_0}^{(n+1)}(x) = \frac{1}{n!} \int_{x_0}^{x} (x - t)^n f^{(n+1)}(t)dt$$

$$= f^{(n+1)}(\xi) \int_{x_0}^{x} \frac{(x - t)^n}{n!}dt = f^{(n+1)}(\xi)\frac{(x - x_0)^{n+!}}{(n + 1)!},$$

proving the theorem. $\qquad \square$

**Example 29.15. A.** For $x \mapsto \ln(1 + x), 0 \le x \le \frac{1}{10}$, we find using the integral form of the remainder

$$\left| R_{\ln(1+\cdot),0}^{(2)}(x) \right| = \left| \int_{0}^{x} (x - t) \left( \frac{d^2}{dt^2} \ln(1 + t) \right) dt \right|$$

$$= \left| -\int_{0}^{x} (x - t) \frac{1}{1 + t^2}dt \right| \le \int_{0}^{x} (x - t)dt$$

$$= \frac{x^2}{2},$$

which implies for $0 \le x \le \frac{1}{10}$ that $| \ln(1 + x) - x | \le \frac{x^2}{2} \le \frac{1}{100}$.
**B.** For $\sin : [0, 2\pi] \to \mathbb{R}$ we find

$$R_{\sin,0}^{(2n+3)}(x) = (-1)^{n+1}\frac{\cos \xi}{(2n + 3)!}x^{2n+3} \quad \text{for } \xi \in [0, 2\pi]$$

and therefore

$$\left| \sin x - T_{\sin,0}^{(2n+1)}(x) \right| \le \frac{|x|^{2n+3}}{(2n + 3)!}$$

Therefore for $n = 1$ and $|x| \leq \frac{1}{10}$ we have already

$$\left| \sin x - T_{\sin,0}^{(3)}(x) \right| \leq \left( \frac{1}{10} \right)^5 \cdot \frac{1}{120}.$$

Finally we consider the Taylor formula as $n$ goes to infinity.

**Definition 29.16.** *Let $f : (a,b) \to \mathbb{R}$ be an arbitrarily often differentiable function and $x_0 \in (a,b)$. We call*

$$T_{f,x_0}(x) := \sum_{k=0}^{\infty} \frac{f^{(k)}(x_0)}{k!}(x - x_0)^k \tag{29.23}$$

*the **Taylor series** (or **Taylor expansion**) of $f$ about $x_0$.*

**Remark 29.17.** So far $T_{f,x_0}(x)$ is a formal power series which does not necessarily converge for all $x \in (a,b)$. Moreover, if $T_{f,x_0}(x)$ converges, the limit does not have to be $f(x)$. In fact the Talyor series $T_{f,x_0}(x)$ converges to $f(x)$ if and only if

$$\lim_{n \to \infty} R_{f,x_0}^{(n+1)}(x) = 0.$$

**Example 29.18.** Consider $f : \mathbb{R} \to \mathbb{R}$ defined by

$$f(x) = \begin{cases} e^{-\frac{1}{x^2}}, & x \neq 0 \\ 0, & x = 0 \end{cases}.$$

We claim that $f \in C^{\infty}(\mathbb{R})$ and $f^{(n)}(0) = 0$ for all $n \in \mathbb{N}_0$. This implies that $T_f(x) = 0$ for all $x \in \mathbb{R}$, in particular $T_f(x)$ converges for all $x \in \mathbb{R}$ but for $x \neq 0$ we have $T_f(x) \neq f(x)$. To prove our claim we show the existence of polynomials $p_n$ such that

$$f^{(n)}(x) = \begin{cases} p_n(\frac{1}{x})e^{-\frac{1}{x^2}}, & x \neq 0 \\ 0, & x = 0 \end{cases}.$$

The case $n = 0$ is clear, just take $p_0 = 1$.
Now, for $x \neq 0$ we have

$$f^{(n+1)}(x) = \frac{d}{dx} f^{(n)}(x) = \frac{d}{dx}\left( p_n \left( \frac{1}{x} \right) e^{-\frac{1}{x^2}} \right)$$

$$= \left( -p_n' \left( \frac{1}{x} \right) \frac{1}{x^2} + 2p_n \left( \frac{1}{x} \right) \frac{1}{x^3} \right) e^{-\frac{1}{x^2}},$$

thus

$$p_{n+1}(t) := -p'_n(t)t^2 + 2p_n(t)t^3.$$

But for $x = 0$

$$f^{(n+1)}(0) = \lim_{x \to 0} \frac{f^{(n)}(x) - f^{(n)}(0)}{x}$$

$$= \lim_{x \to 0} \frac{p_n(\frac{1}{x})e^{-\frac{1}{x^2}}}{x}$$

$$= \lim_{r \to \infty} r p_n(r)e^{-r^2} = 0.$$

Note that if $T_{f,x_0}(x)$ converges to $f(x)$ for some $x \neq x_0$, then in the interval $[x_0 - \rho, x_0 + \rho], 0 < \rho < |x - x_0|$, the convergence is uniform.

We will encounter Taylor series (and power series) later on when discussing functions of several real variables and most of all when treating complex-valued functions of a complex variable.

We have introduced the exponential function now as a convergent power series and we may ask whether we can prove the functional equation for exp, i.e.

$$\exp(x + y) = \exp(x)\exp(y), \tag{29.24}$$

without using the fact that exp satisfies the initial value problem $u' = u, u(0) = 1$, compare with Lemma 9.7. The right hand side of (29.24) is the product of two power series and we first want to discuss products of infinite series.

Let $(a_n)_{n \in \mathbb{N}_0}$ and $(b_n)_{n \in \mathbb{N}_0}$ be two sequences of real numbers and $A := \sum_{n=0}^{\infty} a_n$, $B := \sum_{m=0}^{\infty} b_m$ the corresponding series which we assume to converge. The aim is to find conditions under which we can represent $\left(\sum_{n=0}^{\infty} a_n\right)\left(\sum_{m=0}^{\infty} b_m\right)$ as a series converging to $A \cdot B$.

For two partial sums we have

$$\left(\sum_{n=0}^{N} a_n\right)\left(\sum_{m=0}^{M} b_m\right) = \sum_{n,m} a_n b_m$$

where on the right hand side we form all products on $a_n b_m$, $0 \leq n \leq N$ and $0 \leq m \leq M$, and add them up. However we cannot proceed in the same way with infinitely many terms. We set

$$c_n := a_n b_0 + a_{n-1}b_1 + \cdots + a_0 b_n = \sum_{k=0}^{n} a_{n-k} b_k \tag{29.25}$$

and give

**Definition 29.19.** *Let $\sum_{n=0}^{\infty} a_n$ and $\sum_{m=0}^{\infty} b_m$ be two series of real numbers and define $c_n$ by (29.25). The **Cauchy product** of these series is given by*

$$\sum_{n=0}^{\infty} c_n := \sum_{n=0}^{\infty} \left( \sum_{k=0}^{n} a_{n-k} b_k \right). \tag{29.26}$$

**Remark 29.20.** So far the definition does not include a statement about convergence. Thus $\sum_{n=0}^{\infty} c_n$ stands for the sequence of partial sums $\sum_{n=0}^{N} \left( \sum_{k=0}^{n} a_{n-k} b_k \right)$.

**Theorem 29.21.** *Let $A := \lim_{n \to \infty} A_n$, $A_n := \sum_{k=0}^{n} a_k$, and $B := \lim_{m \to \infty} B_m$, $B_m := \sum_{l=0}^{m} b_l$. If $\sum_{k=0}^{\infty} a_k$ converges absolutely and $\sum_{k=0}^{\infty} b_k$ converges, then their Cauchy product converges to $A \cdot B$, i.e.*

$$A \cdot B = \sum_{n=0}^{\infty} c_n = \lim_{N \to \infty} \sum_{n=0}^{N} \left( \sum_{k=0}^{n} a_{n-k} b_k \right). \tag{29.27}$$

*In the case where $\sum_{l=0}^{\infty} b_l$ converges absolutely, then $\sum_{n=0}^{\infty} c_n$ is also absolutely convergent.*

*Proof.* We may write

$$\sum_{k=0}^{n} c_k = a_0 b_0 + (a_0 b_1 + a_1 b_0) + \cdots + (a_0 b_n + \cdots + a_n b_0)$$

$$= a_0 B_n + a_1 B_{n-1} + \cdots + a_n B_0$$

$$= \sum_{k=0}^{n} a_{n-k}(B_k - B) + B \sum_{k=0}^{n} a_k.$$

By assumption $\sum_{k=0}^{\infty} a_k = A$, and hence we are done if we can prove that

$$\lim_{n \to \infty} \sum_{k=0}^{n} a_{n-k}(B_k - B) = 0.$$

Given $\epsilon > 0$ we can find $N(\epsilon)$ such that for $k \geq N(\epsilon)$ we have $|B_k - B| < \epsilon$. For $n > N(\epsilon)$ it follows that

$$\left| \sum_{k=0}^{n} a_{n-k}(B_k - B) \right| \leq \sum_{k=0}^{N} |a_{n-k}||B_k - B| + \sum_{k=N+1}^{n} |a_{n-k}||B_k - B|$$

$$\leq \max_{k \leq N} |B_k - B| \sum_{k=0}^{N} |a_{n-k}| + \epsilon \sum_{k=N+1}^{n} |a_{n-k}|$$

$$\leq \max_{k \leq N} |B_k - B| \sum_{k=0}^{N} |a_{n-k}| + \epsilon \sum_{k=0}^{\infty} |a_k|.$$

For $n \to \infty$ it follows that $a_n \to 0$ since $\sum_{n=0}^{\infty} a_n$ converges. Therefore it follows that for every fixed $N$

$$\lim_{n \to \infty} \sum_{k=0}^{N} |a_{n-k}| = 0.$$

Hence

$$0 \leq \limsup_{n \to \infty} \left| \sum_{k=0}^{n} a_{n-k}(B_k - B) \right| \leq \epsilon \sum_{k=0}^{\infty} |a_k|,$$

implying the convergence of $\sum_{n=0}^{\infty} c_n$ to $A \cdot B$. Now suppose that both series converge absolutely. Then we get

$$\sum_{n=0}^{M} |c_n| = \sum_{n=0}^{M} \left| \sum_{k=0}^{n} a_{n-k}b_k \right| \leq \sum_{n=0}^{M} \sum_{k=0}^{n} |a_{n-k}||b_k|$$

$$= \left( \sum_{n=0}^{M} |a_n| \right) \left( \sum_{k=0}^{M} |b_k| \right) \leq \left( \sum_{n=0}^{\infty} |a_n| \right) \left( \sum_{k=0}^{\infty} |b_k| \right)$$

implying the absolute convergence of $\sum_{n=0}^{\infty} c_n$. $\qquad\square$

We now apply this result to exp in order to prove its functional equation.

**Proposition 29.22.** *For $x, y \in \mathbb{R}$ the relation (29.24) holds, i.e.*

$$\exp(x) \exp(y) = \exp(x + y).$$

*Proof.* We know that $\exp(x) = \sum_{n=0}^{\infty} \frac{x^n}{n!}$ and $\exp(y) = \sum_{k=0}^{\infty} \frac{y^k}{k!}$ and both series converge absolutely. Therefore, by Theorem 29.21 we find

$$\exp(x)\exp(y) = \sum_{n=0}^{\infty}\left(\sum_{k=0}^{n} \frac{x^{n-k}}{(n-k)!}\frac{y^k}{k!}\right).$$

Using the binomial theorem we get

$$\sum_{n=0}^{\infty}\left(\sum_{k=0}^{n} \frac{x^{n-k}}{(n-k)!}\frac{y^k}{k!}\right) = \sum_{n=0}^{\infty}\frac{1}{n!}\left(\sum_{k=0}^{n}\binom{n}{k}x^{n-k}y^k\right)$$

$$= \sum_{n=0}^{\infty}\frac{1}{n!}(x+y)^n = \exp(x+y).$$

$\square$

## Problems

1. For $n \in \mathbb{N}_0$ consider the functions $g_n(x) = \frac{x^4}{(1+x^4)^n}$ defined on $\mathbb{R}$. Prove that

$$\sum_{n=0}^{\infty} g_n(x) = \begin{cases} 1+x^4, & x \neq 0 \\ 0, & x = 0. \end{cases}$$

Why does this series not converge uniformly?

2. Prove that the following series converge absolutely and uniformly in the given domain.

a) $\sum_{k=1}^{\infty} \frac{\sin kx}{k^\alpha}, \alpha > 1, x \in \mathbb{R}$;

b) $\sum_{n=1}^{\infty} \frac{x^n}{n^{\frac{3}{2}}}, -1 \leq x \leq 1$;

c) $\sum_{n=1}^{\infty} \frac{1}{n^2+r^2}, r \in \mathbb{R}$.

3. For $\alpha \in \mathbb{R}$ define

$$\binom{\alpha}{n} := \prod_{k=1}^{n} \frac{\alpha-k+1}{k}.$$

Prove that for $\alpha \in \mathbb{N}$ this is a binomial coefficient. Let $g_\alpha : (-1,1) \to \mathbb{R}$, $g_\alpha(x) = (1+x)^\alpha$. Show that $g_\alpha^{(k)}(0) = k!\binom{\alpha}{k}$ and find the Taylor polynomial $T_{g_\alpha,0}^{(n)}(x)$.

4. Suppose that $\sum_{k=0}^{\infty} a_k x^k$ and $\sum_{k=0}^{\infty} b_k x^k$ converge absolutely and uniformly on $[-c, c]$. Prove that then $\sum_{k=0}^{\infty}(a_k + b_k)x^k$ and $\sum_{k=0}^{\infty}(\lambda a_k)x^k$, $\lambda \in \mathbb{R}$, converge also in $[-c, c]$ absolutely and uniformly and that we have

$$\sum_{k=0}^{\infty}(a_k + b_k)x^k = \sum_{k=0}^{\infty} a_k x^k + \sum_{k=0}^{\infty} b_k x^k,$$

$$\sum_{k=0}^{\infty}(\lambda a_k)x^k = \lambda \sum_{k=0}^{\infty} a_k x^k.$$

5. Given that $e^x = \sum_{k=0}^{\infty} \frac{x^k}{k!}$ find the Taylor series of sinh and cosh.

6. Find the Taylor series about 0 of

$$f(x) = \frac{1}{2} \ln \frac{1+x}{1-x}, \, |x| < 1.$$

7. For $l \in \mathbb{N}$, we define the **Bessel function** of order $l$ by

$$J_l(x) := \sum_{n=0}^{\infty} \frac{(-1)^n \left(\frac{x}{2}\right)^{2n+l}}{n!(n+l)!}. \tag{29.28}$$

Prove that $J_l$ converges uniformly and absolutely on every compact interval in $\mathbb{R}$. Now prove that $J_l$ solves

$$x^2 J_l''(x) + x J_l'(x) + (x^2 - l^2)J_l(x) = 0.$$

Note that we can write $J_l(x)$ as

$$J_l(x) = \sum_{n=0}^{\infty}(-1)^n \frac{1}{n!(n+l)!2^{2n+l}} x^{2n+l} = \frac{x^l}{2^l} \sum_{n=0}^{\infty}(-1)^n \frac{x^{2n}}{n!(n+l)!2^{2n}}.$$

8. Justify

$$\frac{1}{1+t^2} = \sum_{l=0}^{\infty}(-1)^l t^{2l}, \, |t| < 1,$$

and by using the identity

$$\arctan x = \int_0^x \frac{1}{1+t^2} dt$$

find $T_{\arctan,0}(x)$.

9. Use the result of Problem 8 to show

$$\frac{\pi}{6} = \frac{1}{\sqrt{3}} \sum_{n=0}^{\infty} \frac{1}{2n+1} \left(-\frac{1}{3}\right)^n.$$

10. Prove **Abel's convergence theorem**: if the series $\sum_{k=0}^{\infty} a_k$ converges then

$$\sum_{k=0}^{\infty} a_k = \lim_{\substack{x \to 1 \\ x < 1}} \sum_{k=0}^{\infty} a_k x^k.$$

11. Use Abel's convergence theorem to show
   a) $\ln 2 = \sum_{l=1}^{\infty} (-1)^{l+1} \frac{1}{l} = \sum_{l=1}^{\infty} \frac{1}{(2l-1)2l}$.
   b) $\frac{\pi}{4} = \sum_{k=0}^{\infty} (-1)^k \frac{1}{2k+1}$

12. Show the following inequalities for $x > 0$
   a) $x - \frac{x^2}{2} + \frac{x^3}{3} - \frac{x^4}{4} < \ln(1+x) < x - \frac{x^2}{2} + \frac{x^3}{3}$;
   b) $1 + \frac{x}{2} - \frac{x^2}{8} < \sqrt{1+x} < 1 + \frac{x}{2} - \frac{x^2}{8} + \frac{x^3}{16}$.

13. By using the Cauchy product prove
   a) $\left(\frac{1}{1-x}\right)^2 = \sum_{n=0}^{\infty} (n+1)x^n, |x| < 1$;
   b)

$$\frac{\cos x}{1-x} = 1 + x + \left(1 - \frac{1}{2!}\right) x^2 + \left(1 - \frac{1}{2!}\right) x^3 + \left(1 - \frac{1}{2!} + \frac{1}{4!}\right) x^4$$

$$+ \left(1 - \frac{1}{2!} + \frac{1}{4!}\right) x^5 + \cdots$$

$$= \sum_{n=0}^{\infty} \left( \sum_{k=0}^{n} (-1)^k \frac{1}{(2k)!} \right) (x^{2n} + x^{2n+1}), |x| < 1.$$

14. Let $f, g : (-1, 1) \to \mathbb{R}$ have convergent Taylor expansions $f(x) = \sum_{k=0}^{\infty} \frac{f^{(k)}(0)}{k!} x^k$ and $g(0) = \sum_{k=0}^{\infty} \frac{g^{(k)}(0)}{k!} x^k$. Assuming that $f \cdot g$ also has a convergent Taylor expansion in $(-1, 1)$ prove

$$(f \cdot g)(x) = \sum_{k=0}^{\infty} \left( \sum_{l=0}^{k} \frac{f^{(l)}(0)}{l!} \frac{g^{(k-l)}(0)}{(k-l)!} \right) x^k.$$

426

# 30 Infinite Products and the Gauss Integral

Given a sequence $(c_n)_{n \in \mathbb{N}}$ of real numbers $c_k \neq 0$. For $N \in \mathbb{N}$ we use the notation

$$P_N := \prod_{k=1}^{N} c_k = c_1 \cdot \ldots \cdot c_N. \tag{30.1}$$

We want to study the convergence of the sequence $(P_N)_{N \in \mathbb{N}}$.

**Definition 30.1.** *Given a sequence $(c_n)_{n \in \mathbb{N}}$ of real numbers $c_k \neq 0$. We call the sequence $(P_N)_{N \in \mathbb{N}}$ the* **infinite product** *of $(c_n)_{n \in \mathbb{N}}$ and denote it by*

$$\prod_{k=1}^{\infty} c_k. \tag{30.2}$$

Note that as in the case of a series we may also consider $\prod_{k=m_0}^{\infty} c_k$ with its obvious definition. So $\prod_{k=1}^{\infty} c_k$ is just a further symbol for $(P_N)_{N \in \mathbb{N}} = \left(\prod_{k=1}^{N} c_k\right)_{N \in \mathbb{N}}$ and we are interested in conditions under which $\left(\prod_{k=1}^{N} c_k\right)_{N \in \mathbb{N}}$ converges.

**Definition 30.2.** *We say that the infinite product $\prod_{k=1}^{\infty} c_k$* **converges** *to $P \neq 0$ if the sequence $(P_N)_{N \in \mathbb{N}}$ converges to $P$. In this case we write for the limit $P$*

$$P = \prod_{k=1}^{\infty} c_k. \tag{30.3}$$

*If $\lim\limits_{N \to \infty} \prod_{k=1}^{N} c_k = 0$ we say that $\prod_{k=1}^{\infty} c_k$* **divergent to 0.**

**Remark 30.3.** If $\prod_{k=1}^{\infty} c_k$ converges then it follows that

$$\lim_{N \to \infty} c_N = \lim_{N \to \infty} \frac{P_N}{P_{N-1}} = \frac{\lim_{N \to \infty} P_N}{\lim_{N \to \infty} P_{N-1}} = 1.$$

This condition is however not sufficient for the convergence of an infinite product as is seen by

$$c_k := 1 + \frac{1}{k} = \frac{k+1}{k} \to 1$$

but

$$P_N = \frac{2}{1} \cdot \frac{3}{2} \cdot \ldots \cdot \frac{N+1}{N} = N + 1 \to \infty.$$

427

**Example 30.4.** The infinite product $\prod_{k=2}^{\infty}\left(1-\frac{1}{k^2}\right)$ converges to $\frac{1}{2}$. Indeed we have

$$P_N = \prod_{k=2}^{N}\left(1-\frac{1}{k^2}\right) = \left(1-\frac{1}{2^2}\right)\left(1-\frac{1}{3^2}\right)\cdot\,\ldots\,\cdot\left(1-\frac{1}{N^2}\right)$$
$$= \frac{2^2-1}{2^2}\frac{3^2-1}{3^2}\cdot\,\ldots\,\cdot\frac{N^2-1}{N^2}$$
$$= \frac{(2-1)(2+1)}{2^2}\frac{(3-1)(3+1)}{3^2}\cdot\,\ldots\,\cdot\frac{(N-1)(N+1)}{N^2}$$
$$= \frac{1}{2}\frac{N+1}{N}.$$

The latter follows easily by induction: for $N=2$ we have $\frac{(2-1)(2+1)}{2^2}=\frac{3}{4}$. Furthermore

$$\frac{(2-1)(2+1)}{2^2}\frac{(3-1)(3+1)}{3^2}\cdot\,\ldots\,\cdot\frac{(N-1)(N+1)}{N^2}\frac{N(N+2)}{(N+1)^2}$$
$$= \frac{1}{2}\frac{N+1}{N}\frac{N(N+2)}{(N+1)^2} = \frac{1}{2}\frac{N+2}{N+1}.$$

Thus for $N\to\infty$ we find

$$\prod_{k=2}^{\infty}\left(1-\frac{1}{k^2}\right) = \lim_{N\to\infty}P_N = \lim_{N\to\infty}\frac{1}{2}\left(\frac{N+1}{N}\right) = \frac{1}{2}.$$

Since for the convergence of $\prod_{k=1}^{\infty}c_k$ it is necessary that $\lim_{k\to\infty}c_k=1$ we may introduce $a_k := c_k - 1$, i.e. $c_k = 1 + a_k$, and consider $\prod_{k=1}^{\infty}(1+a_k)$. Clearly we have now the necessary condition $\lim_{k\to\infty}a_k = 0$ and $a_k = -1$ is excluded. Suppose that $a_k > -1$, i.e. $c_k > 0$. Then we find

$$\ln P_N = \ln\prod_{k=1}^{N}(1+a_k) = \sum_{k=1}^{N}\ln(1+a_k),$$

or

$$P_N = \exp\left(\sum_{k=1}^{N}\ln(1+a_k)\right).$$

Since exp is continuous the convergence of $\sum_{k=1}^{\infty}\ln(1+a_k)$ will imply the convergence of $\prod_{k=1}^{\infty}c_k = \prod_{k=1}^{\infty}(1+a_k)$. Conversely, if $\prod_{k=1}^{\infty}(1+a_k)$ converges then $\sum_{k=1}^{\infty}\ln(1+a_k)$ converges too.
Thus we have proved

**Lemma 30.5.** *Let $(a_k)_{k \in \mathbb{N}}$ be a sequence of real numbers where $a_k > -1$ and set $c_k = 1 + a_k$. Then the convergence of $\prod_{k=1}^{\infty} c_k$ is equivalent to the convergence of $\sum_{k=1}^{\infty} \ln(1 + a_k)$.*

**Remark 30.6.** As in the case of series we can sharpen Lemma 30.5 slightly by assuming that $a_k > -1$ for all $k \geq N_0$. Note that if $a_k \leq -1$ for some finite values of $k$, $k \leq N_0$, then for these $k$ the terms $\ln(1 + a_k)$ are not defined. We find however the equivalence of the convergence of $\prod_{k>N_0}^{\infty} c_k$ and $\sum_{k=N_0+1}^{\infty} \ln(1 + a_k)$ and the convergence of this series also implies the convergence of $\prod_{k=1}^{\infty} c_k$.

The **Cauchy criterion for infinite products** is as follows:

**Proposition 30.7.** *The infinite product $\prod_{k=1}^{\infty} c_k$ converges if and only if for every $\epsilon > 0$ there exists $N = N(\epsilon) \in \mathbb{N}$ such that $n > m > N(\epsilon)$ implies*

$$\left| \prod_{k=m+1}^{n} c_k - 1 \right| < \epsilon.$$

*Proof.* We assume first that $\prod_{k=1}^{\infty} c_k$ converges to $c \neq 0$. The Cauchy criterion applied to the convergent sequence $\left( \prod_{k=1}^{N} c_k \right)_{N \in \mathbb{N}}$ states: for every $\epsilon > 0$ and $\eta > 0$ there exists $N(\eta, \epsilon)$ such that $n > m > N(\eta, \epsilon)$ implies

$$\left| \prod_{k=1}^{n} c_k - \prod_{k=1}^{m} c_k \right| < \eta \epsilon$$

or

$$\left| \prod_{k=m+1}^{n} c_k - 1 \right| < \frac{\eta}{\left| \prod_{k=1}^{m} c_k \right|} \epsilon.$$

Since $\lim_{m \to \infty} \prod_{k=1}^{m} c_k = c \neq 0$ it follows that for $m \geq N_0$ we have $\left| \prod_{k=1}^{m} c_k \right| \geq \frac{|c|}{2} \neq 0$. Hence for $n > m > \max(N_0, N(\epsilon))$ we have with $\eta = \frac{2}{|c|}$

$$\left| \prod_{k=m+1}^{n} c_k - 1 \right| < \frac{\eta}{\left| \prod_{k=1}^{m} c_k \right|} \epsilon \leq \epsilon.$$

Now we prove the converse. First we note that for $\epsilon = \frac{1}{2}$ there exists $N_1 \in \mathbb{N}$ such that $n > m > N_1$ implies

$$\left| \prod_{k=m}^{n} c_k - 1 \right| < \frac{1}{2}$$

which yields

$$\frac{1}{2} < \left| \prod_{k=m}^{n} c_k \right| < \frac{3}{2}, \tag{30.4}$$

and in particular $c_l \neq 0$ for $l > N_1$. Now let $N > N_1$ fixed. For every $0 < \epsilon < \frac{1}{2}$ there exists by assumption $N(\epsilon) > N$ such that $n > m > N(\epsilon)$ implies

$$\left| \frac{\prod_{k=N}^{n} c_k}{\prod_{k=N}^{m} c_k} - 1 \right| = \left| \prod_{k=m}^{n} c_k - 1 \right|$$

$$= |c_{m+1} \cdot c_{m+2} \cdot \ldots \cdot c_n - 1| < \frac{2}{3} \epsilon,$$

or

$$\left| \prod_{k=N}^{n} c_k - \prod_{k=m}^{m} c_k \right| < \left| \prod_{N=k}^{m} c_k \right| \cdot \frac{2}{3} \epsilon < \epsilon,$$

where we used (30.4). Thus $\left( \prod_{k=1}^{N} c_k \right)_{N \in \mathbb{N}}$ is a Cauchy sequence in $\mathbb{R}$ and therefore convergent. $\qquad\square$

**Definition 30.8.** *The product $\prod_{k=1}^{\infty} c_k = \prod_{k=1}^{\infty}(1 + a_k)$ is called **absolutely convergent** if $\prod_{k=1}^{\infty}(1 + |a_k|)$ converges.*

**Proposition 30.9.** *If $\prod_{k=1}^{\infty}(1 + a_k)$ converges absolutely then it converges.*

*Proof.* We aim to apply the Cauchy criterion, and for this we note that for $a_1, \ldots, a_n \in \mathbb{R}$ the following holds:

$$|(1 + a_1)(1 + a_2) \cdot \ldots \cdot (1 + a_n) - 1| \leq (1 + |a_1|)(1 + |a_2|) \cdot \ldots \cdot (1 + |a_n|) - 1.$$

Indeed, for $n = 1$ we have

$$|(1 + a_1) - 1| = |a_1| = (1 + |a_1|) - 1.$$

430

Moreover,

$$|(1 + a_1)(1 + a_2) \cdot \ldots \cdot (1 + a_n)(1 + a_{n+1}) - 1|$$
$$= |(1 + a_1)(1 + a_2) \cdot \ldots \cdot (1 + a_n + a_{n+1} + a_n a_{n+1}) - 1|$$
$$\leq (1 + |a_1|)(1 + |a_2|) \cdot \ldots \cdot (1 + |a_n + a_{n+1} + a_n a_{n+1}|) - 1,$$

but $(1 + |a_n + a_{n+1} + a_n a_{n+1}|) \leq (1 + |a_n|)(1 + |a_{n+1}|)$. $\qquad\square$

**Proposition 30.10.** *The product $\prod_{k=1}^{\infty}(1 + a_k)$ converges absolutely if and only if the series $\sum_{k=1}^{\infty} a_k$ converges absolutely.*

*Proof.* Since

$$|a_1| + \cdots + |a_n| \leq (1 + |a_1|)(1 + |a_2|) \cdot \ldots \cdot (1 + |a_n|)$$

it follows that the absolute convergence of $\prod_{k=0}^{\infty}(1 + a_k)$ implies the absolute convergence of $\sum_{k=1}^{\infty} a_k$. On the other hand, for $x \geq 0$ we have $e^x \geq 1 + x$, which yields

$$(1 + |a_1|)(1 + |a_2|) \cdot \ldots \cdot (1 + |a_n|) \leq e^{|a_1| + \cdots + |a_n|}.$$

Now, if $\sum_{k=1}^{\infty} |a_k|$ converges, then $\left(\prod_{k=1}^{N}(1 + |a_k|)\right)_{N \in \mathbb{N}}$ must converge as it is an increasing sequence which is bounded. $\qquad\square$

**Example 30.11.** The product $\prod_{k=2}^{\infty}\left(1 + \frac{(-1)^k}{k}\right)$ converges since

$$\prod_{k=2}^{2n}\left(1 + \frac{(-1)^k}{k}\right) = \frac{3}{2} \cdot \frac{2}{3} \cdot \frac{5}{4} \cdot \frac{4}{5} \cdot \ldots \cdot \left(1 + \frac{1}{2n}\right) = 1 + \frac{1}{2n} \to 1$$

and

$$\prod_{k=2}^{2n-1}\left(1 + \frac{(-1)^k}{k}\right) = \frac{3}{2} \cdot \frac{2}{3} \cdot \frac{5}{4} \cdot \frac{4}{5} \cdot \ldots \cdot \frac{2n-1}{2n-2} \frac{2n-2}{2n-1} = 1.$$

However we already know that $\prod_{k=2}^{\infty}\left(1 + \frac{1}{k}\right)$ does not converge, hence $\prod_{k=2}^{\infty}\left(1 + \frac{(-1)^k}{k}\right)$ does not converge absolutely.

We are interested in finding the value of Wallis' product, i.e. we want to prove

$$\prod_{n=1}^{\infty} \frac{4n^2}{4n^2 - 1} = \frac{\pi}{2}. \tag{30.5}$$

We start by considering

$$A_m := \int_0^{\frac{\pi}{2}} \sin^m x\, dx \qquad (30.6)$$

and claim

$$A_m = \frac{m-1}{m} A_{m-2}, \quad \text{for } m \geq 2. \qquad (30.7)$$

Clearly we have $A_0 = \frac{\pi}{2}$ and $A_1 = 1$. In order to prove (30.7) note that

$$\int_0^{\frac{\pi}{2}} \sin^m x\, dx = \int_0^{\frac{\pi}{2}} \sin^{m-1} x \sin x\, dx$$

$$= \sin^{m-1} x(-\cos x)\Big|_0^{\frac{\pi}{2}} - \int_0^{\frac{\pi}{2}} \frac{d}{dx}(\sin^{m-1} x)(-\cos x)\, dx$$

$$= (m-1)\int_0^{\frac{\pi}{2}} \sin^{m-2} x \cos^2 x\, dx$$

$$= (m-1)\int_0^{\frac{\pi}{2}} \sin^{m-2} x(1 - \sin^2 x)\, dx$$

$$= (m-1)\int_0^{\frac{\pi}{2}} \sin^{m-2} x\, dx - (m-1)\int_0^{\frac{\pi}{2}} \sin^m x\, dx,$$

or

$$A_m = (m-1)A_{m-2} - (m-1)A_m$$

which implies (30.7). Using (30.7) we find

$$A_{2n} = \frac{(2n-1)}{2n}\frac{(2n-3)}{(2n-2)} \cdot \ldots \cdot \frac{3}{4} \cdot \frac{1}{2} \cdot \frac{\pi}{2} \qquad (30.8)$$

and

$$A_{2n+1} = \frac{2n}{(2n+1)}\frac{(2n-2)}{(2n-1)} \cdot \ldots \cdot \frac{4}{5} \cdot \frac{2}{3}. \qquad (30.9)$$

For $x \in [0, \frac{\pi}{2}]$, i.e. $0 \leq \sin x \leq 1$ we have

$$0 \leq \sin^{(2m+2)} x \leq \sin^{(2m+1)} x \leq \sin^{2m} x \leq 1$$

implying that

$$A_{2m+2} \leq A_{2m+1} \leq A_{2m}. \qquad (30.10)$$

Since

$$\lim_{m\to\infty} \frac{A_{2m+2}}{A_{2m}} = \lim_{m\to\infty} \frac{2m+1}{2m+2} = 1$$

we also get by (30.10)

$$\lim_{m\to\infty} \frac{A_{2m+1}}{A_{2m}} = 1.$$

Finally we find

$$\frac{A_{2m+1}}{A_{2m}} = \frac{2m \cdot 2m \cdot \ldots \cdot 4 \cdot 2 \cdot 2}{(2m+1)(2m-1) \cdot \ldots \cdot 3 \cdot 3 \cdot 1} \cdot \frac{2}{\pi},$$

i.e.

$$1 = \lim_{m\to\infty} \frac{A_{2m+1}}{A_{2m}} = \frac{2}{\pi} \lim_{m\to\infty} \frac{2m \cdot 2m \cdot \ldots \cdot 4 \cdot 2 \cdot 2}{(2m+1)(2m-1) \cdot \ldots \cdot 3 \cdot 3 \cdot 1}$$

or

$$\frac{\pi}{2} = \lim_{m\to\infty} \prod_{n=1}^{m} \frac{4n^2}{4n^2 - 1} = \prod_{n=1}^{\infty} \frac{4n^2}{4n^2 - 1}.$$

Thus we have proved

**Theorem 30.12 (Wallis' Product).** *The following holds:*

$$\prod_{n=1}^{\infty} \frac{4n^2}{4n^2 - 1} = \frac{\pi}{2}. \tag{30.11}$$

We want to use (30.11) to prove that

$$\int_{-\infty}^{\infty} e^{-x^2} \, dx = \sqrt{\pi}. \tag{30.12}$$

For this we will study the $\Gamma$-function a bit more closely. We know that on $(0, \infty)$ the $\Gamma$-function is logarithmic convex, i.e. for $0 < \lambda < 1$ and $x, y \in (0, \infty)$ we have

$$\Gamma(\lambda x + (1 - \lambda)y) \leq \Gamma(x)^{\lambda} \Gamma(y)^{1-\lambda}. \tag{30.13}$$

Further we know that $\Gamma(x + 1) = x\Gamma(x)$ implying

$$\Gamma(x + n) = \Gamma(x)x(x + 1) \cdot \ldots \cdot (x + n - 1), \quad x > 0, n \in \mathbb{N}. \tag{30.14}$$

433

Since $n + x = (1 - x)n + x(n + 1)$ the logarithmic convexity of $\Gamma$ implies for $0 < x < 1$

$$\Gamma(x + n) \leq \Gamma(n)^{1-x}\Gamma(n + 1)^x = \Gamma(n)^{1-x}\Gamma(n)^x n^x = (n - 1)!n^x. \quad (30.15)$$

Using $n + 1 = x(n + x) + (1 - x)(n + 1 + x)$ we derive in a similar way for $0 < x < 1$

$$n! = \Gamma(n + 1) \leq \Gamma(n + x)^x\Gamma(n + 1 + x)^{1-x} = \Gamma(n + x)(n + x)^{1-x}. \quad (30.16)$$

Combining (30.15) and (30.16) gives

$$n!(x + n)^{x-1} \leq \Gamma(n + x) \leq (n - 1)!n^x,$$

or with (30.14)

$$a_n(x) := \frac{n!(x + n)^{x-1}}{x(x + 1) \cdot \ldots \cdot (x + n - 1)} \leq \Gamma(x)$$

$$\leq \frac{(n - 1)!n^x}{x(x + 1) \cdot \ldots \cdot (x + n - 1)} =: b_n(x).$$

Note that

$$\frac{b_n(x)}{a_n(x)} = \frac{(n - 1)!n^x}{n!(n + x)^{x-1}} = \frac{(n - 1)!(n + x)n^x}{n!(n + x)^x} = \frac{(n + x)n^x}{n(n + x)}$$

which implies

$$\lim_{n \to \infty} \frac{b_n(x)}{a_n(x)} = \lim_{n \to \infty} \frac{a_n(x)}{b_n(x)} = 1.$$

Thus

$$\frac{a_n(x)}{b_n(x)} \leq \frac{\Gamma(x)}{b_n(x)} \leq 1$$

and for $n \to \infty$ we find for $0 < x < 1$ that $\dfrac{\Gamma(x)}{\lim\limits_{n \to \infty} b_n(x)} \leq 1$ or

$$\Gamma(x) = \lim_{n \to \infty} \frac{(n - 1)!n^x}{x(x + 1) \cdot \ldots \cdot (x + n - 1)}. \quad (30.17)$$

Hence we have proved a product representation for $\Gamma(x)$ provided that $0 < x < 1$. More generally we have

434

**Theorem 30.13.** *For all $x > 0$*

$$\Gamma(x) = \lim_{n \to \infty} \frac{n! n^x}{x(x+1) \cdot \ldots \cdot (x+n)} \qquad (30.18)$$

*holds.*

*Proof.* Since $\lim_{n \to \infty} \dfrac{n}{x+n} = 1$ we deduce (30.18) for $0 < x < 1$ from (30.17), and for $x = 1$ we find

$$\frac{n! n}{1(1+1) \cdot \ldots \cdot (1+n)} = \frac{n! n}{(n+1)!} = \frac{n}{n+1}$$

which yields

$$\Gamma(1) = 1 = \lim_{n \to \infty} \frac{n}{n+1}$$

i.e. (30.18) holds for $0 < x \le 1$. Now, if (30.18) holds for $x \in (0, \infty)$ then it holds also for $y := x + 1$ since

$$\Gamma(y) = \Gamma(x+1) = x\Gamma(x) = \lim_{n \to \infty} \frac{n! n^x}{(x+1) \cdot \ldots \cdot (x+n)}$$

$$= \lim_{n \to \infty} \frac{n! n^{y-1}}{y(y+1) \cdot \ldots \cdot (y+n-1)}$$

$$= \lim_{n \to \infty} \frac{n! n^y}{y(y+1) \cdot \ldots \cdot (y+n-1)(y+n)},$$

where we used $\lim_{n \to \infty} \dfrac{n}{y+n} = 1$, and the theorem is proved. $\qquad\square$

In Lemma 28.25 we have already proved

$$\int_{-\infty}^{\infty} e^{-x^2} dx = \Gamma\left(\frac{1}{2}\right)$$

or equivalently

$$\int_{0}^{\infty} e^{-x^2} dx = \frac{1}{2}\Gamma\left(\frac{1}{2}\right).$$

We note further that by (30.18) we find

$$\Gamma\left(\frac{1}{2}\right) = \lim_{n \to \infty} \frac{n! \sqrt{n}}{\frac{1}{2}(1+\frac{1}{2})(2+\frac{1}{2}) \cdot \ldots \cdot (n+\frac{1}{2})}$$

and

$$\Gamma\left(\frac{1}{2}\right) = \lim_{n \to \infty} \frac{n!\sqrt{n}}{(1 - \frac{1}{2})(2 - \frac{1}{2}) \cdot \ldots \cdot (n - \frac{1}{2})(n + \frac{1}{2})}$$

implying by Theorem 30.12 that

$$\Gamma\left(\frac{1}{2}\right)^2 = \lim_{n \to \infty} \frac{2n}{n + \frac{1}{2}} \frac{(n!)^2}{(1 - \frac{1}{4})(4 - \frac{1}{4}) \cdot \ldots \cdot (n^2 - \frac{1}{4})}$$

$$= 2 \lim_{n \to \infty} \frac{n}{n + \frac{1}{2}} \lim_{n \to \infty} \prod_{k=1}^{n} \frac{n^2}{k^2 - \frac{1}{4}} = \pi,$$

or $\Gamma\left(\frac{1}{2}\right) = \sqrt{\pi}$ which yields

**Theorem 30.14.** *The following holds:*

$$\int_{-\infty}^{\infty} e^{-x^2} dx = \sqrt{\pi}. \tag{30.19}$$

Next we want to study infinite products of functions. For this let $I \subset \mathbb{R}$ be an interval and for $k \in \mathbb{N}$ let $u_k : I \to \mathbb{R}$ be a function. In addition we assume $u_k(x) \neq 0$ for all $x \in I$ and $k \in \mathbb{N}$. With $v_k = 1 - u_k$ we set for $x \in I$

$$\prod_{k=1}^{\infty} u_k(x) = \prod_{k=1}^{\infty}(1 + v_k(x)). \tag{30.20}$$

When the product in (30.20) converges for every $x \in I$ a new function $u : I \to \mathbb{R}$ is defined by

$$u(x) = \prod_{k=1}^{\infty} u_k(x) = \lim_{N \to \infty} \prod_{k=1}^{N} u_k(x), \tag{30.21}$$

or

$$u(x) = \prod_{k=1}^{\infty}(1 + v_k(x)) = \lim_{N \to \infty} \prod_{k=1}^{N}(1 + v_k(x)). \tag{30.22}$$

Thus $u$ is the **pointwise limit** of $\left(\prod_{k=1}^{N} u_k\right)_{N \in \mathbb{N}} = \left(\prod_{k=1}^{N}(1 + v_k)\right)_{N \in \mathbb{N}}$. We say that $u$ has the **product representation** or **product expansion** (30.21) or (30.22). Clearly, in order to check pointwise convergence of $\prod_{k=1}^{\infty}(1 + v_k(x))$ we can use our previous criteria.

**Example 30.15.** The product $\prod_{k=1}^{\infty}\left(1 - \frac{x^2}{k^2}\right)$ converges for every $x \in \mathbb{R}$ such that $x^2 \neq m^2, m \in \mathbb{N}$. Indeed we have absolute convergence since with $a_k = -\frac{x^2}{k^2}$ we have that $\sum_{k=1}^{\infty} |a_k| = x^2 \sum_{k=1}^{\infty} \frac{1}{k^2}$ converges. But now we can extend $\prod_{k=1}^{\infty}\left(1 - \frac{x^2}{k^2}\right)$ to all $x \in \mathbb{R}$ by defining it to be 0 for $x^2 = m^2, m \in \mathbb{N}$. Thus $\prod_{k=1}^{\infty}\left(1 - \frac{x^2}{k^2}\right)$ defines a function on $\mathbb{R}$. Now it is even easier to see that $\prod_{k=1}^{\infty}\left(1 + \frac{x^2}{k^2}\right)$ converges for all $x \in \mathbb{R}$.

**Definition 30.16.** *We call $\prod_{k=1}^{\infty}(1 + v_k)$ **uniformly convergent** to $u$ if the sequence $\left(\prod_{k=1}^{N}(1 + v_k)\right)_{N \in \mathbb{N}}$ converges on $I$ uniformly to $u$.*

The Cauchy criterion extends to uniform convergence in the usual way, i.e. we have

**Theorem 30.17.** *The product $\prod_{k=1}^{\infty}(1 + v_k)$ converges on $I$ uniformly if for every $\epsilon > 0$ there exists $N(\epsilon) \in \mathbb{N}$ such that $n \geq N(\epsilon)$ and $m \in \mathbb{N}$ implies that for all $x \in I$*

$$\left| \prod_{k=n+1}^{n+m} (1 + v_k(x)) - 1 \right| < \epsilon. \tag{30.23}$$

**Theorem 30.18.** *Suppose that the series $\sum_{k=1}^{\infty} |v_k|$ converges uniformly on $I$ and that all functions $v_k$ are continuous. Then $\prod_{k=1}^{\infty}(1 + v_k)$ converges uniformly to a continuous function.*

*Proof.* We know that the convergence of $\sum_{k=1}^{\infty} |v_k(x)|$ implies the convergence of $\prod_{k=1}^{\infty}(1 + v_k(x))$, see Propositions 30.10 and 30.9. Thus we can define a function on $I$ by

$$u(x) := \prod_{k=1}^{\infty}(1 + v_k(x)).$$

Furthermore, there exists $M \in \mathbb{N}$ such that for all $x \in I$ and all $l \in \mathbb{N}$ the following holds

$$|v_{m+1}(x)| + |v_{m+2}(x)| + \cdots + |v_{m+k}(x)| < 1, \tag{30.24}$$

which is again a direct consequence of the uniform convergence of $\sum_{k=1}^{\infty} |v_k|$. Consider now the product

$$w_M(x) := \prod_{k=M+1}^{\infty} (1 + v_k(x)) \tag{30.25}$$

with $p_{n,M}(x) := \prod_{k=M+1}^{n}(1 + v_k(x)), n \geq M + 1$, we find with $p_{M,M}(x) = 0$

$$w_M(x) = \sum_{k=M+1}^{\infty} (p_{k,M}(x) - p_{k-1,M}(x))$$

$$= \sum_{k=M+1}^{\infty} p_{k,M}(x)v_{k+1}(x).$$

We claim that the series $\sum_{k=M+1}^{\infty} p_{k,M}(x)v_{k+1}(x)$ is uniformly convergent on $I$. Indeed, for $k \geq M + 1$ we have

$$|p_{k,M}(x)| \leq (1 + |v_{M+1}(x)|)(1 + |v_{M+2}(x)|) \cdot \ldots \cdot (1 + |v_{M+k}(x)|)$$
$$\leq e^{|v_{M+1}| + \cdots + |v_{M+k}(x)|} < e,$$

where we used (30.24) and the fact that $1 + y \leq e^y$ for $y \geq 1$. Thus we find

$$\sum_{k=M+1}^{\infty} |p_{k,M}(x)v_{k+1}(x)| \leq \sum_{k=M+1}^{\infty} |v_{k+1}(x)|$$

and the uniform convergence of $\sum_{k=1}^{\infty} |v_k(x)|$ together with the Weierstrass M-test yields the uniform and absolute convergence of $\sum_{k=M+1}^{\infty} p_{k,M}(x)v_k(x)$, hence $w_M$ is a continuous function and the convergence in (30.25) is uniform. Multiplying $w_M(x)$ by $\prod_{k=1}^{M}(1 + v_k(x))$ does not change the continuity nor the uniform convergence and the result follows. $\qquad\square$

Finally we want to establish a result for the derivative of an infinite product of differentiable functions.

**Theorem 30.19.** *Suppose that $v_k : I \to \mathbb{R}$, $k \in \mathbb{N}$, is differentiable and that both $\sum_{k=1}^{\infty} |v_k|$ and $\sum_{k=1}^{\infty} |v_k'|$ converge uniformly. Then the product $u(x) := \prod_{k=1}^{\infty}(1 + v_k(x))$ is differentiable and we have*

$$\frac{u'(x)}{u(x)} = \frac{(\prod_{k=1}^{\infty}(1 + v_k(x)))'}{\prod_{k=1}^{\infty}(1 + v_k(x))} = \sum_{k=1}^{\infty} \frac{v_k(x)}{1 + v_k(x)}. \qquad (30.26)$$

*Proof.* First note that for a finite product we have if $g = g_1 \cdot \ldots \cdot g_M$ that

$$\frac{d}{dx}\ln(g_1 \cdot \ldots \cdot g_M) = \frac{(g_1 \cdot \ldots \cdot g_M)'}{g_1 \cdot \ldots \cdot g_M}$$

438

and

$$\frac{d}{dx}\ln(g_1 \cdot \ldots \cdot g_M) = \frac{d}{dx}\sum_{j=1}^{M}\ln g_j = \sum_{j=1}^{M}\frac{g_j'}{g_j},$$

i.e.

$$\frac{(g_1 \cdot \ldots \cdot g_M)'}{g_1 \cdot \ldots \cdot g_M} = \sum_{j=1}^{M}\frac{g_j'}{g_j} \tag{30.27}$$

provided $g_j \neq 0$ and $g_j > 0$. (The case $g < 0$ and some $g_j < 0$ can also be treated by looking at $|g|$.) Now, the uniform convergence of $\sum_{k=1}^{\infty}|v_k(x)|$ implies the existence of $M \in \mathbb{N}$ such that for all $x \in I$ we have

$$\sum_{k=M+1}^{\infty}|v_k(x)| < \frac{1}{2},$$

in particular we have $|v_k(x)| < \frac{1}{2}$ for $k \geq M+1$. This implies by Proposition 30.10 the convergence of $\prod_{k=M+1}^{\infty}(1 + v_k(x))$ and hence by Lemma 30.5 the convergence of $\sum_{k=M+1}^{\infty}\ln(1 + v_k(x))$. Since $|v_k(x)| < \frac{1}{2}$ for $k \geq M+1$ it follows that for these $k$ we have $\left|\frac{1}{1+v_k(x)}\right| < 2$ and consequently, since

$$\sum_{k=M+1}^{\infty}\frac{d}{dx}(\ln(1 + v_k(x))) = \sum_{k=M+1}^{\infty}\frac{v_k'(x)}{1 + v_k(x)},$$

the uniform convergence of $\sum_{k=1}^{\infty}|v_k'|$ implies the absolute and uniform convergence of $\sum_{k=|M+1}^{\infty}\frac{v_k'}{1+v_k}$. By Theorem 26.19 we now have

$$\frac{d}{dx}\sum_{k=M+1}^{\infty}\ln(1 + v_k) = \frac{d}{dx}\sum_{k=M+1}^{\infty}\frac{v_k'}{1 + v_k}.$$

From

$$\prod_{k=M+1}^{N}(1 + v_k) = \exp\sum_{k=M+1}^{N}\ln(1 + v_k)$$

we derive

$$\frac{\frac{d}{dx}\prod_{k=M+1}^{N}(1 + v_k)}{\prod_{k=M+1}^{N}(1 + v_k)} = \sum_{k=M+1}^{N}\frac{v_k'}{1 + v_k}$$

and we are now allowed to pass to the limit to obtain

$$\frac{\frac{d}{dx} \prod_{k=M+1}^{\infty} (1+v_k)}{\prod_{k=M+1}^{\infty} (1+v_k)} = \sum_{k=M+1}^{\infty} \frac{v'_k}{1+v_k}.$$

Using (30.27) for $(1+v_1) \cdot \ldots \cdot (1+v_M)$ we eventually get

$$\frac{\frac{d}{dx} \prod_{k=1}^{\infty} (1+v_k)}{\prod_{k=1}^{\infty} (1+v_k)} = \sum_{k=1}^{\infty} \frac{v'_k}{1+v_k}.$$

$\square$

**Example 30.20.** For $x \in (-1,1)$ we know that $F(x) = x \prod_{k=1}^{\infty} \left(1 - \frac{x^2}{k^2}\right)$ converges uniformly, since $\left| \sum_{k=1}^{\infty} \frac{x^2}{k^2} \right| \le |x|^2 \sum_{k=1}^{\infty} \frac{1}{k^2}$. Further we have

$$\sum_{k=1}^{\infty} \left| \left(\frac{x^2}{k^2}\right)' \right| = 2|x| \sum_{k=1}^{\infty} \frac{1}{k^2}$$

and therefore we find for $x \neq 0$

$$\frac{F'(x)}{F(x)} = \frac{1}{x} + \sum_{k=1}^{\infty} \frac{2x}{x^2 - k^2}.$$

# Problems

1. Prove

    a) $\prod_{k=2}^{\infty} \frac{k^3-1}{k^3+1} = \frac{2}{3}$;

    b) $\prod_{l=1}^{\infty} \left(1 + \frac{1}{l(l+2)}\right) = 2$.

2. Let $a_k \geq 0$, $a_k \neq 1$ for $k \in \mathbb{N}$. Show: the convergence of $\prod_{k=1}^{\infty} (1-a_k)$ is equivalent to the convergence of $\sum_{k=1}^{\infty} a_k$.

3. Suppose that $\sum_{k=1}^{\infty} a_k$ converges.

    a) Prove that $\prod_{k=1}^{\infty} (1+a_k)$ converges if and only if $\sum_{k=1}^{\infty} a_k^2$ converges.

    b) Prove that if $\sum_{k=1}^{\infty} a_k^2$ diverges then $\prod_{k=1}^{\infty} (1+a_k)$ diverges.

4. Prove that if the infinite product $\prod_{k=1}^{\infty}(1+a_k)$ converges absolutely then we can rearrange its factors and the corresponding product converges to the same limit.

5.  a) For $|x| < 1$ prove that $\prod_{k=1}^{\infty}(1 + x^{2^k}) = \frac{1}{1-x}$.

   b) Show that for $x \neq 2^k(\frac{\pi}{2} + l\pi), k \in \mathbb{N}, l \in \mathbb{Z}$, the following holds:

$$\prod_{j=1}^{\infty} \cos \frac{x}{2^j} = \frac{\sin x}{x},$$

and derive $\prod_{j=1}^{\infty} \cos \frac{\pi}{2^{j+1}} = \frac{2}{\pi}$.

441

# 31 More on the Γ-Function

In this chapter we want to discuss more properties of the Γ-function. Although the Γ-function is one of the most important functions in mathematics, typical undergraduate courses do not have enough time to treat the Γ-function in detail. We therefore view this chapter as an addition to the basics that are usually met.

In Theorem 28.29 we have seen that the Γ-function is logarithmic convex. A classical result due to H. Bohr and J. Mollerup gives an interesting characterisation of the Γ-function using logarithmic convexity.

**Theorem 31.1.** *Let $G : (0, \infty) \to \mathbb{R}$ be a positive, logarithmic convex function satisfying the functional equation of the Γ-function, i.e. $G(x + 1) = xG(x)$, and the normalisation $G(1) = 1$. Then $G$ is the Γ-function.*

*Proof.* From the functional equation and the normalisation we deduce for $n \in \mathbb{N}$ the fact that $G(n + 1) = n!$. Now, for $0 < x \leq 1$ we find

$$n + x = (1 - x)n + x(n + 1),$$

i.e. $n + x$ is a convex combination of $n$ and $n + 1$. Since $G$ is logarithmic convex we find

$$\ln G(n + x) \leq (1 - x) \ln G(n) + x \ln G(n + 1),$$

or

$$G(n + x) \leq G(n)^{(1-x)} G(n + 1)^x = ((n - 1)!)^{(1-x)} (n!)^x = n! n^{(x-1)},$$

i.e.

$$G(n + x) \leq n! n^{(x-1)}. \tag{31.1}$$

Moreover,

$$n + 1 = x(n + x) + (1 - x)(n + x + 1),$$

i.e. $n + 1$ is a convex combination of $n + x$ and $n + x + 1$, and this gives

$$
\begin{aligned}
n! = G(n + 1) &\leq G(n + x)^x G(n + x + 1)^{(1-x)} \\
&= G(n + x)^x (G(n + x)(n + x))^{1-x} \\
&= G(n + x)(n + x)^{(1-x)},
\end{aligned}
$$

443

thus we have

$$n! = G(n+1) \leq G(n+x)(n+x)^{(1-x)}. \tag{31.2}$$

Combining (31.1) and (31.2) we find

$$n!(n+x)^{(x-1)} \leq G(n+x) \leq n!n^{(x-1)}. \tag{31.3}$$

Observing that

$$G(n+x) = x(x+1) \cdot \ldots \cdot (x+n-1)G(x),$$

it follows from (31.3) and the fact that $0 < x \leq 1$ that

$$\frac{n!n^x}{x(x+1) \cdot \ldots \cdot (x+n)} \leq \frac{n!(n+x)^x}{x(x+1) \cdot \ldots \cdot (x+n)} \leq G(x)$$

$$\leq \frac{n!n^x}{x(x+1) \cdot \ldots \cdot (x+n-1)n} \leq \frac{n!n^x}{x(x+1) \cdot \ldots \cdot (x+n)},$$

and by Theorem 30.13 we get for $n \to \infty$

$$\Gamma(x) \leq G(x) \leq \Gamma(x), \tag{31.4}$$

i.e. $G(x) = \Gamma(x)$ for $0 < x < 1$. But now $G(x+1) = xG(x)$ and $\Gamma(x+1) = x\Gamma(x)$ imply $G(x) = \Gamma(x)$ for all $x > 0$. $\qquad \square$

In the proof of the above result, Theorem 30.13 was quite important. Using this result we can also give a representation of the $\Gamma$-function as an infinite product involving the exponential function. Recall that the existence of the **Euler constant**

$$\gamma := \lim_{N \to \infty} \left( \sum_{k=1}^{N} \frac{1}{k} - \ln N \right) \tag{31.5}$$

was proved in Theorem 18.24. Denote by $\Gamma_n$ the function

$$\Gamma_n(x) := \frac{n^x n!}{x(x+1) \cdot \ldots \cdot (x+n)}, \tag{31.6}$$

and then Theorem 30.13 reads as

$$\Gamma(x) = \lim_{n \to \infty} \Gamma_n(x). \tag{31.7}$$

From (31.6) we deduce

$$\Gamma_n(x+1) = x\Gamma_n(x)\frac{n}{x+n+1}, \tag{31.8}$$

or

$$\Gamma_n(x) = \frac{1}{x}\frac{x+n+1}{n}\Gamma_n(x+1). \tag{31.9}$$

Next we observe that

$$\frac{k}{x+k} = \frac{1}{1+\frac{x}{k}}$$

and

$$e^{x(\ln n - 1 - \frac{1}{2} - \cdots - \frac{1}{n})} = n^x e^{-\frac{x}{1}} e^{-\frac{x}{2}} \cdots \cdot e^{-\frac{x}{n}},$$

which implies

$$\Gamma_n(x) = e^{x(\ln n - 1 - \frac{1}{2} - \cdots - \frac{1}{n})}\frac{1}{x} \cdot \frac{e^{\frac{x}{1}}}{\left(1+\frac{x}{1}\right)} \cdots \cdot \frac{e^{\frac{x}{n}}}{\left(1+\frac{x}{n}\right)}. \tag{31.10}$$

Now we pass to the limit $n \to \infty$ and using (31.5) we arrive at

**Theorem 31.2.** *For $x > 0$ the $\Gamma$-function has the **Weierstrass product representation***

$$\Gamma(x) = \frac{e^{-\gamma x}}{x} \prod_{k=1}^{\infty} \frac{e^{\frac{x}{k}}}{1+\frac{x}{k}}. \tag{31.11}$$

From (31.10) we can immediately deduce

$$\frac{1}{\Gamma_n(x)} = xe^{x(1+\frac{1}{2}+\cdots+\frac{1}{n}-\ln n)}\left(1+\frac{x}{1}\right)e^{-\frac{x}{1}} \cdot \left(1+\frac{x}{2}\right)e^{-\frac{x}{2}} \cdots \cdot \left(1+\frac{x}{n}\right)e^{-\frac{x}{n}}$$

and passing to the limit $n \to \infty$ yields

**Corollary 31.3.** *For $x > 0$ we have*

$$\frac{1}{\Gamma(x)} = xe^{\gamma x} \prod_{k=1}^{\infty}\left(1+\frac{x}{k}\right)e^{-\frac{x}{k}}. \tag{31.12}$$

The Weierstrass product representation allows us to prove

**Theorem 31.4.** *On $(0,\infty)$ the $\Gamma$-function is arbitrarily often differentiable.*

*Proof.* Since for $x > 0$ we have $\Gamma(x) > 0$ it follows that $\Gamma$ is differentiable if and only if $\ln \Gamma$ is differentiable. From (31.11) we derive

$$\ln \Gamma(x) = -\gamma x - \ln x + \sum_{k=1}^{\infty} \left( \frac{x}{k} - \ln \left( 1 + \frac{x}{k} \right) \right), \tag{31.13}$$

and we know that the series $\sum_{k=1}^{\infty} \left( \frac{x}{k} - \ln \left( 1 + \frac{x}{k} \right) \right)$ converges pointwise. Further we note that

$$\sum_{k=1}^{\infty} \left( \frac{x}{k} - \ln \left( 1 + \frac{x}{k} \right) \right)' = \sum_{k=1}^{\infty} \left( \frac{1}{k} - \frac{1}{k+x} \right) = \sum_{k=1}^{\infty} \frac{x}{k(k+x)},$$

and this series converges uniformly on compact intervals in $(0, \infty)$. Consequently we have

$$(\ln \Gamma(x))' = \frac{\Gamma'(x)}{\Gamma(x)} = -\gamma - \frac{1}{x} + \sum_{k=1}^{\infty} \frac{x}{k(k+x)} = -\gamma - \frac{1}{x} + \sum_{k=1}^{\infty} \left( \frac{1}{k} - \frac{!}{k+x} \right),$$
$$\tag{31.14}$$

which now yields also that $\Gamma$ is arbitrarily often differentiable on $(0, \infty)$. Indeed we find for $l \geq 2$

$$\frac{d^{l-1}}{dx^{l-1}} \left( \frac{\Gamma'(x)}{\Gamma(x)} \right) = \sum_{k=0}^{\infty} \frac{(-1)^l (l-1)!}{(x+k)^l}. \tag{31.15}$$

$\square$

It is worth noting that

$$(\ln \Gamma)''(x) = \sum_{k=0}^{\infty} \frac{1}{(k+x)^2}, \tag{31.16}$$

which in particular confirms that $\ln \Gamma$ is convex, which is of course already known to us.

We want to study the asymptotic behaviour of the $\Gamma$-function. Since on $\mathbb{N}$ it coincides with the factorials we expect rapid growth. More precisely, recalling

$$\left( 1 + \frac{1}{k} \right)^k < e < \left( 1 + \frac{1}{k} \right)^{k+1}, \tag{31.17}$$

see Problem 10 in Chapter 17, we find when multiplying these inequalities for $k = 1, \ldots, (n-1)$

$$\frac{n^{n-1}}{(n-1)!} < e^{n-1} < \frac{n^n}{(n-1)!}, \tag{31.18}$$

or

$$en^n e^{-n} < n! < en^{n+1}e^{-n}, \tag{31.19}$$

which suggests

$$\Gamma(x) = x^{x-\frac{1}{2}}e^{-x}e^{\vartheta(x)}. \tag{31.20}$$

Using methods not yet at our disposal the **general Stirling formula** can be proved.

**Theorem 31.5.** *For $x > 0$ we have*

$$\Gamma(x) = \sqrt{2\pi}x^{x-\frac{1}{2}}e^{-x}e^{\vartheta(x)} \tag{31.21}$$

*where*

$$\vartheta(x) = \int_0^\infty \left(\frac{1}{e^t - 1} - \frac{1}{t} + \frac{1}{2}\right)e^{-xt}dt. \tag{31.22}$$

A proof of Theorem 31.5 is given in R. Beals and R. Wong [1]. Here we give a proof of the Stirling formula for the factorial, or equivalently for $\Gamma(n+1)$. As preparation we prove

**Lemma 31.6.** *For $k \in \mathbb{N}$ we find $\xi_k \in [k, k+1]$ such that*

$$\int_k^{k+1} \ln x\, dx = \frac{1}{2}(\ln k + \ln(k+1)) + \frac{1}{12\xi_k^2}. \tag{31.23}$$

*Proof.* Let $g(x) = \frac{x(1-x)}{2} \geq 0$ on $[0, 1]$ and set $g_k : [k, k+1] \to \mathbb{R}$, $g_k(x) = g(x-k) = \frac{(x-k)(1+k-x)}{2}$. We have $g_k(x) \geq 0$, $g_k'(x) = -x + \frac{2k+1}{2}$ and $g_k''(x) = -1$, and therefore

$$\int_k^{k+1} \ln x\, dx = -\int_k^{k+1} g_k''(x)\ln x\, dx$$

$$= -g_k'(x)\ln x\Big|_k^{k+1} + \int_k^{k+1} g_k'(x)(\ln x)'dx$$

$$= -g_k'(x)\ln x\Big|_k^{k+1} + g_k(x)(\ln x)'\Big|_k^{k+1} - \int_k^{k+1} g_k(x)(\ln x)''dx.$$

447

Now, since

$$-g_k'(x) \ln x \Big|_k^{k+1} = \frac{1}{2}(\ln(k+1) + \ln k)$$

and

$$g_k(x)(\ln x)' \Big|_k^{k+1} = 0,$$

we find

$$
\begin{aligned}
\int_k^{k+1} \ln x \, dx &= \frac{1}{2}(\ln(k+1) + \ln k) + \int_k^{k+1} \frac{1}{x^2} g_k(x) \, dx \\
&= \frac{1}{2}(\ln(k+1) + \ln k) + \frac{1}{\xi_k^2} \int_k^{k+1} g_k(x) \, dx \\
&= \frac{1}{2}(\ln(k+1) + \ln k) + \frac{1}{12\xi_k^2}, \xi_k \in [k, k+1].
\end{aligned}
$$

$\square$

Now we sum (31.23) from $k = 1$ to $n - 1$ and we get

$$\int_1^n \ln x \, dx = \sum_{k=1}^n \ln k - \frac{1}{2} \ln n + \frac{1}{2} \sum_{k=1}^{n-1} \frac{1}{\xi_k^2}.$$

Since $\int_1^n \ln x \, dx = n \ln n - n + 1$ we find further

$$\sum_{k=1}^n \ln k = \left(n + \frac{1}{2}\right) \ln n - n + \eta_n, \tag{31.24}$$

where

$$\eta_n = 1 - \frac{1}{12} \sum_{k=1}^{n-1} \frac{1}{\xi_k^2}. \tag{31.25}$$

But $e^{\sum_{k=1}^n \ln k} = n!$, and therefore

$$n! = n^{\left(n + \frac{1}{2}\right)} e^{-n} c_n, c_n = e^{\eta_n}. \tag{31.26}$$

For $k \leq \xi_k \leq k + 1$ we have $\frac{1}{\xi_k^2} \leq \frac{1}{k^2}$ which yields

$$\eta := \lim_{n \to \infty} \eta_n = 1 - \frac{1}{12} \sum_{k=1}^{\infty} \frac{1}{\xi_k^2}$$

exists, hence

$$c = \lim_{n \to \infty} c_n = e^\eta.$$

We want to find $c$. We must have

$$c = \lim_{n \to \infty} \frac{c_n^2}{c_{2n}},$$

and by (31.26) we have

$$\frac{c_n^2}{c_{2n}} = \frac{(n!)\sqrt{2n}(2n)^{2n}}{n^{2n+1}(2n)!} = \sqrt{2}\frac{2^n(n!)^2}{\sqrt{n}(2n)!}.$$

Recall Wallis' product, Theorem 30.12, i.e.

$$2\prod_{k=1}^{\infty} \frac{4k^2}{4k^2 - 1} = \pi. \tag{31.27}$$

Note that $4k^2 - 1 = (2k - 1)(2k + 1)$ and hence

$$\prod_{k=1}^{N} \frac{4k^2}{4k^2 - 1} = \frac{2 \cdot 2 \cdot 4 \cdot 4 \cdot \ldots \cdot 2N \cdot 2N}{1 \cdot 3 \cdot 3 \cdot 5 \cdot \ldots \cdot (2N - 1) \cdot (2N + 1)}$$

which gives

$$\left(2\prod_{k=1}^{N} \frac{4k^2}{4k^2 - 1}\right)^{\frac{1}{2}} = \sqrt{2}\frac{2 \cdot 4 \cdot \ldots \cdot 2N}{3 \cdot 5 \cdot \ldots \cdot (2N - 1) \cdot \sqrt{2N + 1}}$$

$$= \frac{1}{\sqrt{N + \frac{1}{2}}}\frac{2^2 \cdot 4^2 \cdot \ldots \cdot (2N)^2}{2 \cdot 3 \cdot 4 \cdot 5 \cdot \ldots \cdot (2N - 1)(2N)}$$

$$= \frac{1}{\sqrt{N + \frac{1}{2}}}\frac{2^{2N}(N!)^2}{(2N)!},$$

which yields

$$c = \lim_{N \to \infty} \frac{2^{2N}(N!)^2}{\sqrt{N}(2N)!} = \sqrt{2\pi}, \tag{31.28}$$

implying

$$\lim_{n \to \infty} \frac{n!}{\sqrt{2\pi}n^{(n+\frac{1}{2})}e^{-n}} = 1.$$

Thus we have proved

**Theorem 31.7 (Stirling formula).** *The following holds*

$$\lim_{n\to\infty} \frac{n!}{\sqrt{2\pi}n^{\left(n+\frac{1}{2}\right)}e^{-n}} = 1. \tag{31.29}$$

Further it follows that

**Corollary 31.8.** *For $n \geq 2$ we have*

$$\sqrt{2\pi}n^{\left(n+\frac{1}{2}\right)}e^{-n} < n! < \sqrt{2\pi}n^{\left(n+\frac{1}{2}\right)}e^{-n}e^{\frac{1}{12(n-1)}}. \tag{31.30}$$

*Proof.* We only need to observe that

$$0 < \eta_n - \eta = \frac{1}{12}\sum_{k=n}^{\infty}\frac{1}{\xi_k^2} \leq \frac{1}{12}\sum_{k=n}^{\infty}\frac{1}{k^2} < \frac{1}{12}\int_{n-1}^{\infty}\frac{dx}{x^2} = \frac{1}{12(n-1)}$$

and now we have to use (31.26). $\square$

In Problem 11 in Chapter 28 we have seen that the improper integral

$$B(x,y) = \int_0^1 t^{x-1}(1-t)^{y-1}dt \tag{31.31}$$

converges for $x > 0$ and $y > 0$.

**Definition 31.9.** *The function $B : (0,\infty) \times (0,\infty) \to \mathbb{R}$, $(x,y) \mapsto B(x,y)$, is called **(Euler's) beta-function**.*

Our aim is to relate the beta-function to the $\Gamma$-function.
First we note for $x, y > 0$

$$B(x+1,y) = \int_0^1 t^x(1-t)^{y-1}dt = \int_0^1 (1-t)^{x+y-1}\left(\frac{t}{1-t}\right)^x dt. \tag{31.32}$$

**Lemma 31.10.** *For $x, y > 0$ we have*

$$B(x+1,y) = \frac{x}{x+y}B(x,y). \tag{31.33}$$

450

*Proof.* For $0 < \epsilon, \eta < \frac{1}{2}$ we find using integration by parts

$$\int_\epsilon^{1-\eta} (1-t)^{x+y-1} \left(\frac{t}{1-t}\right)^x dt$$

$$= -\frac{(1-t)^{x+y}}{x+y} \left(\frac{t}{1-t}\right)^x \Big|_\epsilon^{1-\eta} + \int_\epsilon^{1-\eta} \frac{x}{x+y}(1-t)^{x+y} \left(\frac{t}{1-t}\right)^{x-1} \frac{1}{(1-t)^2} dt$$

$$= \frac{(1-\epsilon)^y \epsilon^x - \eta^y (1-\eta)^x}{x+y} + \frac{x}{x+y} \int_\epsilon^{1-\eta} t^{x-1}(1-t)^{y-1} dt.$$

For $\epsilon$ and $\eta$ tending to 0 we get

$$B(x+1, y) = \frac{x}{x+y} B(x, y).$$

$\square$

Now we can prove

**Theorem 31.11.** *For $x, y > 0$ the following holds*

$$B(x, y) = \frac{\Gamma(x)\Gamma(y)}{\Gamma(x+y)}. \tag{31.34}$$

*Proof.* For $y > 0$ fixed we consider the function

$$f(x) := B(x, y)\Gamma(x+y).$$

With (31.33) we find

$$f(x+1) = B(x+1, y)\Gamma(x+1+y)$$

$$= \frac{x}{x+y} B(x, y)(x+y)\Gamma(x+y)$$

$$= xB(x, y)\Gamma(x+y) = xf(x),$$

thus $f$ satisfies the functional equation of the Γ-function. Further, for $y > 0$ fixed the function $x \mapsto \Gamma(x+y)$ and $x \mapsto B(x, y)$ are logarithmic convex. For the Γ-function this is trivial, in the case of the beta-function we only need to note that $x \mapsto t^{x-1}(1-t)^{y-1}$ is logarithmic convex and hence the integral defining $B(x, y)$ is a pointwise limit of logarithmic convex functions. By Problem 12 c) in Chapter 28 it follows that $x \mapsto B(x, y)$ is logarithmic convex. Finally, Problem 12 a) in Chapter 28 shows that $x \mapsto B(x, y)\Gamma(x+y)$ is logarithmic convex. Since both results hold also for $g(x) := \frac{f(x)}{f(1)}$ and

451

$g(1) = 1$ we deduce by Theorem 31.1 that the function $\frac{f(x)}{f(1)}$ is the $\Gamma$-function, i.e. we have

$$f(1)\Gamma(x) = B(x,y)\Gamma(x+y).$$

In order to find $f(1)$ we note that

$$B(1,y) = \int_0^1 (1-t)^{y-1}dt = \frac{1}{y}$$

and therefore

$$f(1) = B(1,y)\Gamma(1+y) = \frac{1}{y}y\Gamma(y) = \Gamma(y)$$

and it follows that

$$B(x,y) = \frac{\Gamma(x)\Gamma(y)}{\Gamma(x+y)}.$$

$\square$

Calculating $B(x,x)$ we find

$$B(x,x) = \int_0^1 t^{x-1}(1-t)^{x-1}dt, \quad x > 0$$

$$= 2\int_0^{\frac{1}{2}} (t(1-t))^{x-1}dt,$$

where we used that $t \mapsto t(1-t)^{x-1}$ is symmetric with respect to the axis $t_0 = \frac{1}{2}$. Using the substitution $s = 4t(1-t)$ we obtain

$$B(x,x) = 2\int_0^1 s^{x-1}(1-s)^{-\frac{1}{2}} \cdot 2^{-2x}ds$$

$$= 2^{1-2x}\int_0^1 s^{x-1}(1-s)^{-\frac{1}{2}}ds = 2^{1-2x}B(x,\frac{1}{2}).$$

Now we apply (31.34) to find

$$\frac{\Gamma(x)^2}{\Gamma(2x)} = B(x,x) = 2^{1-2x}\frac{\Gamma(x)\Gamma\left(\frac{1}{2}\right)}{\Gamma\left(x+\frac{1}{2}\right)}$$

or using $\Gamma\left(\frac{1}{2}\right) = \sqrt{\pi}$ we arrive at the **Legrendre duplication formula** for the $\Gamma$-function:

452

**Theorem 31.12.** *For $x > 0$ we have*

$$\Gamma(2x) = \frac{2^{2x-1}}{\sqrt{\pi}} \Gamma(x)\Gamma\left(x + \frac{1}{2}\right). \tag{31.35}$$

We close our theoretical considerations by proving an interesting relation between the Γ-function and the sine-function. For this we first extend Γ to a larger domain. Let $x > 0$ and $n \in \mathbb{N}$. Iterating the functional equation of the Γ-function we get

$$\Gamma(x + n) = (x + n - 1)(x + n - 2) \cdot \ldots \cdot (x + 1)x\Gamma(x), \tag{31.36}$$

which allows us to define Γ for all $x$, $x > -n$, but $x \neq 0, -1, \ldots, -n$ by

$$\Gamma(x) := \frac{\Gamma(x + n)}{(x + n - 1)(x + n - 2) \cdot \ldots \cdot (x + 1)x}.$$

Thus we can extend Γ to $\mathbb{R} \backslash \{-\mathbb{N}_0\}$, $-\mathbb{N}_0 := \{k | -k \in \mathbb{N}_0\}$ and the functional equation of Γ also holds for this extension.
We now consider the function

$$\varphi(x) := \Gamma(x)\Gamma(1 - x) \sin \pi x \tag{31.37}$$

which is defined for all $x \in \mathbb{R} \backslash \mathbb{Z}$. For such a value of $x$ we find

$$\varphi(x + 1) = \Gamma(x + 1)\Gamma(1 - x - 1) \sin(\pi(x + 1))$$
$$= x\Gamma(x) \cdot \frac{\Gamma(1 - x)}{-x}(-\sin \pi x) = \varphi(x),$$

where we used $\Gamma(1 - x) = -x\Gamma(-x)$. Thus $\varphi$ is a function with period 1. Our next aim is to extend $\varphi$ to $\mathbb{R}$. Applying (31.35) for $x = \frac{1}{2}$ we find

$$\Gamma\left(\frac{x}{2}\right)\Gamma\left(\frac{x + 1}{2}\right) = c_0 2^{-x}\Gamma(x), \tag{31.38}$$

where $c_0 = \frac{\sqrt{\pi}}{2}$. Replacing $x$ by $1 - x$ in (31.38) we get

$$\Gamma\left(\frac{1 - x}{2}\right)\Gamma\left(1 - \frac{x}{2}\right) = c_0 2^{x-1}\Gamma(1 - x), \tag{31.39}$$

and we find

$$\varphi\left(\frac{x}{2}\right)\varphi\left(\frac{x+1}{2}\right) = \Gamma\left(\frac{x}{2}\right)\Gamma\left(1-\frac{x}{2}\right)\sin\frac{\pi x}{2}\Gamma\left(\frac{x+1}{2}\right)\Gamma\left(\frac{1-x}{2}\right)\cos\frac{\pi x}{2}$$

$$= \frac{c_0^2}{4}\Gamma(x)\Gamma(1-x)\sin\pi x = \frac{c_0^2}{4}\varphi(x),$$

or

$$\varphi\left(\frac{x}{2}\right)\varphi\left(\frac{x+1}{2}\right) = \frac{c_0^2}{4}\varphi(x) = \frac{\pi}{16}\varphi(x). \qquad (31.40)$$

For $x \in \mathbb{R} \setminus \mathbb{Z}$ the function $\varphi$ is arbitrarily often differentiable since the sine and the $\Gamma$-functions are. The functional equation of the $\Gamma$-function yields

$$\varphi(x) = \frac{\Gamma(1+x)}{x}\Gamma(1-x)\sin\pi x$$

$$= \Gamma(1+x)\Gamma(1-x)\frac{\sin\pi x}{x}$$

$$= \Gamma(1+x)\Gamma(1-x)\sum_{k=0}^{\infty}(-1)^k\frac{\pi^{2k+1}x^{2k}}{(2k+1)!},$$

and the series on the right hand side converges for all $x \in \mathbb{R}$. Moreover, as $x \to 0$ the right hand side tends to $\pi$ and is indeed an arbitrarily often differentiable function. Thus the function

$$\tilde{\varphi}(x) := \begin{cases} \varphi(x), & x \in \mathbb{R} \setminus \mathbb{Z} \\ \pi, & x \in \mathbb{Z} \end{cases}$$

has period 1 and is on $\mathbb{R}$ arbitrarily often differentiable, and further (31.40) holds for all $x \in \mathbb{R}$.

Now we claim that $\tilde{\varphi}$ is constant. Denote by $g$ the function

$$g(x) := \frac{d^2}{dx^2}\ln\tilde{\varphi}(x), \quad 0 \le x \le 1.$$

Clearly $g$ has period 1 and by (31.40) we find

$$\ln\left(\varphi\left(\frac{x}{2}\right)\varphi\left(\frac{x+1}{2}\right)\right) = \ln\left(\frac{\pi}{16}\varphi(x)\right)$$

or

$$\ln\varphi\left(\frac{x}{2}\right) + \ln\varphi\left(\frac{x+1}{2}\right) = \ln\frac{\pi}{16} + \ln\varphi(x),$$

which yields

$$\frac{1}{4}g\left(\frac{x}{2}\right) + \frac{1}{4}g\left(\frac{x+1}{2}\right) = g(x). \tag{31.41}$$

On $[0,1]$ the function $g$ is continuous, hence bounded, say $|g(x)| \le M$ on $[0,1]$, which implies by (31.41)

$$|g(x)| \le \frac{1}{4}\left|g\left(\frac{x}{2}\right)\right| + \frac{1}{4}\left|g\left(\frac{x+1}{2}\right)\right| \le \frac{M}{2} \tag{31.42}$$

and iterating (31.42) $N$-times we find

$$|g(x)| \le \frac{M}{2^N}, \tag{31.43}$$

which due to the periodicity of $g$ extends to all $x \in \mathbb{R}$, thus we must have $g(x) = 0$ for all $x \in \mathbb{R}$. Hence $\frac{d^2}{dx^2}\ln\tilde{\varphi}(x)$ must be a linear function and periodic, i.e. it must be a constant. Consequently $\tilde{\varphi}$ must be constant, but $\tilde{\varphi}(0) = \pi$. Thus by (31.37) we have proved

**Theorem 31.13.** *For $x \in \mathbb{R} \setminus \mathbb{Z}$*

$$\Gamma(x)\Gamma(1-x) = \frac{\pi}{\sin \pi x}. \tag{31.44}$$

Writing (31.44) as

$$\sin \pi x = \frac{1}{\pi}\frac{1}{\Gamma(x)\Gamma(1-x)}$$

and using again $\Gamma(1-x) = -x\Gamma(-x)$ we obtain

$$\sin \pi x = \frac{\pi}{-x\Gamma(x)\Gamma(-x)}.$$

If we note that the Weiersrtass product representation extends to $x \in \mathbb{R} \setminus \mathbb{Z}$ we find by Theorem 31.2 the following **product representation of the sine function:**

**Theorem 31.14.** *For $x \in \mathbb{R}$ the following holds*

$$\sin \pi x = \pi x \prod_{k=1}^{\infty}\left(1 - \frac{x^2}{k^2}\right) \tag{31.45}$$

**Remark 31.15.** From our derivation we can only conclude that (31.45) holds for $x \in \mathbb{R} \setminus \mathbb{Z}$. But for $x \in \mathbb{Z}$, one term in $\prod_{k=1}^{\infty} \left(1 - \frac{x^2}{k^2}\right)$ vanishes as does $\sin \pi x$, hence we can extend (31.45) to $\mathbb{R}$.

When turning to complex-valued functions of a complex variable and introducing meromorphic functions we will return to the $\Gamma$-function and related functions. In fact many of the formulae proved here will show their full power in the complex setting.

# Problems

1. Show that
$$\Gamma\left(n + \frac{1}{2}\right) = \frac{(2n)!\sqrt{\pi}}{4^n n!}, n \in \mathbb{N}.$$

2. Let $\alpha > -1$ and $f_\alpha : (0, \infty) \to \mathbb{R}$, $f_\alpha(t) = t^\alpha$. Prove that
$$F_\alpha(s) := \int_0^\infty t^\alpha e^{-st} dt = \frac{\Gamma(\alpha + 1)}{s^{\alpha+1}}.$$

3. Prove that
$$\Gamma(x) := \int_0^1 \left(\ln \frac{1}{t}\right)^{x-1} dt$$

and derive
$$\int_0^1 \left(\ln \frac{1}{t}\right)^{\frac{1}{2}} dt = \frac{\sqrt{\pi}}{2},$$

as well as
$$\int_0^1 \left(\ln \frac{1}{t}\right)^{-\frac{1}{2}} dt = \sqrt{\pi}.$$

4. Prove that $\Gamma'(1) = -\gamma$, where $\gamma$ is the Euler constant.

5. The function $\psi(x) := \frac{d}{dx} \ln \Gamma(x) = \frac{\Gamma'(x)}{\Gamma(x)}$ is often called the **digamma-function**. Prove:

   a) $\psi(x) - \psi(1) = -\sum_{k=0}^{\infty} \left(\frac{1}{x+k} - \frac{1}{k+1}\right)$;

   b) $\psi(x + n) = \frac{1}{x} + \cdots + \frac{1}{x+n-1} + \psi(x)$.

456

6. For the Beta-function derive the representation

$$B(x, y) = \int_0^\infty \frac{s^{x-1}}{(1+s)^{x+y}} ds.$$

Hint: use the substitution $t = \frac{s}{1+s}$ in the definition of $B(x, y)$.

7. Find

$$\int_0^\infty \frac{x^5}{(1+x)^7} dx.$$

8. Prove the following product representation of the Beta-function:

$$B(x, y) = \frac{x+y}{xy} \prod_{n=1}^\infty \frac{\left(1 + \frac{x+y}{n}\right)}{\left(1 + \frac{x}{n}\right)\left(1 + \frac{y}{n}\right)}.$$

# 32 Selected Topics on Functions of a Real Variable

We have discussed in much detail continuous functions, differentiable functions (of a certain order including arbitrarily often differentiable functions), integrable functions etc. In particular we could clarify some of their relations, for example that functions differentiable on an open set are continuous, continuous functions on a compact interval are integrable, etc. Maybe most striking was the fundamental theorem of calculus in the form that if $f : [a, b] \to \mathbb{R}$ is continuous then the function $F : [a, b] \to \mathbb{R}$ defined by

$$F(x) := \int_a^x f(t)dt$$

is differentiable and $F'(x) = f(x)$.

However we also have important function classes for which these results do not apply: a monotone function need not be continuous, but one-sided limits exist, see Problem 6 in Chapter 20, or a bounded monotone function on $[a, b]$ is Riemann integrable but we should not expect that

$$G(x) := \int_a^x g(t)dt$$

is differentiable as the example $g : [-1, 1] \to \mathbb{R}$, $g|_{[-1,0]} = 0$, $g|_{(0,1]} = 1$ with corresponding $G$ given by $G|_{[-1,0]} = 0$ and $G(x) = x$ for $x \in (0, 1]$ shows. Thus for handling monotone functions we require an extension of our theory. It turns out that a much better understanding of point sets in $\mathbb{R}$ is needed. In this chapter we want to give some first ideas of the topic "Theory of Real Variables". Only after we have introduced the Lebesgue measure and the Lebesgue integral can we deal with this topic in more detail.

Recall that a set $A$ is called countable if it is the bijective image of $\mathbb{N}$. If $A$ is finite or countable we call $A$ denumerable. In $\mathbb{R}$ we have finite and countable subsets, for example $\mathbb{N}, \mathbb{Z}$ or $\mathbb{Q}$, and non-countable subsets, for example $\mathbb{R}$ or $\mathbb{R} \setminus \mathbb{Q}$. Moreover, in $\mathbb{R}$ we have some topological notions: we have open and closed intervals, in fact open and closed sets, or compact sets. We now want to add a further notion of "smallness".

**Definition 32.1.** *A set $A \subset \mathbb{R}$ is called a **null set** if for every $\epsilon > 0$ there exists a denumerable number of bounded intervals $I_n$ with end points $a_n < b_n$*

459

*such that*

$$A \subset \bigcup_{n \in \mathbb{N}} I_n \quad and \quad \sum_{n=1}^{\infty} (b_n - a_n) \leq \epsilon. \tag{32.1}$$

**Remark 32.2.** There is no need to be more restrictive in the choice of $I_n$, i.e. we may allow open, closed or half-open intervals.

**Lemma 32.3. A.** *If $A' \subset A$ and $A \subset \mathbb{R}$ is a null set, then $A'$ is a null set too.*
**B.** *Every denumerable set $A \subset \mathbb{R}$ is a null set.*

*Proof.* **A.** This is trivial since $A' \subset \bigcup_{n \in \mathbb{N}} I_n$ provided $A \subset \bigcup_{n \in \mathbb{N}} I_n$.
**B.** Let $A = \{a_\nu | \nu \in \mathbb{N}\}$ be denumerable subset of $\mathbb{R}$. (If $A$ is finite with $m$ elements we set $a_{m+j} = a_1$ for $j \in \mathbb{N}$.) Given $\epsilon > 0$, choose $I_\nu = (-\epsilon 2^{-\nu-1} + a_\nu, a_\nu + \epsilon 2^{-\nu-1})$ which yields $b_\nu - a_\nu = \epsilon 2^{-\nu}$ and consequently $A \subset \bigcup_{\nu \in \mathbb{N}} I_\nu$ as well as

$$\sum_{\nu=1}^{\infty} (b_\nu - a_\nu) = \sum_{\nu=1}^{\infty} \epsilon 2^{-\nu} = \epsilon \sum_{\nu=1}^{\infty} 2^{-\nu} = \epsilon.$$

$\square$

Before proceeding further, we briefly consider the idea of how to measure "length". We have no problem in accepting that the length of the bounded interval $[a, b] \subset \mathbb{R}$ is given by $\lambda^{(1)}([a, b]) = b - a$. We may next ask how to determine the "length" or "size" of an arbitrary subset $A \subset \mathbb{R}$. For simplicity we assume that $A$ is bounded. Reasonable properties for a function measuring "length" would include for $A, A_j \subset \mathbb{R}$ the following:

$$\lambda^{(1)}(\emptyset) = 0, \tag{32.2}$$

i.e. the empty set has no length;

$$\lambda^{(1)}(A) \geq 0, \tag{32.3}$$

i.e. length is non-negative;

$$\lambda^{(1)}(A_1 \cup A_2) = \lambda^{(1)}(A_1) \cup \lambda^{(1)}(A_2) \quad for \quad A_1 \cap A_2 = \emptyset, \tag{32.4}$$

or more naturally and more generally

$$\lambda^{(1)} \left( \bigcup_{j=1}^{\infty} A_j \right) = \sum_{j=1}^{\infty} \lambda^{(1)}(A_j) \quad for \quad A_j \cap A_l = \emptyset \quad if \quad j \neq l. \tag{32.5}$$

Moreover, with $c + A = \{c + x | x \in A\}, c \in \mathbb{R}$,

$$\lambda^{(1)}(c + A) = \lambda^{(1)}(A), \tag{32.6}$$

i.e. length is invariant under translations.

Suppose we can define a mapping $\lambda^{(1)}$ with these properties. We want to calculate for $A := [0,1] \cap \mathbb{Q}$ and $B := [0,1] \setminus \mathbb{Q}$ the length $\lambda^{(1)}(A)$ and $\lambda^{(1)}(B)$. Since $1 = \lambda^{(1)}(A \cup B) = \lambda^{(1)}(A) + \lambda^{(1)}(B)$ we only need to find $\lambda^{(1)}(A)$. Since $\mathbb{Q}$ is countable we know that $A = [0,1] \cap \mathbb{Q}$ is countable. Let $\tau : \mathbb{N} \to A$ be a fixed bijective mapping (an enumeration of $A$) and put $A_j = \tau(\{j\})$. Clearly $A_j \cap A_k = \emptyset$ for $j \neq k$ and $A = \cup_{j \in \mathbb{N}} A_j$. Therefore we have

$$\lambda^{(1)}(A) = \sum_{j=1}^{\infty} \lambda^{(1)}(A_j). \tag{32.7}$$

Each set $A_j$ consists of a single point and hence by translation invariance we must have $\lambda^{(1)}(A_j) = \alpha$ for all $j \in \mathbb{N}$. If $\alpha \neq 0$ then $\lambda^{(1)}(A) = \infty$ which is a contradiction to $\lambda^{(1)}(A) \leq 1$. Thus $\alpha = 0$ and therefore $\lambda^{(1)}(A) = 0$ implying that $\lambda^{(1)}(B) = 1$. It follows that the infinite set $A$ must have "length" zero and if we take away this infinite set from $[0,1]$, the length remains unchanged. So far the results might be surprising but they are consistent. However it turns out that we cannot define on all bounded subsets of $\mathbb{R}$ a mapping $\lambda^{(1)}$ with the properties listed above. We will see later that we can construct $\lambda^{(1)}$, the **one-dimensional Lebesgue measure** on a large family of sets, the **Borel sets** $\mathcal{B}$ and with the normalisation $\lambda^{(1)}([0,1]) = 1$, $\lambda^{(1)}$ is even uniquely defined. All open and closed subsets of $\mathbb{R}$ belong to $\mathcal{B}$ as do all countable sets. (Unfortunately not every subset of a null set will belong to $\mathcal{B}$ which will cause a few problems later.) At the moment it is sufficient to accept that for all countable, all closed and all open sets of $\mathbb{R}$ we can define "length" which is finite for bounded sets (if defined) and zero for countable sets. Moreover, if $A \subset \mathbb{R}$ is a Borel set then $A^{\complement}$, its complement, is a Borel set too. If $I$ is a bounded interval with end points $a < b$ then $\lambda^{(1)}(I) = b - a$. If $A \subset [a,b]$ is a Borel set then $[a,b] \setminus A$ is a Borel set and $\lambda^{(1)}([a,b] \setminus A) = (b - a) - \lambda^{(1)}(A)$.

Now we want to discuss a compact set which is not denumerable but nonetheless has "length" zero. This is one of the interesting properties of the famous Cantor set. We start by setting

$$C_0 := [0,1]. \tag{32.8}$$

From $C_0$ we take away the open interval $\left(\frac{1}{3}, \frac{2}{3}\right)$ to obtain

$$C_1 := [0,1] \setminus \left(\frac{1}{3}, \frac{2}{3}\right) = \left[\frac{0}{3}, \frac{1}{3}\right] \cup \left[\frac{2}{3}, \frac{3}{3}\right]. \tag{32.9}$$

In the next step we take away from $\left[\frac{0}{3}, \frac{1}{3}\right]$ and $\left[\frac{2}{3}, \frac{3}{3}\right]$ the open "middle interval" of length $\frac{1}{9}$, i.e.

$$C_2 := \left(\left[\frac{0}{3}, \frac{1}{3}\right] \setminus \left(\frac{1}{9}, \frac{2}{9}\right)\right) \cup \left(\left[\frac{2}{3}, \frac{3}{3}\right] \setminus \left(\frac{7}{9}, \frac{8}{9}\right)\right) \tag{32.10}$$

$$= \left[\frac{0}{9}, \frac{1}{9}\right] \cup \left[\frac{2}{9}, \frac{3}{9}\right] \cup \left[\frac{6}{9}, \frac{7}{9}\right] \cup \left[\frac{8}{9}, \frac{9}{9}\right].$$

We continue this process. Clearly $C_N$ consists of $2^N$ disjoint closed intervals $C_{N,j}, j = 1, \ldots, 2^N$, each of length $\frac{1}{3^N}$. From $C_N$ we move to $C_{N+1}$ by taking away from each interval $C_{N,j}$ the open "middle interval" of length $\frac{1}{3^{N+1}}$ We define the **Cantor set** $C$ by

$$C := \bigcap_{N=0}^{\infty} C_N. \tag{32.11}$$

So what can we say about $C$? First, since each set $C_N$ is closed by Lemma 19.7 it follows that $C$ is closed too. Moreover, since $C \subset [0,1]$, the Cantor set is bounded, hence by the Heine-Borel theorem, Theorem 20.26, it is compact. Further, in the $N^{\text{th}}$ step we get from $C_N$ to $C_{N+1}$ by removing $2^N$ open intervals of length $\frac{1}{3^{N+1}}$. The total length of the removed intervals add up to

$$\sum_{N=0}^{\infty} \frac{2^N}{3^{N+1}} = \frac{1}{3} \sum_{N=0}^{\infty} \left(\frac{2}{3}\right)^N = \frac{1}{3} \frac{1}{1 - \frac{2}{3}} = 1. \tag{32.12}$$

This implies however that

$$\lambda^{(1)}(C) = 0. \tag{32.13}$$

Finally we observe that $C$ is not denumerable. For this we use first Theorem 18.33 which implies that every $x \in [0,1]$ has a ternary or 3-adic representation

$$x = \sum_{n=1}^{\infty} a_n 3^{-n}, \quad a_n \in \{0,1,2\}. \tag{32.14}$$

A different way to write $x$ in this representation is

$$x = 0.a_1a_2a_3\cdots, a_n \in \{0,1,2\}, \tag{32.15}$$

and of course we identify

$$x = 0.00\ldots01000 \text{ (1 is in position } k) \tag{32.16}$$

and

$$y = 0.00\ldots00222\ldots \text{ ( first 2 is in position } k+1). \tag{32.17}$$

Using this identification, in $C_1$ we only find elements with first digit in the ternary representation being either 0 or 2. In $C_2$ we only find elements belonging to $C_1$ and with the second digit being either 0 or 2, and in $C_N$ we only have elements from $C_{N-1}$ with $N^{\text{th}}$ digit either 0 or 2. Thus $x \in C$ implies

$$x = \sum_{n=1}^{\infty} a_n 3^{-n}, \ a_n \in \{0,2\}. \tag{32.18}$$

Conversely, every $x$ with a representation (32.18) must belong to $C$. Now we can use the proof of Theorem 18.35 to show that $C$ is not denumerable. We only have to restrict $A_{k-l}$ in that proof to 0 or 2. Eventually, we have now proved

**Theorem 32.4.** *The Cantor set is a compact, denumerable null set.*

This result tells us that sets being large when judged by their cardinality still can be small with respect to "length" or measure. Having these considerations in mind we return to monotone functions. In the following we consider monotone functions $f$ defined on a compact interval $[a,b]$ which are bounded. If $f$ is monotone decreasing then $-f$ is monotone increasing and hence when investigating the "smoothness" or "regularity" of a monotone function we can confine ourselves to increasing functions. Let $f : [a,b] \to \mathbb{R}$ be a bounded increasing function. Since $f$ is real-valued and increasing we have of course $f(a) \leq f(x) \leq f(b) < \infty$, i.e. $f$ is bounded, however sometimes we prefer to emphasise in this chapter the boundedness of $f$. From Problem 6 in Chapter 20 we know that for $x_0 \in (a,b)$

$$f(x_0+) := \lim_{\substack{x \to x_0 \\ x > x_0}} f(x) = \inf\{f(x) | x_0 < x \leq b\} \tag{32.19}$$

and

$$f(x_0-) := \lim_{\substack{x \to x_0 \\ x < x_0}} f(x) = \sup\{f(x) | a \le x < x_0\} \tag{32.20}$$

exist and the following must hold

$$f(x_0-) \le f(x_0) \le f(x_0+). \tag{32.21}$$

We call

$$[f](x_0) := f(x_0+) - f(x_0-) \ge 0 \tag{32.22}$$

the **jump** of $f$ at $x_0$. In part b) of Problem 6 in Chapter 20 we have proved that $f$ can only have finitely many jumps larger than a given $\eta > 0$. Indeed, since $f$ is bounded, there exists $n_0 \in \mathbb{N}$ such that

$$n_0 \eta \ge f(b) - f(a),$$

implying that an upper bound for the number of jumps of size larger than $\eta$ is the largest $n \in \mathbb{N}$ such that $n\eta \le f(b) - f(a)$. This implies also, again see Problem 6 in Chapter 20, that $f$ can have at most countable many jumps, i.e. outside a countable set $f$ is continuous.

**Lemma 32.5.** *Let* $f : [a, b] \to \mathbb{R}$ *be a bounded increasing function. For* $a = x_0 < x_1 < \cdots < x_n < x_{n+1} = b$ *we have*

$$(f(a+) - f(a)) + \sum_{k=1}^{n} [f](x_k) + (f(b) - f(b-)) \le f(b) - f(a). \tag{32.23}$$

*Proof.* Let $y_k \in (x_k, x_{k+1}), k = 0, \dots, n$. It follows that

$$\begin{aligned} f(x_k+) - f(x_k-) &\le f(y_k) - f(y_{k-1}), \\ f(a+) - f(a) &\le f(y_1) - f(a), \\ f(b) - f(b-) &\le f(b) - f(y_n), \end{aligned}$$

and adding these inequalities yields (32.23). $\qquad \square$

Suppose that $f$ has countable jumps occurring at $x_j$, $j \in \mathbb{N}$, $a < x_1 < \cdots < x_{j-1} < x_j, x_j < b$. For $N \in \mathbb{N}$ denote by $S_N := \left\{ x_1^N, \dots, x_{k(N)}^N \right\}$ the finite subset of $\{x_j | j \in \mathbb{N}\}$ corresponding to jumps of size larger than $\frac{1}{N}$. Clearly we have $S_N \subset S_{N+1}$ and $\cup_{N \in \mathbb{N}} S_N = \{x_j | j \in \mathbb{N}\}$. For $S_N$ inequality (32.23) holds and the sequence $\left(\sum_{j=1}^{k(N)}[f](x_j^N)\right)_{N\in\mathbb{N}}$ is increasing. Since this sequence is also bounded it converges and in the limit we obtain

$$(f(a+) - f(a)) + \sum_{k=1}^{\infty}[f](x_k) + (f(b) - f(b-)) \leq f(b) - f(a). \qquad (32.24)$$

**Definition 32.6.** *Let* $f : [a, b] \to \mathbb{R}$ *be a bounded increasing function. We define the corresponding* **jump function** $s_f : [a, b] \to \mathbb{R}$ *by*

$$s_f(x) := \begin{cases} 0, & x = a \\ (f(a+) - f(a)) + \sum_{y<x}[f](y) + (f(x) - f(x-)), & 0 < x \leq b. \end{cases} \qquad (32.25)$$

Note that since $f$ has at most countable many jumps, say $x_1 < x_2 < \cdots < x_j < \cdots$ the sum in (32.25) stands for

$$\sum_{x_j<x}[f](x_j).$$

Of interest is now

**Theorem 32.7.** *Let* $f : [a, b] \to \mathbb{R}$ *be a bounded increasing function and* $s_j$ *its jump function. The function* $\varphi_f : [a, b] \to \mathbb{R}$ *defined by*

$$\varphi_f := f(x) - s_f(x) \qquad (32.26)$$

*is increasing and continuous.*

*Proof.* Let $a \leq x < y \leq b$. We apply (32.24) to the interval $[x, y]$ and obtain

$$s_f(y) - s_f(x) = f(y) - f(x), \qquad (32.27)$$

which implies $\varphi_f(x) - \varphi_f(y) \geq 0$, i.e. $\varphi_f$ is increasing. Further, passing in (32.27) to the limit $y \to x$ we find

$$s_f(x+) - s_f(x) \leq f(x+) - f(x),$$

but the definition of $\varphi_f$ implies

$$f(x+) - f(x) \leq s_f(y) - s_f(x),$$

which gives for $y \to x$

$$f(x+) - f(x) \leq s_f(x+) - s_f(x),$$

or $f(x+) - f(x) = s_f(x+) - s_f(x)$, i.e. $\varphi_f(x+) = \varphi_f(x)$. Analogously we may prove $\varphi_f(x-) = \varphi(x)$. $\qquad\square$

The jump function $s_f$ is the pointwise and monotone limit of the sequence $(S_N)_{N \in \mathbb{N}}$, $S_N(x) = \sum_{x_j^N < x} [f](x_j)$, which is an increasing step function on $[a, b]$. Thus every monotone increasing function is the sum of continuous increasing functions and a monotone limit of step functions.

Let $f : [a, b] \to \mathbb{R}$ be a bounded function and $a = x_0 < x_1 < \cdots < x_{n-1} < x_n = b$ be a finite partition $Z$ of $[a, b]$ for which we write as before $Z(x_0, \ldots, x_n)$. We can now form

$$V_Z(f) := \sum_{k=0}^{n-1} |f(x_{k+1}) - f(x_k)|. \tag{32.28}$$

**Definition 32.8.** *Let $f :: [a, b] \to \mathbb{R}$ be a function.*
**A.** *By*

$$V(f) := \sup_Z V_Z(f) \tag{32.29}$$

*we denote the **total variation** of $f$, where the supremum is taken over all (finite) partitions of $[a, b]$.*
**B.** *We call $f$ a function of **bounded variation** if $V(f) < \infty$. The set of all functions of bounded variation on $[a, b]$ is denoted by $BV([a, b])$.*

**Remark 32.9. A.** Sometimes it is helpful to emphasise the interval $[a, b]$, and then we write

$$V_a^b(f) := V(f), \quad f : [a, b] \to \mathbb{R}.$$

**B.** If $f \in BV([a, b])$ then $f|_{[c,d]} \in BV([c, d])$, $a \le c < d \le b$.
**C.** Some authors prefer to speak of functions of finite variation, but the symbol $BV$ is now widely used and therefore we prefer to call them functions of bounded variation.

**Lemma 32.10.** *A function of bounded variation is bounded.*

*Proof.* Let $x \in [a, b]$. Then $a \le x \le b$ is a partition of $[a, b]$ and therefore

$$|f(x) - f(a)| + |f(b) - f(x)| \le V(f)$$

which implies

$$|f(x)| \le |f(a)| + V(f).$$

$\square$

**Proposition 32.11.** *If $f : [a, b] \to \mathbb{R}$ is monotone then $f$ belongs to $BV([a, b])$.*

*Proof.* Since $f \in BV([a, b])$ if and only if $-f \in BV([a, b])$ we may assume that $f$ is increasing. For $f$ increasing we have $f(x_{k+1}) - f(x_k) \geq 0$ for any two points $x_k < x_{k+1}$. Hence for every partition $Z$ of $[a, b]$ we find

$$0 \leq V_Z(f) = \sum_{k=0}^{n-1} (f(x_{k+1}) - f(x_k)) = f(b) - f(a),$$

implying that $\sup_Z V_Z(f)$ is finite. $\qquad\square$

**Proposition 32.12.** *A Lipschitz continuous function $f : [a, b] \to \mathbb{R}$ belongs to $BV([a, b])$.*

*Proof.* For some $\kappa \geq 0$ we have for all $x, y \in [a, b]$

$$|f(x) - f(y)| \leq \kappa|x - y|, \tag{32.30}$$

thus for a partition $Z(x_0, \ldots, x_n)$ of $[a, b]$ we find

$$V_Z(f) = \sum_{k=0}^{n-1} |f(x_{k+1} - f(x_k)| \leq \kappa \sum_{k=0}^{n-1} (x_{k+1} - x_k) = \kappa(b - a)$$

which yields

$$V(f) \leq \kappa(b - a).$$

$\qquad\square$

**Example 32.13.** The continuous function $f : [0, 1] \to \mathbb{R}$ defined by

$$f(x) = \begin{cases} 0, & x = 0 \\ x \sin \frac{1}{x}, & x \in (0, 1] \end{cases}$$

is not of bounded variation. To see this, consider the partition $x_0 = 0, x_j = \frac{2}{(2n-2j+1)\pi}, x_n = 1, 1 \leq j \leq n - 1$. With this partition we find for $k = 1, \ldots, n - 2$ that

$$|f(x_{k+1}) - f(x_k)| \geq 2x_k$$

and further we note that

$$\lim_{n \to \infty} \frac{4}{\pi} \sum_{j=1}^{n-2} \frac{1}{2n - 2j + 1} = \frac{4}{\pi} \lim_{n \to \infty} \sum_{k=1}^{n-2} \frac{1}{2k + 1} = \infty,$$

hence $f$ is not of bounded variation.

**Theorem 32.14.** *The set $BV([a,b])$ with the natural pointwise operations forms an algebra. In particular for $f, g \in BV([a,b])$ and $\lambda \in \mathbb{R}$ we have $f + g, \lambda f, f \cdot g \in BV([a,b])$.*

*Proof.* Clearly we need only to prove that $f + g$ and $f \cdot g$ belong to $BV([a,b])$ if $f, g \in BV([a,b])$. For this let $Z(x_0, \ldots, x_n)$ be a partition of $[a,b]$. Since

$$\sum_{k=1}^{n} |f(x_k) + g(x_k) - f(x_{k-1}) - g(x_{k-1})|$$
$$\leq \sum_{k=1}^{n} |f(x_k) - f(x_{k-1})| + \sum_{k=1}^{n} |g(x_k) - g(x_{k-1})|$$

we conclude first that

$$V_Z(f + g) \leq V_Z(f) + V_z(g),$$

and then by taking the supremum over all partitions of $[a,b]$ we get

$$V(f + g) \leq V(f) + V(g).$$

Furthermore we have

$$|f(x_k)g(x_k) - f(x_{k-1}g(x_{k-1})|$$
$$\leq |f(x_k)g(x_k) - f(x_{k-1})g(x_k)| + |f(x_{k-1}g(x_k) - f(x_{k-1})g(x_{k-1})|$$
$$\leq ||g||_\infty |f(x_k) - f(x_{k-1})| + ||f||_\infty |g(x_k) - g(x_{k-1})|,$$

implying

$$V(f \cdot g) \leq ||g||_\infty V(f) + ||f||_\infty V(g).$$

$\square$

The next result gives a surprising characterisation of functions of bounded variation.

**Theorem 32.15.** *For $f \in BV([a,b])$ there exists two monotone increasing functions $g, h : [a,b] \to \mathbb{R}$ such that $f = g - h$.*

*Proof.* Given $f \in BV([a,b])$ we define $v_f : [a,b] \to \mathbb{R}$ by $v_f(x) = V_a^x(f)$. Clearly we have $v_f(a) = 0, v_f(b) = V(f)$ and $v_f(x) \leq v_f(y)$ for $x < y$, i.e. $v_f$ is increasing. We define $g := v_f$ and $h := v_f - f$ and find immediately that $f = g - h$. It remains to prove that $h$ is increasing, i.e. that $f - v_f$ is decreasing. For $a \le x < y \le b$ we have

$$f(y) - f(x) \le V_x^y(f) = v_f(y) - v_f(x),$$

implying

$$-h(y) = f(y) - v_f(y) \le f(x) - v_f(x) = -h(x),$$

i.e. $-h$ is decreasing as claimed.   □

**Corollary 32.16.** *The following holds*

$$BV([a,b]) = \{g - h | g, h : [a,b] \to \mathbb{R} \text{ are increasing }\}. \tag{32.31}$$

An immediate consequence of Corollary 32.16 is that $f \in BV([a,b])$ has at most countably many jump discontinuities and further that the limits

$$f(x_0+) = \lim_{\substack{x \to x_0 \\ x > x_0}} f(x), \quad f(x_0-) = \lim_{\substack{x \to x_0 \\ x < x_0}} f(x) \tag{32.32}$$

exist for every $x_0 \in (a,b)$, so do the limits $f(a+)$ and $f(b-)$.

We have now the following situation: $BV([a,b])$ is a vector space, in fact an algebra, and every element in $BV([a,b])$ is Riemann integrable. However certain results that we have considered for continuous, integrable functions do not hold, for example the fundamental theorem, or rules such as integration by parts, since for this result we need differentiability. A natural question is to which extent can we "rescue" these results, i.e. can we find an extension of our theory of integration which will allow us to prove these results perhaps with some generalised interpretation? It turns out that we can achieve this however we will need the Lebesgue measure and we will take up this problem in Volume 3.

We know that $BV([a,b]) \cup C([a,b])$ is a subset of all Riemann integrable functions. The following result gives a characterisation of a Riemann integrable function a proof of which we will give in Volume 3.

**Theorem 32.17.** *A bounded function $f : [a,b] \to \mathbb{R}$ is Riemann integrable if and only if the set*

$$D_s(f) := \{x \in [a,b] | f \text{ is not continuous at } x\}$$

*is a null set.*

# Problems

1. Define $f : [0,1] \to \mathbb{R}$ by $f(0) = 0$ and $f(x) = x \cos \frac{\pi}{x}$ for $0 < x \leq 1$ and prove that $f$ is continuous but not of bounded variation. Hint: consider the partition $0 < \frac{1}{2k} < \frac{1}{2k-1} < \cdots < \frac{1}{3} < \frac{1}{2} < 1$.

2. Show that if $f, g \in BV([a,b])$ then $g^+, g^-, |g|, \max(f,g)$ and $\min(f,g)$ all belong to $B([a,b])$ too. Hint: prove first that $|g| \in BV([a,b])$ and use the fact that $BV([a,b])$ is a vector space. Recall the representation of max and min using $|\cdot|$.

3. Suppose that $g \in BV([a,b])$ and $\inf |g| > 0$. Prove that $\frac{1}{g} \in BV([a,b])$.

4. Let $f \in C([a,b])$ and $F(x) := \int_a^x f(t)dt, x \in [a,b]$. Show that $F \in BV([a,b])$ and $V(F) = \int_a^b |f(t)|dt$.

5. We call $f : [a,b] \to \mathbb{R}$, $a < b$, **absolutely continuous** if for every $\epsilon > 0$ there exists $\delta > 0$ such that for all $m \in \mathbb{N}$ and any choice of pairwise disjoint open intervals $(a_j, b_j) \subset [a,b]$, $j = 1, \ldots, m$, the estimate $\sum_{j=1}^m (b_j - a_j) < \delta$ implies $\sum_{j=1}^m |f(b_j) - f(a_j)| < \epsilon$.

   a) Prove that every absolutely continuous function is continuous.

   b) Prove that every Lipschitz continuous function is absolutely continuous.

   c) Prove that an absolutely continuous function is of bounded variation.

6. Show that the absolutely continuous functions on $[a,b]$ form an algebra.

7. Let $f \in C([a,b]) \cup BV([a,b])$ and prove that $F : [a,b] \to \mathbb{R}$, $F(x) := \int_a^x f(t)dt$, is absolutely continuous.

# Appendices

The material collected in the following appendices is additional, it is either material which students should have learnt by now and therefore a reminder or it is material which will be taught in more detail elsewhere. In the latter case the material introduced will be brief; omitting proofs and examples. In some of the appendices however we handle in more detail additional aspects of material treated within the main text, or we provide proofs of results that are only cited in the main text.

# Appendix I: Elementary Aspects of Mathematical Logic

Before we consider elementary concepts of logic, let us make some general remarks. A fair summary of our knowledge of the foundations of knowledge could be: at the beginning there was no beginning. This perhaps bizarre statement reflects the fact that there is no point zero to start with, maybe the central insight of philosophy. This has of course impacted on how we think about mathematics. However we must start somewhere. By experience and taking into account the historical development of the subject the best way to start is by taking certain facts for granted and then to investigate the consequences and the nature or essence of these facts. However this may result in severe changes of what we initially took for granted.

A major problem is that we have to use our everyday language to formulate these facts and the objectives we are interested in, but our everyday language however is not precise.

The beginner in mathematics normally encounters these problems when learning how to use mathematical logic and when learning naïve set theory. A suggestion to help in learning these topics is: make a start and then return to these problems occasionally.

The first basic idea we need is that of a (mathematical) statement. We define it as follows:

A **statement** is a sentence which is either true or false.

Mathematics is concerned with deriving new correct (mathematical) statements from given ones. We usually denote a statement by $p, q$ or $r$. Every statement creates a new statement, its **negation** $\neg p$ (read: not $p$):

$\neg p$ is true if $p$ is false and it is false if $p$ is true.

We may present this definition in an easy way by using a **truth table**

| $p$ | $\neg p$ |
|---|---|
| T | F |
| F | T |

Table A.I.1

473

Given several statements we may try to combine them to get compound statements. There are three basic ways to combine two statements $p$ and $q$:

**Conjunction**: $p \wedge q$  (read: $p$ and $q$);

**Disjunction**: $p \vee q$  (read: $p$ or $q$);

**Implication**: $p \implies q$  (read: $p$ implies $q$).

Our main task is to decide, i.e. to define, when these new statements are true and when they are false. Here are the truth tables for conjunction, disjunction and implication. Conjunction:

| $p$ | $q$ | $p \wedge q$ |
|---|---|---|
| T | T | T |
| T | F | F |
| F | T | F |
| F | F | F |

Table A.I.2

Thus the conjunction $p \wedge q$ is true if and only if both $p$ and $q$ are true. Disjunction:

| $p$ | $q$ | $p \vee q$ |
|---|---|---|
| T | T | T |
| T | F | T |
| F | T | T |
| F | F | F |

Table A.I.3

Hence the disjunction $p \vee q$ is true when at least one of $p$ and $q$ is true.

474

Implication:

| $p$ | $q$ | $p \implies q$ |
|---|---|---|
| T | T | T |
| T | F | F |
| F | T | T |
| F | F | T |

Table A.I.4

In the case of implication we have a surprising result: we would expect an implication to be true when both $p$ and $q$ are true, i.e. when the **premise** $p$ and the **conclusion** $q$ are true. However, we can also define that whatever the conclusion is, true or false, it may be derived from a false premise, i.e. $p$ being false also leads to a true statement $p \implies q$ independent of $q$. There are two further formulations related to the implication: we call $p$ a **sufficient condition** for $q$ (read: $p$ is sufficient for $q$), and we call $q$ a **necessary condition** for $p$ (read: $q$ is necessary for $p$). We next want to introduce a further compound statement, but one might have different views on its place in the system. We are speaking about the **equivalence** of two statements $p$ and $q$ for which we write

$$p \iff q \quad \text{(read: } p \text{ is equivalent to } q\text{)},$$

and which we define by

| $p$ | $q$ | $p \iff q$ |
|---|---|---|
| T | T | T |
| T | F | F |
| F | T | F |
| F | F | T |

Table A.I.5

Thus $p$ is equivalent to $q$ if both are true or both are false, but this is not really what we mean when saying that $p$ is equivalent to $q$. What we really mean is the following

$$(p \implies q) \wedge (q \implies p), \tag{A.I.1}$$

475

i.e. $p$ implies $q$ and $q$ implies $p$. Taking (A.I.1) as the definition for equivalence, then we may introduce a new notation, namely

$$p \Longleftrightarrow q \text{ if and only if } (p \implies q) \wedge (q \implies p). \qquad \text{(A.I.2)}$$

Note that the truth tables Tables A.I.1 and Tables A.I.3 imply

$$(p \wedge q) \Longleftrightarrow (q \wedge p) \qquad \text{(A.I.3)}$$

and

$$(p \vee q) \Longleftrightarrow (q \vee p). \qquad \text{(A.I.4)}$$

We want to study some of these compound statements in more detail. We start with the negation of negation, i.e. $\neg(\neg p)$ with truth table

| $p$ | $\neg p$ | $\neg(\neg p)$ |
|-----|----------|----------------|
| T   | F        | T              |
| F   | T        | F              |

Table A.I.6

Thus $\neg(\neg p)$ is equivalent to $p$, i.e. $\neg(\neg p)$ is true if $p$ is true and false if $p$ is false. Next we consider $p \wedge (\neg p)$:

| $p$ | $\neg p$ | $p \wedge (\neg p)$ |
|-----|----------|---------------------|
| T   | F        | F                   |
| F   | T        | F                   |

Table A.I.7

This statement is always false. Conversely, when looking at $p \vee (\neg p)$ we find

| $p$ | $\neg p$ | $p \vee (\neg p)$ |
|-----|----------|-------------------|
| T   | F        | T                 |
| F   | T        | T                 |

Table A.I.8

476

Thus the statement $p \vee (\neg p)$ is always true, i.e. given any statement, either $p$ or $\neg p$ is true, there is no other possibility. This fact is called the **law of the excluded middle** or *tertium non datur*. But note: the law of the excluded middle depends on the fact that any statement is only allowed to be true or false. As soon as we allow a third option we cannot prove the law of the excluded middle. When we take the negation of $p \vee (\neg p)$ we get the statement $\neg(p \vee (\neg p))$ which is always false. Therefore having Table A.I.7 in mind we find

$$(\neg(p \vee (\neg p))) \iff (p \wedge (\neg p)). \qquad (A.I.5)$$

We clearly apply the fact that two compound statements are equivalent if they have identical truth tables.

Of interest are the two laws dealing with negation of conjunction and disjunctions. They are called **de Morgan's laws** and they state

$$(\neg(p \wedge q)) \iff ((\neg p) \vee (\neg q)) \qquad (A.I.6)$$

and

$$(\neg(p \vee q)) \iff ((\neg p) \wedge (\neg q)). \qquad (A.I.7)$$

Note that (A.I.5) follows from (A.I.7) with $\neg p$ instead of $q$. A further important conclusion we can make from the negation of the implication is

$$(\neg(p \implies q)) \iff (p \wedge (\neg q)). \qquad (A.I.8)$$

Thus instead of proving that $p$ does not imply $q$, we may prove that $p$ and $\neg q$ are true. Since (A.I.8) implies

$$(p \implies q) \iff ((\neg p) \vee q) \qquad (A.I.9)$$

instead of proving that $p$ implies $q$ we may prove that either $\neg p$ or $q$ is true. In addition we have

$$(p \implies q) \iff (\neg q \implies (\neg p)), \qquad (A.I.10)$$

i.e. instead of proving $p$ implies $q$ we may prove that $\neg q$ implies $\neg p$. The equivalence (A.I.10) is known as **contra-position** and a proof using $\neg q \implies \neg p$ instead of $p \implies q$ is called a proof by contra-position. Combining (A.I.8) with the law of the excluded middle we obtain a very powerful method for proving statements: *reductio ad absurdum* or **proof by contradiction**. Here is the method:

Suppose we want to prove that $p \implies q$. Instead we assume that $\neg q$ is true and $p$ is true. Now we try to construct a contradiction to the statement $(\neg q) \wedge p$, i.e. we prove that $(\neg q) \wedge p$ is false. This implies by (A.I.8) that $\neg(p \implies q)$ is false too. Hence by the law of the excluded middle $p \implies q$ is true.

The implication has a further very useful property

$$((p \implies q) \wedge (q \implies r)) \implies (p \implies r), \qquad \text{(A.I.11)}$$

i.e. if $p$ implies $q$ and $q$ implies $r$, then $p$ must imply $r$. In fact, most if not all proofs rely on a finite number of applications of (A.I.11).

The following considerations are more involved and often cause some problems to begin with. We must learn to work with statements which include **quantifiers**. To explain this in more detail we need to consider some set theory.

Let $X$ be a non-empty set. Often we need to consider for each $x \in X$ a statement $p$ which depends on $x$. For this we write $p(x)$, for example if $X = \mathbb{N}$ the statement could be:

$$p(n): \quad n \text{ is a prime number.}$$

(Note that we do not interpret $p(x)$ as the value of a mapping at $x \in X$. The co-domain of such a function must be (a subset of) the set of all statements and this is a construction we wish to avoid.)

Now given a set $X \neq \emptyset$ and a family of statements $p(x), x \in X$. An **all-statement** is a statement of the type

$$\text{for all } x \in X \text{ the statement } p(x) \text{ is true.} \qquad \text{(A.I.12)}$$

For example we may consider

for all $n \in \mathbb{N}$ it is true that $n \geq 0$,

here $X = \mathbb{N}$ and $p(n)$ is the statement that $n \geq 0$.
An **existence-statement** is a statement of the type

$$\text{there exists } x \in X \text{ such that } p(x) \text{ is true.} \qquad \text{(A.I.13)}$$

For example we may consider

there exists $z \in \mathbb{Z}$ such that $z \leq 0$,

where now $X = \mathbb{Z}$ and $p(z)$ is the statement $z \leq 0$. For all-statements and existence statements a new notation is introduced. For (A.I.12) we write

$$\forall x \in X : p(x), \qquad (\text{A.I.14})$$

and for (A.I.13) we write

$$\exists x \in X : p(x) \qquad (\text{A.I.15})$$

The symbol "$\forall$" is called the **all-quantifier** and the symbol "$\exists$" is called the **existence-quantifier**. Next we may form compound statements involving quantifiers, for example

$$\forall x \in \mathbb{R} : (\exists n \in \mathbb{N} : n \geq x), \qquad (\text{A.I.16})$$

for which we may also write

$$\forall x \in \mathbb{R} \ \exists n \in \mathbb{N} : n \geq x.$$

Another example is

$$\exists M \in \mathbb{R} : (\forall x \in \mathbb{R} : |\sin x| \leq M), \qquad (\text{A.I.17})$$

for which we often write

$$\exists M \in \mathbb{R} \ \forall x \in \mathbb{R} : |\sin x| \leq M.$$

The rules for negation of statements involving quantifiers are

$$\neg(\forall x \in X : p(x)) \iff (\exists x \in X : \neg p(x)) \qquad (\text{A.I.18})$$

and

$$\neg(\exists x \in X : p(x)) \iff (\forall x \in X : \neg p(x)). \qquad (\text{A.I.19})$$

Thus the negation of (A.I.17) is

$$\neg(\exists M \in \mathbb{R} \ (\forall x \in \mathbb{R} : |\sin x| \leq M))$$

$$\iff \forall M \in \mathbb{R} : \neg(\forall x \in \mathbb{R} : |\sin x| \leq M)$$

$$\iff \forall M \in \mathbb{R} \ (\exists x \in \mathbb{R} : |\sin x| > M),$$

and since (A.I.17) is true, just take $M = 1$, the last statement is of course false.

**Note**: symbols such as $\neg, \wedge, \vee, \implies, \iff, \forall, \exists$ have their meaning in a formal language or in a formal mathematical context. They are not abbreviations. In our course, wherever possible, we try to avoid using these symbols. Clearly, we do not and cannot avoid the ideas of negations, conjunctions, disjunctions, implications, equivalences, all-statements or existence-statements. We believe however that to begin with it is better to use the longhand approach, thus for (A.I.17) we write:

there exists $m \in \mathbb{R}$ such that for all $x \in \mathbb{R}$ it follows that $|\sin x| \leq M$,

whereas the negation of this statement reads as

for all $M \in \mathbb{R}$ there exists $x \in \mathbb{R}$ such that $|\sin x| > M$.

# Appendix II: Sets and Mappings. A Collection of Formulae

In this appendix we give a collection of formulae on set operations and properties of mappings which every mathematics student should eventually know and be able to work with. (In compiling this list we followed closely J. Dieudonné [2].) Many of these formulae have already been used and some of them have been proved in Part 1, partly in the solved exercises. At the end of this appendix we will pick up some of the principal ideas of the proofs of these statements.

**Elementary Operations for Sets**

$$X \setminus X = \emptyset \text{ and } X \setminus \emptyset = X; \tag{A.II.1}$$

$$X \cup X = X \text{ and } X \cap X = X; \tag{A.II.2}$$

$$X \cup Y = Y \cup X \text{ and } X \cap Y = Y \cap X; \tag{A.II.3}$$

The statements $X \subset Y, X \cup Y = Y, X \cap Y = X$ are equivalent; (A.II.4)

The statements $X \subset X \cup Y$ and $X \cap Y \subset X$ are equivalent; (A.II.5)

$$X \subset Z \text{ and } Y \subset Z \text{ if and only if } X \cup Y \subset Z; \tag{A.II.6}$$

$$Z \subset X \text{ and } Z \subset Y \text{ if and only if } Z \subset X \cap Y; \tag{A.II.7}$$

$$X \cup (Y \cup Z) = (X \cup Y) \cup Z, \text{ i.e. } X \cup Y \cup Z \text{ makes sense}; \tag{A.II.8}$$

$$X \cap (Y \cap Z) = (X \cap Y) \cap Z, \text{ i.e. } X \cap Y \cap Z \text{ makes sense}; \tag{A.II.9}$$

$$X \cup (Y \cap Z) = (X \cup Y) \cap (X \cap Z) \text{ and } X \cap (Y \cup Z) = (X \cap Y) \cup (X \cap Z); \tag{A.II.10}$$

if $X \subset E$ and $Y \subset E$, then

$$(X^{\mathsf{C}})^{\mathsf{C}} = X, (X \cup Y)^{\mathsf{C}} = X^{\mathsf{C}} \cap Y^{\mathsf{C}}, (X \cap Y)^{\mathsf{C}} = X^{\mathsf{C}} \cup Y^{\mathsf{C}}; \tag{A.II.11}$$

$$X \subset Y \subset E \text{ is equivalent to } Y^{\mathsf{C}} \subset X^{\mathsf{C}}; \tag{A.II.12}$$

if $X \subset E$ and $Y \subset E$ then $X \cap Y = \emptyset$ if and only if $X \subset Y^{\mathsf{C}}$; (A.II.13)

if $X \subset E$ and $Y \subset E$ then $X \cup Y = E$ if and only if $X^{\mathsf{C}} \subset Y$, and $Y^{\mathsf{C}} \subset X$; (A.II.14)

$$X \times Y = \emptyset \text{ if and only if } X = \emptyset \text{ or } Y = \emptyset; \tag{A.II.15}$$

if $X \times Y \neq \emptyset$ then $X' \times Y' \subset X \times Y$ if and only if $X' \subset X$ and $Y' \subset Y$;
$$\text{(A.II.16)}$$
$$(X \times Y) \cup (X' \times Y) = (X \cup X') \times Y; \qquad \text{(A.II.17)}$$
$$(X \times Y) \cap (X' \times Y') = (X \cap X') \times (Y \cap Y'); \qquad \text{(A.II.18)}$$
$$(X \times Y) \times Z := X \times Y \times Z. \qquad \text{(A.II.19)}$$

**Mappings**

For $Z := X \times Y$ we define

$$pr_1 : \quad \begin{array}{c} Z \to X \\ (x,y) \mapsto x \end{array} \quad \text{and} \quad pr_2 : \quad \begin{array}{c} Z \to Y \\ (x,y) \mapsto y \end{array}$$

For a mapping $F : X \to Y$ we denote by

$$F(A) = \{y \in Y \,|\, y = F(x) \text{ and } x \in A \subset X\} \subset Y$$

the **image** of $A \subset X$, and by

$$F^{-1}(A') = \{x \in X \,|\, y = F(x) \text{ and } y \in A' \subset Y\} \subset X$$

the **pre-image** of $A' \subset Y$. Further we write

$$\Gamma(F) = \{(x, F(x)) \,|\, x \in X\}$$

for the **graph** of $F$. We will write $F^{-1}(y)$ for $F^{-1}(\{y\})$.

$$F(A) = pr_2(\Gamma(F) \cap (A \times Y)); \qquad \text{(A.II.20)}$$

$$A \neq \emptyset \text{ if and only if } F(A) \neq \emptyset; \qquad \text{(A.II.21)}$$

$$F(\{x\}) = \{F(x)\} \text{ for all } x \in X; \qquad \text{(A.II.22)}$$

$$A \subset B \text{ implies } F(A) \subset F(B); \qquad \text{(A.II.23)}$$

$$F(A \cap B) \subset F(A) \cap F(B); \qquad \text{(A.II.24)}$$

$$F(A \cup B) = F(A) \cup F(B); \qquad \text{(A.II.25)}$$

$$F^{-1}(A') = pr_1(\Gamma(F) \cap (X \times A')); \qquad \text{(A.II.26)}$$

$$F^{-1}(A') = F^{-1}(A' \cap F(X)); \qquad \text{(A.II.27)}$$

$$F^{-1}(\emptyset) = \emptyset \qquad \text{(A.II.28)}$$

(but note: $F^{-1}(A') = \emptyset$ does not imply $A' = \emptyset$);

$$A' \subset B' \text{ implies } F^{-1}(A') \subset F^{-1}(B'); \qquad (\text{A.II.29})$$

$$F^{-1}(A' \cap B') = F^{-1}(A') \cap F^{-1}(B'); \qquad (\text{A.II.30})$$

$$F^{-1}(A' \cup B') = F^{-1}(A') \cup F^{-1}(B'); \qquad (\text{A.II.31})$$

$$F^{-1}(A' \setminus B') = F^{-1}(A') \setminus F^{-1}(B') \text{ if } B' \subset A'; \qquad (\text{A.II.32})$$

$$F(F^{-1}(A')) = A' \cap F(X) \text{ for } A' \subset Y; \qquad (\text{A.II.33})$$

$$A \subset F^{-1}(F(A)) \text{ for } A \subset X; \qquad (\text{A.II.34})$$

$$pr_1^{-1}(A) = A \times Y \text{ for } A \subset X; \qquad (\text{A.II.35})$$

$$pr_2^{-1}(A') = X \times A' \text{ for } A' \subset Y; \qquad (\text{A.II.36})$$

$$C \subset pr_1(C) \times pr_2(C) \text{ for } C \subset X \times Y. \qquad (\text{A.II.37})$$

If $F : X \to Y$ and $G : Y \to Z$ we define the **composition** $H := G \circ F$ by

$$H : X \to Z$$
$$x \mapsto H(x) = G(F(x)).$$

$$H(A) = G(F(A)) \text{ for } A \subset X; \qquad (\text{A.II.38})$$

$$H^{-1}(A'') = F^{-1}(G^{-1}(A'')) \text{ for } A'' \subset Z; \qquad (\text{A.II.39})$$

if $F$ and $G$ are injective (surjective, bijective) then

$$H = G \circ F \text{ is injective (surjective, bijective)}; \qquad (\text{A.II.40})$$

if $F : X \to Y$ is bijective we denote its inverse mapping by $F^{-1} : Y \to X$
$$\qquad (\text{A.II.41})$$

(this does not cause any trouble with the notation for the pre-image because in this case the pre-image of one point is either a set containing exactly one point or it is empty.) For a bijective mapping we have

$$F \circ F^{-1} = id_Y$$
$$F^{-1} \circ F = id_X$$

where $id_Y$ is the identity on $Y$ and $id_X$ is the identity on $X$, respectively.

**Families of Sets**

In the following $I$ and $J$ are arbitrary index sets and $(A_i)_{i \in I}$ and $(B_j)_{j \in J}$ are families of sets. We define the union and the intersection of such families by:

$$\bigcup_{i \in I} A_i := \{x | x \in A_i \text{ for some } i \in I\};$$

$$\bigcap_{i \in I} A_i := \{x | x \in A_i \text{ for all } i \in I\}.$$

Clearly if $I = \{1, 2\}$ then

$$\bigcup_{i \in I} A_i = A_1 \cup A_2 \quad \text{and} \quad \bigcap_{i \in I} A_i = A_1 \cap A_2$$

with the obvious generalisation to a finite index set $I$.

$$\left(\bigcup_{i \in I} A_i\right)^{\complement} = \bigcap_{i \in I} A_i^{\complement}; \quad \left(\bigcap_{i \in I} A_i\right)^{\complement} = \bigcup_{i \in I} A_i^{\complement}; \qquad (A.II.42)$$

$$\left(\bigcup_{i \in I} A_i\right) \cap \left(\bigcup_{j \in J} B_j\right) = \bigcup_{(i,j) \in I \times J} (A_i \cap B_j); \qquad (A.II.43)$$

$$\left(\bigcap_{i \in I} A_i\right) \cup \left(\bigcap_{j \in J} B_j\right) = \bigcap_{(i,j) \in I \times J} (A_j \cup B_j); \qquad (A.II.44)$$

Let $F : X \to Y$ be a mapping and $(A_i)_{i \in I}$ a collection of subsets of $X$ and $(A_j')_{j \in J}$ a collection of subsets of Y.

$$F\left(\bigcup_{i \in I} A_i\right) = \bigcup_{i \in I} F(A_i); \qquad (A.II.45)$$

$$F^{-1}\left(\bigcup_{j \in J} A_j'\right) = \bigcup_{j \in J} F^{-1}(A_j'); \qquad (A.II.46)$$

$$F^{-1}\left(\bigcap_{j \in J} A_j'\right) = \bigcap_{j \in J} F^{-1}(A_j'). \qquad (A.II.47)$$

484

If $B \subset X$ is a subset and $(A_i)_{i \in I}$ is a collection of subsets of $X$, i.e. $A_i \subset X$, then we call $(A_i)_{i \in I}$ a **covering** of $B$ if $B \subset \bigcup_{i \in I} A_i$.

## Denumerable Sets

Let $X$ be any set. We call $X$ **denumerable** if it consists either of finitely many elements or if there is a bijective mapping $f : \mathbb{N} \to X$. If we only have the latter case then we call $X$ **countable**

$$\text{every subset of a denumerable set is denumerable;} \qquad \text{(A.II.48)}$$

$$\text{the sets } \mathbb{N}, \mathbb{Z} \text{ and } \mathbb{Q} \text{ are countable;} \qquad \text{(A.II.49)}$$

$$\text{if } X_1, \dots X_k, k \in \mathbb{N} \text{ are countable, then}$$

$$X_1 \times \cdots \times X_k = \prod_{j=1}^{k} X_k \text{ is countable too;} \qquad \text{(A.II.50)}$$

$$\text{the union of denumerable many denumerable sets is denumerable}$$

and

$$\text{the union of countable many countable sets is countable} \qquad \text{(A.II.51)}$$

i.e. if $(X_j)_{j \in \mathbb{N}}$ is a family of countable sets, then

$$\bigcup_{j \in \mathbb{N}} X_i$$

is countable. (Note that instead of $\mathbb{N}$ we may take any countable index set).

Next we want to give some hints on how to prove (in principle) statements about sets and mappings when starting with the basics. There is a natural correspondence between certain logical operations and set theoretical operations. Let us introduce the following statements

$$p : \quad x \in X$$
$$q : \quad x \in Y$$

then

$$x \in X \cap Y \iff p \wedge q$$
$$x \in X \cup Y \iff p \vee q$$
$$x \notin X \iff \neg p$$

and if $X \subset Z$, $Z$ fixed, we have

$$x \in X^{\complement} \iff \neg p.$$

Further, if for some index set $J$, sets $X_j, j \in J$, are given and if

$$p_j : x \in X_j$$

then

$$x \in \bigcap_{j \in J} X_j \implies \forall j \in J : p_j$$

and

$$x \in \bigcup_{j \in J} X_j \iff \exists j \in J : p_j.$$

Now we may use truth tables to prove compound statements when finitely many statements are involved. For example in order to prove the second statement of (A.II.10), i.e.

$$X \cap (Y \cup Z) = (X \cap Y) \cup (X \cap Z)$$

we can look at

| $x \in X$ | $x \in Y$ | $x \in Z$ | $(x \in X) \cap (x \in Y \vee x \in Z)$ | $(x \in X \wedge x \in Y) \vee (x \in X \wedge x \in Z)$ |
|---|---|---|---|---|
| T | T | T | T | T |
| T | T | F | T | T |
| T | F | T | T | T |
| T | F | F | F | F |
| F | T | T | F | F |
| F | T | F | F | F |
| F | F | T | F | F |
| F | F | F | F | F |

Table A.II.1

Since the last two columns coincide the two statements are equivalent, however

$$(x \in X) \wedge (x \in Y \vee x \in Z) \iff x \in X \cap (Y \cup Z)$$

and

$$(x \in X \wedge x \in Y) \vee (x \in X \wedge x \in Z) \iff x \in (X \cap Y) \cup (X \cup Z).$$

486

Note: all statements about relations of sets given in our collection are statements involving quantifiers, for example the above statement (A.II.10) is equivalent to

$$\forall x \in X \cup Y \cup Z : ((x \in X \cap (Y \cup Z)) \iff (x \in (X \cap Y) \cup (X \cap Z))).$$

In our proof we only considered the equivalence for a single $x$, but since $x$ was arbitrary this means that we proved it for all $x \in X \cup Y \cup Z$.

Although the method of truth tables will always provide a proof as long as only finitely many statements are involved, it could be quite a time consuming process to check all cases. For example to prove

$$(X_1 \times Y_1) \cap (X_2 \times Y_2) = (X_1 \cap X_2) \times (Y_1 \cap Y_2) \qquad \text{(A.II.52)}$$

one would have to complete a truth table with 16 rows. However, a short and transparent proof is obtained by using step by step basic definitions and simple rules for handling logical statements:

$$(x,y) \in (X_1 \times Y_1) \cap (X_2 \times Y_2)$$

$$\iff (x,y) \in (X_1 \times Y_1) \wedge (x,y) \in (X_2 \times Y_2)$$
$$\iff x \in X_1 \wedge y \in Y_2 \wedge x \in X_2 \wedge y \in Y_2$$
$$\iff x \in (X_1 \times X_2) \wedge y \in (Y_1 \cap Y_2)$$
$$\iff (x,y) \in (X_1 \cap X_2 \times (Y_1 \cap Y_2).$$

Since the pair $(x,y)$ is arbitrary the statement (A.II.52) (which of course is (A.II.18)) is proved. Similarly we can prove statements with quantifiers, for example the first statement in (A.II.42):

$$\left( \bigcup_{i \in I} A_i \right)^{\complement} = \bigcap_{i \in I} A_i^{\complement}.$$

We have

$$x \in \left( \bigcup_{i \in I} A_i \right)^{\complement} \iff x \notin \bigcup_{i \in I} A_i$$

$$\iff \neg(\exists i \in I : x \in A_i)$$
$$\iff \forall i \in I : \neg(x \in A_i)$$
$$\iff \forall i \in I : x \in A_i^C$$
$$\iff x \in \bigcap_{i \in I} A_i^C.$$

The proofs for the statements listed above involving mappings are reduced to statements for sets. For example the meaning of (A.II.24) is

$$y \in F(A \cap B) \implies y \in F(A) \cap F(B)$$

and in more detail

$$y \in F(A \cap B) \text{ means } y \in \{\tilde{y} \in Y | \exists x \in A \cap B : F(x) = \tilde{y}\},$$

$$y \in F(A) \text{ means } y \in \{\tilde{y} \in Y | \exists x' \in A : F(x') = \tilde{y}\},$$
$$y \in F(B) \text{ means } y \in \{\tilde{y} \in Y | \exists x'' \in B : F(x'') = \tilde{y}\}.$$

Thus $F(A \cap B) \subset F(A) \cap F(B)$ is the statement

$$\{\tilde{y} \in Y | \exists x \in A \cap B : F(x) = \tilde{y}\} \subset$$

$$\{y \in Y | \exists x' \in A : F(x') = \tilde{y}\} \cap \{y \in Y | \exists x'' \in A : F(x'') = \tilde{y}\}.$$

The proofs for statements involving unions or intersections of arbitrary families of sets are similar but they will need quantifiers. Let us prove (A.II.46)

$$F^{-1}\left(\bigcup_{j \in J} A_j'\right) = \bigcup_{j \in J} F^{-1}(A_j').$$

First note that this statement says

$$x \in F^{-1}\left(\bigcup_{j \in J} A_j'\right) \iff x \in \bigcup_{j \in J} F^{-1}(A_j').$$

Now, $x \in F^{-1}\left(\bigcup_{j \in J} A_j'\right)$ means

$$x \in \{\tilde{x} \in X | F(\tilde{x}) \in \bigcup_{j \in J} A_j'\}$$

488

which is equivalent to

$$x \in \{\tilde{x} \in X | \exists j \in J : F(\tilde{x}) \in A'_j\},$$

but the meaning of $x \in \bigcup_{j \in J} F^{-1}(A'_j)$ is nothing but

$$x \in \{\tilde{x} \in X | \exists j \in J : F(\tilde{x}) \in A'_j\}$$

and the statement is proved.

As mentioned at the beginning of this appendix, we only want to indicate the principle strategies on how to prove the statements listed. The reader is encouraged to prove some of the other statements as an exercise.

# Appendix III: The Peano Axioms

As we have stated previously, when starting to think about the foundations of knowledge, in our case the foundations of mathematics, we must come to the conclusion that "at the beginning there was no beginning". To make a start the **axiomatic method** in mathematics as is accepted nowadays by all mathematicians suggests to use a system of **axioms**; statements we accept as true without giving any justification or proof, as a starting point and draw conclusions from these. Of course, a system of axioms should satisfy certain conditions, for example it should not lead to (obvious) contradictions, axioms must be "reasonable" statements etc. In Euclid's geometry such an approach had already been indicated, however he still partly tried to justify axioms or relate the content of axioms to experience. Nowadays, systems of axioms are seen to be completely independent of "exterior experiences". The mystery is that such a method is extremely successful to provide the most powerful tools for science, engineering, economics etc, i.e. real world problems. As E. Wigner put it, we have some "Unreasonable Effectiveness of Mathematics in Natural Sciences".

For beginners in mathematics this method might seem unusual and requires some time to be understood and appreciated. Therefore looking back at Part 1 we can see that we have not used the axiomatic approach to its full extent. It is possible to introduce the natural numbers by a system of axioms in such a way that a beginner should follow. Historically, this approach to the natural numbers was one of the first axiomatic theories. Thus we dedicate this appendix to an axiomatic introduction of the natural numbers. The system of axioms in question are the **Peano Axioms**.

**P.A.1**

> 1 is a natural number.

**P.A.2**

> For every natural number $n$ there exists a unique natural number called the **successor** of $n$ which is denoted by $n'$.

**P.A.3**

> $n' \neq 1$ for all natural numbers $n$.

**P.A.4**

If $n' = m'$ then $n = m$.

**P.A.5 (Axiom of Induction)**

Let $M$ be a subset of the natural numbers such that:

- $1 \in M$;
- if $n \in M$ then $n' \in M$.

Then $M$ is the set of all natural numbers.

Of course, we denote as before the set of all natural numbers by $\mathbb{N}$ and further $2 := 1', 3 := 2', 4 := 3'$ etc.

Here are some consequences of the Peano axioms:

**Proposition A.III.1. A.** *For $n, m \in \mathbb{N}$ it follows that $n \neq m$ implies $n' \neq m'$.*
***B.** For $n \in \mathbb{N}$ we have $n' \neq n$.*
***C.** If $n \neq 1$, $n \in \mathbb{N}$, then there exists a unique $m \in \mathbb{N}$ such that $n = m'$.*

Before we prove this proposition, let us consider some interpretations. P.A.2 states (by its uniqueness property) that if $n = m$ then $n' = m'$. Now part A of the proposition says that two distinct natural numbers have two distinct successors. Part B tells us that $n$ is never its own successor, and part C states that every natural number $n \neq 1$ is indeed a successor of another natural number.

*Proof of Proposition A.III.1.* **A.** Suppose that $n' = m'$. Then by P.A.4 it follows that $n = m$, which is a contradiction, hence $n' \neq m'$.
**B.** Let $M$ be the set of all $n \in \mathbb{N}$ with $n \neq n'$, i.e. $M = \{n \in \mathbb{N} | n \neq n'\}$. By P.A.1 and P.A.3 we have $1' \neq 1$, implying $1 \in M$. Further if $n \in M$, i.e. $n' \neq n$, then by part A it follows that $(n')' \neq n'$, hence $n' \in M$. Now P.A.5 implies $M = \mathbb{N}$.
**C.** Let $M$ be the set containing 1 and all $n \in \mathbb{N}$ such that there is $m \in \mathbb{N}$ with $n = m'$, i.e.

$$M = \{1\} \cup \{n \in \mathbb{N} \setminus \{1\} | \exists m \in \mathbb{N} : n = m'\}.$$

Clearly $1 \in M$. Furthermore, if $n \in M$ then for $m = n$ we find $n' = m'$, i.e. $n' \in M$. Now by P.A.5 we conclude that $M = \mathbb{N}$. $\qquad\square$

So far we have only defined a set $\mathbb{N}$ of natural numbers. Clearly we want to add natural numbers together as we are used to. We achieve this by introducing on $\mathbb{N}$ a binary operation which we call **addition**.

**Theorem A.III.2.** *For every pair of natural numbers $(n, m)$ there exists a unique natural number denoted by $add(n, m)$ such that*

$$add(n, 1) = n' \text{ for every } n \in \mathbb{N}; \qquad (\text{A.III.1})$$

*and*

$$add(n, m') = (add(n, m))' \text{ for all } (n, m) \in \mathbb{N} \times \mathbb{N}. \qquad (\text{A.III.2})$$

Let us now try to understand how to proceed. First we introduce axiomatically a set, called the natural numbers, denoted by $\mathbb{N}$. We then introduce a mapping from $\mathbb{N} \times \mathbb{N}$ to $\mathbb{N}$

$$add : \mathbb{N} \times \mathbb{N} \to \mathbb{N} \qquad (\text{A.III.3})$$

by the two properties (A.III.1) and (A.III.2). Of course we have to prove that such a mapping exists and is unique. This is what the above theorem considers, however we do not give the proof here. Once the theorem is proved, i.e we know there is such a binary operation *add* we can start to study its properties. For simplicity we write from now on

$$n + m := add(n, m) \qquad (\text{A.III.4})$$

and the task is to prove using P.A.1-P.A.5 and Theorem A.III.2 only properties such as

$$(k + m) + n = k + (m + n) \text{ associativity,}$$

or

$$n + m = m + n \text{ commutativity.}$$

E. Landau in [7] gives a very systematical way of introducing $\mathbb{N}$, addition and the extension from $\mathbb{N}$ to $\mathbb{Z}$ as well as from $\mathbb{Z}$ to $\mathbb{Q}$.

Finally we want to discuss how **mathematical induction** relates to the Peano axioms. Recall that mathematical induction works as follows: suppose that for $n \in \mathbb{N}$ a statement $A(n)$ is given. If $A(1)$ is true and if $A(n)$ always implies $A(n + 1)$ then $A(n)$ is true for all $n \in \mathbb{N}$.

Denote by $M$ the set of all natural numbers such that $A(n)$ is true, i.e.

$$M := \{n \in \mathbb{N} | A(n) \text{ is true}\}.$$

We have to prove

$$(A(1) \wedge (A(n) \implies A(n+1))) \implies M = \mathbb{N}.$$

Since $1 \in M$ by assumption and since $n+1 = n'$ we know that $n \in M$ implies $n' \in M$. Hence by P.A.5 it follows that $M = \mathbb{N}$.

Thus introducing $\mathbb{N}$ via the Peano axioms in an axiomatic way we can deduce that mathematical induction is providing what we want.

A final remark: in Chapter 3 we have formulated the principle of mathematical induction for a more general starting point, say $k \in \mathbb{Z}$. Of course we can use the above argument to justify this formulation. We only need to make a change of the enumeration index.

# Appendix IV: Results from Elementary Geometry

Here we recollect some basic results from elementary geometry for reference purposes. Typically students will have already met these results.

We first consider straight lines. Let $g_1$ and $g_2$ be two parallel lines in the plane and $h$ a straight line transverse to both $g_1$ and $g_2$, see Figure A.IV.1 below.

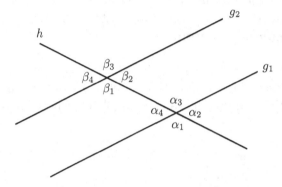

*Figure A.IV.1*

The following relations hold for the above angles:

$$\alpha_1 + \alpha_2 = \pi; \tag{A.IV.1}$$

$$\alpha_1 = \alpha_3 \text{ and } \alpha_2 = \alpha_4; \tag{A.IV.2}$$

$$\alpha_1 = \beta_1, \alpha_2 = \beta_2, \alpha_3 = \beta_3, \alpha_4 = \beta_4; \tag{A.IV.3}$$

$$\alpha_1 = \beta_3, \alpha_2 = \beta_4, \alpha_3 = \beta_1, \alpha_4 = \beta_2. \tag{A.IV.4}$$

Now, let $ABC$ be a triangle in the plane, see Figure A.IV.2.

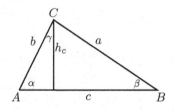

*Figure A.IV.2*

495

Note that
$$\alpha + \beta + \gamma = \pi, \tag{A.IV.5}$$
and for the area of $ABC$ we have
$$area(ABC) = \frac{1}{2}h_c \cdot c \tag{A.IV.6}$$
where $h_c$ is the height from $C$ to $AB$. Clearly we have
$$\frac{1}{2}h_c c = \frac{1}{2}h_b b = \frac{1}{2}h_a a = area(ABC),$$
where $h_b$ and $h_a$ denote the heights from $B$ to the side $AC$ and $A$ to the side $BC$ respectively. In the case of a right angled triangle $ABC$, see Figure A.IV.3 we have **Pythagoras' theorem**
$$a^2 + b^2 = c^2. \tag{A.IV.7}$$

(Note that there is a slight abuse of notation here: $a, b, c$ denote the sides in $ABC$, whereas in (A.IV.7) we use the same symbols to denote the length of these sides.)

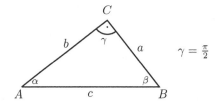

*Figure A.IV.3*

Note that we use the "continental" way to indicate an angle of size $\frac{\pi}{2}$ i.e:

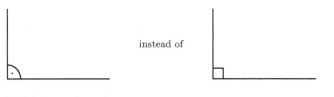

*Figure A.IV.4*

496

Now let $C_r(\mathbf{O})$ be a circle of radius $r$ and centre $\mathbf{O}$, i.e:

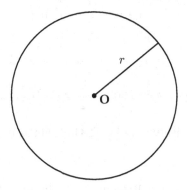

Figure A.IV.5

Its area is given by

$$area(C_r(\mathbf{O})) = \pi r^2 \qquad \text{(A.IV.8)}$$

and its circumference $\partial C_r(\mathbf{O})$ has length

$$length(\partial C_r(\mathbf{O})) = 2\pi r. \qquad \text{(A.IV.9)}$$

There are two scales to measure the size of an angle in the unit circle, i.e. in $C_1(\mathbf{O})$, (these are degrees and radians). An angle is measured as a fraction of $360°$, i.e. by definition we say that the full circle forms an angle of $360°$ and the size of $\alpha$ is just a corresponding fraction, for example a right angle has size $90°$. Or, a better way to do this is to take the length of the segment $\overgroup{AB}$ as a measure of $\alpha$, see Figure A.IV.6.

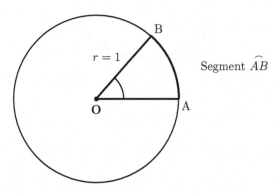

Figure A.IV.6

By the segment $\overset{\frown}{AB}$ we mean the arc joining $A$ and $B$, i.e. on $\partial C_r(\mathbf{O})$. Often we say that it is **measured by the arc length**. This definition of the size of an angle implies the following correspondence:

$$\frac{\pi}{6} \cong 30°, \quad \frac{\pi}{4} \cong 45°, \quad \frac{\pi}{3} \cong 60°, \quad \frac{\pi}{2} \cong 90°,$$

$$\frac{3\pi}{4} \cong 135°, \quad \pi \cong 180°, \quad \frac{3\pi}{2} \cong 270°, \quad 2\pi \cong 360°.$$

For a circle $C_r(\mathbf{O})$, see Figure A.IV.7, the length of the arc $\overset{\frown}{AB}$ with angle $\alpha$ is given by

$$length(\overset{\frown}{AB}) = r\alpha \quad (\alpha \text{ measured by the arc length}) \qquad \text{(A.IV.10)}$$

and the area of the sector $\mathbf{O}\overset{\frown}{AB}$ is given by

$$area(\mathbf{O}\overset{\frown}{AB}) = \frac{r^2\alpha}{2} \quad (\alpha \text{ measured by the arc length}). \qquad \text{(A.IV.11)}$$

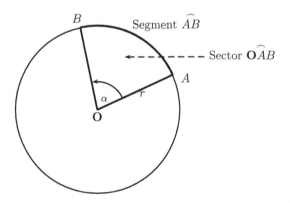

*Figure A.IV.7*

# Appendix V: Trigonometric and Hyperbolic Functions

Trigonometric and hyperbolic functions play an important role in many areas of mathematics. Here we collect some of the most useful formulae for these functions.

## A. Trigonometric Functions

### 1. Symmetries

$$\sin(-x) = -\sin x, \quad \sin(x + 2\pi) = \sin x \qquad \text{(A.V.1)}$$

$$\cos(-x) = \cos x, \quad \cos(x + 2\pi) = \cos x \qquad \text{(A.V.2)}$$

$$\tan(-x) = -\tan x, \quad \tan(x + \pi) = \tan x \qquad \text{(A.V.3)}$$

$$\cot(-x) = -\cot x, \quad \cot(x + \pi) = \cot x \qquad \text{(A.V.4)}$$

### 2. Addition Theorems

$$\sin(x \pm y) = \sin x \cos y \pm \cos x \sin y \qquad \text{(A.V.5)}$$

$$\cos(x \pm y) = \cos x \cos y \mp \sin x \sin y \qquad \text{(A.V.6)}$$

$$\tan(x \pm y) = \frac{\tan x \pm \tan y}{1 \mp \tan x \tan y} \qquad \text{(A.V.7)}$$

$$\cot(x \pm y) = \frac{\cot x \cot y \mp 1}{\cot y \pm \cot x} \qquad \text{(A.V.8)}$$

### 3. Consequences of the Addition Theorems

$$\sin(\frac{\pi}{2} + x) = \cos x, \quad \sin(\pi + x) = -\sin x \qquad \text{(A.V.9)}$$

$$\cos(\frac{\pi}{2} + x) = -\sin x, \quad \cos(\pi + x) = -\cos x \qquad \text{(A.V.10)}$$

$$\tan(\frac{\pi}{2} \pm x) = \mp \cot x \qquad \text{(A.V.11)}$$

$$\cot(\frac{\pi}{2} \pm x) = \mp \tan x \qquad \text{(A.V.12)}$$

### 4. Double Arguments (Double angle formulae)

$$\sin 2x = 2 \sin x \cos x \qquad \text{(A.V.13)}$$

499

$$\cos 2x = \cos^2 x - \sin^2 x \qquad (A.V.14)$$

$$\tan 2x = \frac{2 \tan x}{1 - \tan^2 x} \qquad (A.V.15)$$

$$\cot 2x = \frac{\cot^2 x - 1}{2 \cot x} \qquad (A.V.16)$$

## 5. Half Arguments (Half angle formulae)

$$\sin \frac{x}{2} = \begin{cases} \sqrt{\frac{1}{2}(1 - \cos x)}, & 0 \le x \le \pi \\ -\sqrt{\frac{1}{2}(1 - \cos x)}, & \pi \le x \le 2\pi \end{cases} \qquad (A.V.17)$$

$$\cos \frac{x}{2} = \begin{cases} \sqrt{\frac{1}{2}(1 + \cos x)}, & -\pi \le x \le \pi \\ -\sqrt{\frac{1}{2}(1 + \cos x)}, & \pi \le x \le 3\pi \end{cases} \qquad (A.V.18)$$

$$\tan \frac{x}{2} = \frac{\sin x}{1 + \cos x} = \frac{1 - \cos x}{\sin x} \qquad (A.V.19)$$

$$\cot \frac{x}{2} = \frac{\sin x}{1 - \cos x} = \frac{1 + \cos x}{\sin x} \qquad (A.V.20)$$

## 6. Sums

$$\sin x \pm \sin y = 2 \sin \frac{x \pm y}{2} \cos \frac{x \mp y}{2} \qquad (A.V.21)$$

$$\cos x + \cos y = 2 \cos \frac{x + y}{2} \cos \frac{x - y}{2} \qquad (A.V.22)$$

$$\cos x - \cos y = 2 \sin \frac{x + y}{2} \sin \frac{y - x}{2} \qquad (A.V.23)$$

$$\cos x \pm \sin x = \sqrt{2} \sin(\frac{\pi}{4} \pm x) \qquad (A.V.24)$$

$$\tan x \pm \tan y = \frac{\sin(x \pm y)}{\cos x \cos y} \qquad (A.V.25)$$

$$\cot x \pm \cot y = \pm \frac{\sin(z \pm y)}{\sin x \sin y} \qquad (A.V.26)$$

$$\tan x + \cot y = \frac{\cos(x - y)}{\cos x \sin y} \qquad (A.V.27)$$

$$\cot x - \tan y = \frac{\cos(x + y)}{\sin x \cos y} \qquad (A.V.28)$$

## 7. Products

$$\sin x \sin y = \frac{1}{2}(\cos(x-y) - \cos(x+y)) \tag{A.V.29}$$

$$\cos x \cos y = \frac{1}{2}(\cos(x-y) + \cos(x+y)) \tag{A.V.30}$$

$$\sin x \cos y = \frac{1}{2}(\sin(x-y) + \sin(x+y)) \tag{A.V.31}$$

$$\tan x \tan y = \frac{\tan x + \tan y}{\cot x + \cot y} \tag{A.V.32}$$

$$\cot x \cot y = \frac{\cot x + \cot y}{\tan x + \tan y} \tag{A.V.33}$$

$$\tan x \cot y = \frac{\tan x + \cot y}{\cot x + \tan y} \tag{A.V.34}$$

## 8. Squares

$$\sin^2 x + \cos^2 x = 1 \tag{A.V.35}$$

$$\sin^2 x = \frac{\tan^2 x}{1 + \tan^2 x} = \frac{1}{1 + \cot^2 x} \tag{A.V.36}$$

$$\cos^2 x = \frac{1}{1 + \tan^2 x} = \frac{\cot^2 x}{1 + \cot^2 x} \tag{A.V.37}$$

$$\sin^2 \frac{x}{2} = \frac{1}{2}(1 - \cos x) \tag{A.V.38}$$

$$\cos^2 \frac{x}{2} = \frac{1}{2}(1 + \cos x) \tag{A.V.39}$$

$$\tan^2 x = \frac{\sin^2 x}{1 - \sin^2 x} = \frac{1 - \cos^2 x}{\cos^2 x} \tag{A.V.40}$$

$$\cot^2 x = \frac{\cos^2 x}{1 - \cos^2 x} = \frac{1 - \sin^2 x}{\sin^2 x} \tag{A.V.41}$$

501

### 9. Useful Values

| $x$ | $0$ | $\frac{\pi}{6}$ | $\frac{\pi}{4}$ | $\frac{\pi}{3}$ | $\frac{\pi}{2}$ | $\frac{2\pi}{3}$ | $\frac{3\pi}{4}$ | $\frac{5\pi}{6}$ | $\pi$ |
|---|---|---|---|---|---|---|---|---|---|
| | $0°$ | $30°$ | $45°$ | $60°$ | $90°$ | $120°$ | $135°$ | $150°$ | $180°$ |
| $\sin x$ | $0$ | $\frac{1}{2}$ | $\frac{1}{2}\sqrt{2}$ | $\frac{1}{2}\sqrt{3}$ | $1$ | $\frac{1}{2}\sqrt{3}$ | $\frac{1}{2}\sqrt{2}$ | $\frac{1}{2}$ | $0$ |
| $\cos x$ | $1$ | $\frac{1}{2}\sqrt{3}$ | $\frac{1}{2}\sqrt{2}$ | $\frac{1}{2}$ | $0$ | $-\frac{1}{2}$ | $-\frac{1}{2}\sqrt{2}$ | $-\frac{1}{2}\sqrt{3}$ | $-1$ |
| $\tan x$ | $0$ | $\frac{1}{3}\sqrt{3}$ | $1$ | $\sqrt{3}$ | - | $-\sqrt{3}$ | $-1$ | $-\frac{1}{3}\sqrt{3}$ | $0$ |
| $\cot x$ | - | $\sqrt{3}$ | $1$ | $\frac{1}{3}\sqrt{3}$ | $0$ | $-\frac{1}{3}\sqrt{3}$ | $-1$ | $-\sqrt{3}$ | - |

## B. Hyperbolic Functions

### 1. Symmetries

$$\sinh(-x) = -\sinh x \qquad\qquad \text{(A.V.42)}$$

$$\cosh(-x) = \cosh x \qquad\qquad \text{(A.V.43)}$$

$$\tanh(-x) = -\tanh x \qquad\qquad \text{(A.V.44)}$$

$$\coth(-x) = -\coth x \qquad\qquad \text{(A.V.45)}$$

### 2. Addition Theorems

$$\sinh(x \pm y) = \sinh x \cosh y \pm \cosh x \sinh y \qquad\qquad \text{(A.V.46)}$$

$$\cosh(x \pm y) = \cosh x \cosh y \pm \sinh x \sinh y \qquad\qquad \text{(A.V.47)}$$

$$\tanh(x \pm y) = \frac{\tanh x \pm \tanh y}{1 \pm \tanh x \tanh y} \qquad\qquad \text{(A.V.48)}$$

$$\coth(x \pm y) = \frac{1 \pm \coth x \coth y}{\coth x \pm \coth y} \qquad\qquad \text{(A.V.49)}$$

### 3. Double Arguments

$$\sinh 2x = 2 \sinh x \cosh x \qquad\qquad \text{(A.V.50)}$$

$$\cosh 2x = \sinh^2 x + \cosh^2 x \qquad\qquad \text{(A.V.51)}$$

$$\tanh 2x = \frac{2 \tanh x}{1 + \tanh^2 x} \qquad\qquad \text{(A.V.52)}$$

$$\coth 2x = \frac{1 + \coth^2 x}{2 \coth x} \qquad\qquad \text{(A.V.53)}$$

## 4. Half Arguments

$$\sinh \frac{x}{2} = \begin{cases} \sqrt{\frac{1}{2}(\cosh x - 1)}, & x \geq 0 \\ -\sqrt{\frac{1}{2}(\cosh x - 1)}, & x < 0 \end{cases} \tag{A.V.54}$$

$$\cosh \frac{x}{2} = \sqrt{\frac{1}{2}(\cosh x + 1)} \tag{A.V.55}$$

$$\tanh \frac{x}{2} = \frac{\cosh x - 1}{\sinh x} = \frac{\sinh x}{\cosh x + 1} \tag{A.V.56}$$

$$\coth \frac{x}{2} = \frac{\sinh x}{\cosh x - 1} = \frac{\cosh x + 1}{\sinh x} \tag{A.V.57}$$

## 5. Sums

$$\sinh x \pm \sinh y = 2 \sinh \frac{1}{2}(x \pm y) \cosh \frac{1}{2}(x \mp y) \tag{A.V.58}$$

$$\cosh x + \cosh y = 2 \cosh \frac{1}{2}(x + y) \cosh \frac{1}{2}(x - y) \tag{A.V.59}$$

$$\cosh x - \cosh y = 2 \sinh \frac{1}{2}(x + y) \sinh \frac{1}{2}(x - y) \tag{A.V.60}$$

$$\tanh x \pm \tanh y = \frac{\sinh(x \pm y)}{\cosh x \cosh y} \tag{A.V.61}$$

## 6. Squares

$$\cosh^2 x - \sinh^2 x = 1 \tag{A.V.62}$$

$$\sinh^2 x = \cosh^2 x - 1 = \frac{\tanh^2 x}{1 - \tanh^2 x} = \frac{1}{\coth^2 x - 1} \tag{A.V.63}$$

$$\cosh^2 x = \sinh^2 x + 1 = \frac{1}{1 - \tanh^2 x} = \frac{\coth^2 x}{\coth^2 x - 1} \tag{A.V.64}$$

$$\tanh^2 x = \frac{\sinh^2 x}{\sinh^2 x + 1} = \frac{\cosh^2 x - 1}{\cosh^2 x} = \frac{1}{\coth^2 x} \tag{A.V.65}$$

$$\coth^2 x = \frac{\sinh^2 x + 1}{\sinh^2 x} = \frac{\cosh^2 x}{\cosh^2 x - 1} = \frac{1}{\tanh^2 x} \tag{A.V.66}$$

Note that we will see the relationship between hyperbolic and trigonometric functions when we consider complex arguments later in this course.

# Appendix VI: More on the Completeness of $\mathbb{R}$

In this appendix we want to discuss in more detail some aspects of the **Axiom of Completeness** which as we recall (see Chapter 17) is: In $\mathbb{R}$ every Cauchy sequence has a limit. This axiom was needed to prove many central results including:

- the Bolzano-Weierstrass theorem (Theorem 17.6);

- every increasing (decreasing) sequence bounded from above (below) converges (Theorem 17.14);

- the principle of nested intervals (Theorem 17.15);

- every set bounded from above (below) has a least (greatest) upper (lower) bound (Theorem 19.14).

Without these results we cannot prove many others, hence the completeness of $\mathbb{R}$ is key for our theory. Nonetheless there are at least two problems with the axiom of completeness. Firstly, it looks quite artificial, an ad hoc requirement which turns out to be useful. Secondly, while we may suppose the axiom to hold, we have given no proof so far that an Archimedian ordered field which is complete exists.

First we want to discuss an equivalent way of introducing the completeness of $\mathbb{R}$ by choosing a different axiom as a starting point.

**Axiom A**

> Every non-empty set of real numbers bounded from above has a least upper bound.

Clearly this axiom is equivalent to

**Axiom A′**

> Every non-empty set of real numbers bounded from below has a greatest lower bound.

The first consequence of Axiom A is

**Theorem A.VI.1.** *An increasing sequence $(x_n)_{n \in \mathbb{N}}$, $x_n \in \mathbb{R}$, which is bounded from above converges to the least upper bound $x$ of the set $\{x_n | n \in \mathbb{N}\}$.*

*Proof.* Let $x$ be the least upper bound of $\{x_n | n \in \mathbb{N}\}$. Given $\epsilon > 0$ there exists $N \in \mathbb{N}$ such that $x - \frac{\epsilon}{2} < x_N < x$. Since $(x_n)_{n \in \mathbb{N}}$ is increasing it follows for all $n \geq N$ that $x - \frac{\epsilon}{2} \leq x_N \leq x_n \leq x$, or for all $n \geq N$ we have $0 \leq x - x_n \leq \frac{\epsilon}{2}$, i.e. $|x_n - x| < \epsilon$, implying the convergence of $(x_n)_{n \in \mathbb{N}}$ to $x$. $\qquad\square$

**Corollary A.VI.2.** *A decreasing sequence $(x_n)_{n \in \mathbb{N}}, x_n \in \mathbb{R}$, which is bounded from below converges to the greatest lower bound $x$ of the set $\{x_n | n \in \mathbb{N}\}$.*

**Theorem A.VI.3.** *If Axiom A holds every Cauchy sequence in $\mathbb{R}$ converges.*

*Proof.* Let $(x_n)_{n \in \mathbb{N}}$ be a Cauchy sequence. We know by Proposition 17.3.B that $(x_n)_{n \in \mathbb{N}}$ is bounded. We consider the sets $A_k := \{x_l | l \geq k\}$ which are bounded and $A_{k+1} \subset A_k$, $A_1 = \{x_n | n \in \mathbb{N}\}$. Each of the sets $A_k$ has a greatest lower bound $c_k$ and the sequence $(c_k)_{k \in \mathbb{N}}$ is increasing, i.e. $c_k \leq c_{k+1}$ for $k \in \mathbb{N}$, and bounded from above. By Theorem A.VI.1 this sequence has a limit $c$, $c = \lim_{k \to \infty} c_k$. We claim now that a subsequence of $(x_n)_{n \in \mathbb{N}}$ converges to $c$. Given $\epsilon > 0$ there exists $N \in \mathbb{N}$ such that for $m \geq N$ it follows that $0 < c_m - c < \frac{\epsilon}{2}$. Since $c_m$ is the greatest lower bound of $A_m$ there exists $k_m \geq m$ such that $0 < x_{k_m} - c_m < \frac{\epsilon}{2}$. For the subsequence $(c_{k_m})_{m \in \mathbb{N}}$ the following holds

$$|x_{k_m} - c| \leq |x_{k_m} - c_m| + |c_m - c| < \epsilon,$$

i.e. $(x_{k_m})_{m \in \mathbb{N}}$ converges to $c$. Now Lemma 17.10.B implies the result. $\qquad\square$

Theorem A.VI.3 implies the equivalence of Axiom A (or Axiom A$'$) with the Axiom of Completeness, and arguably Axiom A is more natural to accept. It is possible to prove the equivalence of other statements to the Axiom of Completeness, but we do not want to go into further detail.

The following material is very mathematically advanced and might be skipped in a first reading.

Our goal is to sketch how to construct $\mathbb{R}$. Let us start with the following problem: given $\mathbb{N}$ as a set characterised by the Peano axioms, see Appendix III, can we construct the ring $\mathbb{Z}$? We have of course an idea of what $\mathbb{Z}$ shall constitute of and this will give us hints for our formal construction. Note that every integer $z \in \mathbb{Z}$ is the difference between two natural numbers $m, n \in \mathbb{N}$, i.e. $z = n - m$. The problem is that in $\mathbb{N}$ the operations "$-$" is not yet defined. Moreover, the representation is not unique: $0 = n - n$ for all $n \in \mathbb{N}$, or $n = n + m - m$ for all $m \in \mathbb{N}$. A way forward is to use pairs of natural numbers. On $\mathbb{N} \times \mathbb{N}$ we define the relation

$$(n, m) \backsim_{\mathbb{Z}} (n', m') \quad \text{if and only if} \quad n + m' = m + n'. \qquad (\text{A.VI.1})$$

This definition is of course inspired by the fact that $n + m' = m + n'$ is equivalent to $n - m = n' - m'$, if "$-$" is defined in the usual way. It is easy to see that on $\mathbb{N} \times \mathbb{N}$ the relation "$\backsim_{\mathbb{Z}}$" is an equivalence relation. Indeed, $(n, n) \backsim_{\mathbb{Z}} (n, n)$ is trivial, and since $n + m' = m + n'$ if and only if $n' + m = m' + n$, we also have the symmetry $(n, m) \backsim_{\mathbb{Z}} (n', m')$ if and only if $(n', m') \backsim_{\mathbb{Z}} (n, m)$. Moreover, if $(n, m) \backsim_{\mathbb{Z}} (n', m')$ and $(n', m') \backsim_{\mathbb{Z}} (n'', m'')$ it follows that $n + m' = m + n'$ and $n' + m'' = n'' + m'$ and therefore $n + m' + n' + m'' = m + n' + n'' + m'$ and the arithmetic rules in $\mathbb{N}$ yield $n + m'' = m + n''$ or $(n, m) \backsim_{\mathbb{Z}} (n'', m'')$. We denote now by $\mathbb{Z} := \mathbb{N} \times \mathbb{N}/\backsim_{\mathbb{Z}}$ the family of all equivalence classes and introduce on $\mathbb{Z}$ the operations

$$[(n, m)] \oplus [(n', m')] := [(n + n', m + m')]$$

and

$$[(n, m)] \odot [(n', m')] := [(nn' + mm', nm' + mn')].$$

First we can prove that these definitions are independent of the representatives chosen. Moreover we can identify $n \in \mathbb{N}$ with $[n + m, m], m \in \mathbb{N}$, and we may define $0 := [n, n]$, as we may set $-n$ for $[m, n + m]$. It takes some work, but it is not difficult to see that $\mathbb{N} \times \mathbb{N}/\backsim_{\mathbb{Z}}$ with the operations $\oplus$ and $\odot$ forms a ring and we will use the standard notations from now on, i.e. $0, 1, n, -n, n + m, n - m$ when working in $\mathbb{Z}$. We do not want to go much further into the details since we will do so when passing from $\mathbb{Z}$ to $\mathbb{Q}$ for which we employ a similar construction.
On $\mathbb{Z} \times \mathbb{N}$ we define the relation

$$(k, m) \backsim_{\mathbb{Q}} (l, n) \quad \text{if and only if} \quad nk = ml. \qquad (\text{A.VI.2})$$

Again it is easy to see that $\backsim_{\mathbb{Q}}$ is an equivalence relation:

$$(k, m) \backsim_{\mathbb{Q}} (k, m) \text{ is trivial}$$

and since $kn = ml$ if and only if $lm = nk$ the symmetry relation

$$(k, m) \backsim_{\mathbb{Q}} (l, n) \quad \text{if and only if} \quad (l, n) \backsim_{\mathbb{Q}} (k, m)$$

follows. Moreover, if $(k, m) \backsim_{\mathbb{Q}} (l, n)$ and $(l, n) \backsim_{\mathbb{Q}} (p, q)$ we have $kn = lm$ and $lp = qn$ which yields $knlp = qnlm$ or $kp = qm$, i.e. $(k, m) \backsim (p, q)$. We denote by $\mathbb{Q}$ the family of all equivalence classes, i.e.

$$\mathbb{Q} := \mathbb{Z} \times \mathbb{N}/\backsim_{\mathbb{Q}},$$

and for $[(k, m)] \in \mathbb{Z} \times \mathbb{N}/\backsim_{\mathbb{Q}}$ we will soon write again $\frac{k}{m}$.

Next we want to define the "usual" algebraic operations on $\mathbb{Q}$, and again we take guidance from our previous knowledge about the rationals. The rules we know are

$$\frac{k}{m} + \frac{l}{n} = \frac{nk + lm}{nm}$$

and

$$\frac{k}{m} \cdot \frac{l}{n} = \frac{k \cdot l}{m \cdot n},$$

therefore we define

$$[(k, m)] \oplus [(l, m)] := [(nk + lm, nm)] \qquad \text{(A.VI.3)}$$

and

$$[(k, m)] \odot [(l, n)] := [(kl, mn)]. \qquad \text{(A.VI.4)}$$

Note that $mn = m + \cdots + m$ ($n$ summands), so we need only addition in $\mathbb{N}$ (which we get from the Peano axioms) to define $\oplus$ and $\odot$. First we need to prove that our definitions are independent of the choice of representatives. So let $(k, m) \backsim_{\mathbb{Q}} (k', m')$ and $(l, n) \backsim_{\mathbb{Q}} (l', n')$. We find

$$[(k, m)] \oplus [(l, n)] = [(nk + lm, nm)]$$

and

$$[(k', m')] \oplus [(l', n')] = [(n'k' + l'm', n'm')].$$

However we have $km' = k'm$ and $ln' = l'n$ and therefore

$$nkn'm' + lmn'm' = n'k'nm + l'm'nm$$
$$= n'(k'm)n + (l'n)mm' = n'(km')n + (ln')mm',$$

implying

$$(nk + lm)n'm' = (n'k' + l'm')nm,$$

or

$$(nk + lm, nm) \backsim_{\mathbb{Q}} (n'k' + l'm', n'm'),$$

508

i.e.

$$[(nk + lm, nm)] = [(n'k' + l'm', n'm')].$$

Analogously we can prove that (A.VI.4) is independent of the representatives. The next task is to verify the field axioms for $(\mathbb{Q}, \oplus, \odot)$. For example we find

$$[(k, m)] \oplus [(l, n)] = [(l, n)] \oplus [(k, m)]$$

since

$$[(k, m)] \oplus [(l, n)] = [(kn + lm, nm)]$$

and

$$[(l, n) \oplus (k, m)] = [(lm + kn, mn)].$$

For $n \in \mathbb{Z}$ we identify $[(n, 1)]$ with $n$, and since $(0, 1) \backsim_{\mathbb{Q}} (0, m)$ for all $m \in \mathbb{N}$ we can represent 0 by any pair of the type $(0, m)$. Further, for $n \in \mathbb{N}$ we can identify $[n, n]$ with 1, indeed we get

$$[(n, n) \cdot (k, m)] = [nk, nm]$$

but $(nk, nm) \backsim_{\mathbb{Q}} (k, m)$ as $nkm = nmk$, and further

$$[(0, 1)] \odot [(l, n)] = [(0, n)] = [(0, 1)].$$

For $[(m, n)] \neq [(0, 1)]$ we can form its inverse of multiplication by $[(n, m)]$:

$$[(n, m)] \odot [(m, n)] = [(mn, mn)].$$

Thus, along these lines it is possible to prove that $(\mathbb{Q}, \oplus, \odot)$ is a field and we can consider this field as a model of the rational numbers. We can also introduce an order relation $\circleddash$ on $(\mathbb{Q}, \oplus, \odot)$ by

$$[(k, m)] \circleddash [(l, n)]$$

if and only if $kn \leq lm$. Again it is possible to prove that the definition is independent from the choice of representatives and that typical properties of an order relation hold. For example we know that $kn \leq lm$ and $lq \leq np$, i.e. $knlq \leq lmnp$, implies $kq \leq mp$ and therefore

$$[(k, m)] \circleddash [(l, n)] \quad \text{and} \quad [(l, n)] \circleddash [(p, q)] \quad \text{implies} \quad [(k, m)] \leq [(p, q)].$$

Moreover we find

$$[(0, 1)] \circleddash [(k, m)] \tag{A.VI.5}$$

509

if and only if $k \in \mathbb{N}_0$ since (A.VI.5) means $0 \cdot m \leq k$. In particular we have $[(0,1)] \leq [(n,n)]$ for $n \in \mathbb{N}$.

The principle should be now clear: the natural numbers $\mathbb{N}$ and addition in $\mathbb{N}$ we introduce using the Peano axioms, and then we can construct the ring $\mathbb{Z}$ and the ordered field $\mathbb{Q}$ using appropriate equivalence relations. This is now our basic idea to pass from $\mathbb{Q}$ to $\mathbb{R}$: on the set of all Cauchy sequences of rational numbers we will introduce an equivalence relation "$\backsim_{\mathbb{R}}$" and on the corresponding equivalence classes we can implement the structure of a complete ordered field in which Archimedes' axiom holds. Of course this field will become $\mathbb{R}$.

We denote by $\mathcal{C}$ the set of all Cauchy sequences of rational numbers. Hence $(x_n)_{n \in \mathbb{N}} \in \mathcal{C}$ if $x_n \in \mathbb{Q}$ and for every $\epsilon \in \mathbb{Q}, \epsilon > 0$, there exists $N = N(\epsilon) \in \mathbb{N}$ such that $n, m \geq N(\epsilon)$ implies $|x_n - x_m| < \epsilon$. Two Cauchy sequences $(x_n)_{n \in \mathbb{N}}$, $(y_n)_{n \in \mathbb{N}} \in \mathcal{C}$ are said to be equivalent if their difference tends to $0 \in \mathbb{Q}$:

$$(x_n)_{n \in \mathbb{N}} \backsim_{\mathbb{R}} (y_n)_{n \in \mathbb{N}} \text{ if and only if } \lim_{n \to \infty} (x_n - y_n) = 0. \tag{A.VI.6}$$

First we claim that "$\backsim_{\mathbb{R}}$" is an equivalence relation on $\mathcal{C}$. Clearly, $\lim_{n \to \infty} (x_n - x_n) = 0$ for every sequence $(x_n)_{n \in \mathbb{N}}$ and therefore $(x_n)_{n \in \mathbb{N}} \backsim (x_n)_{n \in \mathbb{N}}$. Moreover, since $\lim_{n \to \infty} (x_n - y_n) = 0$ if and only if $\lim_{n \to \infty} (y_n - x_n) = 0$ it follows that $(x_n)_{n \in \mathbb{N}} \backsim (y_n)_{n \in \mathbb{N}}$ if and only if $(y_n)_{n \in \mathbb{N}} \backsim (x_n)_{n \in \mathbb{N}}$, i.e. the relation $\backsim_{\mathbb{R}}$ is symmetric. Finally we observe that $\lim_{n \to \infty} (x_n - y_n) = 0$ and $\lim_{n \to \infty} (y_n - z_n) = 0$ implies

$$\lim_{n \to \infty} (x_n - z_n) = \lim_{n \to \infty} (x_n - y_n + y_n - z_n) = \lim_{n \to \infty} (x_n - y_n) + \lim_{n \to \infty} (y_n - z_n) = 0,$$

i.e. $(x_n)_{n \in \mathbb{N}} \backsim_{\mathbb{R}} (y_n)_{n \in \mathbb{N}}$ and $(y_n)_{n \in \mathbb{N}} \backsim_{\mathbb{R}} (z_n)_{n \in \mathbb{N}}$ implies $(x_n)_{n \in \mathbb{N}} \backsim_{\mathbb{R}} (z_n)_{n \in \mathbb{N}}$. Hence we have proved that "$\backsim_{\mathbb{R}}$" is an equivalence relation on $\mathcal{C}$. Now we consider

$$\mathbb{R} := \mathcal{C}/\backsim_{\mathbb{R}}, \tag{A.VI.7}$$

the set of all equivalence classes of Cauchy sequence of rational numbers. On $\mathcal{C}/\backsim_{\mathbb{R}}$ we introduce the following two operators:

$$[(x_n)_{n \in \mathbb{N}}] \oplus [(y_n)_{n \in \mathbb{N}}] := [(x_n + y_n)_{n \in \mathbb{N}}] \tag{A.VI.8}$$

and

$$[(x_n)_{n \in \mathbb{N}}] \odot [(y_n)_{n \in \mathbb{N}}] := [(x_n y_n)_{n \in \mathbb{N}}]. \tag{A.VI.9}$$

First we need to show that these definitions are independent of the choice of representatives. If $(x_n)_{n\in\mathbb{N}} \backsim_\mathbb{R} (x'_n)_{n\in\mathbb{N}}$ and $(y_n)_{n\in\mathbb{N}} \backsim_\mathbb{R} (y'_n)_{n\in\mathbb{N}}$ then it follows immediately that $(x_n + y_n)_{n\in\mathbb{N}} \backsim_\mathbb{R} (x'_n + y'_n)_{n\in\mathbb{N}}$ since

$$\lim_{n\to\infty} (x_n + y_n - (x'_n + y'_n)) = \lim_{n\to\infty} (x_n - x'_n) + \lim_{n\to\infty} (y_n - y'_n) = 0.$$

Furthermore we know that $(y_n)_{n\in\mathbb{N}}$ and $(x'_n)_{n\in\mathbb{N}}$ are bounded and

$$x_n y_n - x'_n y'_n = (x_n - x'_n)y_n + x'_n(y_n - y'_n)$$

implies now that $\lim_{n\to\infty} (x_n y_n - x'_n y'_n) = 0$, i.e. $(x_n y_n)_{n\in\mathbb{N}} \backsim_\mathbb{R} (x'_n y'_n)_{n\in\mathbb{N}}$.

Next we claim that $(\mathcal{C}/\backsim_\mathbb{R}, \oplus, \odot)$ is a field. We will check only some of the axioms and the reader is invited to check the remaining ones. For the addition $\oplus$ we find for example

$$[(x_n)_{n\in\mathbb{N}}] \oplus [(y_n)_{n\in\mathbb{N}}] = [(x_n + y_n)_{n\in\mathbb{N}}]$$
$$= [(y_n + x_n)_{n\in\mathbb{N}}] = [(y_n)_{n\in\mathbb{N}}] \oplus [(x_n)_{n\in\mathbb{N}}].$$

Further, with $[0] := [(c_n)_{n\in\mathbb{N}}], c_n = 0$ for all $n \in \mathbb{N}$,

$$[(x_n)_{n\in\mathbb{N}}] \oplus [0] = [(x_n + c_n)_{n\in\mathbb{N}}] = [(x_n)_{n\in\mathbb{N}}]$$

or

$$[(x_n)_{n\in\mathbb{N}}] \oplus [(-x_n)_{n\in\mathbb{N}}] = [(x_n - x_n)_{n\in\mathbb{N}}] = [0].$$

For the multiplication $\odot$ we have for example with $[e] = [(e_n)_{n\in\mathbb{N}}], e_n = 1$ for $n \in \mathbb{N}$, that

$$[(x_n)_{n\in\mathbb{N}}] \odot [e] = [(x_n e_n)_{n\in\mathbb{N}}] = [(x_n)_{n\in\mathbb{N}}].$$

More delicate is to prove that if $[(x_n)_{n\in\mathbb{N}}] \neq [0]$, then we can find an inverse with respect to the multiplication. We observe that if $[(x_n)_{n\in\mathbb{N}}] \neq [0]$ there exists $\delta \in \mathbb{Q}, \delta > 0$, and $N(\delta) \in \mathbb{N}$ such that $|x_n| \geq \delta$ for all $n \geq N(\delta)$. If this is not the case then $(x_n)_{n\in\mathbb{N}}$ has a subsequence $(x_{n_k})_{k\in\mathbb{N}}$ converging to zero, and by Lemma 17.10.B we conclude that $(x_n)_{n\in\mathbb{N}}$ must converge to zero, i.e. $[(x_n)_{n\in\mathbb{N}}] = [0]$, which is a contradiction. (Note that the proof of Lemma 17.10.B works for Cauchy sequences in $\mathbb{Q}$.) For $(x_n)_{n\in\mathbb{N}} \in \mathcal{C}$ not equivalent to $(c_n)_{n\in\mathbb{N}}, c_n = 0$ for all $n \in \mathbb{N}$, we define

$$\tilde{x}_n := \begin{cases} x_n^{-1}, & n \geq N(\delta) \\ 0, n < N(\delta) \end{cases}$$

511

where $\delta$ and $N(\delta)$ are as before. We find

$$[(x_n)_{n\in\mathbb{N}}] \odot [(\tilde{x}_n)_{n\in\mathbb{N}}] = [(x_n \cdot \tilde{x}_n)_{n\in\mathbb{N}}]$$

where

$$x_n\tilde{x}_n = \begin{cases} 1, & n \geq N(\delta) \\ 0, & n < N(\delta), \end{cases}$$

which implies that $(x_n\tilde{x}_n)_{n\in\mathbb{N}} \backsim (e_n)_{n\in\mathbb{N}}$, $e_n = 1$ for $n \in \mathbb{N}$. The remaining axioms, in particular the associative laws and the distributivity law are proved in a straightforward way along the lines as indicated above.

We want to define an order structure on $\mathcal{C}/\backsim_\mathbb{R}$. We call $(x_n)_{n\in\mathbb{N}} \in \mathcal{C}$ **positive** if there exists $\delta \in \mathbb{Q}, \delta > 0$, and $N(\delta) \in \mathbb{N}$ such that $x_n \geq \delta$ for all $n \geq N(\delta)$. Again, our first task before looking at $\mathcal{C}/\backsim_\mathbb{R}$ is to prove that the definition is independent of the representative. For this let $(x_n)_{n\in\mathbb{N}} \in \mathcal{C}$ be positive and $(x'_n)_{n\in\mathbb{N}} \in \mathcal{C}$ be equivalent to $(x_n)_{n\in\mathbb{N}}$. Then there exists $\delta \in \mathbb{Q}, \delta > 0$, and $N(\delta) \in \mathbb{N}$ such that $x_n \geq \delta$ for $n \geq N_\delta$. Further, since $\lim_{n\to\infty}(x_n - x'_n) = 0$ we can find $\tilde{N}(\delta) \in \mathbb{N}$ such that $|x_n - x'_n| < \frac{\delta}{2}$ for all $n \geq \tilde{N}(\delta)$. This however implies for $n \geq \max(N(\delta), \tilde{N}(\delta))$ that $x'_n > x_n - \frac{\delta}{2} \geq \frac{\delta}{2}$ and hence $(x'_n)_{n\in\mathbb{N}}$ is positive too. With $[0] = [(c_n)_{n\in\mathbb{N}}], c_n = 0$ for all $n \in \mathbb{N}$, we define

$$[(x_n)_{n\in\mathbb{N}}] \geqslant [0] \tag{A.VI.10}$$

if and only if $(x_n)_{n\in\mathbb{N}}$ is positive. The claim is that $\left(\mathcal{C}/\backsim_\mathbb{R}, \oplus, \odot, \geqslant\right)$ is an ordered field. Again we will verify only some of the axioms and leave the rest to the reader.

For example, if $[(x_n)_{n\in\mathbb{N}}] \geqslant [0]$ and $[(y_n)_{n\in\mathbb{N}}] \geqslant [0]$ then we can find $\delta \in \mathbb{Q}$ and $N(\delta) \in \mathbb{N}$ such that $x_n \geq \delta$ and $y_n \geq \delta$ for $n \geq N(\delta)$, implying that $x_n + y_n \geq 2\delta$ for $n \geq N(\delta)$, hence $[(x_n)_{n\in\mathbb{N}}] \oplus [(y_n)_{n\in\mathbb{N}}] \geqslant [0]$, and further we find $x_n y_n \geq \delta^2$, i.e. $[(x_n)_{n\in\mathbb{N}}] \odot [(y_n)_{n\in\mathbb{N}}] \geqslant [0]$. It is a bit more tricky to show that one and only one of the statements

$$[(x_n)_{n\in\mathbb{N}}] = [0], \quad [(x_n)_{n\in\mathbb{N}}] \geqslant [0] \text{ and } [(x_n)_{n\in\mathbb{N}}] \neq [0], \quad [0] \geqslant [(x_n)_{n\in\mathbb{N}}] \text{ and } [(x_n)_{n\in\mathbb{N}}] \neq [0]$$

holds. Let $[(x_n)_{n\in\mathbb{N}}] \neq [0]$. We claim $[(|x_n|)_{n\in\mathbb{N}}] \geqslant [0]$. If this is not the case, then there exists a subsequence $(x_{n_k})_{k\in\mathbb{N}}$ of $(x_n)_{n\in\mathbb{N}}$ such that $|x_{n_k}| < \frac{1}{k}$ implying by Lemma 17.10.B that $(x_n)_{n\in\mathbb{N}} \backsim_\mathbb{R} (c_n)_{n\in\mathbb{N}}, c_n = 0$ for all $n \in \mathbb{N}$. Now, we know $|x_n| \geq \delta > 0$ for $n \geq N(\delta)$ and $(x_n)_{n\in\mathbb{N}}$ is a Cauchy sequence. Thus there exists $\tilde{N}(\delta) \in \mathbb{N}$ such that $n, m \geq \tilde{N}(\delta)$ implies $|x_m - x_n| < \frac{\delta}{2}$. For $m_0 \geq \max(N(\delta), \tilde{N}(\delta))$ it follows from $|x_n| \geq \delta$ that either $x_{m_0} \geq \delta$

or $-x_{m_0} \geq \delta$. In the first case we get $x_{m_0} - x_n \leq |x_{m_0} - x_n| < \frac{\delta}{2}$ or $x_n > x_{m_0} - \frac{\delta}{2} \geq \frac{\delta}{2}$, and in the second case we find $-x_n > -x_{m_0} - \frac{\delta}{2} \geq \frac{\delta}{2}$, proving that $[(|x_n|)_{n\in\mathbb{N}}]$ is indeed positive. Therefore either $[(x_n)_{n\in\mathbb{N}}] \ominus [0]$ or $[0] \ominus [(x_n)_{n\in\mathbb{N}}]$.

Thus we have already constructed an ordered field $(\mathcal{C}/\!\!\sim_{\mathbb{R}}, \oplus, \odot, \ominus)$. We now want to embed $\mathbb{Q}$ into $\mathcal{C}/\!\!\sim_{\mathbb{R}}$ while preserving all structures. For $q \in \mathbb{Q}$ we can form the class $[q]$ by defining $[q] := [(x_n)_{n\in\mathbb{N}}], x_n = q$ for all $n \in \mathbb{N}$. Consider

$$j : \mathbb{Q} \to \mathcal{C}/\!\!\sim_{\mathbb{R}}, \quad j(q) := [q]. \tag{A.VI.11}$$

Clearly, $q \neq q'$ implies $j(q) \neq j(q')$, i.e. $j$ is an injective mapping. Moreover the following hold (we leave the proofs for the reader):

$$j(q_1 + q_2) = [q_1] \oplus [q_2];$$
$$j(q_1 \cdot q_2) = [q_1] \odot [q_2];$$
$$q_1 \geq q_2 \text{ implies } [q_2] \ominus [q_2];$$
$$j^{-1}([q_1] \oplus [q_2]) = j^{-1}([q_1]) + j^{-1}([q_2]);$$
$$j^{-1}([q_1] \odot [q_2]) = j^{-1}([q_1]) j^{-1}([q_2]);$$
$$[q_1] \ominus [q_2] \text{ implies } j^{-1}([q_1]) \geq j^{-1}([q_2]).$$

These results show that $j(\mathbb{Q})$ is a subset of $\mathcal{C}/\!\!\sim_{\mathbb{R}}$ which is a subfield and respects the order relation, i.e. $j(\mathbb{Q})$ is in all structures isomorphic to $\mathbb{Q}$. With some further effort one can see that for $[(x_n)_{n\in\mathbb{N}}] \in \mathcal{C}/\!\!\sim_{\mathbb{R}}, [(x_n)_{n\in\mathbb{N}}] \ominus [0]$, there exists $[q] \in j(\mathbb{Q})$ such that $[(x_n)_{n\in\mathbb{N}}] \ominus [q] \ominus [0]$ for which we can of course write $[(x_n)_{n\in\mathbb{N}}] \ominus j(q) \ominus j(0)$. We can now introduce as usual the notation $\ominus$, $\oslash$, and $\olessthan$. Moreover we can define the absolute value on $\mathcal{C}/\!\!\sim_{\mathbb{R}}$ by

$$||(x_n)_{n\in\mathbb{N}}|| := \begin{cases} [(x_n)_{n\in\mathbb{N}}], & \text{if } [(x_n)_{n\in\mathbb{R}}] \ominus [0] \\ -[(x_n)_{n\in\mathbb{N}}], & \text{if } [(x_n)_{n\in\mathbb{N}}] \oslash [0], \end{cases} \tag{A.VI.12}$$

where $-[(x_n)_{n\in\mathbb{N}}]$ denotes the inverse of $[(x_n)_{n\in\mathbb{N}}]$ with respect to the addition $\oplus$. It is not difficult to see that

$$||(x_n)_{n\in\mathbb{N}}|| = [(|x_n|)_{n\in\mathbb{N}}] \tag{A.VI.13}$$

and in particular

$$|j(q)| = j(|q|) \tag{A.VI.14}$$

as well as

$$|j^{-1}([q])| = j^{-1}([|q|]) \tag{A.VI.15}$$

hold. It remains to prove that $\mathcal{C}/\backsim_{\mathbb{R}}$ is complete. In order to simplify notation from now on we often write $\mathbb{R}$ for $\mathcal{C}/\backsim_{\mathbb{R}}$ and $x \in \mathbb{R}$ for elements in $\mathcal{C}/\backsim_{\mathbb{R}}$. But we still make a distinction between $\mathbb{Q}$ and $j(\mathbb{Q}) \subset \mathbb{R}$. We also will use the easier notation $+$ for $\oplus$, $\cdot$ for $\odot$, $\geq$ for $\geqslant$, etc.

Using the absolute value as defined by (A.VI.12) we can now define convergence in $\mathbb{R}$ as we are used to: $(x_n)_{n\in\mathbb{N}}, x_n \in \mathbb{R}$, converges to $x \in \mathbb{R}$ if for every $\epsilon > 0, \epsilon \in \mathbb{R}$, there exists $N = N(\epsilon) \in \mathbb{N}$ such that $n \geq N(\epsilon)$ implies $|x_n - x| < \epsilon$. Further, $(x_n)_{n\in\mathbb{N}}, x_n \in \mathbb{R}$, is a Cauchy sequence in $\mathbb{R}$ if for every $\epsilon > 0, \epsilon \in \mathbb{R}$, there exists $N(\epsilon) \in \mathbb{N}$ such that $n, m \geq N(\epsilon)$ yields $|x_n - x_m| < \epsilon$.

We prove the completeness of $\mathbb{R}$ in three steps. First we prove that $(q_n)_{n\in\mathbb{N}}$, $q_n \in \mathbb{Q}$, is a Cauchy sequence in $\mathbb{Q}$ if and only if $(j(q_n))_{n\in\mathbb{N}}$ is a Cauchy sequence in $\mathbb{R} = \mathcal{C}/\backsim_{\mathbb{R}}$. Then we show that every Cauchy sequence $(j(q_n))_{n\in\mathbb{N}}$, $q_n \in \mathbb{Q}$, has a limit in $\mathbb{R}$. Eventually we will prove that every Cauchy sequence in $\mathbb{R}$ has a limit.

**Theorem A.VI.4.** *The sequence $(q_n)_{n\in\mathbb{N}}, q_n \in \mathbb{N}$, is a Cauchy sequence in $\mathbb{Q}$ if and only if the sequence $(j(q_n))_{n\in\mathbb{N}}$ is a Cauchy sequence in $\mathbb{R}$.*

*Proof.* Let $(q_n)_{n\in\mathbb{N}}, q_n \in \mathbb{N}$, be a Cauchy sequence in $\mathbb{Q}$ and $\epsilon > 0, \epsilon \in \mathbb{R}$. Then there exists $\epsilon' \in j(\mathbb{Q})$ such that $0 < \epsilon' < \epsilon$. Since $(q_n)_{n\in\mathbb{N}}$ is a Cauchy sequence in $\mathbb{Q}$, for $j^{-1}(\epsilon') > 0$ there exists $N(\epsilon)$ such that $n, m \geq N(\epsilon)$ implies $|q_n - q_m| < j^{-1}(\epsilon')$ and we conclude

$$
\begin{aligned}
|[q_n] - [q_m]| &= |j(q_n) - j(q_m)| \\
&= |j(q_n - q_m)| = j(|q_n - q_m|) < j(j^{-1}(\epsilon')) = \epsilon' < \epsilon,
\end{aligned}
$$

i.e. $(j(q_n))_{n\in\mathbb{N}}$ is a Cauchy sequence in $\mathbb{R}$.

Now let $(j(q_n))_{n\in\mathbb{N}}, q_n \in \mathbb{Q}$, be a Cauchy sequence in $\mathbb{R}$. Hence, given $\epsilon \in \mathbb{Q}, \epsilon > 0$, we can find $N(\epsilon) \in \mathbb{N}$ such that $n, m \geq N(\epsilon)$ implies $|j(q_n) - j(q_m)| < j(\epsilon)$, which yields

$$
\begin{aligned}
|q_n - q_m| &= |j^{-1}(q_n) - j^{-1}(q_m)| = |j^{-1}(q_n - q_m)| \\
&= j^{-1}(|q_n - q_m|) < j^{-1}(j(\epsilon)) = \epsilon.
\end{aligned}
$$

$\square$

**Theorem A.VI.5.** *Every Cauchy sequence $(j(q_n))_{n\in\mathbb{N}}, q_n \in \mathbb{Q}$, converges to a limit $x \in \mathbb{R}$.*

*Proof.* We have to prove the existence of $x \in \mathbb{R}$ such that (as a limit in $\mathbb{R}$) $\lim_{n\to\infty} j(q_n) = x$. Since $(j(q_n))_{n\in\mathbb{N}}$ is a Cauchy sequence in $\mathbb{R}$ it follows that $(q_n)_{n\in\mathbb{N}}$ is a Cauchy sequence in $\mathbb{Q}$. Consequently $(q_n)_{n\in\mathbb{N}}$ defines an element in $\mathbb{R}$ ($= \mathcal{C}/\sim_\mathbb{R}$), and this element we denote by $x$ and we claim $\lim_{n\to\infty} j(q_n) = x$. Given $\epsilon > 0, \epsilon \in \mathbb{R}$, we can find as before $\epsilon' > 0, \epsilon' \in j(\mathbb{Q})$, such that $0 < \epsilon' < \epsilon$. Since $(q_n)_{n\in\mathbb{N}}$ is a Cauchy sequence in $\mathbb{Q}$ there exists $N \in \mathbb{N}$ such that for $n, m \geq N$ we have

$$|q_n - q_m| < \frac{j^{-1}(\epsilon')}{2}.$$

For $m \in \mathbb{N}$ fixed consider the sequence $(y_n^{(m)})_{n\in\mathbb{N}}$ where

$$y_n^{(m)} := j^{-1}(\epsilon') - |q_n - q_m|.$$

With $(q_n)_{n\in\mathbb{N}}$ also $(q_n - q_m)_{n\in\mathbb{N}}$ is a Cauchy sequence, hence $(y_n^{(m)})_{n\in\mathbb{N}}$ is a Cauchy sequence in $\mathbb{Q}$. Moreover we have

$$y_n^{(m)} = j^{-1}(\epsilon') - |q_n - q_m| > j^{-1}(\epsilon') - \frac{j^{-1}(\epsilon')}{2} = \frac{j^{-1}(\epsilon')}{2} > 0,$$

i.e. $(y_n^{(m)})_{n\in\mathbb{N}}$ is a sequence of positive numbers, implying that

$$[(y_n^{(m)})_{n\in\mathbb{N}}] = [(j^{-1}(\epsilon') - |q_n - q_m|)_{n\in\mathbb{N}}] > 0,$$

hence

$$[(q_n - q_m)_{n\in\mathbb{N}}] < [j^{-1}(\epsilon')] = \epsilon',$$

or

$$|j(q_m) - x| = |[q_m] - x| = |[q_m] - [(q_n)_{n\in\mathbb{N}}]|$$
$$= |[(q_m - q_n)_{n\in\mathbb{N}}]| = [|(q_m - q_n)_{n\in\mathbb{N}}|] < \epsilon' < \epsilon.$$

Since $m \geq N$ was arbitrary it follows that $\lim_{m\to\infty} j(q_m) = x$. $\square$

**Corollary A.VI.6.** *For every $x \in \mathbb{R}$ and $\epsilon > 0$ there exists $x' \in j(\mathbb{Q})$ such that $|x' - x| < \epsilon$.*

*Proof.* Let $x \in \mathbb{R}$, i.e. $x = [(q_n)_{n\in\mathbb{N}}]$ for a Cauchy sequence $(q_n)_{n\in\mathbb{N}}, q_n \in \mathbb{Q}$. By Theorem A.VI.5 we have $\lim_{n\to\infty} j(q_n) = x$, so given $\epsilon > 0$ we choose $N(\epsilon) \in \mathbb{N}$ such that $|j(q_n) - x| < \epsilon$ for $n \geq N(\epsilon)$ and it follows that $|x' - x| < \epsilon$ for $x' = j(q_{N(\epsilon)+1})$. $\square$

Eventually we can prove

**Theorem A.VI.7.** *In $\mathbb{R}$ every Cauchy sequence converges.*

*Proof.* Let $(x_n)_{n\in\mathbb{R}}$ be a Cauchy sequence in $\mathbb{R}$. We have to prove the existence of $x \in \mathbb{R}$ such that $\lim_{n\to\infty} x_n = x$. Let $(\epsilon_n)_{n\in\mathbb{N}}, \epsilon_n > 0$, be a sequence in $\mathbb{R}$ such that $\lim_{n\to\infty} \epsilon_n = 0$. (Any sequence $(\eta_n)_{n\in\mathbb{N}}, \eta_n \in \mathbb{Q}$, such that $\eta_n > 0$ and $\lim_{n\to\infty} \eta_n = 0$ will induce on $\mathbb{R}$ such a sequence by $\epsilon_n := j(\eta_n)$.)
For $n \in \mathbb{N}$ there exists $q_n \in \mathbb{Q}$ such that

$$|j(q_n) - x_n| < \epsilon_n.$$

We claim that $(j(q_n))_{n\in\mathbb{N}}$ is a Cauchy sequence in $\mathbb{R}$. For $n, m \in \mathbb{N}$ we find

$$|j(q_n) - j(q_m)| \leq |j(q_n) - x_n| + |x_n - x_m| + |j(q_m) - x_m|$$
$$\leq \epsilon_n + \epsilon_m + |x_n - x_m|.$$

Since $\lim_{n\to\infty} \epsilon_n = 0$ and $(x_n)_{n\in\mathbb{N}}$ is a Cauchy sequence, given $\epsilon > 0$ we can find $N(\epsilon) \in \mathbb{N}$ such that $n, m \geq N(\epsilon)$ implies

$$\epsilon_n < \frac{\epsilon}{3}, \quad \epsilon_m < \frac{\epsilon}{3}, \quad |x_n - x_m| < \frac{\epsilon}{3},$$

or $|j(q_n) - j(q_m)| < \epsilon$, i.e. $(j(q_n))_{n\in\mathbb{N}}$ is a Cauchy sequence in $\mathbb{R}$. By Theorem A.VI.4 we know that $(q_n)_{n\in\mathbb{N}}$ must be a Cauchy sequence in $\mathbb{Q}$. We define $x := [(q_n)_{n\in\mathbb{N}}]$ and show that $\lim_{n\to\infty} x_n = x$. From Theorem A.VI.5 we deduce that $\lim_{n\to\infty} j(q_n) = x$ and therefore

$$|x_n - x| \leq |x_n - q_n| + |q_n - x| < \epsilon_n + |q_n - x|.$$

Given $\epsilon > 0$, since $\lim_{n\to\infty} \epsilon_n = 0$ and $\lim_{n\to\infty} q_n = x$, we can find $N \in \mathbb{N}$ such that $n \geq N$ yields $\epsilon_n < \frac{\epsilon}{2}$ and $|q_n - x| < \frac{\epsilon}{2}$, or for $n \geq N$

$$|x_n - x| < \frac{\epsilon}{2} + \frac{\epsilon}{2} = \epsilon,$$

which implies $\lim_{n\to\infty} x_n = x$. $\square$

Our presentation follows that in K. Endl and W. Luh [3]. However we have left some of the details to the reader (including the fact that $\mathbb{R}$ as constructed is an Archimedian field). The reason why the proof is so long is partly because we have a lot of structure on $\mathbb{Q}$ which needs to be transferred to $\mathbb{R}$: algebraic structures, order structure and convergence (topological structure).

Finally we want to mention a further possibility of constructing $\mathbb{R}$ from $\mathbb{Q}$. Let $A, B \subset \mathbb{R}$ be two non-empty sets such that $A \cup B = \mathbb{R}$. In addition we require for all $a \in A$ and $b \in B$ that $a < b$. We call such a pair of subsets of $\mathbb{R}$ a **Dedekind cut** and denote it by $(A|B)$. Further we call any $s \in \mathbb{R}$ a separating number for $(A|B)$ if for all $a \in A$ and $b \in B$ we have $a \leq s \leq b$. Equivalent to our axiom of completeness is:

**Axiom D**

> Every Dedekind cut has exactly one separating number.

As it stands Axiom D is "artificial" as our Axiom of Completeness. However we may introduce Dedekind cuts first in $\mathbb{Q}$ and then prove that with the help of these cuts we can construct $\mathbb{R}$. In W. Rudin [11] this construction is given in detail.

Axiom A and Axiom D use the order structure of $\mathbb{R}$ (or $\mathbb{Q}$) and look more natural than the Axiom of Completeness using Cauchy sequences. However the construction using Cauchy sequences extends to many more situations where no order structure is given but just a metric.

# Appendix VII: Limes Superior and Limes Inferior

The concepts of limes superior and limes inferior are difficult ones, and students will have to spend time to understand these ideas and how to work with them. While we have proved some basic properties of lim sup and lim inf in the main text or in the solved problems, we believe that it is of some benefit to students to have a more detailed list of properties of lim sup and lim inf. We will not give proofs, however many detailed proofs can be found in R. L. Schilling [12].

In the following $(a_n)_{n \in \mathbb{N}}$ and $(b_n)_{n \in \mathbb{N}}$ are sequences of real numbers and $\lambda > 0$ is a fixed real number.

$$\limsup_{n \to \infty} a_n = -\liminf_{n \to \infty}(-a_n), \; \liminf_{n \to \infty} a_n = -\limsup_{n \to \infty}(-a_n); \qquad \text{(A.VII.1)}$$

$$\liminf_{n \to \infty} a_n \leq \limsup_{n \to \infty} a_n; \qquad \text{(A.VII.2)}$$

if $a' \in \mathbb{R}$ is an accumulation point of $(a_n)_{n \in \mathbb{N}}$ then $\qquad \text{(A.VII.3)}$

$$\liminf_{n \to \infty} a_n \leq a' \leq \limsup_{n \to \infty} a_n,$$

and $\liminf_{n \to \infty} a_n$ as well as $\limsup_{n \to \infty} a_n$ are accumulation points of $(a_n)_{n \in \mathbb{N}}$;

if $\lim_{n \to \infty} = a \in \mathbb{R}$ exists then $\limsup_{n \to \infty} a_n = \liminf_{n \to \infty} a_n = \lim_{n \to \infty} a_n; \quad \text{(A.VII.4)}$

$$\limsup_{n \to \infty}(a_n + b_n) \leq \limsup_{n \to \infty} a_n + \limsup_{n \to \infty} b_n; \qquad \text{(A.VII.5)}$$

$$\liminf_{n \to \infty}(a_n + b_n) \geq \liminf_{n \to \infty} a_n + \liminf_{n \to \infty} b_n; \qquad \text{(A.VII.6)}$$

$$\liminf_{n \to \infty}(a_n + b_n) \leq \liminf_{n \to \infty} a_n + \limsup_{n \to \infty} b_n \leq \limsup_{n \to \infty}(a_n + b_n); \qquad \text{(A.VII.7)}$$

if $\lim_{n \to \infty} b_n = b \in \mathbb{R}$ exists then $\qquad \text{(A.VII.8)}$

$$\limsup_{n \to \infty}(a_n + b_n) = \limsup_{n \to \infty} a_n + \lim_{n \to \infty} b_n,$$

$$\liminf_{n \to \infty}(a_n + b_n) = \liminf_{n \to \infty} a_n + \lim_{n \to \infty} b_n;$$

if $a_n \geq 0$ and $b_n \geq 0$ for all $n \in \mathbb{N}$ then $\hspace{3cm}$ (A.VII.9)

$$\limsup_{n \to \infty}(a_n b_n) \leq \left(\limsup_{n \to \infty} a_n\right)\left(\limsup_{n \to \infty} b_n\right),$$

$$\liminf_{n \to \infty}(a_n b_n) \geq \left(\liminf_{n \to \infty} a_n\right)\left(\liminf_{n \to \infty} b_n\right);$$

if $a_n \geq 0$ and $b_n \geq 0$ for all $n \in \mathbb{N}$ then $\hspace{3cm}$ (A.VII.10)

$$\liminf_{n \to \infty}(a_n b_n) \leq \left(\liminf_{n \to \infty} a_n\right)\left(\limsup_{n \to \infty} b_n\right) \leq \limsup_{n \to \infty}(a_n b_n);$$

if $a_n \geq 0$ and $b_n \geq 0$ for all $n \in \mathbb{N}$ and $\lim\limits_{n \to \infty} b_n = b \in \mathbb{R}$ exists, then we have

$$\hspace{9cm} \text{(A.VII.11)}$$

$$\limsup_{n \to \infty}(a_n b_n) = \left(\limsup_{n \to \infty} a_n\right)\left(\lim_{n \to \infty} b_n\right),$$

$$\liminf_{n \to \infty}(a_n b_n) = \left(\liminf_{n \to \infty} a_n\right)\left(\lim_{n \to \infty} b_n\right),$$

in particular for $\lambda > 0$ it follows

$$\limsup_{n \to \infty}(\lambda a_n) = \lambda \limsup_{n \to \infty} a_n, \ \liminf_{n \to \infty}(\lambda a_n) = \lambda \liminf_{n \to \infty} a_n;$$

$$\limsup_{n \to \infty}|a_n| = 0 \text{ implies } \lim_{n \to \infty} a_n = 0; \hspace{2cm} \text{(A.VII.12)}$$

$$\limsup_{n \to \infty} a_n = \infty \text{ if and only if } \liminf \frac{1}{a_n} = 0, \hspace{1cm} \text{(A.VII.13)}$$

$$\liminf_{n \to \infty} a_n = \infty \text{ if and only if } \limsup_{n \to \infty} \frac{1}{a_n} = 0;$$

if $0 < \liminf\limits_{n \to \infty} a_n \leq \limsup\limits_{n \to \infty} a_n < \infty$ then $\hspace{2cm}$ (A.VII.14)

$$\limsup_{n \to \infty} \frac{1}{a_n} = \frac{1}{\liminf_{n \to \infty} a_n},$$

$$\liminf_{n \to \infty} \frac{1}{a_n} = \frac{1}{\limsup_{n \to \infty} a_n};$$

if $(a_n)_{n \in \mathbb{N}}$ is bounded then $\hspace{5cm}$ (A.VII.15)

$$\liminf_{n\to\infty} a_n \leq \liminf_{n\to\infty} \frac{a_1 + \cdots + a_n}{n} \leq \limsup_{n\to\infty} \frac{a_1 + \cdots + a_n}{n} \leq \limsup_{n\to\infty} a_n;$$

if $(a_n)_{n\in\mathbb{N}}$ is bounded and $a_n > 0$ then $\qquad$ (A.VII.16)

$$\liminf_{n\to\infty} a_n \leq \liminf_{n\to\infty} \sqrt[n]{a_1 \cdot \ldots \cdot a_n} \leq \limsup_{n\to\infty} \sqrt[n]{a_1 \cdot \ldots \cdot a_n} \leq \limsup_{n\to\infty} a_n.$$

A proof of (A.VII.15) and (A.VII.16) is given for example in H. Heuser, [5, Section 28].

# Appendix VIII: Connected Sets in $\mathbb{R}$

In this appendix we provide proofs for Theorem 19.25 and Theorem 19.27.

Recall that Theorem 19.25 states that a non-empty subset of $\mathbb{R}$ is connected if and only if it is an interval.

*Proof of Theorem 19.25.* Suppose that $A \subset \mathbb{R}$ is not an interval. It follows that there exist $a < b < c$ such that $a, c \in A$ and $b \notin A$. Define $\mathcal{O}_1 := (-\infty, b)$ and $\mathcal{O}_2 = (b, \infty)$. Clearly $\mathcal{O}_1 \cap \mathcal{O}_2 = \emptyset$ and both sets are open. Moreover $A \cap \mathcal{O}_1$ and $A \cap \mathcal{O}_2$ are non-empty and $A \subset \mathcal{O}_1 \cup \mathcal{O}_2$. Thus $A$ has a non-trivial splitting and is therefore not connected, and we have proved that for $A$ to be connected it is necessary that $A$ is an interval.

Next we prove that $[a, b] \subset \mathbb{R}$ is connected. Suppose that $[a, b]$ is not connected and that $\{\mathcal{O}_1, \mathcal{O}_2\}$ is a non-trivial splitting of $[a, b]$ with $a \in \mathcal{O}_1$. Define

$$c := \sup\{x \in \mathbb{R} | [a, x] \subset \mathcal{O}_1 \cap [a, b]\}.$$

If $c < b$ and $c \in \mathcal{O}_1$ then there exists $\eta > 0$ such that $[c - \eta, c + \eta] \subset \mathcal{O}_1 \cap [a, b]$ and $[a, c + \eta] \subset \mathcal{O}_1 \cap [a, b]$ which is a contradiction. Consequently $c \in \mathcal{O}_2 \cap [a, b]$ and $[c - \delta, c + \delta] \subset \mathcal{O}_2 \cap [a, b]$ for some $\delta > 0$. But now we find for $c - \delta \leq x$ that $[a, x]$ is not a subset of $\mathcal{O}_1 \cap [a, b]$ which again contradicts the definition of $c$. Hence $c = b$, $[a, b) \subset \mathcal{O}_1 \cap [a, b]$ and $\{b\} = \mathcal{O}_2 \cap [a, b]$. But $\mathcal{O}_2$ is open and therefore either $\mathcal{O}_2 \cap [a, b]$ is empty or contains more than one point. Now any open interval $(a, b)$ has the representation

$$(a, b) = \bigcup_{m \geq m_0} \left[a + \frac{1}{m}, b - \frac{1}{m}\right], \mathbb{R} = \bigcup_{m \in \mathbb{N}} [-m, m],$$

$$(-\infty, b) = \bigcup_{m \in \mathbb{N}} \left[-m, b - \frac{1}{m}\right], (a, \infty) = \bigcup_{m \in \mathbb{N}} \left[a - \frac{1}{m}, m\right],$$

and a half-open interval is of the type $I \cup \{c\}$ where $I$ is an open interval with $c$ being an end point. Thus if we can prove that the union of intersecting connecting sets is connected and noting that $\{a\}$, $a \in \mathbb{R}$, is trivially connected we are done.

Our claim is: let $A_j \subset \mathbb{R}$, $j \in J \neq \emptyset$, be a family of connected sets such that $\bigcap_{j \in J} A_j \neq \emptyset$. Then $\bigcup_{j \in J} A_j$ is connected too.

Suppose $\bigcup_{j \in J} A_j$ is not connected and let $\{\mathcal{O}_1, \mathcal{O}_2\}$ be a non-trivial splitting of $\bigcup_{j \in J} A_j$ such that $\mathcal{O}_1 \cap \bigcap_{j \in J} A_j \neq \emptyset$. Since $A_j$ is connected it follows that

$A_j \cap \mathcal{O}_2 = \emptyset$ for all $j \in A$, implying $\mathcal{O}_2 \cap \bigcap_{j \in J} A_j = \emptyset$ which is a contradiction. Hence $\bigcup_{j \in J} A_j$ does not have a non-trivial splitting and therefore it is connected. $\qquad \square$

Next we prove Theorem 19.27 which states that every open set in $\mathbb{R}$ is a denumerable union of disjoint open intervals.

*Proof of Theorem 19.27.* Let $A \subset \mathbb{R}$ be open and $x \in A$. Then there exists $\delta > 0$ such that $(x, x+\delta) \subset A$ and $(x-\delta, x) \subset A$. Let $b := \sup\{y | (x, y) \subset A\}$ and $a := \inf\{z | (z, x) \subset A\}$. Clearly $a < x < b$ and $I_x := (a, b)$ is an open interval containing $x$. We claim $I_x \subset A$. Take $w \in I_x$, and assume $x < w < b$, the case $a < w < x$ goes analogously. The definition of $b$ implies the existence of $y > w$ such that $(x, y) \subset A$ but $w \in (x, y)$, so $w \in A$. Next we prove that $b \notin A$ (the fact that $a \notin A$ goes analogously). Suppose $b \in A$. In this case there would exist some $\epsilon > 0$ such that $(b - \epsilon, b + \epsilon) \subset A$, hence $(x, b + \epsilon) \subset A$ contradicting the definition of $b$. We consider now $(I_x)_{x \in A}$. Each $x \in A$ belongs to some of these intervals, for example $I_x$, and each $I_x$ is contained in $A$, thus $A = \bigcup_{x \in A} I_x$. We want to prove that either $I_{x_1} \cap I_{x_2} = \emptyset$ or $x_1 = x_2$. Let $I_{x_1}$ and $I_{x_2}$, $x_1, x_2 \in A$, say $I_{x_1} = (\alpha_1, \beta_1)$ and $I_{x_2} = (\alpha_2, \beta_2)$, and suppose $x \in (\alpha_1, \beta_1) \cap (\alpha_2, \beta_2)$. In this case it follows that $\alpha_2 < \beta_1$ and $\alpha_1 < \beta_2$. But $\alpha_2 \notin A$ hence $\alpha_2 \notin (\alpha_1, \beta_1)$ and therefore $\alpha_2 \leq \alpha_1$. Since $\alpha_1 \notin A$ and hence $\alpha_2 \notin (\alpha_2, \beta_2)$ we have $\alpha_1 \leq \alpha_2$, i.e. $\alpha_1 = \alpha_2$. Similarly we can prove $\beta_1 = \beta_2$ to get $(\alpha_1, \beta_1) = (\alpha_2, \beta_2)$. Thus if $I_{x_1} \cap I_{x_2} \neq \emptyset$ then $I_{x_1} = I_{x_2}$. So we have already proved that $A$ is the union of disjoint open intervals. By Theorem 19.11 each of these intervals must contain a rational number. But the rational numbers are countable and no rational number can belong to two of these intervals. Hence we have at most countably many open intervals, the union of which is $A$. $\qquad \square$

# Solutions to Problems of Part 1

## Chapter 1

1. The set $\{\phi\}$ is not empty. It contains one element, the empty set, i.e. $\phi \in \{\phi\}$.

2.     a) For a real number $x$ belonging to the set
$\{x \in \mathbb{R} \,|\, x^2 = 16 \text{ and } 2x + 3 = 12\}$ two conditions must be satisfied: $x^2 = 16$ and $2x + 3 = 12$. The first condition implies that $x = 4$ or $x = -4$, however the second condition implies that $x = \frac{9}{2}$. Hence we cannot satisfy both conditions so the set is empty.

    b) For a rational number to belong to the set
$\{x \in \mathbb{Q} \,|\, x^2 = 9 \text{ and } 3x - 6 = 3\}$ the following two conditions must be satisfied: $x^2 = 9$ and $3x - 6 = 3$. The first condition gives $x = 3$ or $x = -3$ while the second one leads to $x = 3$. Hence $\{x \in \mathbb{Q} \,|\, x^2 = 9 \ a \ 3x - 6 = 3\} = \{3\} \neq \phi$.

    c) It is clear that the set $\{x \in \mathbb{R} \,|\, x \neq x\}$ is empty. There is no real number not equal to itself.

    d) The condition $x^2 = \frac{1}{4}$ implies that $x = \frac{1}{2}$ or $x = -\frac{1}{2}$, both are not integers, hence the set $\{x \in \mathbb{Z} \,|\, x^2 = \frac{1}{4}\}$ is empty.

    e) Since $x^2 = \frac{1}{4}$ implies that $x = \frac{1}{2}$ or $x = -\frac{1}{2}$ and they are both rational numbers, it follows that $\{x \in \mathbb{Q} \,|\, x^2 = \frac{1}{4}\} = \{\frac{1}{2}, -\frac{1}{2}\}$, hence the set is a non-empty set.
Note that this is different to problem d): In both problems we have to deal with the same condition $x^2 = \frac{1}{4}$. However, we seek integers in problem d) while in problem e) we seek rational numbers.

3.     a) Since every element in $A$ is an odd integer we have $A \subset B$.

    b) 9 is not a prime number, therefore $A$ is not a subset of $C$: there is (at least) one element in $A$ which does not belong to $C$.

    c) Every number belonging to $C$ is an odd integer, hence belonging to $B$, then $C \subset B$.

4. The set $\mathbb{Z} \setminus M$ consists of all integers $x \in \mathbb{Z}$ that do not belong to $M$, i.e. in order to belong to $\mathbb{Z} \setminus M$ a number $x$ must be an integer and $x < 5$. Therefore we have

$$\mathbb{Z} \setminus M = \{x \in \mathbb{Z} \,|\, x < 5\} = \{x \in \mathbb{Z} \,|\, x \leq 4\}.$$

5. The set $R = \{k \in \mathbb{N} \,|\, k^2 \leq 10\}$ consists of all the numbers $1, 2$ and $3$, i.e. $R = \{1, 2, 3\}$. Consequently we have

$$B \setminus R = \{1, 2, 3, 4, 5, 6\} \setminus \{1, 2, 3\} = \{4, 5, 6\}.$$

6.     a) The condition $5x + 7 = 13$ implies $x = \frac{6}{5} \notin \mathbb{Z}$ which gives

$$\{x \in \mathbb{Z} \,|\, 5x + 7 = 13\} = \phi.$$

b) This is the same condition as in a), but now we seek rational solutions and $\frac{6}{5} \in \mathbb{Q}$. Therefore we have

$$\{x \in \mathbb{Q} \mid 5x + 7 = 13\} = \left\{\frac{6}{5}\right\}.$$

c) Now we have to handle the inequality $5x + 7 \leq 13$ which is equal to $x \leq \frac{6}{5}$. However, only integer solutions are allowed, which leads to

$$\{x \in \mathbb{Z} \mid 5x + 7 \leq 13\} = \{x \in \mathbb{Z} \mid x \leq 1\}.$$

7.  a)

$$
\frac{-7}{3}\left(\frac{27}{8} - \frac{18}{5}\right) = \frac{-7}{3}\left(\frac{27 \cdot 5 - 18 \cdot 8}{8 \cdot 5}\right)
$$
$$
= \frac{-7}{3}\left(\frac{135 - 144}{40}\right)
$$
$$
= \frac{-7}{3}\left(-\frac{9}{40}\right) = \frac{7}{3} \cdot \frac{9}{40} = \frac{7 \cdot 3}{40} = \frac{21}{40}.
$$

b)

$$
\frac{\frac{3}{4} + \frac{7}{12}}{\frac{2}{19} - \frac{1}{7}} = \frac{\frac{3 \cdot 3}{3 \cdot 4} + \frac{7}{12}}{\frac{2 \cdot 7}{19 \cdot 7} - \frac{19}{19 \cdot 7}} = \frac{\frac{9+7}{12}}{-\frac{5}{133}}
$$
$$
= -\frac{\frac{16}{12}}{\frac{5}{133}} = \frac{\frac{4}{3}}{\frac{5}{133}} = \frac{4}{3} \cdot \frac{133}{5} = \frac{532}{15}.
$$

c)

$$
\frac{4^2 - 3^3}{5^2 + 19} = \frac{16 - 27}{25 + 19} = -\frac{11}{44} = -\frac{1}{4}.
$$

8.  a)

$$
\frac{3a + 4(a+b)^2 - 6a(\frac{1}{2}+b) - 2b(a+2b)}{\frac{1}{2}(a+b)}
$$
$$
= \frac{3a + 4(a^2 + 2ab + b^2) - 3a - 6ab - 2ab - 4b^2}{\frac{1}{2}(a+b)}
$$
$$
= \frac{4a^2 + 8ab + 4b^2 - 8ab - 4b^2}{\frac{1}{2}(a+b)}
$$
$$
= \frac{4a^2}{\frac{1}{2}(a+b)} = \frac{8a^2}{a+b}.
$$

b) We need to prove that

$$
\frac{1}{2}(a^2 - 3b^2 - c^2 - 2ab + 4bc) = \frac{1}{4}(a+b-c)(2a-6b+2c).
$$

526

Now

$$\frac{1}{4}(a + b - c)(2a - 6b + 2c) = \frac{1}{2}(a + b - c)(a - 3b + c)$$
$$= \frac{1}{2}(a^2 + ab - ac - 3ab - 3b^2 + 3bc + ac + bc - c^2)$$
$$= \frac{1}{2}(a^2 - 3b^2 - c^2 - 2ab + 4bc)$$

and therefore the result is proved.

c)

$$\frac{a - b}{a + b} + \frac{4ab}{(a + b)^2} - \frac{a + b}{a - b}$$
$$= \frac{(a - b)(a + b)(a - b)}{(a + b)^2(a - b)} + \frac{4ab(a - b)}{(a + b)^2(a - b)} - \frac{(a + b)(a + b)^2}{(a + b)^2(a - b)}$$
$$= \frac{(a^2 - b^2)(a - b) + 4a^2b - 4ab^2 - (a + b)^3}{(a + b)^2(a - b)}$$
$$= \frac{a^3 - ab^2 - a^2b + b^3 + 4a^2b - 4ab^2 - a^3 - 3a^2b - 3ab^2 - b^3}{(a + b)^2(a - b)}$$
$$= -\frac{8ab^2}{(a + b)^2(a - b)}.$$

d)

$$\frac{x^3 - y^3}{y - x} - y^4x^2\left(\frac{1}{y^3x} - \frac{x}{y} + \frac{y}{x}\right)$$
$$= -(x^2 + xy + y^2) - yx + y^3x^3 - y^5x$$
$$= x^3y^3 - x^2 - 2xy - xy^5 - y^2.$$

9.

$$\frac{\frac{1}{9}\left(\frac{8}{11} - \frac{2}{9}\right)\left(\frac{12}{5} - \frac{6}{7}\right)}{\frac{8}{3}\left(\frac{3}{4} - \frac{7}{2}\right)}$$
$$= \frac{\frac{1}{9}\left(\frac{72}{99} - \frac{22}{99}\right)\left(\frac{84}{35} - \frac{30}{35}\right)}{\frac{8}{3} \cdot \left(-\frac{11}{4}\right)}$$
$$= -\frac{\frac{1}{9} \cdot \frac{50}{99} \cdot \frac{54}{35}}{\frac{82}{12}} = -\frac{\frac{1 \cdot 50 \cdot 54}{9 \cdot 99 \cdot 35}}{\frac{22}{3}}$$
$$= -\frac{\frac{10 \cdot 6}{9 \cdot 11 \cdot 7}}{\frac{22}{3}} = -\frac{\frac{20}{3 \cdot 11 \cdot 7}}{\frac{22}{3}} = -\frac{\frac{20}{11 \cdot 7}}{\frac{22}{1}}$$
$$= -\frac{20}{77 \cdot 22} = -\frac{10}{77 \cdot 11} = -\frac{10}{847}.$$

527

10.    a)

$$\left(\frac{2}{3}\right)^3 - \left(\frac{1}{2}\right)^4 + 5\left(\frac{8}{9}\right)$$
$$= \frac{8}{27} - \frac{1}{16} + \frac{40}{9}$$
$$= \frac{8\cdot 16 - 27 + 40\cdot 3\cdot 16}{27\cdot 16}$$
$$= \frac{2021}{432}.$$

b)

$$\frac{\left(\frac{2}{5}\right)^3 - \left(\frac{3}{8}\right)^2}{\frac{19}{40}} = \frac{\frac{8}{125} - \frac{9}{64}}{\frac{19}{40}}$$
$$= \frac{\frac{512-1125}{8000}}{\frac{19}{40}} = -\frac{\frac{613}{200}}{19}$$
$$= -\frac{613}{3800}.$$

11.    a)

$$\frac{(a+b)^3 - (b-a)^2(b+a)}{4ab}$$
$$= \frac{a^3 + 3a^2b + 3ab^2 + b^3 - (b^2 - 2ab + a^2)(b+a)}{4ab}$$
$$= \frac{a^3 + 3a^2b + 3ab^2 + b^3 - b^3 + a^2b + ab^2 - a^3}{4ab}$$
$$= \frac{4a^2b + 4ab^2}{4ab} = a+b;$$

b)

$$\frac{\left(\frac{a}{b}\right)^3 - \left(\frac{b}{a}\right)^4}{a^2b^3}$$
$$= \frac{\frac{a^3}{b^3} - \frac{b^4}{a^4}}{a^2b^3} = \frac{a^7 - b^7}{a^6b^6}.$$

12. a) $\sqrt{625} = 25$; b) $\sqrt{\frac{225}{49}} = \frac{15}{7}$; c) $\sqrt{\frac{a^4b^6}{(a+b)^2}} = \frac{a^2b^3}{a+b}$.

13.    a) First we observe that
$$3x - 12 \geq -7$$
is equivalent to
$$3x \geq 5$$
or
$$x \geq \frac{5}{3}.$$
Hence every $x \in \mathbb{R}$ with $x \geq \frac{5}{3}$ solves the above inequality.

$\frac{5}{3}$

528

b) Note that

$$\frac{7}{4} + \frac{2}{5}x \le \frac{3}{8}x$$

is equivalent to

$$\frac{2}{5}x - \frac{3}{8}x \le -\frac{7}{4},$$

i.e.

$$\frac{1}{40}x \le -\frac{7}{4}$$

or

$$x \le -70.$$

Thus every $x \in \mathbb{R}$ satisfying $x \le -70$ solves this inequality.

$$-70$$

c) In order to have $(x - 3)(x + 4) \ge 0$ we must have that either

$$(x - 3) \ge 0 \quad \text{and} \quad (x + 4) \ge 0$$

or

$$(x - 3) \le 0 \quad \text{and} \quad (x + 4) \le 0$$

is true.
The first pair of inequalities imply

$$x \ge 3 \quad \text{and} \quad x \ge -4$$

hence whenever $x \ge 3$ then $(x - 3)(x + 4) \ge 0$.
The second pair of inequalities give

$$x \le 3 \quad \text{and} \quad x \le -4$$

which yields that for all $x \le -4$ we have $(x - 3)(x + 4) \ge 0$.

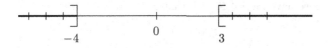

$$-4 \qquad\qquad 0 \qquad\qquad 3$$

14. The term $x^{y^z}$ is not well defined. For $x = 2, y = 3, z = 2$ we have

$$x^y = 2^3 = 8 \quad \text{and therefore} \quad (x^y)^z = 8^2 = 64$$

however, since $y^z = 3^2 = 9$, it follows that

$$x^{(y^z)} = 2^9 = 512.$$

Thus $(x^y)^z \ne x^{(y^z)}$ and therefore the brackets are needed.

529

15.    a) Using that $b^{-1} = \frac{1}{b}, d^{-1} = \frac{1}{d}$ we have

$$\frac{1}{b} + \frac{1}{d} = b^{-1} + d^{-1} = (b^{-1} + d^{-1})\frac{b \cdot d}{b \cdot d} = (b^{-1}(b \cdot d) + d^{-1}(b \cdot d))\frac{1}{b \cdot d}$$

$$= ((b^{-1} \cdot b) \cdot d + (d^{-1} \cdot d)b)\frac{1}{b \cdot d} = (d + b)\frac{1}{b \cdot d} = \frac{d + b}{b \cdot d}.$$

   b) We first show that $(x^{-1})^{-1} = x$ for $x \neq 0$. Since $(x^{-1})^{-1}x^{-1} = 1$ and $x \cdot x^{-1} = 1$ it follows that

$$(x^{-1})^{-1} \cdot x^{-1} = x \cdot x^{-1}$$

or

$$(x^{-1})^{-1}x^{-1} \cdot x = x \cdot x^{-1} \cdot x,$$

i.e.

$$(x^{-1})^{-1} = x$$

and using fractions we get $\frac{1}{\frac{1}{x}} = x$. Now we find

$$\frac{\frac{a}{b}}{\frac{c}{d}} = \frac{a}{b} \cdot \left(\frac{c}{d}\right)^{-1} = \frac{a}{b}\left(c \cdot \frac{1}{d}\right)^{-1} = \frac{a}{b} \cdot c^{-1}\left(\frac{1}{d}\right)^{-1} = \frac{a}{b} \cdot \frac{1}{c} \cdot d = \frac{ad}{bc}.$$

16.    a) A straightforward calculation gives

$$a\left(x + \frac{b}{2a}\right)^2 - \frac{b^2}{4a} + c$$

$$= a\left(x^2 + 2\frac{bx}{2a} + \frac{b^2}{4a^2}\right) - \frac{b^2}{4a} + c$$

$$= ax^2 + bx + \frac{b^2}{4a} - \frac{b^2}{4a} + c$$

$$= ax^2 + bx + c,$$

therefore the equivalence is established.

   b) By part (a), we have for $x \in \mathbb{R}$ such that $ax^2 + bx + c = 0$ that

$$a\left(x + \frac{b}{2a}\right)^2 = \frac{b^2}{4a} - c$$

or

$$\left(x + \frac{b}{2a}\right)^2 = \frac{1}{4a^2}(b^2 - 4ac).$$

By assumption $b^2 - 4ac \geq 0$, therefore we can take the square root on the right hand side to get

$$\sqrt{\frac{1}{4a^2}(b^2 - 4ac)} = \frac{1}{2a}\sqrt{b^2 - 4ac}.$$

Now we wish to take the square root on the left hand side. If $x + \frac{b}{2a} \geq 0$ we have no problem to find

$$x + \frac{b}{2a} = \frac{1}{2a}\sqrt{b^2 - 4ac},$$

or

$$x = -\frac{b}{2a} + \frac{1}{2a}\sqrt{b^2 - 4ac}.$$

If $x + \frac{b}{2a} \leq 0$, we know that $-\left(x + \frac{b}{2a}\right) = -x - \frac{b}{2a} \geq 0$, however $\left(x + \frac{b}{2a}\right)^2 = \left(-x - \frac{b}{2a}\right)^2$. Thus we have

$$\left(-x - \frac{b}{2a}\right)^2 = \frac{1}{4a^2}(b^2 - 4ac)$$

or

$$-x - \frac{b}{2a} = \frac{1}{2a}\sqrt{b^2 - 4ac},$$

implying

$$x = -\frac{b}{2a} - \frac{1}{2a}\sqrt{b^2 - 4ac}.$$

Thus so long as $b^2 - 4ac \geq 0$ we find that the **solutions of the quadratic equation** $ax^2 + bx + c = 0$ are

$$x_1 = -\frac{b}{2a} + \frac{1}{2a}\sqrt{b^2 - 4ac}$$

and

$$x_2 = -\frac{b}{2a} - \frac{1}{2a}\sqrt{b^2 - 4ac}.$$

If $b^2 = 4ac$ we have only one solution $x_1 = x_2 = -\frac{b}{2a}$.

## Chapter 2

1. Since $A^C = \{x \in X \mid x \notin A\}$ we have $A^C = \{e, f, g, h, i\}$. The set $A \cap C$ is given by those elements belonging to both the sets $A$ and $C$, hence,

$$A \cap C = \{c, d\}.$$

Now we find

$$(A \cap C)^C = \{a, b, e, f, g, h, i\}.$$

The set $B \setminus C$ consists of every $x \in X$ which belongs to $B$ but does not belong to $C$, so we find

$$B \setminus C = \{b, h\}.$$

Finally, since $A \cup B = \{a, b, c, d, f, h\}$, we have

$$(A \cup B)^C = \{e, g, i\}.$$

531

2.  a) From the definition we know

$$B_4(2) = \{x \in \mathbb{R} \,|\, |x - 2| < 4\} = \{x \in \mathbb{R} \,|\, -4 < x - 2 < 4\} = \{x \in \mathbb{R} \,|\, -2 < x < 6\}$$

and analogously

$$B_3(8) = \{x \in \mathbb{R} \,|\, |x - 8| < 3\} = \{x \in \mathbb{R} \,|\, 5 < x < 11\}.$$

Thus for $x \in B_4(2) \cap B_3(8)$ the two sets of inequalities

$$-2 < x < 6 \quad \text{and} \quad 5 < x < 11$$

must be true, i.e. $x$ must satisfy $5 < x < 6$, so

$$B_4(2) \cap B_3(8) = \{x \in \mathbb{R} \,|\, 5 \le x < 6\}.$$

Here is the graphical solution to the problem:

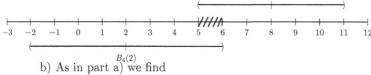

b) As in part a) we find

$$B_2(5) = \{x \in \mathbb{R} \,|\, 3 < x < 7\}$$

and

$$B_7(-2) = \{x \in \mathbb{R} \,|\, -9 < x < 5\}$$

implying that

$$B_2(5) \cap B_7(-2) = \{x \in \mathbb{R} \,|\, 3 \le x < 5\}$$

and therefore we have

$$(B_2(5) \cap B_7(-2))^{\complement} = \{x \in \mathbb{R} \,|\, x < 3 \text{ or } x \ge 5\}.$$

The graphical solution to the problem is the following:

c) We have

$$\left(-3, \frac{3}{2}\right) \cup \left[-\frac{1}{4}, \frac{7}{3}\right] = \left\{x \in \mathbb{R} \,\middle|\, -3 < x < \frac{3}{2} \text{ or } -\frac{1}{4} < x < \frac{7}{3}\right\}$$

$$= \left\{x \in \mathbb{R} \,\middle|\, -3 < x < \frac{7}{3}\right\},$$

which yields

$$\left(\left(-3, \frac{3}{2}\right) \cup \left[-\frac{1}{4}, \frac{7}{3}\right]\right)^{\complement} = \left\{x \in \mathbb{R} \,\middle|\, x \le -3 \text{ or } x \ge \frac{7}{3}\right\}.$$

Graphically we have:

532

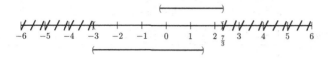

d) Since

$$\left[-2, \frac{7}{3}\right) = \left\{x \in \mathbb{R} \,\middle|\, -2 \le x < \frac{7}{3}\right\}$$

and further

$$\left[\frac{3}{5}, \frac{15}{4}\right] = \left\{x \in \mathbb{R} \,\middle|\, \frac{3}{5} \le x \le \frac{15}{4}\right\}$$

we find

$$\left[-2, \frac{7}{3}\right) \cap \left[\frac{3}{5}, \frac{15}{4}\right] = \left[\frac{3}{5}, \frac{7}{3}\right)$$

which we can also see from the following:

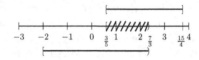

3.  a) It is true that $x \in A \cap B$ implies $x \in A$ and $x \in B$, hence

$$x \in A \cap B \implies x \in A$$

or we can write $A \cap B \subset A$.
On the other hand $x \in A$ implies $x \in A$ or $x \in B$, i.e.

$$x \in A \implies x \in A \cup B,$$

or we can write $A \subset A \cup B$.

b) Let $x \in (A \setminus B) \cap B$. Then $x \in A \setminus B$ and $x \in B$, or

$$(x \in A \wedge x \notin B) \wedge x \in B$$

which implies that $x \notin B$ and $x \in B$ which does not hold for any $x$, i.e. $(A \backslash B) \cap B = \phi$.

c) The statement $x \in B \setminus A$ means that $x \in B$ and $x \notin A$, which is equivalent to $x \in B$ and $x \in A^C$ which is the statement that $x \in B \cap A^C$.

4.  a) The following holds:

$$x \in (A \cap B) \cup C$$
$$\iff (x \in A \cap B) \vee (x \in C)$$
$$\iff ((x \in A) \wedge (x \in B)) \vee (x \in C)$$
$$\iff ((x \in A) \vee (x \in C)) \wedge ((x \in B) \vee (x \in C))$$
$$\iff x \in (A \cup C) \cap (B \cup C).$$

533

b) We use a truth table to prove this equality:

| $x \in A$ | $x \in A^{\complement}$ | $x \in B$ | $x \in B^{\complement}$ | $x \in A \cup B$ | $x \in (A \cup B)^{\complement}$ | $x \in A^{\complement} \cap B^{\complement}$ |
|---|---|---|---|---|---|---|
| $T$ | $F$ | $T$ | $F$ | $T$ | $F$ | $F$ |
| $T$ | $F$ | $F$ | $T$ | $T$ | $F$ | $F$ |
| $F$ | $T$ | $T$ | $F$ | $T$ | $F$ | $F$ |
| $F$ | $T$ | $F$ | $T$ | $F$ | $T$ | $T$ |

Since the last two columns coincide it follows that

$$(A \cup B)^{\complement} = A^{\complement} \cap B^{\complement}.$$

5. We prove

$$\text{a)} \implies \text{b)} \implies \text{c)} \implies \text{d)} \implies \text{e)} \implies \text{f)} \implies \text{a)}.$$

Now $A \subset B$ means $x \in A \implies x \in B$. Therefore $x \in A$ implies $x \in A$ and $x \in B$, hence $x \in A \cap B$, i.e. $A \subset A \cap B$. The inclusion $A \cap B \subset A$ has already been proved (see Problem 3 a)). Thus $A \subset B$ implies $A \cap B = A$, i.e. a) implies b).
Next observe that $A \cap B = A$ is equivalent to $(A \cap B)^{\complement} = A^{\complement}$, or $A^{\complement} \cup B^{\complement} = A^{\complement}$. Thus $x \in B^{\complement}$ must imply $x \in A^{\complement}$, $B^{\complement} \subset A^{\complement}$, proving b) implies c).
Suppose now that $B^{\complement} \subset A^{\complement}$. We have already proved that $A \subset B$ implies $A \cap B = A$. Thus $B^{\complement} \subset A^{\complement}$ implies $A^{\complement} \cap B^{\complement} = B^{\complement}$. Taking the complement on both sides gives $(A^{\complement} \cap B^{\complement})^{\complement} = B$, but $(A^{\complement} \cap B^{\complement})^{\complement} = A \cup B$, so we find $A \cup B = B$ and therefore c) $\implies$ d).
If $A \cup B = B$ then $A^{\complement} \cup B = A^{\complement} \cup A \cup B = X$, since $A^{\complement} \cup A = X$. Therefore we have proved that d) implies e).
In order to prove that e) implies f) we only need to take complements: if $B \cup A^{\complement} = X$ then $(B \cup A^{\complement})^{\complement} = X^{\complement}$ or $B^{\complement} \cap A = \phi$.
Finally, we show that $A \cap B^{\complement} = \phi$ implies $A \subset B$. Note that $A \cap B^{\complement} = \phi$ means that $x \in A$ and $x \in B^{\complement}$ cannot hold, i.e. $x \in A$ implies $x \in B$ which is $A \subset B$ of course, i.e. we have proved that f) implies a).

6. We have:

a)
$$\left| -\frac{5}{8} \right| = \frac{5}{8};$$

b)
$$\left| \frac{11}{3} - 3 \right| = \left| \frac{11 - 9}{3} \right| = \left| \frac{2}{3} \right| = \frac{2}{3};$$

c)
$$\left| \frac{7}{9} - \frac{12}{5} \right| = \left| \frac{35 - 108}{45} \right| = \left| -\frac{73}{45} \right| = \frac{73}{45};$$

d)
$$\left| \, |-3| - |-5| \, \right| = |3 - 5| = |-2| = 2;$$

e) $\sqrt{a^2} = |a|$;
(note that $a$ can be non-positive since $a \in \mathbb{R}$).

534

7. For $\varepsilon > 0$ and $a, b \in \mathbb{R}$ we have by (2.9):

$$|ab| = \left| \sqrt{2\varepsilon}a \frac{1}{\sqrt{2\varepsilon}}b \right|$$

$$\leq \frac{(\sqrt{2\varepsilon}a)^2}{2} + \frac{\left( \frac{1}{\sqrt{2\varepsilon}}b \right)^2}{2} = \varepsilon a^2 + \frac{1}{4\varepsilon}b^2.$$

In order to prove

$$\min\{a, b\} = \frac{1}{2}\left( a + b - |a - b| \right),$$

note that $a \leq b$ implies $a = \min\{a, b\}$ as well as $a - b \leq 0$ or $|a - b| = b - a$ implying

$$\frac{1}{2}\left( a + b - |a - b| \right) = \frac{1}{2}\left( a + b - (b - a) \right) = \frac{1}{2}2a = a.$$

However, if $b \leq a$ we have $b = \min\{a, b\}$ and $a - b \geq 0$ which gives $|a - b| = a - b$ and therefore

$$\frac{1}{2}\left( a + b - |a - b| \right) = \frac{1}{2}\left( a + b - (a - b) \right) = \frac{1}{2}2b = b.$$

Since $a > 0$, we deduce that

$$a^2 + 1 \geq 2a,$$

or

$$a^2 - 2a + 1 = (a - 1)^2 \geq 0,$$

which is clearly correct. Every step is an equivalent formulation of the previous one, hence we have the equivalence of

$$a + \frac{1}{a} \geq 0 \quad \text{and} \quad (a - 1)^2 \geq 0.$$

8. We may use the triangle inequality (2.11). Therefore we can easily see that

$$|a - c| = |a - b + b - c| \leq |a - b| + |b - c|.$$

For the second estimate we use the converse triangle inequality, i.e. (2.55), which states for $\alpha, \beta \in \mathbb{R}$, that
$||\alpha| - |\beta|| \leq |\alpha - \beta|$. Now with $\alpha = |a - b|$ and $\beta = c$ we find

$$\left| |a - b| - |c| \right| \leq \left| |a - b| - c \right|$$

and the triangle inequality gives:

$$\left| |a - b| - c \right| \leq |a - b| + |c| \leq |a| + |b| + |c|.$$

9.     a) For $x \in \mathbb{R}$
$$8x - 11 > -24x + 89$$
is equivalent to
$$32x > 100$$
or
$$x > \frac{25}{8}.$$
Thus $8x - 11 > -24x + 89$ holds for all $x > \frac{25}{8}$.

b) We have to satisfy two inequalities:
$$-3 \leq 7x - 2 \quad \text{and} \quad 7x - 2 < 6x + 5.$$
The first yields:
$$-\frac{1}{7} \leq x;$$
and the second:
$$x < 7.$$
We must now be careful, we only seek integer solutions of the system
$$-\frac{1}{7} \leq x < 7,$$
namely $x_1 = 0, x_2 = 1, x_3 = 2, x_4 = 3, x_5 = 4, x_6 = 5, x_7 = 6$. For this we may write: the solution set is given by $\{0, 1, 2, 3, 4, 5, 6\}$.

c) We discuss the following four cases:
(i) $x - 3 \geq 0$ and $x + 3 \geq 0$, i.e. $x \geq 3$ and $x \geq -3$, which implies $x \geq 3$;
(ii) $x - 3 \geq 0$ and $x + 3 \leq 0$, i.e. $x \geq 3$ and $x \leq -3$, which cannot happen;
(iii) $x - 3 \leq 0$ and $x + 3 \geq 0$, i.e. $x \leq 3$ and $x \geq -3$ which means $x \in [-3, 3]$;
(iv) $x - 3 \leq 0$ and $x + 3 \leq 0$, i.e. $x \leq 3$ and $x \leq -3$ which implies $x \leq -3$.
We now consider each case:
In case (i) $|x - 3| \leq |x + 3|$ is equivalent to
$$x - 3 \leq x + 3$$
which holds for all $x$, hence for $x \geq 3$ the inequality holds.
In case (ii) $|x - 3| \leq |x + 3|$ can never hold.
In case (iii) $|x - 3| \leq |x + 3|$ is equivalent to
$$-(x - 3) \leq x + 3 \quad \text{or} \quad -x + 3 \leq x + 3,$$
which can only hold for $x \geq 0$, then for $x \in [0, 3]$ the inequality has a solution.
In case (iv) $|x - 3| \leq |x + 3|$ is equivalent to
$$-(x - 3) \leq -(x + 3) \quad \text{or} \quad -x + 3 \leq -x - 3$$
which never holds.
Thus the inequality $|x - 3| \leq |x + 3|$ holds for all $x \geq 0$.

10.     a) Note that
$$2x + 6(2 - x) \geq 8 - 2x$$

is equivalent to
$$2x + 12 - 6x \geq 8 - 2x$$

or
$$-4x + 12 \geq 8 - 2x,$$

i.e.
$$-2x \geq -4$$

or $x \leq 2$. Thus the inequality is solved by every $x \in \mathbb{R}$, $x \leq 2$.

b) First note that
$$x^2 + 2x - 10 < 3x + 2$$

is equivalent to
$$x^2 - x - 12 < 0.$$

We now factorise the left hand side noting that $x^2 - x - 12 = (x-4)(x+3)$. (We find this factorisation by determining the roots of the quadratic equation $x^2 - x - 12 = 0$.) The condition $(x - 4)(x + 3) < 0$ is fulfilled either if $x - 4 > 0$ and $x + 3 < 0$ or if $x - 4 < 0$ and $x + 3 > 0$.
In the first case we have:
$$x > 4 \quad \text{and} \quad x < -3,$$

and in this case there is no solution.
The second case holds if
$$x < 4 \quad \text{and} \quad x > -3,$$

implying that every $x \in (-3, 4)$ solves this inequality.

**Chapter 3**

1.      a) For $k = 0$ we have
$$0^3 + 1^3 + 2^3 = 1 + 8 = 9$$

which is divisible by 9. If the statement holds for $k$, then we find for $k + 1$ that
$$(k + 1)^3 + (k + 2)^3 + (k + 3)^3$$
$$= (k + 1)^3 + (k + 2)^3 + k^3 + 3 \cdot 3k^2 + 3 \cdot 3^2 k + 3^3$$
$$= (k^3 + (k + 1)^3 + (k + 2)^3) + 9(k^2 + 3k + 3).$$

Now the first term $k^3 + (k + 1)^3 + (k + 2)^3$ as well as the second term is divisible by 9 and the result follows by mathematical induction.
b) For $n = 0$ we find
$$\frac{0^5}{5} + \frac{0^4}{2} + \frac{0^3}{3} - \frac{0}{30} = 0 \in \mathbb{Z}.$$

Suppose now that

$$\frac{n^5}{5} + \frac{n^4}{2} + \frac{n^3}{3} - \frac{n}{30}$$

is an integer. We need to show that

$$\frac{(n+1)^5}{5} + \frac{(n+1)^4}{2} + \frac{(n+1)^3}{3} - \frac{n+1}{30}$$

is an integer too. Expanding all terms we arrive at

$$\frac{n^5 + 5n^4 + 10n^3 + 10n^2 + 5n + 1}{5} + \frac{n^4 + 4n^3 + 6n^2 + 4n + 1}{2}$$
$$+ \frac{n^3 + 3n^2 + 3n + 1}{3} - \frac{n+1}{30}$$
$$= \frac{n^5}{5} + \frac{n^4}{2} + \frac{n^3}{3} - \frac{n}{30}$$
$$+ (n^4 + 2n^3 + 2n^2 + n) + (2n^3 + 3n^2 + 2n)$$
$$+ (n^2 + n) + \frac{6 + 15 + 10}{30} - \frac{1}{30}.$$

Now by our assumption

$$\frac{n^5}{5} + \frac{n^4}{2} + \frac{n^3}{3} - \frac{n}{30}$$

is an integer. Moreover $(n^4 + 2n^3 + 2n^2 + n)$, $(2n^3 + 3n^2 + 2n)$, $(n^2 + n)$ and $\frac{31}{30} - \frac{1}{30}$ are integers. Therefore the result follows by mathematical induction.

2.    a) For $n = 1$ we have $x^1 - y^1 = 1 \cdot (x - y)$. For $n \in \mathbb{N}$ suppose that the following hold

$$x^n - y^n = (x - y)Q_n(x, y).$$

We need to show that for $x^{n+1} - y^{n+1}$ we have a similar factorisation. Since

$$x^{n+1} - y^{n+1} = xx^n - yy^n = xx^n - xy^n + xy^n - yy^n$$
$$= x(x^n - y^n) + (x - y)y^n$$
$$= x(x - y)Q_n(x, y) + (x - y)y^n$$
$$= (x - y)(xQ_n(x, y) + y^n)$$

we have a factorisation as required with $Q_{n+1}(x, y) = xQ_n(x, y) + y^n$, and the result follows.

b) For $n = 1$, the statement reduces to

$$(1 - 1)x + y^1 \geq 1x^{n-1}y$$

or $y = y$ which is of course correct. Now for $n \in \mathbb{N}$ fixed suppose that we have

$$(*) \qquad (n - 1)x^n + y^n \geq nx^{n-1}y.$$

538

We want to prove that

$$nx^{n+1} + y^{n+1} \geq (n+1)x^n y.$$

Since $x > 0$ we may multiply $(*)$ by $x$ to obtain

$$(n-1)x^{n+1} + y^n x > nx^n y,$$

then adding $x^{n+1}$ and subtracting $y^n x$ yields

$$nx^{n+1} \geq nx^n y + x^{n+1} - y^n x,$$

and adding $y^{n+1}$ leads to:

$$nx^{n+1} + y^{n+1} \geq nx^n y + x^{n+1} - y^n x + y^{n+1}$$
$$= (n+1)x^n y + x^{n+1} - y^n x + y^{n+1} - x^n y.$$

Thus we need to show that

$$x^{n+1} - y^n x + y^{n+1} - x^n y \geq 0.$$

Note that

$$x^{n+1} - y^n x + y^{n+1} - x^n \cdot y = x^n(x-y) + y^n(y-x)$$
$$= (x^n - y^n)(x-y).$$

Now if $x > y$ then $x - y > 0$ as well as $x^n - y^n > 0$. However if $x < y$ then $x^n - y^n < 0$. In both cases we find that $(x^n - y^n)(x-y) \geq 0$ and the inequality follows from mathematical induction.

3.  a)

$$\sum_{j=-2}^{2} \frac{1}{2^j} = \frac{1}{2^{-2}} + \frac{1}{2^{-1}} + \frac{1}{2^0} + \frac{1}{2^1} + \frac{1}{2^2}$$

$$= 2^2 + 2 + 1 + \frac{1}{2} + \frac{1}{4} = 7\frac{3}{4}.$$

b)

$$\sum_{k=2}^{5}(a^k - a^{k-2}) = a^2 - a^{2-2} + a^3 - a^{3-2} + a^4 - a^{4-2} + a^5 - a^{5-2}$$

$$= a^2 - 1 + a^3 - a + a^4 - a^2 + a^5 - a^3$$

$$= a^5 + a^4 - a - 1.$$

539

c)

$$\sum_{l=1}^{6}(-1)^l\frac{l+1}{l} = (-1)^1\frac{1+1}{1} + (-1)^2\frac{2+1}{2} + (-1)^3\frac{3+1}{3}$$

$$+ (-1)^4\frac{4+1}{4} + (-1)^5\frac{5+1}{5} + (-1)^6\frac{6+1}{6}$$

$$= -2 + \frac{3}{2} - \frac{4}{3} + \frac{5}{4} - \frac{6}{5} + \frac{7}{6}$$

$$= \left(-2 - \frac{4}{3} - \frac{6}{5}\right) + \left(\frac{3}{2} + \frac{5}{4} + \frac{7}{6}\right)$$

$$= -\frac{68}{15} + \frac{47}{12} = \frac{-272+235}{60} = -\frac{37}{60}.$$

4.  a) In both cases our formal proof will use induction. However before giving the formal proofs let us rewrite the statement as

$$\lambda(a_1 + \ldots + a_N) = (\lambda a_1 + \ldots + \lambda a_N)$$

and

$$(a_1 + \ldots + a_N) + (b_1 + \ldots + b_N) = (a_1 + b_1) + \ldots + (a_N + b_N),$$

thus we get a feeling for the content of these statements: the first is an extension of the law of distributivity, the second follows as an extension of the commutativity of addition. Here are the formal proofs:
for $N = 1$ we obviously have

$$\lambda\sum_{j=1}^{1}a_j = \lambda a_1 = \sum_{j=1}^{1}(\lambda a_j).$$

Now if

$$\lambda\sum_{j=1}^{N}a_j = \sum_{j=1}^{N}(\lambda a_j)$$

then it follows that

$$\lambda\sum_{j=1}^{N+1}a_j = \lambda\left(\sum_{j=1}^{N}a_j + a_{N+1}\right)$$

$$= \lambda\sum_{j=1}^{N}a_j + \lambda a_{N+1}$$

$$= \sum_{j=1}^{N}(\lambda a_j) + \lambda a_{N+1} = \sum_{j=1}^{N+1}(\lambda a_j).$$

Further for $N = 1$ we have:

$$\sum_{j=1}^{1}a_j + \sum_{j=1}^{1}b_j = a_1 + b_1 = \sum_{j=1}^{1}(a_j + b_j).$$

540

If

$$\sum_{j=1}^{N} a_j + \sum_{j=1}^{N} b_j = \sum_{j=1}^{N} (a_j + b_j)$$

holds then we find

$$\sum_{j=1}^{N+1} a_j + \sum_{j=1}^{N+1} b_j = \sum_{j=1}^{N} a_j + a_{N+1} + \sum_{j=1}^{N} b_j + b_{N+1}$$

$$= \sum_{j=1}^{N} (a_j + b_j) + (a_{N+1} + b_{N+1})$$

$$= \sum_{j=1}^{N+1} (a_j + b_j).$$

Hence both statements follow by mathematical induction.

b) Applying the results of part a) we note that:

$$(x - y) \sum_{k=0}^{5} x^k y^{5-k} = \sum_{k=0}^{5} x^{k+1} y^{5-k} - \sum_{k=0}^{5} x^k y^{6-k}$$

$$= xy^5 + x^2 y^4 + x^3 y^3 + x^4 y^2 + x^5 y + x^6$$

$$- y^6 - xy^5 - x^2 y^4 - x^3 y^3 - x^4 y^2 - x^5 y$$

$$= x^6 - y^6.$$

5. We prove each of the following identities by mathematical induction:

a) For $n = 1$ we have

$$\sum_{k=1}^{1} \frac{1}{(2k-1)(2k+1)} = \frac{1}{(2-1)(2+1)} = \frac{1}{3} = \frac{1}{2 \cdot 1 + 1}.$$

Now if

$$\sum_{k=1}^{n} \frac{1}{(2k-1)(2k+1)} = \frac{n}{2n+1}$$

then it follows that

$$\sum_{k=1}^{n+1} \frac{1}{(2k-1)(2k+1)} = \sum_{k=1}^{n} \frac{1}{(2k-1)(2k+1)} + \frac{1}{(2(n+1)-1)(2(n+1)+1)}$$

$$= \frac{n}{2n+1} + \frac{1}{(2n+1)(2n+3)}$$

$$= \frac{n(2n+3)+1}{(2n+1)(2n+3)} = \frac{(2n+1)(n+1)}{(2n+1)(2n+3)} = \frac{n+1}{2n+3},$$

which proves the statement.

541

b) For $k = 1$ we find

$$\sum_{n=1}^{1} n \cdot n! = 1 \cdot 1! = 1 = (1+1)! - 1.$$

Next we observe that

$$\begin{aligned}
\sum_{n=1}^{k+1} n \cdot n! &= \sum_{n=1}^{k} nn! + (k+1)(k+1)! \\
&= (k+1)! - 1 + (k+1)(k+1)! \\
&= (k+2)(k+1)! - 1 = (k+2)! - 1
\end{aligned}$$

proving the assertion.

c) For $m = 1$ it follows that

$$\sum_{j=1}^{1}(a + (j-1)d) = a + (1-1)d = a$$

$$= \frac{1}{2}1(2a + (1-1)d).$$

If

$$\sum_{j=1}^{m}(a + (j-1)d) = \frac{1}{2}m(2a + (m-1)d),$$

holds then

$$\begin{aligned}
\sum_{j=1}^{m+1}(a + (j-1)d) &= \sum_{j=1}^{m}(a + (j-1)d) + a + ((m+1) - 1)d \\
&= \frac{1}{2}m(2a + (m-1)d) + a + md \\
&= \frac{1}{2} \cdot 2am + a + \frac{1}{2}m(m-1)d + md \\
&= \frac{1}{2}2a(m+1) + \frac{1}{2}(m+1)md \\
&= \frac{1}{2}(m+1)(2a + md).
\end{aligned}$$

6.   a)

$$\prod_{k=-2}^{2} 2^{-k} = 2^{-(-2)} \cdot 2^{-(-1)} \cdot 2^{-(0)} \cdot 2^{-1} \cdot 2^{-2}$$

$$= 2^2 \cdot 2 \cdot 1 \cdot 2^{-1} \cdot 2^{-2} = 1.$$

542

b)

$$\prod_{j=3}^{6}(j-4) = (3-4)(4-4)(5-4)(6-4) = 0.$$

c)

$$\prod_{j=1}^{5} \frac{j+2}{j+4} = \frac{3}{5} \cdot \frac{4}{6} \cdot \frac{5}{7} \cdot \frac{6}{8} \cdot \frac{7}{9} = \frac{1}{6}.$$

7. Again, as in Problem 4 a), we first rewrite the statement to understand its content:

$$\prod_{j=1}^{N}(\mu a_j) + \prod_{j=1}^{N}(\nu a_j)$$

$$= \mu a_1 \cdot \mu a_2 \cdot \ldots \cdot \mu a_N + \nu a_1 \cdot \nu a_2 \cdot \ldots \cdot \nu a_N$$

$$= \mu^N a_1 \cdot \ldots \cdot a_N + \nu^N a_1 \cdot \ldots \cdot a_N$$

$$= (\mu^N + \nu^N)a_1 \cdot \ldots \cdot a_N = (\mu^N + \nu^N)\prod_{j=1}^{N} a_j.$$

Here is the formal proof by induction:
for $N = 1$ we have

$$\prod_{j=1}^{1} \mu a_j + \prod_{j=1}^{1} \nu a_j = \mu a_1 + \nu a_1 = (\mu + \nu)a_1 = (\mu + \nu)\prod_{j=1}^{1} a_j.$$

Next we observe that

$$\prod_{j=1}^{N+1} \mu a_j + \prod_{j=1}^{N+1} \nu a_j = \left(\prod_{j=1}^{N} \mu a_j\right)\mu a_{N+1} + \left(\prod_{j=1}^{N} \nu a_j\right)\nu a_{N+1}$$

$$= \mu^N \left(\prod_{j=1}^{N} a_j\right)\mu a_{N+1} + \nu^N \left(\prod_{j=1}^{N} a_j\right)\nu a_{N+1}$$

$$= \mu^{N+1} \prod_{j=1}^{N+1} a_j + \nu^{N+1} \prod_{j=1}^{N+1} a_j.$$

8.  a)

$$7! = 1 \cdot 2 \cdot 3 \cdot 4 \cdot 5 \cdot 6 \cdot 7 = 5040$$

and

$$\frac{63!}{60!} = \frac{60! \, 61 \cdot 62 \cdot 63}{60!} = 61 \cdot 62 \cdot 63 = 238,266.$$

b)

$$\frac{(n+1)! - n!}{n} = \frac{(n+1)n! - n!}{n} = \frac{((n+1)-1)n!}{n} = n!$$

c)
$$\frac{(n+1)!}{(n-1)!} = \frac{(n-1)! \cdot n \cdot (n+1)}{(n-1)!} = n(n+1).$$

9.    a) For $n = 2$ we find

$$\prod_{k=1}^{2} \frac{2k-1}{2k} = \frac{1}{2} \cdot \frac{3}{4} = \frac{3}{8} = \frac{1}{2^4}\binom{4}{2} = \frac{6}{16}.$$

Now we want to show that the statement for $n$ implies that for $n+1$:

$$\prod_{k=1}^{n+1} \frac{2k-1}{2k} = \left(\prod_{k=1}^{n} \frac{2k-1}{2k}\right) \frac{2(n+1)-1}{2(n+1)}$$

$$= \frac{1}{2^{2n}}\binom{2n}{n} \frac{2n+1}{2n+2}$$

$$= \frac{1}{2^{2(n+1)}} \frac{(2n)!}{n!n!} \frac{2(2n+1)}{n+1}.$$

Thus it remains to prove that:

$$\frac{(2n)!}{n!n!} \frac{2(2n+1)}{n+1} = \binom{2(n+1)}{n+1} = \frac{(2(n+1))!}{(n+1)!(n+1)!}.$$

Note that

$$\frac{(2n)!2(2n+1)}{n!n!(n+1)} = \frac{2(2n+1)!}{(n+1)!n!}$$

and

$$\frac{(2(n+1))!}{(n+1)!(n+1)!} = \frac{(2n+1)!(2n+2)}{(n+1)!(n+1)!} = \frac{2(2n+1)!(n+1)}{(n+1)!(n+1)!} = \frac{2(2n+1)!}{(n+1)!n!}$$

and the identity is now proved.

b) Since

$$\prod_{k=1}^{1-1}\left(1+\frac{1}{k}\right)^k = \prod_{k=1}^{0}\left(1+\frac{1}{k}\right)^k = 1$$

the statement is true for $n = 1$. Now under the assumption that the statement holds for $n$ we get for $n+1$ that:

$$\prod_{k=1}^{n}\left(1+\frac{1}{k}\right)^k = \left(\prod_{k=1}^{n-1}\left(1+\frac{1}{k}\right)^k\right)\left(1+\frac{1}{n}\right)^n$$

$$= \frac{n^n}{n!}\left(1+\frac{1}{n}\right)^n = \frac{n^n}{n!}\left(\frac{n+1}{n}\right)^n = \frac{(n+1)^n}{n!}$$

$$= \frac{(n+1)^n(n+1)}{n!(n+1)} = \frac{(n+1)^{n+1}}{(n+1)!},$$

and the assertion is proved.

544

10.    a)

$$(5x^2 + 3y)^4 = \sum_{k=0}^{4} \binom{4}{k} (5x^2)^{4-k} (3y)^k$$

$$= \binom{4}{0}(5x^2)^4 + \binom{4}{1}(5x^2)^3(3y) + \binom{4}{2}(5x^2)^2(3y)^2 + \binom{4}{3}(5x^2)(3y)^3 + \binom{4}{4}(3y)^4$$

$$= 625x^8 + 1500x^6 y + 1350x^4 y^2 + 540x^2 y^3 + 81y^4.$$

b)

$$(x - y)^n = \sum_{k=0}^{n} \binom{n}{k} x^{n-k}(-y)^k$$

$$= \sum_{k=0}^{n} (-1)^k \binom{n}{k} x^{n-k} y^k.$$

11.    a) By definition we have

$$\binom{n}{k} = \frac{n!}{(n-k)!k!} = \frac{n(n-1)\cdot\ldots\cdot(n-k+1)(n-k)\cdot\ldots\cdot 2\cdot 1}{((n-k)(n-k-1)\cdot\ldots\cdot 2\cdot 1)(1\cdot 2\cdot\ldots\cdot k)}$$

$$= \frac{n(n-1)\cdot\ldots\cdot(n-k+1)}{1\cdot 2\cdot\ldots\cdot k}.$$

b) Using the definition for $\binom{\frac{1}{2}}{k}$ we find

$$\binom{\frac{1}{2}}{k} = \frac{\frac{1}{2}(\frac{1}{2}-1)(\frac{1}{2}-2)\cdot\ldots\cdot(\frac{1}{2}-k+1)}{1\cdot 2\cdot\ldots\cdot k}$$

$$= \frac{\frac{1}{2^k}(1(1-2)(1-4)\cdot\ldots\cdot(1-2k+2))}{1\cdot 2\cdot\ldots\cdot k}$$

$$= (-1)^{k-1}\frac{1(2-1)(4-1)\cdot\ldots\cdot(2k-2-1)}{2^k 1\cdot 2\cdot\ldots\cdot k}$$

$$= (-1)^{k-1}\frac{1\cdot 3\cdot\ldots\cdot(2k-3)}{2\cdot 4\cdot\ldots\cdot(2k)}.$$

12. We use mathematical induction:

a) Since by assumption $p \geq 2$, we have the correct statement $p \geq 1$ for $k = 1$. Now suppose that $p^k > k$. We want to prove $p^{k+1} > k+1$. Since

$$pp^k > pk$$

and

$$kp \geq 2k = k+1,$$

it follows that

$$p^{k+1} > k+1.$$

545

b) For $k = 1$ it is true that $p > 1$ for $p \geq 3$ and also for $k = 2$ we have $p^2 > k^2$ since $p \geq 3$. (Note that $p \geq 2$ is not sufficient to get the strict inequality.) Assume $p^k > k^2$ and $k \geq 2$. Multiplying by $p$ yields:

$$p^{k+1} > pk^2 \geq 3k^2$$

and it remains to prove that $3k^2 \geq (k+1)^2$ which is equivalent to

$$3k^2 \geq k^2 + 2k + 1$$

or

$$2k^2 \geq 2k + 1$$

which is equivalent to

$$k^2 + (k-1)^2 \geq 2,$$

which holds since $k \geq 2$. Thus by mathematical induction the statement holds for all $k \geq 2$. The case $k = 1$ has already been proved.

c) Note that for $k = 2, 3$ and $4$ the statement is false. For $k = 5$ we have:

$$2^5 = 32 > 25 = 5^2.$$

If we multiply $2^k > k^2$ by 2 we find

$$2^{k+1} > 2k^2$$

and the proof reduces to show that $2k^2 \geq (k+1)^2$ or $k^2 \geq 2k + 1$ which follows from $(k-1)^2 \geq 0$.

13.    a) For $N = 1$ we have:

$$\sum_{j=1}^{1} \frac{1}{\sqrt{j}} = 1 \leq 2\sqrt{1} = 2.$$

Now under the assumption that the statement holds for $N$ we find for $N + 1$ that:

$$\sum_{j=1}^{N+1} \frac{1}{\sqrt{j}} = \sum_{j=1}^{N} \frac{1}{\sqrt{j}} + \frac{1}{\sqrt{N+1}} \leq 2\sqrt{N} + \frac{1}{\sqrt{N+1}}$$

and it remains to show that

$$2\sqrt{N} + \frac{1}{\sqrt{N+1}} \leq 2\sqrt{N+1},$$

which is equivalent to

$$\frac{1}{\sqrt{N+1}} \leq 2\left(\sqrt{N+1} - \sqrt{N}\right).$$

Now multiplying this inequality by $\left(\sqrt{N+1}+\sqrt{N}\right)$ gives the equivalent statement:

$$\frac{\sqrt{N+1}+\sqrt{N}}{\sqrt{N+1}} \leq 2(\sqrt{N+1}-\sqrt{N})(\sqrt{N+1}+\sqrt{N}) = 2$$

or

$$1+\sqrt{\frac{N}{N+1}} \leq 2,$$

i.e. we need to justify the equivalent statement

$$\sqrt{\frac{N}{N+1}} \leq 1$$

which follows from $\frac{N}{N+1} < 1$.

b) For $k = 1$ we find:

$$\prod_{m=1}^{1}(2m)! = 2! = 2 \geq ((1+1)!)^1 = 2.$$

Suppose that

$$\prod_{m=1}^{k}(2m)! \geq ((k+1)!)^k.$$

For $k+1$ it follows that

$$\prod_{m=1}^{k+1}(2m)! = \left(\prod_{m=1}^{k}(2m)!\right)(2(k+1))!$$
$$\geq ((k+1)!)^k(2(k+1))!$$

our problem is to prove

$$((k+1)!)^k(2(k+1))! \geq ((k+2)!)^{k+1}$$

which is equivalent to

$$(2(k+1))! \geq \frac{((k+2)!)^{k+1}}{((k+1)!)^k}$$
$$= \frac{((k+1)!(k+2))^k(k+2)!}{((k+1)!)^k}$$
$$= (k+2)^k(k+2)!$$

However note:

$$(2(k+1))! = (k+2)!(k+2+1) \cdot \ldots \cdot (k+2+k)$$

547

or more formally

$$(2(k+1))! = (k+2)! \prod_{j=1}^{k} (k+2+j).$$

Since for $j = 1, \ldots, k$ we have $k + 2 + j \geq k + 2$, it follows that

$$\prod_{j=1}^{k} (k+2+j) \geq (k+2)^k.$$

Hence we conclude

$$(2(k+1))! \geq (k+2)^k (k+2)!$$

14. We first prove $(**)$ for $n = 2^k$. For this we use mathematical induction. The case $k = 1$, i.e. $n = 2$ follows from

$$(a_1 + a_2)^2 - 4a_1 a_2 = (a_1 - a_2)^2 \geq 0$$

or

$$\sqrt{a_1 a_2} \leq \frac{a_1 + a_2}{2}.$$

Now suppose that $(**)$ holds for $n = 2^{k-1}$, i.e.

$$(a_1 \cdot \ldots \cdot a_{2^{k-1}})^{\frac{1}{2^{k-1}}} \leq \frac{a_1 + \ldots + a_{2^{k-1}}}{2^{k-1}}.$$

However we also have for the "next" $2^{k-1}$ terms the following estimate:

$$(a_{2^{k-1}+1} \cdot \ldots \cdot a_{2^k})^{\frac{1}{2^{k-1}}} \leq \frac{a_{2^{k-1}+1} + \cdots + a_{2^k}}{2^{k-1}},$$

or equivalently

$$a_1 \cdot \ldots \cdot a_{2^{k-1}} \leq \left( \frac{a_1 + \ldots + a_{2^{k-1}}}{2^{k-1}} \right)^{2^{k-1}}$$

and

$$a_{2^{k-1}+1} \cdot \ldots \cdot a_{2^k} \leq \left( \frac{a_{2^{k-1}+1} + \ldots + a_{2^k}}{2^{k-1}} \right)^{2^{k-1}},$$

which gives

$$a_1 \cdot \ldots \cdot a_{2^k} \leq \left( \left( \frac{a_1 + \ldots + a_{2^{k-1}}}{2^{k-1}} \right) \left( \frac{a_{2^{k-1}+1} + \ldots + a_{2^k}}{2^{k-1}} \right) \right)^{2^{k-1}}$$

or

$$(a_1 \cdot \ldots \cdot a_{2^k})^{\frac{1}{2^{k-1}}} \leq \frac{(a_1 + \ldots + a_{2^{k-1}})(a_{2^{k-1}+1} + \ldots + a_{2^k})}{2^{k-1} \cdot 2^{k-1}}.$$

The case $k = 1$ also gives:

$$(a_1 + \ldots + a_{2^{k-1}})(a_{2^{k-1}+1} + \ldots + a_{2^k}) \leq \frac{1}{4}(a_1 + \ldots + a_{2^k})^2,$$

implying

$$(a_1 \cdot \ldots \cdot a_{2^k})^{\frac{1}{2^{k-1}}} \leq \frac{(a_1 + \ldots + a_{2^k})^2}{2^k \cdot 2^k}$$

or

$$(a_1 \cdot \ldots \cdot a_{2^k})^{\frac{1}{2^k}} \leq \frac{a_1 + \ldots + a_{2^k}}{2^k}.$$

Now let $n$ be any integer. Choose a $k \in \mathbb{N}$ such that $2^k > n$ and introduce

$$a_j := a := \frac{1}{n} \sum_{k=1}^{n} a_k$$

for $n < j < 2^k$. We may now apply the result for $2^k$ when looking at

$$a_1 \cdot \ldots \cdot a_n \cdot a^{2^k - n} \leq \left( \frac{a_1 + \ldots + a_n + a + \ldots + a}{2^k} \right)^{2^k}.$$

Now $\frac{a_j}{2^k} \leq a$ for every $1 \leq j \leq n$ and therefore we have

$$a_1 \cdot \ldots \cdot a_n \cdot a^{2^k - n} \leq a^{2^k}$$

or

$$a_1 \cdot \ldots \cdot a_n \leq a^n,$$

15. First note that $x_n$ and $a_n$ are defined by recursion.
Note that

$$x_n := \frac{1}{2} \left( x_{n-1} + \frac{c}{x_{n-1}} \right), n \in \mathbb{N}, x_0 = 1,$$

and

$$a_n = \frac{c}{x_n}, n \in \mathbb{N} \cup \{0\},$$

implying that

$$(*) \quad x_n = \frac{1}{2}(x_{n-1} + a_{n-1}), n \in \mathbb{N}.$$

All terms are non-negative and

$$x_n^2 = \left( \frac{x_{n-1} + a_{n-1}}{2} \right)^2 \geq x_{n-1} a_{n-1}.$$

hence

$$x_n \geq \frac{x_{n-1} a_{n-1}}{x_n} = \frac{x_{n-1}}{x_n} \frac{c}{x_{n-1}} = \frac{c}{x_n} = a_n.$$

Combining this with $(*)$ we find for $n \in \mathbb{N}$

$$a_n \leq x_{n+1} \leq x_n.$$

Therefore we deduce

$$\frac{x_n}{x_{n+1}} \geq 1$$

and consequently

$$a_{n+1} = \frac{x_n a_n}{x_{n+1}} \geq a_n.$$

Together we now have

$$a_n \leq a_{n+1} \leq x_{n+1} \leq x_n.$$

549

**Chapter 4**

1.   a) By definition we have

$$A \times B = \{(x, y) \,|\, x \in A \text{ and } y \in B\}$$

and

$$B \times A = \{(x, y) \,|\, x \in B \text{ and } y \in A\}.$$

Therefore it follows that

$$A \times B = \{(3,1), (4,1), (5,1), (6,1), (3,2), (4,2), (5,2), (6,2), (3,3), (4,3), (5,3), (6,3)\}$$

and

$$B \times A = \{(1,3), (2,3), (3,3), (1,4), (2,4), (3,4), (1,5), (2,5), (3,5), (1,6), (2,6), (3,6)\}.$$

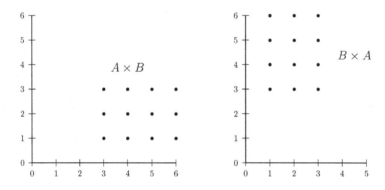

b) We need to prove: if $(k, m) \in \mathbb{N} \times \mathbb{Z}$ then $(k, m) \in \mathbb{R} \times \mathbb{Q}$. Since $k \in \mathbb{N}$ implies $k \in \mathbb{R}$, i.e. $\mathbb{N} \subset \mathbb{R}$, and since $m \in \mathbb{Z}$ implies $m \in \mathbb{Q}$, i.e. $\mathbb{Z} \subset \mathbb{Q}$, $(k, m) \in \mathbb{N} \times \mathbb{Z}$ yields $(k, m) \in \mathbb{R} \times \mathbb{Q}$.

c) First note that $X \cup Y = \{1, 2, 3, 4, 5\}$ and $Y \cup Z = \{3, 4, 5, 6, 7\}$. Now it follows that

$$(X \cup Y) \times Z = \{(1,6), (1,7), (2,6), (3,6), (3,7),$$
$$(4,6), (4,7), (5,6), (5,7)\},$$

$$X \times (Y \cup Z) = \{(1,3), (1,4), (1,5), (1,6), (1,7), (2,3), (2,4), (2,5), (2,6), (2,7), (3,3),$$
$$(3,4), (3,5), (3,6), (3,7)\}.$$

Finally, from

$$X \times Z = \{(1,6), (2,6), (3,6), (1,7), (2,7), (3,7)\}$$

and

$$Y \times Z = \{(3,6), (4,6), (5,6), (3,7), (4,7), (5,7)\}$$

we deduce

$$(X \times Z) \cap (Y \times Z) = \{(3,6), (3,7)\}.$$

550

2.     a) Since $(x, y) \in (A \cup B) \times C$ is equivalent to $x \in A \cup B$ and $y \in C$, i.e. $x \in A$ or $x \in B$ and $y \in C$, we note that this is equivalent to

$$x \in A \text{ and } y \in C \quad \text{or} \quad x \in B \text{ and } y \in C,$$

i.e. $(x, y) \in (A \times C) \cup (B \times C)$.

   b) Note $(x, y) \in (A \times B) \cap (C \times D)$ means

$$(x, y) \in (A \times B) \quad \text{and} \quad (x, y) \in (C \times D),$$

i.e.

$$x \in A \text{ and } y \in B \quad \text{and} \quad x \in C \text{ and } y \in D$$

or

$$x \in A \text{ and } x \in C \quad \text{and} \quad y \in B \text{ and } y \in D,$$

i.e. $x \in A \cap C$ and $y \in B \cap D$ implying that $(x, y) \in (A \cap C) \times (B \cap D)$. However all arguments are reversible, hence we also deduce that $(x, y) \in (A \cap C) \times (B \cap D)$ implies that $(x, y) \in (A \times B) \cap (C \times D)$.

3. Suppose that $X' \times Y' \subset X \times Y$, i.e. $(x, y) \in X' \times Y'$ implies that $(x, y) \in X \times Y$. This means that $x \in X'$ and $y \in Y'$ implies $x \in X$ and $y \in Y$, hence $X' \subset X$ and $Y' \subset Y$. Next if $X' \subset X$ and $Y' \subset Y$, then $(x, y) \in X' \times Y'$ which implies $(x, y) \in X \times Y$.

4. The following hold:

$$\bigcup_{j=1}^{5}(\{j\} \times I_j) = (\{1\} \times [1, 2]) \cup (\{2\} \times [2, 3]) \cup (\{3\} \times [3, 4]) \cup (\{4\} \times [4, 5]) \cup (\{5\} \times [5, 6])$$

and

$$\bigcup_{j=1}^{5}(I_j \times \{j\}) = ([1, 2] \times \{1\}) \cup ([2, 3] \times \{2\}) \cup ([3, 4] \times \{3\}) \cup ([4, 5] \times \{4\}) \cup ([5, 6] \times \{5\}).$$

This gives:

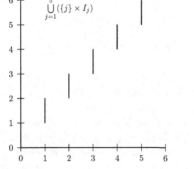 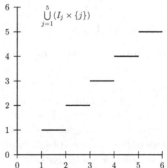

5. We need to prove that $m \equiv n \bmod(p)$ is a reflexive, symmetric and transitive relation on $\mathbb{Z}$. Clearly $m \equiv m \bmod(p)$ since $m - m$ is divisible by $p$. Further if $m - n$ is divisible by $p$ then $n - m$ is divisible by $p$ also, since $m - n = rp$ implies $n - m = (-r)p$. Hence this relation is reflexive and symmetric. Now, suppose $m \equiv n \bmod(p)$ and $n \equiv k \bmod(p)$. We want to prove that $m \equiv k \bmod(p)$. The congruence $m \equiv n \bmod(p)$ stands for $m - n = r_1 p$ and the congruence $n \equiv k \bmod(p)$ stands for $n - k = r_2 p$ with $r_1, r_2 \in \mathbb{Z}$. Now it follows that

$$m - k = (m - n) + (n - k) = r_1 p + r_2 p = (r_1 + r_2)p,$$

i.e. $m \equiv k \bmod(p)$ implying the transitivity, and therefore we have proved that $m \equiv n \bmod(p)$ is an equivalence relation.

6. Again we have to prove that "$\sim$" on $\mathbb{Z} \times \mathbb{N}$ is a reflexive, symmetric and transitive relation. Now for $(k, m) \in \mathbb{Z} \times \mathbb{N}$ we see that $km = mk$ implying $(k, m) \sim (m, k)$, i.e. "$\sim$" is reflexive. Also $kn = lm$ is equivalent to $lm = kn$, i.e. $(k, m) \sim (l, n)$ if and only if $(l, n) \sim (k, m)$ i.e. symmetry is proved. Further, suppose that $(k, m) \sim (l, n)$ and $(l, n) \sim (p, q)$. It follows that $kn = lm$ and $lq = pn$, implying $lqkn = lmpn$. Now $n \in \mathbb{N}$, hence $n \neq 0$ and therefore we find $lqk = lmp$. If $l \neq 0$, then it follows that $qk = mp$ or $(k, m) \sim (p, q)$. However $l = 0$ implies $p = 0$ and $k = 0$, and therefore $qk = 0 = mp$, which proves the transitivity of "$\sim$", i.e. this is an equivalence relation on $\mathbb{Z} \times \mathbb{N}$.

7.    a) Since $\phi$ is the only subset of $\phi$ we find that $\mathcal{P}(\phi) = \{\phi\}$. Note that $\mathcal{P}(\phi) \neq \phi$, the set $\{\phi\}$ contains one element, the set $\phi$.

   b) We have

$$\mathcal{P}(\{1, 2, 3\}) = \{\phi, \{1\}, \{2\}, \{3\}, \{1, 2\}, \{1, 3\}, \{2, 3\}, \{1, 2, 3\}\}.$$

8. We need to add up the number of subsets of $X$ with $0, 1, 2, \ldots, N$ elements. However the number of subsets of $X$ with $k$ elements is $\binom{N}{k}$ by the hint, so we need to find:

$$\sum_{k=0}^{N} \binom{N}{k} = \sum_{k=0}^{N} \binom{N}{k} 1^k 1^{N-k} = (1 + 1)^N = 2^N.$$

Let us give a proof of the following:

**Proposition.** *The number of subsets with $k$ elements of a set with $N$ elements is* $\binom{N}{k}$.

*Proof.* Denote the number of subsets with $k$ elements of the set $X = \{x_1, \ldots, x_N\}$ with $N$ elements by $\nu_{N,k}$. The aim is to prove that $\nu_{N,k} = \binom{N}{k}$. We use mathematical induction, i.e. we assume that the statement holds for $N$ and every $k \leq N$. For $N = 0$ we only have one subset, namely $\phi$, hence $\nu_{0,0} = 1 = \binom{0}{0}$. For $N = 1$ we have one subset with zero elements, namely $\phi$, and one subset with one element, namely $\{x_1\}$. Hence $\nu_{1,0} = 1 = \binom{1}{0}$ and $\nu_{1,1} = 1 = \binom{1}{1}$. Now suppose that the number of subsets with $k$ elements of a set with $N$ elements is $\nu_{N,k} = \binom{N}{k}$. We want to find $\nu_{N+1,k}$. Two cases are trivial:

$$\nu_{N+1,0} = 1 = \binom{N+1}{0} \quad \text{and} \quad \nu_{N+1,N+1} = 1 = \binom{N+1}{N+1}.$$

Thus we may assume $1 \leq k \leq N$. The subsets of $X = \{x_1, \ldots, x_{N+1}\}$ having $k$ elements form two disjoint sets $K_0$ and $K_1$. In $K_0$ we collect all subsets of $X$ with $k$ elements which do not contain $x_{N+1}$, whereas $K_1$ is the family of subsets of $X$ having $k$ elements, one of which is $x_{N+1}$. The number of elements of $K_0$ is by our assumption $\binom{N}{k}$. We are looking for the number of subsets with $k$ elements of a set with $N$ elements. Every set belonging to $K_1$ contains $x_{N+1}$ and $k - 1$ further elements belonging to $\{x_1, \ldots x_n\}$. Thus $K_1$ has by our assumption $\binom{N}{k-1}$ elements. This implies

$$\nu_{N+1,k} = \binom{N}{k} + \binom{N}{k-1} = \binom{N+1}{k},$$

where we used Lemma 3.8 in the last step. $\qquad\qquad\square$

9. The solutions of the quadratic equation $y^2 - 2y + x = 0$ are formally given by

$$y_{1,2} = 1 \pm \sqrt{1 - x},$$

but we are confined to real numbers, hence for $1 > x$ we have two solutions, for $x = 1$ we have one solution and for $x < 1$ we have no solution. Therefore we cannot define a mapping on $\mathbb{R}$ which maps $x$ to the solution of $y^2 - 2y + x = 0$.

10. Let $p(x) = \sum_{j=0}^{k} a_j x^j$ and $q(x) = \sum_{l=0}^{m} b_l x^l$ and suppose that $k \leq m$. Define for $j = k+1, \ldots, m$ the coefficients $a_j := 0$ to get $p(x) = \sum_{j=0}^{m} a_j x^j$. Now we define

$$p(x) + q(x) = \sum_{j=0}^{m} a_j x^j + \sum_{j=0}^{m} b_j x^j = \sum_{j=1}^{m}(a_j + b_j)x^j$$

proving that $p + q$ is a polynomial.

Further we have

$$p(x)q(x) = \left(\sum_{j=1}^{k} a_j x^j\right)\left(\sum_{l=0}^{m} b_l x^l\right)$$

$$= \sum_{j=0}^{k}\sum_{l=0}^{m} a_j b_l x^{j+l} = \sum_{n=1}^{k+m}\left(\sum_{j+l=n} a_j b_l\right) x^n$$

and it follows that $p \cdot q$ is a polynomial.

11.    a) We need to determine the coefficients $b_l$, $0 \le l \le 2n$ given the coefficients $a_{2j}$, $0 \le j \le n$. The only choice is

$$b_l = \begin{cases} a_{2j}, & l = 2j, \ j = 0, \ldots, n \\ 0, & l = 1, 3, \ldots, 2n - 1. \end{cases}$$

With this choice we clearly have:

$$p(x) = \sum_{j=0}^{n} a_{2j} x^{2j} = \sum_{l=0}^{2n} b_l x^l.$$

b) Since for all $j \in \mathbb{N}$ we have $|x|^{2j} = (x^2)^j = x^{2j}$, it follows that $f$ and $p$ have the same domain, namely $\mathbb{R}$, and on $\mathbb{R}$ they coincide:

$$p(x) = \sum_{j=0}^{n} a_{2j} x^{2j} = \sum_{j=0}^{n} a_{2j} |x|^{2j}.$$

c) For $x \ge 0$ we have $|x| = x$ and therefore $|x|^3 = x^3$. However, for $x < 0$ we have $|x| = -x$ and therefore $|x|^3 = (-1)^3 x^3 = -x^3 \ne x^3$. Hence the largest domain where $h$ and $g$ coincide is

$$\mathbb{R}_+ = \{x \in \mathbb{R} \,|\, x \ge 0\}.$$

12.    a) For all $x \in \mathbb{R}$ we know that $x^2 + 7 \ne 0$ and therefore

$$\frac{x^3 - 5x^2 - 17}{x^2 + 7}$$

is defined for all $x \in \mathbb{R}$. Hence we can define a rational function:

$$q_1 : \mathbb{R} \longrightarrow \mathbb{R}$$

$$x \longmapsto q_1(x) = \frac{x^3 - 5x^2 - 17}{x^2 + 7}.$$

b) The term $(x - 3)(x + 4)(2x + 7)^8$ has zeroes for $x = 3$, $x = -4$ and $x = -\frac{7}{2}$. Therefore we can define on $\mathbb{R} \setminus \{3, -4, -\frac{7}{2}\}$ the function $q_2 : \mathbb{R} \setminus \{3, -4, -\frac{7}{2}\} \longrightarrow \mathbb{R}$, $x \longmapsto q_2(x)$, where

$$q_2(x) = \frac{(x - 3)^2 (2x + 7)^5}{(x - 3)(x + 4)(2x + 7)^8}.$$

However, on $\mathbb{R} \setminus \{3, -4, -\frac{7}{2}\}$ we find

$$q_2(x) = \frac{(x - 3)}{(x + 4)(2x + 7)^3}$$

and this term is defined on $\mathbb{R} \setminus \{-4, -\frac{7}{2}\}$. Therefore we may extend $q_2 : \mathbb{R} \setminus \{3, -4, -\frac{7}{2}\} \longrightarrow \mathbb{R}$ to a function $\tilde{q}_2 : \mathbb{R} \setminus \{-4, -\frac{7}{2}\} \longrightarrow \mathbb{R}$ by

$$x \longmapsto \tilde{q}_2(x) = \frac{(x - 3)}{(x + 4)(2x + 7)^3}.$$

c) The term $(x - 4)(x + 2)$ is zero for $x = 4$ and $x = -2$. It follows that on $\mathbb{R} \setminus \{4, -2\}$ we can define the function:

$$q_3 : \mathbb{R} \setminus \{4, -2\} \longrightarrow \mathbb{R}$$

$$x \longmapsto q_3(x) = \frac{x^2 - x - 12}{(x - 4)(x + 2)}.$$

However, for $x = 4$ we have $4^2 - 4 - 12 = 0$, or $x^2 - x - 12 = (x - 4)(x + 3)$. Thus on $\mathbb{R} \setminus \{4, -2\}$ we find

$$\frac{x^2 - x - 12}{(x - 4)(x + 2)} = \frac{(x - 4)(x + 3)}{(x - 4)(x + 2)} = \frac{x + 3}{x + 2}.$$

Therefore we can extend $q_3$ to a function $\tilde{q}_3 : \mathbb{R} \setminus \{-2\} \longrightarrow \mathbb{R}$ by

$$x \longmapsto \tilde{q}_3(x) = \frac{x + 3}{x + 2}.$$

13. (i)

a) By definition we have $x \in f^{-1}(A \cap B)$ if there exists $y \in A \cap B$ such that $f(x) = y$. Since $y \in A$ it follows that $x \in f^{-1}(A)$ and since $y \in B$ it follows that $x \in f^{-1}(B)$, i.e. $x \in f^{-1}(A) \cap f^{-1}(B)$. We have proved that $f^{-1}(A \cap B) \subset f^{-1}(A) \cap f^{-1}(B)$. Now let $x \in f^{-1}(A) \cap f^{-1}(B)$, i.e. $x \in f^{-1}(A)$ and $x \in f^{-1}(B)$. Hence there exists $y_1 \in A$ such that $f(x) = y_1$ and $y_2 \in B$ such that $f(x) = y_2$. However this implies $y_1 = y_2$ and $y_1 = y_2 \in A \cap B$. Consequently $x \in f^{-1}(A \cap B)$ proving $f^{-1}(A) \cap f^{-1}(B) \subset f^{-1}(A \cap B)$ which now proves $f^{-1}(A \cap B) = f^{-1}(A) \cap f^{-1}(B)$.

b) If $x \in f^{-1}(A \cup B)$ then there exists $y \in A \cup B$ such that $f(x) = y$. Consequently $x \in f^{-1}(A)$ or $x \in f^{-1}(B)$ implying $f^{-1}(A \cup B) \subset f^{-1}(A) \cup f^{-1}(B)$. Now, let $x \in f^{-1}(A) \cup f^{-1}(B)$. Then there exists $y \in A \cup B$ such that $f(x) = y$ implying that $x \in f^{-1}(A \cup B)$ or $f^{-1}(A) \cup f^{-1}(B) \subset f^{-1}(A \cup B)$ proving the assertion.

(ii)

a) For $y \in f(A \cap B)$ there exists $x \in A \cap B$ such that $f(x) = y$, hence $y \in f(A)$ and $y \in f(B)$, i.e. $y \in f(A) \cap f(B)$ and we have proved that $f(A \cap B) \subset f(A) \cap f(B)$. Of course we do not expect equality to hold: take $f : \mathbb{R} \longrightarrow \mathbb{R}$, $x \mapsto f(x) = x^2$, and choose $A = \{1\}$ and $B = \{-1\}$. Then $A \cap B = \phi$ and consequently $f(\phi) = \phi$ while $f(A) = \{1\}$ and $f(B) = 1$, i.e. $f(A) \cap f(B) = \{1\}$.

b) If $y \in f(A \cup B)$ then there exists $x \in A \cup B$ such that $f(x) = y$, thus $y \in f(A)$ or $y \in f(B)$ implying $f(A \cup B) \subset f(A) \cup f(B)$. Now let $y \in f(A) \cup f(B)$ then there exists $x \in A$ or $x \in B$ such that $f(x) = y$, i.e. $x \in A \cup B$ and $f(x) = y$ which yields $y \in f(A \cup B)$. Thus we have proved that $f(A) \cup f(B) \subset f(A \cup B)$.

c) By definition $f(\{x\}) = \{y \in Y \mid y = f(x)\} = \{f(x)\}$.

14. (i)

a) Since $x^2 + 1 \geq 1$ we first note that $f^{-1}(\{y\}) = \phi$ if $y < 1$. In the case where $y = 1$ we deduce $f^{-1}(\{1\}) = \{0\}$, whereas for $y > 1$, $x^2 + 1 = y$ implies $x_{1,2} = \pm\sqrt{y - 1}$, i.e. $f^{-1}(\{y\}) = \{+\sqrt{y - 1}, -\sqrt{y - 1}\}$. This is easier to see in the following figure:

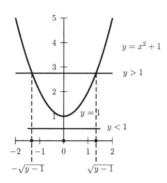

b) Since $\frac{1}{x} \neq 0$ for all $x \in \mathbb{R} \setminus \{0\}$, we deduce that $g^{-1}(\{0\}) = \phi$, whereas for $z \neq 0$ it always follows from $g(x) = z$, i.e. $z = \frac{1}{x}$, that $g^{-1}(\{z\}) = \frac{1}{z}$ for $z \neq 0$.

c) Consider the following figure:

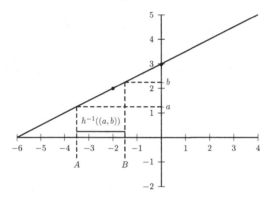

It suggests that $h^{-1}((a, b))$ is the interval $(A, B)$ with $A$ given by $h(A) = a$ and $B$ by $h(B) = b$ which implies $A = 2a-6$ and $B = 2b-6$. Thus $h^{-1}((a, b)) = (2a-6, 2b-6)$.

(ii)

a) Using previous knowledge we find

$$f\left(\left[\frac{1}{4}, 9\right]\right) = \left\{y \in \mathbb{R} \mid y = \sqrt{x},\ x \in \left[\frac{1}{4}, 9\right]\right\} = \left[\frac{1}{2}, 3\right].$$

b) We have

$$g(\{1, 2, 3, 4\}) = \{g(1), g(2), g(3), g(4)\} = \left\{0, \frac{1}{2}, \frac{8}{11}, \frac{5}{6}\right\}.$$

c) We have
$$h(\mathbb{N}) = \{y \in \mathbb{R} \mid y = 2^n, n \in \mathbb{N}\}.$$

556

**Chapter 5**

1.    a) In order for $f_1 : \mathbb{R} \longrightarrow \mathbb{R}_+$ to be injective we must have that $f_1(x) = f_1(y)$ implies $x = y$, i.e. we need to consider the equation

$$|x - 3| + 2 = |y - 3| + 2,$$

or

$$|x - 3| = |y - 3|.$$

For every real number $a \in \mathbb{R} \setminus \{0\}$ where $x = 3 + a$ and $y = 3 - a$ it follows that

$$|x - 3| = |3 + a - 3| = |a| = |3 - a - 3| = |y - 3|,$$

but $x = 3 + a \neq 3 - a = y$ for $a \neq 0$. Hence $f_1$ is not injective and therefore it cannot be bijective.

In order for $f_1 : \mathbb{R} \longrightarrow \mathbb{R}_+$ to be surjective, we need to find for every $b \geq 0$ some $x \in \mathbb{R}$ such that $f_1(x) = b$. Since $f_1(x) = |x - 3| + 2 \geq 2$ for all $x \in \mathbb{R}$, we cannot find any $x \in \mathbb{R}$ such that $|x - 3| + 2 = b$ for $0 \leq b < 2$. Hence $f_1 : \mathbb{R} \longrightarrow \mathbb{R}_+$ is also not surjective.

The graph of $f_1$ is as follows:

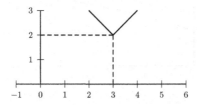

b) We first test $f_2 : [1, \infty) \longrightarrow (0, 2]$ for injectivity. Given $x, y \in [1, \infty)$ and suppose that $f_2(x) = f_2(y)$, i.e.

$$\frac{2}{x} = \frac{2}{y} \quad \text{or} \quad 2y = 2x.$$

This implies that $x = y$ and therefore $f_2$ is injective. Now let $b \in (0, 2]$ and consider the equation

$$b = f_2(x) = \frac{2}{x}.$$

This equation has the unique solution $x = \frac{2}{b}$ and for $0 < b \leq 2$ it follows that $1 \leq x < \infty$. Thus $f_2$ is surjective and with the previous result it follows that $f_2$ is bijective.

The graph of $f_2$ is as follows:

557

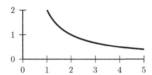

c) Again we start by checking the injectivity of $f_3 : [-2, 7] \longrightarrow [0, 3]$. For $x, y \in [-2, 7]$ we find the condition

$$\sqrt{x + 2} = \sqrt{y + 2} \quad \text{or} \quad x + 2 = y + 2$$

implying $x = y$, i.e. $f_3$ is injective. Now let $b \in [0, 3]$ and consider the equation

$$\sqrt{x + 2} = b \quad \text{or} \quad x + 2 = b^2.$$

This yields that $x = b^2 - 2$ and for $b \in [0, 3]$ we have $-2 \le b^2 - 2 \le 7$. Thus $f_3$ is surjective,

$$f_3(b^2 - 2) = \sqrt{b^2 - 2 + 2} = \sqrt{b^2} = b.$$

Consequently, since $f_3$ has already been proved to be injective it follows that $f_3$ is bijective. Here is the graph of $f_3$:

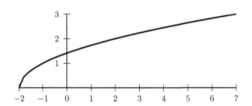

2.    a) For $p_1 = 2$ and $q_1 = 3$ we have $g\left(\frac{p_1}{q_1}\right) = p_1 + q_1 = 5$. Further for $p_2 = 4$ and $q_2 = 1$ we have $g\left(\frac{p_2}{q_2}\right) = p_2 + q_2 = 5$, and therefore $g$ is not injective and therefore it is not bijective either. Clearly, $g$ is surjective: given $k \in \mathbb{Z}$ then we take $p = k - 1$ and $q = 1$ to find $g\left(\frac{k-1}{1}\right) = g\left(\frac{p}{q}\right) = p + q = k - 1 + 1 = k$.

b) Note that $r$ maps pairs of real numbers into pairs of real numbers. Let $(x_1, y_1)$ and $(x_2, y_2)$ be two pairs of real numbers and suppose that

$$r(x_1, y_1) = r(x_2, y_2), \text{ i.e. } (y_1, x_1) = (y_2, x_2).$$

The equality $(y_1, x_1) = (y_2, x_2)$ means that $y_1 = y_2$ and $x_1 = x_2$, hence the pairs $(x_1, y_1)$ and $(x_2, y_2)$ are equal implying the injectivity of $r$. The surjectivity of $r$ is straightforward: given the pair $(a, b) \in \mathbb{R} \times \mathbb{R}$ then for $(x, y) := (b, a)$ it follows that

$$r(x, y) = r(b, a) = (a, b).$$

Thus $r$ is injective and surjective, hence it is bijective.

558

3.     a) Not taking into account domain and range (co-domain) problems, formally $(f \circ g)(x)$ is given by

$$(f \circ g)(x) = f(g(x))$$
$$= 5(g(x))^2 - 2(g(x)) + 1$$
$$= 5\left(\sqrt{5+x}\right)^2 - 2\sqrt{5+x} + 1$$
$$= 5x - 2\sqrt{5+x} + 26.$$

Since $g$ is defined for all $x \geq -5$ and since $f$ is defined for all real numbers it follows that $f \circ g$ is defined on $[5, \infty)$ and therefore $(f \circ g)(x) = 5x - 2\sqrt{5+x} + 26$ holds for $x \in [-5, \infty)$.

b) For $D_1 = \mathbb{R}$ we can define $f \circ h$ by

$$(f \circ h)(x) = f\left(\sqrt{x^4 + 2}\right) = |\sqrt{x^4 + 2} + 3| - 2.$$

The same holds for $D_2 = \mathbb{R}$ and $h \circ f$:

$$(h \circ f)(x) = h(|x+3| - 2) = \sqrt{(|x+3| - 2)^4 + 2}.$$

Note that we can define $f \circ h : \mathbb{R} \longrightarrow \mathbb{R}$ as well as $h \circ f : \mathbb{R} \longrightarrow \mathbb{R}$, but of course $f \circ h \neq h \circ f$.

c) Here we have to be more careful since the range of $h$ is $R(h) = [-1, \infty)$ but $\sqrt{\cdot}$ is not defined for non-positive numbers. However for $x \geq -1$ and $x \leq -3$ it follows that $|x+2| - 1 \geq 0$. Hence on
$D = \{x \in \mathbb{R} \mid x \leq -3 \text{ or } x \geq -1\} = \mathbb{R} \setminus (-3, -1)$ we can define

$$(f \circ h)(x) = f(h(x)) = f(|x+2| - 1) = \sqrt{|x+2| - 1}.$$

4. Let $f_1 : D_1 \longrightarrow F_1$ and $f_2 : D_2 \longrightarrow F_2$ be two injective mappings such that $f_1(D_1) = D_2$. Then $f_2 \circ f_1 : D_1 \longrightarrow F_2$ is defined and for $x, y \in D_1$ the equality

$$(f_2 \circ f_1)(x) = (f_2 \circ f_1)(y), \quad \text{i.e.} \quad f_2(f_1(x)) = f_2(f_1(y)),$$

implies $f_1(x) = f_1(y)$ by the injectivity of $f_2$, and now the injectivity of $f_1$ implies $x = y$, i.e. $f_2 \circ f_1$ is injective.
Now let $g_1 : D_1 \longrightarrow F_1$ and $g_2 : D_2 \longrightarrow F_2$ be two surjective mappings such that $g_1(D_1) = D_2(= F_1)$. Then $g_2 \circ g_1 : D_1 \longrightarrow F_2$ is defined and for $b \in F_2$ the surjectivity of $g_2$ implies the existence of $a \in D_2$ such that $g_2(a) = b$. Since $g_1$ is surjective we know that $g_1(D_1) = F_1 = D_2$, thus given $a \in D_2 = F_1$ we find $x \in D_1$ such that $g_1(x) = a$ implying that

$$(g_2 \circ g_1)(x) = g_2(g_1(x)) = g_2(a) = b,$$

i.e. $g_2 \circ g_1$ is surjective.
Now if $f_1 : D_1 \longrightarrow F_1$ and $f_2 : D_2 \longrightarrow F_2$ are injective and surjective, i.e. bijective, and if $f_1(D_1) = D_2(= F_1)$, then $f_2 \circ f_1$ is also injective and surjective, i.e. it is bijective.

559

5. Since all mappings belonging to $Aut(X)$ are bijective their composition is always defined.

(i) Since the composition of mappings is associative the statement $(f \circ g) \circ h = f \circ (g \circ h)$ follows immediately.

(ii) Clearly $f \circ g$ maps $X$ to $X$. We need to prove that it is bijective. By Problem 4 we know however that the composition of bijective mappings is bijective, hence $f \circ g \in Aut(X)$.

(iii) The map $id_X : X \longrightarrow X$, $x \mapsto id_X(x) = x$, belongs to $Aut(X)$ and the following holds for $f \in Aut(X)$:

$$(f \circ id_X)(x) = f(id_X(x)) = f(x) = id_X(f(x)) = (id_X \circ f)(x).$$

(iv) Since $f \in Aut(X)$ it is bijective and with $k_f := f^{-1}$ we find

$$f \circ f^{-1} = f^{-1} \circ f = id_X.$$

6.     a) Let $f : X \longrightarrow Y$ be injective. Then $f : X \longrightarrow R(f)$ is bijective and $f^{-1} : R(f) \longrightarrow X$ exists. Define $g : Y \longrightarrow X$ by

$$g(y) := \begin{cases} f^{-1}(y) & \text{for } y \in R(f) \\ x_0 \in X & \text{for } y \in Y \setminus R(f). \end{cases}$$

Then we find for $x \in X$ since $f(x) = y \in R(f)$ that

$$(g \circ f)(x) = g(f(x)) = g(y) = f^{-1}(y) = (f^{-1} \circ f)(x) = x,$$

i.e. $g \circ f = id_X$.
Conversely, suppose that there exists a mapping $g : Y \longrightarrow X$ such that $g \circ f = id_X$. For $x, y \in X$ with $f(x) = f(y)$ it follows that

$$x = g(f(x)) = g(f(y)) = y,$$

i.e. $x = y$ and $f$ must be injective.

b) Suppose now that $f : X \longrightarrow Y$ is surjective. For $y \in Y$ choose $x_y \in X$ such that $f(x_y) = y$. This defines a mapping

$$h : Y \longrightarrow X$$

$$y \mapsto x_y.$$

For this mapping we find $f \circ h : Y \longrightarrow Y$ and

$$(f \circ h)(y) = f(h(y)) = f(x_y) = y,$$

i.e. $f \circ h = id_Y$.
Conversely if there exists a mapping $h : Y \longrightarrow X$ such that $f \circ h = id_Y$ it follows for any $b \in Y$ that

$$b = id_Y(b) = (f \circ h)(b) = f(h(b)).$$

Thus given $b \in Y$ there exists $x := h(b) \in X$ such that $f(x) = f(h(b)) = b$, i.e. $f$ is surjective.

7. We already know (or can easily check) that
$f : (0, \infty) \longrightarrow (0, \infty)$ is bijective, hence $f^{-1}$ exists. The claim is that $f \circ f = id$.
However for $x \in (0, \infty)$ we find

$$f(f(x)) = f\left(\frac{1}{x}\right) = \frac{1}{\frac{1}{x}} = x = id_{(0,\infty)}(x).$$

8. First note that the range of $h$ is a subset of $\mathbb{R}_+ \setminus \{0\}$, and therefore we can define
$f \circ h$ and $g \circ h$ on a suitable domain, namely the domain of $h$. By definition we
have

$$(f + g) \circ h = f \circ h + g \circ h,$$
$$(f \cdot g) \circ h = (f \circ h) \cdot (g \circ h),$$
$$\frac{1}{g} \circ h = \frac{1}{g \circ h},$$

and therefore it follows that

$$((f + g) \circ h)(x) = f(h(x)) + g(h(x))$$
$$= \frac{1}{h(x)} + \sqrt{h(x)} + |h(x) - 2|$$
$$= \frac{1}{x^2 + 1} + \sqrt{x^2 + 2} + x^2,$$

$$((f \cdot g) \circ h)(x) = f(h(x)) \cdot g(h(x))$$
$$= \frac{1}{h(x)} \left(\sqrt{h(x)} + |h(x) - 2|\right)$$
$$= \frac{1}{\sqrt{x^2 + 1}} + \frac{x^2}{x^2 + 1},$$

$$\left(\left(\frac{1}{g}\right) \circ h\right)(x) = \frac{1}{g(h(x))} = \frac{1}{\sqrt{h(x)} + |h(x) - 2|}$$
$$= \frac{1}{\sqrt{x^2 + 2} + x^2}.$$

9. For every real number $a \in \mathbb{R}$ we have

$$\frac{|a| + a}{2} \geq 0 \quad \text{and} \quad \frac{|a| - a}{2} \geq 0.$$

Indeed, if $a \geq 0$ then $|a| = a$ and

$$\frac{|a| + a}{2} = \frac{2a}{2} = a \geq 0$$

561

and

$$\frac{|a| - a}{2} = \frac{a - a}{2} = 0.$$

But if $a < 0$ then $|a| = -a$ and we find

$$\frac{|a| + a}{2} = \frac{-a + a}{2} = 0$$

as well as

$$\frac{|a| - a}{2} = \frac{-2a}{2} = -a > 0 \text{ since } a < 0.$$

Since for $x \in X$ by definition we have $f(x) \in \mathbb{R}$ it follows that

$$f^+(x) = \frac{|f(x)| + f(x)}{2} \geq 0 \quad \text{and} \quad f^-(x) = \frac{|f(x)| - f(x)}{2} \geq 0.$$

We call $f^+$ the positive part and $f^-$ the negative part of $f$. Note that the negative part of $f$ is a non-negative function.
Now it follows that

$$f^+(x) - f^-(x) = \frac{|f(x)| + f(x)}{2} - \frac{|f(x)| - f(x)}{2} = f(x)$$

and

$$f^+(x) + f^-(x) = \frac{|f(x)| + f(x)}{2} + \frac{|f(x)| - f(x)}{2} = |f(x)|.$$

10.　　a) We need to solve the equation

$$y = \frac{1}{1 + x^2};$$

or $1 + x^2 = \frac{1}{y}$, i.e. $x^2 = \frac{1}{y} - 1$.
Since $y \in (0, 1]$ it follows that $\frac{1}{y} - 1 \geq 0$. Hence

$$x = \sqrt{\frac{1}{y} - 1} = \sqrt{\frac{1 - y}{y}}.$$

Thus we find the inverse function of $f_1$ to be:

$$f_1^{-1} : (0, 1] \longrightarrow [0, \infty)$$

$$y \mapsto \sqrt{\frac{1 - y}{y}}.$$

b) We first sketch $f_2$:

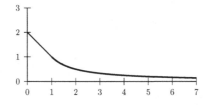

562

In order to find $f_2^{-1}$ we need to solve the equation

$$f_2(x) = y.$$

For $0 \le x \le 1$ we find

$$-x + 2 = y, \quad \text{i.e.} \quad x = 2 - y.$$

For $1 < x < \infty$ we have

$$\frac{1}{x} = y \quad \text{or} \quad x = \frac{1}{y},$$

and therefore we obtain

$$f_2^{-1} : (0, 2] \longrightarrow [0, \infty)$$

$$y \mapsto \begin{cases} \frac{1}{y}, & y \in (0, 1) \\ 2 - y, & y \in [1, 2]. \end{cases}$$

c) Now the equation we have to solve is given by

$$f_3(n) = q \quad \text{or} \quad \frac{1}{n^3} = q$$

which yields

$$n = \frac{1}{\sqrt[3]{q}} = q^{-\frac{1}{3}}.$$

Thus $f_3^{-1} : \{q \mid q = \frac{1}{n^3} \text{ and } n \in \mathbb{N}\} \longrightarrow \mathbb{N}, \ q \mapsto q^{-\frac{1}{3}}$.
Note that

$$\left(\frac{1}{n^3}\right)^{-\frac{1}{3}} = (n^3)^{\frac{1}{3}} = n,$$

therefore $f_3^{-1}$ has the desired properties.

11.    a) First consider the figure below of the unit disc $\overline{B_1(0)}$

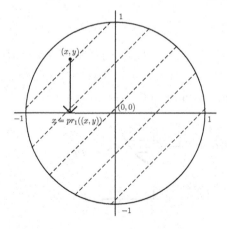

For $(x, y) \in \overline{B_1(0)}$ we find that $pr_1((x, y)) = x$. Denote the set $\left\{(x_0, y) \mid -\sqrt{1 - x_0^2} \leq y \leq \sqrt{1 - x_0^2}\right\}$ by $A(x_0)$ for $x_0 \in [-1, 1]$. Then we find $pr_1(A(x_0)) = x_0$.

Now for the circle $S^1 = \{(x, y) \in \mathbb{R}^2 \mid x^2 + y^2 = 1\}$ we find again that $pr_1((x, y)) = x$, see the following figure:

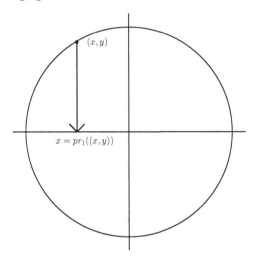

For $x_0 \in [-1, 1]$ and $(x_0, y) \in S^1$ we find with $y = \pm\sqrt{1 - x_0^2}$ that only the points $\left(x_0, \pm\sqrt{1 - x_0^2}\right)$ are mapped to $x_0$ by $pr_1$. In both cases we have however $pr_1(\overline{B_1(0)}) = pr_1(S^1) = [-1, 1]$.

b) We may rewrite $R(g)$ as

$$R(g) = \{(x, y) \mid x \in [0, 1] \text{ and } g(x) = x^2 + 1\}$$
$$= \{(x, x^2 + 1) \mid x \in [0, 1]\}.$$

This implies that

$$pr_2(R(g)) = \{x^2 + 1 \mid x \in [0, 1]\}$$
$$= [1, 2],$$

i.e. we are dealing with the following situation:

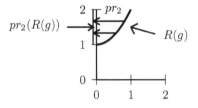

564

12. First we look at $pr_1 : X \times Y \longrightarrow X$, $(x,y) \mapsto x$. Now, by the very definition of the pre-image we have for $A \subset X$

$$pr_1^{-1}(A) = \{(x,y) \in X \times Y \mid x \in A\}$$
$$= \{(x,y) \mid x \in A, \ y \in Y\}$$
$$= A \times Y.$$

Analogously we find for $pr_2 : X \times Y \longrightarrow Y$, $(x,y) \mapsto y$, that for $B \subset Y$ the following holds

$$pr_2^{-1}(B) = \{(x,y) \in X \times Y \mid y \in B\}$$
$$= \{(x,y) \mid x \in X, \ y \in B\}$$
$$= X \times B.$$

13. Suppose that $j : \mathbb{N} \longrightarrow \mathbb{R}$ is injective. Then $j : \mathbb{N} \longrightarrow j(\mathbb{N})$ is surjective and injective, hence bijective, implying that $j(\mathbb{N})$ is countable as it is a bijective image of $\mathbb{N}$. Now consider the mapping $j : \mathbb{N} \longrightarrow \{1\} \cup \{2k \mid k \in \mathbb{N}\}$ with

$$j(n) := \begin{cases} 1, & \text{for } n \text{ being odd} \\ 2n & \text{for } n \text{ being even.} \end{cases}$$

Clearly $j$ is not injective but $j(\mathbb{N})$ is countable. Indeed we know that $\{2k \mid k \in \mathbb{N}\}$ is countable and the union of a countable set with a finite set is again countable.

14. We have to prove that '$\sim$' is symmetric, reflexive and transitive.
If $f, g \in M(D; \mathbb{R})$ and $f \sim g$ then there exists a finite set $A_{f,g} = \{x_1, \ldots, x_m\} \subset D$ such that $f(x) = g(x)$ for $x \in D \setminus A_{f,g}$. But for $x \in D \setminus A_{f,g}$ we also have $g(x) = f(x)$, i.e. $f \sim g$ implies that $g \sim f$ and '$\sim$' is symmetric. Since $f(x) = f(x)$ for all $x \in D$ and by definition the empty set is finite it follows with $A_{f,f} = \phi$ that $f(x) = f(x)$ for all $x \in D \setminus A_{f,f}$, i.e. '$\sim$' is reflexive.
Finally, if $f, g, h \in M(D; \mathbb{R})$ and $f \sim g$ as well as $g \sim h$, we find sets $A_{f,g}$ and $A_{g,h}$ such that

$$f(x) = g(x) \text{ for } x \in D \setminus A_{f,g} \quad \text{and} \quad g(x) = h(x) \text{ for } x \in D \setminus A_{g,h}.$$

Now $A_{f,h} := A_{f,g} \cup A_{g,h}$ is a finite set and for
$x \in D \setminus A_{f,h} = D \setminus (A_{f,g} \cup A_{g,h})$ we have

$$f(x) = g(x) = h(x),$$

i.e. $f(x) = h(x)$ for $x \in D \setminus A_{f,h}$ implying the transitivity of '$\sim$'. Therefore it follows that '$\sim$' is an equivalence relation.

15. The mapping $J$ is injective: if $((x,y), z) \neq ((x',y'), z')$ then either $z \neq z'$ or $(x,y) \neq (x', y')$. Hence at least one of the statements $z \neq z'$, $x \neq x'$, $y \neq y'$ is true which implies that $(x,y,z) \neq (x', y', z')$. The mapping $J$ is surjective: given $(x,y,z) \in X \times Y \times Z$, then $((x,y), z) \in (X \times Y) \times Z$ and $J(((x,y), z)) = (x,y,z)$. Hence $J$ is bijective.

565

**Chapter 6**

1. Firstly, a general remark: in order to calculate limits using $(6.18)-(6.20)$ we assume that all the relevant assumptions hold. However, while doing these calculations it is important that we can justify that all steps are correct.

a)

$$\lim_{x\to\frac{3}{4}} \left(\tfrac{5}{3}x^2 - \tfrac{7}{12}x\right) \underset{(6.18)}{=} \lim_{x\to\frac{3}{4}} \tfrac{5}{3}x^2 - \lim_{x\to\frac{3}{4}} \tfrac{7}{12}x$$

$$\underset{(6.19)}{=} \left(\lim_{x\to\frac{3}{4}} \tfrac{5}{3}\right)\left(\lim_{x\to\frac{3}{4}} x^2\right) - \left(\lim_{x\to\frac{3}{4}} \tfrac{7}{12}\right)\left(\lim_{x\to\frac{3}{4}} x\right)$$

$$= \frac{5}{3}\cdot\left(\frac{3}{4}\right)^2 - \frac{7}{12}\cdot\left(\frac{3}{4}\right) = \frac{5}{3}\cdot\frac{9}{16} - \frac{7}{12}\cdot\frac{3}{4} = \frac{5\cdot3}{16} - \frac{7}{16} = \frac{1}{2}.$$

b) First note that for $x \neq 1$

$$\frac{1-x^2}{1-x} = \frac{(1-x)(1+x)}{1-x} = 1+x$$

and therefore

$$\lim_{x\to1} \frac{1-x^2}{1-x} = \lim_{x\to1}(1+x) \underset{(6.18)}{=} \lim_{x\to1} 1 + \lim_{x\to1} x = 2.$$

c)

$$\lim_{x\to3} \frac{x^3 - 4x^2 + 7x - 13}{-\frac{7}{5}x^2 + \frac{1}{1+x^2}}$$

$$\underset{(6.20)}{=} \frac{\lim_{x\to3}(x^3 - 4x^2 + 7x - 13)}{\lim_{x\to3}\left(-\frac{7}{5}x^2 + \frac{1}{1+x^2}\right)}$$

$$\underset{(6.18),(6.20)}{=} \frac{\lim_{x\to3} x^3 - \lim_{x\to3} 4x^2 + \lim_{x\to3} 7x - \lim_{x\to3} 13}{\lim_{x\to3}\left(-\frac{7}{5}x^2\right) + \dfrac{1}{\lim_{x\to3}(1+x^2)}}$$

$$= \frac{3^3 - 4\cdot3^2 + 7\cdot3 - 13}{-\frac{7}{5}\cdot3^2 + \frac{1}{1+3^2}}$$

$$= \frac{27 - 36 + 21 - 13}{-\frac{63}{5} + \frac{1}{10}}$$

$$= \frac{10(48-49)}{-126+1} = \frac{-10}{-125} = \frac{2}{25}.$$

Note that since $\lim_{x\to3}\left(-\dfrac{7}{5}x^2 + \dfrac{1}{1+x^2}\right) = \dfrac{-125}{10} \neq 0$ we may apply $(6.20)$.

2. The remark made at the beginning of the solution of Problem 1 also applies here.

   a)

   $$\lim_{x \to 4} \frac{x^2 - 2x + 5}{x - 2} = \frac{\lim_{x \to 4} (x^2 - 2x + 5)}{\lim_{x \to 4} (x - 2)}$$

   $$= \frac{\lim_{x \to 4} x^2 - \lim_{x \to 4} 2x + \lim_{x \to 4} 5}{\lim_{x \to 4} x - \lim_{x \to 4} 2}$$

   $$= \frac{16 - 8 + 5}{4 - 2} = \frac{13}{2},$$

   and we need to note that $\lim_{x \to 4} (x - 2) = 2 \neq 0$.

   b)

   $$\lim_{x \to -3} \frac{x^2 - 9}{(x + 5)(x + 3)} = \lim_{x \to -3} \frac{(x - 3)(x + 3)}{(x + 5)(x + 3)}$$

   $$= \lim_{x \to -3} \frac{x - 3}{x + 5} = \frac{\lim_{x \to -3} (x - 3)}{\lim_{x \to -3} (x + 5)}$$

   $$= \frac{-6}{2} = -3.$$

   We need to note that for $x \neq -3$ we have $\frac{x^2 - 9}{(x+5)(x+3)} = \frac{x-3}{x+5}$, and that $\lim_{x \to -3} (x+5) = 2 \neq 0$.

3. For $x \neq 3$ we have

   $$\lim_{x \to 3} f(x) = \lim_{x \to 3} (x^3 - 22) = 27 - 22 = 5$$

   and since $5 = \lim_{x \to 3} f(x) \neq f(3) = 17$, it follows that $f$ is not continuous at $x = 3$.

4.    a) Since $h$ is bounded we know that $|h(x)| \leq M$ for some $M \geq 0$ therefore we find that $|xh(x)| \leq M|x|$. Therefore it remains to prove that $\lim_{x \to 0} (M|x|) = 0$ (using the assumption in the question) which is equivalent to $\lim_{x \to 0} |x| = 0$.

   We must satisfy the definition of the limit of a function: given $\epsilon > 0$ we chose $\delta = \epsilon$ to find for $|x| < \delta$ that

   $$||x| - 0| = |x| < \delta = \epsilon \text{ which implies } \lim_{x \to 0} |x| = 0.$$

   Now we sketch the proof of the assumption:

   $|f(x)| \leq g(x)$ for all $x \in (a, b)$ and $\lim_{x \to c} g(x) = 0$, $c \in (a, b)$, implies $\lim_{x \to c} f(x) = 0$.

   We know that for $\epsilon > 0$ there exists $\delta > 0$ such that $0 < |x - c| < \delta$ implies $|g(x)| = g(x) < \epsilon$. Therefore for $\epsilon > 0$ given we find with the same $\delta > 0$ for $0 < |x - c| < \delta$ that

   $$|f(x) - 0| = |f(x)| \leq g(x) < \epsilon,$$

567

i.e. $\lim\limits_{x\to 0} f(x) = 0$.

b) For the function $f$ we find the estimate

$$|f(x)| \le |x| \left| \sin \frac{1}{x} \right| \le |x| \text{ for } x \ne 0$$
$$|f(0)| = 0 = |0| \text{ for } x = 0.$$

Therefore it follows that

$$|f(x)| \le |x| \text{ for all } x \in \mathbb{R}$$

and applying part a) in particular that $\lim\limits_{x\to 0} |x| = 0$ it follows that $\lim\limits_{x\to 0} f(x) = 0$.

5. Consider the following

$$\frac{f(x) - f(x_0)}{x - x_0} = \frac{\frac{3}{4}x^2 - 2 - \left(\frac{3}{4}\left(-\frac{1}{2}\right)^2 - 2\right)}{x - \left(-\frac{1}{2}\right)}$$

$$= \frac{3}{4}\left(\frac{x^2 - \frac{1}{4}}{x + \frac{1}{2}}\right) = \frac{3}{4}\frac{\left(x - \frac{1}{2}\right)\left(x + \frac{1}{2}\right)}{x + \frac{1}{2}}$$

$$= \frac{3}{4}\left(x - \frac{1}{2}\right).$$

Recall $f'(x) = \lim\limits_{x\to x_0} \dfrac{f(x) - f(x_0)}{x - x_0}$. Therefore for the limit we now find

$$\lim_{x\to -\frac{1}{2}} \frac{f(x) - f\left(-\frac{1}{2}\right)}{x - \left(-\frac{1}{2}\right)} = \lim_{x\to -\frac{1}{2}} \frac{f(x) - f\left(-\frac{1}{2}\right)}{x + \frac{1}{2}}$$

$$= \lim_{x\to -\frac{1}{2}} \frac{3}{4}\left(x - \frac{1}{2}\right) = \frac{3}{4}\left(-\frac{1}{2} - \frac{1}{2}\right) = \frac{3}{4}(-1) = -\frac{3}{4}$$

thus $f'\left(-\frac{1}{2}\right) = -\frac{3}{4}$.

6. First let us sketch the graph of $\chi_{[0,1]} : \mathbb{R} \longrightarrow \mathbb{R}$, where

$$\chi_{[0,1]}(x) = \begin{cases} 1, & x \in [0,1] \\ 0, & x \notin [0,1]. \end{cases}$$

568

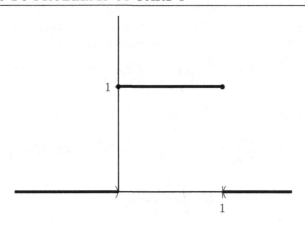

Now, for $x_0 < 0$ we find for $x \in \mathbb{R}$ such that $|x - x_0| < \delta$ and $\delta < |x_0|$, in particular $x < x_0 + \delta < 0$, that

$$\frac{\chi_{[0,1]}(x) - \chi_{[0,1]}(x_0)}{x - x_0} = \frac{0 - 0}{x - x_0} = 0,$$

implying $\chi'_{[0,1]}(x_0) = 0$. In a similar way we find that $\chi_{[0,1]}$ is differentiable for $0 < x_0 < 1$: for $x$ close to $x_0$ and $x \in (0,1)$ we find

$$\frac{\chi_{[0,1]}(x) - \chi_{[0,1]}(x_0)}{x - x_0} = 0$$

which gives $\chi'_{[0,1]}(x_0) = 0$. Moreover, for $x_0 > 1$ and $1 < x$ it follows once again that

$$\frac{\chi_{[0,1]}(x) - \chi_{[0,1]}(x_0)}{x - x_0} = 0$$

hence $\chi'_{[0,1]}(x_0) = 0$.
Before we investigate the case $x_0 = 0$ or $x_0 = 1$, we make the following observation: in order for

$$\lim_{x \to x_0} \frac{\chi_{[0,1]}(x) - \chi_{[0,1]}(x_0)}{x - x_0}$$

to exist it is necessary that for all $0 < \delta \le \delta_0$ the function

$$x \mapsto \frac{\chi_{[0,1]}(x) - \chi_{[0,1]}(x_0)}{x - x_0}$$

is bounded on $0 < |x - x_0| < \delta$.
Suppose that

$$\lim_{x \to x_0} \frac{\chi_{[0,1]}(x) - \chi_{[0,1]}(x_0)}{x - x_0} = a$$

569

for some $a \in \mathbb{R}$. Then for $\epsilon = 1$ there exists $\tilde{\delta} > 0$ such that $0 < |x - x_0| < \tilde{\delta}$ implies

$$\left| \frac{\chi_{[0,1]}(x) - \chi_{[0,1]}(x_0)}{x - x_0} - a \right| < 1.$$

Thus for $0 < |x - x_0| < \tilde{\delta}$ it follows that

$$\left| \frac{\chi_{[0,1]}(x) - \chi_{[0,1]}(x_0)}{x - x_0} \right| - |a| \leq \left| \frac{\chi_{[0,1]}(x) - \chi_{[0,1]}(x_0)}{x - x_0} - a \right| < 1,$$

or

$$\left| \frac{\chi_{[0,1]}(x) - \chi_{[0,1]}(x_0)}{x - x_0} \right| < 1 + |a|.$$

Now, for $x_0 = 0$ we find with $0 < |x - x_0| = |x| < 1$ that

$$\frac{\chi_{[0,1]}(x) - \chi_{[0,1]}(x_0)}{x - x_0} = \frac{\chi_{[0,1]}(x) - 1}{x}$$

$$= \begin{cases} 0, & 0 < x < 1 \\ -\frac{1}{x}, & -1 < x < 0 \end{cases}$$

which is for $0 < |x| < 1$ unbounded. Suppose that $\left| -\frac{1}{x} \right| = \left| \frac{1}{x} \right| \leq c$ for some $c$ and every $0 < |x| < 1$, then it would follow that $0 < \frac{1}{c} \leq x$ for all $x \in (0,1)$ implying $\frac{1}{c} = 0$ which is a contradiction.
For $x_0 = 1$ we have to consider

$$\frac{\chi_{[0,1]}(x) - \chi_{[0,1]}(1)}{x - 1}$$

$$= \begin{cases} 0, & 0 < x < 1 \\ -\frac{1}{x-1}, & 1 < x \end{cases}$$

and this is again an unbounded function for $0 < |x - 1| < 1$. Thus neither $\chi'_{[0,1]}(0)$ nor $\chi'_{[0,1]}(1)$ exist.

7. Again it is helpful to sketch the graph of $g$, say on $[0, 3]$

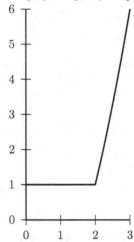

In order to have differentiability at $x_0 = 2$ we need to investigate the limit as $x \mapsto 2$
for

$$\frac{g(x) - g(2)}{x - 2} = \begin{cases} 0, & \text{for } x \leq 2 \\ \frac{x^2 - 4}{x - 2} = x + 2, & \text{for } x > 2. \end{cases}$$

Suppose that

$$\lim_{x \to 2} \frac{g(x) - g(2)}{x - 2} = a$$

for some $a \in \mathbb{R}$.

Then for all $\epsilon > 0$ there exists $\delta > 0$ such that $0 < |x-2| < \delta$ implies $\left| \frac{g(x)-g(2)}{x-2} - a \right| < \epsilon$. In particular for $\epsilon = \frac{1}{2}$ there exists some $\delta > 0$ such that $-\delta + 2 < x < 2 + \delta$ implies

$$\left| \frac{g(x) - g(2)}{x - 2} - a \right| < \frac{1}{2}.$$

For $-\delta + 2 < x < 2$ we have $|a| < \frac{1}{2}$, where for $2 < x < 2 + \delta$ we find that

$$|x + 2 - a| < \frac{1}{2}$$

or

$$|x + 2| - \frac{1}{2} < |a|$$

where we used the converse triangle inequality

$$|x + 2 - a| \geq |x + 2| - |a|.$$

We may assume $\delta < 1$ to see that

$$x + 2 - \frac{1}{2} < |a|$$

or

$$x + \frac{3}{2} < |a|$$

but $x > 1$, implies $|a| > \frac{3}{2}$ which is a contradiction. Thus $g$ is not differentiable at $x_0 = 2$.

However $g$ is continuous at $x_0 = 2$. For this we need to prove that for every $\epsilon > 0$ there exists $\delta > 0$ such that $0 < |x - 2| < \delta$ implies $|g(x) - 1| < \epsilon$. Now for $-\delta + 2 < x < 2$ we have $|g(x) - 1| = 0$, hence every $\delta > 0$ will work. Whereas for $2 < x < \delta + 2$ we find

$$|g(x) - 1| = |x^2 - 3 - 1| = |x^2 - 4| = |x + 2||x - 2|$$

and since we may assume without loss of generality that $\delta < 1$ we find $|x - 2| < \delta < 1$ implies $|x| \leq 3$ and therefore

$$|g(x) - 1| = |x + 2||x - 2| \leq 4|x - 2|.$$

Thus for $\delta = \frac{\epsilon}{4}$ we find $0 < |x - 2| < \delta$ implies

$$|g(x) - 1| \leq 4|x - 2| < 4 \cdot \delta = 4 \cdot \frac{\epsilon}{4} = \epsilon$$

proving the continuity of $g$ at $x_0 = 2$.

8. In the following we make use of (6.36), (6.37), (6.38), (6.40) and (6.42).

   a)

$$\frac{d}{dx} f(x) = \frac{d}{dx} \left( \frac{7}{5} x^2 - \frac{2}{x^3} \right)$$

$$= 2 \cdot \frac{7}{5} x - 2(-3) \frac{1}{x^4} = \frac{14}{5} x + \frac{6}{x^4}.$$

   b)

$$\frac{d}{dt} \left( \frac{t^7 + 12t^3 - 2}{t^5} \right) = \frac{d}{dt} \left( (t^7 + 12t^3 - 2) \cdot \frac{1}{t^5} \right)$$

$$\overset{(6.38)}{=} \frac{d}{dt} (t^7 + 12t^3 - 2) \frac{1}{t^5} + (t^7 + 12t^3 - 2) \frac{d}{dt} \left( \frac{1}{t^5} \right)$$

$$= (7t^6 + 12 \cdot 3t^2) \frac{1}{t^5} + (t^7 + 12t^3 - 2) \left( -5 \cdot \frac{1}{t^6} \right)$$

$$= 2t - \frac{24}{t^3} + \frac{10}{t^6}.$$

   c)

$$\frac{d}{ds} h(s) = \frac{d}{ds} \left( \sum_{j=1}^{M} j s^{-j} \right)$$

$$= \sum_{j=1}^{M} j \frac{d}{ds} (s^{-j}) = \sum_{j=1}^{M} (-j^2) s^{-j-1}.$$

572

9. The proof that $\chi_{\mathbb{R}_+}$ is not differentiable at $x_0 = 0$ follows in the same way as the proof that $\chi_{[0,1]}$ is not differentiable at $x_0 = 0$, see Problem 6.
In order to investigate the differentiability of $h : \mathbb{R} \longrightarrow \mathbb{R}$, $x \mapsto x^2 f(x) = x^2 \chi_{\mathbb{R}_+}(x)$ we must consider the limit
$$\lim_{x \to 0} \frac{h(x) - h(0)}{x - 0}.$$

Note that
$$\left| \frac{h(x) - h(0)}{x - 0} \right| = \left| \frac{x^2 \chi_{\mathbb{R}_+}(x) - 0}{x - 0} \right|$$
$$= \frac{x^2 \chi_{\mathbb{R}_+}(x)}{|x|} \leq |x|.$$

Therefore, given $\epsilon > 0$ we find for $\delta = \epsilon$ that $0 < |x - 0| = |x| < \delta$ implies
$$\left| \frac{h(x) - h(0)}{x - 0} - 0 \right| = \left| \frac{h(x) - h(0)}{x - 0} \right| \leq |x| < \delta = \epsilon,$$

i.e. $h$ is differentiable at $x_0 = 0$ and $h'(0) = 0$.

## Chapter 7

1. The case $k = 1$ is known: $h_1(x) = \sqrt{x}$ and $\frac{d}{dx} h_1(x) = \frac{1}{2\sqrt{x}} = \frac{1}{2}\sqrt{x^{-1}}$. Now for $k = 2n$, $n \in \mathbb{N}$, being even we have $h_k(x) = h_{2n}(x) = \sqrt{x^{2n}} = x^n$ and therefore
$$\frac{d}{dx} h_k(x) = n x^{n-1} = \frac{k}{2} x^{\frac{k}{2}-1} = \frac{k}{2}\sqrt{x^{k-2}}.$$

Whereas for $k = 2n + 1$, $n \in \mathbb{N}$, being odd we find
$$h_k(x) = h_{2n+1}(x) = \sqrt{x^{2n+1}} = x^n \sqrt{x}$$

which gives
$$\frac{d}{dx} h_k(x) = \frac{d}{dx}(x^n \sqrt{x}) = n x^{n-1}\sqrt{x} + x^n \frac{1}{2\sqrt{x}}$$
$$= \left(n + \frac{1}{2}\right) x^{n-\frac{1}{2}} = \frac{k}{2} x^{\frac{k}{2}-1} = \frac{k}{2}\sqrt{x^{k-2}}.$$

Thus we have for all $k \in \mathbb{N}$
$$\frac{d}{dx}\sqrt{x^k} = \frac{k}{2}\sqrt{x^{k-2}}.$$

2.     i)
$$\frac{d}{dx} f(x) = -\frac{k}{2}(1 + x^2)^{-\frac{k}{2}-1}(2x)$$
$$= -kx(1 + x^2)^{\frac{-k-2}{2}} = \frac{-kx}{(1 + x^2)^{\frac{k+2}{2}}};$$

ii)

$$\frac{d}{dy}g(y) = \frac{1}{2\sqrt{1+\frac{1}{y^4}}} \cdot \frac{d}{dy}\left(\frac{1}{y^4}\right)$$

$$= \frac{1}{2\sqrt{1+\frac{1}{y^4}}}\left(-4\frac{1}{y^5}\right)$$

$$= -\frac{2}{y^5\sqrt{1+\frac{1}{y^4}}} = \frac{-2}{y^3\sqrt{y^4+1}};$$

iii)

$$\frac{d}{dz}\sqrt{\frac{z^4}{1+z^2}} = \frac{d}{dz}\frac{z^2}{\sqrt{1+z^2}}$$

$$= \frac{2z}{\sqrt{1+z^2}} + z^2\frac{d}{dz}(1+z^2)^{-\frac{1}{2}}$$

$$= \frac{2z}{\sqrt{1+z^2}} + z^2\left(-\frac{1}{2}2z(1+z^2)^{-\frac{3}{2}}\right)$$

$$= \frac{2z(1+z^2) - z^3}{(1+z^2)^{\frac{3}{2}}} = \frac{z^3+2z}{(1+z^2)^{\frac{3}{2}}}.$$

3. i) Using the quotient rule we find

$$\frac{d}{du}\left(\frac{3u^5 - 7u^9}{1+u^6+u^8}\right)$$

$$= \frac{\left(\frac{d}{du}(3u^5 - 7u^9)\right)(1+u^6+u^8) - (3u^5 - 7u^9)\frac{d}{du}(1+u^6+u^8)}{(1+u^6+u^8)^2}$$

$$= \frac{(15u^2 - 63u^8)(1+u^6+u^8) - (3u^5 - 7u^9)(6u^5+8u^7)}{(1+u^6+u^8)^2}$$

$$= \frac{15u^2 - 48u^8 - 3u^{10} - 24u^{12} - 21u^{14} - 7u^{16}}{(1+u^6+u^8)^2};$$

ii) By the quotient rule it follows that

$$\frac{d}{dv}\left(\frac{(1+v^2)^{\frac{1}{2}}}{(5+v^2)^{\frac{7}{2}}}\right) = \frac{\left(\frac{d}{dv}(1+v^2)^{\frac{1}{2}}\right)(5+v^2)^{\frac{7}{2}} - (1+v^2)^{\frac{1}{2}}\frac{d}{dv}(5+v^2)^{\frac{7}{2}}}{(5+v^2)^7}$$

$$= \frac{v(1+v^2)^{-\frac{1}{2}}(5+v^2)^{\frac{7}{2}} - (1+v^2)^{\frac{1}{2}}(7\cdot v(5+v^2)^{\frac{5}{2}})}{(5+v^2)^7}$$

$$= \frac{v(5+v^2) - 7v(1+v^2)}{(1+v^2)^{\frac{1}{2}}(5+v^2)^{\frac{9}{2}}} = \frac{-2v(1+3v^2)}{(1+v^2)^{\frac{1}{2}}(5+v^2)^{\frac{9}{2}}};$$

574

iii) Again, our main tool is the quotient rule:

$$\frac{d}{dz}h(z) = \frac{d}{dz}\left(\frac{\sqrt{z^5} - 2z^4}{12 + z^2(1+z^3)}\right)$$

$$= \frac{\left(\frac{d}{dz}(\sqrt{z^5} - 2z^4)\right)(12 + z^2(1+z^3)) - (\sqrt{z^5} - 2z^4)\frac{d}{dz}(12 + z^2(1+z^3))}{(12 + z^2(1+z^3))^2}$$

$$= \frac{\left(\frac{5}{2}\sqrt{z^3} - 8z^3\right)(12 + z^2 + z^5) - (\sqrt{z^5} - 2z^4)(2z + 5z^4)}{(12 + z^2 + z^5)^2}$$

$$= \frac{(5z^5 + 5z^2 + 60)\sqrt{z^3} - 16z^8 - 16z^5 - 192z^3 + 2(10z^8 + 4z^5 + \sqrt{z^3}(-2z^2 - 5z^5))}{2(12 + z^2 + z^5)^2}$$

$$= \frac{\sqrt{z^3}(5z^5 + 5z^2 + 60 - 4z^2 - 10z^5) - 16z^8 - 16z^5 - 192z^3 + 20z^8 + 8z^5}{2(12 + z^2 + z^5)^2}$$

$$= \frac{(-5z^5 + z^2 + 60)\sqrt{z^3} + 4z^8 - 8z^5 - 192z^3}{2(12 + z^2 + z^5)^2}.$$

4. For $f^{-1}$ we have

$$\frac{d}{dy}(f^{-1})(y) = \frac{1}{f'(f^{-1}(y))}.$$

Since $f'(x) = kx^{k-1}$ and $f^{-1}(y) = y^{\frac{1}{k}}$, we find

$$\frac{d}{dy}(f^{-1})(y) = \frac{1}{k(f^{-1}(y))^{k-1}} = \frac{1}{ky^{\frac{k-1}{k}}}$$

$$= \frac{1}{k}y^{\frac{1}{k}-1}.$$

Note that for $k = 2$ we know that the inverse of $x \mapsto x^2$, $x > 0$ is $y \mapsto \sqrt{y} = y^{\frac{1}{2}}$ and $\frac{d}{dy}\left(y^{\frac{1}{2}}\right) = \frac{1}{2}y^{-\frac{1}{2}} = \frac{1}{2\sqrt{y}}$, as we already know.

5. In all three sub-problems we use the result of Problem 4, namely that

$$\frac{d}{dx}\left(x^{\frac{1}{k}}\right) = \frac{d}{dx}\sqrt[k]{x} = \frac{1}{k}x^{\frac{1}{k}-1} = \frac{1}{k}x^{\frac{1-k}{k}} = \frac{1}{k}\frac{1}{\sqrt[k]{x^{k-1}}}.$$

i)

$$\frac{d}{ds}f(s) = \frac{d}{ds}(1 + s^2)^{\frac{1}{k}}$$

$$= 2s\frac{1}{k}(1 + s^2)^{\frac{1}{k}-1} = \frac{2s}{k}(1 + s^2)^{\frac{1}{k}-1}$$

$$\left(= \frac{2s}{k(1 + s^2)^{1-\frac{1}{k}}}\right).$$

ii)

$$\frac{d}{dt}g(t) = \frac{d}{dt}\left(\frac{\sqrt{1+t^4}}{\sqrt[5]{1+t^6+t^8}}\right)$$

$$= \frac{\left(\frac{d}{dt}\sqrt{1+t^4}\right)\left(\sqrt[5]{1+t^6+t^8}\right) - \sqrt{1+t^4}\frac{d}{dt}\left(\sqrt[5]{1+t^6+t^8}\right)}{\left(\sqrt[5]{1+t^6+t^8}\right)^2}$$

$$= \frac{\frac{2t^3(1+t^6+t^8)^{\frac{1}{5}}}{(1+t^4)^{\frac{1}{2}}} - \frac{(1+t^4)^{\frac{1}{2}}(6t^5+8t^7)}{5(1+t^6+t^8)^{\frac{4}{5}}}}{(1+t^6+t^8)^{\frac{2}{5}}}$$

$$= \frac{10t^3(1+t^6+t^8) - (1+t^4)(6t^5+8t^7)}{5(1+t^4)^{\frac{1}{2}}(1+t^6+t^8)^{\frac{6}{5}}}$$

$$= \frac{2t^{11}+4t^9-8t^7-6t^5+10t^3}{5(1+t^4)^{\frac{1}{2}}(1+t^6+t^8)^{\frac{6}{5}}}.$$

iii)

$$\frac{d}{du}\left(\frac{u^{\frac{1}{7}}}{\sqrt{\frac{1+u^2}{1+u^4}}}\right) = \frac{d}{du}\left(\frac{u^{\frac{1}{7}}(1+u^4)^{\frac{1}{2}}}{(1+u^2)^{\frac{1}{2}}}\right)$$

$$= \frac{\frac{d}{du}(u^{\frac{1}{7}}(1+u^4)^{\frac{1}{2}})(1+u^2)^{\frac{1}{2}} - u^{\frac{1}{7}}(1+u^4)^{\frac{1}{2}}\frac{d}{du}(1+u^2)^{\frac{1}{2}}}{1+u^2}$$

$$= \frac{\left(\frac{1}{7}u^{-\frac{6}{7}}(1+u^4)^{\frac{1}{2}} + u^{\frac{1}{7}}2u^3(1+u^4)^{-\frac{1}{2}}\right)(1+u^2)^{\frac{1}{2}} - u^{\frac{1}{7}}(1+u^4)^{\frac{1}{2}}u(1+u^2)^{-\frac{1}{2}}}{1+u^2}$$

$$= \frac{(1+u^4+14u^4)(1+u^2) - 7u^2(1+u^4)}{7u^{\frac{6}{7}}(1+u^4)^{\frac{1}{2}}(1+u^2)^{\frac{3}{2}}}$$

$$= \frac{8u^6+15u^4-6u^2+1}{7u^{\frac{6}{7}}(1+u^4)^{\frac{1}{2}}(1+u^2)^{\frac{3}{2}}}.$$

6.    i) By the chain rule we find

$$\frac{d}{dx}\left(x^{\frac{l}{k}}\right) = \frac{d}{dx}\left(\left(x^{\frac{1}{k}}\right)^l\right)$$

$$= l\left(x^{\frac{1}{k}}\right)^{l-1}\cdot\frac{1}{k}x^{\frac{1}{k}-1}$$

$$= \frac{l}{k}x^{\frac{l-1}{k}+\frac{1}{k}-1}$$

$$= \frac{l}{k}x^{\frac{l}{k}-1}.$$

576

ii)

$$\frac{d}{ds} g(s) = \frac{d}{ds} \left( \frac{(1+s^2)^{-\frac{3}{2}}}{(1+s^4)^5} \right)$$

$$= \frac{d}{ds} ((1+s^2)^{-\frac{3}{2}} (1+s^4)^{-5})$$

$$= -\frac{3}{2} \cdot 2s(1+s^2)^{-\frac{5}{2}} (1+s^4)^{-5} + (1+s^2)^{-\frac{3}{2}} (-5 \cdot 4s^3 (1+s^4)^{-6})$$

$$= (1+s^2)^{-\frac{5}{2}} (1+s^4)^{-6} (-3s(1+s^4) + (1+s^2)(-20s^3))$$

$$= \frac{-(20s^5 + 3s^4 + 20s^3 + 3s)}{(1+s^2)^{\frac{5}{2}} (1+s^4)^6}.$$

7. A straightforward calculation using the chain rule and then the quotient rule gives

$$\frac{d}{dx} \sqrt{\frac{p(x)}{q(x)} - 2} = \frac{1}{2\sqrt{\frac{p(x)}{q(x)} - 2}} \frac{d}{dx} \left( \frac{p(x)}{q(x)} - 2 \right)$$

$$= \frac{p'(x)q(x) - p(x)q'(x)}{2q^2(x)\sqrt{\frac{p(x)}{q(x)} - 2}}$$

$$= \frac{p'(x)q(x) - p(x)q'(x)}{2q(x)^{\frac{3}{2}} \sqrt{p(x) - 2q(x)}}.$$

8. Again, we just apply the chain rule to find

$$\frac{dg(t)}{dt} = \frac{d}{dt} \sqrt{(t^2 - 1)(2t + 3)^{\frac{1}{2}}}$$

$$= \frac{1}{2\sqrt{(t^2 - 1)(2t + 3)^{\frac{1}{2}}}} \frac{d}{dt} \left( (t^2 - 1)(2t + 3)^{\frac{1}{2}} \right)$$

$$= \frac{1}{2\sqrt{(t^2 - 1)(2t + 3)^{\frac{1}{2}}}} \left( 2t(2t + 3)^{\frac{1}{2}} + (t^2 - 1)2 \cdot \frac{1}{2}(2t + 3)^{-\frac{1}{2}} \right)$$

$$= \frac{1}{2\sqrt{(t^2 - 1)(2t + 3)^{\frac{1}{2}}}} \cdot \frac{2t(2t + 3) + t^2 - 1}{(2t + 3)^{\frac{1}{2}}}$$

$$= \frac{2t(2t + 3) + t^2 - 1}{2\sqrt{(t^2 - 1)(2t + 3)^{\frac{3}{2}}}} = \frac{5t^2 + 6t - 1}{2\sqrt{(t^2 - 1)(2t + 3)^{\frac{3}{2}}}}.$$

9. Since $(h \circ f)^{-1} = f^{-1} \circ h^{-1}$ we have to apply the chain rule to $f^{-1} \circ h^{-1}$, thus

$$\frac{d}{dz} (f^{-1} \circ h^{-1})(z) = \left( \frac{d}{dy} f^{-1} \right) (h^{-1}(z)) \left( \frac{dh^{-1}}{dz} \right) (z)$$

577

with $f(x) = y$, $h(y) = z$, i.e. $z = h(f(x))$. Now using (7.7) we find

$$\left(\frac{dh^{-1}}{dz}\right)(z) = \frac{1}{h'(y)} = \frac{1}{h'(h^{-1}(z))};$$

and further

$$\left(\frac{d}{dy}f^{-1}\right)(h^{-1}(z)) = \frac{1}{f'(f^{-1}(h^{-1}(z)))} = \frac{1}{f'((h \circ f)^{-1}(z))},$$

which gives

$$\frac{d}{dz}((h \circ f)^{-1})(z) = \frac{1}{f'((h \circ f)^{-1}(z))} \frac{1}{h'(h^{-1}(z))}.$$

10. First note that $\frac{d}{dx}p(x) = \sum_{k=1}^{m} k a_k x^{k-1}$. Now using the chain rule we find

i)

$$\frac{d}{dx}p(u(x)) = p'(u(x))u'(x)$$

$$= u'(x) \sum_{k=1}^{m} k a_k (u(x))^{k-1};$$

ii)

$$\frac{d}{dx}u(p(x)) = u'(p(x))p'(x)$$

$$= u\left(\sum_{k=0}^{m} a_k x^k\right) \sum_{k=1}^{m} k a_k x^{k-1};$$

iii)

$$\frac{d}{dx}\frac{1}{u(p(x))} = \frac{-1}{u(p(x))^2}\frac{d}{dx}u(p(x))$$

$$= \frac{-u'(p(x))}{u(p(x))^2}p'(x) = -\frac{u'\left(\sum_{k=0}^{m} a_k x^k\right)}{\left(u\left(\sum_{k=0}^{m} a_k x^k\right)\right)^2} \cdot \sum_{k=1}^{m} k a_k x^{k-1}.$$

**Chapter 8**

1.    a) This follows straightforward from the definition of the composition of mappings and the boundedness of $g$. For $x \in D_1$ set $y := f(x)$ and observe

$$|(g \circ f)(x)| = |g(f(x))| = |g(y)| \leq M.$$

Thus $g \circ f$ is bounded with bound $M$.

b) Since $|f(x)| = |(x-1)^2| = (x-1)^2 \le x^2 + 2|x| + 1$ we have $|f(x)| \le 9$ for $x \in (1,2)$. We can in fact improve the bound: for $x \in (1,2)$ it follows that $x - 1 \in (0,1)$ and therefore $(x-1)^2 \le 1$, i.e. a sharper bound for $f$ on $(1,2)$ is 1, i.e. $|f(x)| \le 1$ for $x \in (1,2)$.

The function $g \circ f$ is given by

$$(g \circ f)(x) = \frac{1}{(x-1)^2}, \quad x \in (1,2).$$

We claim that this function is unbounded. For this suppose that there exists $M \ge 0$ such that

$$(*) \qquad \left| \frac{1}{(x-1)^2} \right| = \frac{1}{(x-1)^2} \le M \text{ for all } x \in (1,2).$$

Now take $x_n = 1 + \frac{1}{n}$, $n \in \mathbb{N} \setminus \{1\}$. It follows that $x_n \in (1,2)$ and

$$\frac{1}{(x_n - 1)^2} = \frac{1}{\left(1 + \frac{1}{n} - 1\right)^2} = \frac{1}{\frac{1}{n^2}} = n^2$$

and $(*)$ implies that

$$n^2 \le M \quad \text{for all } n \ge 2$$

which of course is a contradiction. Thus $g \circ f$ is unbounded on $(1,2)$.

c) We may choose $a = 0$ and $b = 1$ and consider the function $f : (0,1) \longrightarrow \mathbb{R}$, $x \mapsto f(x) = \frac{1}{x}$. This function is unbounded on $(0,1)$. Indeed as in part b) suppose that for some $M \ge 0$ we have $\left|\frac{1}{x}\right| = \frac{1}{x} \le M$ for all $x \in (0,1)$. Then for $x_n = \frac{1}{n}$, $n \in \mathbb{N} \setminus \{1\}$, we would deduce $n \le M$ for $n \ge 2$ which is a contradiction. However for $x \in [a_1, b_1] \subset (0,1)$ we find

$$\frac{1}{x} \le \frac{1}{a_1},$$

thus $f|_{[a_1,b_1]}$ is bounded by $\frac{1}{a_1}$.

2. Note that we need to find a bound, not necessarily the best bound, i.e. the smallest bound for $f$. Thus we may use rather crude estimates as long as we achieve our goal.

Therefore let $p$ be a polynomial of degree $k \in \mathbb{N}_0$. It follows with $p(x) = \sum_{j=0}^{k} a_j x^j$

and

$$c_0 := \max\{|a_j| \mid j \in \{0, 1, \ldots, k\}\}$$

that

$$|p(x)| = \left| \sum_{j=0}^{k} a_j x^j \right| \le \sum_{j=0}^{k} |a_j||x|^j$$

$$(*) \qquad \le c_0 \sum_{j=0}^{k} |x|^j.$$

Now we claim that for all $x \in \mathbb{R}$

$$|x| \leq (1 + x^2)^{\frac{1}{2}}$$

which follows immediately from $x^2 \leq 1 + x^2$. Thus by $(*)$ we get

$$|p(x)| \leq c_0 \sum_{j=0}^{k} |x|^j \leq c_0 \sum_{j=0}^{k} (1 + x^2)^{\frac{j}{2}}$$

$$\leq (k+1)c_0 (1 + x^2)^{\frac{k}{2}}.$$

Consequently we find

$$\frac{|p(x)|}{(1 + x^2)^n} \leq (k+1)c_0 \frac{(1 + |x|^2)^{\frac{k}{2}}}{(1 + |x|^2)^n} = (k+1)c_0 \left(1 + |x|^2\right)^{\frac{k}{2} - n}.$$

For $\frac{k}{2} - n \leq 0$, i.e. $n \geq \frac{k}{2}$, the right hand side is bounded since for any $l \geq 0$ the function $x \mapsto (1 + x^2)^{-l}$ is bounded. The latter statement follows from $(1 + x^2)^{-1} \leq 1$, which is equivalent to $1 \leq 1 + x^2$.

3.    a) First note that

$$\frac{d}{dx}\left(\frac{x^3 + 2x - 5}{x - 1}\right) = \frac{(3x^2 + 2)(x - 1) - (x^3 + 2x - 5) \cdot 1}{(x - 1)^2}$$

$$= \frac{2x^3 - 3x^2 + 3}{(x - 1)^2}$$

and now it follows that

$$\frac{d^2}{dx^2}\left(\frac{x^3 + 2x - 5}{x - 1}\right) = \frac{d}{dx}\left(\frac{2x^3 - 3x^2 + 3}{(x - 1)^2}\right)$$

$$= \frac{(6x^2 - 6x)(x - 1)^2 - (2x^3 - 2x^2 + 3)2(x - 1)}{(x - 1)^4}$$

$$= \frac{2x^4 - 10x^3 + 14x^2 - 12x + 6}{(x - 1)^4}.$$

b) It might be easier to write $(t^4 + 1)^{\frac{1}{2}}$ instead of $\sqrt{t^4 + 1}$. Now we find

$$\frac{d}{dt}\sqrt{t^4 + 1} = \frac{d}{dt}(t^4 + 1)^{\frac{1}{2}} = 4t^3 \cdot \frac{1}{2}(t^4 + 1)^{-\frac{1}{2}}$$

$$= 2t^3(t^4 + 1)^{-\frac{1}{2}},$$

$$\frac{d^2}{dt^2}\sqrt{t^4 + 1} = \frac{d}{dt}(2t^3(t^4 + 1)^{-\frac{1}{2}})$$

$$= 6t^2(t^4 + 1)^{-\frac{1}{2}} + 2t^3\left(-\frac{1}{2} \cdot 4t^3(t^4 + 1)^{-\frac{3}{2}}\right)$$

$$= \frac{6t^2(t^4 + 1) - 4t^6}{(t^4 + 1)^{\frac{3}{2}}} = (2t^6 + 6t^2)(t^4 + 1)^{-\frac{3}{2}}$$

580

and therefore we get

$$\frac{d^3}{dt^3}\sqrt{t^4+1} = \frac{d}{dt}\left((2t^6+6t^2)(t^4+1)^{-\frac{3}{2}}\right)$$

$$= (12t^5+12t)(t^4+1)^{-\frac{3}{2}} + (2t^6+6t^2)\left(-\frac{3}{2}\cdot 4t^3\cdot(t^4+1)^{-\frac{5}{2}}\right)$$

$$= \frac{(12t^5+12t)(t^4+1)}{(t^4+1)^{\frac{5}{2}}} + \frac{-12t^9-36t^5}{(t^4+1)^{-\frac{5}{2}}}$$

$$= \frac{-12t^5+12t}{(t^4+1)^{\frac{5}{2}}}.$$

c) We first want to investigate the differentiability of $s \mapsto |s|^5$. For $s < 0$ this is just the function $s \mapsto -s^5$ with derivative $-5s^4$, where for $s > 0$ it is the function $s \mapsto s^5$ with derivative $5s^4$. Now for $s = 0$ we find

$$\frac{|s|^5-0}{s-0} = \frac{|s|^5}{s} = \begin{cases} s^4, & s > 0 \\ 0, & s = 0 \\ -|s|^4, & s < 0 \end{cases}$$

implying that $s \mapsto |s|^5$ is differentiable and

$$\frac{d}{ds}\left(|s|^5\right) = \begin{cases} 5s^4, & s > 0 \\ 0, & s = 0 \\ -5s^4, & s < 0. \end{cases}$$

Now it follows with $g(s) = |s|^5$ that

$$\frac{d}{ds}\left(\frac{|s|^5}{s^2+4}\right) = \frac{g'(s)(s^2+4) - g(s)2s}{(s^2+4)^2}$$

$$= \frac{g'(s)s^2 + 4g'(s) - 2s|s|^5}{(s^2+4)^2}$$

$$= \begin{cases} \frac{5s^4\cdot s^2 + 4\cdot 5s^4 - 2ss^5}{(s^2+4)^2}, & s > 0 \\ 0, & s = 0 \\ \frac{-5s^4 s^2 - 4\cdot 5s^4 + 2ss^5}{(s^2+4)^2}, & s < 0 \end{cases}$$

$$= \begin{cases} \frac{3s^6+20s^4}{(s^2+4)^2}, & s > 0 \\ 0, & s = 0 \\ \frac{-3s^6-20s^4}{(s^2+4)^2}, & s < 0. \end{cases}$$

In order to find the second derivative of $h(s) := \frac{|s|^5}{s^2+4}$ we need to find the limit

$$\lim_{s\to 0}\frac{h'(s) - h'(0)}{s-0}.$$

Now for $s \geq 0$ we have

$$\frac{h'(s) - h'(0)}{s-0} = \frac{3s^5 + 20s^3}{(s^2+4)^2}$$

581

and for $s \leq 0$ we have

$$\frac{h'(s) - h'(0)}{s - 0} = \frac{-3s^5 - 20s^3}{(s^2 + 4)^2}.$$

and so for $s \in \mathbb{R}$ we have

$$\left| \frac{h'(s) - h'(0)}{s - 0} \right| \leq 3s^5 + 20s^3$$

implying that

$$\lim_{s \to 0} \frac{h'(s) - h'(0)}{s - 0} = 0,$$

thus $s \mapsto \frac{|s|^5}{s^2 + 4}$ has a second derivative at $s = 0$ and this second derivative at 0 is 0.

4. With $f(x) = u^2(x) + 1$ and $g(x) = (v^2(x) + 1)^{-1}$ we find

$$\frac{d^2}{dx^2} \left( (u^2(x) + 1)(v^2(x) + 1)^{-1} \right) = \frac{d^2}{dx^2}(f(x)g(x))$$
$$= f''(x)g(x) + 2f'(x)g(x) + f(x)g''(x).$$

Next we note

$$f'(x) = 2u'(x)u(x) \text{ a } f''(x) = 2u''(x)u(x) + 2(u'(x))^2$$

as well as

$$g'(x) = -\frac{2v'(x)v(x)}{(v^2(x) + 1)^2} = -2v'(x)v(x)(v^2(x) + 1)^{-2}$$

and

$$g''(x) = (-2v'(x)v(x))'(v^2(x) + 1)^{-2} - 2v'(x)v(x)((v^2(x) + 1)^{-2})'$$
$$= (-2v''(x)v(x) - 2v'(x)^2)(v^2(x) + 1)^{-2}$$
$$- 2v'(x)v(x)(-2v'(x)v(x))(-2(v^2(x) + 1)^{-3})$$
$$= \frac{(-2v''(x)v(x) - 2v'(x)^2)(v^2(x) + 1) - 8v'(x)^2v^2(x)}{(v^2(x) + 1)^3}$$
$$= \frac{-2v''(x)v^3(x) - 2v''(x)v(x) - 10v'(x)^2v(x)^2 - 2v'(x)^2}{(v^2(x) + 1)^3}.$$

Therefore we find

$$\frac{d^2}{dx^2}((u^2(x) + 1)(v^2(x) + 1)^{-1})$$
$$= \frac{2u''(x)u(x) + 2u'(x)^2}{(v^2(x) + 1)} + 2\frac{2u'(x)v(x)(-2v'(x)v(x))}{(v^2(x) + 1)^2}$$
$$+ \frac{(u^2(x) + 1)(-2v''(x)v^3(x) - 2v''(x)v(x) - 10v'(x)^2v(x)^2 - 2v'(x)^2)}{(v^2(x) + 1)^3}$$
$$= \frac{Q(u, v)(x)}{(v^2(x) + 1)^3},$$

582

where

$$Q(u,v)(x) = (2u''(x)u(x) + 2(u'(x))^2)(v^2(x) + 1)^2$$
$$- 8u'(x)v'(x)u(x)v(x)(v^2(x) + 1)$$
$$+ (u^2(x) + 1)(-2v''(x)v^3(x) - 2v''(x)v(x) - 10v'(x)^2v(x)^2 - 2v'(x)^2).$$

5. By the chain rule we find

$$\frac{d}{dx}(g \circ f)(x) = g'(f(x))f'(x)$$

and therefore

$$\frac{d^2}{dx^2}(g \circ f)(x) = \frac{d}{dx}(g'(f(x))f'(x))$$
$$= \left(\frac{d}{dx}(g'(f(x)))\right)f'(x) + g'(f(x))f''(x)$$
$$= g''(f(x))f'(x)^2 + g'(f(x))f''(x).$$

For $h(t) = (1 + f^2(t))^{-\frac{1}{2}}$ we find with $g(s) = (1 + s^2)^{-\frac{1}{2}}$ that $h(t) = (g \circ f)(t)$ and therefore we may apply the above formula. For this note that

$$g'(s) = -\frac{s}{(1 + s^2)^{\frac{3}{2}}} = -s(1 + s^2)^{-\frac{3}{2}}$$

and

$$g''(s) = \frac{2s^2 - 1}{(1 + s^2)^{\frac{5}{2}}} = (2s^2 - 1)(1 + s^2)^{-\frac{5}{2}}.$$

Now we set

$$\frac{d^2}{dt^2}\left((1 + f^2(t))^{-\frac{1}{2}}\right)$$
$$= (2f^2(t) - 1)(1 + f^2(t))^{-\frac{5}{2}} \cdot f'(t)^2 + f''(t)(-5(1 + f(t))^2)^{-\frac{3}{2}}$$
$$= \frac{(2f^2(t) - 1)f'(t)^2 - 5f''(t)\left(1 + f'(t)^2\right)}{(1 + f(t)^2)^{\frac{5}{2}}}.$$

6. First we observe

$$\frac{1}{(u^2(x) + 2)^2} = \frac{1}{\left(\left(\frac{x^2}{\sqrt{1+x^2}}\right)^2 + 2\right)^2}$$
$$= \frac{1}{\left(\frac{x^4}{1+x^2} + \frac{2(1+x^2)}{1+x^2}\right)^2}$$
$$= \frac{(1 + x^2)^2}{x^4 + 2x^2 + 2} = \frac{x^4 + 2x^2 + 1}{x^4 + 2x^2 + 2}$$
$$= 1 - \frac{1}{x^4 + 2x^2 + 2}.$$

This implies immediately that

$$\frac{d^2}{dx^2}\left(\frac{1}{(u^2(x)+2)^2}\right) = -\frac{d^2}{dx^2}(x^4 + 2x^2 + 2)^{-1}$$

with $g(y) = \frac{1}{y}$ and $f(x) = x^4 + 2x^2 + 2$ we find $g'(y) = -\frac{1}{y^2}$, $g''(y) = \frac{2}{y^3}$, $f'(x) = 4x^3 + 4x$, and $f''(x) = 12x^2 + 4$. Thus it follows

$$\frac{d^2}{dx^2}\left(\frac{1}{(u^2(x)+2)^2}\right) = g''(f(x))f'(x)^2 + g'(f(x))f''(x)$$

$$= -\frac{1}{(x^4+2x^2+2)^3}\cdot(4x^3+4x)^2 - \frac{1}{(x^4+2x^2+2)^2}(12x^2+4)$$

$$= \frac{-(4x^3+4x)^2 - (12x^2+4)(x^4+2x^2+2)}{(x^4+2x^2+2)^3}$$

$$= \frac{-28x^6 - 60x^4 - 48x^2 - 8}{(x^4+2x^2+2)^3}.$$

7. We prove

$$\frac{d^n}{dx^n}\left(\frac{1}{1+x^2}\right) = \frac{p_n(x)}{(1+x^2)^{n+1}}$$

by induction. For $n = 0$ we have $p_0(x) = 1$. Now we calculate

$$\frac{d^{n+1}}{dx^{n+1}}\left(\frac{1}{1+x^2}\right) = \frac{d}{dx}\left(\frac{p_n(x)}{(1+x^2)^{n+1}}\right)$$

where we used the induction hypothesis. It follows that

$$\frac{d}{dx}\left(\frac{p_n(x)}{(1+x^2)^{n+1}}\right) = \frac{p_n'(x)(1+x^2)^{n+1} - p_n(x)(2(n+1)x(1+x^2)^n)}{(1+x^2)^{2n+2}}$$

$$= \frac{p_n'(x)(1+x^2) - p_n(x)(2(n+1)x)}{(1+x^2)^{n+2}} = \frac{p_{n+1}(x)}{(1+x^2)^{n+2}}$$

with

$$p_{n+1}(x) = p_n'(x)(1+x^2) - 2(n+1)xp_n(x).$$

The degree of $p_n(x)$ is at most $n$ and that of $p_n'(x)$ is at most $n-1$, therefore the degree of $p_{n+1}(x)$ is at most $n+1$.

Now the estimate follows using Problem 2

$$\left|\frac{d^n}{dx^n}\left(\frac{1}{1+x^2}\right)\right| = \left|\frac{p_n(x)}{(1+x^2)^{n+1}}\right| \leq \frac{c_n(1+|x^2|)^{\frac{n}{2}}}{(1+x^2)^{n+1}}$$

$$= \frac{c_n}{(1+x^2)^{\frac{n+2}{2}}}.$$

8.     a) By the definition of the absolute value, we know that $|x^3| \geq 0$ for all $x \in \mathbb{R}$ and $|x|^3 = 0$ if and only if $x = 0$ implying that $f(x) = |x|^3$ has a local minimum at $x_0 = 0$.

(Note that we did not use differential calculus as it is not necessary or helpful here.)

    b) We first find $g'(s)$:

$$g(s) = (s^2 - 2s)(2 + 3s^2)^{-1},$$

therefore

$$\begin{aligned}
g'(s) &= (2s - 2)(2 + 3s^2)^{-1} + (s^2 - 2s)(-1(2 + 3s^2)^{-2}6s) \\
&= \frac{(2s - 2)(2 + 3s^2) - 6s(s^2 - 2s)}{(2 + 3s^2)^2} \\
&= \frac{6s^2 + 4s - 4}{(2 + 3s^2)^2}.
\end{aligned}$$

Therefore the condition $g'(s) = 0$ is equivalent to

$$2(3s^2 + 2s - 2) = 0,$$

i.e. we have to solve the quadratic equation

$$3s^2 + 2s - 2 = 0$$

which gives

$$s_{1,2} = -\frac{1}{3} \pm \frac{1}{3}\sqrt{7}.$$

In order to decide whether we have a local extreme value at $s_1$ or $s_2$, and when we do in order to find what type it is we make use of $g''(s)$.

$$\begin{aligned}
g''(s) &= \frac{d}{ds}((6s^2 + 4s - 4)(2 + 3s^2)^{-2}) \\
&= (12s + 4)(2 + 3s^2)^{-2} + (6s^2 + 4s - 4)(-2(2 + 3s^2)^{-3}(6s)) \\
&= \frac{(12x + 4)(2 + 3x^2) - 12x(6x^2 + 4x - 4)}{(2 + 3x^2)^3} \\
&= \frac{-36s^3 - 36s^2 - 24s + 8}{(2 + 3s^2)^3}.
\end{aligned}$$

Now we need to determine whether $g''(s_1)\,(g''(s_2))$ is strictly positive or strictly negative.

**But** we do not need to calculate the exact value of $g''(s_1)\,(g''(s_2))$. Therefore we only need to look at the sign of the polynomial - $36s^3 - 36s^2 - 24s + 8$ at $s_1$ and $s_2$. Note that

$$s_1 = -\frac{1}{3} + \frac{1}{3}\sqrt{7}, \ s_1^2 = \frac{8}{9} - \frac{2}{9}\sqrt{7}, \ x_1^3 = -\frac{22}{27} + \frac{10}{27}\sqrt{7}$$

while

$$s_2 = -\frac{1}{3} - \frac{1}{3}\sqrt{7}, \quad x_2^2 = \frac{8}{9} + \frac{2}{9}, \quad s_2^3 = -\frac{22}{27} - \frac{10}{27}\sqrt{7}.$$

Therefore we find

$$g''(s_1) = -36(s_1^3 + s_1^2) - 24s_1 + 8$$

$$= -36\left(-\frac{22}{27} + \frac{10}{27}\sqrt{7} + \frac{8}{9} - \frac{2}{9}\sqrt{7}\right) + 8 - 8\sqrt{7} + 8$$

$$= -36\left(\frac{2}{27} + \frac{4}{27}\sqrt{7}\right) - 8\sqrt{7} + 16$$

$$= -\frac{8}{3} - \frac{16}{3}\sqrt{7} - 8\sqrt{7} + 16$$

$$= \frac{40}{3} - \frac{40}{3}\sqrt{7} = \frac{40}{3}(1 - \sqrt{7}) < 0,$$

implying that $g$ has a local maximum at $s_1 = -\frac{1}{3} + \frac{1}{3}\sqrt{7}$. For $g''(s_2)$ we find

$$g''(s_2) = -36(s_2^3 + s_2^2) - 24s_2 + 8$$

$$= -36\left(-\frac{22}{27} - \frac{10}{27}\sqrt{7} + \frac{8}{9} + \frac{2}{9}\sqrt{7}\right) - 24\left(-\frac{1}{3} - \frac{1}{3}\sqrt{7}\right) + 8$$

$$= -36\left(\frac{2}{27} - \frac{4}{27}\sqrt{7}\right) + 8 + 8\sqrt{7} + 8$$

$$= 36\left(\frac{4}{27}\sqrt{7} - \frac{2}{27}\right) + 8\sqrt{7} + 16 > 0,$$

implying that $g$ has a local minimum at $s_2 = -\frac{1}{3} - \frac{1}{3}\sqrt{7}$.

c) For the first derivative of $h$ we find

$$h'(u) = \frac{-2u^2 - u + 1}{\sqrt{1 - u^2}}$$

which has zeroes for $u_1 = \frac{1}{2}$ and $u_2 = -1$, but $-1 \notin (-1, 1)$. Thus $h$ may only have a local extreme value at $u_1 = \frac{1}{2}$.

Now

$$h''(u) = \frac{2u^3 - 3u - 1}{(1 - u^2)^{\frac{3}{2}}}$$

and for $u_0 = \frac{1}{2}$ we find

$$h''\left(\frac{1}{2}\right) = \frac{2 \cdot \frac{1}{2^3} - 3 \cdot \frac{1}{2} - 1}{\left(1 - \left(\frac{1}{2}\right)^2\right)^{\frac{3}{2}}} = \frac{-9}{4\left(\frac{3}{4}\right)^{\frac{3}{2}}} < 0.$$

Therefore $h$ has a local maximum at $u_0 = \frac{1}{2}$.

586

9.  a) We may take, for example $f : (-1, 1) \longrightarrow \mathbb{R}$, $f(t) = \frac{1}{1+t}$. It follows that $(f \circ g)(x) = \frac{1}{1+x^2}$ and we find immediately that

$$\left| \frac{1}{1+x^2} \right| = \frac{1}{1+x^2} \leq 1 \text{ for all } x \in (-1, 1).$$

Since $(f \circ g)(0) = 1$ it follows that $f \circ g$ has a maximum at $x = 0$.

b) The function $f$ has a local maximum at $x_0$ if for some $\epsilon > 0$ it follows for $x \in (-\epsilon + x_0, x_0 + \epsilon)$ that $f(x) \leq f(x_0)$. This implies for all $x \in (-\epsilon + x_0, x_0 + \epsilon)$ that

$$h(x_0 + c) = f(x_0 + c - c) = f(x_0) \geq f(x) = f(x + c - c) = h(x + c),$$

i.e.

$$h(x_0 + c) \geq h(y) \text{ for all } y \in (-\epsilon + x_0 + c, x_0 + c + \epsilon),$$

and with $y_0 := x_0 + c \in (-\epsilon + x_0 + c, x_0 + c + \epsilon)$ we have

$$h(y) \leq h(y_0) \text{ for all } y \in (-\epsilon + x_0 + c, x_0 + c + \epsilon),$$

implying that $h$ has a maximum at $y_0 = x_0 + c$.
Note that in the case where $f$ is twice differentiable we may use calculus. First note that we know $f'(x_0) = 0$ and $f''(x_0) < 0$. However $h'(x) = f'(x - c)$. Thus $h'(x_0 + c) = 0$ and since $h''(x) = f''(x - c)$ we also know that $h''(x_0 + c) < 0$ implying that $h$ has a local maximum at $x_0 + c$.

10.  a) By the mean value theorem we have

$$|\sin x - \sin y| = |\sin' \xi||x - y| = |\cos \xi||x - y| \leq |x - y|$$

and for $y = 0$ we find

$$|\sin x| \leq |x|.$$

b) We apply the mean value theorem in the form

$$|f(x) - f(y)| \leq M|x - y|$$

where $|f'(z)| \leq M$ for all $z$, $f : [x, y] \to \mathbb{R}$.
Thus in this case, we have

$$|g'(z)| = \frac{1}{2\sqrt{z}} \leq \frac{1}{2} \text{ for } z \in [1, 2],$$

therefore with $x = \frac{11}{10}$ and $y = \frac{10}{10} = 1$ :

$$\left| \sqrt{\frac{11}{10}} - 1 \right| \leq \frac{1}{2} \frac{1}{10} = \frac{1}{20}$$

587

or

$$-\frac{1}{20} + 1 \leq \sqrt{\frac{11}{10}} \leq 1 + \frac{1}{20},$$

i.e.

$$\frac{19}{20} \leq \sqrt{\frac{11}{10}} \leq \frac{21}{20}.$$

11.    a) We first note that each of the functions is increasing.

$$\chi_n : \mathbb{R} \to R$$
$$x \mapsto \chi_n(x) = \chi_{[n,\infty)}(x)$$

Indeed,
if $x < y < n$ then $\chi_n(x) = \chi(y)$,
if $x < n \leq y$ then $\chi_n(x) = 0 < 1 = \chi_n(y)$,
if $n \leq x < y$ then $\chi_n(x) = \chi_n(y)$.
Since the sum of increasing functions is increasing ($g(x) \leq g(y)$ and $f(x) \leq f(y)$ implies $g(x) + f(x) \leq g(y) + f(y)$), it follows that $X_N$ is increasing.

Here is the graph of $X_5$

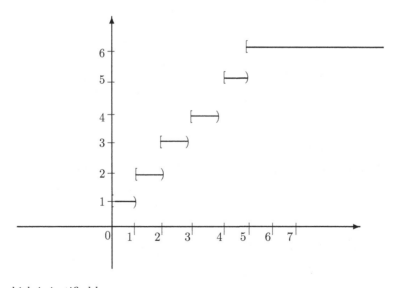

which is justified by:

$$\chi_0(x) = \qquad 1 \text{ for all } x \geq 0$$

$$\chi_1(x) = \begin{cases} 0 \text{ for all } & x < 1 \\ 1 \text{ for all } & x \geq 1 \end{cases}$$

$$\chi_2(x) = \begin{cases} 0 \text{ for all } & x < 2 \\ 1 \text{ for all } & x \geq 2 \end{cases}$$

$$\chi_3(x) = \begin{cases} 0 \text{ for all } & x < 3 \\ 1 \text{ for all } & x \geq 3 \end{cases}$$

$$\chi_4(x) = \begin{cases} 0 \text{ for all } & x < 4 \\ 1 \text{ for all } & x \geq 4 \end{cases}$$

$$\chi_5(x) = \begin{cases} 0 \text{ for all } & x < 5 \\ 1 \text{ for all } & x \geq 5 \end{cases}$$

therefore for all

$$0 < x < 1: \quad \sum_{n=0}^{5} \chi_n(x) = 1$$

$$1 \leq x < 2: \quad \sum_{n=0}^{5} \chi_n(x) = 2$$

$$2 \leq x < 3: \quad \sum_{n=0}^{5} \chi_n(x) = 3$$

$$3 \leq x < 4: \quad \sum_{n=0}^{5} \chi_n(x) = 4$$

$$4 \leq x < 5: \quad \sum_{n=0}^{5} \chi_n(x) = 5$$

$$5 \leq x \quad: \quad \sum_{n=0}^{5} \chi_n(x) = 6.$$

b) We need to consider the sign of $f'_a$. Note that

$$f'(x) = \frac{d}{dx}(x(1 + ax^2)^{-1})$$
$$= (1 + ax^2)^{-1} + x(-1(1 + ax^2))^{-2}(2ax))$$
$$= \frac{1 + ax^2 - 2ax^2}{(1 + ax^2)^2} = \frac{1 - ax^2}{(1 + ax^2)^2}.$$

Since $(1 + ax^2)^2 > 0$ for all $x \in \mathbb{R}$ we only need to look at $1 - ax^2$. For $x < -\frac{1}{\sqrt{a}}$ or $x > \frac{1}{\sqrt{a}}$, it follows that $1 - ax^2 < 0$, hence in $(\frac{1}{\sqrt{a}}, \infty)$ the function is strictly

589

decreasing. For $x \in \left(0, \frac{1}{\sqrt{a}}\right)$ we have $1 - ax^2 > 0$ and therefore in $\left(0, \frac{1}{\sqrt{a}}\right)$ the function is strictly increasing.

12.  a) Let $t_1, t_2 \in (c, d)$ such that $t_1 < t_2$. Since $g$ is increasing it follows that $g(t_1) \le g(t_2)$. Now the fact that $f$ is also increasing gives $f(g(t_1)) \le f(g(t_2))$, i.e. $f \circ g$ is increasing.

b) Using the chain rule we find

$$(f \circ g)'(x) = f'(g(x))g'(x)$$

and

$$(g \circ f)'(x) = g'(f(x))f'(x).$$

In both cases, if $f'$ and $g'$ are positive functions, or if $f'$ and $g'$ are negative functions we have that $(f \circ g)'$ and $(g \circ f)'$ are non-negative functions, hence increasing.

13. By the mean value theorem applied to $h = g - f$ there exists $\xi \in (a, b)$ such that

$$h(x) - h(a) = h'(\xi)(x - a) = (g'(\xi) - f'(\xi))(x - a) > 0.$$

However $h(a) = g(a) - f(a) = 0$ and therefore $g(x) - f(x) = h(x) > 0$ for all $x \in (a, b)$, i.e. $f(x) < g(x)$ for all $x \in (a, b)$.

**Chapter 9**

1.  a) Recall that $\lim\limits_{x \to \infty} f(x) = \infty$ means that for all $M > 0$ there exists $N \in \mathbb{N}$ such that $x > N$ implies $f(x) > M$. Given $M > 0$ we have to find $N \in \mathbb{N}$ such that $x > N$ implies $x^5 - 5 > M$, or $x^5 > M + 5$. Hence for $N := [\sqrt{M + 5}] + 1$ it follows for $x > N = [\sqrt{M + 5}] + 1$ that

$$x^2 - 5 > \left([\sqrt{M + 5}] + 1\right)^2 - 5$$
$$= [M + 5] + 2\sqrt{M + 5} - 4$$
$$= M + 4 + 2\sqrt{M + 5} - 4 > M.$$

b) Let us rewrite $p(x)$ as

$$p(x) = a_k x^k \left(1 + \sum_{l=0}^{k-1} \frac{a_l}{a_k} x^{l-k}\right)$$

which is correct for say $x > 1$. Now for $0 \le l < k$ there exists $N_l$ such that for $x > N_l$

$$\left|\frac{a_l}{a_k}\right| x^{l-k} < \frac{1}{2k}.$$

Indeed, this is equivalent to

$$x^{k-l} > 2k \left|\frac{a_l}{a_k}\right|$$

for $x > N_l$, and this follows from Example 9.10. Therefore we see for $N_k :=$ $\max\{N_0, \ldots, N_{k-1}\}$ that $x > N_k$ implies

$$1 + \sum_{l=0}^{k-1} \frac{a_l}{a_k} x^{l-k} \geq 1 - \sum_{l=0}^{k-1} \left|\frac{a_l}{a_k}\right| x^{l-k}$$

$$\geq 1 - k \cdot \frac{1}{2k} = \frac{1}{2}.$$

This now implies for $x > N_k$

$$p(x) \geq \frac{a_k}{2} x^k.$$

Again using Example 9.10 we deduce that given $M > 0$ there exists $\tilde{N} \in \mathbb{N}$ such that $x > \tilde{N}$ it follows that $\frac{a_k}{2} x^k > M$. Hence for $N = \max\{\tilde{N}, N_k\}$ it follows that $x > N$ implies $p(x) > M$ or

$$\lim_{x \to \infty} p(x) = \infty.$$

c) Note that

$$\frac{1 + a + ax^2}{1 + x^2} = \frac{1}{1 + x^2} + \frac{a(1 + x^2)}{1 + x^2} = \frac{1}{1 + x^2} + a,$$

therefore for $\epsilon > 0$ we have to find $N \in \mathbb{N}$ such that $x > N$ implies that

$$\left|\frac{1 + a + ax^2}{1 + x^2} - a\right| = \left|\frac{1}{1 + x^2} + a - a\right| = \frac{1}{1 + x^2} < \epsilon.$$

We can now continue as in Example 9.9 and take $N = N(\epsilon) = \left[\frac{1}{\epsilon}\right] + 1$, and see for $x > N(\epsilon)$ that

$$\left|\frac{1 + a + ax^2}{1 + x^2} - a\right| = \frac{1}{1 + x^2} < \frac{1}{x} < \frac{1}{\left[\frac{1}{\epsilon}\right] + 1} < \epsilon.$$

2.   a) Lemma 9.11.B says that for $a > 0$ and $n \in \mathbb{N}_0$

$$(1 + a)^n \geq 1 + na + \frac{n(n - 1)}{2} a^2.$$

For $n \geq 2$ it follows that $\frac{n(n-1)}{2} a^2 > 0$ implies

$$(*) \quad (1 + a)^n > 1 + na$$

for $a > 0$ and $n \geq 2$.

   b) We apply $(*)$ to see for $n \geq 2$

$$\left(1 + \frac{1}{n^2 - 1}\right)^n > 1 + n\frac{1}{n^2 - 1} = 1 + \frac{n}{n^2 - 1}$$

and it remains to prove that for $n \geq 2$ it follows that $\frac{n}{n^2-1} \geq \frac{1}{n}$ which is equivalent to $\frac{n^2}{n^2-1} \geq 1$ and this of course is correct.

591

3. From the definition we find

$$a^{x+y} = \exp((x+y)\ln a) = \exp(x\ln a + y\ln a)$$
$$= \exp(x\ln a)\exp(y\ln a) = a^x a^y,$$

as well as

$$a^0 = \exp(0\ln a) = \exp(0) = 1.$$

4.     a) By the chain rule we find

$$\frac{d}{dx}\exp(-\sqrt{x^2+1}) = \left(\frac{d}{dx}(-\sqrt{x^2+1})\right)(\exp')(-\sqrt{x^2+1})$$
$$= \frac{-x}{\sqrt{x^2+1}}\exp(-\sqrt{x^2+1}).$$

b) Again we use the chain rule to get

$$\frac{d}{du}\exp(-\log_a(1+u^2)) = \left(\frac{d}{du}(-\log_a(1+u^2))\right)(\exp')(-\log_a(1+u^2))$$
$$= \frac{-2u}{(\ln a)(1+u^2)}\exp(-\log_a(1+u^2)).$$

c) First we find

$$\frac{d}{dt}\left(\exp\left(-\frac{1}{1+t^2}\right)\right) = \left(\frac{d}{dt}\left(-\frac{1}{1+t^2}\right)\right)(\exp')\left(-\frac{1}{1+t^2}\right)$$
$$= \frac{2t}{(1+t^2)^2}\exp\left(-\frac{1}{1+t^2}\right)$$

and now it follows that

$$\frac{d^2}{dt^2}\left(\exp\left(-\frac{1}{1+t^2}\right)\right) = \frac{d}{dt}\left(\frac{2t}{(1+t^2)^2}\exp\left(-\frac{1}{1+t^2}\right)\right)$$
$$= \left(\frac{d}{dt}\left(\frac{2t}{(1+t^2)^2}\right)\right)\exp\left(-\frac{1}{1+t^2}\right) + \frac{2t}{(1+t^2)^2}\frac{d}{dt}\left(\exp\left(-\frac{1}{1+t^2}\right)\right)$$
$$= \frac{2-6t^2}{(1+t^2)^3}\exp\left(-\frac{1}{1+t^2}\right) + \frac{2t}{(1+t^2)^2}\cdot\frac{2t}{(1+t^2)^2}\exp\left(-\frac{1}{1+t^2}\right)$$
$$= \frac{2-6t^4}{(1+t^2)^4}\exp\left(-\frac{1}{1+t^2}\right).$$

5. The case $n = 0$ is straightforward, just take $p_0(x) = 1$. Now suppose that

$$\frac{d^n}{dx^n}e^{-x^2} = p_n(x)e^{-x^2}$$

592

with $p_n(x)$ of degree $n$. It follows that

$$\frac{d^{n+1}}{dx^{n+1}}e^{-x^2} = \frac{d}{dx}\left(p_n(x)e^{-x^2}\right)$$
$$= p_n'(x)e^{-x^2} - 2xp_n(x)e^{-x^2}$$
$$= (p_n'(x) - 2xp_n(x))e^{-x^2} = p_{n+1}(x)e^{-x^2}.$$

The polynomial $p_{n+1}(x) := p_n'(x) - 2xp_n(x)$ has degree at most $n+1$ since the degree of $p_n'(x)$ is at most $n-1$ and that of $-2xp(x)$ is at most $n+1$.

6.    a) By the chain rule we find

$$\frac{d}{ds}\ln(\sqrt{s^4+1} - s^2) = \left(\frac{d}{ds}(\sqrt{s^4+1} - s^2)\right)(\ln')(\sqrt{s^4+1} - s^2)$$
$$= \frac{2s^3 - 2s\sqrt{s^4+1}}{\sqrt{s^4+1}}\frac{1}{\sqrt{s^4+1} - s^2}$$
$$= \frac{2s^3 - 2s\sqrt{s^4+1}}{s^4 - s^2\sqrt{s^4+1} + 1}.$$

b) Once again by the chain rule we find

$$\frac{d}{dx}(\ln(a^x)) = \left(\frac{d}{dx}a^x\right)(\ln')(a^x)$$
$$= (\ln a)a^x\frac{1}{a^x} = \ln a.$$

Note that the derivative is constant.

c) First note that

$$\frac{d}{dy}\ln((y^2+1)^{-k}) = \left(\frac{d}{dy}(y^2+1)^{-k}\right)(\ln')\left((y^2+1)^{-k}\right)$$
$$= \frac{-2yk}{(y^2+1)^{k+1}}\cdot\frac{1}{\frac{1}{(y^2+1)^k}} = \frac{-2yk}{(y^2+1)}$$

and it follows now that

$$\frac{d^2}{dy^2}\ln\left((y^2+1)^{-k}\right) = \frac{d}{dy}\left(\frac{-2ky}{y^2+1}\right)$$
$$= \frac{-2k(y^2+1) - (-2ky)(2y)}{(y^2+1)^2}$$
$$= \frac{2ky^2 - 2k}{(y^2+1)^2}.$$

7.    a) We can use Lemma 9.14 in the following way:

$$\lim_{x\to\infty} \frac{x}{\exp(ax)} = \frac{1}{a}\lim_{x\to\infty}\frac{ax}{\exp(ax)}$$

$$= \frac{1}{a}\lim_{y\to\infty}\frac{y}{\exp(y)} = \infty.$$

Here we used the fact that $x\to\infty$ if and only if $y = \frac{x}{a}\to\infty$.

b) For $n\in\mathbb{N}$ we have

$$\exp(ax) = \exp\left(\frac{ax}{n}+\cdots+\frac{ax}{n}\right)$$

$$= \exp\left(\frac{ax}{n}\right)\cdot\ldots\cdot\exp\left(\frac{ax}{n}\right) \quad (n\text{-terms})$$

and therefore it follows that

$$\lim_{x\to\infty}\frac{x^n}{\exp(ax)} = \lim_{x\to\infty}\left(\frac{x}{\exp\left(\frac{ax}{n}\right)}\cdot\ldots\cdot\frac{x}{\exp\left(\frac{ax}{n}\right)}\right)$$

$$= \lim_{x\to\infty}\left(\frac{x}{\exp\left(\frac{ax}{n}\right)}\right)\cdot\ldots\cdot\lim_{x\to\infty}\left(\frac{x}{\exp\left(\frac{ax}{n}\right)}\right) = \infty.$$

The following is important to note: we have not yet proved that if $\lim_{x\to\infty} f(x) = \infty$ and $\lim_{x\to\infty} g(x) = \infty$ then it follows that $\lim_{x\to\infty}(f(x)g(x)) = \left(\lim_{x\to\infty}f(x)\right)\left(\lim_{x\to\infty}g(x)\right) = \infty$. Suppose that $\lim_{x\to\infty} f(x) = \infty$ and $\lim_{x\to\infty} g(x) = \infty$. Given $M > 0$ there exists $N$ such that for $x > N$ we have $f(x) > \sqrt{M}$ and $g(x) > \sqrt{M}$. Therefore for $x > N$ it follows that $f(x)\cdot g(x) > \sqrt{M}\sqrt{M} = M$, i.e. we have proved that $\lim_{x\to\infty} f(x)g(x) = \infty$. Finally we use the convention that $(+\infty)\cdot(+\infty) = +\infty$.

8. Firstly we can use the considerations of Problem 1 b). Thus we first write for $x\neq 0$

$$p(x) = b_m x^m\left(1 + \sum_{k=0}^{m-1}\frac{b_k}{b_m}x^{k-m}\right).$$

If $m$ is even we find further for $K = \ln M$, $K > 0$, i.e. $M > 1$ given there exists $N\in\mathbb{N}$ such that $x > N$ implies

$$p(-x) = b_m x^m\left(1 + \sum_{k=0}^{m-1}\frac{b_k}{b_m}(-1)^{k-m}x^{k-m}\right)$$

$$\geq \ln M.$$

Now it follows for $x > N$ that

$$\exp(p(-x)) \geq M$$

implying for $m$ even it follows that $\lim_{x \to -\infty} \exp(p(x)) = \infty$.

Now let $m$ be odd. First we prove that

$$\lim_{x \to -\infty} \exp(x^m) = 0$$

which follows from

$$\lim_{x \to -\infty} \exp(x^m) = \lim_{y \to \infty} \frac{1}{\exp(y^m)} = 0.$$

Now suppose that we can prove that there exists some $N \in \mathbb{N}$ such that $x < -N$ implies

$$(*) \quad p(x) \leq cx^m$$

with $c > 0$ independent of $N$. In this case we would have

$$0 \leq \exp(p(x)) \leq \exp c x^m$$

and therefore

$$0 \leq \lim_{x \to -\infty} \exp(p(x)) \leq \lim_{x \to -\infty} \exp(cx^m) = 0,$$

i.e.

$$\lim_{x \to -\infty} \exp p(x) = 0.$$

In order to prove $(*)$, note that for $x < 0$

$$\frac{1}{b_m} \frac{p(x)}{x^m} = 1 + \sum_{k=0}^{m-1} \frac{b_k}{b_m} x^{k-m}$$

and we are done if we can show that

$$1 + \sum_{k=0}^{m-1} \frac{b_k}{b_m} x^{k-m} \leq \tilde{c}, \ \tilde{c} > 0.$$

Now note that for $x \leq -1$

$$1 + \sum_{k=0}^{m-1} \frac{b_k}{b_m} x^{k-m} \leq 1 + \sum_{k=0}^{m-1} \frac{|b_k|}{|b_m|} |x|^{k-m}$$

$$= 1 + \sum_{k=0}^{m-1} \frac{|b_k|}{|b_m|},$$

and then the result follows.

9. We again use the fact that for $x \neq 0$

$$(*) \quad p(x) = a_n x^n \left( 1 + \sum_{k=0}^{n-1} \frac{a_k}{a_n} x^{k-n} \right)$$

and the result shown in Problem 1 b) that for large $x$

$$1 + \sum_{k=0}^{n-1} \frac{a_k}{a_n} x^{k-n} \geq \frac{1}{2}.$$

Hence for $x \geq R$ we have that $\ln p(x)$ is defined. Now we can investigate

$$\lim_{x \to \infty} \frac{\ln p(x)}{x}.$$

With $(*)$ it follows for $x \geq R$ that

$$0 \leq \frac{\ln p(x)}{x} = \frac{\ln a_n x^n \left(1 + \sum_{k=0}^{n-1} \frac{a_k}{a_n} x^{k-n}\right)}{x}$$

$$= \frac{\ln a_n x^n}{x} + \frac{\ln\left(1 + \sum_{k=0}^{n-1} \frac{a_k}{a_n} x^{k-n}\right)}{x}$$

$$\leq \frac{\ln a_n x^n}{x} + \frac{\ln\left(1 + \sum_{k=0}^{n-1} \frac{|a_k|}{|a_n|} R^{k-n}\right)}{x}.$$

Clearly $\lim_{x \to \infty} \dfrac{\ln\left(1 + \sum_{k=0}^{n-1} \frac{|a_k|}{|a_n|} R^{k-n}\right)}{x} = 0.$

Thus we want to prove

$$\lim_{x \to \infty} \frac{\ln a_n x^n}{x} = 0,$$

but

$$\frac{\ln(a_n x^n)}{x} = \frac{\ln a_n}{x} + \frac{n \ln x}{x}$$

and Theorem 9.16 gives the result.

10.     a) First note that for $x, y > 0$

$$(xy)^{\frac{1}{2}} \leq \frac{x+y}{2}.$$

This estimate is equivalent to

$$4xy \leq (x+y)^2 = x^2 + 2xy + y^2$$

or

$$2xy \leq x^2 + y^2$$

which is correct since $0 \leq (x-y)^2 = x^2 - 2xy + y^2$.
Now the monotonicity of ln gives

$$\ln(xy)^{\frac{1}{2}} \leq \ln\left(\frac{x+y}{2}\right)$$

but

$$\ln(xy)^{\frac{1}{2}} = \frac{1}{2}\ln(xy) = \frac{\ln x + \ln y}{2}.$$

A function satisfying $g\left(\frac{x+y}{2}\right) \leq \frac{g(x)+g(y)}{2}$ is called **convex in the sense of J. Jensen** or **mid-point convex**.

b) The mean value theorem gives

$$|\ln x - \ln y| = |\ln' \xi||x - y|$$

for some $y \leq \xi \leq x$. Now $\ln' \xi = \frac{1}{\xi}$ and by assumption $|x - y| = 1$. Therefore we have

$$|\ln x - \ln y| = \frac{1}{\xi}.$$

Since ln is monotone increasing we have $\ln x - \ln y > 0$, i.e. $\ln x - \ln y = |\ln x - \ln y|$ and further $\frac{1}{x} < \frac{1}{\xi} < \frac{1}{y}$ implying

$$\frac{1}{x} \leq \ln x - \ln y \leq \frac{1}{y}.$$

11. The logarithmic derivative of $v$ is given by $\frac{v'}{v}$. Thus we have

$$\frac{v'(x)}{v(x)} = 1, \ v(0) = 1,$$

or $v'(x) = v(x)$ and $v(0) = 1$. Thus it follows that $v(x) = \exp x$.

**Chapter 10**

1.  a) For $x \in \mathbb{R}$ we have

$$(f \circ g)(-x) = f(g(-x)) = f(g(x)) = (f \circ g)(x),$$

therefore $f \circ g$ is an even function.

b) For $x \in \mathbb{R}$ we find

$$(f \circ g)(-x) = f(g(-x)) = f(-g(x))$$
$$= -f(g(x)) = -(f \circ g)(x),$$

hence $f \circ g$ is an odd function.

c) Let $c = \min\{|a|, b\}$. Then $-\frac{c}{2}, \frac{c}{2} \in (a, b)$ and $f\left(-\frac{c}{2}\right) = f\left(\frac{c}{2}\right)$. Therefore $f|_{(a,b)}$ is not injective and therefore it does not have an inverse function.

597

2.    a) Let $f$ be an even function and note that

$$\frac{f(-y+h) - f(-y)}{h} = \frac{f(y-h) - f(y)}{h} = -\frac{f(y-h) - f(y)}{-h}$$

and for $h \to 0$ we find

$$f'(-y) = \lim_{h \to 0} \frac{f(-y+h) - f(-y)}{h} = -\lim_{h \to 0} \frac{f(y-h) - f(y)}{-h} = -f'(y)$$

implying that $f'$ is an odd function.
Now if $f$ is an odd function we have

$$\frac{f(-y+h) - f(-y)}{h} = \frac{-f(y-h) + f(y)}{h} = \frac{f(y-h) - f(y)}{-h}$$

and in the limit we have

$$f'(-y) = \lim_{h \to 0} \frac{f(-y+h) - f(-y)}{h} = \lim_{h \to 0} \frac{f(y-h) - f(y)}{-h} = f'(y),$$

i.e. $f'$ is even.
Thus by iteration if $f$ is an even $C^k$ function then all derivatives $f^{(l)}$ with $l \leq k$ and $l$ even are even functions and all derivatives $f^{(l)}$ with $l \leq k$ and $l$ odd are odd functions.
    b) We define

$$g(x) = \begin{cases} f(x), & x \geq 0 \\ f(-x), & x \leq 0 \end{cases}$$

and

$$h(x) = \begin{cases} f(x), & x > 0 \\ 0, & x = 0 \\ -f(-x), & x < 0. \end{cases}$$

Clearly $g$ is even and $h$ is odd. Note that we obtain $g$ by reflecting $f$ in the $y$-axis, where $h$ is obtained by a point reflection of $f_{[0,\infty)}$ at $x_0 = 0$.

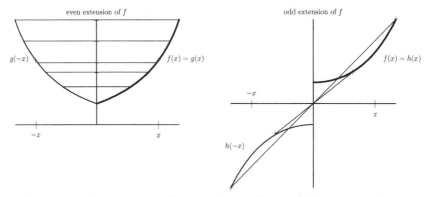
even extension of $f$                    odd extension of $f$

3.    a) This limit does not exist. Suppose that it does and that it is equal to $a$, i.e. for all $\epsilon > 0$ there exists $N(\epsilon) \in \mathbb{N}$ such that $x > N(\epsilon)$ implies that $|\sin x - a| < \epsilon$.

For $\epsilon = \frac{1}{2}$ take $k > N(\epsilon)$ implying $x_k := 2\pi k > N(\epsilon)$ and $y_k := \frac{\pi}{2} + 2\pi k > N(\epsilon)$ implying

$$1 = |\sin x_k - \sin y_k| = |\sin x_k - a + a - \sin y_k|$$

$$\leq |\sin x_k - a| + |\sin y_k - a| < \frac{\epsilon}{2} + \frac{\epsilon}{2} = \frac{1}{2} + \frac{1}{2} = 1,$$

which is a contradiction.

b) Since $|(\sin x)^k| \leq 1$ for all $x \in \mathbb{R}$ and $k \in \mathbb{N}$ it follows that

$$\left| \frac{(\sin x)^k}{x} - 0 \right| \leq \frac{1}{x}, \ x > 0.$$

Now, given $\epsilon > 0$ choose $N(\epsilon) = \left[\frac{1}{\epsilon}\right] + 1$ to find for $x > N(\epsilon)$, i.e. $\frac{1}{x} < \frac{1}{\left[\frac{1}{\epsilon}\right]+1} < \epsilon$, that

$$\left| \frac{(\sin x)^k}{x} - 0 \right| = \left| \frac{(\sin x)^k}{x} \right| < \frac{1}{x} < \epsilon.$$

4. We use Figure 10.2 in the following.

a) First note that $\sin \frac{\pi}{4} = \cos \frac{\pi}{4}$ and since $1 = \sin^2 \frac{\pi}{4} + \cos^2 \frac{\pi}{4} = 2\sin^2 \frac{\pi}{4}$ it follows that $\sin \frac{\pi}{4} = \cos \frac{\pi}{4} = \frac{\sqrt{2}}{2}$.
By (10.10) we now find

$$\cos \frac{\pi}{4} - \cos 0 = \frac{1}{2}\sqrt{2} - 1 = -2 \left( \sin \frac{\pi}{8} \right)^2$$

or

$$\left( \sin \frac{\pi}{8} \right)^2 = \frac{1}{2} \left( 1 - \frac{1}{2}\sqrt{2} \right),$$

i.e. $\sin \frac{\pi}{8} = \sqrt{\frac{1}{2} - \frac{1}{4}\sqrt{2}}$.

b) By looking at the figure below and using elementary geometry we deduce that $\sin \frac{\pi}{6} = \frac{1}{2}$

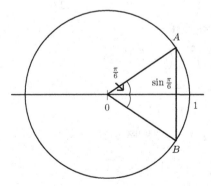

$OBA$ is an equilateral triangle, hence $2 \sin \frac{\pi}{6} = 1$.

Since $\cos^2 \frac{\pi}{6} = 1 - \sin^2 \frac{\pi}{6} = 1 - \frac{1}{4} = \frac{3}{4}$ we find that

$$\cos \frac{\pi}{6} = \frac{1}{2}\sqrt{3}.$$

c) First note that

$$\sin \frac{\pi}{3} = \sin\left(\frac{\pi}{6} + \frac{\pi}{6}\right) = \sin \frac{\pi}{6} \cos \frac{\pi}{6} + \cos \frac{\pi}{6} \sin \frac{\pi}{6}$$
$$= 2 \sin \frac{\pi}{6} \cos \frac{\pi}{6} = \frac{1}{2}\sqrt{3}$$

and

$$\cos \frac{\pi}{3} = \cos\left(\frac{\pi}{6} + \frac{\pi}{6}\right) = \cos \frac{\pi}{6} \cos \frac{\pi}{6} - \sin \frac{\pi}{6} \sin \frac{\pi}{6}$$
$$= \left(\frac{1}{2}\sqrt{3}\right)\left(\frac{1}{2}\sqrt{3}\right) - \frac{1}{2}\cdot\frac{1}{2} = \frac{1}{2}.$$

Therefore we find

$$\tan \frac{\pi}{3} = \frac{\sin \frac{\pi}{3}}{\cos \frac{\pi}{3}} = \frac{\frac{1}{2}\sqrt{3}}{\frac{1}{2}} = \sqrt{3}.$$

d) Since $\sin \frac{\pi}{6} = \frac{1}{2}$ we find

$$\frac{1}{2} = \sin \frac{\pi}{6} = \sin\left(\frac{\pi}{12} + \frac{\pi}{12}\right) = \sin \frac{\pi}{12} \cos \frac{\pi}{12} + \cos \frac{\pi}{12} \sin \frac{\pi}{12}$$
$$= 2\sin \frac{\pi}{12} \cos \frac{\pi}{12}$$

or

$$\sin \frac{\pi}{12} \cos \frac{\pi}{12} = \frac{1}{4},$$

which yields

$$(*) \qquad \frac{1}{\sin \frac{\pi}{12}} = 4\cos \frac{\pi}{12}.$$

Further we find that

$$\frac{1}{2}\sqrt{3} = \cos \frac{\pi}{6} = \cos\left(\frac{\pi}{12} + \frac{\pi}{12}\right) = \left(\cos \frac{\pi}{12}\right)^2 - \left(\sin \frac{\pi}{12}\right)^2.$$

Since $\left(\cos \frac{\pi}{12}\right)^2 + \left(\sin \frac{\pi}{12}\right)^2 = 1$ it follows that

$$\frac{1}{2}\sqrt{3} = \left(\cos \frac{\pi}{12}\right)^2 + \left(\cos \frac{\pi}{12}\right)^2 - 1$$

or

$$\cos \frac{\pi}{12} = \sqrt{\frac{1}{2} + \frac{1}{4}\sqrt{3}}.$$

Finally, since $\cot \frac{\pi}{12} = \frac{\cos \frac{\pi}{12}}{\sin \frac{\pi}{12}} = 4\cos^2 \frac{\pi}{12}$, where we used $(*)$, we find

$$\cot \frac{\pi}{12} = 4\left(\frac{1}{2} + \frac{1}{4}\sqrt{3}\right) = 2 + \sqrt{3}.$$

5. The problem is equivalent to finding the value $x$ such that

   a) $\sin x = \frac{\sqrt{3}}{2}$,

   b) $\cos x = -\frac{1}{2}\sqrt{2}$,

   c) $\tan x = \frac{1}{\sqrt{3}}$,

   d) $\cot x = -\sqrt{3}$.

   a) $\frac{\sqrt{3}}{2} = \cos\frac{\pi}{6} = \sin\left(\frac{\pi}{6} + \frac{\pi}{2}\right) = \sin\frac{2\pi}{3}$, therefore $x = \frac{2\pi}{3}$, note that we have used part b) of Problem 1.

   b) Since $\cos(\pi + x) = -\cos x$ and $\cos\frac{\pi}{4} = \frac{1}{2}\sqrt{2}$ we find $\cos\frac{5\pi}{4} = -\frac{1}{2}\sqrt{2}$, i.e. $x = \frac{5\pi}{4}$.

   c) Since $\tan x = \frac{\sin x}{\cos x}$ and $\cos\frac{\pi}{6} = \frac{1}{2}\sqrt{3}$, $\sin\frac{\pi}{6} = \frac{1}{2}$ we find $\tan\frac{\pi}{6} = \frac{1}{\sqrt{3}}$.

   d) Now $\cot x = \frac{1}{\tan x}$ and $\tan x = -\tan(-x)$, so we know that $\tan\left(-\frac{\pi}{6}\right) = -\frac{1}{\sqrt{3}}$ or $\cot\left(-\frac{\pi}{6}\right) = -\sqrt{3}$. Since in addition $\cot(x + \pi) = \cot x$, and it is preferred to consider $\cot$ on $(0, \pi)$ we take $x = \frac{5\pi}{6}$ as a solution i.e. $\cot\frac{5\pi}{6} = -\sqrt{3}$.

6.  a) By the mean value theorem for some $\xi$ between $x$ and $y$ we get

$$|\sin x - \sin y| = |\sin'\xi||x - y| = |\cos\xi||x - y| \le |x - y|.$$

   b) Again the mean value theorem yields for some $\xi$ between $x$ and $y$

$$|\tan x - \tan y| = |\tan'\xi||x - y| = \frac{1}{|\cos^2\xi|}|x - y|$$

and since that in $[-a, a]$ the function $\cos$ has its minimum at $a$ we find

$$|\tan x - \tan y| \le \frac{1}{\cos^2 a}|x - y|.$$

   c) For the first statement we use a short induction argument. For $n = 1$ we have $|\sin x| \le |\sin x|$. Now suppose that $|\sin nx| \le n|\sin x|$. Using the addition theorems we find

$$\sin((n + 1)x) = (\sin nx)\cos x + \sin x \cos nx$$

or

$$|\sin(n + 1)x| \le |\sin nx||\cos x| + |\sin x||\cos nx|$$
$$\le n|\sin nx| + |\sin x| = (n + 1)|\sin x|.$$

The statement $|\sin ax| \le a|\sin x|$ for all $a > 0$ and all $x \in \mathbb{R}$ is not correct. Take $a = \frac{1}{2}$ and $x = \pi$, then the claim is

$$1 = \sin\frac{\pi}{2} \le \frac{1}{2}|\sin\pi| = 0,$$

which of course is not correct.

601

7. Let $x \in \mathbb{R}$. It follows that

$$(f \circ g)(x + a) = f(g(x + a)) = f(g(x)) = (f \circ g)(x),$$

thus $f \circ g$ has period $a$.

The function $g \circ f : \mathbb{R} \longrightarrow \mathbb{R}$ does not have to be periodic. Consider the periodic function $g = \sin$ and the function $f : \mathbb{R} \longrightarrow \mathbb{R}$, $x \mapsto \frac{1}{1+x^2}$. Note that the range of $f$ is $(0, 1]$. On $[0, 1]$ the function $\sin$ is strictly increasing. Now, let $a > 0$. It follows for $x \in \mathbb{R}$ that $x + a \neq x$ and therefore $\frac{1}{1+(x+a)^2} \neq \frac{1}{1+x^2}$. Thus either $\sin\left(\frac{1}{1+(x+a)^2}\right) < \sin\left(\frac{1}{1+x^2}\right)$ or $\sin\left(\frac{1}{1+x^2}\right) < \sin\left(\frac{1}{1+(x+a)^2}\right)$, but we never have equality, hence $x \mapsto \sin\left(\frac{1}{1+x^2}\right)$ is not periodic.

8.  a)

$$\frac{d}{dx} \cos(\ln(1 + x^2))$$

$$= \left(\frac{d}{dx} \ln(1 + x^2)\right) (\cos')(\ln(1 + x^2))$$

$$= -\frac{2x}{1 + x^2} \sin(\ln(1 + x^2)).$$

b)

$$\frac{d}{dt} \left(\frac{\sin(\tan t)}{\sqrt{1 - \cos^4 t}}\right)$$

$$= \frac{\left(\frac{d}{dt} \sin(\tan t)\right) \sqrt{1 - \cos^4 t} - \sin(\tan t) \frac{d}{dt} \sqrt{1 - \cos^4 t}}{1 - \cos^4 t}$$

$$= \frac{\frac{1}{\cos^2 t} \cos(\tan t) \sqrt{1 - \cos^4 t} - \sin(\tan t) \cdot (4 \cos^3 t)(\sin t) \frac{1}{2}(1 - \cos^4 t)^{-\frac{1}{2}}}{1 - \cos^4 t}$$

$$= \frac{(\cos(\tan t))(\cos t)(1 - \cos^4 t) - 2 \sin(\tan t) \cos^3 t \sin t}{(1 - \cos^4 t)^{\frac{3}{2}}}$$

$$= \frac{-(\cos(\tan t))(\cos^5 t) - 2(\sin(\tan t))(\sin t)(\cos^3 t) + (\cos(\tan t))(\cos t)}{(1 - \cos^4 t)^{\frac{3}{2}}}.$$

c)

$$\frac{d}{ds} \arcsin\left(\sqrt{1 + \cos s}\right)$$

$$= \left(\frac{d}{ds} \sqrt{1 + \cos s}\right) (\arcsin')\left(\sqrt{1 + \cos s}\right)$$

$$= \frac{-\sin s}{2\sqrt{1 + \cos s}} \frac{1}{\sqrt{1 - \left(\sqrt{1 + \cos s}\right)^2}}$$

$$= \frac{-\sin s}{2\sqrt{1 + \cos s}\sqrt{\cos s}}.$$

d)

$$\frac{d}{du}\left(\arctan\left(e^{-u^2}\cot u\right)\right)$$

$$= \frac{d}{du}\left(e^{-u^2}\cot u\right)(\arctan')\left(e^{-u^2}\cot u\right)$$

$$= \left(-2ue^{-u^2}\cot u + e^{-u^2}\frac{1}{\sin^2 u}\right)\frac{1}{1+\left(e^{-u^2}\cot u\right)^2}$$

$$= \frac{(-2u(\cot u)\sin^2 u + 1)e^{-u^2}}{1+e^{-2u^2}\cot^2 u}.$$

9. We first find $\sum_{j=1}^{n}\cos jt$. For $j \geq 1$ it follows that

$$\cos jt \cdot \sin\frac{t}{2} = \frac{1}{2}\left(\sin\left(j+\frac{1}{2}\right)t - \sin\left(j-\frac{1}{2}\right)t\right)$$

and therefore

$$\left(\sum_{j=1}^{n}\cos jt\right)\sin\frac{t}{2}$$

$$= \frac{1}{2}\sum_{j=1}^{n}\left(\sin\left(j+\frac{1}{2}\right)t - \sin\left(j-\frac{1}{2}\right)t\right)$$

$$= \frac{1}{2}\left(\sin\left(n+\frac{1}{2}\right)t - \sin\frac{t}{2}\right)$$

or for $t \neq 0$

$$\sum_{j=1}^{n}\cos jt = \frac{1}{2}\left(\frac{\sin\left(\frac{2n+1}{2}t\right)}{\sin\frac{t}{2}} - 1\right),$$

i.e. for $t \neq 0$ we have

$$C_n(t) = \frac{1}{2} + \sum_{j=1}^{n}\cos jt = \frac{1}{2}\frac{\sin\left(\frac{2n+1}{2}t\right)}{\sin\frac{t}{2}}.$$

Thus we find for $t \neq 0$ that

$$C_n(2t) = \frac{1}{2}\frac{\sin(2n+1)t}{\sin t},$$

which yields for $t \neq 0$ that $C_n(2t) = \frac{1}{2}D_n(t)$. For $t = 0$ we find $C_n(0) = \frac{1}{2} + \sum_{j=1}^{n}(\cos j0) = \frac{2n+1}{2}$, hence

$$C_n(2t) = \frac{1}{2}D_n(t)$$

603

for all $t \in \left[-\frac{\pi}{2}, \frac{\pi}{2}\right]$. Since $t \mapsto C_n(2t)$ is arbitrarily often differentiable on $\left(-\frac{\pi}{2}, \frac{\pi}{2}\right)$ it follows that $D_n(.)$ is arbitrarily often differentiable on $\left(-\frac{\pi}{2}, \frac{\pi}{2}\right)$.

**Chapter 11**

1. The interior points are $(-1, 2) \cup (5, 6)$ and the boundary is $\{-1\} \cup \{2\} \cup \{3\} \cup \{4\} \cup \{5\} \cup \{6\}$, i.e.

$$\partial D = \{-1, 2, 3, 4, 5, 6\}.$$

To prove this we first look at the following sketch:

For $-1 < x < 2$, i.e. $x \in (-1, 2)$ we take $\varepsilon_1 = \frac{1}{2}\min\{|x + 1|, |x - 2|\}$ to find that $(-\varepsilon_1 + x, x + \varepsilon_1) \subset (-1, 2) \subset D$, and for $x \in (5, 6)$ we find with $\varepsilon_2 := \frac{1}{2}\min\{|x - 5|, |x - 6|\}$ that $(-\varepsilon_2 + x, x + \varepsilon_2) \subset (5, 6) \subset D$. The points $-1, 2, 3, 4, 5, 6$ are not internal points: every interval with centre $x \in \{-1, 2, 3, 4, 5, 6\}$ will contain points not belonging to $D$. This is almost the proof that $\partial D = \{-1, 2, 3, 4, 5, 6\}$. For $-1, 3, 4, 5, 6 \in D$ it follows immediately that these points belong to $\partial D$. For 2 we need to argue more carefully: if $(-\varepsilon + 2, 2 + \varepsilon)$, $0 < \varepsilon < 1$, is an open inteval with centre 2, then $(-\varepsilon + 2, 2) \subset D$. Hence $(-\varepsilon + 2, 2 + \varepsilon)$, $0 < \varepsilon < 1$, always contains points in $D$ and $D^{\complement}$. However this also holds when we do not use the restriction $\varepsilon < 1$.

2.    a) In order for $\sqrt{(x^2 - 1)(x^2 + 4x)}$ to be defined we need to have that $(x^2 - 1)(x^2 + 4x) \geq 0$, i.e. $(x^2 - 1) \leq 0$ and $(x^2 + 4x) \leq 0$ or $(x^2 - 1) \geq 0$ and $(x^2 + 4x) \geq 0$. Now $x^2 - 1 \leq 0$ if $x \in [-1, 1]$ and $x^2 + 4x = x(x + 4) \leq 0$ if either $x \leq 0$ and $x \geq -4$ or $x \geq 0$ and $x \leq -4$. Moreover $x^2 - 1 \geq 0$ if $x \in \mathbb{R} \setminus (-1, 1)$ and $x(x + 4) \geq 0$ if either $x \geq 0$ and $x \geq -4$ or $x \leq 0$ and $x \leq -4$. Together we find that $(x^2 - 1)(x^2 + 4x) \geq 0$ if

$$x \in \mathbb{R} \setminus (-4, 1) \cup [-1, 0].$$

b) First note that we need to have $x^3 + 4x^2 - 5x \neq 0$, or $x(x^2 + 4x - 5) \neq 0$, i.e. $x \in \mathbb{R} \setminus \{-5, 0, 1\}$. Next note that cos is defined on all of $\mathbb{R}$ but ln is only defined on $(0, \infty)$. Hence we need $\arctan x > 0$ implying $x > 0$. Therefore we have

$$D = \mathbb{R}_+ \setminus \{0, 1\}.$$

c) As in part a) we note that the condition is that $(\sinh x)(1 - x^4)$ is non-negative. Now for $x \geq 0$ we have $\sinh x \geq 0$ and for $x < 0$ we have

$\sinh x < 0$. Further for $x \in [-1, 1]$ we have $\left(1 - x^4\right) \geq 0$ and in $\mathbb{R} \setminus [-1, 1]$ we find that $\left(1 - x^4\right) < 0$. Therefore it follows that

$$D = (-\infty, -1] \cup [0, 1].$$

d) The definition of cot gives

$$\cot(\arcsin x) = \frac{\cos(\arcsin x)}{\sin(\arcsin x)} = \frac{\cos(\arcsin x)}{x}$$

which implies

$$D = [-1, 1] \setminus \{0\}.$$

3.    a) Since $\cos \pi = -1$, hence $1 + \cos \pi = 0$, and $x^2 - 2x + 1$ has a zero when $x = 1$, we may therefore use the rules of l'Hospital:

$$\lim_{x \to 1} \frac{1 + \cos \pi x}{x^2 - 2x + 1} = \lim_{x \to 1} \frac{-\pi \sin \pi x}{2x - 2}$$

and since $\sin \pi = 0$ as is $2x - 2$ for $x = 1$ we use the rules again to find

$$\lim_{x \to 1} \frac{1 + \cos \pi x}{x^2 - 2x + 1} = \lim_{x \to 1} \frac{-\pi \sin \pi x}{2x - 2} = \lim_{x \to 1} \frac{-\pi^2 \cos \pi x}{2} = \frac{\pi^2}{2}.$$

b) Note that for $t = 0$ we have $\cos 3t = \cos 2t = 1$, however $\ln 1 = 0$. Applying the rules of l'Hospital gives

$$\lim_{t \to 0} \frac{\ln(\cos 3t)}{\ln(\cos 2t)} = \lim_{t \to 0} \frac{\frac{-3 \sin 3t}{\cos 3t}}{\frac{-2 \sin 2t}{\cos 2t}}$$
$$= \lim_{t \to 0} \frac{3 \sin 3t \cos 2t}{2 \sin 2t \cos 3t}$$
$$= \frac{3}{2} \lim_{t \to 0} \frac{\cos 2t}{\cos 3t} \lim_{t \to 0} \frac{\sin 3t}{\sin 2t}$$
$$= \frac{3}{2} \lim_{t \to 0} \frac{\sin 3t}{\sin 2t} = \frac{3}{2} \lim_{t \to 0} \frac{3 \cos 3t}{2 \cos 2t} = \frac{3}{2} \cdot \frac{3}{2} = \frac{9}{4},$$

where we used the rules once again in the final step.

c) It follows that

$$\lim_{y \to \infty} \frac{3y^2 - y + 5}{5y^2 - 6y - 3} = \lim_{y \to \infty} \frac{6y - 1}{10y + 6} = \lim_{y \to \infty} \frac{6}{10} = \frac{3}{5}.$$

d) We first rewrite the term $\frac{1}{\sin^2 u} - \frac{1}{u^2}$ as:

$$\frac{1}{\sin^2 u} - \frac{1}{u^2} = \frac{u^2 - \sin^2 u}{u^2 \sin^2 u} = \frac{u^2 - \sin^2 u}{u^4} \cdot \frac{u^2}{\sin^2 u}.$$

605

Next we note that

$$\lim_{u \to 0} \frac{u^2}{\sin^2 u} = \left( \frac{1}{\displaystyle\lim_{u \to 0} \frac{\sin u}{u}} \right)^2 = 1$$

and therefore it remains to find $\displaystyle\lim_{u \to 0} \frac{u^2 - \sin^2 u}{u^4}$. We have

$$\begin{aligned}
\lim_{u \to 0} \frac{u^2 - \sin^2 u}{u^4} &= \lim_{u \to 0} \frac{2u - 2\sin u \cos u}{4u^3} \\
&= \lim_{u \to 0} \frac{2u - \sin 2u}{4u^3} \\
&= \lim_{u \to 0} \frac{2 - 2\cos 2u}{12u^2} \\
&= \lim_{u \to 0} \frac{4\sin 2u}{24u} = \frac{1}{3} \lim_{u \to 0} \frac{\sin 2u}{2u} = \frac{1}{3}.
\end{aligned}$$

4.     a) We claim that $x \mapsto f(x) = x^2$ is the asymptote for $g$ as $x \to \infty$. To show this we must prove that

$$\lim_{x \to \infty} \frac{\ln\left(1 + x^2 + e^{x^2}\right)}{x^2} = 1.$$

Using the de l'Hospital rules we find

$$\begin{aligned}
\lim_{x \to \infty} \frac{\ln\left(1 + x^2 + e^{x^2}\right)}{x^2} &= \lim_{x \to \infty} \frac{2x + 2xe^{x^2}}{2x\left(1 + x^2 + e^{x^2}\right)} \\
&= \lim_{x \to \infty} \frac{1 + e^{x^2}}{1 + x^2 + e^{x^2}} = \lim_{x \to \infty} \frac{e^{x^2}\left(2xe^{-x^2} + 1\right)}{e^{x^2}\left(e^{-x^2} + x^2 e^{-x^2} + 1\right)} \\
&= \lim_{x \to \infty} \frac{2xe^{-x^2} + 1}{e^{-x^2} + x^2 e^{-x^2} + 1} = 1.
\end{aligned}$$

b) Since $h$ is even we only need to consider the case $t \to \infty$. We guess, since $\displaystyle\lim_{t \to \infty} \frac{1}{1 + t^2} = 0$, that the asymptote is the function $t \mapsto f(t) = 1$.
Now we have

$$\begin{aligned}
\lim_{t \to \infty} \frac{e^{-\frac{1}{1+t^2}}}{1} &= \lim_{t \to \infty} e^{-\frac{1}{1+t^2}} \\
&= e^{-\lim_{t \to \infty} \frac{1}{1+t^2}} = e^0 = 1.
\end{aligned}$$

5.     a) In order for $f_1(x) = \frac{2x^2 + 12x - 2}{15(x^2 - 1)^{\frac{3}{2}}}$ to be defined we have to have $x^2 - 1 > 0$. If $x^2 - 1 = 0$ then we would be dividing by 0, note that the numerator is not 0 for

606

$x = \pm 1$. If $x^2 - 1 < 0$ the square root is not defined. Hence the maximal domain $D_1$ of $f_1$ is $D_1 = \mathbb{R} \setminus [-1, 1]$.

On this domain $f_1$ is neither even nor odd and $f_1(x) = 0$ if and only if $x^2 + 6x - 1 = 0$, i.e. for

$$x_{1,2} = -3 \pm \sqrt{9+1} = -3 \pm \sqrt{10}.$$

Since $3 < \sqrt{10} < 4$ the only root is $x_0 := -3 - \sqrt{10}$, the number $-3 + \sqrt{10}$ does not belong to $D_1$.

For $x > 1$ we have $2x^2 + 12x - 2 > 0$ and therefore

$$\lim_{\substack{x \to 1 \\ x > 0}} \frac{2x^2 + 12x - 2}{15(x^2 - 1)^{\frac{1}{2}}} = +\infty.$$

For $x < -3 - \sqrt{10}$ we know that $2x^2 + 12x - 2 > 0$ but for $-3 - \sqrt{10} < x < -1$ we have $2x^2 + 12x - 2 < 0$ which implies that

$$\lim_{\substack{x \to -1 \\ x < -1}} \frac{2x^2 + 12x - 2}{15(x^2 - 1)^{\frac{1}{2}}} = -\infty.$$

We claim that for $x \to \infty$ the asymptote is $x \mapsto \frac{2}{15}x$. Indeed we find

$$\frac{\frac{2x^2 + 12x - 2}{15(x^2-1)^{\frac{1}{2}}}}{\frac{2}{15}x} = \frac{x^2 + 6x - 1}{x \cdot x \left(1 - \frac{1}{x^2}\right)^{\frac{1}{2}}}$$

and for $x \to \infty$ we have

$$\lim_{x \to \infty} \frac{x^2 + 6x - 1}{x^2 \left(1 - \frac{1}{x^2}\right)^{\frac{1}{2}}} = \lim_{x \to \infty} \frac{1 + \frac{6}{x} - \frac{1}{x}}{\left(1 - \frac{1}{x^2}\right)^{\frac{1}{2}}} = 1.$$

The asymptote for $x \to -\infty$ is the function $x \mapsto -\frac{2}{15}x$, since for $x < -1$

$$\frac{\frac{2x^2 + 12x - 2}{15(x^2-1)^{\frac{1}{2}}}}{-\frac{2}{15}x} = \frac{x^2 + 6x - 1}{-x(-x)\left(1 - \frac{1}{x^2}\right)^{\frac{1}{2}}} = \frac{x^2 + 6x - 1}{x^2 \left(1 - \frac{1}{x^2}\right)^{\frac{1}{2}}}$$

and the result follows as before.

To find local extreme values we consider

$$\frac{d}{dx}\left(\frac{2x^2 + 12x - 2}{15(x^2 - 1)^{\frac{1}{2}}}\right) = \frac{2x^3 - 2x - 12}{15(x^2 - 1)^{\frac{3}{2}}}.$$

The condition $\frac{df_1}{dx}(x) = 0$ is equivalent to

$$x^3 - x - 6 = 0.$$

One zero is easy to find: $x = 2$, indeed we have

$$2^3 - 2 - 6 = 0.$$

607

Now we look at

$$\left(x^3 - x - 6\right) : (x - 2) = x^2 + 2x + 3.$$

$$\underline{x^3 - 2x^2}$$
$$2x^2 - x$$
$$\underline{2x^2 - 4x}$$
$$3x - 6.$$

Thus we have

$$\left(x^3 - x - 6\right) = (x - 2)\left(x^2 + 2x + 3\right).$$

The zeroes of $x^2 + 2x + 3$ are $x_{1,2} = -1 \pm \sqrt{1 - 3}$, hence they are not real and therefore only at $x_0 = 2$ may we have local extreme values.

We know that $\lim\limits_{\substack{x \to 1 \\ x > 1}} f_1(x) = +\infty$ and $\lim\limits_{x \to \infty} f_1(x) = +\infty$, therefore we have a local minimum at $x_0 = 2$.

This can be checked by looking at $f_1''(2)$:

$$f_1''(x) = \frac{d}{dx}\left(\frac{2x^3 - 2x - 12}{15\left(x^2 - 1\right)^{\frac{1}{2}}}\right) = \frac{-2x^2 + 12x + 2}{15\left(x^2 - 1\right)^{\frac{5}{2}}}$$

and

$$f_1''(2) = \frac{-2 \cdot 4 + 12 \cdot 2 + 2}{15\sqrt{3}} = \frac{18}{15\sqrt{3}} = \frac{6}{5\sqrt{3}} > 0.$$

Note that $f(2) = \frac{2}{\sqrt{3}}$. Finally we sketch the graph of $f_1$

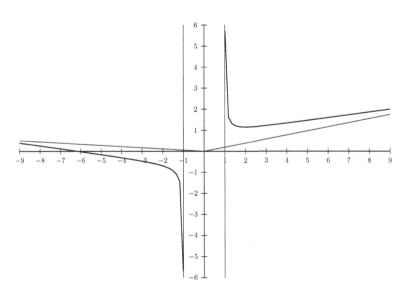

608

b) First we note that $\frac{s^2}{1+s^4} \in [0,1)$ for all $s \in \mathbb{R}$ and therefore $s \mapsto \tan\frac{s^2}{1+s^4}$ is defined on $\mathbb{R}$. Secondly $s \mapsto \frac{s^2}{1+s^4}$ is even implying that $s \mapsto \tan\frac{s^2}{1+s^4}$ is even. Consequently we can restrict our considerations to $s \in \mathbb{R}_+$. For $s_0 = 0$ it follows that $\frac{s_0^2}{1+s_0^4} = 0$, i.e. $\tan\frac{s_0^2}{1+s_0^4} = 0$ and $s_0$ is the only zero in $\mathbb{R}_+$. Also, since $\lim_{s\to\infty}\frac{s^2}{1+s^4} = 0$ it follows that $\lim_{s\to\infty}\tan\frac{s^2}{1+s^4} = 0$. Next we search for local extreme values and therefore we consider

$$\frac{d}{ds}\tan\left(\frac{s^2}{1+s^4}\right) = \frac{d}{ds}\left(\frac{s^2}{1+s^4}\right)\cdot\frac{1}{\cos^2\left(\frac{s^2}{1+s^4}\right)}$$

$$= \frac{2s - 2s^5}{(1+s^4)}\frac{1}{\cos^2\left(\frac{s^2}{1+s^4}\right)}.$$

Thus $f_2'(s_0) = 0$ implies that $2s_0 - 2s_0^5 = 0$, or since

$$2s_0 - 2s_0^5 = 2s_0\left(1 - s_0^4\right) = 2s_0\left(1 - s_0^2\right)\left(1 + s_0^2\right)$$

the real zeroes are $s_0 = 0$, $s_1 = 1$, $s_2 = -1$, and due to symmetry we only need to investigate the function at $s_0 = 0$ and $s_1 = 1$.
We know that $\tan t \geq 0$ for $t \in [0, \frac{\pi}{2})$ therefore the function $s \mapsto \tan\frac{s^2}{1+s^4}$ must have a minimum at $s_0 = 0$. Further, for $s \to \infty$ we know that $\lim_{s\to\infty}\tan\frac{s^2}{1+s^4} = 0$, hence we have a maximum at $x_1 = 1$, and by symmetry also at $s_2 = -1$. The value at $x_1 = 1$ is $\tan\frac{1}{2}$.

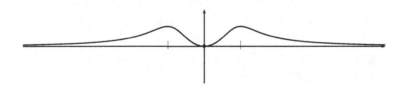

c) Since arsinh is defined on $\mathbb{R}$ and so is $t \mapsto 1 - e^{-t^2}$, the maximal domain of $f_3$ is $D_3 = \mathbb{R}$. Further $t \mapsto 1 - e^{-t^2}$ is even, therefore $f_3$ is even too, hence we need to consider $f_3$ only on $[0, \infty)$.
The only zero of arsinh is 0, and $1 - e^{-t^2} = 0$ only for $t = 0$, therefore $f_3$ has a zero at 0. For $t \to \infty$ it follows that $-e^{-t^2}$ tends to 0, hence

$$\lim_{t\to\infty}\operatorname{arsinh}\left(1 - e^{-t^2}\right) = \operatorname{arsinh}(1) = \ln\left(1 + \sqrt{2}\right).$$

We now look at the first derivative of $f_3$ to investigate local extreme values:

$$\frac{d}{dt} \operatorname{arsinh}\left(1 - e^{-t^2}\right) = \frac{d}{dt}\left(1 - e^{-t^2}\right)(\operatorname{arsinh}')\left(1 - e^{-t^2}\right)$$

$$= 2te^{-t^2}\frac{1}{\sqrt{1 + \left(1 - e^{-t^2}\right)^2}} > 0 \text{ for } t > 0,$$

hence on $[0, \infty)$ arsinh is strictly monotone increasing and the only local extreme value is at 0, which is equal to

$$\operatorname{arsinh}\left(1 - e^{-0^2}\right) = \operatorname{arsinh} 0 = 0.$$

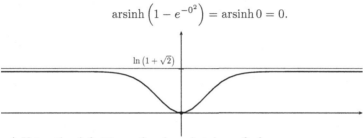

6.  a) Using the definitions of cosh and sinh we find

$$\cosh^2 x - \sinh^2 x = \left(\frac{e^x + e^{-x}}{2}\right)^2 - \left(\frac{e^x - e^{-x}}{2}\right)^2$$

$$= \frac{1}{4}\left(e^{2x} + 2 + e^{-2x} - e^{2x} + 2 - e^{-2x}\right) = 1.$$

b) Since $\coth x = \frac{\cosh x}{\sinh x}$ we have

$$\frac{1}{\coth^2 x - 1} = \frac{1}{\frac{\cosh^2 x}{\sinh^2 x} - 1} = \frac{1}{\frac{\cosh^2 x}{\sinh^2 x} - \frac{\sinh^2 x}{\sinh^2 x}}$$

$$= \frac{\sinh^2 x}{\cosh^2 x - \sinh^2 x} = \sinh^2 x,$$

where we used part a).

c) We start with

$$\sinh x \cosh y + \cosh x \sinh y = \frac{e^x - e^{-x}}{2}\frac{e^y + e^{-y}}{2} + \frac{e^x + e^{-x}}{2}\frac{e^y - e^{-y}}{2}$$

$$= \frac{1}{4}\left(e^{x+y} - e^{-x+y} + e^{x-y} - e^{-(x+y)} + e^{x+y} + e^{-x+y} - e^{x-y} - e^{-(x+y)}\right)$$

$$= \frac{1}{4}\left(2e^{x+y} - 2e^{-(x+y)}\right) = \frac{e^{x+y} - e^{-(x+y)}}{2} = \sinh(x + y).$$

Now for $-y$ instead of $y$ we find

$$\sinh(x - y) = \sinh x \cosh(-y) + \cosh x \sinh(-y)$$

$$= \sinh x \cosh y - \cosh x \sinh y$$

610

where we used the fact that cosh is an even function and sinh is an odd function.

d) Recall that $\tanh x = \frac{\sinh x}{\cosh x}$ and therefore

$$\frac{\tanh x - \tanh y}{1 - \tanh x \tanh y} = \frac{\frac{\sinh x}{\cosh x} - \frac{\sinh y}{\cosh y}}{1 - \frac{\sinh x}{\cosh x}\frac{\sinh y}{\cosh y}}$$

$$= \frac{\frac{\sinh x \cosh y - \sinh y \cosh x}{\cosh x \cosh y}}{\frac{\cosh x \cosh y - \sinh x \sinh y}{\cosh x \cosh y}}$$

$$= \frac{\sinh x \cosh x - \sinh y \cosh x}{\cosh x \cosh y - \sinh x \sinh y} = \frac{\sinh(x - y)}{\cosh(x - y)} = \tanh(x - y).$$

**Chapter 12**

1.    a) We use the formula

$$(*) \quad S(f, Z_n, \xi) = \sum_{k=1}^{n} f(\xi_k)(t_k - t_{k-1})$$

and therefore with $t_k = 1 + \frac{k-1}{n}$, $k = 1, \ldots, n+1$, we have

$$t_k - t_{k-1} = 1 + \frac{k-1}{n} - 1 - \frac{k-2}{n} = \frac{1}{n}$$

and

$$\xi_k = 1 + \frac{k-1}{n} + \frac{1}{2n} = 1 + \frac{2k-1}{2n}$$

which gives

$$S(f, Z_n, \xi) = \sum_{k=1}^{n} f(\xi_k)(t_k - t_{k-1})$$

$$= \frac{1}{n} \sum_{k=1}^{n} f\left(1 + \frac{2k-1}{2n}\right)$$

$$\frac{2}{n} \sum_{k=1}^{n} \left(1 + \frac{2k-1}{2n}\right)^2 - \frac{1}{n} \sum_{k=1}^{n} \left(1 + \frac{2k-1}{2n}\right).$$

Now we note that

$$\frac{2}{n} \sum_{k=1}^{n} \left(1 + \frac{2k-1}{2}\right)^2 = \frac{1}{2n^3} \sum_{k=1}^{n} (2n + 2k - 1)^2$$

$$= \frac{1}{2n^3} \left(\sum_{k=1}^{n} (4n^2 - 4n + 1) + \sum_{k=1}^{n} (8n - 4)k + \sum_{k=1}^{n} 4k^2\right)$$

$$= \frac{4n^2 - 4n + 1}{2n^2} + \frac{(8n - 4)n(n + 1)}{4n^3} + \frac{2n(n + 1)(2n + 1)}{6n^3}$$

$$= \frac{14}{3} - \frac{2}{3n^2}$$

611

and

$$\frac{1}{n}\sum_{k=1}^{n}\left(1+\frac{2k-1}{2n}\right)=\frac{1}{n}\sum_{k=1}^{n}1+\frac{1}{n}\sum_{k=1}^{n}\frac{k}{n}+\frac{1}{n}\sum_{k=1}^{n}\left(-\frac{1}{n}\right)$$

$$=\frac{1}{n}\cdot n+\frac{1}{n}\frac{n(n+1)}{2n}+\frac{1}{n}\left(-\frac{1}{n}n\right)$$

$$=\frac{3}{2}-\frac{1}{2n}$$

therefore it follows that

$$S(f,Z_n,\xi)=\frac{14}{3}-\frac{2}{3n^2}-\frac{3}{2}+\frac{1}{2n}$$

$$=\frac{19}{6}-\frac{2}{3n^2}+\frac{1}{2n}.$$

Note that

$$\int_1^2 (2t-t)\mathrm{d}t=\frac{2}{3}t^3-\frac{t^2}{2}\Big|_1^2=\frac{2}{3}\cdot 8-\frac{4}{2}-\frac{2}{3}+\frac{1}{2}$$

$$=\frac{32-12-4+3}{6}=\frac{19}{6},$$

and the larger the $n$ the closer $\frac{1}{2n}-\frac{2}{3n^2}$ is to 0.

b) Again we wish to use $*$, therefore we note that

$$t_k=\frac{a(m^2-(k-1)^2)+(k-1)^2 b}{m^2},\quad k=1,\ldots,m+1$$

$$t_{k+1}-t_k=\frac{2k+1}{m^2}(b-a),k=1,\ldots,m$$

$$\xi_k=\frac{2}{3}t_k+\frac{1}{3}t_{k+1}$$

$$=\frac{1}{3m^2}\left(\left(3m^2-3k^2+4k-2\right)a+\left(3k^2-4k+2\right)b\right),\quad k=1,\ldots,m$$

and we find

$$S(h,Z_m,\xi)=\sum_{k=1}^{m}h(\xi_k)(t_{k+1}-t_k)$$

$$=\sum_{k=1}^{m}\left(\frac{1}{1+\xi_k^2}\frac{2k+1}{m^2}(b-a)\right)$$

$$=9m^2(b-a)\sum_{k=1}^{m}\frac{2k+1}{9m^4+((3m^2-3k^2+4k-2)a+(3k^2-4k+2)b)^2}$$

an expression we do not wish to simplify further.

2. From the definition we find

$$S(g|_{[a,t_k]}, Z_n|_{[a,t_k]}, \xi|_{[a,t_k]})$$
$$= \sum_{j=1}^{k} g(\xi_j)(t_j - t_{j-1})$$

and

$$S(g|_{[t_k,b]}, Z_n|_{[t_k,b]}, \xi|_{[t_k,b]})$$
$$= \sum_{j=k+1}^{n} g(\xi_j)(t_j - t_{j-1})$$

which implies

$$S(g, Z_n, \xi) = \sum_{j=1}^{n} g(\xi_j)(t_j - t_{j-1})$$
$$= \sum_{j=1}^{k} g(\xi_j)(t_j - t_{j-1}) + \sum_{j=k+1}^{n} g(\xi_j)(t_j - t_{j-1})$$
$$= S(g|_{[a,t_k]}, Z_n|_{[a,t_k]}, \xi|_{[a,t_k]}) + S(g|_{[t_k,b]}, Z_n|_{[t_k,b]}, \xi|_{[t_k,b]}).$$

It is clear from the following figure that such a result is true.

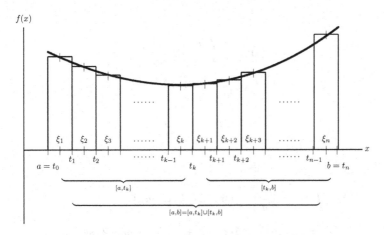

3.      a) The figure in the problem shows the graph of $x \mapsto |x|$ on $[-2, 1]$ and indicates the two triangles. They are right angled triangles and by denoting the distance between two points $A_1$, $A_2$ in the plane by $l(A_1, A_2)$ we find

$$\text{area}(ABC) = \frac{1}{2}l(A, B)l(A, C) = \frac{1}{2} \cdot 2 \cdot 2 = 2$$

613

and

$$\text{area}(BDE) = \frac{1}{2}l(B,D)l(D,E) = \frac{1}{2} \cdot 1 \cdot 1 = \frac{1}{2}$$

therefore we find

$$\int_{-2}^{1} |x|dx = \frac{5}{2}.$$

b) The figure in the problem shows the graph of $r \mapsto g(r) = \sqrt{R^2 - r^2}$, $r \in [-R, R]$, and the area of the upper disc with radius $R$ is $\frac{1}{2}\pi R^2$, hence we have

$$\int_{-R}^{R} \sqrt{R^2 - r^2}dr = \frac{1}{2}\pi R^2.$$

4.   a)

$$F'(x) = \frac{d}{dx}(\ln(\cosh x))$$

$$= \left(\frac{d}{dx}\cosh x\right)(\ln')(\cosh x)$$

$$= \sinh x \frac{1}{\cosh x} = \tanh x,$$

therefore

$$\int \tanh x dx = \ln \cosh x.$$

b)

$$F'(s) = \frac{d}{ds}\left(\frac{a^s}{\ln a}\right) = \frac{d}{ds}\left(\frac{e^{s\ln a}}{\ln a}\right)$$

$$= \frac{\ln a}{\ln a}e^{s\ln a} = a^s,$$

i.e.

$$\int a^s ds = \frac{a^s}{\ln a}.$$

c)

$$F'(u) = \frac{d}{du}\left(\frac{e^u}{26}(\sin 5u - 5\cos 5u)\right)$$

$$= \frac{e^u}{26}(\sin 5u - 5\cos 5u) + \frac{e^u}{26}(5\cos 5u + 25\sin 5u)$$

$$= \frac{e^u \sin 5u}{26},$$

and we find

$$\int \frac{e^u \sin 5u}{26}du = \frac{e^u}{26}(\sin 5u - 5\cos 5u).$$

614

d)

$$F'(r) = \frac{d}{dr}\left(-\frac{1}{2}\cos\left(r^2 + 4r - 6\right)\right)$$
$$= -\frac{1}{2}(2r + 4)\left(-\sin\left(r^2 + 4r - 6\right)\right)$$
$$= (r + 2)\sin\left(r^2 + 4r - 6\right)$$

which gives

$$\int (r + 2)\sin\left(r^2 + 4r - 6\right)\,dr = -\frac{1}{2}\cos\left(r^2 + 4r - 6\right).$$

## Chapter 13

1. We have

$$\int_0^1 \sum_{k=1}^n (1 + k^2)x^{\frac{1}{k^2}}\,dx = \sum_{k=1}^n \int_0^1 (1 + k^2)x^{\frac{1}{k^2}}\,dx$$

$$= \sum_{k=1}^n (1 + k^2)\int_0^1 x^{\frac{1}{k^2}}\,dx$$

$$= \sum_{k=1}^n (1 + k^2)\frac{1}{\left(1 + \frac{1}{k^2}\right)}x^{1+\frac{1}{k^2}}\Bigg|_0^1$$

$$= \sum_{k=1}^n (1 + k^2)\frac{1}{\frac{k^2+1}{k^2}}x^{1+\frac{1}{k^2}}\Bigg|_0^1$$

$$= \sum_{k=1}^n k^2 = \frac{n(n + 1)(2n + 1)}{6}.$$

2.   a) We know that $|f| = f^+ + f^-$ and $f^+, f^-$ are integrable. Hence

$$\int_a^b |f(x)|\,dx = \int_a^b \left(f^+(x) + f^-(x)\right)\,dx = \int_a^b f^+(x)\,dx + \int_a^b f^-(x)\,dx.$$

b) The function $t \mapsto M$ is integrable $|f(t)| \leq M$, therefore by monotonicity we have

$$\int_a^b |f(t)|\,dt \leq \int_a^b M\,dt = M(b - a).$$

Note that the monotonicity of the integral, i.e. the fact that $f \leq g$ implies $\int_a^b f(t)\,dt \leq \int_a^b g(t)\,dt$ is a simple consequence of the fact that the integral of a non-negative function is non-negative: $f \leq g$ if and only if $g - f \geq 0$, hence $\int_a^b (g - f)(t)\,dt \geq 0$ implying $\int_a^b f(t)\,dt \leq \int_a^b f(t)\,dt.$

c) If $h \geq 0$ is integrable on $[a, b]$ then $\int_a^b h(t)\mathrm{d}t \geq 0$. Now $f'(x) \geq 0$ on $[-1, 0]$ implies that $f$ is monotone increasing on $[-1, 0]$ and since $f(-1) = 0$ it follows that $f(x) \geq 0$ for all $x \in [-1, 0]$ implying $\int_{-1}^0 f(x)\mathrm{d}x \geq 0$.

3. The function $x \mapsto \left(1 + \frac{1}{1+x^2}\right) \sin x^3$ is odd:

$$\left(1 + \frac{1}{1 + (-x)^2}\right) \sin \left((-x)^3\right) = \left(1 + \frac{1}{1+x^2}\right) \sin \left(-(x^3)\right)$$
$$= -\left(1 + \frac{1}{1+x^2}\right) \sin x^3.$$

Now note that for every odd function $f : [-a, a] \longrightarrow \mathbb{R}$ we have $\int_{-a}^a f(t)\mathrm{d}t = 0$. Indeed

$$\int_{-a}^a f(t)\mathrm{d}t = \int_0^a f(t)\mathrm{d}t + \int_{-a}^0 f(t)\mathrm{d}t$$
$$= \int_0^a f(t)\mathrm{d}t - \int_{-a}^0 f(-t)\mathrm{d}t$$
$$= \int_0^a f(t)\mathrm{d}t - \int_a^0 f(s) \cdot (-1)\mathrm{d}s$$
$$= \int_0^a f(t)\mathrm{d}t + \int_a^0 f(s)\mathrm{d}s = \int_0^a f(t)\mathrm{d}t - \int_0^a f(s)\mathrm{d}s = 0.$$

4. Denote the **Dirichlet kernel** by $D_n$:

$$D_n(t) := \begin{cases} \frac{\sin(2n+1)t}{\sin t}, & t \neq 0, \ t \in \left[-\frac{\pi}{2}, \frac{\pi}{2}\right] \\ 2n + 1, & t = 0. \end{cases}$$

By Problem 3 in Chapter 9 we have

$$\frac{1}{2}D_n(t) = \frac{1}{2} + \sum_{k=1}^n \cos(2kt)$$

implying

$$\frac{2}{\pi}\int_0^{\frac{\pi}{2}} D_n(t)\mathrm{d}t = \frac{2}{\pi}\int_0^{\frac{\pi}{2}}\left(1 + 2\sum_{k=1}^n \cos 2kt\right)\mathrm{d}t$$

$$= 1 + 2\sum_{k=1}^n \int_0^{\frac{\pi}{2}} \cos 2kt\,\mathrm{d}t$$

$$= 1 + \sum_{k=1}^n \frac{1}{k}\int_0^{k\pi} \cos s\,\mathrm{d}s$$

$$= 1 + \sum_{k=1}^n \frac{1}{k}\sin s\Big|_0^{k\pi} = 1.$$

5. For $\alpha \neq 0$ and $\alpha t = s$ we have

$$\int_a^b f(\alpha t)\mathrm{d}t = \frac{1}{\alpha}\int_{\alpha a}^{\alpha b} f(s)\mathrm{d}s$$

and further with $r = \alpha t + \beta$ we have

$$\int_a^b f(\alpha t + \beta)\mathrm{d}t = \frac{1}{\alpha}\int_{\alpha a + \beta}^{\alpha b + \beta} f(s)\mathrm{d}s.$$

6.    a) We have

$$\int_0^{\frac{\pi}{4}} \vartheta \cos\vartheta\,\mathrm{d}\vartheta = \vartheta\sin\vartheta\Big|_0^{\frac{\pi}{4}} - \int_0^{\frac{\pi}{4}} \sin\vartheta\,\mathrm{d}\vartheta$$

$$= \vartheta\sin\vartheta\Big|_0^{\frac{\pi}{4}} + \cos\vartheta\Big|_0^{\frac{\pi}{4}}$$

$$= \frac{\pi}{4}\frac{1}{2}\sqrt{2} + \frac{1}{2}\sqrt{2} - 1 = \frac{1}{2}\sqrt{2}\left(1 + \frac{\pi}{4}\right) - 1.$$

b) We have

$$\int_{\frac{1}{2}}^2 x\ln(2x+1)\mathrm{d}x = \frac{x^2}{2}\ln(2x+1)\Big|_{\frac{1}{2}}^2 - \int \frac{x^2}{2}\frac{2}{2x+1}\mathrm{d}x$$

$$= 2\ln 5 - \frac{1}{8}\ln 2 - \int_{\frac{1}{2}}^2 \frac{x^2}{2x+1}\mathrm{d}x,$$

and we know

$$(*)\quad \frac{x^2}{2x+1} = \frac{1}{2}x - \frac{1}{4} + \frac{1}{8x+4},$$

617

this is obtained by division:

$$x^2 : 2x + 1 = \frac{1}{2}x - \frac{1}{4}$$

$$x^2 + \frac{1}{2}x$$

$$\overline{\quad -\frac{1}{2}x \quad}$$

$$-\frac{1}{2}x - \frac{1}{4}$$

$$\overline{\quad +\frac{1}{4} \quad}$$

giving (*). Thus we find

$$\int_{\frac{1}{2}}^{2} \frac{x^2}{2x+1}\mathrm{d}x = \int_{\frac{1}{2}}^{2} \left( \frac{1}{2}x - \frac{1}{4} + \frac{1}{8x+4} \right) \mathrm{d}x$$

$$= \frac{1}{4}x^2 \bigg|_{\frac{1}{2}}^{2} - \frac{1}{4}x \bigg|_{\frac{1}{2}}^{2} + \frac{1}{8}\ln(8x+4) \bigg|_{\frac{1}{2}}^{2}$$

$$= \frac{9}{16} + \frac{1}{8}\ln 20 - \frac{1}{8}\ln 8,$$

which implies

$$\int_{\frac{1}{2}}^{2} x\ln(2x+1)\mathrm{d}x = 2\ln 5 - \frac{1}{8}\ln 2 - \frac{9}{16} - \frac{1}{8}\ln 20 + \frac{1}{8}\ln 8.$$

c) We have

$$\int_{0}^{\frac{1}{m}} s\sinh ms\,\mathrm{d}s = s\frac{\cosh(ms)}{m} \bigg|_{0}^{\frac{1}{m}} - \frac{1}{m}\int_{0}^{\frac{1}{m}} \cosh ms\,\mathrm{d}s$$

$$= \frac{\cosh 1}{m^2} - \frac{1}{m^2}\sinh ms \bigg|_{0}^{\frac{1}{m}}$$

$$= \frac{\cosh 1 - \sinh 1}{m^2} = \frac{e + e^{-1} - \left(e - e^{-1}\right)}{2m^2}$$

$$= \frac{1}{m^2 e}.$$

d) We have

$$\int_{1}^{3} \frac{\ln t}{\sqrt{t}}\mathrm{d}t = \int_{1}^{3} t^{-\frac{1}{2}}\ln t\,\mathrm{d}t = 2t^{\frac{1}{2}}\ln t \bigg|_{1}^{3} - \int_{1}^{3} 2t^{\frac{1}{2}}\frac{1}{t}\mathrm{d}t$$

$$= 2\sqrt{3}\ln 3 - 2\int_{1}^{3} t^{-\frac{1}{2}}\mathrm{d}t$$

$$= 2\sqrt{3}\ln 3 - 4t^{\frac{1}{2}} \bigg|_{1}^{3} = 4 + 2\sqrt{3}(\ln 3 - 2).$$

618

e) We have

$$\int_0^\pi e^{2r} \sin 3r dr = \frac{1}{2} e^{2r} \sin 3r \Big|_0^\pi - \frac{3}{2} \int_0^\pi e^{2r} \cos 3r dr$$

$$= -\frac{3}{2} \left( \frac{1}{2} e^{2r} \cos 3r \Big|_0^\pi + \frac{3}{2} \int_0^\pi e^{2r} \sin 3r dr \right)$$

or

$$\left( 1 + \frac{9}{4} \right) \int_0^\pi e^{2r} \sin 3r dr = -\frac{3}{4} \left( -e^{2\pi} - 1 \right)$$

which gives

$$\int_0^\pi e^{2r} \sin 3r dr = \frac{4}{13} \cdot \frac{3}{4} \left( e^{2\pi} + 1 \right) = \frac{3 \left( e^{2\pi} + 1 \right)}{13}.$$

7. Performing integration by parts twice gives

$$\frac{1}{\pi} \int_{-\pi}^\pi (\cos nx)(\cos mx) dx = \frac{1}{\pi m} (\cos nx)(\sin mx) \Big|_{-\pi}^\pi + \frac{1}{\pi} \int_{-\pi}^\pi \frac{n}{m} (\sin nx)(\sin mx) dx$$

$$= \frac{1}{\pi} \frac{n}{m} \int_{-\pi}^\pi (\sin nx)(\sin mx) dx$$

$$= -\frac{1}{\pi} \frac{n}{m^2} (\sin nx)(\cos mx) \Big|_{-\pi}^\pi + \frac{1}{\pi} \frac{n^2}{m^2} \int_{-\pi}^\pi (\cos nx)(\cos mx) dx,$$

or

$$(*) \qquad \frac{1}{\pi} \left( 1 - \frac{n^2}{m^2} \right) \int_{-\pi}^\pi \cos nx \cos mx dx = 0.$$

For $n \neq m$ it follows from $(*)$ that

$$\frac{1}{\pi} \int_{-\pi}^\pi (\cos nx)(\cos mx) dx = 0.$$

In the case where $n = m$ we find by noting

$$\frac{1}{\pi} \int_{-\pi}^\pi \cos^2 (nx) dx = \frac{1}{\pi} \int_{-\pi}^\pi \sin^2 (nx) dx$$

that

$$2 = \frac{1}{\pi} \int_{-\pi}^\pi \left( \cos^2 nx + \sin^2 nx \right) dx = \frac{2}{\pi} \int_{-\pi}^\pi \cos^2 nx dx,$$

which gives

$$\frac{1}{\pi} \int_{-\pi}^\pi \cos^2 nx dx = 1.$$

619

Furthermore, for $n, m \in \mathbb{N}$ we have

$$\frac{1}{\pi} \int_{-\pi}^{\pi} (\sin nx)(\cos mx) \mathrm{d}x = \frac{1}{\pi m} \sin nx \sin mx - \frac{1}{\pi} \frac{n}{m} \int_{-\pi}^{\pi} (\cos nx)(\sin mx) \mathrm{d}x$$

$$= -\frac{1}{\pi} \frac{n}{m} \int_{-\pi}^{\pi} (\cos nx)(\sin mx) \mathrm{d}x$$

$$= -\frac{n}{\pi m^2} (\cos nx)(\cos mx) \Big|_{-\pi}^{\pi} - \frac{1}{\pi} \frac{n^2}{m^2} \int_{-\pi}^{\pi} (\sin nx)(\cos mx) \mathrm{d}x$$

$$= -\frac{1}{\pi} \frac{n^2}{m^2} \int_{-\pi}^{\pi} (\sin nx)(\cos mx) \mathrm{d}x.$$

Since $\frac{n^2}{m^2} > 0$ it follows that

$$\frac{1}{\pi} \int_{-\pi}^{\pi} (\sin nx)(\cos mx) \mathrm{d}x = 0.$$

8.    a) Integration by parts gives

$$\int x^2 e^{\lambda x} \mathrm{d}x = \frac{x^2 e^{\lambda x}}{\lambda} - \frac{2}{\lambda} \int x e^{\lambda x} \mathrm{d}x$$

$$= \frac{x^2 e^{\lambda x}}{\lambda} - \frac{2x e^{\lambda x}}{\lambda^2} + \frac{2}{\lambda^2} \int e^{\lambda x} \mathrm{d}x$$

$$= \left( \frac{x^2}{\lambda} - \frac{2x}{\lambda^2} + \frac{2}{\lambda^3} \right) e^{\lambda x}.$$

b) First, it is clear that the case $a = 0$ gives

$$\int \frac{\mathrm{d}t}{at^2 + bt + c} = \int \frac{\mathrm{d}t}{bt + c} = \frac{1}{b} \ln|bt + c|.$$

Therefore let $a \neq 0$ and set $D = b^2 - 4ac$. We consider the three cases $D > 0$, $D = 0$, and $D < 0$.
**D>0.** It follows that

$$at^2 + bt + c = a \left( \left( t + \frac{b}{2a} \right)^2 - \left( \frac{\sqrt{D}}{2a} \right)^2 \right)$$

$$= a \left( t + \frac{b + \sqrt{D}}{2a} \right) \left( t + \frac{b - \sqrt{D}}{2a} \right).$$

Over any interval which does not contain $t_1, t_2, t_{1,2} = \frac{-b \pm \sqrt{D}}{2a}$ the function $t \mapsto$

$\frac{1}{at^2+bt+c}$ is integrable and we have

$$\frac{1}{at^2 + bt + c} = \frac{1}{a\left(t + \frac{b+\sqrt{D}}{2a}\right)\left(t + \frac{b-\sqrt{D}}{2a}\right)}$$

$$= \frac{1}{a}\left(\frac{\frac{a}{\sqrt{D}}}{t + \frac{b-\sqrt{D}}{2a}} - \frac{\frac{a}{\sqrt{D}}}{t + \frac{b+\sqrt{D}}{2a}}\right)$$

$$= \frac{1}{\sqrt{D}}\left(\frac{1}{t + \frac{b-\sqrt{D}}{2a}} - \frac{1}{t + \frac{b+\sqrt{D}}{2a}}\right),$$

which implies

$$\int \frac{1}{at^2 + bt + c}\,dt = \frac{1}{\sqrt{D}}\int \frac{1}{t+\frac{b-\sqrt{D}}{2a}}\,dt - \frac{1}{\sqrt{D}}\int \frac{1}{t+\frac{b+\sqrt{D}}{2a}}\,dt$$

$$= \frac{1}{\sqrt{D}}\ln\left|\frac{2at + b - \sqrt{D}}{2at - b + \sqrt{D}}\right|.$$

**D=0.** In this case we have

$$at^2 + bt + c = a\left(t + \frac{b}{2a}\right)^2$$

and consequently

$$\int \frac{dt}{at^2 + bt + c} = \frac{1}{a}\int \frac{dt}{\left(x+\frac{b}{2a}\right)^2} = -\frac{1}{a\left(x + \frac{b}{2a}\right)} = \frac{-2}{2ax + b}.$$

**D<0.** Now we find

$$at^2 + bt + c = a\left(\left(t + \frac{b}{2a}\right)^2 + \left(\frac{\sqrt{|D|}}{2a}\right)^2\right)$$

which has no real zero and therefore

$$\int \frac{dt}{at^2 + bt + c} = \frac{1}{a}\int \frac{dt}{\left(t+\frac{b}{2a}\right)^2 + \left(\frac{\sqrt{|D|}}{2a}\right)^2}.$$

Since $\int \frac{dx}{x^2 + \alpha^2} = \frac{1}{\alpha}\arctan\frac{x}{\alpha}$ it follows by the substitution $t + \frac{b}{2a} = y$ that

$$\int \frac{dt}{at^2 + bt + c} = \frac{1}{a}\frac{2a}{\sqrt{|D|}}\arctan\left(\frac{t + \frac{b}{2a}}{\frac{\sqrt{|D|}}{2a}}\right)$$

$$= \frac{2}{\sqrt{|D|}}\arctan\left(\frac{2at + b}{\sqrt{|D|}}\right).$$

9. We have

$$\int_0^a g(t)\mathrm{d}t = \int_c^{a+c} g(s-c)\mathrm{d}s = \int_c^{a+c} g(s)\mathrm{d}s.$$

10. In general, we will first determine a primitive and then evaluate it at the boundary points of the integral

a) With the change of variable $y = \ln x$ we find that $\frac{dy}{dx} = \frac{1}{x}$, $x = e^y$, and consequently

$$\int \frac{\mathrm{d}x}{x(\ln x)^3} = \int \frac{\mathrm{d}y}{y^3} = \frac{-1}{2y^2} = \frac{-1}{2(\ln x)^2}$$

which gives

$$\int_e^{e^2} \frac{\mathrm{d}x}{x(\ln x)^3} = \frac{-1}{2(\ln x)^2}\Big|_e^{e^2} = \frac{-1}{2(\ln e^2)^2} + \frac{1}{2(\ln e)^2}$$
$$= -\frac{1}{8} + \frac{1}{2} = \frac{3}{8}.$$

b) With the change of variable $\tan\frac{t}{2} = s$ we find that $\cos t = \cos^2\frac{t}{2} - \sin^2\frac{t}{2} = \frac{1-s^2}{1+s^2}$ and $\frac{dt}{ds} = \frac{2}{1+s^2}$, thus we find

$$\int \frac{\mathrm{d}t}{5 + 3\cos t} = \int \frac{2}{5 + 3\frac{1-s^2}{1+s^2}} \frac{\mathrm{d}s}{1+s^2}$$
$$= \int \frac{2}{5 + 5s^2 + 3 - 3s^2}\mathrm{d}s = \int \frac{1}{s^2 + 4}\mathrm{d}s$$
$$= \frac{1}{2}\arctan\frac{s}{2} = \frac{1}{2}\arctan\left(\frac{1}{2}\tan\frac{t}{2}\right).$$

Thus we find

$$\int_{\frac{\pi}{3}}^{\frac{\pi}{2}} \frac{\mathrm{d}t}{5 + \cos t} = \frac{1}{2}\arctan\left(\frac{1}{2}\tan\frac{t}{2}\right)\Big|_{\frac{\pi}{3}}^{\frac{\pi}{2}}$$
$$= \frac{1}{2}\arctan\left(\frac{1}{2}\tan\frac{\pi}{4}\right) - \frac{1}{2}\arctan\left(\frac{1}{2}\tan\frac{\pi}{6}\right)$$
$$= \frac{1}{2}\arctan\frac{1}{2} - \frac{1}{2}\arctan\frac{1}{2\sqrt{3}}.$$

c) With $\arcsin y^2 = \nu$ we find that $\frac{d\nu}{dy} = \frac{2y}{(1-y^4)^{\frac{1}{2}}}$ giving

$$\int \frac{y\arcsin y^2}{\sqrt{1-y^4}}\mathrm{d}y = \frac{1}{2}\int \nu\mathrm{d}\nu = \frac{\nu^2}{4}$$
$$= \frac{1}{4}\left(\arcsin y^2\right)^2$$

and therefore

$$\int_0^{\frac{1}{\sqrt{2}}} \frac{y \arcsin y^2}{\sqrt{1-y^4}}\,dy = \frac{1}{4}\left(\arcsin y^2\right)^2 \Big|_0^{\frac{1}{\sqrt{2}}}$$

$$= \frac{1}{4}\left(\arcsin \frac{1}{2}\right)^2 = \frac{1}{4}\left(\frac{\pi}{6}\right)^2 = \frac{\pi^2}{144}.$$

d) We have

$$\int \frac{ds}{\sqrt{5-4s-s^2}} = \int \frac{ds}{\sqrt{9-(s+2)^2}} = \frac{1}{3}\int \frac{ds}{\sqrt{1-\left(\frac{s+2}{3}\right)^2}}$$

and the change of variable $\frac{s+2}{3} = y$ gives

$$\int \frac{ds}{\sqrt{5-4s-s^2}} = \int \frac{1}{\sqrt{1-y^2}}\,dy = \arcsin y = \arcsin\left(\frac{s+2}{3}\right)$$

implying

$$\int_{\frac{1}{2}}^1 \frac{ds}{\sqrt{5-4s-s^2}} = \arcsin\left(\frac{s+2}{3}\right)\Big|_{\frac{1}{2}}^1 = \arcsin 1 - \arcsin\frac{5}{3}$$

$$= \frac{\pi}{2} - \arcsin\frac{5}{3}.$$

e) With $x = \sinh t$ we find $\frac{dx}{dt} = \cosh t$ and further

$$\int \frac{dx}{(1+x^2)^{\frac{3}{2}}} = \int \frac{\cosh t\,dt}{(1+\sinh^2 t)^{\frac{3}{2}}} = \int \frac{\cosh t}{(\cosh t)^3}\,dt$$

$$= \tanh t = \frac{\sinh t}{\sqrt{1+\sinh^2 t}} = \frac{x}{\sqrt{1+x^2}}.$$

Therefore we get

$$\int_1^4 \frac{dx}{(1+x^2)^{\frac{3}{2}}} = \frac{x}{\sqrt{1+x^2}}\Big|_1^4 = \frac{4}{\sqrt{17}} - \frac{1}{\sqrt{2}}.$$

11.    a) To find a primitive of $t \mapsto 3^{\sqrt{2t+1}}$ we use the change of variable $y = \sqrt{2t+1}$, i.e. $2x+1 = y^2$ and $\frac{dx}{dy} = y$ to get

$$\int 3^{\sqrt{2t+1}}\,dt = \int 3^y y\,dy,$$

now we use integration by parts:

$$\int 3^y y\,dy = \frac{y \cdot 3^y}{\ln 3} - \int \frac{3^y}{\ln 3}\,dy$$

$$= \frac{y \cdot 3^y}{\ln 3} - \frac{e^y}{(\ln 3)^2}.$$

623

Thus we arrive at

$$\int_0^4 3^{\sqrt{2t+1}} dt = \left( \frac{\sqrt{2t+1} \, 3^{\sqrt{2t+1}}}{\ln 3} - \frac{e^{\sqrt{2t+1}}}{(\ln 3)^2} \right) \Big|_0^4$$

$$= \frac{3 \cdot 3^3}{\ln 3} - \frac{1 \cdot 3^1}{\ln 3} - \frac{e^3}{(\ln 3)^2} + \frac{e}{(\ln 3)^2}$$

$$= \frac{240}{\ln 3} - \frac{(e^3 + e)}{(\ln 3)^2}.$$

b) The substitution $x = \pi - y$ gives

$$I = \int_0^\pi \frac{x \sin x}{1 + \cos^2 x} dx = \int_0^\pi \frac{(\pi - y) \sin y}{1 + \cos^2 y} dy$$

$$= \pi \int_0^\pi \frac{\sin y}{1 + \cos^2 y} dy - \int_0^\pi \frac{y \sin y}{1 + \cos^2 y} dy$$

$$= -\pi \int_0^\pi \frac{(\cos')(y)}{1 + \cos^2 y} dy - I$$

$$= -\pi \arctan(\cos y) |_0^\pi - I$$

$$= \frac{\pi^2}{2} - I,$$

or

$$I = \frac{\pi^2}{4}.$$

12. First we note

$$x^4 - x = x \left( x^3 - 1 \right) = x(x - 1) \left( x^2 + x + 1 \right).$$

We wish to find $a, b, c, d \in \mathbb{R}$ such that

$$\frac{x+1}{x^4 - x} = \frac{x+1}{x(x-1)(x^2+x+1)} = \frac{a}{x} + \frac{b}{x-1} + \frac{cx+d}{x^2+x+1}$$

$$= \frac{x^3(a+b+c) + x^2(b+d-c) + x(b-d) - a}{x(x-1)(x^2+x+1)},$$

i.e.

$$a + b + c = 0$$
$$a + d - c = 0$$
$$b - d = 1$$
$$- a = 1$$

which gives $a = -1$, $b = \frac{2}{3}$, $c = \frac{1}{3}$, $d = -\frac{1}{3}$. Hence we have

$$\int \frac{x+1}{x^4 - x} dx = -\int \frac{1}{x} dx + \frac{2}{3} \int \frac{1}{x-1} dx + \frac{1}{3} \int \frac{x-1}{x^2+x+1} dx$$

$$= -\ln|x| + \frac{2}{3}\ln|x-1| + \frac{1}{3} \int \frac{x-1}{x^2+x+1} dx,$$

624

and further

$$\int \frac{x-1}{x^2+x+1}dx = \frac{1}{2}\int \frac{2x-2}{x^2+x+1}dx$$

$$= \frac{1}{2}\int \frac{2x+1}{x^2+x+1}dx - \frac{1}{2}\int \frac{3}{x^2+x+1}dx$$

$$= \frac{1}{2}\ln\left(x^2+x+1\right) - \frac{3}{2}\left(\frac{2}{\sqrt{3}}\arctan\left(\frac{2x+1}{\sqrt{3}}\right)\right)$$

$$= \frac{1}{2}\ln\left(x^2+x+1\right) - \sqrt{3}\arctan\left(\frac{2x+1}{\sqrt{3}}\right),$$

where we use the solution to Problem 8 b) for $D < 0$.

13. This formula follows by iterating integration by parts:

$$\int_a^b f(t)g^{(3)}(t)dt = fg''\big|_a^b - \int_a^b f'g^{(2)}dt$$

$$= fg''\big|_a^b - \left(f'g'\big|_a^b - \int_a^b f''(t)g'(t)dt\right)$$

$$= fg''\big|_a^b - f'g'\big|_a^b + \int_a^b f''(t)g'(t)dt$$

$$= fg''\big|_a^b - f'g'\big|_a^b + f''g\big|_a^b - \int_a^b f^{(3)}(t)g(t)dt.$$

14. Note that $\frac{d}{ds}\sqrt{g(s)} = \frac{g'(s)}{2\sqrt{g(s)}}$ and therefore we find

$$\int \frac{g'(s)}{\sqrt{g(s)}}ds = 2\sqrt{g(s)}.$$

In the case of $\int_{\frac{\pi}{6}}^{\frac{\pi}{2}} \frac{\cos r}{\sqrt{\sin r}}dr$ we now find

$$\int_{\frac{\pi}{6}}^{\frac{\pi}{2}} \frac{\cos r}{\sqrt{\sin r}}dr = 2\sqrt{\sin r}\Big|_{\frac{\pi}{6}}^{\frac{\pi}{2}}$$

$$= 2 - \sqrt{2}.$$

15. Integration by parts gives

$$\int_{-\pi}^{\pi} f(t)\cos nt\,dt = \frac{1}{n}(\sin nt)f(t)\Big|_{-\pi}^{\pi} - \frac{1}{n}\int_{-\pi}^{\pi} f'(t)\sin nt\,dt$$

which implies

$$\left|\int_{-\pi}^{\pi} f(t)\cos nt\,dt\right| = \frac{1}{n}\left|\int_{-\pi}^{\pi} f'(t)\sin nt\,dt\right|$$

$$\leq \frac{1}{n}\int_{-\pi}^{\pi} M\,dt = \frac{2\pi M}{n},$$

625

where we used the fact that $f'(t) \sin nt \leq |f'(t) \sin nt| \leq M$.

16. We have

$$
\begin{aligned}
\int_1^x t^{-n} \mathrm{d}t &= \left. \frac{1}{-n+1} t^{-n+1} \right|_1^x \\
&= \frac{1}{n-1} - \frac{1}{n-1} \frac{1}{x^{n-1}}
\end{aligned}
$$

which implies

$$
\lim_{x \to \infty} \int_1^x t^{-n} \mathrm{d}t = \frac{1}{n-1} - \lim_{x \to \infty} \frac{1}{n-1} \frac{1}{x^{n-1}} = \frac{1}{n-1}.
$$

# Solutions to Problems of Part 2

## Chapter 14

1. For $x, y \in \mathbb{R}$ it follows that $\frac{x}{2}, \frac{y}{2}$ and $\frac{x+y}{2}$ belong to $\mathbb{R}$ too and $x < y$ implies $\frac{x}{2} < \frac{y}{2}$. Therefore we find $x = \frac{x}{2} + \frac{x}{2} < \frac{x}{2} + \frac{y}{2} = \frac{x+y}{2} = \frac{x}{2} + \frac{y}{2} < \frac{y}{2} + \frac{y}{2} = y$.

2.     a) By definition $x < 0$ if and only if $0 > x$ and this is equivalent to $0 - x > 0$ or $-x > 0$.

    b) For $x > 0$ it follows that $x^2 > 0$ from (14.12). If $x < 0$ then $-x > 0$, hence $(-x)^2 > 0$ which gives $x^2 > 0$.

    c) Since $y - x > 0$ and $-a > 0$ it follows that $-a(y - x) > 0$ or $ax - ay > 0$ which implies $ax > ay$.

3. Given $x, y \in \mathbb{Q}$, $x > 0, y > 0$ we need to find an $n \in \mathbb{N}$ such that $nx > y$. Let $x = \frac{p_1}{q_1}$ and $y = \frac{p_2}{q_2}$ with $p_1, p_2, q_1, q_2 \in \mathbb{N}$. With $r = q_1 q_2, p = p_1 q_2, q = q_1 p_2$ we find $x = \frac{p}{r}$ and $y = \frac{q}{r}$. Now $p \geq 1$ and therefore $p \cdot q \geq q$ and $r \geq 1$ implies $q \cdot r \geq q$ or $q \geq \frac{q}{r}$. Then with $\tilde{n} = r \cdot q$ we obtain

$$\tilde{n}x = rq\frac{p}{r} = pq \geq q \geq \frac{q}{r} = y$$

and therefore with $n = \tilde{n} + 1$ we have $nx > y$.

4. Suppose that for $a = \frac{p}{q} \in \mathbb{Q}$, $p \in \mathbb{Z}$ and $q \in \mathbb{N}$ having no common divisor, we have $a^2 = 3$, i.e. $\frac{p^2}{q^2} = 3$ or $p^2 = 3q^2$. Then 3 must divide $p$, say $p = 3r$. Therefore we have $9r^2 = 3q^2$ or $3r^2 = q^2$ and it follows that 3 also divides $q$ which is a contradiction.

5. Taking $x = \frac{1}{n}$ in Bernoulli's inequality we find

$$(1 + \frac{1}{n})^n \geq 1 + n\frac{1}{n} = 2$$

or

$$\left(\frac{n}{n} + \frac{1}{n}\right)^n = \frac{(n+1)^n}{n^n} \geq 2$$

implying

$$(*) \qquad 2n^n \leq (n+1)^n.$$

We now use induction to prove

$$n! \leq 2\left(\frac{n}{2}\right)^n.$$

For $n = 1$ we have

$$1! \leq 2\left(\frac{1}{2}\right)^1 = 1$$

627

which is correct. Next we find using $(*)$ that

$$
\begin{aligned}
(n+1)! &= n!(n+1) \leq 2\left(\frac{n}{2}\right)^n (n+1) \\
&= \frac{1}{2^n} \cdot 2n^n(n+1) \\
&\leq \frac{1}{2^n}(n+1)^n(n+1) = \frac{2}{2^{n+1}}(n+1)^{n+1} \\
&= 2\left(\frac{n+1}{2}\right)^{n+1}
\end{aligned}
$$

proving the assertion.

6. First note that for $x_k = x$ for all $k = 1, \ldots, n$ we find Bernoulli's inequality. For $n = 1$ the statement is trivial. Now if it holds for $n$ then we consider the case $n+1$

$$
\begin{aligned}
\prod_{k=1}^{n+1}(1+x_k) &= \prod_{k=1}^{n}(1+x_k)(1+x_{n+1}) \\
&= \prod_{k=1}^{n}(1+x_k) + x_{n+1}\prod_{n=1}^{n}(1+x_k) \\
&\geq 1 + \sum_{k=1}^{n}x_k + x_{n+1} = 1 + \sum_{k=1}^{n+1}x_k.
\end{aligned}
$$

7. Suppose that

$$
(*) \qquad (a_1 \cdot \ldots \cdot a_n)^{\frac{1}{n}} \leq \frac{a_1 + \cdots + a_n}{n}
$$

holds and that $n \geq 2$. If $0 < x \leq 1 - \frac{1}{n}$ then $x^n > 0 \geq 1 + n(x-1)$ or with $x = 1+y$ we have $(1+y)^n \geq 1 + ny$.

Thus we need to prove the case $n \geq 2$ and $x > 1 - \frac{1}{n}$. In this case $1 + n(x-1) > 0$ and we may apply $(*)$ to $a_1 = 1 + n(x-1)$ and $a_2 = \cdots = a_n = 1$ to find

$$
\begin{aligned}
x^n &= \left(\frac{1 + n(x-1) + 1 + \cdots + 1}{n}\right)^n \\
&\geq (1 + n(x-1)) \cdot 1 \cdot \ldots \cdot 1 = 1 + n(x-1)
\end{aligned}
$$

or again with $x = 1 + y : (1+y)^n \geq 1 + ny$.

8. We take $a_1 = \cdots = a_n = 1 + \frac{x}{n}$ and $a_{n+1} = \cdots = a_m = 1$. Note that $a_n > 0$ is equivalent to $-x < n$. Now we find

$$
\begin{aligned}
G_m &= (a_1 \cdot \ldots \cdot a_m)^{\frac{1}{m}} = (1 + \frac{x}{n})^{\frac{n}{m}} \\
&\leq \frac{1}{m}\sum_{j=1}^{m}a_j = \frac{1}{m}(n(1 + \frac{x}{n}) + (m-n)1) \\
&= 1 + \frac{x}{m},
\end{aligned}
$$

which implies

$$(1 + \frac{x}{n})^n \le (1 + \frac{x}{m})^m \text{ for } -x < n < m.$$

9. The Cauchy-Schwarz inequality yields

$$\left| \sum_{k=1}^{n} a_k \right| \le \sum_{k=1}^{n} |a_k| = \sum_{k=1}^{n} 1 \cdot |a_k|$$

$$\le \left( \sum_{k=1}^{n} 1^2 \right)^{\frac{1}{2}} \left( \sum_{k=1}^{n} a_k^2 \right)^{\frac{1}{2}} = \sqrt{n} \left( \sum_{k=1}^{n} a_k^2 \right)^{\frac{1}{2}}.$$

Now, from the above calculations we derive

$$\frac{1}{\sqrt{n}} \sum_{k=1}^{n} |a_k| \le \left( \sum_{k=1}^{n} a_k^2 \right)^{\frac{1}{2}},$$

and since

$$\sum_{k=1}^{n} a_k^2 \le n \cdot \max \{a_1^2, \ldots, a_n^2\}$$

$$= n \cdot \max \{|a_1|^2, \ldots, |a_n|^2\} = n \cdot \max \{|a_1|, \ldots, |a_n|\}^2$$

we find

$$\left( \sum_{k=1}^{n} a_k^2 \right)^{\frac{1}{2}} \le \sqrt{n} \max \{|a_1|, \ldots, |a_n|\}.$$

**Chapter 15**

1. Let $g : \mathbb{N} \to M$ be a bijective mapping which must exist since $M$ is countable. Define $a_k := f(g(k)), k \in \mathbb{N}$ then $\{a_k | a_k = f(g(k))\} = \{f(m) | m \in M\}$.

2.    a) We need to prove that for every $\epsilon > 0$ there exists $N(\epsilon) \in \mathbb{N}$ such that $n > N(\epsilon)$ implies $|a_n - a| < \epsilon$. For $n \ge M$ it follows however that

$$|a_n - a| = |a - a| = 0 < \epsilon$$

implying $\lim_{n \to \infty} a_n = a$.

   b) Since $\lim_{n \to \infty} a_n = a$, for $\epsilon > 0$ there exists $N_1(\epsilon) \in \mathbb{N}$ such that $n \ge N_1(\epsilon)$ implies $|a_n - a| < \epsilon$. Therefore, for $n \ge N(\epsilon) = \max\{N_1(\epsilon), M\}$ we have $|b_n - a| = |a_n - a| < \epsilon$.

Both problems show that the first $M$ ($M$ could be very large) elements of a sequence do not have any affect on the limit.

629

3. Let $|b_n| \leq M$ for all $n \geq k$ and given $\epsilon > 0$ choose $N(\epsilon) \in \mathbb{N}$ such that $n \geq N(\epsilon)$ implies $|a_n| < \frac{\epsilon}{M}$. It follows for $n \geq N(\epsilon)$ that

$$|b_n a_n| = |b_n||a_n| \leq M|a_n| < M\frac{\epsilon}{M} = \epsilon,$$

i.e.

$$\lim_{n \to \infty} |b_n a_n| = 0.$$

4.　　a) Since by assumption $a_n - a \leq c_n - a \leq b_n - a$ we find for all $n \geq k$ that

$$|c_n - a| \leq \max\{|a_n - a|, |b_n - a|\}.$$

Now, since $\lim_{n \to \infty} a_n = \lim_{n \to \infty} b_n = a$, given $\epsilon > 0$ there exists $N_1(\epsilon), N_2(\epsilon) \in \mathbb{N}$ such that $n \geq N_1(\epsilon)$ implies $|a_n - a| < \epsilon$ and $n \geq N_2(\epsilon)$ implies $|b_n - a| < \epsilon$. Hence for $n \geq \max\{N_1(\epsilon), N_2(\epsilon)\}$, we deduce that $\max\{|a_n - a|, |b_n - a|\} < \epsilon$ which implies for $n \geq \max\{N_1(\epsilon), N_2(\epsilon)\}$ that $|c_n - a| < \epsilon$.

　　b) We may look at $(c_n)_{n \in \mathbb{N}}, c_n = (-1)^n$, which is a divergent sequence as we know by Example 15.5.C. However $-1 \leq c_n \leq 1$ for all $n \in \mathbb{N}$ and the sequence $(a_n)_{n \in \mathbb{N}}, a_n = -1$ for all $n \in \mathbb{N}$ has limit $a = -1$. Also the sequence $(b_n)_{n \in \mathbb{N}}, b_n = 1$ for all $n \in \mathbb{N}$ has limit $b = 1$. Thus $a < b, a_n \leq c_n \leq b_n$ but $(c_n)_{n \in \mathbb{N}}$ does not converge.

5.　　a) Here we will use the converse triangle inequality

$$||x| - |y|| \leq |x - y| \text{ for } x, y \in \mathbb{R}.$$

Since $\lim_{n \to \infty} a_n = a$ yields for $\epsilon > 0$ the existence of $N(\epsilon) \in \mathbb{N}$ such that $n \geq N(\epsilon)$ implies $|a_n - a| < \epsilon$, for these $n$, i.e. $n \geq N(\epsilon)$, it follows that

$$||a_n| - |a|| \leq |a_n - a| < \epsilon,$$

i.e.

$$\lim_{n \to \infty} |a_n| = |a|.$$

Since $\lim_{n \to \infty} a_n = a$ is equivalent to $\lim_{n \to \infty} (a_n - a) = 0$, we deduce that $\lim_{n \to \infty} a_n = a$ implies $\lim_{n \to \infty} |a_n - a| = 0$. That $\lim_{n \to \infty} |a_n - a| = 0$ implies $\lim_{n \to \infty} a_n = a$ follows from the definition.

　　b) Given $\epsilon > 0$ there exists $N(\epsilon) \in \mathbb{N}$ such that $n \geq N(\epsilon)$ implies $\mu_n < \epsilon$ or $|a_n - a| \leq \mu_n < \epsilon$, implying $\lim_{n \to \infty} a_n = a$.

6. We know (compare with Lemma 2.7) that

$$\max\{a, b\} = \frac{1}{2}(a + b + |a - b|)$$

and

$$\min\{a, b\} = \frac{1}{2}(a + b - |a - b|).$$

Therefore we find

$$\max\{a_n, b_n\} = \frac{1}{2}(a_n + b_n + |a_n - b_n|)$$

and

$$\min\{a_n, b_n\} = \frac{1}{2}(a_n + b_n - |a_n - b_n|),$$

or with $c_n := a_n + b_n$, $d_n := a_n - b_n$

$$\max\{a_n, b_n\} = \frac{1}{2}(c_n + |d_n|)$$

and

$$\min\{a_n, b_n\} = \frac{1}{2}(c_n - |d_n|).$$

Since $\lim_{n\to\infty} c_n = a + b$ and $\lim_{n\to\infty} |d_n| = |a - b|$ where for the last limit we have used Problem 5 a), it follows that

$$\lim_{n\to\infty} \max\{a_n, b_n\} = \frac{1}{2}(a + b + |a - b|) = \max\{a, b\}$$

and

$$\lim_{n\to\infty} \min\{a_n, b_n\} = \frac{1}{2}(a + b - |a - b|) = \min\{a, b\}.$$

7.   a) First we observe that $\left|\frac{5}{n+6} - 0\right| = \frac{5}{n+6} < \frac{5}{n}$. Now, given $\epsilon > 0$ we choose $N(\epsilon) \in \mathbb{N}$, $N(\epsilon) > \frac{5}{\epsilon}$, to find for $n \geq N(\epsilon)$ that

$$\left|\frac{5}{n+6} - 0\right| = \frac{5}{n+6} < \frac{5}{n} < \epsilon,$$

which implies $\lim_{n\to\infty} \frac{5}{n+6} = 0$.

   b) We have

$$\left|\frac{4n}{3n+2} - \frac{4}{3}\right| = \left|\frac{3\cdot 4n - 4(3n+2)}{3(3n+2)}\right|$$

$$= \left|\frac{-8}{9n+6}\right| = \frac{8}{9n+6} < \frac{8}{9n}.$$

Therefore, if $\frac{8}{9n} < \frac{1}{1000}$, i.e. $n > \frac{8000}{9}$, it follows that

$$\left|\frac{4n}{3n+2} - \frac{4}{3}\right| < \frac{1}{1000}.$$

8.   a) Given $\epsilon > 0$ for $n > \left[\frac{1}{\epsilon}\right] + 1$ it follows that $\frac{1}{n} < \epsilon$. For these $n$ we find $\frac{1}{n^k} \leq \frac{1}{n} < \epsilon$, i.e. $\lim_{n\to\infty} \frac{1}{n^k} = 0$.

b) Since $\lim\limits_{n\to\infty} \dfrac{1}{n} = 0$, given $\epsilon > 0$ we find $N(\epsilon) \in \mathbb{N}$ such that $n \geq N(\epsilon)$ implies $\frac{1}{n} < \epsilon^k$, implying for these $n$ that

$$\left| \frac{1}{n^{\frac{1}{k}}} - 0 \right| = \frac{1}{n^{\frac{1}{k}}} < \epsilon,$$

which proves $\lim\limits_{n\to\infty} \dfrac{1}{n^{\frac{1}{k}}} = 0$.

9.    a)

$$\lim_{n\to\infty} \frac{(n+1)^2 - n^2}{n} = \lim_{n\to\infty} \frac{n^2 + 2n + 1 - n^2}{n}$$

$$= \lim_{n\to\infty} \frac{2n+1}{n} = \lim_{n\to\infty} \left(2 + \frac{1}{n}\right) = 2.$$

b)

$$\lim_{n\to\infty} (\sqrt{n+1} - \sqrt{n}) = \lim_{n\to\infty} \frac{(\sqrt{n+1} - \sqrt{n})(\sqrt{n+1} + \sqrt{n})}{\sqrt{n+1} + \sqrt{n}}$$

$$= \lim_{n\to\infty} \frac{n+1-n}{\sqrt{n+1} + \sqrt{n}} = \lim_{n\to\infty} \frac{1}{\sqrt{n+1} + \sqrt{n}} = 0,$$

where the latter follows from $\frac{1}{\sqrt{n+1}+\sqrt{n}} \leq \frac{1}{\sqrt{n}}$ and $\lim\limits_{n\to\infty} \dfrac{1}{\sqrt{n}} = 0$.

c)

$$\lim_{n\to\infty} \frac{\sum_{j=1}^{n} j}{n^2} = \lim_{n\to\infty} \frac{\frac{n(n+1)}{2}}{n^2}$$

$$= \lim_{n\to\infty} \frac{n^2 + n}{2n^2} = \lim_{n\to\infty} \left(\frac{1}{2} + \frac{1}{2n}\right) = \frac{1}{2}.$$

d)

$$\lim_{n\to\infty} \frac{\sum_{j=1}^{n} j^2}{n^3} = \lim_{n\to\infty} \frac{\frac{n(n+1)(2n+1)}{6}}{n^3}$$

$$= \lim_{n\to\infty} \frac{2n^3 + 3n^2 + n}{6n^3} = \lim_{n\to\infty} \left(\frac{1}{3} + \frac{1}{2n} + \frac{1}{6n^2}\right) = \frac{1}{3}.$$

e)

$$\lim_{n\to\infty} \frac{1 + 2 \cdot 3^n}{5 + 4 \cdot 3^n} = \lim_{n\to\infty} \left(\frac{\frac{1}{3^n} + 2}{\frac{5}{3^n} + 4}\right)$$

$$= \lim_{n\to\infty} \left(\frac{\left(\frac{1}{3}\right)^n + 2}{5\left(\frac{1}{3}\right)^n + 4}\right)$$

$$= \frac{\lim\limits_{n\to\infty} \left(\left(\frac{1}{3}\right)^n + 2\right)}{\lim\limits_{n\to\infty} \left(5\left(\frac{1}{3}\right)^n + 4\right)} = \frac{2}{4} = \frac{1}{2}.$$

632

f)

$$\lim_{n\to\infty} \frac{n+4^n}{5^n} = \lim_{n\to\infty} \frac{n}{5^n} + \lim_{n\to\infty} \left(\frac{4}{5}\right)^n = 0$$

since $\lim_{n\to\infty} q^n = 0$ for $|q| < 1$ and $\frac{n}{5^n} \le \frac{n}{2^n}$ and $\lim_{n\to\infty} \frac{n}{2^n} = 0$ by Example 15.5.E.

10. The case $a = 1$ is trivial. Let $a > 1$ then $a^{\frac{1}{n}} = 1 + b_n$ for some $b_n > 0$ and further

$$a = (1+b_n)^n = \sum_{k=0}^{n} \binom{n}{k} b_n^k > 1 + \binom{n}{1} b_n$$

$$= 1 + nb_n$$

or

$$0 < b_n = \frac{a-1}{n},$$

implying $\lim_{n\to\infty} b_n = 0$. Therefore

$$\lim_{n\to\infty} \sqrt[n]{a} = \lim_{n\to\infty} a^{\frac{1}{n}} = \lim_{n\to\infty} (1+b_n) = 1.$$

11. We first note that

$$\prod_{j=1}^{n}\left(1 - \frac{1}{j+1}\right) = \prod_{j=1}^{n}\left(\frac{j+1-1}{j+1}\right) = \prod_{j=1}^{n}\left(\frac{j}{j+1}\right)$$

$$= \frac{\prod_{j=1}^{n} j}{\prod_{j=1}^{n}(j+1)} = \frac{1\cdot 2\cdot\ldots\cdot n}{2\cdot 3\cdot\ldots\cdot n\cdot(n+1)} = \frac{1}{n+1},$$

and now we conclude that

$$\lim_{n\to\infty} \prod_{j=1}^{n}\left(1 - \frac{1}{j+1}\right) = \lim_{n\to\infty} \frac{1}{n+1} = 0.$$

12. Since every polynomial has at most a finite number of real zeroes and since we are only interested in limits we may assume in the following that for $\nu \ge K$ none of the polynomials under consideration has a root. Next we consider

$$\sum_{k=0}^{n} a_k \nu^k = a_n \nu^n + \nu^n \sum_{k=0}^{n-1} a_k \nu^{k-n}$$

and

$$\sum_{l=0}^{m} b_l \nu^l = b_m \nu^l + \nu^m \sum_{k=0}^{m-1} b_l \nu^{l-m}$$

which gives

$$\frac{\sum_{k=0}^{n} a_k \nu^k}{\sum_{l=0}^{m} b_l \nu^l} = \frac{a_n \nu^n + \nu^n \sum_{k=0}^{n-1} a_k \nu^{k-n}}{b_m \nu^m + \nu^m \sum_{k=0}^{m-1} b_l \nu^{l-m}}.$$

If $n = m$ then it follows that

$$\frac{\sum_{k=0}^{n} a_k \nu^k}{\sum_{l=0}^{n} b_l \nu^l} = \frac{a_n + \sum_{k=0}^{n-1} a_k \nu^{k-n}}{b_n + \sum_{k=0}^{n-1} b_l^{l-n}}$$

and since $\lim_{\nu \to \infty} \nu^{k-n} = 0$ for $k < n$, we deduce for $a_n \neq 0, b_m \neq 0$ that

$$\lim_{\nu \to \infty} \frac{\sum_{k=0}^{n} a_\nu \nu^k}{\sum_{l=0}^{n} b_l \nu^l} = \frac{a_n}{b_n}.$$

Now if $m > n$ we find

$$\frac{\sum_{k=0}^{n} a_k \nu^k}{\sum_{l=0}^{m} b_l \nu^l} = \frac{a_n + \sum_{k=0}^{n-1} a_k \nu^{k-n}}{\nu^{m-n} \left( b_m + \sum_{k=0}^{m-1} b_l \nu^{l-m} \right)}$$

$$= \frac{1}{\nu^{m-n}} \cdot \frac{a_n + \sum_{k=0}^{n-1} a_k \nu^{k-n}}{b_m + \sum_{k=0}^{m-1} b_l \nu^{l-m}},$$

and we note

$$\lim_{\nu \to \infty} \frac{a_n + \sum_{k=0}^{n-1} a_k \nu^{k-n}}{b_m + \sum_{k=0}^{m-1} b_l \nu^{l-m}} = \frac{a_n}{b_m}$$

as well as $\lim_{\nu \to \infty} \frac{1}{\nu^{m-n}} = 0$ for $m > n$, hence

$$\lim_{\nu \to \infty} \frac{\sum_{k=0}^{n} a_k \nu^k}{\sum_{l=0}^{m} b_l \nu^l} = 0 \text{ for } m > n.$$

In the case where $n > m$ we find

$$\frac{\sum_{k=0}^{n} a_k \nu^k}{\sum_{l=0}^{m} b_l \nu^l} = \nu^{n-m} \left( \frac{a_n + \sum_{k=0}^{n-1} a_k \nu^{k-n}}{b_m + \sum_{l=0}^{m-1} b_l \nu^{l-m}} \right).$$

We know that

$$\lim_{\nu \to \infty} \left( \frac{a_n + \sum_{k=0}^{n-1} a_k \nu^{k-n}}{b + \sum_{l=0}^{m-1} b_l \nu^{l-m}} \right) = \frac{a_n}{b_m}$$

and $a_n \neq 0$ by assumption. Thus there exists $N_0 \in \mathbb{N}$ such that $\nu \geq N_0$ implies

$$-\left| \frac{a_n}{2b_m} \right| \leq \frac{a_n + \sum_{k=0}^{n-1} a_k \nu^{k-n}}{b_m + \sum_{l=0}^{m-1} b_l \nu^{l-m}} - \frac{a_n}{b_m} \leq \left| \frac{a_n}{2b_m} \right|$$

If $\frac{a_n}{b_m} \geq 0$ then we find

$$\frac{a_n}{2b_m} \leq \frac{a_n + \sum_{k=0}^{n-1} a_k \nu^{k-n}}{b_m + \sum_{l=0}^{m-1} b_l \nu^{l-m}} \leq \frac{3a_n}{2b_m}$$

634

implying

$$\frac{a_n}{2b_m}\nu^{n-m} \le \frac{\sum_{k=0}^{n} a_k \nu^k}{\sum_{l=0}^{m} b_l \nu^l} \le \frac{3a_n}{2b_m}\nu^{n-m}$$

which yields that $\frac{\sum_{k=0}^{n} a_k \nu^k}{\sum_{l=0}^{m} b_l \nu^l}$ is unbounded and therefore must diverge to $+\infty$.
If $\frac{a_n}{b_m} < 0$ then we consider

$$\frac{-\sum_{k=0}^{n} a_k \nu^k}{\sum_{l=0}^{m} b_l \nu^l}$$

which must tend to $+\infty$ as $\nu \to +\infty$, and hence

$$\frac{\sum_{k=0}^{n} a_k \nu^k}{\sum_{l=0}^{m} b_l \nu^l}$$

tends to $-\infty$ as $k \to \infty$.

13. With $b_n := a_n - a$ we have to prove that if $\lim_{j \to \infty} b_j = 0$ then $\lim_{n \to \infty} \frac{\sum_{j=1}^{n} b_j}{n} = 0$. For $m < n$ we have

$$\frac{b_1 + \cdots + b_n}{n} = \frac{b_1 + \cdots + b_m}{n} + \frac{b_{m+1} + b_{m+2} + \cdots + b_n}{n}$$

and therefore

$$\left| \frac{b_1 + \cdots + b_n}{n} \right| \le \frac{|b_1 + \cdots + b_m|}{n} + \frac{|b_{m+1}| + \cdots + |b_n|}{n}.$$

Since $\lim_{j \to \infty} b_j = 0$, given $\epsilon > 0$ we can find $m \in \mathbb{N}$ such that $|b_j| < \frac{\epsilon}{2}$ for $j > m$, which implies

$$\frac{|b_{m+1}| + \cdots + |b_n|}{n} \le \frac{n-m}{n}\frac{\epsilon}{2} < \frac{\epsilon}{2}.$$

Now we can choose $N$ such that for $n > N > m$ it follows that

$$\frac{|b_1 + \cdots + b_m|}{n} < \frac{\epsilon}{2}$$

which eventually yields for $n > N > m$ that

$$\frac{|b_1 + \cdots + b_n|}{n} < \epsilon,$$

i.e.

$$\lim_{n \to \infty} \frac{|b_1 + \cdots + b_n|}{n} = 0.$$

14. Given $x_0$ we take $x_n = x_0 + \frac{1}{n}$ and we deduce that for $\epsilon > 0$ there exists $\delta > 0$ such that $|x_n - x_0| = \frac{1}{n} < \delta$ then

$$\left| \frac{f(x_n) - f(x_0)}{x_n - x_0} - A \right| < \epsilon.$$

However

$$\frac{f(x_n) - f(x_0)}{x_n - x_0} - A = \frac{f(x_0 + \frac{1}{n}) - f(x_0)}{\frac{1}{n}} - A$$

$$= n(f(x_0 + \frac{1}{n}) - f(x_0)) - A.$$

Thus given $\epsilon > 0$ we can find $N(\epsilon) \in \mathbb{N}$ such that $n > N(\epsilon) \geq \frac{1}{\delta}$ implies

$$|n(f(x_0 + \frac{1}{n}) - f(x_0)) - A| < \epsilon,$$

i.e.

$$\lim_{n \to \infty} (n(f(x_0 + \frac{1}{n}) - f(x_0)) - A) = 0.$$

**Chapter 16**

1. Given a sequence of partial sums $(s_n)_{n \in \mathbb{N}}$ we find the corresponding sequence $(a_n)_{n \in \mathbb{N}}$ with $s_n = \sum_{k=1}^{n} a_k$ by $a_n = s_n - s_{n-1}$, hence

$$\begin{aligned} a_n &= \frac{n(n+1)(2n+1)}{6} - \frac{(n-1)n(2(n-1)+1)}{6} \\ &= \frac{n}{6}(2n^2 + 3n + 1 - (n-1)(2n-1)) \\ &= \frac{n}{6}(2n^2 + 3n + 1 - 2n^2 + 2n + n - 1) \\ &= \frac{n}{6} \cdot 6n = n^2. \end{aligned}$$

Indeed we already know that $\sum_{k=1}^{n} k^2 = \frac{n(n+1)(2n+1)}{6}$.

2. From $a_n \leq b_n$ we deduce $\sum_{k=1}^{n} a_n \leq \sum_{k=1}^{n} b_k$ which implies

$$\sum_{k=1}^{\infty} a_k = \lim_{n \to \infty} \sum_{k=1}^{n} a_n \leq \lim_{n \to \infty} \sum_{k=1}^{n} b_k = \sum_{k=1}^{\infty} b_k.$$

3.

$$\begin{aligned} \sum_{k=1}^{\infty} \frac{1}{4k^2 - 1} &= \lim_{n \to \infty} \sum_{k=1}^{n} \frac{1}{4k^2 - 1} \\ &= \lim_{n \to \infty} \frac{1}{2} \sum_{k=1}^{n} \left( \frac{1}{2k - 1} - \frac{1}{2k + 1} \right) \\ &= \lim_{n \to \infty} \frac{1}{2} \left( 1 + \sum_{k=2}^{n} \frac{1}{2k - 1} - \sum_{k=1}^{n-1} \frac{1}{2k + 1} - \frac{1}{2n + 1} \right) \\ &= \frac{1}{2} \lim_{n \to \infty} \left( 1 + \sum_{k=1}^{n-1} \frac{1}{2k + 1} - \sum_{k=1}^{n-1} \frac{1}{2k + 1} - \frac{1}{2n + 1} \right) \\ &= \frac{1}{2} \lim_{n \to \infty} \left( 1 - \frac{1}{2n + 1} \right) = \frac{1}{2}. \end{aligned}$$

4.  a) This is the geometric series with $q = -\frac{1}{5}$. Since $|q| < 1$ we have

$$\sum_{k=0}^{\infty} \frac{(-1)^k}{5^k} = \frac{1}{1 - \left(-\frac{1}{5}\right)} = \frac{1}{1 + \frac{1}{5}} = \frac{5}{6}.$$

b) Note that $e^{-nx} = (e^{-x})^n$ and for $x < 0$ we know that $0 < e^{-x} < 1$, hence

$$\sum_{n=0}^{\infty} e^{-nx} = \frac{1}{1 - e^{-x}}.$$

c) We have

$$\sum_{k=2}^{\infty} \left(\frac{4}{7}\right)^k = \sum_{k=0}^{\infty} \left(\frac{4}{7}\right)^k - 1 - \frac{4}{7}$$

$$= \frac{1}{1 - \frac{4}{7}} - \frac{11}{7} = \frac{7}{3} - \frac{11}{7} = \frac{16}{21}.$$

5.  The condition for the convergence of the geometric series is $\left|\frac{1}{y-2}\right| < 1$ or $1 < |y-2|$. Thus if $y > 3$ or $y < 1$ the series $\sum_{k=0}^{\infty} \frac{1}{(y-2)^k}$ converges with limit

$$\sum_{k=0}^{\infty} \frac{1}{(y-2)^k} = \frac{1}{1 - (y-2)} = \frac{1}{3 - y}.$$

6.  We only need to note that

$$s_n := \sum_{k=1}^{n} (a_k - a_{k-1}) = (a_1 - a_0) + (a_2 - a_1) + \cdots + (a_n - a_{n-1}) = a_n - a_0$$

and

$$\tilde{s}_n := \sum_{k=1}^{n} (a_k - a_{k+1}) = (a_1 - a_2) + (a_2 - a_3) + \cdots + (a_n - a_{n+1}) = a_1 - a_{n+1}$$

Now we find

$$\sum_{k=1}^{\infty} (a_k - a_{k-1}) = \lim_{n \to \infty} s_n = \lim_{n \to \infty} (a_n - a_0) = \lim_{n \to \infty} a_n - a_0$$

and

$$\sum_{k=1}^{\infty} (a_{k+1} - a_k) = \lim_{n \to \infty} \tilde{s}_n = \lim_{n \to \infty} (a_1 - a_{n+1}) = a_1 - \lim_{n \to \infty} a_n.$$

637

7.    a) We know that

$$\sum_{k=0}^{\infty} \frac{1}{2^k} = \frac{1}{1-\frac{1}{2}} = 2$$

and

$$\sum_{k=0}^{\infty} \frac{(-1)^k}{3^k} = \sum_{k=0}^{\infty} \left(-\frac{1}{3}\right)^k = \frac{1}{1-\frac{1}{3}} = \frac{3}{4}$$

which implies

$$\sum_{k=0}^{\infty} \left(\frac{1}{2^k} + \frac{(-1)^k}{3^k}\right) = \frac{11}{4}.$$

b) First we note that

$$\ln\left(1 - \frac{1}{k^2}\right) = \ln\left(\frac{k^2-1}{k^2}\right)$$
$$= \ln(k+1) + \ln(k-1) - 2\ln k$$
$$= (\ln(k+1) - \ln k) - (\ln k - \ln(k-1)).$$

Thus

$$\sum_{k=2}^{\infty} \ln\left(1 - \frac{1}{k^2}\right) = \sum_{k=2}^{\infty}((\ln(k+1) - \ln k) - (\ln k - \ln(k-1)))$$

is a telescopic series with respect to $a_k = \ln k - \ln(k-1)$. Therefore, by using a straightforward modification of Problem 6 we get

$$\sum_{k=2}^{\infty} \ln\left(1 - \frac{1}{k^2}\right) = \lim_{k\to\infty}(\ln(k+1) - \ln k) - (\ln 2 - \ln 1)$$
$$= \lim_{k\to\infty} \ln\left(1 + \frac{1}{k}\right) + \ln\frac{1}{2} = \ln\frac{1}{2},$$

where we used the assumption that $\lim_{k\to\infty} \ln\left(1 + \frac{1}{k}\right) = 0$.

c) The series $\sum_{k=1}^{\infty} \frac{1}{(2k-1)^2}$ sums up all reciprocals of squares of odd natural numbers, $\sum_{k=1}^{\infty} \frac{1}{(2k)^2}$ sums up all reciprocals of squares of even natural numbers, thus

$$\sum_{k=1}^{\infty} \frac{1}{k^2} = \sum_{k=1}^{\infty} \frac{1}{(2k)^2} + \sum_{k=1}^{\infty} \frac{1}{(2k-1)^2}$$
$$= \frac{1}{4}\sum_{k=1}^{\infty} \frac{1}{k^2} + \sum_{k=1}^{\infty} \frac{1}{(2k-1)^2}$$

or

$$\frac{3}{4}\sum_{k=1}^{\infty} \frac{1}{k^2} = \sum_{k=1}^{\infty} \frac{1}{(2k-1)^2},$$

i.e.

$$\sum_{k=1}^{\infty} \frac{1}{(2k-1)^2} = \frac{3A}{4}.$$

8.    a) For $a_n = \frac{n^3+2n^2-2}{15n^2+n}$ it follows that

$$
\begin{aligned}
a_n &= \frac{n^3\left(1+\frac{2}{n}-\frac{2}{n^3}\right)}{n^2(15+\frac{1}{n})} \\
&= n\left(\frac{1+\frac{2}{n}-\frac{2}{n^3}}{15+\frac{1}{n}}\right) \\
&\geq \frac{n}{16},
\end{aligned}
$$

where we used that $1+\frac{2}{n}-\frac{2}{n^3} \geq 1$ and $15+\frac{1}{n} \leq 16$. Now, given $K \in \mathbb{R}$ we only have to choose $N > 16K$ to have that $n \geq N$ implies

$$a_n \geq \frac{n}{16} \geq \frac{N}{16} > K.$$

b) Using the hint we find

$$0 < \sin\frac{1}{n} \leq \frac{1}{n} \text{ or } n \leq \frac{1}{\sin\frac{1}{n}}.$$

Now, given $K > 0$ choose $N \in \mathbb{N}$ such that $N > K$. Then it follows for $n \geq N$ that

$$\frac{1}{\sin\frac{1}{n}} \geq n \geq N \geq K,$$

hence $\lim\limits_{n\to\infty}\left(\frac{1}{\sin\frac{1}{n}}\right) = +\infty$.

9.    a) Take $a_n = n^2$ and $b_n = \frac{1}{n}$, then $\lim\limits_{n\to\infty} a_n = \lim\limits_{n\to\infty} n^2 = \infty$ and $\lim\limits_{n\to\infty} b_n = \lim\limits_{n\to\infty}\frac{1}{n} = 0$. Moreover $a_n \cdot b_n = n^2 \cdot \frac{1}{n} = n$ and $\lim\limits_{n\to\infty}(a_n b_n) = \lim\limits_{n\to\infty} n = +\infty$.

b) Take $a_n = -n^2$ and $b_n = \frac{1}{n}$ and we find $\lim\limits_{n\to\infty} a_n = -\infty$, $\lim\limits_{n\to\infty} b_n = 0$ and $\lim\limits_{n\to\infty}(a_n b_n) = \lim\limits_{n\to\infty}\left(-n^2\frac{1}{n}\right) = \lim\limits_{n\to\infty}(-n) = -\infty$.

c) Take $a_n = n$ and $b_n = \frac{c}{n}$, then $\lim\limits_{n\to\infty} a_n = \infty$, $\lim\limits_{n\to\infty} b_n = 0$ and $\lim\limits_{n\to\infty}(a_n b_n) = \lim\limits_{n\to\infty}\left(n \cdot \frac{c}{n}\right) = c$.

## Chapter 17

1.    a) Since

$$
\begin{aligned}
s_{2n} - s_n &= \sum_{j=n+1}^{2n} \frac{1}{j} \\
&= \frac{1}{n+1} + \cdots + \frac{1}{2n} > \frac{n}{2n} = \frac{1}{2}
\end{aligned}
$$

the first part follows in a straightforward way. However now the second part is trivial: $(s_n)_{n \in \mathbb{N}}$ is not a Cauchy sequence, hence has no limit.

b) We will prove that

$$(*) \qquad 0 < s_{n+m} - s_n < \frac{1}{n+1}$$

which implies that $(s_n)_{n \in \mathbb{N}}$ is a Cauchy sequence: given $\epsilon > 0$ take $N > \frac{1}{\epsilon}$ to find for $n \geq N$ that $|s_{n+m} - s_n| < \epsilon$ compare with Remark 17.2.

Next we prove $(*)$. If $m$ is even we find

$$s_{n+m} - s_n = \left( \frac{1}{n+1} - \frac{1}{n+2} \right) + \left( \frac{1}{n+3} - \frac{1}{n+4} \right) + \cdots + \left( \frac{1}{n+m-1} - \frac{1}{n+m} \right) > 0$$

and if $m$ is odd we have

$$s_{n+m} - s_n = \left( \frac{1}{n+1} - \frac{1}{n+2} \right) + \left( \frac{1}{n+3} - \frac{1}{n+4} \right) - \cdots$$
$$- \left( \frac{1}{n+m-1} - \frac{1}{n+m} \right) + \frac{1}{n+m} > 0$$

which proves the lower bound in $(*)$.

To show the upper bound we note that for $m$ even that

$$s_{n+m} - s_n = \frac{1}{n+1} - \left( \frac{1}{n+2} - \frac{1}{n+3} \right) - \left( \frac{1}{n+4} - \frac{1}{n+5} \right) - \cdots - \frac{1}{n+m} > \frac{1}{n+1}$$

and for $m$ odd that

$$s_{n+m} - s_n = \frac{1}{n+1} - \left( \frac{1}{n+2} - \frac{1}{n+3} \right) - \cdots - \left( \frac{1}{n+m-1} - \frac{1}{n+m} \right) > \frac{1}{n+1}$$

and therefore $(*)$ is proved.

Note that in 1b) a lot of cancellation happens when calculating $s_{n+m} - s_n$, whereas while calculating 1a) there is no cancellation; we just add up the strictly positive terms.

2. We know that for $\epsilon > 0$ there exists $N \in \mathbb{N}$ such that $n \geq N$ implies $2^{-N+1} < \epsilon$. For $n \geq N$ and $m \in \mathbb{N}$ it follows that

$$|a_n - a_{n+m}| = \left| \sum_{j=0}^{m-1} (a_{n+j+1} - a_{n+j}) \right|$$

$$\leq \sum_{j=0}^{m-1} |a_{n+j+1} - a_{n+j}|$$

$$\leq \sum_{j=1}^{m-1} 2^{-n-j} \leq 2^{-n} \sum_{j=0}^{\infty} 2^{-j} = 2^{-n+1} < \epsilon,$$

proving that $(a_n)_{n \in \mathbb{N}}$ is a Cauchy sequence.

3. Since $(a_n)_{n \in \mathbb{N}}$ and $(b_n)_{n \in \mathbb{N}}$ converge they are bounded, i.e. $-A \leq a_n \leq A$ and $-B \leq b_n \leq B$ for some $A > 0$ and $B > 0$. It then follows that

$$-A \leq a_n \leq c_n \leq b_n \leq B$$

implying that the sequence $(c_n)_{n \in \mathbb{N}}$ is bounded. By the Bolzano-Weierstrass theorem it must contain a convergent subsequence.

4. Since $0 < \frac{\sqrt{n}}{n+1}$ for all $n \in \mathbb{N}$, once we know that $\left(\frac{\sqrt{n}}{n+1}\right)_{n \in \mathbb{N}}$ is decreasing we can deduce that it converges. In order to prove that $\left(\frac{\sqrt{n}}{n+1}\right)_{n \in \mathbb{N}}$ is decreasing we need to prove

$$\frac{\sqrt{n}}{n+1} \geq \frac{\sqrt{n+1}}{n+2} \quad \text{for all } n \in \mathbb{N}.$$

This is equivalent to $\frac{n+2}{n+1} \geq \frac{\sqrt{n+1}}{\sqrt{n}}$ which is equivalent to $\sqrt{n}(n+2) \geq \sqrt{n+1}(n+1)$ or

$$n(n+2)^2 \geq (n+1)^3,$$

i.e.

$$n^3 + 4n + 4n \geq n^3 + 3n^2 + 3n + 1$$

which is a correct statement.

5. First we prove by induction for $k \in \mathbb{N}$ that $k! \geq 2^{k-1}$. For $k = 1$ we have $1 \geq 1$ which is clearly true. Now we observe that

$$(k+1)! = k!(k+1) \geq 2^{k-1}(k+1) \geq 2 \cdot 2^{k-1} = 2^k = 2^{(k+1)-1},$$

and the result follows. Next we find

$$\sum_{k=0}^{n} \frac{1}{k!} = 1 + \sum_{k=1}^{n} \frac{1}{k!} \leq 1 + \sum_{k=1}^{n} 2^{-k+1}$$

$$= 1 + 2 \sum_{k=1}^{n} 2^{-k} = 1 + 2\left(\sum_{k=0}^{n} 2^{-k} - 1\right)$$

$$= 1 + 2\left(1 - 2\left(\frac{1}{2}\right)^n\right) < 3,$$

where we used Theorem 16.4. Thus $\left(\sum_{k=0}^{n} \frac{1}{k!}\right)_{n \in \mathbb{N}}$ is a monotone increasing sequence which is bounded from above.

6. We have to prove that for every $K > 0$ there exists $N \in \mathbb{N}$ such that $n \geq N$ implies $a_n \geq K$. Suppose that there exists $K_0 > 0$ such that for infinitely many $n_l, l \in \mathbb{N}, a_{n_l} < K_0$. The subsequence $(a_{n_l})_{l \in \mathbb{N}}$ satisfies $0 \leq a_{n_l} \leq K_0$, i.e. it is bounded. Consequently it must have a convergent subsequence, implying that $(a_n)_{n \in \mathbb{N}}$ must have an accumulation point which is a contradiction.

7. We define
$$b_n := \begin{cases} -2, & n = 3k+1, k \in \mathbb{N}_0 \\ -\frac{1}{3}, & n = 3k+2, k \in \mathbb{N}_0 \\ 17, & n = 3k, k \in \mathbb{N}_0. \end{cases}$$

Since $-3 < -2 < -\frac{1}{3} < 17 < 19$ we have $-3 \le a_n \le 19$ for all $n \in \mathbb{N}$. Further the subsequence $(a_{3k+1})_{k\in\mathbb{N}_0}$ converges to $-2$, the subsequence $(a_{3k+2})_{k\in\mathbb{N}_0}$ converges to $-\frac{1}{3}$ and the subsequence $(a_{3k})_{k\in\mathbb{N}_0}$ converges to $17$.

8. Once we know that $(x_n)_{n\ge0}$ converges to a limit $\lim_{n\to\infty} x_n = x > 0$ we find

$$x = \frac{k-1}{k} \frac{x^k + a}{x^{k-1}} \text{ or } kx^k = (k-1)x^k + a$$

i.e. $x^k = a$ or $x = a^{\frac{1}{k}} = \sqrt[k]{a}$.

Now we prove that $(x_n)_{n\ge0}$ is bounded from below and decreasing, hence it must converge. We need the following steps:

i) We prove $x_n > 0$. We know that $x_0 > 0$ and $a > 0$. Since

$$x_{n+1} = \frac{(k-1)x_n^k + a}{kx_n^{k-1}},$$

we find $x_{n+1} > 0$ if $x_n > 0$, thus by induction $x_n > 0$ for all $n$.

ii) Since $x_n > 0$ it follows that $1 - \frac{x_n^k - a}{kx_n^k} \ge 0$, note that this is equivalent to $kx_n^k - x_n^k + a \ge 0$. Thus $-\frac{x_n^k - a}{kx_n^k} \ge -1$.

iii) Using Bernoulli's inequality we get

$$\left( x_n - \frac{x_n^k - a}{kx_n^{k-1}} \right)^k = \left( x_n \left( 1 - \frac{x_n^k - a}{kx_n^k} \right) \right)^k$$

$$= x_n^k \left( 1 - \frac{x_n^k - a}{kx_n^k} \right)^k$$

$$\ge x_n^k \left( 1 + k \left( -\frac{x_n^k - a}{kx_n^k} \right) \right)$$

$$= x_n^k \left( 1 - 1 + \frac{a}{x_n^k} \right) = a.$$

iv) Since $x_{n+1} = x_n - \frac{x_n^k - a}{kx_n^{k-1}}$ we deduce that $x_n^k \ge a$ for all $n \in \mathbb{N}$.

v) Since $x_n \ge 0$ and $\frac{x_n^k - a}{kx_n^{k-1}} \ge 0$ it follows that

$$x_{n+1} = x_n - \frac{x_n^k - a}{kx_n^{k-1}} \le x_n.$$

Thus $(x_n)_{n\in\mathbb{N}}$ is decreasing and bounded below by $a^{\frac{1}{k}}$. Therefore $\lim_{n\to\infty} x_n$ exists and it must be $a^{\frac{1}{k}}$.

9. By the binomial theorem we find for $n > 1$ that

$$
\begin{aligned}
a_n &= \left(1 + \frac{1}{n}\right)^n = \sum_{k=0}^{n} \binom{n}{k} \frac{1}{n^k} \\
&= 1 + n\frac{1}{n} + \frac{n(n-1)}{2!}\frac{1}{n^2} + \cdots + \frac{n(n-1)(n-2)\cdots 1}{n!}\frac{1}{n^n} \\
(*) \quad &= 1 + 1 + \frac{1}{2!}\left(1 - \frac{1}{n}\right) + \cdots + \frac{1}{n!}\left(1 - \frac{1}{n}\right)\cdots\left(1 - \frac{n-1}{n}\right) \\
&\leq \sum_{j=0}^{n} \frac{1}{j!} \leq \sum_{j=0}^{\infty} \frac{1}{j!},
\end{aligned}
$$

and by Problem 5 the series $\sum_{j=0}^{\infty} \frac{1}{j!}$ converges. Moreover, for $k > n$ we have

$$
\left(1 + \frac{1}{k}\right)^k > 1 + 1 + \frac{1}{2!}\left(1 - \frac{1}{k}\right) + \cdots + \frac{1}{n!}\left(1 - \frac{1}{k}\right)\left(1 - \frac{2}{k}\right)\cdots\left(1 - \frac{n}{k}\right),
$$

where we used the calculations made above leading to $(*)$. For $n$ fixed and $k \to \infty$ we obtain

$$
\lim_{k \to \infty} \left(1 + \frac{1}{k}\right)^k \geq \sum_{j=0}^{n} \frac{1}{j!}
$$

and therefore

$$
\lim_{k \to \infty} \left(1 + \frac{1}{k}\right)^k \geq \lim_{n \to \infty} \sum_{j=0}^{n} \frac{1}{j!} = \sum_{j=0}^{\infty} \frac{1}{j!}.
$$

Thus we have proved that

$$
\sum_{j=0}^{\infty} \frac{1}{j!} \leq \lim_{n \to \infty} \left(1 + \frac{1}{n}\right)^n \leq \sum_{j=0}^{\infty} \frac{1}{j!},
$$

i.e.

$$
e = \lim_{n \to \infty} \left(1 + \frac{1}{n}\right)^n = \sum_{j=0}^{\infty} \frac{1}{j!}
$$

10. Since $a_n = \left(1 + \frac{1}{n}\right)^n < \left(1 + \frac{1}{n}\right)^{n+1} = b_n$ the intervals $[a_n, b_n]$ are bounded, closed and non-empty. We are done if we can prove that $(a_n)_{n \in \mathbb{N}}$ is monotone increasing and that $(b_n)_{n \in \mathbb{N}}$ is monotone decreasing. We then have $a_1 \leq a_n \leq b_n \leq b_1$ and in addition

$$
\begin{aligned}
b_n - a_n &= \left(1 + \frac{1}{n}\right)^{n+1} - \left(1 + \frac{1}{n}\right)^n \\
&= \left(1 + \frac{1}{n}\right)^n \frac{1}{n} = a_n \frac{1}{n} \leq b_1 \frac{1}{n} = \frac{4}{n},
\end{aligned}
$$

643

hence $b_n - a_n \to 0$ as $n \to \infty$.

For $n \geq 2$ Bernoulli's inequality yields

$$\left(1 - \frac{1}{n^2}\right)^n > (1 - n)\frac{1}{n^2} = 1 - \frac{1}{n},$$

and therefore

$$\frac{n-1}{n} < \left(\frac{n^2-1}{n^2}\right)^n = \frac{(n+1)^n(n-1)^n}{n^n n^n}$$

or

$$\left(\frac{n}{n-1}\right)^{n-1} < \left(\frac{n+1}{n}\right)^n,$$

i.e. for $n \geq 2$

$$a_{n-1} = \left(1 + \frac{1}{n-1}\right)^{n-1} = \left(\frac{n}{n-1}\right)^{n-1} < \left(\frac{n+1}{n}\right)^n = \left(1 + \frac{1}{n}\right)^n = a_n.$$

Therefore $(a_n)_{n \in \mathbb{N}}$ is strictly monotone increasing. Once again, using Bernoulli's inequality we find for $n \geq 2$

$$\left(1 + \frac{1}{n^2-1}\right)^n > 1 + \frac{n}{n^2-1} > 1 + \frac{n}{n^2} = 1 + \frac{1}{n},$$

and further

$$1 + \frac{1}{n} < \left(\frac{n^2}{n^2-1}\right)^n = \left(\frac{n}{n-1}\right)^n \left(\frac{n+1}{n}\right)^{-n}$$

$$= \left(1 + \frac{1}{n-1}\right)^n \left(1 + \frac{1}{n}\right)^{-n},$$

which yields

$$b_n = \left(1 + \frac{1}{n}\right)^{n+1} < \left(1 + \frac{1}{n-1}\right)^n = b_{n-1}.$$

Now we see that $(b_n)_{n \in \mathbb{N}}$ is strictly monotone decreasing and the result follows, recall that $\lim\limits_{n \to \infty} \left(1 + \frac{1}{n}\right)^n = e$.

## Chapter 18

1. This problem helps to understand how to handle the Cauchy criterion better. With $l := n + m$ and $j = n + 1$ it follows that

$$\sum_{k=1}^{m} \left(\frac{1}{2}\right)^{n+k} = \sum_{i=j}^{l} \left(\frac{1}{2}\right)^i.$$

Thus we know that for $\epsilon > 0$ there exists $N \in \mathbb{N}$ such that $l \geq j \geq N$ implies $\sum_{i=j}^{l} \left(\frac{1}{2}\right)^i = \sum_{i=j}^{l} 2^{-i} < \epsilon$. Therefore the Cauchy criterion is satisfied.

2. Given $\epsilon > 0$ there exists $m \in \mathbb{N}$ such that $k > m$ and $l \geq 1$ implies

$$|s_{k+l} - s_k| = a_{k+1} + \cdots + a_{k+l} < \frac{\epsilon}{2}.$$

Now take $n \geq 2m$ then for $k = [\frac{1}{2}n]$ it follows that $k \geq m$ and therefore

$$a_{k+1} + \cdots + a_n < \frac{\epsilon}{2}.$$

Since the sequence is decreasing and $a_j \geq 0$, we find

$$(n-k)a_n = (n - [\tfrac{1}{2}n])a_n < \frac{\epsilon}{2},$$

but $n - [\frac{1}{2}n] \geq \frac{n}{2}$ implying

$$\frac{n}{2}a_n < \frac{\epsilon}{2} \quad \text{or} \quad na_n < \epsilon.$$

Thus we have proved: given $\epsilon > 0$ there exists $M = 2m \in \mathbb{N}$ such that $n > M = 2m$ yields $na_n = |na_n| < \epsilon$, i.e. $\lim\limits_{n \to \infty} na_n = 0$.

3. Since $a_n \geq 0$ and $a_{j+1} \leq a_j$ it follows for $s = \sum_{n=1}^{\infty} a_n < \infty$ that

$$
\begin{aligned}
s &\geq a_1 + a_2 + (a_3 + a_4) + (a_5 + a_6 + a_7 + a_8) + \cdots + (a_{2^{n-1}+1} + \cdots + a_{2^n}) \\
&\geq \frac{1}{2}a_1 + a_2 + 2a_4 + 4a_8 + \cdots + 2^{n-1}a_{2^n},
\end{aligned}
$$

or

$$a_1 + 2a_2 + \cdots + 2^n a_{2^n} \leq 2s.$$

This implies the convergence of $\sum_{n=1}^{\infty} 2^n a_{2^n}$ by Theorem 18.4. Now suppose $\sum_{n=1}^{\infty} 2^n a_{2^n}$ converges, then we find for $2^k \geq n$

$$
\begin{aligned}
a_1 + a_2 + \cdots + a_n &\leq a_1 + (a_2 + a_3) + (a_4 + \cdots a_7) + \cdots + (a_{2^k} + \cdots + a_{2^{k+1}-1}) \\
&\leq a_1 + 2a_2 + 4a_4 + \cdots + 2^k a_{2^k},
\end{aligned}
$$

and again Theorem 18.4 gives the result.

Note that the statement of this problem is sometimes called **Cauchy's condensation theorem**.

4.　　a) For the "condensed" series we find

$$\sum_{n=1}^{\infty} 2^n \cdot \frac{1}{(2^n)^\alpha} = \sum_{n=1}^{\infty} \left(\frac{1}{2^{\alpha-1}}\right)^n$$

and therefore for $\alpha > 1$ the series $\sum_{n=1}^{\infty} \frac{1}{n^\alpha}$ converges and for $\alpha \leq 1$ the series $\sum_{n=1}^{\infty} \frac{1}{n^\alpha}$ diverges.

　　b) We note that

$$2^n \frac{1}{2^n (\ln 2^n)^\alpha} = \frac{1}{n^\alpha (\ln 2)^\alpha}$$

and consequently $\sum_{n=2}^{\infty} \frac{1}{n(\ln n)^\alpha}$ converges for $\alpha > 1$ and diverges for $\alpha \leq 1$.

5.     a) We need to consider the sequence $\left(\frac{1}{n^\alpha}\right)_{n\in\mathbb{N}}$. Clearly for $\alpha \in \mathbb{R}$ we have $\frac{1}{n^\alpha} \geq 0$. Moreover, for $\alpha \geq 0$ the sequence is decreasing, while for $\alpha < 0$ it is not decreasing, but only for $\alpha > 0$ we have $\lim_{n\to\infty} \frac{1}{n^\alpha} = 0$. Hence for $\alpha > 0$ the series $\sum_{n=1}^\infty \frac{(-1)^{n+1}}{n^\alpha}$ converges by Leibniz's test, for $\alpha = 0$ the series reduces to $\sum_{n=1}^\infty (-1)^{k+1}$ which we know to be divergent and for $\alpha < 0$ it follows that $\lim_{n\to\infty} \frac{1}{n^\alpha} = +\infty$ therefore $\sum_{n=1}^\infty \frac{(-1)^{k+1}}{n^\alpha}$ diverges in this case.

b) Clearly $\frac{1}{2n-1} \geq 0$ for $n \geq 1$, $\lim_{n\to\infty} \frac{1}{2n-1} = 0$ and $\frac{1}{2(n+1)-1} = \frac{1}{2n+1} < \frac{1}{2n-1}$, i.e. the sequence $\left(\frac{1}{2n-1}\right)_{n\in\mathbb{N}}$ is decreasing and again Leibniz's test gives the convergence of $\sum_{n=1}^\infty \frac{(-1)^{n+1}}{2n-1}$.

c) For $n \geq 2$ the term $\frac{1}{n\ln n}$ is defined and positive. Since $0 < \frac{1}{n\ln n} < \frac{1}{n}$ we deduce that $\lim_{n\to\infty} \frac{1}{n\ln n} = 0$. Moreover we claim that

$$\frac{1}{(n+1)\ln(n+1)} < \frac{1}{n\ln n}$$

which follows from $\frac{1}{n+1} < \frac{1}{n}$ and $\frac{1}{\ln(n+1)} < \frac{1}{\ln n}$. Hence by Leibniz's criterion we know the convergence of $\sum_{n=2}^\infty \frac{(-1)^n}{n\ln n}$.

6. From the assumption it follows that

$$\sum_{k=1}^n a_k \leq \sum_{k=1}^n b_k$$

and that for $K \geq 0$ there exists $N \in \mathbb{N}$ such that $n \geq N$ implies $\sum_{k=1}^n a_k \geq K$. This however gives that $n \geq N$ implies $\sum_{k=1}^n b_k \geq K$, i.e. $\sum_{k=1}^\infty b_k$ diverges.

7.     a)

$$\left|\frac{(-1)^k k^2}{k^4 + 2k}\right| \leq \frac{k^2}{k^4 + 2k} \leq \frac{1}{k^2}$$

and $\sum_{k=1}^\infty \frac{1}{k^2}$ converges, hence $\sum_{k=1}^\infty \frac{(-1)^k k^2}{k^4+2k}$ converges.

b)

$$\frac{k!}{k^k} = \frac{1 \cdot \ldots \cdot k}{k \cdot \ldots \cdot k} = \frac{1}{k} \cdot \frac{2}{k} \cdot \left(\frac{3}{k} \cdot \ldots \cdot \frac{k}{k}\right) \leq \frac{2}{k^2}$$

and since $\sum_{k=1}^\infty \frac{2}{k^2}$ converges we deduce that $\sum_{k=1}^\infty \frac{k!}{k^k}$ converges.

c) First we claim that $\ln(n + 1) \leq n$ for all $n \in \mathbb{N}$. Indeed for $n = 1$ we have $\ln 2 \leq 2$ since $\ln e = 1$ and $\ln$ is a monotone function. Further $n + 1 \leq e^n$ is equivalent to $\ln(n+1) \leq n$ and if $n+1 \leq e^n$ then we get $n+2 \leq e^n +1 \leq e^n +e^n = 2e^n \leq ee^n = e^{n+1}$, thus we know that $\ln(n+1) \leq n$ for all $n \in \mathbb{N}$ and it follows that

$$\frac{\ln(n+1)}{3n^3 + 7} \leq \frac{\ln(n+1)}{3n^3} \leq \frac{1}{3n^2}$$

646

and the convergence of $\sum_{n=1}^{\infty} \frac{1}{3n^2}$ implies the result.

d) Since $|\sin x| \leq |x|$ we deduce $\left|\sin \frac{1}{n^3}\right| \leq \frac{1}{n^3}$ implying the convergence of $\sum_{n=1}^{\infty} \sin \frac{1}{n^3}$ since $\sum_{n=1}^{\infty} \frac{1}{n^3}$

e) For all $x \in \mathbb{R}$ we have $\left|\frac{\cos kx}{1+k^2}\right| \leq \frac{1}{1+k^2}$ and since $\sum_{k=1}^{\infty} \frac{1}{1+k^2}$ converges it follows that $\sum_{k=1}^{\infty} \frac{\cos kx}{1+k^2}$ converges for all $x \in \mathbb{R}$

f) For $x \leq 0$ we know that $e^{mx} \leq 1$ and therefore $\frac{e^{mx}}{m^4} \leq \frac{1}{m^4}$ and the convergence of $\sum_{m=1}^{\infty} \frac{1}{m^4}$ implies the convergence of $\sum_{m=1}^{\infty} \frac{e^{mx}}{m^4}$ for $x \leq 0$. However for $x > 0$ the series $\sum_{m=1}^{\infty} \frac{e^{mx}}{m^4}$ diverges since $\lim\limits_{m\to\infty} \frac{e^{mx}}{m^4} \neq 0$. Indeed, applying the de l'Hospital rules four times to the function $t \mapsto \frac{e^{tx}}{t^4}$ we find

$$\lim_{t\to\infty} \frac{e^{tx}}{t^4} = \lim_{t\to\infty} \frac{x^4 e^{tx}}{4!} = \lim_{t\to\infty} \frac{x^4}{4!} e^{tx} = \infty.$$

Therefore there exists $N \in \mathbb{N}$ such that $t \geq N$ implies $\frac{e^{tx}}{t^4} \geq 1$, hence for $m \in \mathbb{N}$ it follows that $m \geq N$ implies $\frac{e^{mx}}{m^4} \geq 1$.

g)

$$\sum_{l=1}^{\infty} \frac{x^2}{l^2+x^2} = x^2 \sum_{l=1}^{\infty} \frac{1}{l^2+x^2} \leq x^2 \sum_{l=1}^{\infty} \frac{1}{l^2} < \infty,$$

i.e. the series converges for all $x \in \mathbb{R}$.

h) Since

$$\frac{n+5}{(2n+1)\sqrt{n+3}} \geq \frac{n+5}{2n\sqrt{n}} \geq \frac{1}{2\sqrt{n}}$$

and

$$\sum_{n=1}^{\infty} \frac{1}{2\sqrt{n}}$$

diverges, the series $\sum_{n=1}^{\infty} \frac{n+5}{(2n+1)\sqrt{n+3}}$ diverges too.

8. For $n \in \mathbb{N}$ we know both the Cauchy-Schwarz inequality (compare with Corollary 14.3)

$$(*) \quad \left|\sum_{k=1}^{n} a_k b_k\right| \leq \sum_{k=1}^{n} |a_k b_k| \leq \left(\sum_{k=1}^{n} a_k^2\right)^{\frac{1}{2}} \left(\sum_{k=1}^{n} b_k^2\right)^{\frac{1}{2}}$$

and the Minkowski inequality (compare with Lemma 14.5)

$$(**) \quad \left(\sum_{k=1}^{n} |a_k+b_k|^2\right)^{\frac{1}{2}} \leq \left(\sum_{k=1}^{n} a_k^2\right)^{\frac{1}{2}} + \left(\sum_{k=1}^{n} b_k^2\right)^{\frac{1}{2}}.$$

Now, since by the assumptions that $\sum_{k=1}^{\infty} a_k^2 < \infty$ and $\sum_{k=1}^{\infty} b_k^2 < \infty$ we may take the limit as $n \to \infty$ of $(*)$ and the right hand side of $(**)$ to obtain

$$\left|\sum_{k=1}^{n} a_k b_k\right| \leq \sum_{k=1}^{n} |a_k b_k| \leq \left(\sum_{k=1}^{\infty} a_k^2\right)^{\frac{1}{2}} \left(\sum_{k=1}^{\infty} b_k^2\right)^{\frac{1}{2}}$$

and

$$\left(\sum_{k=1}^{n}|a_k+b_k|^2\right)^{\frac{1}{2}} \le \left(\sum_{k=1}^{\infty}a_k^2\right)^{\frac{1}{2}} + \left(\sum_{k=1}^{\infty}b_k^2\right)^{\frac{1}{2}}.$$

Now we see that the partial sums $\sum_{k=1}^{n}|a_kb_k|$ and $\left(\sum_{k=1}^{n}|a_k+b_k|^2\right)^{\frac{1}{2}}$ are monotone and bounded, hence

$$\sum_{k=1}^{\infty}|a_kb_k| \le \left(\sum_{k=1}^{\infty}a_k^2\right)^{\frac{1}{2}} \left(\sum_{k=1}^{\infty}b_k^2\right)^{\frac{1}{2}}$$

and

$$\left(\sum_{k=1}^{\infty}|a_k+b_k|^2\right)^{\frac{1}{2}} \le \left(\sum_{k=1}^{\infty}a_k^2\right)^{\frac{1}{2}} + \left(\sum_{k=1}^{\infty}b_k^2\right)^{\frac{1}{2}}.$$

Finally we note that the (absolute) convergence of $\sum_{k=1}^{\infty}|a_kb_k|$ implies the convergence of $\sum_{k=1}^{\infty}a_kb_k$.

9. Since for every sequence $(a_k)_{k\in\mathbb{N}}$ the term $\frac{|a_k|}{1+|a_k|}$ is bounded by 1 and $\sum_{k=1}^{\infty}\frac{1}{2^k}$ converges, it follows immediately that $\sum_{k=1}^{\infty}\frac{1}{2^k}\frac{|a_k|}{1+|a_k|}$ converges. The inequality

$$\sum_{k=1}^{\infty}\frac{1}{2^k}\frac{|a_k+b_k|}{1+|a_k+b_k|} \le \sum_{k=1}^{\infty}\frac{1}{2^k}\frac{|a_k|}{1+|a_k|} + \sum_{k=1}^{\infty}\frac{1}{2^k}\frac{|b_k|}{1+|b_k|}$$

follows once we have proved for $a,b \in \mathbb{R}$

$$\frac{|a+b|}{1+|a+b|} \le \frac{|a|}{1+|a|} + \frac{|b|}{1+|b|}.$$

However a straightforward calculation first shows

$$\frac{|a+b|}{1+|a+b|} \le \frac{|a|+|b|}{1+|a|+|b|}$$

and a further calculation shows

$$\frac{|a|+|b|}{1+|a|+|b|} \le \frac{|a|}{1+|a|} + \frac{|b|}{1+|b|}.$$

(The reader is encouraged to do these simple calculations.)

10.     a) We use the ratio test here:

$$\frac{(n+1)^6 e^{-(n+1)^2}}{n^6 e^{-n^2}} = \left(\frac{n+1}{n}\right)^6 e^{-2n-1} \le \left(\frac{n+1}{n}\right)^6 e^{-2n}.$$

Since $\lim_{n\to\infty}\left(\frac{n+1}{n}\right)^6 = 1$ and $\lim_{n\to\infty}e^{-2n} = 0$, it follows that

$$\lim_{n\to\infty}\frac{(n+1)^6 e^{-(n+1)^2}}{n^6 e^{-n^2}} = 0$$

648

implying the convergence of $\sum_{n=1}^{\infty} n^6 e^{-n^2}$.

b) Note that
$$\frac{4n^2 + 15n - 3}{n^2(n+1)^{\frac{3}{2}}} \le \frac{4n^2 + 15n^2}{n^2 n^{\frac{3}{2}}} = \frac{19}{n^{\frac{3}{2}}}$$

and the series $\sum_{n=1}^{\infty} \frac{1}{n^{\frac{3}{2}}}$ converges implying the convergence of $\sum_{n=1}^{\infty} \frac{4n^2 + 15n - 3}{n^2(n+1)^{\frac{3}{2}}}$.

c) For $x \in \mathbb{R}$ fixed we find
$$\frac{\frac{x^{k+1}}{(k+1)!}}{\frac{x^k}{k!}} = x\frac{k!}{(k+1)!} = \frac{x}{k+1}.$$

Thus $\lim\limits_{k \to \infty} \left| \frac{\frac{x^{k+1}}{(k+1)!}}{\frac{x^k}{k!}} \right| = 0$ and therefore $\sum_{k=0}^{\infty} \frac{x^k}{k!}$ converges for every $x \in \mathbb{R}$.

d) For every $x \in \mathbb{R}$ it follows that
$$\left| \frac{\frac{(-1)^{k+1}x^{2k+2}}{(2(k+1))!}}{\frac{(-1)^k x^k}{(2k)!}} \right| = x^2 \frac{(2k)!}{(2(k+1))!}$$

$$= x^2 \frac{(2k)!}{(2k)!(2k+1)(2k+2)} = \frac{x^2}{(2k+1)(2k+2)}.$$

Since $\lim\limits_{k \to \infty} \left( \frac{x^2}{(2k+1)(2k+2)} \right) = 0$ we deduce that the series $\sum_{k=0}^{\infty} (-1)^k \frac{x^{2k}}{(2k)!}$ converges for every $x \in \mathbb{R}$.

11. The condition $\left| \frac{a_{n+1}}{a_n} \right| \ge \lambda > 1$ implies $|a_{n+1}| > |a_n|$, i.e. $(|a_n|)_{n \in \mathbb{N}}$ is a strictly increasing sequence of non-negative numbers, hence it cannot converge to 0, but this implies that $(a_n)_{n \in \mathbb{N}}$ itself cannot converge to 0.
We now study the two examples

a)
$$\left| \frac{\frac{(-1)^{n+1} 3^{n+1}}{(n+1)^4}}{\frac{(-1)^n 3^n}{n^4}} \right| = 3 \left( \frac{n+1}{n} \right)^4 \ge 3,$$

implying the divergence of $\sum_{n=1}^{\infty} \frac{(-1)^n 3^n}{n^4}$.

b) We find
$$\frac{a_{n+1}}{a_n} = \frac{(n+1)\sqrt{n+1}}{n\sqrt{n}} \frac{(n+3)\sqrt{4n+15}}{(n+4)\sqrt{4n+19}}$$

$$= \frac{(n+1)(n+3)}{(n+4)n} \frac{\sqrt{n+1}\sqrt{4n+15}}{\sqrt{n}\sqrt{4n+19}}$$

$$= \frac{n^2 + 4n + 4}{n^2 + 4n} \sqrt{\frac{4n^2 + 19n + 15}{4n^2 + 19n}} > 1,$$

649

implying the divergence of $\sum_{n=1}^{\infty} \frac{n^{\frac{3}{2}}}{(n+3)\sqrt{4n+15}}$.

12.    a) From $|a_n|^{\frac{1}{n}} \geq 1$, i.e. $|a_n| \geq 1$, we deduce that $(a_n)_{n\in\mathbb{N}}$ cannot have the limit 0, hence $\sum_{n=1}^{\infty} a_n$ must diverge.

b) First we note that for $\epsilon := \frac{1-a}{2} > 0$ there exists $N \in \mathbb{N}$ such that $n \geq N$ implies

$$||a_n|^{\frac{1}{n}} - a| < \epsilon = \frac{1-a}{2}$$

or

$$|a_n|^{\frac{1}{n}} < a + \frac{1-a}{2} = \frac{a+1}{2} < 1.$$

This implies the convergence of $\sum_{n=N}^{\infty} |a_n|$ by Theorem 18.18 and therefore $\sum_{n=1}^{\infty} |a_n|$ converges too.

13. From $\left|\frac{a_{n+1}}{a_n}\right| \leq 1 - \frac{a}{n}$ for $n \geq N$ we deduce that

$$n|a_{n+1}| \leq n|a_n| - a|a_n|$$

or

$$(a-1)|a_n| \leq (n-1)|a_n| - n|a_{n+1}|.$$

Since $a > 1$ we find

$$0 < (n-1)|a_n| - a|a_{n+1}|,$$

or

$$(n-1)|a_n| > n|a_{n+1}|.$$

Hence the sequence $(n|a_{n+1}|)_{n\in\mathbb{N}}$ is strictly monotone and decreasing and bounded from below by 0, implying its convergence. Therefore we deduce that the series $\sum_{n=1}^{\infty}((n-1)|a_n| - n|a_{n+1}|)$ which is a telescopic series converges, compare with Chapter 16, Problem 6, and this implies, see Chapter 16, Problem 6 again, that $\sum_{n=1}^{\infty} |a_n|$ converges. Note that we can also prove: if for all $n \geq N$ we have $\frac{a_{n+1}}{a_n} \geq 1 - \frac{1}{n}$ then $\sum_{n=1}^{\infty} a_n$ diverges.

14. Note that if

$$\lim_{n\to\infty} n\left(1 - \left|\frac{a_{n+1}}{a_n}\right|\right) > 1$$

then there exists $N \in \mathbb{N}$ such that $n \geq N$ implies for some $a$

$$n\left(1 - \left|\frac{a_{n+1}}{a_n}\right|\right) \geq a > 1.$$

Now for $a_n = \left(\frac{1\cdot4\cdot\ldots\cdot(3n-2)}{3\cdot9\cdot\ldots\cdot3n}\right)^2$ we find

$$n\left(1 - \left|\frac{a_{n+1}}{a_n}\right|\right) = n\left(1 - \left(\frac{3n+1}{3(n+1)}\right)^2\right)$$

$$= n\left(\frac{12n+8}{9n^2+18n+9}\right) = \frac{12n^2+8n}{9n^2+18n+9}$$

implying $\lim\limits_{n\to\infty} n \left(1 - \left|\dfrac{a_{n+1}}{a_n}\right|\right) = \dfrac{4}{3} > 1$ and therefore by Raabe's criterion the series converges. Note that

$$\lim_{n\to\infty} \left|\frac{a_{n+1}}{a_n}\right| = \lim_{n\to\infty} \frac{9n^2 + 6n + 1}{9n^2 + 18n + 9} = 1,$$

therefore the ratio test cannot give the result.

15.    a) We note that

$$\int_2^N \frac{1}{x(\ln x)^\alpha} dx = \int_{\ln 2}^{\ln N} \frac{1}{e^y y^\alpha} e^y dy$$

$$= \int_{\ln 2}^{\ln N} \frac{1}{y^\alpha} dy = \frac{1}{1-\alpha} y^{1-\alpha} \Big|_{\ln 2}^{\ln N},$$

therefore, for $\alpha > 1$ it follows that

$$\lim_{N\to\infty} \int_2^N \frac{1}{x(\ln x)^\alpha} dx = \frac{(\ln 2)^{1-\alpha}}{\alpha - 1}$$

and the series $\sum_{k=2}^\infty \frac{1}{k(\ln k)^\alpha}$ converges. On the other hand, for $\alpha < 1$, it follows that

$$\lim_{N\to\infty} \int_2^N \frac{1}{x(\ln x)^\alpha} dx = +\infty$$

implying the divergence of the corresponding series.
For $\alpha = 1$ we have to note in the above calculation that

$$\int_2^N \frac{1}{x \ln x} dx = \int_{\ln 2}^{\ln N} \frac{1}{y} dy = \ln N - \ln 2$$

which yields $\lim_{N\to\infty} \int_0^N \frac{1}{x \ln x} dx = \infty$ and again we get the divergence of the corresponding series. Also compare with Problem 4 b).

    b) Since

$$\int_1^N xe^{-x^2} dx = \int_1^N \left(-\frac{1}{2}\right) \frac{d}{dx} e^{-x^2} dx = -\frac{1}{2} e^{-x^2} \Big|_1^N$$

we find $\lim\limits_{N\to\infty} \int_1^N xe^{-x^2} dx = \dfrac{1}{2e}$ and hence the series $\sum_{l=1}^\infty le^{-l^2}$ converges.

    c) We have

$$\int_2^N \frac{\ln x}{x} dx = \ln(\ln x)\Big|_2^N = \ln(\ln N) - \ln(\ln 2)$$

implying $\lim\limits_{N\to\infty} \int_2^N \dfrac{\ln x}{x} dx = \infty$ and therefore the divergence of $\sum_{k=2}^\infty \frac{\ln k}{k}$.

d) Integration by parts yields

$$\int_2^N \frac{\ln x}{x^2} dx = \frac{1}{x}\ln x - \frac{1}{x}\Big|_2^N = -\frac{1+\ln x}{x}\Big|_2^N$$

and we conclude that

$$\lim_{n\to\infty} \int_2^N \frac{\ln x}{x^2} dx = \frac{1+\ln 2}{2}$$

and $\sum_{k=2}^{\infty} \frac{\ln k}{k^2}$ converges.

16. For our purpose we may assume that $a_n \neq 0$ for all $n \in \mathbb{N}$. Let $a_n^+ = \begin{cases} a_n, & a_n > 0 \\ 0, & a_n < 0 \end{cases}$

and $a_n^- = \begin{cases} -a_n, & a_n < 0 \\ 0, & a_n > 0 \end{cases}$ then $a_n = a_n^+ - a_n^-$. Since $\sum_{n=1}^{\infty} a_n$ converges the convergence of $\sum_{n=1}^{\infty} a_n^+$ or $\sum_{n=1}^{\infty} a_n^-$ implies the convergence of $\sum_{n=1}^{\infty} |a_n| = \sum_{n=1}^{\infty} (a_n^+ + a_n^-)$, hence both series $\sum_{n=1}^{\infty} a_n^+$ and $\sum_{n=1}^{\infty} a_n^-$ diverge, i.e.

$$\lim_{n\to\infty} \sum_{n=1}^N a_n^+ = \lim_{n\to\infty} \sum_{n=1}^{\infty} a_n^- = \infty.$$

However, since $\sum_{n=1}^{\infty} a_n$ converges it follows that $\lim_{n\to\infty} a_n = 0$ implying that $\lim_{n\to\infty} a_n^+ = 0$ and $\lim_{n\to\infty} a_n^- = 0$.

Given $A \in \mathbb{R}$ and denote by $(b_n)_{n\in\mathbb{N}}$ the subsequence of all positive elements of $(a_n)_{n\in\mathbb{N}}$ and by $(c_n)_{n\in\mathbb{N}}$ the subsequence of all negative elements of $(a_n)_{n\in\mathbb{N}}$. Choose $n_0$ to be the smallest index such that

$$\sum_{k=1}^{n_0} b_k > A,$$

next choose $n_1$ to be the smallest index such that

$$\sum_{k=1}^{n_0} b_k - \sum_{k=1}^{n_1} |c_k| < A$$

and continue to choose $n_2$ such that

$$\sum_{k=1}^{n_0} b_k - \sum_{k=1}^{n_1} |c_k| + \sum_{k=n_0+1}^{n_2} b_k > A,$$

and now continue with this process. We eventually obtain a series

$$(*) \quad b_1 + \cdots + b_{n_0} - |c_1| - \cdots - |c_{n_1}| + b_{n_0+1} + \cdots + b_{n_2} - |c_{n_1+1}| \cdots$$

which is a rearrangement of $\sum_{k=1}^{\infty} a_k$. Moreover

$$0 \le S - \left( \sum_{k=1}^{n_0} b_k - \sum_{k=1}^{n_1} |c_k| + \cdots - \sum_{k=n_{2l-1}+1}^{n_{2l+1}} |c_k| \right) < \sum_{k=n_{2l-1}+1}^{n_{2l+2}} b_k$$

and

$$0 \le \sum_{k=1}^{n_0} b_k - \sum_{k=1}^{n_1} |c_k| + \cdots + \sum_{k=n_{2l-2}+1}^{n_{2l}} b_k - S < \sum_{k=n_{2l-1}+1}^{n_{2l+1}} |c_k|.$$

Since $\lim_{k \to \infty} b_k = \lim_{k \to \infty} c_k = 0$, it follows that the rearranged series converges to $S$.

17. We could start by looking at $\frac{1}{7} = \sum_{n=-k}^{\infty} a_n b^{-n}$ and to try to find the numbers $k$ and $n$ for a given $b$. However there is a more systematic suggestion for any $b$ and $0 < x < 1$:
Clearly

$$x = \sum_{n=1}^{\infty} a_n b^{-n}, \quad 0 \le a_n < b, b \in \mathbb{N}, b \ge 2.$$

Then $a_1$ is the largest integer such that $a_1 b^{-1} \le x$. If $a_1, \ldots, a_{n-1}$ are already known, then $a_n$ is the largest integer such that

$$a_n b^{-n} \le x - (a_1 b^{-1} + \cdots + a_{n-1} b^{-n+1})$$

or

$$a_n \le x b^n - a_1 b^{n-1} - \cdots - a_{n-1} b.$$

We set

$$y_n := x b^n - a_1 b^{n-1} - \cdots - a_{n-1} b$$

and find the following algorithm:

$$\begin{aligned}
y_1 &:= xb \\
a_1 &:= [y_1] \\
y_{n+1} &:= (y_n - a_n)b \\
a_{n+1} &:= [y_{n+1}].
\end{aligned}$$

We now solve (i)-(iii).

(i) $b = 2$

| $n$ | 1 | 2 | 3 | 4 |
|-----|---|---|---|---|
| $y_n$ | $\frac{2}{7}$ | $\frac{4}{7}$ | $\frac{8}{7}$ | $\frac{2}{7}$ |
| $a_n$ | 0 | 0 | 1 | 0 |

653

We may stop now as the results start to repeat, therefore we have the following periodic expansion

$$\frac{1}{7} = 0.001001\ldots \qquad (b = 2)$$

(ii) The case $b = 7$ is trivial

$$\frac{1}{7} = 0.1 \qquad (b = 7)$$

(iii) $b = 10$

| $n$ | 1 | 2 | 3 | 4 | 5 | 6 | 7 |
|---|---|---|---|---|---|---|---|
| $y_n$ | $\frac{10}{7}$ | $\frac{30}{7}$ | $\frac{20}{7}$ | $\frac{60}{7}$ | $\frac{40}{7}$ | $\frac{50}{7}$ | $\frac{10}{7}$ |
| $a_n$ | 1 | 4 | 2 | 8 | 5 | 7 | 1 |

and again the result is the periodic expansion

$$\frac{1}{7} = 0.142857142857\ldots$$

It is straightforward to see that for every rational number $\frac{l}{k}, 0 < l < k$, and for every $b \geq 2$ the b-adic fraction representation of $\frac{l}{k}$ is periodic.
Using induction we can prove

$$y_n = \frac{p_n}{n}, \quad p_n \in \mathbb{N}_0, 0 \leq p_n < nb.$$

However there are only finitely many possibilities for $y_n$, thus there exists two numbers $r, s \in \mathbb{N}$ such that

$$y_{r+s} = y_r$$

implying $a_{n+s} = a_n$ for all $n \geq r$.

18. In the case where we have a bijective mapping $f : (a, b) \to (0, 1)$, then $(a, b)$ is not countable, hence $D$, a set containing $(a, b)$ is not countable. Thus we need to construct $f$. This is easily done, for example $f : (a, b) \to (0, 1), t \mapsto f(t), f(t) = \frac{t}{b-a} - \frac{a}{b-a}$ is strictly monotone since $f'(t) = \frac{1}{b-a} > 0, f(a) = 0, f(b) = 1$, hence $f$ maps $(a, b)$ bijectively to $(0, 1)$.

**Chapter 19**

1. By definition a set is closed if and only if it contains all of its accumulation points. We claim that $b$ is an accumulation point of $[a, b)$. Since $b \notin [a, b)$ this implies that $[a, b)$ is not closed. The sequence $(b_n)_{n \in \mathbb{N}}, b_n := b - \frac{b-a}{2n}$, consists of elements belonging to $[a, b)$ since $a < b - \frac{b-a}{2n} < b$, and $\lim_{n \to \infty} b_n = b$. Hence $b$ is an accumulation point of $[a, b)$ and therefore $[a, b)$ is not closed. In order for $[a, b)$ to be open the set $[a, b)^{\complement} = (-\infty, a) \cup [b, \infty)$ must be closed. However $a$ is an accumulation point of $[a, b)^{\complement}$ not belonging to the set, so $[a, b)^{\complement}$ is not closed and therefore $[a, b)$ is not open.

2. The set $\mathbb{Q} \subset \mathbb{R}$ cannot be closed since we know that for example $\sqrt{2} \in \mathbb{R}$ is an accumulation point of $\mathbb{Q}$ not belonging to $\mathbb{Q}$. However $\mathbb{Q} \subset \mathbb{R}$ cannot be open either. If it was open then $\mathbb{Q}^{\complement}$ must be closed. The sequence $\left( \frac{\sqrt{2}}{n} \right)_{n \in \mathbb{N}}$ is a sequence in $\mathbb{Q}^{\complement}$ with accumulation point $0 \in \mathbb{Q}$, so $\mathbb{Q}^{\complement}$ is not closed and consequently $\mathbb{Q}$ is not open in $\mathbb{R}$.

3. It follows that $\{a_\nu | \nu \in \mathbb{R}\}$ is the complement of the set $(-\infty, a_1) \cup (a_1, a_2) \cup (a_2, a_3) \cup \cdots$, i.e. we have

$$\{a_\nu | \nu \in \mathbb{R}\} = \left( \bigcup_{\nu=1}^{\infty} (a_{\nu-1}, a_\nu) \right)^{\complement}$$

with $a_0 := -\infty$. Since $\bigcup_{\nu=1}^{\infty} (a_{\nu-1}, a_\nu)$ is open as it is a union of open sets it follows that $\{a_\nu | \nu \in \mathbb{R}\}$ is closed.

4. We claim that with $B_\nu = \left[ \frac{1}{\nu+2}, 1 - \frac{1}{\nu+2} \right]$ we have

$$\bigcup_{\nu \in \mathbb{N}} B_\nu = (0, 1).$$

Clearly we have $B_\nu \subset (0, 1)$ for all $\nu \in \mathbb{N}$, which implies $\bigcup_{\nu \in \mathbb{N}} B_\nu \subset (0, 1)$. Next we prove that $(0, 1) \subset \bigcup_{\nu \in \mathbb{N}} B_\nu$. For $x \in (0, 1)$ we need to show the existence of $\nu_0 \in \mathbb{N}$ such that $x \in \left[ \frac{1}{\nu_0+2}, 1 - \frac{1}{\nu_0+2} \right]$. However $\lim_{\nu \to \infty} \frac{1}{\nu+2} = 0$ and $\lim_{\nu \to \infty} \left( 1 - \frac{1}{\nu+2} \right) = 1$, implying that for $\nu$ large enough $\frac{1}{\nu+2} < x < 1 - \frac{1}{\nu+2}$.

5. In general $\{a_\nu | \nu \in \mathbb{N}\}$ is not closed. A closed set must contain all of its accumulation points and if $a$ is not an element of $\{a_\nu | \nu \in \mathbb{N}\}$ then this set is not closed. However, for a converging sequence $(a_\nu)_{\nu \in \mathbb{N}}$ with limit $a$ the set $\{a_\nu | \nu \in \mathbb{N}\} \cup \{a\}$ is closed since it contains all of its accumulation points.

6. Since $A$ and $B$ are bounded with some $K_1, K_2 > 0$ we have $-K_1 \le a \le K_1$ for all $a \in A$ and $-K_2 \le b \le K_2$ for all $b \in B$. Consequently, for all $a \in A$ and $b \in B$ it follows that

$$-(K_1 + K_2) \le a + b \le (K_1 + K_2),$$

i.e. $A + B$ is bounded.

7.  a) First we prove that $(-3, 2) \cup (4, 6)$ is open. Since the open interval $(-3, 2)$ and $(4, 6)$ are open it follows that their union is open. Clearly $(-3, 2) \cup (4, 6) \subset M$. There are only three points, $\{4\}, \{6\}$ and $\{10\}$, not belonging to $(-3, 2) \cup (4, 6)$ but to $M$. For none of these points exists an open interval containing the point and belonging entirely to $M$. Hence $(-3, 2) \cup (4, 6)$ is the largest open set contained in $M$.

The set $[-3, 2] \cup [4, 6] \cup \{10\}$ is closed since it is a finite union of the closed sets $[-3, 2], [4, 6]$ and $\{10\}$, note that $\{10\} = ((-\infty, 10) \cup (10, \infty))^{\complement}$ and $(-\infty, 10)$ as well as $(10, \infty)$ are open. Clearly $M \subset [-3, 2] \cup [4, 6] \cup \{10\}$. There are only two points, $\{-3\}$ and $\{2\}$, belonging to $[-3, 2] \cup [4, 6] \cup \{10\}$ not belonging to $M$. However both

655

are accumulation points of $M : -3 = \lim_{n \to \infty} (-3 + \frac{1}{n})$ and $-3 + \frac{1}{n} \in M$ for all $n \in \mathbb{N}$, and $2 = \lim_{n \to \infty} (2 - \frac{1}{n})$ and $2 - \frac{1}{n} \in M$ for all $n \in \mathbb{N}$. Hence $[-3, 2] \cup [4, 6] \cup \{10\}$ is the largest closed set containing $M$.

b) Since $0 \in (-\frac{1}{n}, \frac{1}{n})$ for every $n \in \mathbb{N}$ it follows that $0 \in \bigcap_{n \in \mathbb{N}} (-\frac{1}{n}, \frac{1}{n})$, i.e. $\{0\} \subset \bigcap_{n \in \mathbb{N}} (-\frac{1}{n}, \frac{1}{n})$. Suppose $a \in \bigcap_{n \in \mathbb{N}} (-\frac{1}{n}, \frac{1}{n})$ then $a \in (-\frac{1}{n}, \frac{1}{n})$ for all $n \in \mathbb{N}$. If $a \neq 0$ then there exists $n \in \mathbb{N}$ such that $a < -\frac{1}{n}$ or $a > \frac{1}{n}$, implying $a \notin (-\frac{1}{n}, \frac{1}{n})$. Thus $\bigcap_{n \in \mathbb{N}} (-\frac{1}{n}, \frac{1}{n}) \subset \{0\}$ and together with the first part we have $\bigcap_{n \in \mathbb{N}} (-\frac{1}{n}, \frac{1}{n}) = \{0\}$. Each of the sets $(-\frac{1}{n}, \frac{1}{n})$ is open as it is an open interval. However $\{0\}$ is not open since it does not contain an entire open interval $(-\varepsilon, \varepsilon), \varepsilon > 0$.

8.    a) We claim $G = \{y \in \mathbb{R} | y = \frac{1}{x}, x \geq \frac{1}{2}\} = (0, 2]$. Indeed, $z \in G$ implies $z = \frac{1}{x}$ for some $x \geq \frac{1}{2}$. On $[\frac{1}{2}, \infty)$ the function $x \mapsto \frac{1}{x}$ is strictly decreasing, strictly positive and tends to 0 for $x$ tending to $\infty$, hence $G = (0, 2]$, $\inf G = 0$ and $\sup G = 2$. Since $2 \in G$ we have $2 = \max G (= \sup G)$, but $0 \notin G$ and therefore $G$ has no minimum.

b) Consider the sequence

$$a_n := \begin{cases} 3, & n = 1 \\ 0, & n = 2 \\ \frac{5}{2} - \frac{1}{n}, & n = 3k \\ \frac{n}{n+1}, & n = 3k + 1 \\ \frac{1}{2} + \frac{1}{n}, & n = 3k + 2 \end{cases}$$

It holds

$$\lim_{k \to \infty} a_{3k} = \lim_{k \to \infty} \left( \frac{5}{2} - \frac{1}{3k} \right) = \frac{5}{2}$$

$$\lim_{k \to \infty} a_{3k+1} = \lim_{k \to \infty} \left( \frac{3k+1}{3k+2} \right) = 1$$

$$\lim_{k \to \infty} a_{3k+2} = \lim_{k \to \infty} \left( \frac{1}{2} + \frac{1}{3k+2} \right) = \frac{1}{2}$$

and these are obviously all convergent subsequence of $(a_n)_{n \in \mathbb{N}}$. Hence $(a_n)_{n \in \mathbb{N}}$ has 3 accumulation points. Further

$$0 \le \frac{3}{2} \le \frac{5}{2} - \frac{1}{n} \le \frac{5}{2} \le 3,$$

$$0 \le \frac{n}{n+1} \le 1 \le 3$$

and

$$0 \le \frac{1}{2} \le \frac{1}{2} + \frac{1}{n} \le \frac{3}{2} \le 3$$

implying that $\sup\{a_n | n \in \mathbb{N}\} = a_1 = 3$ and $\inf\{a_n | n \in \mathbb{N}\} = a_2 = 0$. Note that in our case the supremum is a maximum and the infimum is a minimum.

Of course we expect

$$\limsup_{n \to \infty} a_n = \lim_{k \to \infty} a_{3k} = \frac{5}{2}$$

and

$$\liminf_{n \to \infty} a_n = \lim_{k \to \infty} a_{3k+2} = \frac{1}{2}.$$

Here comes the proof:

For $n \ge 3$:

$$\sup\{a_\ell | \ell \ge h\} = \begin{cases} \frac{5}{2} - \frac{1}{n}, & n = 3k \\ \frac{5}{2} - \frac{1}{n+1}, & n = 3k+1 \\ \frac{5}{2} - \frac{1}{n+2}, & n = 3k+1 \end{cases}$$

which implies

$$\lim_{n \to \infty} (\sup\{a_\ell | \ell \ge n\}) = \frac{5}{2},$$

i.e.

$$\limsup_{n \to \infty} a_n = \frac{5}{2}.$$

Moreover we have for $n \ge 3$

$$\inf\{a_\ell | \ell \ge n\} = \begin{cases} \frac{1}{2} + \frac{1}{n+2}, & n = 3k \\ \frac{1}{2} + \frac{1}{n+1}, & n = 3k+1 \\ \frac{1}{2} + \frac{1}{n}, & n = 3k+2 \end{cases}$$

implying

$$\lim_{n\to\infty} (\inf\{a_\ell | \ell \geq n\}) = \frac{1}{2}$$

or

$$\liminf_{n\to\infty} a_n = \frac{1}{2}.$$

9. Consider the following table:

| | $\sup\{a_n | n \in \mathbb{N}\}$ | $\inf\{a_n | n \in \mathbb{N}\}$ | $\limsup_{n\to\infty} a_n$ | $\liminf_{n\to\infty} |a_n$ |
|---|---|---|---|---|
| $a_n = 2 - \frac{n-1}{10}$ | $2$ | $-\infty$ | $-\infty$ | $-\infty$ |
| $a_n = \frac{(-1)^{n-1}}{n+1}$ | $\frac{1}{2}$ | $-\frac{1}{3}$ | $0$ | $0$ |
| $a_n = \frac{2}{3}\left(1 - \frac{1}{10^n}\right)$ | $\frac{2}{3}$ | $\frac{3}{5}$ | $\frac{2}{3}$ | $\frac{2}{3}$ |

Since each sequence converges or diverges to $-\infty$, the lim inf and lim sup is in each case the limit.

10. Once Corollary 19.23 is at our disposal, the problem is trivial. Here we provide a solution using the same idea as in the first part of Example 19.17. Clearly $a$ is an upper bound of $\{a_n | n \in \mathbb{N}\}$ and therefore $\sup\{a_n | n \in \mathbb{N}\} \leq a$. Given $\epsilon > 0$ we can find $N(\epsilon)$ such that $a - a_n < \epsilon$ for all $n \geq N(\epsilon)$, note that since $a_n \leq a$ we do not need to use the absolute value in this estimate. Thus for $n \geq N(\epsilon)$ we have $a - \epsilon < a_n$, implying that $a - \epsilon$ cannot be an upper bound.

11. We may assume that $a$ is finite, for $a = +\infty$ the statement is trivial. Suppose that for some $\epsilon > 0$ there exists infinitely many $a_{n_l}, l \in \mathbb{N}$, such that $a_{n_l} \geq a + \epsilon$. Then all accumulation points of the sequence $(a_{n_l})_{l\in\mathbb{N}}$ are greater or equal to $a + \epsilon > a$. Hence $(a_n)_{n\in\mathbb{N}}$ has a subsequence converging to a point larger than its limit superior which is of course a contradiction.

12. Let $a = \limsup_{n\to\infty} a_n$ and $b = \limsup_{n\to\infty} b_n$. For $\epsilon > 0$ we have $a_n < a + \epsilon$ and $b_n < b + \epsilon$ for all but finitely many $n \in \mathbb{N}$. This implies $\lambda a_n < \lambda a + \lambda \epsilon$ for all but finitely many $n \in \mathbb{N}$ and $a_n + b_n < a + b + 2\epsilon$ for all but finitely many $n \in \mathbb{N}$, which gives a) and b) respectively.

Now we apply b) to the sequences $(a_n + b_n)_{n\in\mathbb{N}}$ and $(b_n)_{n\in\mathbb{N}}$ to find with (19.20)

$$\limsup_{n\to\infty} a_n = \limsup_{n\to\infty}(a_n + b_n - b_n)$$
$$\leq \limsup_{n\to\infty}(a_n + b_n) + \limsup_{n\to\infty}(-b_n)$$
$$= \limsup_{n\to\infty}(a_n + b_n) - \liminf_{n\to\infty} b_n,$$

which yields

$$\limsup_{n\to\infty}(a_n + b_n) \geq \limsup_{n\to\infty} a_n + \liminf_{n\to\infty} b_n,$$

proving c). Now d) follows since

$$\limsup_{n\to\infty} b_n = \liminf_{n\to\infty} b_n = \lim_{n\to\infty} b_n$$

and combining b) and c) we find

$$\limsup_{n\to\infty} a_n + \lim_{n\to\infty} b_n \le \limsup_{n\to\infty}(a_n + b_n) \le \limsup_{n\to\infty} a_n + \lim_{n\to\infty} b_n.$$

13. We need to find two non-empty open sets $O_1$ and $O_2$ such that $O_1 \cap O_2 = \emptyset$ and $A \subset O_1 \cup O_2$. The two open intervals $\left(-\frac{1}{2}, \frac{3}{2}\right)$ and $\left(\frac{7}{4}, 5\right)$ will suffice. Clearly we have $\left(-\frac{1}{2}, \frac{3}{2}\right) \cap \left(\frac{7}{4}, 5\right) = \emptyset$ and further we find $[0, 1] \subset \left(-\frac{1}{2}, \frac{3}{2}\right)$ and $\{2\} \cup (3, 4) \subset \left(\frac{7}{4}, 5\right)$ implying $[0, 1] \cup \{2\} \cup (3, 4) \subset \left(-\frac{1}{2}, \frac{3}{2}\right) \cup \left(\frac{7}{4}, 5\right)$.

## Chapter 20

1. This is merely a reformulation, but a helpful one, of Theorem 20.2.(ii) by replacing $f(x)$ by $f(\lim_{n\to\infty} x_n)$. Let us add a remark. It is important that

$$\lim_{n\to\infty} f(x_n) = f(\lim_{n\to\infty} x_n)$$

holds for all sequences converging to $x \in [a, b], x_n \in [a, b]$. Consider the function

$$g(x) = \begin{cases} 1, & x \ge 0 \\ -1, & x < 0. \end{cases}$$

The graph of $g$ is given in the figure below.

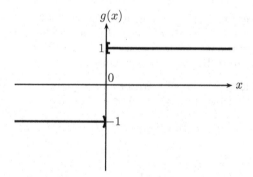

For the sequence $\left(\frac{1}{2n}\right)_{n\in\mathbb{N}}$ it holds that $\lim_{n\to\infty} \frac{1}{2n} = 0$ and $\lim_{n\to\infty} g\left(\frac{1}{2n}\right) = 1 = g(0)$. For the sequence $\left(\frac{-1}{2n+1}\right)_{n\in(N)}$ it holds that $\lim_{n\to\infty} g\left(\frac{-1}{2n+1}\right) = -1 \ne g(0)$. Obviously, $g$ is discontinuous.

659

2. Suppose that $f$ is continuous and $V \subset \mathbb{R}$ is open. We have to prove that $f^{-1}(V)$ is open. Take $y_0 = f(x_0) \in V$. Since $V$ is open there exists $\epsilon > 0$ such that $(y_0 - \epsilon, y_0 + \epsilon) \subset V$. By continuity of $f$, we can find $\delta > 0$ such that $(x_0 - \delta, x_0 + \delta) \subset D$ and $|f(x) - f(x_0)| < \epsilon$, thus $(x_0 - \delta, x_0 + \delta) \subset f^{-1}(V)$ proving that $f^{-1}(V)$ is open.

Conversely, suppose the pre-image $f^{-1}(V)$ of every open set $V \subset \mathbb{R}$ is open. Take $x_0 \in D$ and set $y_0 = f(x_0)$. The interval $(y_0 - \epsilon, y_0 + \delta) \subset \mathbb{R}$ is open and consequently $f^{-1}((y_0 - \epsilon, y_0 + \epsilon))$ is open too and $x_0 \in f^{-1}((y_0 - \epsilon, y_0 + \epsilon))$. Thus there exists $\delta > 0$ such that $(x_0 - \delta, x_0 + \delta) \subset f^{-1}((y_0 - \epsilon, y_0 + \epsilon))$ and $f((x_0 - \delta, x_0 + \delta)) \subset (y_0 - \epsilon, y_0 + \epsilon)$. In other words: For $x_0 \in D$, given $\epsilon > 0$ exists $\delta > 0$ such that $|x - x_0| < \delta, x \in D$, implies $|f(x) - f(x_0)| < \epsilon$, i.e. $f$ is continuous at all $x_0 \in D$.

3. The function $f : D \mapsto \mathbb{R}$ has at $x_0$ the limit $a$ from the right if $x \in D, 0 < |x - x_0| < \delta$ and $x > x_0$ implies $|f(x) - a| < \epsilon$, i.e. $0 < x - x_0 < \delta$ implies $|f(x) - a| < \epsilon$.

Analogously we find that $f : D \mapsto \mathbb{R}$ has at $x_0$ the limit $b$ from the left if for every $\epsilon > 0$ exists $\delta > 0$ such that $0 < x_0 - x < \delta$ implies $|f(x) - b| < \epsilon$.

(Note that we have taken for granted the assumption that there exists a sequence $(x_k)_{k \in \mathbb{N}}, x_k \in D, x_k \neq x_0$, converging to $x_0$.)

4.  a) By Theorem 18.30 we know that every real number can be approximated by rational numbers and by Theorem 18.35 the real numbers are not countable. Thus, given $x \in [0, 1]$ we can find a sequence of rational numbers $(q_n)_{n \in \mathbb{N}}, q_n \in [0, 1]$, converging to $x$, and we can find a sequence of irrational numbers $(r_n)_{n \in \mathbb{N}}, r_n \in [0, 1]$, converging to $x$ too. However

$$\lim_{n \to \infty} \chi_{[0,1] \cap \mathbb{Q}}(q_n) = 1 \neq 0 = \lim_{n \to \infty} \chi_{[0,1] \cap \mathbb{Q}}(r_n),$$

implying that $X_{[0,1] \cap \mathbb{Q}}$ is at all $x \in [0, 1]$ discontinuous.

b) For $x \neq 0$ we can argue as in part a). However for $x = 0$ we find that every sequence, whether consisting only of rational points, only of irrational points, or both rational and irrational points converging to 0, will be mapped by $f$ onto a sequence converging to 0. Hence $f$ is continuous at 0.

5. Let $M$ be a bound for $g$, i.e. $|g(x)| \leq M$ for all $x \in [0, 1]$. Given $\epsilon > 0$, take $\delta = \frac{\epsilon}{M}$ to find for $x \in [0, 1]$ such that $|x| = |x - 0| < \delta = \frac{\epsilon}{M}$ that

$$|f(x) - f(0)| = |xg(x)| \leq M|x| < \delta M = \epsilon,$$

i.e. $f$ is continuous at 0.

6.  a) Suppose that $f$ is increasing, the decreasing case goes analogously. Since $x_\nu > x_0$ and $\lim_{\nu \to \infty} x_\nu = x_0$ it follows that $x_0 < x_\nu \leq x_N, x_N := \max\{x_\nu | \nu \in \mathbb{N}\}$. Consequently $(f(x_\nu))_{\nu \in \mathbb{N}}$ is bounded from below by $f(x_0)$ and from above by $f(x_N)$. By the Bolzano-Weierstrass theorem $(f(x_\nu))_{\nu \in \mathbb{N}}$ has at least one converging subsequence.

Let $\left(f(x_{\nu_l^1})\right)_{l \in \mathbb{N}}$ and $\left(f(x_{\nu_k^2})\right)_{k \in \mathbb{N}}$ be two converging subsequences. We want to show that they have the same limit. For $\nu_l^1$ there exists $\nu_{k(l)}^2$ such that $x_{\nu_l^1} \geq x_{\nu_{k(l)}^2}$ and for $\nu_{k(l)}^2$ exists $\nu_{m(k)}^1$ such that $x_{\nu_{k(l)}^2} \geq x_{\nu_{m(k)}^1}$ implying that

$$f(x_{\nu_l^1}) \geq f(x_{\nu_{k(l)}^2}) \geq f(x_{\nu_{m(k)}^1})$$

and therefore $\lim_{l \to \infty} f(x_{\nu_l^1}) = \lim_{k \to \infty} f(x_{\nu_k^2})$. Thus all subsequences of $(f(x_\nu))_{\nu \in \mathbb{N}}$ converge to the same limit.

b) We suppose again that $f$ is monotone increasing. Let $x_0 \in I$ and $(x_\nu)_{\nu \in \mathbb{N}}$, $x_\nu \in I$, $x_\nu > x_0$, be a sequence converging from the right to $x_0$ and $(y_\nu)_{\nu \in \mathbb{N}}, y_\nu \in I, y_\nu < x_0$, be a sequence converging from the left to $x_0$. Part a) implies that $(f(x_\nu))_{\nu \in \mathbb{N}}$ has a limit $f(x_0+)$ and $(f(y_\nu))_{\nu \in \mathbb{N}}$ has a limit $f(x_0-)$. By monotonicity, if $f(x_0+) = f(x_0-)$ then both must coincide with $f(x_0)$. Denote by $D(f, I)$ the set of all points of discontinuity which $f$ has in $I$. It follows that

$$D(f, I) = \{x \in I | f(x-) < f(x+)\}.$$

For every $x \in D(f, I)$ exists a rational number $r(x)$ such that $f(x-) < r(x) < f(x+)$. The mapping $x \to r(x)$ maps $D(f, I)$ injectively to $\mathbb{Q}$, hence $D(f, I)$ must be denumerable.

7. Given $f : I \to \mathbb{R}$ a monotone function and denote by $D(f, I)$ the denumerable set of its discontinuities, which we write as a monotone sequence $x_1 < x_2 < x_3 \cdots$ Now we define $h : I \to \mathbb{R}$ as follows

$$h(x) = \begin{cases} f(x), & x \in I \backslash D(f, I) \\ f(x_k+), & x_k \in D(f, I) \end{cases}$$

Clearly $f \big|_{I \backslash D(f,I)} = h \big|_{I \backslash D(f,I)}$, so $f$ and $h$ coincide outside a countable set and further $\lim_{x \to x_k, x < x_k} h(x) = f(x_k-)$ exists. Finally we have

$$\lim_{x \to x_k, x > x_k} h(x) = \lim_{x \to x_k, x > x_k} f(x) = f(x_k+) = h(x),$$

hence $f$ is continuous from the right.

Note that càdlàg function are most important when investigating certain stochastic procedures, e.g. Lévy processes or more generally Feller processes.

8.     a) Since

$$\varphi(x) = \frac{1}{2}(f(x) + g(x) + |f(x) - g(x)|)$$

and

$$\psi(x) = \frac{1}{2}(f(x) + g(x) - |f(x) - g(x)|)$$

the result follows immediately since in both cases on the right hand side we have continuous functions. (Recall $f \pm g$ are continuous as is $|h|$ for $h$ continuous).

b) Clearly

$$f_+(x) = \max(f(x), 0)$$

and
$$f_-(x) = -\min(f(x), 0)$$

This implies immediately with part a) that for a continuous function $f$ the function $f_+$ and $f_-$ are continuous. Now

$$f_+(x) - f_-(x) = \max(f(x), 0) + \min(f(x), 0)$$

if $f(x) \geq 0$ then $f_+(x) = f(x)$ and $f_-(x) = 0$, and if $f(x) \leq 0$ then $-f_-(x) = f(x) = \min(f(x), 0)$, and $f_+(x) = 0$. Thus in each case we get

$$f_+(x) - f_-(x) = f(x).$$

This decomposition also implies that if $f_+$ and $f_-$ are continuous then $f$ is continuous.

Finally we note

$$|f(x)| = \begin{cases} f(x), & f(x) \geq 0 \\ -f(x), & f(x) \leq 0 \end{cases} = \begin{cases} f_+(x), f(x) \geq 0 \\ f_-(x), f(x) \leq 0 \end{cases}$$

but $f_+(x) = 0$ if $f_-(x) \neq 0$ and $f_-(x) = 0$ if $f_+(x) \neq 0$ and therefore

$$|f(x)| = f_+(x) + f_-(x).$$

Note that both the positive and the negative part of a function are non-negative functions.

9. Let $x \in [a, b]$. There exists a sequence of rational numbers $(q_\nu)_{\nu \in \mathbb{N}}, q_\nu \in (a, b)$, converging to $x$, i.e. $\lim_{\nu \to \infty} q_\nu = x$. Consequently, we have $f(q_\nu) = g(q_\nu)$ and hence, by continuity of $f$ and $g$

$$f(x) = \lim_{\nu \to \infty} f(q_\nu) = \lim_{\nu \to \infty} g(q_\nu) = g(x).$$

10. The even extension of $f$ is given by $f_e : [-a, a] \to \mathbb{R}$, where

$$f_e(x) = \begin{cases} f(x), & x \in [0, a] \\ f(-x), & x \in [-a, 0]. \end{cases}$$

If $x > 0$ then $f_e(x) = f(x)$ and $f_e$ is continuous at $x$. If $x < 0$ then $f_e(x) = (f_0(-id))(x)$ and the continuity follows. Since $\lim_{x \to 0, x > 0} f_e(x) = f(0) = \lim_{x \to 0, x < 0} f_e(x)$, the continuity is also proven for $x = 0$.

Now the odd extension of $f - f(0)$ is given by $g : [-a, a] \to \mathbb{R}$

$$g(x) := \begin{cases} f(x) - f(0), & x \in [0, a] \\ -f(-x) + f(0), & x \in [-a, 0] \end{cases}$$

662

and the continuity of $g$ follows with arguments similar to those given above. For $x \in (0, a]$ the function $x \mapsto g(x) = f(x) - f(0)$ is continuous since $f$ is. For $x \in [-a, 0)$ the function $x \mapsto g(x) = -f(-x) + f(0)$ is continuous as composition and sum of continuous functions.

For $x = 0$ we find

$$\lim_{x \to 0, x>0} g(x) = \lim_{x \to 0, x>0} (f(x) - f(0)) = 0 = \lim_{x \to 0, x<0} (-f(-x) + f(0)) = \lim_{x \to 0, x<0} g(x)$$

hence $g$ is also continuous at $x = 0$.

11.     a) For $f, g : D \to \mathbb{R}$ continuous and $\lambda \in \mathbb{R}$ we know that $f + g$ and $\lambda f$ as well as $f \cdot g$ are continuous, so $C(D)$ is an $\mathbb{R}$-algebra.

b) for $f, g \in C(D)$ and $\lambda, \mu \in \mathbb{R}$ we find

$$A_{op}(\lambda f + \mu g)(x) = a(x)(\lambda f + \mu g)(x)$$

$$= \lambda a(x) f(x) + \mu a(x) g(x)$$

$$= (\lambda A_{op} f)(x) + (\mu A_{op} g)(x)$$

$$= (\lambda A_{op} f + \mu A_{op} g)(x)$$

so $A_{op}(\lambda f + \mu g) = \lambda A_{op} f + \mu A_{op} g$ proving the linearity of $A_{op} : C(D) \to C(D)$.

12.     a) Consider $h : D \to \mathbb{R}, h(D) \subset D$, and let $x_0 \in D$ be a fixed point of $h$, i.e. $h(x_0) = x_0$. In this case, the graph of $h$ must intersect the line $g(x) = x$ at $x_0$, see

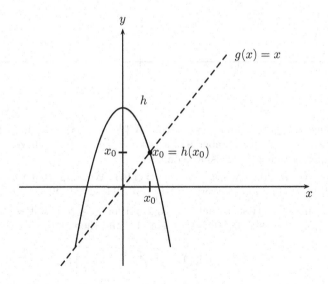

b) This follows from a simple algebraic consideration:

$$g(x_0) = h(x_0) + x_0 - a$$

implies $h(x_0) - a = 0$. Conversely $h(x_0) = a$ implies

$$g(x_0) = h(x_0) + x_0 - a = x_0.$$

13.    a) We apply Theorem 20.14 to the function $g - f$. For $h(x) := g(x) - f(x)$ defined on $[a, b]$ we have $h(a) = g(a) - f(a) > 0$ and $h(b) = g(b) - f(b) < 0$ implying that for some $x_0 \in (a, b)$ it holds that $h(x_0) = 0$ or $g(x_0) = f(x_0)$.

b) We apply part a) to $g(x) = \sin x$ and $f(x) = \frac{1}{2+\cos^4 x}$ defined on $[\frac{\pi}{2}, \frac{3\pi}{2}]$. Since $\sin \frac{\pi}{2} = 1$ and $\cos \frac{\pi}{2} = 0$ we find $g(\frac{\pi}{2}) = 1 > \frac{1}{2} = f(\frac{\pi}{2})$. But for $x = \frac{3\pi}{2}$ we have $\sin \frac{3\pi}{2} = -1$ and $f(\frac{3\pi}{2}) > 0$, so $g(\frac{3\pi}{2}) < f(\frac{3\pi}{2})$ Hence there exists at least one $\xi \in (\frac{\pi}{2}, \frac{3\pi}{2})$ with $\sin \xi = \frac{1}{2+\cos^4 \xi}$

14.    a) Below is a picture of $A_n = \left(\frac{1}{2}\left(\frac{1}{n} + \frac{1}{n+1}\right), \frac{1}{2}\left(\frac{1}{n} - \frac{1}{n-1}\right)\right)$

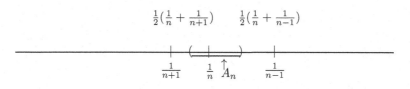

This gives

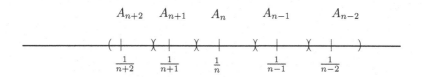

Hence $(A_n)_{n\in\mathbb{N}}$ is an open covering of $\{\frac{1}{n} \mid n \in \mathbb{N}\}$ and $A_n \cap A_m = \emptyset$ for $n \neq m$. Therefore a finite number of the sets, $A_n, n \in \mathbb{N}$, can never cover $\{\frac{1}{n} \mid n \in \mathbb{N}\}$ implying that this set is not compact.

Now we consider the set $\{\frac{1}{n} \mid n \in \mathbb{N}\} \cup \{0\}$. We claim that this set is compact. Let $(A_\nu)_{\nu\in I}$ be an open covering of $\{\frac{1}{n} \mid n \in \mathbb{N}\} \cup \{0\}$. Since $\lim_{n\to\infty} \frac{1}{n} = 0$ there exists $A_{\nu_0}$ and $N \in \mathbb{N}$ such that $0 \in A_{\nu_0}$ and for $k > N$ it follows that $x_k \in A_{\nu_0}$. For $k \leq N - 1$ exists $A_{\nu_k}$ such that $\frac{1}{k} \in A_{\nu_k}$ and therefore

$$A_{\nu_0} \cup A_{\nu_1} \cup \cdots \cup A_{\nu_{N-1}} \supset \{\frac{1}{n} \mid n \in \mathbb{N}\} \cup \{0\},$$

i.e. we have a finite subcovering.

b) We can use the idea developed when proving that $B$ in part a) is compact. Let $(U_j)_{j \in I}$ be an open covering of $C := \{a_k | k \in \mathbb{N}_0\}$. Since $a_0 \in C$ there exists $U_{j_0}$ such that $a_0 \in U_{j_0}$. Further, since $U_{j_0}$ is open there exists an $\epsilon > 0$ such that $(a_0 - \epsilon, a_0 + \epsilon) \subset U_{j_0}$. Now, $\lim_{k \to \infty} a_k = a_0$ implies the existence of $N = N(\epsilon)$ such that $k \geq N$ implies $a_k \in (a_0 - \epsilon, a_0 + \epsilon) \subset U_{j_0}$, note that $a_k \in (a_0 - \epsilon, a_0 + \epsilon)$ is equivalent to $|a_k - a_0| < \epsilon$. For $a_1, \cdots, a_N$ we can find $U_{j_1}, \cdots, U_{j_N}$ such that $a_l \in U_{j_l}$ and consequently $C \subset U_{j_0} \cup U_{j_n} \cup \cdots \cup U_{j_N}$, i.e. we have constructed a finite subcovering of $C$.

15. First we sketch the situation. The set $U_x, x \in [0,1]$, is an open interval with mid point $x$ and of length $\frac{3N}{2}$. See:

The points $0, \frac{1}{N}, \cdots, \frac{N-1}{N}, 1$ give a partition of $(0,1)$ and the distance of two neighbouring points is $\frac{1}{N}$. Therefore, for $\frac{k}{N}$ and $\frac{k+1}{N}$ we find

$$U_{\frac{k}{N}} \cap U_{\frac{k+1}{N}} = \left( \frac{4k+1}{4N}, \frac{4k+3}{4N} \right) \neq \emptyset$$

and

$$(0,1) \subset \bigcup_{k=0}^{N} U_{\frac{k}{N}} = \left( \frac{-3}{4N}, \frac{4N+3}{4N} \right).$$

Thus, $(U_{\frac{k}{N}})_{k=0,\cdots,N}$ is indeed a finite subcovering of $(0,1)$. However, since $(0,1)$ is open it cannot be compact. Finding a finite subcovering for a special open covering is of course not sufficient for compactness.

16. First we note that $\bigcap_{\nu \in I} K_\nu \subset K_{j_0}$ for every $j_0 \in I$. Now, $K_\nu \subset \mathbb{R}$ is compact, hence closed and bounded implying immediately that $\bigcap_{\nu \in I} K_\nu$ is bounded. Further we know that the intersection of an arbitrary family of closed sets is closed, hence $\bigcap_{\nu \in I} K_\nu$ is closed and bounded and therefore compact.

The sets $[-\nu, \nu], \nu \in \mathbb{N}$, are compact but the set $\bigcup_{n \in \mathbb{N}} [-\nu, \nu] = \mathbb{R}$ is not.

17. a) For $x \in K$ exists $\delta_x > 0$ such that for $y \in K$ and $|x - y| < \delta_x$ it follows that $|f(y) - f(x)| < \frac{f(x)}{4}$. The family of intervals $(x - \delta_x, x + \delta_x), x \in K$, forms an open

covering of K and therefore, by compactness, we can find points $x_1, \cdots, x_N \in K$ such that $(x_j - \delta_{x_j}, x_j + \delta_{x_j})_{j=1,\cdots,N}$ forms a finite subcovering of K. On $((x_j - \delta_{x_j}, x_j + \delta_{x_j})$ it holds that $|f(y) - f(x_j)| < \frac{f(x_j)}{4}$ or $-\frac{f(x_j)}{4} < f(y) - f(x_j) < \frac{f(x_j)}{4}$, implying $0 < \frac{3f(x_j)}{4} < f(y)$ which yields

$$0 < \min_{1 \leq j \leq N} \frac{3f(x_j)}{4} < f(y).$$

b) Since $f$ is uniformly continuous on $D$, given $\epsilon = 1$ there exists $\delta > 0$ such that $|x - y| < \delta$ implies $|f(x) - f(y)| < 1$. Since $D$ is bounded we can cover $D$ with a finite number $N$ of intervals of length $2\delta, \delta > 0$, with midpoints $x_j, j = 1, \cdots, N$, belonging to $D$. On $(x_j - \delta, x_j + \delta)$ we have $|f(x) - f(x_j)| < 1$, or $|f(x)| < 1 + |f(x_j)|$ implying $|f(x)| < 1 + \max_{1 \leq j \leq N} |f(x_j)|$ for all $x \in D$, i.e. $f$ is bounded.

18. First we note that $f \big|_{[-a, -a+1]}$ is uniformly continuous as a continuous function on a compact set. Thus, given $\epsilon > 0$ there exists $\delta_1 > 0$ such that $x, y \in [-a, -a + 1]$ and $|x - y| < \delta_1$ it follows that $|\sqrt{x + a} - \sqrt{y + a}| < \epsilon$.

Next we observe that if either $x \geq -a + 1$ or $y \geq -a + 1$ then

$$|\sqrt{x + a} - \sqrt{y + a}| \leq |\sqrt{x + a} + \sqrt{y + a}||\sqrt{x + a} - \sqrt{y + a}| = |x - y|.$$

Thus, given $\epsilon > 0$ choose $\delta = \min(\delta_1, \epsilon)$ to find for all $x, y \in [-a, \infty)$ that $|x - y| < \delta$ implies $|\sqrt{x + a} - \sqrt{y + a}| < \epsilon$, i.e. $f$ is uniformly continuous on $[-a, \infty)$.

19. By uniform continuity of $g$, given $\epsilon > 0$ there exists $\delta > 0$ such that $|f(x) - f(y)| \leq \epsilon$ for all $x, y \in [a, b]$ such that $|x - y| < \delta$. Now let $a = x_0 < x_1 < \cdots < x_n = b$ be a partition of $[a, b]$ such that $|x_k - x_{k-1}| \leq \delta$ for $k = 1, \cdots, n$. We define $\varphi : [a, b] \to \mathbb{R}$ as follows: on $[x_k, x_{k-1}]$ we set

$$\varphi \big|_{[x_{k-1}, x_k]}(x) = \frac{f(x_k) - f(x_{k-1})}{x_k - x_{k-1}} x + \frac{f(x_{k-1})x_k - f(x_k)x_{k-1}}{x_k - x_{k-1}},$$

i.e. the graph of $\varphi \big|_{[x_{k-1}, x_k]}$ is the line segment connecting $(x_{k-1}, f(x_{k-1}))$ with $(x_k, f(x_k))$. Clearly, $\varphi$ is piecewise linear. By assumption, we have for $x, y \in [x_{k-1}, x_k]$ that $|f(x) - f(y)| \leq \epsilon$ or $f([x_{k-1}, x_k]) \subset [\gamma_k - \epsilon, \gamma_k]$ where $\gamma_k := \sup\{f(x)|x \in [x_{k-1}, x_k]\}$, but we have also, by construction, that $|f(x) - \varphi(x)| \leq \epsilon$ for all $x \in [a, b]$.

20. Given $\varepsilon > 0$ we take $\delta = \frac{\varepsilon}{\kappa}$ to find for all $x, y \in D$ with $|x - y| < \delta$ that

$$|f(x) - f(y)| \leq \kappa |x - y| \leq \kappa \delta = \kappa \frac{\varepsilon}{\kappa} = \varepsilon$$

which implies the uniform continuity of $f$.

21.     a) By definition $f : D \to \mathbb{R}$ is uniformly continuous if for every $\epsilon > 0$ and all $x, y \in D$ there exists $\delta > 0$ such that $|x - y| < \delta$ implies $|f(x) - f(y)| < \epsilon$. Thus if

we restrict $x, y$ to $D\prime \subset D$, given $\epsilon > 0$ we may still work with the same $\delta > 0$ to get the uniform continuity of $f|_{D\prime}$.

b) Since $\lim\limits_{x \to a, x > a} g(x) = A$ exists, we can define $\tilde{g} : [a, b] \to \mathbb{R}$ by

$$\tilde{g} := \begin{cases} g(x), & x \in (a, b] \\ A, & x = a. \end{cases}$$

By construction, $\tilde{g}$ is continuous on the compact interval $[a, b]$, hence $\tilde{g}$ is uniformly continuous. Now, the result follows from part a).

## Chapter 21

1. If $S_{x_0, x}(t) = at + b$ then we must have $S_{x_0, x}(x_0) = f(x_0)$ and $S_{x_0, x}(x) = f(x)$ which yields $f(x_0) = ax_0 + b$ and $f(x) = ax + b$, or

$$a = \frac{f(x) - f(x_0)}{x - x_0}$$

and

$$b = \frac{f(x_0)x - f(x)x_0}{x - x_0},$$

i.e.

$$S_{x_0, x}(t) = \frac{f(x) - f(x_0)}{x - x_0}t + \frac{f(x_0)x - f(x)x_0}{x - x_0}.$$

The tangent line through $(x_0, f(x_0))$ is given by $g_{x_0}(t) = \alpha t + \beta$ with $g_{x_0}(x_0) = f(x_0)$ and $g'_{x_0}(x_0) = f'(x_0)$. This implies $g_{x_0}(t) = f'(x_0)t + f(x_0) - f'(x_0)x_0 = f'(x_0)(t - x_0) + f(x_0)$. Now we find

$$S_{x_0, x}(t) - g_{x_0}(t) = \frac{f(x) - f(x_0)}{x - x_0}t + \frac{f(x_0)x - f(x)x_0}{x - x_0} - f'(x_0)t - f(x) + f'(x_0)x_0$$

$$= \left( \frac{f(x) - f(x_0)}{x - x_0} - f'(x_0) \right) t + x_0 \left( f'(x_0) - \frac{f(x) - f(x_0)}{x - x_0} \right),$$

which implies for $t \in \mathbb{R}$ fixed that

$$\lim_{x \to x_0} (S_{x_0, x}(t) - g_{x_0}(t)) = 0$$

or $\lim\limits_{x \to x_0} S_{x_0, x}(t) = g_{x_0}(t)$.

2. We use mathematical induction. For $k = 1$ the statement is just the well known Leibniz's rule

$$(f \cdot g)'(x) = f'(x)g(x) + f(x)g'(x)$$

Now suppose that

$$\frac{d^k}{dx^k}(f \cdot g)(x) = \sum_{l=0}^{k} \binom{k}{l} f^{(k-l)}(x)g^{(l)}(x)$$

667

and consider

$$(*) \quad \frac{d^{k+1}}{dx^{k+1}}(f \cdot g)(x) = \frac{d}{dx} \sum_{l=0}^{k} \binom{k}{l} f^{(k-l)}(x) g^{(l)}(x)$$

$$= \sum_{l=0}^{k} \binom{k}{l} \frac{d}{dx}(f^{(k-l)}(x) \cdot g^{(l)}(x))$$

$$= \sum_{l=0}^{k} \binom{k}{l} \left\{ f^{(k+1-l)}(x) g^{(l)}(x) + f^{(k-l)}(x) g^{(l+1)}(x) \right\}.$$

The last term we now handle is the analogous term in the proof of the binomial theorem, Theorem 3.9.

$$\sum_{l=0}^{k} \binom{k}{l} f^{(k+1-l)}(x) g^{(l)} + \sum_{l=0}^{k} \binom{k}{l} f^{(k-l)}(x) g^{(l+1)}(x)$$

$$= \quad f^{(k+1)}(x) g(x) + \sum_{l=1}^{k} \binom{k}{l} f^{(k+1-l)}(x) g^{(l)}(x)$$

$$+ \quad \sum_{l=0}^{k-1} \binom{k}{l} f^{(k-l)}(x) g^{(l+1)}(x) + f(x) g^{(k+1)}(x)$$

$$= \quad f^{(k+1)}(x) g(x) + \sum_{l=1}^{k} \binom{k}{l} f^{(k+1-l)}(x) g^{(l)}(x)$$

$$+ \quad \sum_{l=1}^{k} \binom{k}{l-1} f^{(k-(l-1))}(x) g^{(l)}(x) + f(x) g^{(k+1)}(x)$$

$$= \quad \binom{k+1}{0} f^{(k+1)}(x) g(x) + \sum_{l=1}^{k} \left( \binom{k}{l} + \binom{k}{l+1} \right) f^{(k+1-l)}(x) g^{(l)}(x)$$

$$+ \quad \binom{k+1}{k+1} f(x) g^{(l)}(x)$$

$$(**) \quad = \quad \sum_{l=0}^{k+1} \binom{k+1}{l} f^{(k+1-l)}(x) g^{(l)}(x)$$

where we used Lemma 3.8, i.e. $\binom{k+1}{l} = \binom{k}{l-1} + \binom{k}{l}$. Thus the general Leibniz's rule is proved by combining $(*)$ and $(**)$.

3. We need to prove for $f, g \in C^k(I)$ and $\lambda, \mu \in \mathbb{R}$ that $\lambda f + \mu g \in C^k(I)$ and $f \cdot g \in C^k(I)$. Now the linearity of the derivative, i.e. $\frac{d}{dx}(\lambda f + \mu g) = \lambda \frac{df}{dx} + \mu \frac{dg}{dx}$ implies immediately the linearity of higher derivative, $l \leq k$

$$\frac{d^l}{dx^l}(\lambda f + \mu g) = \frac{d}{dx}\left( \frac{d^{l-1}}{dx^{l-1}}(\lambda f + \mu g) \right)$$

668

$$= \frac{d}{dx}\left(\lambda\frac{d^{l-1}}{dx^{l-1}}f + \mu\frac{d^{l-1}}{dx^{l-1}}g\right)$$

$$= \lambda\frac{d^l f}{dx^l} + \mu\frac{d^l g}{dx^l},$$

where as the general Leibniz's rule, see Problem 2, yields for $l \leq j$

$$\frac{d^l}{dx^l}(f \cdot g) = \sum_{j=0}^{l}\binom{l}{j}f^{(l-j)}g^{(j)}$$

and it follows $\frac{d^l}{dx^l}(fg) \in C^{k-l}(I)$.

4. For $x \neq 0$ and $x \neq 1$ the function is obviously differentiable. For being also differentiable at $x_0 = 0$ and $x_1 = 1$ the function must be at these points continuous and the right and the left derivative, i.e.

$$\lim_{x\to\tilde{x},x>\tilde{x}}\frac{f(x) - f(\tilde{x})}{x - \tilde{x}} \quad \text{and} \quad \lim_{x\to\tilde{x},x<\tilde{x}}\frac{f(x) - f(\tilde{x})}{x - \tilde{x}}$$

must exist and coincide, $\tilde{x} \in \{x_0, x_1\}$.

This yields

$$a \cdot 0 + b = c \cdot 0^2 + d \cdot 0$$

$$a = 2c \cdot 0 + d$$

$$c + d = 1 - \frac{1}{1}$$

$$2c \cdot 1 + d = \frac{1}{1^2}$$

or $b = 0, a = d, c + d = 0, 2c + d = 1$ i.e. $a = -1, b = 0, c = 1, d = -1$.

5. Using the law of the logarithms

$$\ln(a \cdot b) = \ln a + \ln b$$

we find first

$$\ln \prod_{k=1}^{n} f_k(x) = \sum_{k=1}^{n}\ln f_k(x),$$

and consequently

$$\left(\ln \prod_{k=1}^{n} f_k(x)\right)' = \left(\sum_{k=1}^{n}\ln f_k(x)\right)'$$

which yields

$$\frac{(\prod_{k=1}^{n} f_k(x))'}{\prod_{k=1}^{n} f_k(x)} = \sum_{k=1}^{n}\frac{f_k'(x)}{f_k(x)},$$

where we used $(\ln g)' = \frac{g'}{g}$.

669

6.    a) In the case that $f$ is a differentiable function at $x_0$, we find

$$\frac{f(x_0 + h) - f(x_0 - h)}{2h} = \frac{f(x_0 + h) - f(x_0) + f(x_0) - f(x_0 - h)}{2h}$$

$$= \frac{1}{2}\frac{f(x_0 + h) - f(x_0)}{h} + \frac{1}{2}\frac{f(x_0 - h) - f(x_0)}{-h}$$

and passing to the limit $h \to 0$ yields

$$\lim_{h \to 0}\frac{f(x_0 + h) - f(x_0 - h)}{2h} = \frac{1}{2}\lim_{h \to 0}\frac{f(x_0 + h) - f(x_0)}{h} + \frac{1}{2}\lim_{h \to 0}\frac{f(x_0 - h) - f(x_0)}{-h}$$

$$= \frac{1}{2}f'(x_0) + \frac{1}{2}f'(x_0) = f'(x_0).$$

b) The function $g : (-a.a) \to \mathbb{R}, g(x) = |x|$, is not differentiable at $x_0 = 0$, compare Example 7.7. However we have for $x_0 = 0$ that $g(0 + h) = |h| = |-h| = g(0 - h)$,
and therefore

$$\frac{g(0 + h) - g(0 - h)}{2h} = 0$$

implying that

$$\lim \frac{g(0 + h) - g(0 - h)}{2h} = 0$$

but $g'(0)$ does not exist.

7. Consider the quotient

$$\frac{f(x) - f(0)}{x - 0} = \frac{x^2 h(x)}{x} = |x|h(x).$$

Now, since $h$ is bounded, i.e. $|h(x)| \leq M$ for all $x$, we deduce that

$$f'(0) = \lim_{x \to 0}\frac{f(x) - f(0)}{x - 0} = \lim_{x \to 0}|x|h(x) = 0,$$

since $||x|h(x)| \leq |x|M$ and $\lim_{x \to 0}|x|M = 0$.

8.    a) For an even function $f : \mathbb{R} \to \mathbb{R}$ we find for all $x_0 \in \mathbb{R}$

$$f'(x_0) = \lim_{x \to x_0}\frac{f(x) - f(x_0)}{x - x_0} = \lim_{x \to x_0}\frac{f(-x) - f(-x_0)}{x - x_0}$$

$$= \lim_{y \to y_0}\frac{f(y) - f(y_0)}{-y - (-y_0)} = \lim_{y \to y_0}\left(-\left(\frac{f(y) - f(y_0)}{y - y_0}\right)\right)$$

$$= -\lim_{y \to y_0}\frac{f(y) - f(y_0)}{y \to y_0} = -f'(y_0) = -f'(-x_0),$$

670

i.e. for all $x_0 \in \mathbb{R}$ we have $f'(x_0) = -f'(-x_0)$ which means that $f'$ is odd. Now, if $g : \mathbb{R} \to \mathbb{R}$ is an odd function it follows for all $x_0 \in \mathbb{R}$ that

$$
\begin{aligned}
g'(x_0) &= \lim_{x \to x_0} \frac{g(x) - g(x_0)}{x - x_0} \\
&= \lim_{x \to x_0} \frac{-g(-x) - (-g(-x_0))}{x - x_0} \\
&= \lim_{y \to y_0} \frac{-g(y) + g(y_0)}{-y - (-y_0)} \\
&= \lim_{y \to y_0} \frac{-(g(y) - g(y_0))}{-(y - y_0)} \\
&= \lim_{y \to y_0} \frac{g(y) - g(y_0)}{y - y_0} = g'(y_0) = g'(-x_0),
\end{aligned}
$$

i.e. $g'$ is even.

b) Since $f$ is $a$-periodic, we have

$$
\frac{d}{dx}\left(f(x) \cdot f(x+a)\right) = \frac{d}{dx} f^2(x) = 2f'(x)f(x)
$$

and

$$
\frac{d}{dx} f(x)f(x+a) = f'(x)f(x+a) + f(x)f'(x+a),
$$
$$
= f'(x)f(x) + f(x)f'(x+a)
$$

or

$$
f'(x)f(x) = f(x)f'(x+a).
$$

By assumption, we have $f(x) \neq 0$, so we deduce for all $x \in \mathbb{R}$, $f'(x) = f'(x+a)$, i.e. $f'$ is $a$-periodic too. Note that the assumption $f(x) = 0$ for $x$ can be reduced when assuming that $f'$ is continuous. In this case it would be sufficient for $f(x) \neq 0$ for all $x \in \mathbb{R}\backslash\mathbb{Q}$, or more generally for a dense set in $\mathbb{R}$, a notion we will discuss later on.

9. For $|x| < 1$ we find

$$
\sum_{k=0}^{N} x^k = \frac{1 - x^{N+1}}{1 - x}
$$

which implies by differentiation

$$
\sum_{k=1}^{N} kx^{k-1} = \frac{1}{(1-x)^2} + \frac{Nx^{N+1} - (N+1)x^N}{(1-x)^2},
$$

and further for $x \neq 0$

$$
\frac{1}{x} \sum_{k=1}^{N} kx^k = \frac{x}{(1-x)^2} + \frac{Nx^{N+2} - (N+1)x^{N+1}}{(1-x)^2},
$$

or

$$\sum_{k=1}^{N} kx^k = \frac{x}{(1-x)^2} + \frac{Nx^{N+2} - (N+1)x^{N+1}}{(1-x)^2},$$

and this identity holds also for $x = 0$. Hence for all $|x| < 1$ we find, since in this case $\lim_{N\to\infty} Nx^{N+l} = 0, l = 1, 2, \cdots$, that

$$\sum_{k=1}^{\infty} kx^k = \lim_{N\to\infty} \sum_{k=1}^{N} kx^k = \frac{x}{(1-x)^2}.$$

10.      a) For $k = 1$ we have

$$\frac{d}{dx}(1+x^2)^{-\frac{1}{2}} = -\frac{1}{2}(2x)(1+x^2)^{-\frac{3}{2}}$$

$$= \frac{-x}{(1+x^2)^{\frac{3}{2}}}$$

and $P_1(x) = -x$. Now suppose that $\frac{d^k}{dx^k}(1+x^2)^{-\frac{1}{2}} = \frac{P_k(x)}{(1+x^2)^{\frac{2k+1}{2}}}$ and that the degree of $P_k$ is less or equal to $k$. We want to prove

$$\frac{d^{k+1}}{dx^{k+1}}(1+x^2)^{-\frac{1}{2}} = \frac{P_{k+1}(x)}{(1+x^2)^{\frac{2(k+1)+1}{2}}} = \frac{P_{k+1}(x)}{(1+x^2)^{\frac{2k+3}{2}}}$$

with a polynomial of degree at most $k + 1$. Note that

$$\frac{d^{k+1}}{dx^{k+1}}(1+x^2)^{-\frac{1}{2}} = \frac{d}{dx}\left(\frac{P_k(x)}{(1+x^2)^{\frac{2k+1}{2}}}\right)$$

$$= \left(\frac{d}{dx}P_k(x)\right)\frac{1}{(1+x^2)^{\frac{2k+1}{2}}} + P_k(x)\frac{d}{dx}\frac{1}{(1+x^2)^{\frac{2k+1}{2}}}$$

$$= \frac{P_k'(x)(1+x^2) + P_k(x)(-\frac{2k+1}{2})(2x)}{(1+x^2)^{\frac{2(k+1)+1}{2}}}$$

$$= \frac{P_k'(x)(1+x^2) - (2k+1)xP_k(x)}{(1+x^2)^{\frac{2k+3}{2}}}.$$

Since $P_k(x)$ has degree less or equal to $k$, $P_k'(x)$ has degree less or equal to $k - 1$ and this implies that $(1 + x^2)P_k'(x)$ as well as $xP_k(x)$ has degree less or equal to $k + 1$.

Now, if $P_k(x)$ has degree less or equal to $k$, then we know that $|P_k(x)| \le C_k(1+x^2)^{\frac{k}{2}}$, which implies

$$|\frac{d^k}{dx^k}(1+x^2)^{-\frac{1}{2}}| \le \frac{|P_k(x)|}{(1+x^2)^{\frac{2k+1}{2}}} \le C_k\frac{(1+x^2)^{\frac{k}{2}}}{(1+x^2)^{\frac{2k+1}{2}}}$$

672

$$= C_k \frac{1}{(1+x^2)^{\frac{k+1}{2}}}.$$

b) Let $f \in C_b^m(\mathbb{R})$. First we note that for $l \le m$, since $|f^{(l)}(x)| \le M_l$ for some $M_l$, it follows that

$$\left| f^{(l)}\left((1+x^2)^{-\frac{1}{2}}\right) \right| \le M_l.$$

Next, with $g(x) = (1+x^2)^{-\frac{1}{2}}$ we find with part a),

$$|g^{(j)}(x)|^{k_j} \le \gamma_{j,k_j} \left( \frac{1}{(1+x^2)^{\frac{1}{2}}} \frac{1}{(1+x^2)^{\frac{j}{2}}} \right)^{k_j}$$

$$\le \gamma_{j,k_j} \frac{1}{(1+x^2)^{\frac{1}{2}}} \frac{1}{(1+x^2)^{\frac{jk_j}{2}}}.$$

Now the Faà di Bruno formula yields

$$|f^{(m)}((1+x^2)^{-\frac{1}{2}})|$$

$$\le \left| \sum c_{m,k_1,\cdots,k_n} f^{(k)}((1+x^2)^{-\frac{1}{2}}) \left( \frac{g^{(1)}(x)}{1!} \right)^{k_1} \cdots \left( \frac{g^{(m)}(x)}{m!} \right)^{k_m} \right.$$

$$\le \sum \tilde{}_{m,k_1,\cdots,k_m} |g^{(1)}(x)^{k_1}| \cdots |g^{(m)}(x)|^{k_m}$$

$$\le C \frac{1}{(1+x^2)^{\frac{1}{2}}} \frac{1}{(1+x^2)^{\sum jk_j}}$$

$$= C \frac{1}{(1+x^2)^{\frac{m+1}{2}}}.$$

**Chapter 22**

1. We may first apply Rolle's theorem to $f$ which yields the existence of $x_0 \in (a,b)$ such that $f'(x_0) = 0$. Now consider $f'|_{[a,x_0]}$ and $f'|_{[x_0,b]}$. Since $f'(a) = 0$ and $f'(b) = 0$, both functions $f'|_{[a,x_0]}$ and $f'|_{[x_0,b]}$ satisfy the assumption of Rolle's theorem. Thus there exists $x_1 \in (a,x_0)$ and $x_2 \in (x_0,b)$, hence $x_1 \ne x_2$, such that $f''(x_1) = f''(x_2) = 0$

2. By our assumption, we find for $x,y \in (a,b), x \ne y$, that

$$\left| \frac{f(x) - f(y)}{x - y} \right| \le c|x - y|^\alpha,$$

which implies that $f$ is differentiable at $x$ and

$$f'(x) = \lim_{y \to x} \frac{f(x) - f(y)}{x - y} = 0$$

for all $x \in (a,b)$ which yields that $f$ must be constant on $(a.b)$. Clearly $f$ is continuous on $[a,b]$, hence $f$ is constant on $[a,b]$.

Remark: The condition $|f(x) - f(y)| \leq C|x - y|^\beta, \beta > 0$, is called the **Hölder condition**. For $\beta = 1$ we recover the Lipschitz condition. The result proved above says that if $f$ is **Hölder continuous**, i.e. satisfies the Hölder condition with exponent (Hölder exponent) $\beta > 1$, then $f$ is constant. For $0 < \beta < 1$ there are non-trivial functions satisfying the Hölder condition.

3. Let $x, y \in (a, b), x < y$. We can find $\xi \in (x, y)$ such that $f(x) - f(y) = f'(\xi)(x - y)$ which gives
$$|f(x) - f(y)| = |f'(\xi)||x - y| \leq M|x - y|,$$
i.e. $f$ is Lipschitz continuous and by Problem 19 in Chapter 20, $f$ is uniformly continuous.

4. We consider the function $y \mapsto \ln(1 + y)$ as the interval $[0, \frac{x}{q}]$ and $[\frac{x}{q}, \frac{x}{p}]$ for $x > 0$ and apply in both intervals the mean value theorem. Thus using for $0 < y_1 < y_2$ the formula
$$\ln(1 + y_2) - \ln(1 + y_1) = \frac{1}{1 + \xi}(y_2 - y_1), \quad \xi \in (y_1, y_2)$$
we find
$$\ln\left(1 + \frac{x}{q}\right) = \ln\left(1 + \frac{x}{q}\right) - \ln 1 = \frac{1}{1 + \xi_0}\frac{x}{q}, 0 < \xi_0 < \frac{x}{q}$$
and
$$\ln\left(1 + \frac{x}{p}\right) - \ln\left(1 + \frac{x}{q}\right) = \frac{1}{1 + \xi_1}\left(\frac{x}{p} - \frac{x}{q}\right), \quad \frac{x}{q} < \xi_1 < \frac{x}{p}.$$
Thus, since $\frac{1}{1+\xi_0} > \frac{1}{1+\xi_1}$ we obtain
$$\frac{\ln\left(1 + \frac{x}{q}\right)}{\frac{x}{q}} > \frac{\ln\left(1 + \frac{x}{p}\right) - \ln\left(1 + \frac{x}{q}\right)}{\frac{x}{p} - \frac{x}{q}},$$
implying
$$\left(\frac{x}{p} - \frac{x}{q}\right)\ln\left(1 + \frac{x}{q}\right) > \frac{x}{q}\left(\ln\left(1 + \frac{x}{p}\right) - \ln\left(1 + \frac{x}{q}\right)\right),$$
which gives
$$\frac{x}{p}\ln\left(1 + \frac{x}{q}\right) > \frac{x}{q}\ln\left(1 + \frac{x}{p}\right)$$
or
$$q\ln\left(1 + \frac{x}{q}\right) > p\ln\left(1 + \frac{x}{p}\right),$$
i.e.
$$\ln\left(1 + \frac{x}{p}\right)^p < \ln\left(1 + \frac{x}{q}\right)^q$$
and this implies of course
$$\left(1 + \frac{x}{p}\right)^p < \left(1 + \frac{x}{q}\right)^q.$$

5.    a) By l'Hospital's rule we find first

$$\lim_{x\to\infty} \frac{e^{\alpha x}}{x} = \lim_{x\to\infty} \frac{\alpha e^{\alpha x}}{1} = \infty,$$

and since $\frac{e^{\alpha x}}{x^\beta} = \left(\frac{e^{\frac{\alpha}{\beta}x}}{x}\right)^\beta$, we deduce

$$\lim_{x\to\infty} \frac{e^{\alpha x}}{x^\beta} = \infty.$$

b) First we note that

$$\lim_{x\to\infty} \frac{\ln x}{x^\alpha} = \lim_{x\to\infty} \frac{\frac{1}{x}}{\alpha x^{\alpha-1}} = \lim_{x\to\infty} \frac{1}{\alpha x^\alpha} = 0$$

and now we note

$$\frac{(\ln x)^\beta}{x^\alpha} = \left(\frac{\ln x}{x^{\frac{\alpha}{\beta}}}\right)^\beta$$

and the result follows.

c) Since $x^x = e^{x\ln x}$ we find by the continuity of exp

$$\lim_{x\to 0} x^x = \lim_{x\to 0} e^{x\ln x} = \exp(\lim_{x\to 0} x\ln x)$$

$$= \exp(0) = 1,$$

where we used

$$\lim_{x\to 0} x\ln x = \lim_{x\to 0} \frac{\ln x}{\frac{1}{x}} = \lim_{x\to 0} \frac{\frac{1}{x}}{-\frac{1}{x^2}} = \lim_{x\to 0}(-x) = 0.$$

6. First note that

$$(8-x)^{\frac{1}{x-7}} = \left(e^{\ln(8-x)}\right)^{\frac{1}{x-7}} = e^{\frac{1}{x-7}\ln(8-x)},$$

and therefore, since

$$\lim_{x\to 7} \frac{\ln(8-x)}{x-7} = \lim_{y\to 0} \frac{\ln(1-y)}{y} = \lim_{y\to 0} \frac{-\frac{1}{1-y}}{1} = -1,$$

we find

$$\lim_{x\to 7}(8-x)^{\frac{1}{x-7}} = e^{-1}.$$

7. We write

$$\left(f\left(\frac{a}{\sqrt{x}}\right)\right)^x = e^{x\ln f\left(\frac{a}{\sqrt{x}}\right)}$$

and note that

$$\lim_{x\to\infty,x>0} \left(f\left(\frac{a}{\sqrt{x}}\right)\right)^x = \lim_{x\to\infty,x>0} e^{x\ln f\left(\frac{a}{\sqrt{x}}\right)}$$

675

$$= exp\left(\lim_{x\to\infty, x>0}\left(x\ln f\left(\frac{a}{\sqrt{x}}\right)\right)\right)$$

and further, by applying l'Hospital's rule twice

$$\lim_{x\to\infty, x>0}\left(x\ln f\left(\frac{a}{\sqrt{x}}\right)\right) = \lim_{y\to 0}\frac{\ln f(a\sqrt{y})}{y}$$

$$= \lim_{y\to 0, y>0}\frac{af'(a\sqrt{y})}{2\sqrt{y}f(a\sqrt{y})}$$

$$= \lim_{y\to 0, y>0}\frac{a^2 f''(a\sqrt{y})}{2f(a\sqrt{y}) + 2a\sqrt{y}f'(a\sqrt{y})} = -\frac{a^2}{2},$$

where we used $f(0) = 1$, $f'(0) = 0$ and $f''(0) = -1$. Hence we arrive at

$$\lim_{x\to\infty, x>0}\left(f\left(\frac{a}{\sqrt{x}}\right)\right)^x = e^{-\frac{a^2}{2}}.$$

8. Let $x \in (a, b)$ and $h > 0$ such that $x + h \in (a, b)$. It follows that $f(x+h) - f(x) \le 0$ and therefore

$$f'(x) = \lim_{h\to 0}\frac{f(x+h) - f(x)}{h} \le 0.$$

9. Clearly for $k = 0$ we have $e^{-at} \ge 0$. Moreover for $k \in \mathbb{N}$ we find

$$\frac{d^k}{dt^k}(e^{-at}) = (-a)^k e^{-at} = (-1)^k a^k e^{-at}$$

implying that $(-1)^k\left((-1)^k a^k e^{-at}\right) = a^k e^{-at} \ge 0$, i.e. $t \mapsto e^{-at}, a > 0$, is completely monotone.

Now, $1 - e^{-at}, a > 0, t \ge 0$, is always non-negative and for $k \in \mathbb{N}$ we find

$$\frac{d^k}{dx^k}\left(1 - e^{-at}\right) = -\frac{d^k}{dx^k}\left(e^{-at}\right)$$

and by the previous result follows that $(-1)^k\frac{d^k}{dx^k}\left(1 - e^{-at}\right) \le 0$.

Finally, for $\alpha = 1$ we have $t \ge 0$ for $t > 0$ and $\frac{d}{dt}(t) = 1 \ge 0$,, as well as $\frac{d^k}{dt^k}(t) = 0$ for $k \ge 2$. Thus $t \mapsto t$ is a Bernstein function. For $0 < \alpha < 1$ we find first $t^\alpha > 0$ for $\alpha > 0$ and $t > 0$ and further

$$\frac{d^k}{dt^k}(t^\alpha) = \alpha(\alpha - 1)(\alpha - 2)\cdots(\alpha - k)t^{\alpha-k}$$

$$= \alpha|\alpha - 1||\alpha - 2|\cdots|\alpha - k|(-1)^{k-1}t^{\alpha-k},$$

and we arrive at

$$(-1)^k\frac{d^k}{dt^k}(t^\alpha) = (-1)\alpha|\alpha - 1||\alpha - 2|\cdots|\alpha - k|t^{\alpha-k} \le 0.$$

10.    a) We have

$$f'(x) = \frac{1}{9} \frac{\frac{1}{3} - x}{(x^2(1-x))^{\frac{1}{3}}}$$

provided $x \neq 0$ and $x \neq 1$. Thus $f'$ vanishes at $x_0 = \frac{1}{3}$ and for $0 < x < \frac{1}{3}$ we have $f'(x) > 0$ and for $\frac{1}{3} < x < 1$ we have $f'(x) < 0$, i.e. approaching from the left of the point $x_0 = \frac{1}{3}$ the function $f$ is strictly increasing, and for $x > \frac{1}{3}$ the function is strictly decreasing, hence we must have a local maximum at $x_0 = \frac{1}{3}$ and the value is $f\left(\frac{1}{3}\right) = \frac{4^{\frac{1}{3}}}{3}$.

For $x = 0$ and $x = 1$ the function $f$ is not differentiable. For $x \in (0, 1)$ we have $f(x) > 0$ and further $f(x) < 0$ for $x < 0$, hence there is no local extreme value at $x = 0$. However for $x = 1$ we find that $f(1) = 0$ and $f(x) > 0$ for $x > 1$ as well as for $x \in (0, 1)$, hence there is a local minimum at $x = 1$, see J. Kaczar and M.T. Nowak [6, p. 298].

b) The function $f$ has only strictly positive values and $\lim_{x \to \infty} f(x) = \lim_{x \to -\infty} f(x) = 0$. For $x = 0$ and $x = 1$ the function is not differentiable. Thus, to find the maximum we have to look at $(-\infty, 0), (0, 1)$ and $(1, \infty)$ for a local maximum and compare with $f(0) = \frac{3}{2}$ and $f(1) = \frac{3}{2}$.

Now for $x < 0$ we find

$$f(x) = \frac{1}{1 - x} + \frac{1}{1 - x + 1} = \frac{1}{1 - x} + \frac{1}{2 - x}$$

and $f'(x) = \frac{1}{(1-x)^2} + \frac{1}{(2-x)^2} > 0$, implying that $f$ is on $(-\infty, 0)$ strictly increasing, hence $f(x) < \frac{3}{2}$ for $x \in (-\infty, 0)$.

For $0 < x < 1$ we find

$$f(x) = \frac{1}{1 + x} + \frac{1}{2 - x}$$

and

$$f'(x) = -\frac{1}{(1 + x)^2} + \frac{1}{(2 - x)^2}$$

implying $f'\left(\frac{1}{2}\right) = 0$. Since

$$f''(x) = \frac{2}{(1 + x)^3} + \frac{2}{(2 - x)^3}$$

and therefore $f''\left(\frac{1}{2}\right) > 0$, we find that $f$ has a local minimum at $\frac{1}{2}$.

Finally, for $x > 1$ we have

$$f(x) = \frac{1}{1 + x} + \frac{1}{1 + x - 1} = \frac{1}{1 + x} + \frac{1}{x}$$

and

$$f'(x) = -\frac{1}{(1 - x)^2} - \frac{1}{x^2}$$

Thus on $(1, \infty)$ the function $f$ is strictly decreasing. It follows that the global maximum of $f$ is $\frac{3}{2}$ and it is attained at two points $x_0 = 0$ and $x_1 = 1$.

11. The tangent line at $x_0$ is the graph of

$$g_{x_0}(t) = g'(x_0)t + g(x_0) - x_0 g'(x_0)$$

and the normal line at $x_0$ is the graph of

$$n_{x_0}(t) = -\frac{1}{g'(x_0)}t + g(x_0) + \frac{x_0}{g'(x_0)}$$

provided $g'(x_0) \neq 0$.
With $g(x) = \sqrt{1-x^2}, x \in (-1,1)$, we find

$$g'(x) = -\frac{x}{\sqrt{1-x^2}} \neq 0$$

for $x \neq 0$. Thus

$$g_{x_0}(t) = \frac{-x_0 t}{\sqrt{1-x_0^2}} + \sqrt{1-x_0^2} + \frac{x_0^2}{\sqrt{1-x_0^2}}$$

$$= \frac{-x_0 t + 1 - x_0^2 + x_0^2}{\sqrt{1-x_0^2}} = \frac{-x_0 t + 1}{\sqrt{1-x_0^2}}$$

and for $x_0 \neq 0$

$$n_{x_0}(t) = \frac{\sqrt{1-x_0^2}}{x_0}t + \sqrt{1-x_0^2} - \frac{x_0\sqrt{1-x_0^2}}{x_0}$$

$$= \frac{\sqrt{1-x_0^2}}{x_0}t.$$

For $x_0 = 0$ the normal line is of course the abscissa. Since $g$ is the upper half circle, we expect the centre of curvature to be for all $x_0 \in (-1,1)$ the origin. In general we have for $c = (c_1, c_2)$

$$c_1 = x_0 - g'(x_0)\frac{1 + g'^2(x_0)}{g''(x_0)}$$

and

$$c_2 = g(x_0) + \frac{1 + g'^2(x_0)}{g''(x_0)}.$$

Since $g''(x_0) = \frac{-1}{(1-x_0^2)^{\frac{3}{2}}}$ we find

$$1 + g'^2(x_0) = 1 + \frac{x_0^2}{1-x_0^2} = \frac{1}{1-x_0^2},$$

$$\frac{1 + g'^2(x_0)}{g''(x_0)} = -\frac{(1-x_0^2)^{\frac{3}{2}}}{1-x_0^2} = -\sqrt{1-x_0^2},$$

678

$$g'(x_0)\frac{1+g'^2(x_0)}{g''(x_0)} = -\frac{x_0}{\sqrt{1-x_0^2}} \cdot \left(-\sqrt{1-x_0^2}\right) = x_0,$$

and consequently

$$c_1 = x_0 - x_0 = 0$$

as well as

$$c_2 = \sqrt{1-x_0^2} - \sqrt{1-x_0^2} = 0.$$

Finally, as radius of curvature we find

$$r = \frac{\left(1+g'^2(x_0)\right)^{\frac{3}{2}}}{|g''(x_0)|} = \frac{1}{(1-x_0^2)^{\frac{3}{2}}} \cdot \frac{(1-x_0^2)^{\frac{3}{2}}}{1} = 1,$$

as we shall expect: the circle of curvature of a circle is the circle itself.

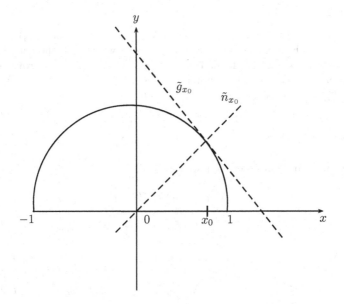

12. For $x \in (0,\infty)$ we have $f'(x) = -\frac{1}{x^2}$ and $f''(x) = \frac{2}{x^3}$ which yields for the normal line that it is the graph of the function

$$n_{x_0}(t) = -\frac{1}{f'(x_0)}t + f(x_0) + \frac{x_0}{f'(x_0)}$$

$$= x_0^2 t + \frac{1}{x_0} - x_0^3.$$

The centre of curvature $c = (c_1, c_2)$ is given by

$$c_1 = x_0 - f'(x_0)\frac{1+f'^2(x_0)}{f''(x_0)}$$

679

and

$$c_2 = f(x_0) + \frac{1 + f'^2(x_0)}{f''(x_0)},$$

and since

$$\frac{1 + f'^2(x_0)}{f''(x_0)} = \frac{1 + x_0^4}{2x_0}$$

we get

$$c_1 = x_0 + \frac{1}{x_0^2} \cdot \frac{x_0^4 + 1}{2x_0} = \frac{2x_0^4 + x_0^4 + 1}{2x_0^3} = \frac{3x_0^4 + 1}{2x_0^4},$$

$$c_2 = \frac{1}{x_0} + \frac{1 + x_0^4}{2x_0} = \frac{3 + x_0^4}{2x_0}.$$

**Chapter 23**

1. For $m = 2$ the statement is $f(\lambda_1 x_1 + \lambda_2 x_2) \leq \lambda_1 f(x_1) + \lambda_2 f(x_2), \lambda_1, \lambda_2 \in [0,1], \lambda_1 + \lambda_2 = 1$. Thus with $\lambda := \lambda_1$ and $\lambda_2 := 1 - \lambda$ we recover the definition of convexity. Now suppose (23.12) holds for some $m \geq 2$. We want to prove that it also holds for $m + 1$. For this, take points $x_1, \cdots, x_{m+1} \in I$ and $\lambda_1, \cdots, \lambda_{m+1} \in [0,1]$ with $\sum_{j=1}^{m+1} \lambda_j = 1$. Since for $\lambda_m + \lambda_{m+1} > 0$

$$\lambda_m x_m + \lambda_{m+1} x_{m+1} = (\lambda_m + \lambda_{m+1}) \left( \frac{\lambda_m}{\lambda_m + \lambda_{m+1}} x_m + \frac{\lambda_{m+1}}{\lambda_m + \lambda_{m+1}} x_{m+1} \right) = \tilde{\lambda}_m \tilde{x}_n,$$

by our induction hypothesis we find

$$f(\lambda_1 x_1 + \cdots + \lambda_{m+1} x_{m+1}) = f(\lambda_1 x_1 + \cdots + \lambda_{m-1} x_{m-1} + \tilde{\lambda}_m \tilde{x}_m)$$

$$\leq \lambda_1 f(x_1) + \cdots + \tilde{\lambda}_m f(\tilde{x}_m)$$

$$= \lambda_1 f(x_1) + \cdots + \lambda_{m-1} f(x_{m-1}) + (\lambda_m + \lambda_{m+1}) f\left( \frac{\lambda_m}{\lambda_m + \lambda_{m+1}} x_m + \frac{\lambda_{m+1}}{\lambda_m + \lambda_{m+1}} x_{m+1} \right)$$

$$\leq \lambda_1 f(x_1) + \cdots + \lambda_{m-1} f(x_{m-1}) + \lambda_m f(x_m) + \lambda_{m+1} f(x_{m+1}),$$

where we used in the last step the convexity of $f$ and the fact that $\frac{\lambda_m}{\lambda_m + \lambda_{m+1}} + \frac{\lambda_{m+1}}{\lambda_m + \lambda_{m+1}} = 1$.

2. We will prove more, namely that if $I$ has end points $a < b$ and $a < a_1, b_1 < b$ then $f\big|_{[a_1,b_1]}$ is Lipschitz continuous.

Choose $\eta > 0$ such that $a < a_1 - \eta$ and $b_1 + \eta < b$. Now choose $x_1, y_1, x_2, y_2 \in I$ such that $x_1 < y_1 < a_1 - \eta$ and $b_1 + \eta < x_2 < y_2$. Take $x, y \in [a_1, b_1]$ and suppose $x < y$ (otherwise change the role of $x$ and $y$ in the following argument). We apply Lemma 23.4 to $x_1, y_1, x$ and then to $y_1, x_1, y$ to find

$$\frac{f(y_1) - f(x_1)}{y_1 - x_1} \leq \frac{f(x) - f(y_1)}{x - y_1}$$

and

$$\frac{f(x) - f(y_1)}{x - y_1} \leq \frac{f(y) - f(x)}{y - x},$$

hence

$$\frac{f(y_1) - f(x_1)}{y_1 - x_1} \leq \frac{f(y) - f(x)}{y - x}.$$

Applying Lemma 23.4 once more first to $x, y, x_2$ and then to $y, x_2, y_2$ we arrive at

$$\frac{f(y) - f(x)}{y - x} \leq \frac{f(y_2) - f(x_2)}{y_2 - x_2}.$$

Thus we get the estimate

$$\left| \frac{f(y) - f(x)}{y - x} \right| \leq \max \left\{ \left| \frac{f(y_1) - f(x_1)}{y_1 - x_1} \right|, \left| \frac{f(y_2) - f(x_2)}{y_2 - x_2} \right| \right\}$$

which implies with

$$L := \max \left\{ \left| \frac{f(y_1) - f(x_1)}{y_1 - x_1} \right|, \left| \frac{f(y_2) - f(x_2)}{y_2 - x_2} \right| \right\}$$

the Lipschitz estimate

$$|f(y) - f(x)| \leq L|y - x|$$

for $x, y \in [a_1, b_1]$.

3. Suppose that $f$ has at $x_0 \in \mathbb{R}$ a local minimum, i.e. for some $\delta > 0$ it follows that $|x - x_0| \leq \delta$ implies $f(x_0) \leq f(x)$. For $x \in \mathbb{R}$ such that $|x - x_0| > \delta$ we note that $\frac{\delta}{|x - x_0|} \in (0, 1)$ and further with

$$y := \frac{\delta}{|x - x_0|} x + \left( 1 - \frac{\delta}{|x - x_0|} \right) x_0$$

we first find $|y - x_0| = \delta$, hence $f(x_0) \leq f(y)$, and using the convexity of $f$ we find

$$f(x_0) \leq f(y) \leq \frac{\delta}{|x - x_0|} f(x) + \left( 1 - \frac{\delta}{|x - x_0|} \right) f(x_0),$$

implying $f(x_0) \leq f(x)$ i.e. $f(x_0)$ is a global minimum of $f$.

4.  a) On $(0, \infty)$ the function $\ln$ is twice continuously differentiable with $\frac{d^2 \ln x}{dx^2} = -\frac{1}{x^2}$. Hence the function $x \mapsto -\ln x$ is convex, i.e. $\ln$ is concave. Using Jensen's inequality we obtain

$$-\ln \left( \frac{x_1}{n} + \cdots + \frac{x_n}{n} \right) \leq -\frac{1}{n} \ln x_1 - \frac{1}{n} \ln x_2 - \cdots - \frac{1}{n} \ln x_n,$$

or

$$\ln \left( \frac{x_1}{n} + \cdots + \frac{x_n}{n} \right) \geq \frac{1}{n} (\ln x_1 + \cdots + \ln x_n),$$

681

i.e.

$$\ln\left(\frac{x_1}{n}+\cdots+\frac{x_n}{n}\right)\geq\frac{1}{n}\ln(x_1\cdots.x_n)$$

which yields

$$\frac{1}{n}\sum_{k=1}^{n}x_k\geq\left(\prod_{k=1}^{n}x_k\right)^{\frac{1}{n}}.$$

b) Since

$$\frac{d^2}{dx^2}(x\ln x)=\frac{1}{x}>0$$

for $x\in(0,\infty)$, we note that $f$ is convex, and consequentially by convexity

$$\frac{x+y}{2}\ln\left(\frac{x+y}{2}\right)\leq\frac{x}{2}\ln x+\frac{y}{2}\ln y,$$

or

$$(x+y)\ln\left(\frac{x+y}{2}\right)\leq x\ln x+y\ln y.$$

5. First we sketch the situation.

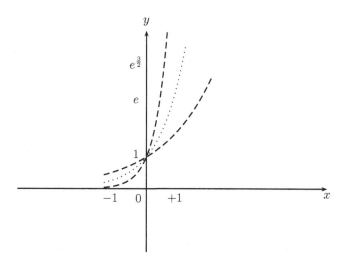

Now, for $x\in[-1,0]$ and $1\leq a\leq\frac{3}{2}$ we have $e^{ax}\leq e^x$ and for $x\in[0,1]$ and $1\leq a\leq\frac{3}{2}$ we find $e^{ax}\leq e^{\frac{3}{2}x}$ implying

$$\sup_{a\in[1,\frac{3}{2}]}f_a(x)=\begin{cases}e^{\frac{3}{2}x}, & x\in[0,1]\\ e^x, & x\in[-1,0].\end{cases}$$

682

6. For $0 \leq \lambda \leq 1$ and $x, y \in \mathbb{R}$ we find using first the monotonicity of $h$ and the convexity of $f$, and then the convexity of $h$.

$$(h \circ f)(\lambda x + (1 - \lambda)y) = h(f(\lambda x + (1 - \lambda)y))$$

$$\leq h(\lambda f(x) + (1 - \lambda)f(y))$$

$$\leq \lambda h\,(f(x)) + (1 - \lambda)h\,(f(y))$$

$$= \lambda(h \circ f)(x) + (1 - \lambda)(h \circ f)(y).$$

7. First we note that $\frac{||x-y||_k}{1=||x-y||_k} \leq 1$ implying that the series converges for all $x, y \in \mathbb{R}^n$. Moreover, from the definition follows that $d(x, y) \geq 0$ for all $x, y \in \mathbb{R}$ and if $d(x, y) = 0$ then $||x - y||_k = 0$ for all $k \in \mathbb{N}$, hence, since $||.||_k$ is a norm, $x = y$. Since for every norm $||x - y|| = ||y - x||$ holds we also find that $d$ is symmetric i.e. $d(x, y) = d(y, x)$.

Moreover, the monotonicity of $f(t) = \frac{1}{1+t}, t \geq 0$, implies

$$f\,(||x - y||_k) \leq f\,(||x - z||_k + ||z - y||_k),$$

and it follows

$$d(x, y) = \sum_{k=1}^{\infty} \frac{1}{2^k} f\,(||x - y||_k)$$

$$\leq \sum_{k=1}^{\infty} \frac{1}{2^k} f\,(||x - z||_k + ||z - y||_k)$$

$$= \sum_{k=1}^{\infty} \frac{1}{2^k} \frac{||x - z||_k + ||z - y||_k}{1 + ||x - z||_k + ||z - y||_k}$$

$$= \sum_{k=1}^{\infty} \frac{1}{2^k} \frac{||x - z||_k}{1 + ||x - z||_k + ||z - y||_k} + \sum_{k=1}^{\infty} \frac{1}{2^k} \frac{||z - y||_k}{1 + ||x - z||_k + ||z - y||_k}$$

$$\leq \sum_{k=1}^{\infty} \frac{1}{2^k} \frac{||x - z||_k}{1 + ||x - z||_k} + \sum_{k=1}^{\infty} \frac{1}{2^k} \frac{||z - y||_k}{1 + ||z - y||_k}$$

$$= d(x, z) + d(z, y).$$

8. For $x, y, z \in \mathbb{R}^n$ we find with $||x|| := ||x||_2$ using $2\sqrt{ab} \leq a + b$ which holds for $a, b \geq 0$, that

$$2(1 + ||y||)(1 + ||z||) = 2 + 2||y|| + 2||z|| + 2||y|| \cdot ||z||$$

$$= (1 + ||y|| + ||z|| + (||y|| + ||z||)) + (1 + 2||y|| \cdot ||z||)$$

$$\geq 1 + ||y|| + ||z|| + 2\sqrt{||y|| \cdot ||z||}$$

$$= 1 + \left(\sqrt{||y||} + \sqrt{||z||}\right)^2$$

683

$$\geq \left(1 + \sqrt{||y + z||}^2\right) = 1 + ||y + z||,$$

where for the last estimate we need $\sqrt{a + b} \leq \sqrt{a} + \sqrt{b}$ for $a, b \geq 0$. Thus with $z = x - y$ we get

$$2(1 + ||y||)(1 + ||x - y||) \geq 1 + ||y + x - y|| = 1 + ||x||,$$

or

$$\frac{1 + ||x||}{1 + ||y||} \leq 2(1 + ||x - y||).$$

9.      a) We first prove that $||.||$ is a norm on $\mathbb{R}^n$. Note that

$$||x|| = ||x||_{(1)} + ||x||_{(2)} \geq 0$$

and $||x|| = 0$ implies $||x||_{(1)} + ||x||_{(2)} = 0$, i.e. $||x||_{(1)} = 0$ and $||x||_{(2)} = 0$ implying $x = 0$.

Moreover, for $\lambda \in \mathbb{R}$ we find with $x \in \mathbb{R}^n$

$$||\lambda x|| = ||\lambda x||_{(1)} + ||\lambda x||_{(2)} = |\lambda| ||x||_{(1)} + |\lambda| ||x||_{(2)} = |\lambda| (||x||_{(1)} + ||x||_{(2)}) = |\lambda| \, ||x||.$$

Finally for $x, y \in \mathbb{R}^n$ we get

$$\begin{aligned}
||x + y|| &= ||x + y||_{(1)} + ||x + y||_{(2)} \\
&\leq ||x||_{(1)} + ||y||_{(1)} + ||x||_{(2)} + ||y||_{(2)} \\
&= ||x||_{(1)} + ||x||_{(2)} + ||y||_{(1)} + ||y||_{(2)} \\
&= ||x|| + ||y||.
\end{aligned}$$

Now we turn to $|||.|||$. Clearly

$$|||x||| = \max(||x||_{(1)}, ||x||_{(2)}) \geq 0$$

and if $|||x||| = 0$ then $||x||_{(1)} = 0$ and $||x||_{(2)} = 0$ implying $x = 0$. For $\lambda \in \mathbb{R}$ and $x \in \mathbb{R}^n$ we have

$$\begin{aligned}
|||\lambda||| &= \max(||\lambda x||_{(1)}, ||\lambda x||_{(2)}) \\
&= \max(|\lambda| \, ||x||_{(1)}, |\lambda| ||x||_{(2)}) \\
&= |\lambda x| \max(||x||_{(1)}, ||x||_{(2)}) = |\lambda| \, |||x|||.
\end{aligned}$$

684

However, for $x, y \in \mathbb{R}^n$ we have

$$
\begin{aligned}
|||x + y||| &= \max(\|x + y\|_{(1)}, \|x + y\|_{(2)}) \\
&\leq \max(\|x\|_{(1)} + \|y\|_{(1)}, \|x\|_{(2)} + \|y\|_{(2)}) \\
&= \frac{1}{2}(\|x\|_{(1)} + \|y\|_{(1)} + \|x\|_{(2)} + \|y\|_{(2)} + |\|x\|_{(1)} + \|y\|_{(1)} - \|x\|_{(2)} - \|y\|_{(2)}|) \\
&= \frac{1}{2}(\|x\|_{(1)} + \|x\|_{(2)}) + \|y\|_{(1)} + \|y\|_{(2)} + |(\|x\|_{(1)} - \|x\|_{(2)}) + (\|y\|_{(1)} - \|y\|_{(2)})| \\
&\leq \frac{1}{2}(\|x\|_{(1)} + \|x\|_{(2)} + (\|x\|_{(1)} - \|x\|_{(2)})) \\
&\quad + \frac{1}{2}(\|y\|_{(1)} + \|y\|_{(2)} + (\|y\|_{(1)} - \|y\|_{(2)})) \\
&= \max(\|x\|_{(1)}, \|x\|_{(2)}) + \max(\|y\|_{(1)}, \|y\|_{(2)}) \\
&= |||x||| + |||y|||,
\end{aligned}
$$

i.e. $|||.|||$ is a norm on $\mathbb{R}^n$.

b) The triangle inequality yields for $y, z \in \mathbb{R}^n$ that

$$
\|z + y\| \leq \|z\| + \|y\| \quad \text{or} \quad \|z + y\| - \|y\| \leq \|z\|,
$$

which gives with $x = z + y$, i.e. $z = x - y$, that

$$
\|x\| - \|y\| \leq \|x - y\|.
$$

Analogously we obtain

$$
-(\|x\| - \|y\|) = \|y\| - \|x\| \leq \|x - y\|
$$

implying

$$
\left| \|x\| - \|y\| \right| \leq \|x - y\|,
$$

where the inequality

$$
\|x\| - \|y\| \leq \left| \|x\| - \|y\| \right|
$$

is obvious.

10. First we recall $(x_k)_{k \in \mathbb{N}}$ converges to $x$ in $\|.\|_p$ if for every $\varepsilon > 0$ there exists $N(\varepsilon)$ such that $k \geq N$ implies $\|x_k - x\|_p = \left( \sum_{j=1}^n |x_k^{(j)} - x^{(j)}|^p \right)^{1/p} < \varepsilon$ which implies immediately that $k \geq N$ yields $\left| x_k^{(j)} - x^{(j)} \right| < \epsilon$ for $j = 1, \ldots, n$, i.e. $\left( x_k^{(j)} \right)_{k \in \mathbb{N}}$ converges to $x^{(j)}$. Conversely, suppose that for every $j = 1, \ldots, n$ the sequence $\left( x_k^{(j)} \right)_{k \in \mathbb{N}}$ converges to $x^{(j)}$. Given $\epsilon > 0$ we can find $N(\epsilon)$ such that for $j = 1, \ldots, n$ we have that $k \geq N(\epsilon)$ implies for all $j = 1, \ldots, n$ that $\left| x_k^{(j)} - x^{(j)} \right| < \frac{\epsilon}{n^{\frac{1}{p}}}$ which gives

$$
\|x_k - x\|_p = \left( \sum_{j=1}^n \left| x_k^{(j)} - x^{(j)} \right|^p \right)^{\frac{1}{p}} < \epsilon.
$$

11. Since $\lim\limits_{k\to\infty} ||x_k - x||_p = 0$ we have that for every $\epsilon > 0$ there exists $N(\epsilon) \in \mathbb{N}$ such that $k \geq N(\epsilon)$ implies $||x_k - x||_p < \frac{\epsilon}{c}$. Consequentially given $\epsilon > 0$ and $N(\epsilon)$ chosen as above we find

$$||x_k - x|| \leq c||x_k - x||_p < c \cdot \frac{\epsilon}{c} = \epsilon,$$

i.e. $\lim\limits_{k\to\infty} ||x_k - x|| = 0$.

## Chapter 24

1. For $x \in \mathbb{R}$ fixed the sequence $(g_n(x))_{n\in\mathbb{N}}$ converges clearly to 0. However $\sup\limits_{x\in\mathbb{R}} |g_{2n}(x)| = \infty$, thus we cannot expect $||g_n - 0||_\infty = ||g_n||_\infty$ converging to 0, and therefore the convergence is not uniform.

2. Once we have proved that $f$ is the pointwise limit of $(f_n)_{n\in\mathbb{N}}$ we have also shown that the convergence cannot be uniform. Each $f_n$ is continuous, $f$ is not. But the uniform limit of continuous function must be continuous.

   Now, for $x = 0$ we have $f_n(0) = \frac{1}{2}$, hence $\lim\limits_{n\to\infty} f_n(0) = \frac{1}{2}$. If $x \neq 0$ then $(nx - 1)^2$ diverges to $+\infty$ and hence $\lim\limits_{n\to\infty} f_n(x) - \lim\limits_{n\to\infty} \frac{1}{1 + (nx - 1)^2} = 0$.

3.   a) We know for $x \neq 1$ that $x^n \to 0$, and $f_n(1) = 0$ for all $n$. Thus we conclude that $(f_n)_{n\in\mathbb{N}}$ converges pointwise to 0. The function $f_n$ attains the maximum on $[0,1]$ and it is attained at $x_n = \frac{n}{n+1}$, since $f'_n(x) = nx^{n-1} - (n+1)x^n$, and $nx^{n-1} - (n+1)x^n = 0$ implies either $x = 0$ or $x = \frac{n}{n+1}$. For $x_n$ we find $f_n(x_n) = \left(\frac{n}{n+1}\right)^n \left(1 - \frac{n}{n+1}\right) = \frac{n^n}{(n+1)^{n+1}}$, and $\lim\limits_{n\to\infty} \frac{n^n}{(n+1)^{n+1}} = \lim\limits_{n\to\infty} \left(\left(\frac{n}{n+1}\right)^n \frac{1}{n+1}\right) = 0$. Thus we have $\lim\limits_{n\to\infty} ||f_n - 0||_\infty = \lim\limits_{n\to\infty} \left(\left(\frac{n}{n+1}\right)^n \frac{1}{n+1}\right) = 0$ implying the uniform convergence of $(f_n)_{n\to\infty}$ to 0.

   b) We notice that

   $$g_n(x) = \frac{x^2}{\frac{1}{n}(1 + nx)} = \frac{x^2}{\frac{1}{n} + x}$$

   and hence the pointwise limit is $g(x) = x$. Moreover we find for $x \in [0,1]$

   $$|g_n(x) - x| = \left|\frac{nx^2}{1 + nx} - x\right| = \left|\frac{nx^2 - (1 + nx)x}{1 + nx}\right|$$

   $$= \frac{x}{1 + nx}$$

   and

   $$\sup_{x\in[0,1]} |g_n(x) - x| = \sup_{x\in[0,1]} \frac{x}{1 + nx} = \frac{1}{1 + n}$$

   note $\frac{d}{dx}\left(\frac{x}{1+nx}\right) = \frac{1}{(1+nx)^2}$, i.e. $x \mapsto \frac{x}{1+nx}$ is monotone increasing, hence on $[0,1]$ we have $\frac{x}{1+nx} \leq \frac{1}{1+n}$. Now we conclude that $(g_n)_{n\in\mathbb{N}}$ converges uniformly on $[0,1]$ to $g(x) = x$.

c) Since arctan is continuous and $\arctan 0 = 0$ we find for each $x \in \mathbb{R}$ that $\lim_{n \to \infty} \left( \arctan \left( \dfrac{4x}{x^2 + n^4} \right) \right) = 0$. The mean value theorem implies

$$|\arctan z - \arctan y| \leq |z - y|$$

since $\left| \frac{d}{dx}(\arctan x) \right| = \frac{1}{1+x^2} \leq 1$. Consequently we have

$$\sup_{x \in \mathbb{R}} \left| \arctan \frac{4x}{x^2 + n^4} - 0 \right| \leq \sup_{x \in \mathbb{R}} \left| \frac{4x}{x^2 + n^4} - 0 \right| \leq \sup_{x \in \mathbb{R}} \frac{4|x|}{x^2 + n^4} = \sup_{x > 0} \frac{4x}{x^2 + n^4}.$$

But $\sup_{x > 0} \dfrac{4x}{x^2 + n^4} = \dfrac{4n^2}{16n^4 + n^4}$ which tends to 0 as $n \to \infty$, thus $(h_n)_{n \in \mathbb{N}}$, $h_n(x) = \arctan \frac{4x}{x^2 + n^4}$, converges on $\mathbb{R}$ uniformly to 0.

d) Since $|\cos a_n x| \leq 1$ and $\lim_{n \to \infty} \dfrac{1}{n^\alpha} = 0$ we find

$$\sup_{x \in \mathbb{R}} |\cos(a_n x) - 0| = \sup_{x \in \mathbb{R}} |\cos(a_n x)| \leq \frac{1}{n^\alpha},$$

and once again we have uniform convergence.

4. Let $x, y \in I, x < y$. It follows that

$$f(x) = \lim_{n \to \infty} f_n(x) \leq \lim f_n(y) = f(y).$$

5. Since for $c_k$ there exists a sequence $(c_{k,n})_{n \in \mathbb{N}}$ of rational numbers $c_{k,n}$ converging to $c_k$, given $\epsilon > 0$ we can find $N(\epsilon) \in \mathbb{N}$ such that $n \geq N(\epsilon)$ implies $|c_k - c_{n,k}| < \frac{\epsilon}{N+1}$. This implies since $0 \leq x \leq 1$

$$|p(x) - p_n(x)| = \left| \sum_{k=0}^{N} (c_k - c_{n,k}) x^k \right|$$

$$\leq \sum_{k=0}^{N} |c_k - c_{n,k}| x^k$$

$$< \sum_{k=0}^{N} \frac{\epsilon}{N+1} = \epsilon,$$

and consequently for $n \geq N(\epsilon)$

$$\|p(.) - p_n(x)\|_\infty < \epsilon,$$

implying the uniform convergence of $(p_n)_{n \in \mathbb{N}}$ to $p$.

6. Let $\epsilon > 0$ be given. There exists $N(\epsilon) \in \mathbb{N}$ such that $n \geq N(\epsilon)$ implies $|f_n(y) - f(y)| \leq \|f_n - f\|_\infty < \frac{\epsilon}{2}$ as well as $|f(x_n) - f(x)| < \frac{\epsilon}{2}$ the latter due to the continuity of $f$ at $x$ and the convergence of $(x_n)_{n \in \mathbb{N}}$ to $x$. Hence for $n \geq N(\epsilon)$ it follows that $|f_n(x_n) - f(x)| \leq |f_n(x_n) - f(x_n)| + |f(x_n) - f(x)| < \frac{\epsilon}{2} + \frac{\epsilon}{2}$, i.e. $\lim_{n \to \infty} f_n(x_n) = f(x)$.

687

7. Let $x \in I$, there exists $\alpha < \beta$ such that $x \in [\alpha, \beta] \subset [a, b]$ and consequently we have
$$g_{\alpha,\beta}(x) = \lim_{n \to \infty} f_n \big|_{[\alpha,\beta]}(x)$$
For any interval $[\alpha', \beta'] \subset I$ such that $x \in [\alpha', \beta']$ it follows $g_{\alpha,\beta}(x) = g_{\alpha',\beta'}(x)$. So we may define $f : I \to \mathbb{R}, f(x) = g_{\alpha,\beta}(x)$ for some $[\alpha, \beta] \subset I, x \in [\alpha, \beta]$. Moreover for every $x \in I$ we have $f(x) = \lim_{n \to \infty} f_n(x)$. The continuity of $f$ at $x$ follows from the uniform convergence of $f_n \big|_{[\alpha,\beta]}$ to $g_{\alpha,\beta}$. Thus $g_{\alpha,\beta}$ is continuous on $[\alpha, \beta]$ and consequently $f$ is continuous for every $x \in I$.

Note that in general we cannot prove the uniform convergence of $f_n$ to $f$ on $(a, b)$.

8. We note first that $g_n(x) = \dfrac{f\left(x + \frac{1}{n}\right) - f(x)}{\frac{1}{n}}$ and now we use the mean value theorem to find
$$g_n(x) - f'(x) = \frac{f\left(x + \frac{1}{n}\right) - f(x)}{\frac{1}{n}} - f'(x)$$
$$= f'(\xi_n) - f'(x)$$
for some $\xi_n \in \left(x_1, x + \frac{1}{n}\right)$, or
$$|g_n(x) - f'(x)| = |f'(\xi_n) - f'(x)|.$$

Now we use the uniform continuity of $f'$: For $\epsilon > 0$ we can find $\delta > 0$ such that $|y - z| < \delta$ implies $|f'(y) - f'(z)| < \epsilon$. For $\delta$ we may find $N \in \mathbb{N}, N = N(\epsilon)$, such that $n \geq N(\epsilon)$ implies $\frac{1}{n} < \delta$, and consequently for $n \geq N(\epsilon)$
$$|g_n(x) - f'(x)| < \epsilon,$$
proving the uniform convergence of $(g_n)_{n \in \mathbb{N}}$ to $f$.

9. Since $f'_n(x) = \frac{1 - n^2 x^2}{(1 + n^2 x^2)^2}$ we find
$$\sup_{x \in [-1,1]} |f_n(x) - 0| = \sup_{x \in [-1,1]} |f_n(x)| = \frac{1}{2n},$$
and we obtain uniform convergence. Now for $x = 0$ we have $f'_n(0) = 1$ for all $n$, whereas for $x \in [-1, 1] \backslash \{0\}$ we have
$$\lim_{n \to \infty} \frac{1 - n^2 x^2}{(1 - n^2 x^2)^2} = 0.$$
Since the pointwise limit is not continuous, the convergence of the derivative cannot be uniform.

## Chapter 25

1. Let $\varphi \in T[a, b]$ be given with respect to the partition $Z_\varphi(x_0, \ldots, x_n)$ and $\psi \in T[a, b]$ with respect to the partition $Z_\psi(t_0, \ldots, t_m)$. Denote the joint partition by $Z = Z_\varphi \cup Z_\psi, Z = Z(y_0, \ldots, y_k)$. For $1 \leq l \leq k$ it follows that $\varphi|_{(y_{l-1}, y_l)} = c_l$ and $\psi|_{(y_{l-1}, y_l)} = d_l$ for some $c_l, d_l \in \mathbb{R}$. Consequently $(\varphi \cdot \psi)|_{(y_{l-1}, y_l)} = c_l \cdot d_l$ and hence $\varphi \cdot \psi \in T[a, b]$.

2. Since $f$ is Riemann integrable, given $\epsilon > 0$ there exists step functions $\varphi, \psi \in T[a, b]$ such that $\varphi \leq f \leq \psi$ and $\int_a^b (\psi - \varphi)dx < \epsilon$. Let $\varphi$ be given with respect to $Z_\varphi$ and $\psi$ with respect to $Z_\psi$. Let $Z = Z_\varphi \cup Z_\psi$ be the joint partition and $\tilde{Z} :=$ $Z \cup \{y_1, \ldots, y_N\}$. With respect to $\tilde{Z}$ we define two step functions

$$\tilde{\varphi}(x) = \begin{cases} c_j, & x \in \{y_1, \ldots, y_N\} \\ \varphi(x), & x \in [a, b] \setminus \{y_1, \ldots, y_N\} \end{cases}$$

and

$$\tilde{\psi}(x) = \begin{cases} c_j, & x \in \{y_1, \ldots, y_N\} \\ \psi(x), & x \in [a, b] \setminus \{y_1, \ldots, y_N\}. \end{cases}$$

It follows that $\tilde{\varphi} \leq \tilde{f} \leq \tilde{\psi}$ and further

$$\int_a^b (\tilde{\psi} - \tilde{\varphi})(x)dx = \int_a^b (\psi - \varphi)(x)dx < \epsilon.$$

The latter equality follows when using $\tilde{Z}$ to calculate both integrals. Note that for a step function represented with respect to $Z = Z(t_0, \ldots, t_m)$ the values at $t_j$, $1 \leq j \leq m$, do not contribute to the integral.

3.  a) Consider the function $g : [a, b] \to \mathbb{R}$ defined by

$$g(x) := \begin{cases} 1, & a \leq x < c < b \\ 0, & c \leq x \leq b. \end{cases}$$

Clearly $g|_{(a,c)} = 1$ and $g|_{(c,b)} = 0$ are continuous and the one-sided limits exist. However at $c$ we have

$$\lim_{\substack{x \to c \\ x > c}} g(x) = 1 \quad \text{and} \quad \lim_{\substack{x \to c \\ x < c}} g(x) = 0,$$

and therefore $g$ is not continuous at $c$.

b) First we note that if $h : (\lambda, \mu) \to \mathbb{R}$ is a bounded continuous function such that $\lim_{\substack{x \to \lambda \\ x > \lambda}} h(x) =: h_\lambda$ and $\lim_{\substack{x \to \mu \\ x < \mu}} h(x) =: h_\mu$ exist, then we can extend $h$ to a continuous function $\bar{h} : [\lambda, \mu] \to \mathbb{R}$ by defining

$$\bar{h}(x) := \begin{cases} h_\lambda, & x = \lambda \\ h(x), & x \in (\lambda, \mu) \\ h_\mu, & x = \mu. \end{cases}$$

Therefore for the piecewise continuous function $f : [a, b] \to \mathbb{R}$ there exists a partition $Z(x_0, \ldots, x_n)$ of $[a, b]$ such that $f|_{(x_{k-1}, x_k)}$ is continuous with continuous extension $\tilde{f}_k : [x_{k-1}, x_k] \to \mathbb{R}$.

689

For $\bar{f}_k$ there exists step functions $\varphi_k, \psi_k \in T[a,b]$ such that $\varphi_k \leq \bar{f}_k \leq \psi_k$ and $\int_{x_{k-1}}^{x_k} (\psi_k - \varphi_k)(x)dx < \frac{\epsilon}{n}$. We define the step functions

$$\varphi(x) := \begin{cases} f(x_j), & x = x_j, j = 0, \ldots, n \\ \varphi_k(x), & x \in (x_{k-1}, x_k), k = 1, \ldots, n \end{cases}$$

and

$$\psi(x) := \begin{cases} f(x_j), & x = x_j, j = 0, \ldots, n \\ \psi_k(x), & x \in (x_{k-1}, x_k), k = 1, \ldots, n. \end{cases}$$

It follows that $\varphi \leq f \leq \psi$ and

$$\int_a^b (\psi - \varphi)(x)dx = \sum_{k=1}^{n} \int_{x_{k-1}}^{x_k} (\psi - \varphi)(x)dx$$

$$= \sum_{k=1}^{n} \int_{x_{k-1}}^{x_k} (\psi_k - \varphi_k)(x)dx < \sum_{k=1}^{n} \frac{\epsilon}{n} = \epsilon,$$

where we used the result of Problem 2.

4. Since $f$ is Riemann integrable, given $\epsilon > 0$ there exists step functions $\varphi, \psi \in T[a,b]$ such that $\varphi \leq f \leq \psi$ and

$$\int_a^b (\psi - \varphi)(x)dx \leq \gamma^2 \epsilon.$$

Since $f \geq \gamma$ we may assume that $\varphi \geq \gamma$. It follows that $\frac{1}{\psi}, \frac{1}{\varphi} \in T[a,b]$ and $\frac{1}{\psi} \leq \frac{1}{f} \leq \frac{1}{\varphi} \leq \frac{1}{\gamma}$. Thus $\frac{1}{f}$ is bounded and further

$$\int_a^b \left( \frac{1}{\varphi} - \frac{1}{\psi} \right)(x)dx = \int_a^b \frac{1}{\varphi(x)\psi(x)}(\psi(x) - \varphi(x))dx$$

$$\leq \frac{1}{\gamma^2} \int_a^b (\psi - \varphi)(x)dx \leq \frac{1}{\gamma^2}\gamma^2 \epsilon = \epsilon,$$

proving the Riemann integrability of $\frac{1}{f}$.

5. By Problem 2 we know that changing a Riemann integrable function at finitely many points will not affect the value of its integral. Thus in general

$$\int_a^b |f(x)|dx = 0 \quad \text{does not imply} \quad f(x) = 0 \quad \text{for all } x \in [a,b].$$

6. Suppose $f \in C([a,b])$ and $\int_a^b |f(x)|dx = 0$. Suppose further that for some $x_0 \in (a,b)$ we have $f(x_0) \neq 0$, say $f(x_0) > 0$, the case $f(x_0) < 0$ is analogous. Since $f$ is continuous there exists $\delta > 0$ such that $(-\delta + x_0, x_0 + \delta) \subset (a,b)$ and $f(x) > \frac{f(x_0)}{2}$ for $x \in \left( -\frac{\delta}{2} + x_0, x_0 + \frac{\delta}{2} \right)$. Consequently

$$0 = \int_a^b |f(x)|dx \geq \int_{-\frac{\delta}{2}+x_0}^{\frac{\delta}{2}+x_0} f(x)dx \geq \int_{-\frac{\delta}{2}+x_0}^{\frac{\delta}{2}+x_0} \frac{f(x_0)}{2}dx = \frac{f(x_0)\delta}{2} > 0,$$

which is a contradiction. It is now easy to show that $||f||_{L^1} = \int_a^b |f(x)|dx$ is a norm on $C([a,b])$. We need to prove:

i) $||f||_{L^1} \geq 0$ and $||f||_{L^1} = 0$ if and only if $f = 0$, i.e. $f$ is constant and has the value 0;

ii) $||\lambda f||_{L^1} = |\lambda| ||f||_{L^1}$;

iii) $||f + g||_{L^1} \leq ||f||_{L^1} + ||g||_{L^1}$.

Clearly $\int_a^b |f(x)|dx \geq 0$ and $\int_a^b |f(x)|dx = 0$ if and only if $f = 0$ has just been proved above. For $\lambda \in \mathbb{R}$ we have

$$||\lambda f||_{L^1} = \int_a^b |\lambda f(x)|dx = |\lambda| \int_a^b |f(x)|dx = |\lambda| ||f||_{L^1},$$

and iii) is Minkowski's inequality.

7.    a) This problem is more of an interpretation of the result given in Theorem 25.24. By definition

$$\lim_{n \to \infty} S_n(f) = \int_a^b f(x)dx$$

if for $\epsilon > 0$ there exists $N \in \mathbb{N}$ such that $n \geq N$ implies

$$(*) \qquad \left| S_n(f) - \int_a^b f(x)dx \right| < \epsilon.$$

By Theorem 25.24 for $\epsilon > 0$ and any partition $Z(x_0, \ldots, x_n)$ with mesh size less than $\delta = \delta(\epsilon)$ and points $\xi_j \in [x_{j-1}, x_j]$ we have

$$(**) \qquad \left| S(f) - \int_a^b f(x)dx \right| < \epsilon$$

where $S(f)$ denotes the Riemann sum for $f$ with respect to $Z$ and $\xi_1, \ldots, \xi_n$. The mesh size of $Z_n$ is

$$x_j^{(n)} - x_{j-1}^{(n)} = a + \frac{j}{n}(b-a) - a - \frac{j-1}{n}(b-a) = \frac{b-a}{n}.$$

Hence, given $\epsilon > 0$ we determine $N \in \mathbb{N}$ such that for $n \geq N$ it follows that $\frac{b-a}{n} < \delta$ and now $(**)$ implies $(*)$.

b) Since

$$S_n(f) = \sum_{j=1}^{n} f(x_j^{(n)}) \frac{b-a}{n}$$

and since

$$\lim_{n \to \infty} S_n(f) = \int_a^b f(x)dx$$

it follows that

$$\lim_{n \to \infty} \left( \frac{1}{n} \sum_{j=1}^{n} f(x_j^{(n)}) \right) = \lim_{n \to \infty} \left( \frac{1}{b-a} S_n(f) \right) = \frac{1}{b-a} \int_a^b f(x)dx = \fint_a^b f(x)dx.$$

691

8. The following is Hölder's inequality for finite sums

$$\sum_{k=1}^{n} |\alpha_k \beta_k| \leq \left( \sum_{k=1}^{n} |\alpha_k|^p \right)^{\frac{1}{p}} \left( \sum_{k=1}^{n} |\beta_k|^q \right)^{\frac{1}{q}}.$$

Now let $Z_n = (x_0^{(n)}, \ldots, x_n^{(n)})$ be a sequence of partitions of $[a,b]$ whose mesh size converges to 0. For the corresponding Riemann sums of $f \cdot g$ we find with $\xi_j \in [x_{j-1}^{(n)}, x_j^{(n)}]$

$$\sum_{j=1}^{n} |f(\xi_j^{(n)}) g(\xi_j^{(n)}) (x_j^{(n)} - x_{j-1}^{(n)})|$$

$$= \sum_{j=1}^{n} |f(\xi_j^{(n)})(x_j^{(n)} - x_{j-1}^{(n)})|^{\frac{1}{p}} |g(x_j^{(n)})|^q (x_j^{(n)} - x_{j-1}^{(n)})|^{\frac{1}{q}}$$

$$\leq \left( \sum_{j=1}^{n} |f(\xi_j^{(n)})|^p (x_j^{(n)} - x_{j-1}^{(n)}) \right)^{\frac{1}{p}} \left( \sum_{j=1}^{n} |g(\xi_j^{(n)})|^q (x_j^{(n)} - x_{j-1}^{(n)}) \right)^{\frac{1}{q}}.$$

Passing to the limit $n \to \infty$ we obtain

$$\int_a^b |f(x)g(x)| dx \leq \left( \int_a^b |f(x)|^p dx \right)^{\frac{1}{p}} \left( \int_a^b |g(x)|^q dx \right)^{\frac{1}{q}}.$$

Note that if we agree that $q = \infty$ is the conjugate of $p = 1$, i.e. $\frac{1}{p} + \frac{1}{q} = 1$, then we find

$$\int_a^b |f(x)g(x)| dx \leq \sup_{x \in [a,b]} |g(x)| \int_a^b |f(x)| dx,$$

or

$$\int_a^b |f(x)g(x)| dx \leq \|f\|_{L^1} \|g\|_\infty.$$

9.   a) We follow the hint and apply Hölder's inequality to $|f|^p$ and 1: so given $p$ and $q$ we take $r = \frac{q}{p}$, and $r'$ such that $\frac{1}{r} + \frac{1}{r'} = 1$, i.e. $r' = \frac{r}{r-1} = \frac{q}{q-p}$, to find

$$\int_a^b |f(x)|^p dx = \int_a^b |f(x)|^p \cdot 1 dx$$

$$\leq \left( \int_a^b |f(x)|^{p \cdot r} dx \right)^{\frac{1}{r}} \left( \int_a^b 1^{r'} dx \right)^{\frac{1}{r'}}$$

$$= \left( \int_a^b |f(x)|^q dx \right)^{\frac{p}{q}} (b-a)^{\frac{q-p}{q}}.$$

b) By the Cauchy-Schwarz inequality we have

$$(*) \qquad \int_a^b |f(x)g(x)|dx \le \left( \int_a^b |f(x)|^2 dx \right)^{\frac{1}{2}} \left( \int_a^b |g(x)|^2 dx \right)^{\frac{1}{2}}.$$

Further for $A, B > 0$, noting that $A \cdot B = \sqrt{2\epsilon}A \cdot \frac{1}{\sqrt{2\epsilon}}B$, we have

$$(**) \qquad AB \le \frac{1}{2}(\sqrt{2\epsilon}A)^2 + \frac{1}{2}\left( \frac{1}{2\epsilon}B \right)^2 = \epsilon A^2 + \frac{1}{4\epsilon}B^2.$$

The result follows by applying $(**)$ to $(*)$ with $A = \left( \int_a^b |f(x)|^2 dx \right)^{\frac{1}{2}}$ and $B = \left( \int_a^b |g(x)|^2 dx \right)^{\frac{1}{2}}$.

10. By the Cauchy-Schwarz inequality we find

$$\left( \int_a^b f(x) \sin kx \, dx \right)^2 \le \left( \int_a^b f^2(x) \, dx \right) \left( \int_a^b \sin^2 kx \, dx \right)$$

and

$$\left( \int_a^b f(x) \cos kx \, dx \right)^2 \le \left( \int_a^b f^2(x) \, dx \right) \left( \int_a^b \cos^2 kx \, dx \right).$$

Adding these inequalities gives

$$\left( \int_a^b f(x) \sin kx \, dx \right)^2 + \left( \int_a^b f(x) \cos kx \, dx \right)^2$$

$$\le \left( \int_a^b f^2(x) \, dx \right) \left( \int_a^b \sin^2 kx \, dx + \int_a^b \cos^2 kx \, dx \right)$$

$$= \left( \int_a^b f^2(x) \, dx \right) \left( \int_a^b (\sin^2 kx + \cos^2 kx) \, dx \right)$$

$$= (b-a) \int_a^b f^2(x)dx,$$

since $\sin^2 kx + \cos^2 kx = 1$.

11. Let $Z = Z(x_0, \dots, x_n)$ be a partition of $[a, b]$ and consider the two Riemannian sums

$$\sum_{j=1}^n h(x_j)(x_j - x_{j-1}) \text{ and } \sum_{j=1}^n f(h(x_{j-1}))(x_j - x_{j-1}).$$

693

We set $\lambda_j := \frac{x_j - x_{j-1}}{b-a}$ and therefore $0 \le \lambda_j \le 1$ and $\lambda_1 + \cdots + \lambda_n = 1$. By Problem 1 in Chapter 23 we find

$$f\left(\frac{1}{b-a}\sum_{j=1}^{n} h(x_j)(x_j - x_{j-1})\right) = f\left(\sum_{j=1}^{n} \lambda_j h(x_j)\right)$$

$$\le \sum_{j=1}^{n} \lambda_j f(h(x_j))$$

$$\frac{1}{b-a}\sum_{j=1}^{n} f(h(x_j))(x_j - x_{j-1}).$$

If we replace now $Z$ by a sequence $(Z_k)_{k\in\mathbb{N}}$ of partitions such that $Z_{k+1}$ is a refinement of $Z_k$ and for the mesh sizes we have $\eta(Z_k) \to 0$ as $k \to \infty$, then it follows from

$$f\left(\frac{1}{b-a}\sum_{j=1}^{n_k} h(x_j)(x_j - x_{j-1})\right) \le \frac{1}{b-a}\sum_{j=1}^{n_k} f(h(x_j))(x_j - x_{j-1})$$

and the continuity of $f$, recall that convex functions on an interval are continuous in the interior, see Corollary 23.6, that

$$f\left(\frac{1}{b-a}\int_a^b h(t)dt\right) \le \frac{1}{b-a}\int_a^b f(h(t))dt.$$

12. The following is the graph of $g_n$

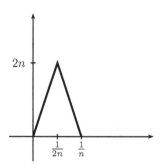

First let us show that $g_n$ is continuous. Clearly $g_n\big|_{[0,\frac{1}{2n})\cup(\frac{1}{2n},\frac{1}{n})\cup(\frac{1}{n},0]}$ is continuous. If $x_0 = \frac{1}{2n}$ we have

$$\lim_{\substack{x \to \frac{1}{2n} \\ x < \frac{1}{2n}}} g_n(x) = \lim_{\substack{x \to \frac{1}{2n} \\ x < \frac{1}{2n}}} 4n^2 x = 2n$$

and

$$\lim_{\substack{x \to \frac{1}{2n} \\ x > \frac{1}{2n}}} g_n(x) = \lim_{\substack{x \to \frac{1}{2n} \\ x > \frac{1}{2n}}} (-4n^2 x + 4n) = 2n,$$

694

i.e. $g_n$ is continuous at $x_0 = \dfrac{1}{2n}$. At $x_0 = \dfrac{1}{n}$ we find

$$\lim_{\substack{x \to \frac{1}{n} \\ x < \frac{1}{n}}} g_n(x) = \lim_{\substack{x \to \frac{1}{n} \\ x < \frac{1}{n}}} (-4n^2x + 4n) = 0$$

and

$$\lim_{\substack{x \to \frac{1}{n} \\ x > \frac{1}{n}}} g_n(x) = \lim_{\substack{x \to \frac{1}{n} \\ x > \frac{1}{n}}} 0 = 0,$$

implying $g_n$ is continuous on $[0,1]$. As indicated in the hint the integral is the area of the triangle $(0,0), (\frac{1}{n},0), (\frac{1}{2n}, 2n)$, hence

$$\int_0^1 g_n(x)dx = \frac{1}{2} \cdot \frac{1}{n} \cdot 2n = 1.$$

thus

$$\lim_{n \to \infty} \int_0^1 g_n(x)dx = 1.$$

Finally we claim: For every $x \in [0,1]$ it holds

$$\lim_{n \to \infty} g_n(x) = 0.$$

For $x = 0$ or $x = 1$ this follows from the definition. Now let $x \in (0,1)$. Since $x > 0$ it follows that for some $N$ it holds $x < \frac{1}{N}$ and now, for $n \geq N$ it follows $g_n(x) = 0$ implying that $\lim_{n\to\infty} g_n(x) = 0$. Since $\int_0^1 0dx = 0$ we find in this case

$$1 = \lim_{n \to \infty} \int_0^1 g_n(x)dx \neq \int_0^1 \lim g_n(x)dx = 0.$$

13. Since $(f_n)_{n\in\mathbb{N}}$ converges uniformly to $f$, given $\epsilon > 0$ there exists $N = N(\epsilon) \in \mathbb{N}$ such that $x \in [a,b]$ and $n \geq N(\epsilon)$ implies $|f_n(x) - f(x)| < \frac{\epsilon}{2(b-a)}$. For any $n$ we have

$$f_n - |f_n - f| \leq f \leq f_n + |f_n - f|.$$

Now, since $f_n$ is Riemann integrable there exists step functions $\varphi_n, \psi_n \in T[a,b]$ such that $\varphi_n \leq f_n \leq \psi_n$ and $\int_a^b (\psi_n - \varphi_n)dx < \frac{\epsilon}{2}$. Hence for $n \geq N(\epsilon)$ we find with the step functions $\varphi_n - \frac{\epsilon}{2(b-a)}$ and $\psi_n + \frac{\epsilon}{2(b-a)}$ that

$$\varphi_n - \frac{\epsilon}{2(b-a)} \leq f \leq \psi_n + \frac{\epsilon}{2(b-a)}$$

and

$$\int_a^b \left( \psi_n + \frac{\epsilon}{2(b-a)} - \varphi_n + \frac{\epsilon}{2(b-a)} \right) dx = \int_a^b (\psi_n - \varphi_n)dx + \frac{\epsilon}{2} < \epsilon.$$

Hence we have proved that the uniform limit of a sequence of Riemann integrable functions is Riemann integrable. Now it follows that

$$\left| \int_a^b f(x)dx - \int_a^b f_n(x)dx \right| = \left| \int_a^b (f(x) - f_n(x))dx \right|$$

$$\leq \int_a^b |f(x) - f_n(x)|dx \leq (b-a)\|f_n - f\|_\infty$$

implying

$$\lim_{n \to \infty} \int_a^b f_n(x)dx = \int_a^b f(x)dx.$$

## Chapter 26

1.  a) Let $F$ be a primitive of $f$, i.e.

$$F(x) = \int_a^x f(t)dt + c.$$

Since $F'(x) = f(x)$ by Theorem 26.1, if $f \in C^k([a,b])$ then $F' \in C^k([a,b])$ and $F$ is $(k+1)$-times continuously differentiable.

b) Recall that if $V$ is a $\mathbb{R}$ vector space a set $W_a = a + W$, $a \in V$ and $W \subset V$ a subspace, is called an affine subspace of $V$. The dimension of $W_a$ is that of $W$. Clearly the constant function $f_c : [a,b] \to \mathbb{R}$, $f_c(x) = c$, forms a one-dimensional subspace of $C([a,b])$, a basis for example is given by $f_1, f_1(x) = 1$. If $f \in C([a,b])$ then the set of all its primitives is given by

$$\left\{ g : [a,b] \to \mathbb{R} \mid g(x) = \int_a^x f(t)dt + f_c, c \in \mathbb{R} \right\}$$

or with $W := \{f_c | c \in \mathbb{R}\} \subset C^1([a,b])$ and $F \in C^1([a,b])$, $F(x) := \int_a^x f(t)dt$, the set of all primitives of $f$ is the affine subspace $F + W \subset C^1([a,b])$.

2. Note that nothing is claimed about the existence of a fixed point. The statement is that if $T$ has a fixed point then the fixed point must belong to $C^\infty([a,b])$.
   Now, by Theorem 26.1 we have that $Tf$ is differentiable and $(Tf)'(x) = e^{-x}f(x)$. This implies that for a fixed point $Tg(x) = g(x)$ that $g$ is in $C^1$, i.e. a continuously differentiable function. Therefore $t \mapsto e^{-t}g(t)$ is a $C^1$ function implying that $Tg$ is a $C^2$ function. By induction it follows that if $g = Tg$ and $g \in C^k([a,b])$ then $g \in C^{k+1}([a,b])$ and therefore a fixed point belongs to $C^\infty([a,b])$.

3.  a) Let $I_1 = [a_1, b_1)$ and $I_2 = [a_2, b_2)$ and assume that $a_1 \leq a_2$. If $b_1 < a_2$ then $I_1 \cup I_2$ is the union of two disjoint intervals. In the case that $a_2 \leq b_1$ then we either have $I_2 \subset I_1$, namely if $b_2 \leq b_1$, hence $I_1 \cup I_2 = I_1$, or, if $b_1 \leq b_2$ then $I_1 \cup I_2 = [a_1, b_2)$. Now, for finitely many right half-open intervals $I_1, \ldots, I_N$, $I_j = [a_j, b_j)$ we proceed by induction. The case $N = 2$ has just been proved. We assume that $I_1 \cup \cdots \cup I_{N-1}$ is the union of mutually disjoint right half-open

intervals with some $b_{j_0}$, $j_0 \leq N - 1$, being the supremum of $I_1 \cup \cdots \cup I_{N-1}$. Now, if $b_{j_0} < a_N$ we are done. If not, for some $j_1 \leq N - 1$ we have $[a_{j_1}, b_{j_1}) \cap I_N \neq \emptyset$. If now $b_N < b_{j_2}$ for some $j_2 \geq j_1$, then

$$I_1 \cup \cdots \cup I_N = I_1 \cup \cdots \cup I_{j_1-1} \cup [a_{j_1}, b_{j_2}) \cup I_{j_2+1} \cup \cdots \cup I_{N-1}.$$

If however $b_j < b_N$ for $j \leq N - 1$ we have $I_1 \cup \cdots \cup I_N = I_1 \cup \cdots \cup I_{j_1-1} \cup [a_{j_1}, b_N)$. For the intersection of $[a_1, b_1) \cap [a_2, b_2), a_1 \leq a_2$ we find for $b_1 < a_2$ that $[a_1, b_1) \cap [a_2, b_2) = \emptyset$, otherwise we find $[a_1, b_1) \cap [a_2, b_2) = [a_2, \min(b_1, b_2))$.

b) Let $I_1 = [a_1, b_1)$ and $I_2 = [a_2, b_2), a_1 \leq a_2$. If $I_1 \cap I_2 = \emptyset$ there is nothing to prove. If $I_1 \cap I_2 \neq \emptyset$ then $I_1 \cup I_2 = [a, b)$ and $I_1 \cap I_2 = [c, d)$ with the following possibilities $[a, b) = [a_1, b_1)$ implying $[c, d) = [a_2, b_2)$, $[a, b) = [a_1, b_2)$ implying that $[c, d) = [a_2, b_1)$ with the convention that $[a_2, b_1) = \emptyset$ if $a_2 = b_1$. In the first case we have

$$\mu(I_1 \cup I_2) + \mu(I_1 \cap I_2) = \int_{a_1}^{b_1} f(t)dt + \int_{a_2}^{b_2} f(t)dt = \mu(I_1) + \mu(I_2)$$

and in the second case we find

$$\mu(I_1 \cup I_2) + \mu(I_1 \cap I_2) = \int_{a_1}^{b_2} f(t)dt + \int_{a_2}^{b_1} f(t)dt$$

$$= \int_{a_1}^{b_1} f(t)dt + \int_{b_1}^{b_2} f(t)dt + \int_{a_2}^{b_1} f(t)dt$$

$$= \int_{a_1}^{b_1} f(t)dt + \int_{a_2}^{b_2} f(t)dt = \mu(I_1) + \mu(I_2).$$

c) Since $\mu_{a_0}(x) = \mu([a_0, x)) = \int_{a_0}^{x} f(t)dt$ the result follows from Theorem 26.1.

4. First we note that for every $x$ we get

$$0 = \int_{-x}^{x} f(t)dt = \int_{0}^{x} (f(-t) + f(t))dt = 0$$

which yields for all $x, y$ that

$$\int_{x}^{y} (f(-t) + f(t))dt = 0.$$

Now we claim that if for a continuous function $g : [a, b] \to \mathbb{R}$ we have for all $\alpha, \beta \in [a, b], \alpha < \beta$ that $\int_{\alpha}^{\beta} g(t)dt = 0$ then $g(t) = 0$ for all $t$. Indeed, take $t_0 \in [a, b]$ and $h > 0$ such that $a \leq t_0 < t < t_0 + h \leq b$, to find by our assumptions and by the mean value theorem

$$0 = \int_{t_0}^{t_0+h} g(t)dt = g(\xi_h)h, \xi_h \in [t_0, t_0 + h].$$

This implies that $g(\xi_h) = 0$ and since $\lim_{h \to 0} \xi_h = t_0$ the continuity of $g$ implies that $g(t_0) = 0$. Therefore we deduce that $f(-x) + f(x) = 0$ for all $x$, i.e. $f(-x) = -f(x)$ which implies that $f$ is odd.

5.   a) We note that

$$y^\rho - x^\rho = \rho \int_x^y t^{\rho-1} dt \le \rho \int_x^y 1 \, dt = \rho(y - x).$$

b) Since

$$\int_x^y \cos t \, dt = \sin y - \sin x$$

and for $-\frac{\pi}{4} \le x < y \le \frac{\pi}{4}$ we have

$$\int_x^y \cos t \, dt \ge \frac{1}{2}\sqrt{2}(y - x)$$

the estimate $(y - x) \le \frac{2}{\sqrt{2}}(\sin y - \sin x)$ follows.

6. For $F$ we have

$$F(x) - F(y) = \int_a^x f(t)dt - \int_a^y f(t)dt$$
$$= \int_y^x f(t)dt$$

which implies

$$|F(x) - F(y)| = \left| \int_y^x f(t)dt \right| \le ||f||_\infty \left| \int_y^x 1 dt \right| = ||f||_\infty |x - y|.$$

7.   a) Since $(f \cdot g)(x) = 0$ for all $x \in [a, b]$, but $f$ and $g$ are not both zero, it follows that $f \perp g$.

b) The product of an odd function and an even function is odd. For any odd function $h : [-a, a] \to \mathbb{R}$ we have, see Proposition 26.7.B, $\int_{-a}^a h(t)dt = 0$.

c) Let $g, h \in C([a, b])$ such that $f \perp g$ and $f \perp h$. For $\lambda, \mu \in \mathbb{R}$ we find

$$\int_a^b f(x)(\lambda g(x) + \mu h(x))dx = \lambda \int_a^b f(x)g(x)dx + \mu \int_a^b f(x)h(x)dx = 0,$$

hence $f \perp (\lambda g + \mu h)$, which implies the required result.

8. We start with

$$0 = \int_a^b f(x)f'(x)dx = \int_a^b \frac{1}{2}\frac{d}{dx}(f^2(x))dx = \frac{1}{2}f^2(b) - \frac{1}{2}f^2(a),$$

or $f^2(b) = f^2(a)$ implying $|f(b)| = |f(a)|$.

698

9. Obviously we have for $f, g \in C_0^1([a, b])$ and $\lambda \in \mathbb{R}$ that

$$\|f'\|_{L^2} \geq 0,$$

$$\|\lambda f'\|_{L^2} = |\lambda| \|f'\|_{L^2},$$

and

$$\|f' + g'\|_{L^2} \leq \|f'\|_{L^2} + \|g'\|_{L^2},$$

since these results hold for all $f, g \in C^1([a, b])$ and $\lambda \in \mathbb{R}$. In order to prove that $\|f'\|_{L^2}$ is a norm we need to show in addition that $\|f'\|_{L^2} = 0$ implies $f(x) = 0$ for all $x \in [a, b]$, i.e. $f$ is the zero element in $C_0^1([a, b])$. By Proposition 26.16 we know Poincaré's inequality:

$$\|f\|_{L^2} \leq \gamma_0 \|f'\|_{L^2}.$$

Thus $\|f'\|_{L^2} = 0$ implies $\|f\|_{L^2} = 0$ or $\int_a^b (f(x))^2 dx = 0$. But by Problem 6 in Chapter 25 we now find that $f(x) = 0$ for all $x \in [a, b]$.

10. Since $G$ is differentiable and

$$G'(x) = \beta'(x)f(\beta(x)) - \alpha'(x)f(\alpha(x))$$

the fact that $f(y) \geq 0$ for all $y$ and $\beta'(x) \geq 0$ whereas $\alpha'(x) \leq 0$, which follows from the fact that $\beta$ is increasing and $\alpha$ is decreasing, we find $G'(x) \geq 0$, hence $G$ is increasing.

11. For $\epsilon > 0$ we can find $N_0 \in \mathbb{N}$ such that for $n, m \geq N_0$ it follows that

$$|f_n(x_0) - f_m(x_0)| < \frac{\epsilon}{2}$$

and

$$|f_n'(t) - f_m'(t)| < \frac{\epsilon}{2(b-a)} \quad \text{for all } t \in [a, b].$$

We now apply the mean value theorem to $f_n - f_m$ to find

$$|(f_n - f_m)(x) - (f_n - f_m)(t)| < \frac{\epsilon|x - t|}{2(b-a)} \leq \frac{\epsilon}{2}$$

for all $x, t \in [a, b]$ and $n, m \geq N_0$. Hence it follows with $t = x_0$ for $n, m \geq N_0$ that

$$|f_n(x) - f_m(x)| \leq |(f_n - f_m)(x) - (f_n - f_m)(x_0)| + |f_n(x_0) - f_m(x_0)| < \epsilon,$$

or $n, m \geq N_0$ implies

$$\|f_n - f_m\|_\infty < \epsilon,$$

i.e. $(f_n)_{n \in \mathbb{N}}$ is a Cauchy sequence with respect to $\|\cdot\|_\infty$. Therefore it has a limit $f$ which is a continuous function $f$. Moreover $f_n$ converges pointwise to $f$. Denote by $f^*$ the uniform limit of $(f_n')_{n \in \mathbb{N}}$. It follows that

$$f_n(x) = f_n(x_0) + \int_{x_0}^x f_n'(t) dt$$

and for $n \to \infty$ we get

$$f(x) = f(x_0) + \int_{x_0}^{x} f^*(t)dt,$$

implying that $f'(x) = f^*(x)$.

12. We know by Theorem 16.4 that

$$S_N(x) = \frac{1 - x^{N+1}}{1 - x}$$

implying that

$$S'_N(x) = \frac{1 - (N+1)x^N + Nx^{N+1}}{(1-x)^2}.$$

Moreover, for $|x| < 1$ we have $\sum_{k=0}^{\infty} x^k = \frac{1}{1-x}$, so we need to prove that for $[a,b] \subset (-1,1)$

$$\sup_{x \in [a,b]} \left| S_N(x) - \frac{1}{1-x} \right| \quad \text{and} \quad \sup_{x \in [a,b]} \left| S'_N(x) - \frac{1}{(1-x)^2} \right|$$

both tend to 0.

Denote by $\kappa_1 := \sup_{x \in [a,b]} \left| \frac{1}{1-x} \right| < \infty$ and $\kappa_2 := \sup_{x \in [a,b]} \left| \frac{1}{(1-x)^2} \right| < \infty$. It follows that

$$\left| S_N(x) - \frac{1}{1-x} \right| = \frac{|x|^{N+1}}{1-x} \leq \kappa_1 \max(|a|^{N+1}, |b|^{N+1})$$

hence

$$\sup_{x \in [a,b]} \left| S_N(x) - \frac{1}{1-x} \right| \leq \kappa_1 \max(|a|^{N+1}, |b|^{N+1})$$

and since $|a| < 1$ and $|b| < 1$ the uniform convergence of $S_N(x)$ to $\frac{1}{1-x}$ is proved. Further

$$\left| S'_N(x) - \frac{1}{(1-x)^2} \right| = \frac{(N+1)|x|^N + N|x|^{N+1}}{(1-x)^2}$$
$$\leq \kappa_2(2N+1) \max(|a|^N, |b|^N),$$

where we used that for $|y| < 1$ we have $|y|^{N+1} < |y|^N$. Since $\lim_{N \to \infty}(2N+1)|y|^N = 0$ for $|y| < 1$ it also follows that $S'_N(x)$ converges uniformly to $\frac{1}{(1-x)^2}$. Therefore we have

$$\frac{1}{(1-x)^2} = \frac{d}{dx} \sum_{k=1}^{\infty} x^k = \frac{d}{dx} \lim_{N \to \infty} S_N(x)$$

$$= \lim_{N \to \infty} \left( \frac{d}{dx} S_N(x) \right) = \lim_{N \to \infty} \sum_{k=1}^{N} kx^{k-1} = \sum_{k=1}^{\infty} kx^{k-1}.$$

For $x = \frac{1}{m}, m \geq 2$, we find

$$\sum_{k=1}^{\infty} k \left( \frac{1}{m} \right)^{k-1} = m \sum_{k=1}^{\infty} \frac{k}{m^k} = \frac{1}{\left( 1 - \frac{1}{m} \right)^2} = \frac{m^2}{m^2 - 1},$$

i.e. $\sum_{k=1}^{\infty} \frac{k}{m^k} = \frac{m}{m^2 - 1}.$

## Chapter 27

1. For $x \in (-\infty, 0) \cup (0, c) \cup (c, \infty)$ the function $u_c$ is clearly differentiable. At $x = 0$ we find

$$\frac{u_c(x) - u_c(0)}{x - 0} = \begin{cases} \frac{x}{4}, & x > 0 \\ 0, & c \leq x \leq 0 \end{cases}$$

implying

$$\lim_{x \to 0} \frac{u_c(x) - u_c(0)}{x - 0} = 0,$$

and at $x = c$ we have

$$\frac{u_c(x) - u_c(c)}{x - c} = \begin{cases} 0, & c \geq x \leq 0 \\ -\frac{(x-c)}{4}, & x \leq c, \end{cases}$$

which yields

$$\lim_{x \to 0} \frac{u_c(x) - u_c(c)}{x - c} = 0,$$

hence $u_c$ is on $\mathbb{R}$ differentiable. Moreover

$$u_c'(c) = \begin{cases} \frac{x}{2}, & x > 0 \\ 0, & c \leq x \leq 0 \\ -\frac{(x-c)}{2}, & x < c, \end{cases}$$

i.e.

$$u_c'(x) = \sqrt{|u_c(x)|}.$$

Next we observe that $u_c(2) = 1$ for all $c < 0$, implying that for all $c < 0$ a solution to

$$v'(x) = \sqrt{|v(x)|}, \quad v(2) = 1$$

is given by $u_c$. Hence we have existence but not uniqueness.

2. The calculation is simple and goes as follows:

$$\begin{aligned} f(x)(\lambda u_1 + \mu u_2)'(x) &+ h(x)(\lambda u_1 + \mu u_2)(x) \\ &= f(x)(\lambda u_1'(x) + \mu u_2'(x)) + h(x)(\lambda u_1(x) + \mu u_2(x)) \\ &= \lambda f(x) u_1'(x) + \mu f(x) u_2'(x) + \lambda h(x) u_1(x) + \mu h(x) u_2(x) \\ &= \lambda (f(x) u_1'(x) + h(x) u_1(x)) + \mu (f(x) u_2'(x) + h(x) u_2(x)) \\ &= 0. \end{aligned}$$

It is important to note that if $u'$ and $u$ only appear linearly in a differential equation then linear combinations of solutions are solutions.

3. First we note that

$$u(a) = u_a e^{-\int_a^a \frac{p_1(t)}{p_0(t)} dt} = u_a e^0 = u_a,$$

i.e. the initial condition is fulfilled. Differentiating $u$ we find

$$u'(x) = \frac{d}{dx} \left( u_a e^{-\int_a^x \frac{p_1(t)}{p_0(t)} dt} \right)$$

$$= u_a \left( \frac{d}{dx} \left( -\int_a^x \frac{p_1(t)}{p_0(t)} dt \right) \right) e^{-\int_a^x \frac{p_1(t)}{p_0(t)} dt}$$

$$= -u_a \frac{p_1(x)}{p_0(x)} e^{-\int_a^x \frac{p_1(t)}{p_0(t)} dt},$$

and it follows that

$$p_0(x)u'(x) + p_1(x)u(x)$$

$$= -p_0(x)u_a \frac{p_1(x)}{p_0(x)} e^{-\int_a^x \frac{p_1(t)}{p_0(t)} dt} + p_1(x)u_a e^{-\int_a^x \frac{p_1(t)}{p_0(t)} dt}$$

$$= u_0 e^{-\int_a^x \frac{p_1(t)}{p_0(t)} dt} (-p_1(x) + p_1(x)) = 0.$$

4.  a) Using the method of separation of variables we find $xu' = 2u$ or $\frac{du}{u} = \frac{2dx}{x}$ which yields

$$\ln |u(x)| = 2 \ln |x| + c$$

with some constant $c$. From here we derive

$$u(x) = kx^2$$

where $k$ is any real number. The initial condition demands $u(1) = k = 3$, so we expect $u(x) = 3x^2$ to be a solution to this initial value problem. Indeed we have $u(1) = 3 \cdot 1^2 = 3$ and $u'(x) = 6x$, hence $xu'(x) = 6x^2 = 2 \cdot u(x)$. Obviously $u$ is defined on the whole real line.

b) From $y'(t) = 2y^2(t)$ we derive $\frac{dy}{y^2} = 2dt$, or

$$-\frac{1}{y} = 2t + c$$

which gives $y(t) = -\frac{1}{2t+c}$. Adjusting the initial value requires

$$y(0) = -\frac{1}{c} = -1,$$

implying that $y(t) = -\frac{1}{2t+1}$ is a candidate for a solution. We find $y(0) = -1$ and further

$$y'(t) = \frac{2}{(2t+1)^2} = 2y^2(t).$$

c) The differential equation $\varphi'(s) = \frac{\varphi(s)}{\tan s}$ leads to

$$\frac{d\varphi}{\varphi} = \frac{ds}{\tan s} = \frac{\cos s}{\sin s} ds = \frac{(\sin s)'}{\sin s} ds$$

or

$$\ln |\varphi| = \ln |\sin s| + c,$$

which yields

$$\varphi(s) = \gamma \sin s$$

for some $\gamma \in \mathbb{R}$. The condition $\varphi(\frac{\pi}{4}) = \frac{\pi}{4}$ implies

$$\frac{\pi}{4} = \gamma \sin \frac{\pi}{4} = \gamma \frac{\sqrt{2}}{2}$$

or

$$\gamma = \frac{\pi}{2\sqrt{2}} = \frac{\pi\sqrt{2}}{4}.$$

For $\varphi(s) = \frac{\pi\sqrt{2}}{4} \sin s$ we find $\varphi'(s) = \frac{\pi\sqrt{2}}{4} \cos s$ which gives

$$\varphi'(s) = \frac{\pi\sqrt{2}}{4} \cos s = \frac{\pi\sqrt{2}}{4} \sin s \left(\frac{\cos s}{\sin s}\right) = \frac{\varphi(s)}{\tan s},$$

as well as $\varphi(\frac{\pi}{4}) = \frac{\pi\sqrt{2}}{4} \sin \frac{\pi}{4} = \frac{\pi\sqrt{2}}{4} \cdot \frac{\sqrt{2}}{2} = \frac{\pi}{4}$. Note that $\varphi$ is defined on $\mathbb{R}$, but the coefficient in the differential equation is not defined for $s = k\pi$, where $\tan s = 0$ and for $s = \frac{\pi}{2} + k \in \mathbb{Z}$ where $\tan s$ is not defined.

d) From $5x^4(r)x'(r) = r \cos r$ we deduce

$$5x^4 dx = r \cos r \, dr$$

or

$$x^5 = \cos r + r \sin r + c,$$

which yields

$$x(r) = (\cos r + r \sin r + c)^{\frac{1}{5}},$$

and $x(\frac{\pi}{2}) = 1$ implies

$$1 = (\cos \frac{\pi}{2} + \frac{\pi}{2} \sin \frac{\pi}{2} + c)^{\frac{1}{5}}$$

which is solved by $c = 1 - \frac{\pi}{2}$. An easy calculation now shows that $x(r) = (\cos r + r \sin r + 1 - \frac{\pi}{2})^{\frac{1}{5}}$ indeed solves the initial value problem: $x(\frac{\pi}{2}) = (\cos \frac{\pi}{2} + \frac{\pi}{2} \sin \frac{\pi}{2} + 1 - \frac{\pi}{2})^{\frac{1}{2}} = 1$, and

$$\frac{d}{dr}(\cos r + r \sin r + 1 - \frac{\pi}{2})^{\frac{1}{5}}$$
$$= \frac{1}{5}(\cos r + r \sin r + 1 - \frac{\pi}{2})^{-\frac{4}{5}}(-\sin r + \sin r + r \cos r)$$
$$= \frac{1}{5}(\cos r + r \sin r + 1 - \frac{\pi}{2})^{-\frac{4}{5}}(r \cos r),$$

703

or

$$5x^4(r)x'(r)$$

$$= 5(\cos r + r\sin r + 1 - \frac{\pi}{2})^{\frac{4}{5}}\frac{1}{5}(\cos r + r\sin r + 1 - \frac{\pi}{2})^{-\frac{4}{5}}(r\cos r)$$

$$= r\cos r.$$

Again the solution is defined for all $r \in \mathbb{R}$.

5.  a)

$$\frac{d}{dx}\int_{\cos x}^{\sqrt{x^2+1}} g(z)dz$$

$$= g(\sqrt{x^2+1})\frac{d}{dx}\sqrt{x^2+1} - g(\cos x)\frac{d}{dx}\cos x$$

$$= g(\sqrt{x^2+1})\frac{x}{\sqrt{x^2+1}} + g(\cos x)\sin x.$$

b)

$$\frac{d}{dx}\int_{v(x)}^{u(x)} \frac{1}{1+t^2}dt = \frac{1}{1+u^2(x)}u'(x) - \frac{1}{1+v^2(x)}v'(x).$$

6. We have

$$\frac{d}{dx}\int_{-u(x)}^{u(x)} h(t)dt = h(u(x))u'(x) - h(-u(x))(-u(x))'$$

$$= h(u(x))u'(x) + h(u(x))(-u(x))'$$

$$= h(u(x))u'(x) - h(u(x))u(x) = 0,$$

thus $x \mapsto \int_{-u(x)}^{u(x)} h(t)dt$ has derivative zero, and therefore it must be constant. We know that for every odd function $h$ we have

$$\int_{-a}^{a} h(t)dt = 0,$$

compare with Proposition 26.7.B, and therefore we must have for all $x \in \mathbb{R}$ that $\int_{-u(x)}^{u(x)} h(t)dt = 0$.

7. Since $u^{2k} \geq 0$ it follows from $u' = \frac{1}{1+u^{2k}}$ that $u$ is strictly monotone increasing, and since $u(0) = 1$ we deduce that on $[0, \infty)$ the function $u$ is positive. Further we have

$$u''(x) = \frac{d}{dx}\frac{1}{1+u^{2k}(x)} = \frac{2ku^{2k-1}(x)u'(x)}{(1+u^{2k}(x))^2}$$

$$= \frac{2ku^{2k-1}(x)}{(1+u^{2k}(x))^3} > 0$$

704

since $u^{2k-1}(x) > 0$ (which follows from $u$ being strictly positive). Hence $u$ is convex. The fact that $u$ is an arbitrarily often differentiable function follows as discussed at the end of Chapter 27: we know that

$$u' = g_1(u), \quad g_1(t) = \frac{1}{1+t^{2k}}$$

and

$$u'' = g_2(u), \quad g_2(t) = \frac{2kt^{2k-1}}{(1+t^{2k})^3}.$$

Now we claim that $u^{(n)} = g_n(u)$ with an arbitrarily often differentiable function $g_n$. For $n = 1$ (and $n = 2$) we know the result. Now if $u^{(n)} = g_n(u)$ then

$$u^{(n+1)} = \frac{d}{dx}g_n(u) = g_n'(u) \cdot u'$$
$$= g_n'(u)g_1(u)$$

implying the result.

## Chapter 28

1.　　a) Denote by $g_\alpha$ the function $g_\alpha : (a, b] \to \mathbb{R}$, $g_\alpha(x) = (x - a)^{-\alpha}$. A primitive of $g_\alpha$ is given by

$$G_\alpha(x) = \begin{cases} \frac{1}{1-\alpha}(x-a)^{1-\alpha}, & \alpha \neq 1 \\ \ln(x-a), & \alpha = 1. \end{cases}$$

Consequently we have

$$\int_{a-\epsilon}^{b} \frac{dx}{(x-a)^\alpha} = \begin{cases} \frac{1}{1-\alpha}\left((b-a)^{1-\alpha} - \epsilon^{1-\alpha}\right) \\ \ln(b-a) - \ln\epsilon, \end{cases}$$

and for $\epsilon \to 0$ we find if $\alpha < 1$, then

$$\lim_{\epsilon \to 0} \int_{a-\epsilon}^{b} \frac{dx}{(x-a)^\alpha} = \frac{1}{1-\alpha}(b-a)^{1-\alpha},$$

however for $\alpha \geq 1$ the limit

$$\lim_{\epsilon \to 0} \int_{a-\epsilon}^{b} \frac{dx}{(x-a)^\alpha}$$

does not exist (as a finite limit).

　　b) We have an unbounded integrand at $x = 0$ and at $x = 2$. Therefore we split the integral accordingly: for $0 < \epsilon < 1$

$$\int_{\epsilon}^{2-\epsilon} \frac{dx}{\sqrt{x(2-x)}} = \int_{\epsilon}^{1} \frac{dx}{\sqrt{x(2-x)}} + \int_{1}^{2-\epsilon} \frac{dx}{\sqrt{x(2-x)}}.$$

For $0 < \epsilon \le x \le 1$ we have $\frac{1}{\sqrt{x(2-x)}} \le \frac{1}{\sqrt{x}}$ and therefore

$$0 \le \int_\epsilon^1 \frac{dx}{\sqrt{x(2-x)}} \le \int_\epsilon^1 \frac{dx}{\sqrt{x}} = 2 - 2\sqrt{\epsilon},$$

implying the convergence of the first integral. For $1 \le x \le 2 - \epsilon$ we find $\frac{1}{\sqrt{x(2-x)}} \le \frac{1}{\sqrt{2-x}}$ which yields

$$0 \le \int_1^{2-\epsilon} \frac{dx}{\sqrt{x(2-x)}} \le \int_1^{2-\epsilon} \frac{dx}{\sqrt{2-x}} = 2 - 2\sqrt{\epsilon}.$$

and hence the second integral converges too, i.e. $\int_0^2 \frac{dx}{\sqrt{x(2-x)}}$ converges.

c) If the integral converges we can split the integral as follows:

$$\int_0^\infty x^\alpha \, dx = \int_0^1 x^\alpha \, dx + \int_1^\infty x^\alpha \, dx.$$

The first integral converges if and only if $\alpha > -1$ but in this case the second integral diverges. Hence $\int_0^\infty x^\alpha \, dx$ will never converge.

d) A primitive of $g(x) = e^{-ax} \cos(wx)$ is the function

$$G(x) = -\frac{e^{-ax}}{a^2 + w^2} (a \cos(wx) - w \sin(wx))$$

and therefore

$$\int_0^R e^{-ax} \cos(wx) dx = G(R) - G(0).$$

Since $\lim_{R \to \infty} G(R) = 0$ we find

$$\int_0^\infty e^{-ax} \cos(wx) dx = \frac{a}{a^2 + w^2}.$$

2. First we note

$$\frac{|f(r)|}{(1+r^2)^{\frac{\alpha}{2}}} \le c_0 \frac{(1+r^2)^{\frac{\beta}{2}}}{(1+r^2)^{\frac{\alpha}{2}}} = c_0 (1+r^2)^{\frac{\beta-\alpha}{2}}.$$

It follows that

$$\int_0^R \frac{|f(r)|}{(1+r^2)^{\frac{\alpha}{2}}} dr \le c_0 \int_0^R \frac{1}{(1+r^2)^{\frac{\alpha-\beta}{2}}} dr$$

$$= c_0 \int_0^1 \frac{1}{(1+r^2)^{\frac{\alpha-\beta}{2}}} dr + c_0 \int_1^R \frac{1}{(1+r^2)^{\frac{\alpha-\beta}{2}}} dr,$$

706

and clearly the first integral on the right hand side exists for all $\alpha$ and $\beta$. If $r \geq 1$ and $\beta < \alpha$ then $\frac{1}{(1+r^2)^{\frac{\alpha-\beta}{2}}} \leq \frac{1}{2^{\frac{\alpha-\beta}{2}} r^{\frac{\alpha-\beta}{2}}}$ and since

$$\lim_{R \to \infty} \int_1^R r^{-\alpha+\beta} dr = \lim_{R \to \infty} \left( \frac{1}{1-\alpha+\beta} r^{1-\alpha+\beta} \Big|_1^R \right)$$

$$= \frac{1}{\alpha-1-\beta} + \lim_{R \to \infty} \frac{R^{1-\alpha+\beta}}{1-\alpha+\beta}$$

exists only for $1-\alpha+\beta < 0$, i.e. $\beta+1 < \alpha$, it follows that for $\beta+1 < \alpha$ the integral $\int_0^\infty \frac{f(r)}{(1+r^2)^{\frac{\alpha}{2}}} dr$ converges absolutely. Now if $f$ is a polynomial of degree $m$ we know that $|f(r)| \leq c_0(1+r^2)^{\frac{m}{2}}$ and therefore for $m+1 < \alpha$ the integral $\int_0^\infty \frac{f(r)}{(1+r^2)^{\frac{\alpha}{2}}} dr$ converges absolutely in this case. In the case where $m+1 \geq \alpha$ the integral must diverge. We may assume that $f(r) \geq 0$ for $r \geq R_0$, otherwise we switch to $-f$. From Example 11.4 we know that

$$\lim_{r \to \infty} \frac{f(r)}{a_m r^m} = 1$$

when $a_m > 0$ is the leading coefficient of $f(r)$. Thus we can find $R_1 \geq R_0$ such that $r \geq R_1$ implies $\left| \frac{f(r)}{a_m r^m} - 1 \right| < \frac{1}{2}$, or $\frac{a_m}{2} r^m \leq f(r)$. Since for $m+1 \geq \alpha$ the integral $\int_{R_1}^\infty \frac{a_m r^m}{2(1+r^2)^{\frac{\alpha}{2}}} dr$ diverges it follows that $\int_0^\infty \frac{f(r)}{(1+r^2)^{\frac{\alpha}{2}}} dr$ diverges.

3. For $k = 0$ we find

$$\int_0^1 (1-x)^\alpha dx = \frac{1}{\alpha+1} = \frac{0!}{\alpha+1}.$$

Assuming that

$$\int_0^1 x^k(1-x)^\alpha dx = \frac{k!}{(\alpha+1)(\alpha+2) \cdot \ldots \cdot (\alpha+k+1)}$$

we find when integrating by parts

$$\int_0^1 x^{k+1}(1-x)^\alpha dx = \int_0^1 x^{k+1} \left( \frac{d}{dx} \left( -\frac{(1-x)^{\alpha+1}}{\alpha+1} \right) \right) dx$$

$$= \frac{k+1}{\alpha+1} \int_0^1 x^k(1-x)^{\alpha+1} dx$$

$$= \frac{(k+1)}{\alpha+1} \cdot \frac{k!}{(\alpha+2)(\alpha+3) \cdot \ldots \cdot (\alpha+k+2)},$$

where we have used that the boundary terms

$$x^{k+1} \left( -\frac{(1-x)^{\alpha+1}}{\alpha+1} \right) \Big|_0^1$$

vanish.

4. The second integral is straightforward since

$$\left| \frac{\sin^2 t}{t^2 + a^2} \right| \leq \frac{1}{t^2 + a^2},$$

implying the absolute convergence of $\int_0^\infty \frac{\sin^2 t}{t^2+a^2} dt$. The first integral we split into two integrals and we consider

$$\int_\epsilon^1 \frac{\ln x}{x^2 + a^2} dx \quad \text{and} \quad \int_1^R \frac{\ln x}{x^2 + a^2} dx.$$

We note that

$$0 \leq \int_\epsilon^1 \frac{-\ln x}{x^2 + a^2} dx \leq - \int_\epsilon^1 \ln x\, dx$$

$$= - (x \ln x - x)\big|_\epsilon^1 = 1 + \epsilon - \epsilon \ln \epsilon,$$

and since $\lim_{\epsilon \to 0}(\epsilon - \epsilon \ln \epsilon) = 0$, compare with the calculation in Example 11.6.C, it follows that the integral converges. The second integral converges since we know that for $x \geq 1$ we have $\ln x \leq c_0 \sqrt{x}$ and therefore

$$\int_1^R \frac{\ln x}{x^2 + a^2} dx \leq \int_1^R \frac{c_0 \sqrt{x}}{x^2 + a^2} dx \leq c_0 \int_1^R x^{-\frac{3}{2}} dx = 2c_0(1 - R^{-\frac{1}{2}})$$

implying the convergence of $\int_0^\infty \frac{\ln x}{x^2+a^2} dx$.

5. For the first part we observe that since $g$ is continuous and $g(0) \neq 0$ for some $\eta > 0$ we have $g(x) \neq 0$ for $x \in (-\eta, \eta)$. We may assume that $g > 0$ in $(-\eta, \eta)$ and consequently there exists $0 < m \leq M$ such that $0 < m \leq g(x) \leq M$ for $x \in \left(-\frac{\eta}{2}, \frac{\eta}{2}\right)$. This implies for $0 < \epsilon < \frac{\eta}{2}$ that

$$\int_\epsilon^{\frac{\eta}{2}} \frac{g(x)}{x} dx \geq m \int_\epsilon^{\frac{\eta}{2}} \frac{1}{x} dx = m(\ln \frac{\eta}{2} - \ln \epsilon)$$

and therefore $\lim_{\epsilon \to 0} \int_\epsilon^{\frac{\eta}{2}} \frac{g(x)}{x} dx$, and hence $\int_0^1 \frac{g(x)}{x} dx$ does not exist. The second integral goes analogously. Note that now we have $x < 0$, and since

$$\int_{-\frac{\eta}{2}}^{-\epsilon} \frac{g(x)}{x} dx = - \int_{-\frac{\eta}{2}}^{-\epsilon} \frac{(-g(x))}{x} dx$$

the estimate

$$\int_{-\frac{\eta}{2}}^{-\epsilon} \frac{(-g(x))}{x} dx \leq -M \int_{-\frac{\eta}{2}}^{-\epsilon} \frac{1}{x} dx$$

yields for $\epsilon \to 0$ the divergence of this integral, and hence the divergence of $\int_{-1}^0 \frac{g(x)}{x} dx$. Since $g$ is even the function $x \mapsto \frac{g(x)}{x}, x \neq 0$, is odd and therefore

$$\int_{-1}^{-\epsilon} \frac{g(x)}{x} dx = - \int_\epsilon^1 \frac{g(x)}{x} dx,$$

708

implying that

$$(*) \qquad \lim_{\epsilon \to 0} \left( \int_{-1}^{-\epsilon} \frac{g(x)}{x} dx + \int_{\epsilon}^{1} \frac{g(x)}{x} dx \right) = 0.$$

Clearly, $(*)$ does not imply $\int_{-1}^{1} \frac{g(x)}{x} dx = 0$ since we know that the latter integral does not exist.

6. Suppose that $\alpha > 1$ and $\lim_{x \to \infty} x^{\alpha} f(x) = c_0$. It follows that there exists $R > 0$ such that $x \geq R$ implies

$$|x^{\alpha} f(x)| - |c_0| \leq |x^{\alpha} f(x) - c_0| < 1,$$

or

$$|f(x)| \leq \frac{1 + |c_0|}{x^{\alpha}},$$

implying for $\alpha > 1$ the convergence of $\int_{1}^{R} |f(x)| dx$ and hence the convergence of $\int_{0}^{\infty} |f(x)| dx$.

Now suppose for $c_0 \neq 0$ and $\alpha \leq 1$ that $\lim_{x \to \infty} x^{\alpha} f(x) = c_0$. We consider the case $c_0 > 0$, the case $c_0 < 0$ goes analogously. The existence of the limit implies $x^{\alpha} f(x) \geq 0$ for $x \geq R_0$, i.e. $f(x) \geq 0$ for $x \geq R_0$, and consequently we can find $R_1 \geq R_0$ such that $x \geq R_1$ implies

$$c_0 - x^{\alpha} f(x) \leq |c_0 - x^{\alpha} f(x)| < \frac{c_0}{2},$$

or for $x \geq R_1$

$$\frac{c_0}{2x^{\alpha}} \leq f(x)$$

implying

$$\int_{R_1}^{\infty} \frac{c_0}{2x^{\alpha}} dx \leq \int_{R_1}^{\infty} f(x) dx,$$

but for $\alpha \leq 1$ the integral on the left hand side diverges. Note that in the second case $c_0 = \infty$ is allowed. Clearly, we can also apply these criteria to continuous functions $f : [a, \infty) \to \mathbb{R}$.

7.  a) Since $\lim_{x \to \infty} \left( x \frac{\ln x}{1 + x} \right) = \infty$, by the second case in Problem 6 the integral diverges.

b) Here we have two boundary points which can cause potential problems and therefore we split the integral as follows:

$$\int_{0}^{\infty} \frac{1 - \cos y}{y^2} dy = \int_{0}^{\pi} \frac{1 - \cos y}{y^2} dy + \int_{\pi}^{\infty} \frac{1 - \cos y}{y^2} dy.$$

Since $\lim_{y \to 0} \frac{1 - \cos y}{y^2} = \frac{1}{2}$ (use the rules of l'Hospital), it turns out that the first integral is a Riemann integral and not an improper integral. For the second integral we observe that

$$\lim_{y \to \infty} \left( y^{\frac{3}{2}} \left( \frac{1 - \cos y}{y^2} \right) \right) = 0,$$

709

and the first part of Problem 6 gives the convergence of the integral.

c) The substitution $t \mapsto -s$ gives

$$\int_{-\infty}^{-1} \frac{e^t}{t} dt = -\int_1^\infty \frac{e^{-s}}{s} ds$$

and we need only to note that $\lim_{s \to \infty} s^2 \left( \frac{e^{-s}}{s} \right) = 0$ to deduce that $\int_{-\infty}^{-1} \frac{e^t}{t} dt$ converges.

8. Following the hint we write

$$\int_0^\infty \left| \frac{\sin x}{x} \right| dx = \sum_{n=0}^\infty \int_{n\pi}^{(n+1)\pi} \left| \frac{\sin x}{x} \right| dx.$$

Taking into account that $\sin k\pi = 0$ as well as $|\sin x| = |\sin(x+\pi)|$, the substitution $x = t + n\pi$ yields

$$\int_{n\pi}^{(n+1)\pi} \left| \frac{\sin x}{x} \right| dx = \int_0^\pi \frac{\sin t}{t + n\pi} dt.$$

Since for $0 \le t \le \pi$ it follows that $\frac{1}{t+n\pi} \ge \frac{1}{(n+1)\pi}$ we find

$$\int_0^\pi \frac{\sin t}{t + n\pi} dt \ge \frac{1}{(n+1)\pi} \int_0^\pi \sin t \, dt = \frac{2}{\pi(n+1)},$$

which implies

$$\int_0^\infty \left| \frac{\sin x}{x} \right| dx \ge \sum_{n=0}^\infty \frac{2}{\pi(n+1)} = \frac{2}{\pi} \sum_{n=1}^\infty \frac{1}{n},$$

and since the series $\sum_{n=1}^\infty \frac{1}{n}$ diverges we have proved that the integral $\int_0^\infty \left| \frac{\sin x}{x} \right| dx$ diverges.

9. Since $\lim_{x \to a} \frac{f(x)}{g(x)} = c_0 > 0$, for $\frac{c_0}{2} > 0$ there exists $\delta > 0$ such that $0 < x - a < \delta$ implies

$$\left| \frac{f(x)}{g(x)} - c_0 \right| < \frac{c_0}{2},$$

or $\frac{c_0}{2} < \frac{f(x)}{g(x)} < \frac{3c_0}{2}$, i.e.

$$\frac{c_0}{2} g(x) \le f(x) \le \frac{3c_0}{2} g(x).$$

This yields

$$\frac{c_0}{2} \int_a^{a+\delta} g(x) dx \le \int_a^{a+\delta} f(x) dx \le \frac{3c_0}{2} \int_a^{a+\delta} g(x) dx,$$

710

implying that $\int_a^b f(x)dx$ exists if and only if $\int_a^b g(x)dx = \int_a^{a+\delta} g(x)dx + \int_{a+\delta}^b f(x)dx$ exists. In the case where $\lim\limits_{x \to a} \dfrac{f(x)}{g(x)} = 0$ we can still find for $\epsilon > 0$ some $\delta > 0$ such that for $0 < x - a < \delta$ it follows that

$$f(x) \leq \epsilon g(x),$$

implying that $\int_a^b f(x)dx$ converges if $\int_a^b g(x)dx = \int_a^{a+\delta} g(x)dx + \int_{a+\delta}^b g(x)dx$ converges. Now, if $\lim\limits_{x \to a} \dfrac{f(x)}{g(x)} = \infty$ then for $R > 0$ there exists $\delta > 0$ such that $0 < x - a < \delta$ implies $\frac{f(x)}{g(x)} \geq R$, or $f(x) \geq Rg(x)$. Therefore the divergence of $\int_a^b g(x)dx$ implies the divergence of $\int_a^b f(x)dx$.

10. For $0 < \epsilon < r < 1$ we find with $-\ln s = u$ that

$$\int_\epsilon^r \frac{dr}{\sqrt{-\ln s}} = \int_{-\ln \epsilon}^{-\ln r} u^{-\frac{1}{2}}(-e^{-u})du = \int_{-\ln r}^{-\ln \epsilon} u^{-\frac{1}{2}}e^{-u}du.$$

Now, as $\epsilon \to 0$ it follows that $-\ln \epsilon \to \infty$ and as $r \to 1$ it follows that $-\ln r \to 0$. Hence for $0 < \epsilon < \alpha < r$ we find

$$\lim_{r \to 1} \int_\alpha^r \frac{ds}{\sqrt{-\ln s}} + \lim_{\epsilon \to 0} \int_\epsilon^\alpha \frac{ds}{\sqrt{-\ln s}}$$

$$= \lim_{r \to 1} \int_{-\ln r}^{-\ln \alpha} u^{-\frac{1}{2}}e^{-u}du + \lim_{\epsilon \to 0} \int_{-\ln \alpha}^{-\ln \epsilon} u^{-\frac{1}{2}}e^{-u}du$$

$$= \int_0^\infty u^{-\frac{1}{2}}e^{-u}du = \Gamma\left(\frac{1}{2}\right).$$

11. For $0 < \epsilon < \frac{1}{2}$ we find

$$\int_\epsilon^{1-\epsilon} t^{x-1}(1-t)^{y-1}dt = \int_\epsilon^{\frac{1}{2}} t^{x-1}(1-t)^{y-1}dt + \int_{\frac{1}{2}}^{1-\epsilon} t^{x-1}(1-t)^{y-1}dt.$$

Since $x > 0$ it follows that $x - 1 > -1$ and consequently, see Example 28.3,

$$\lim_{\epsilon \to 0} \int_\epsilon^{\frac{1}{2}} t^{x-1}(1-t)^{y-1}dt$$

exists. Analogously we deduce, also compare with Problem 1 a), that

$$\lim_{\epsilon \to 0} \int_{\frac{1}{2}}^{1-\epsilon} t^{x-1}(1-t)^{y-1}dt$$

exists, implying the convergence of $B(x, y) = \int_0^1 t^{x-1}(1-t)^{y-1}dt$. Substituting $t$ by $1 - s$ we find

$$B(x, y) = -\int_1^0 (1-s)^{x-1}s^{y-1}ds = \int_0^1 s^{y-1}(1-s)^{x-1}ds = B(y, x).$$

711

This calculation has however a problem: we have not proved the substitution rule for improper integrals. Thus we should start with

$$\int_{\epsilon}^{1-\epsilon} t^{x-1}(1-t)^{y-1}dt = -\int_{1-\epsilon}^{\epsilon}(1-s)^{x-1}s^{y-1}ds = \int_{\epsilon}^{1-\epsilon}s^{y-1}(1-s)^{x-1}ds$$

and pass to the limit.

Finally substituting $t = \sin^2\vartheta$ (and allowing ourselves to use a substitution rule for this particular improper integral) we find for $x = m$ and $y = n$, while noting that for $t = 0$ we have $\vartheta = 0$ ($t = \epsilon, \vartheta = \arcsin\sqrt{\epsilon}$) and for $t = 1$ we have $\vartheta = \frac{\pi}{2}$ ($t = 1-\epsilon, \vartheta = \arcsin\sqrt{1-\epsilon}$), that

$$B(m,n) = \int_0^1 t^{m-1}(1-t)^{n-1}dt$$

$$= 2\int_0^{\frac{\pi}{2}}(\sin^2\vartheta)^{m-1}(\cos^2\vartheta)^{n-1}\cos\vartheta\sin\vartheta d\vartheta$$

$$= 2\int_0^{\frac{\pi}{2}}(\sin\vartheta)^{2m-1}(\cos\vartheta)^{2n-1}d\vartheta,$$

where we used $1 - \sin^2\vartheta = \cos^2\vartheta$ and $\frac{dt}{d\vartheta} = 2\cos\vartheta\sin\vartheta$.

(The more correct calculation would be to first derive

$$\int_{\epsilon}^{1-\epsilon} t^{m-1}(1-t)^{n-1}dt = 2\int_{\arcsin\sqrt{\epsilon}}^{\arcsin\sqrt{1-\epsilon}}(\sin\vartheta)^{2m-1}(\cos\vartheta)^{2n-1}d\vartheta$$

and pass to the limit $\epsilon \to 0$.)

The mapping $(x,y) \mapsto B(x,y)$ is the **(Euler) beta-function** and we will study it, in particular its relation to the $\Gamma$-function, in Chapter 31.

12.　　a) Since the sum of two convex functions is convex we have for $h$ and $g$ being logarithmic convex that

$$\log h + \log g = \log(hg)$$

is convex, i.e. $h \cdot g$ is logarithmic convex.

　　b) We just need to note that the convexity of $\log f$ implies

$$0 \le (\log f)'' = \left(\frac{f'}{g}\right)' = \frac{ff'' - (f')^2}{f^2}.$$

　　c) Since the limit of a sequence of convex functions is convex the continuity of the logarithmic function implies the result.

**Chapter 29**

1. For $x \neq 0$ we have $\frac{1}{1+x^4} < 1$ and consequently

$$\sum_{n=0}^{\infty} g_n(x) = x^4 \sum_{n=0}^{\infty} \frac{1}{(1+x^4)^n} = x^4 \frac{1}{1 - \frac{1}{1+x^4}}$$
$$= 1 + x^4, x \neq 0.$$

However, for $x = 0$ we have $g_n(x) = 0$ for all $n$, thus $\sum_{n=0}^{\infty} g_n(0) = 0$. If follows that

$$\lim_{x \to 0} \sum_{n=0}^{\infty} g_n(x) = 1 \neq \sum_{n=0}^{\infty} g_n(0),$$

i.e. $\sum_{n=0}^{\infty} g_n(x)$ is not continuous for $x \neq 0$. Since all functions $g_n$ are continuous the convergence of $\sum_{n=0}^{\infty} g_n(x)$ cannot be uniform on any interval containing 0.

2.     a) Since $|\sin kx| \leq 1$ for all $x \in \mathbb{R}$ and $k \in \mathbb{N}_0$ it follows that $\left|\frac{\sin kx}{k^\alpha}\right| \leq \frac{1}{k^\alpha}$ and for $\alpha > 1$ the series $\sum_{\alpha=0}^{\infty} \frac{1}{k^\alpha}$ converges, hence $\sum_{\alpha=0}^{\infty} \frac{\sin kx}{k^\alpha}$ converges absolutely and uniformly.

    b) We observe that for $|x| \leq 1$ we have $\left|\frac{x^n}{n^{\frac{3}{2}}}\right| \leq \frac{1}{n^{\frac{3}{2}}}$ and the convergence of $\sum_{n=1}^{\infty} \frac{1}{n^{\frac{3}{2}}}$ implies the absolute and uniform convergence of $\sum_{n=1}^{\infty} \frac{x^n}{n^{\frac{3}{2}}}$ for $|x| \leq 1$.

    c) Note that $\frac{1}{n^2+r^2} \leq \frac{1}{n^2}$ for any $r \in \mathbb{R}$ and since $\sum_{n=1}^{\infty} \frac{1}{n^2} < \infty$ it follows that $\sum_{n=1}^{\infty} \frac{1}{n^2+r^2}$ converges for all $r \in \mathbb{R}$ absolutely and uniformly.

3. For $\alpha = m \in \mathbb{N}, m \geq n$, we find

$$\prod_{k=1}^{n} \frac{m-k+1}{k} = \frac{m!}{n!(m-n)!} = \binom{m}{n}.$$

Now, for $k \in \mathbb{N}_0$ we have

$$g_\alpha^{(k)}(x) = \alpha(\alpha-1) \cdot \ldots \cdot (\alpha - k + 1)(1+x)^{\alpha-k}$$
$$= k! \binom{\alpha}{k}(1+x)^{\alpha-k},$$

i.e. $g_\alpha^{(k)}(0) = k!\binom{\alpha}{k}$. Consequently the $n^{\text{th}}$ Taylor polynomial of $g_\alpha$ about 0 is given by

$$T_{g_\alpha}^{(n)}(0) = \sum_{k=0}^{n} \binom{\alpha}{k} x^k.$$

**Note:** with $c_k = \binom{\alpha}{k}x^k$ we find

$$\left|\frac{a_{k+1}}{a_k}\right| = \left|\frac{\binom{\alpha}{k+1}x^{k+1}}{\binom{\alpha}{k}x^k}\right| = |x|\left|\frac{\alpha-k}{k+1}\right|.$$

Since $\lim\limits_{k\to\infty}\left|\dfrac{\alpha-k}{k+1}\right| = 1$, we find for $\eta$ such that $|x| < \eta < 1$ some $N = N(\eta)$ with the property that $n \geq N(\eta)$ implies $\left|\dfrac{a_{k+1}}{a_k}\right| \leq \eta < 1$. Consequently, the series

$$\sum_{k=0}^{\infty}\binom{\alpha}{k}x^k$$

converges for $|x| < 1$. It takes further effort to prove that $\sum_{k=0}^{\infty}\binom{\alpha}{k}x^k$ is indeed the Taylor series of $g_\alpha$, $|x| < 1$.

4. For $N \in \mathbb{N}$ we have

$$\sum_{k=0}^{N}|(a_k + b_k)x^k| \leq \sum_{k=0}^{N}|a_k||x|^k + \sum_{k=0}^{N}|b_k||x|^k$$
$$\leq \sum_{k=0}^{\infty}|a_k||x|^k + \sum_{k=0}^{\infty}|b_k||x|^k$$

as well as

$$\sum_{k=0}^{N}|(\lambda a_k)x^k| \leq |\lambda|\sum_{k=0}^{N}|a_k||x|^k \leq |\lambda|\sum_{k=0}^{\infty}|a_k||x|^k,$$

which allows us in each case to pass to the limit as $N \to \infty$. Once we have secured absolute and uniform convergence, we may pass in the equalities

$$\sum_{k=0}^{N}(a_k + b_k)x^k = \sum_{k=0}^{N}a_k x^k + \sum_{k=0}^{N}b_k x^k$$

and

$$\sum_{k=0}^{N}(\lambda a_k)x^k = \lambda\sum_{k=0}^{N}a_k x^k$$

to the limit as $N \to \infty$.

5. We note that $\sinh x = \dfrac{e^x - e^{-x}}{2}$ and $\cosh x = \dfrac{e^x + e^{-x}}{2}$, and therefore we find

$$\sinh x = \frac{1}{2}\left(\sum_{k=0}^{\infty}\frac{x^k}{k!} - \sum_{k=0}^{k}(-1)^k\frac{x^k}{k!}\right)$$
$$= \frac{1}{2}\left(\sum_{l=0}^{\infty}\frac{x^{2l}}{(2l)!} + \sum_{m=1}^{\infty}\frac{x^{2m-1}}{(2m-1)!}\right.$$
$$\left. - \sum_{l=0}^{\infty}(-1)^{2l}\frac{x^{2l}}{(2l)!} - \sum_{m=1}^{\infty}(-1)^{2m-1}\frac{x^{2m-1}}{(2m-1)!}\right)$$
$$= \sum_{m=1}^{\infty}\frac{x^{2m-1}}{(2m-1)!},$$

714

and further

$$\cosh x = \frac{1}{2}\left(\sum_{k=0}^{\infty}\frac{x^k}{k!} + \sum_{k=0}^{\infty}(-1)^k\frac{x^k}{k!}\right)$$

$$= \frac{1}{2}\left(\sum_{l=0}^{\infty}\frac{x^{2l}}{(2l)!} + \sum_{m=1}^{\infty}\frac{x^{2m-1}}{(2m-1)!}\right.$$

$$\left. + \sum_{l=0}^{\infty}(-1)^{2l}\frac{x^{2l}}{(2l)!} + \sum_{m=1}^{\infty}(-1)^{2m-1}\frac{x^{2m-1}}{(2m-1)!}\right)$$

$$= \sum_{l=0}^{\infty}\frac{x^{2l}}{(2l)!}.$$

6. We know that for $|x| < 1$ the following holds:

$$\ln(1+x) = \sum_{n=1}^{\infty}(-1)^{n+1}\frac{x^n}{n},$$

which implies for $|x| < 1$ that

$$\frac{1}{2}\ln\frac{1+x}{1-x} = \frac{1}{2}(\ln(1+x) - \ln(1-x))$$

$$= \frac{1}{2}\left(\sum_{n=1}^{\infty}(-1)^{n+1}\frac{x^n}{n} - \sum_{n=1}^{\infty}(-1)^{n+1}\frac{(-x)^n}{n}\right)$$

$$= \frac{1}{2}\left(\sum_{n=1}^{\infty}(-1)^{n+1}\frac{x^n}{n} - \sum_{n=1}^{\infty}(-1)^{2n+1}\frac{x^n}{n}\right)$$

$$= \frac{1}{2}\sum_{n=1}^{\infty}\left((-1)^{n+1} + 1\right)\frac{x^n}{n}$$

$$= \sum_{n=0}^{\infty}\frac{x^{2n+1}}{2n+1}.$$

7. For $x \in \mathbb{R}$ fixed we apply the ratio test to the series representing $J_l(x)$ :

$$\left|\frac{\frac{(-1)^{n+1}(\frac{x}{2})^{l+2(n+1)}}{(n+1)!(n+1+l)!}}{\frac{(-1)^n(\frac{x}{2})^{l+2n}}{n!(n+l)!}}\right| = \frac{\frac{x^l x^2 x^{2n}}{2^2 2^l 2^{2n}(n+1)n!(n+1+l)(n+l)!}}{\frac{x^l x^{2n}}{2^l 2^{2n}n!(n+l)!}} = \frac{x^2}{2^2(n+1)(n+1+l)}.$$

Thus, in order to obtain the convergence of $J_l(x)$ we need to assume that there exists $N \in \mathbb{N}$ such that $n \geq N$ implies

$$\frac{x^2}{2^2(n+1)(n+1+l)} \leq \tau < 1,$$

715

of course $N$ may depend on $x$. Now

$$\frac{x^2}{2^2(n+1)(n+1+l)} \leq \frac{x^2}{2^2 n^2},$$

thus if $\frac{x^2}{2^2 n^2} \leq \frac{1}{4}$ (but any $0 < \tau < 1$ will do instead of $\frac{1}{4}$) then we are done. Now

$$\frac{x^2}{2^2 n^2} \leq \frac{1}{4} \text{ implies } |x| \leq n.$$

Hence, for $N := [x] + 1$ it follows for $n \geq N$ that

$$\frac{x^2}{2^2(n+1)(n+1+l)} \leq \frac{1}{4}$$

implying the convergence of $J_l(x)$. (Note: there is no need to assume $l \in \mathbb{N}_0$).

We can now differentiate $J_l(x)$ term by term to find

$$J_l(x) = \sum_{n=0}^{\infty} \frac{(-1)^n x^{l+2n}}{2^{l+2n} n! (n+l)!}$$

$$J_l'(x) = \sum_{n=0}^{\infty} \frac{(-1)^n (l+2n) x^{l+2n-1}}{2^{l+2n} n! (n+l)!}$$

$$J_l''(x) = \sum_{n=0}^{\infty} \frac{(-1)^n (l+2n)(l+2n-1) x^{l+2n-2}}{2^{l+2n} n! (n+l)!}$$

and therefore we have

$$(x^2 - l^2) J_l(x) = \sum_{n=0}^{\infty} \frac{(-1)^n x^{l+2n+2}}{2^{l+2n} n! (n+l)!} - \sum_{n=0}^{\infty} \frac{(-1)^n l^2 x^{l+2n}}{2^{l+2n} n! (n+l)!}$$

$$x J_l'(x) = \sum_{n=0}^{\infty} \frac{(-1)^n (l+2n) x^{l+2n}}{2^{l+2n} n! (n+l)!}$$

$$x^2 J_l''(x) = \sum_{n=0}^{\infty} \frac{(-1)^n (l+2n)(l+2n-1) x^{l+2n}}{2^{l+2n} n! (n+l)!}$$

716

Now we have to add up these three terms to find

$$x^2 J_l''(x) \quad + x J_l'(x) + (x^2 - l^2) J_l(x)$$

$$= \sum_{n=0}^{\infty} \frac{(-1)^n x^{l+2n+2}}{2^{l+2n} n!(n+l)!}$$

$$+ \sum_{n=0}^{\infty} \frac{(-1)^n \{-l^2 + (l+2n) + (l+2n)(l+2n-1)\} x^{l+2n}}{2^{l+2n} n!(n+l)!}$$

$$= \sum_{n=0}^{\infty} \frac{(-1)^n x^{l+2n+2}}{2^{l+2n} n!(n+l)!} + \sum_{n=0}^{\infty} \frac{(-1)^n (4n(n+l)) x^{l+2n}}{2^{l+2n} n!(n+l)!}$$

$$= \sum_{n=0}^{\infty} \frac{(-1)^n x^{l+2n+2}}{2^{l+2n} n!(n+l)!} + \sum_{n=1}^{\infty} \frac{(-1)^n 4 x^{l+2n}}{2^{l+2n} (n-1)!(n+l-1)!}$$

$$= \sum_{n=1}^{\infty} \frac{(-1)^{n-1} x^{l+2n}}{2^{l+2n-2} (n-1)!(n-1+l)!} + \sum_{n=1}^{\infty} \frac{(-1)^n 4 x^{l+2n}}{2^{l+2n} (n-1)!(n+l-1)!}$$

$$= -\sum_{n=1}^{\infty} \frac{(-1)^n 4 x^{l+2n}}{2^{l+2n} (n-1)!(n-1+l)!} + \sum_{n=1}^{\infty} \frac{(-1)^n 4 x^{l+2n}}{2^{l+2n} (n-1)!(n+l-1)!}$$

$$= 0.$$

8. Since for $|r| < 1$ we have $\displaystyle\sum_{n=0}^{\infty} r^n = \frac{1}{1-r}$ we find with $r = -t^2$

$$\frac{1}{1+t^2} = \sum_{n=0}^{\infty} (-t^2)^n = \sum_{n=0}^{\infty} (-1)^n t^{2n}.$$

For $|x| < 1$ it holds

$$\arctan x = \int_0^x \frac{1}{1+t^2} dt = \int_0^x \sum_{n=0}^{\infty} (-1)^n t^{2n} dt.$$

Since $|x| < 1$ implies $|t| < 1$ the series under the integral sign converges uniformly and therefore we find by changing the order of summation and integration that

$$\arctan x = \sum_{n=0}^{\infty} \int_0^x (-1)^n t^{2n} dt = \sum_{n=0}^{\infty} (-1)^n \frac{x^{2n+1}}{2n+1}.$$

717

9. Since $\tan \frac{\pi}{6} = \frac{\sqrt{3}}{3}$ we have by Problem 8

$$\frac{\pi}{6} = \arctan \frac{\sqrt{3}}{3} = \sum_{n=0}^{\infty} (-1)^n \frac{1}{2n+1} \left( \frac{\sqrt{3}}{3} \right)^{2n+1}$$

$$= \frac{\sqrt{3}}{3} \sum_{n=0}^{\infty} (-1)^n \frac{1}{2n+1} \frac{3^n}{3^{2n}} = \frac{1}{\sqrt{3}} \sum_{n=0}^{\infty} \frac{(-1)^n}{(2n+1)3^n}.$$

10. For $n \in \mathbb{N}_0$ we denote the $n^{\text{th}}$ partial sum of $\sum_{m=0}^{\infty} a_n$ by $S_n := \sum_{k=0}^{n} a_k$ and further

we set $S_{-1} := 0$, implying that $a_n = S_n - S_{n-1}$ and $S := \lim_{n \to \infty} S_n = \sum_{n=0}^{\infty} a_n$. It

follows for $|x| < 1$ that by $g : (-1, 1) \to \mathbb{R}$, $g(x) = \sum_{n=0}^{\infty} a_n x^n$, a function is defined which satisfies

$$g(x) = (1 - x) \sum_{n=0}^{\infty} S_n x^n.$$

Now let $\epsilon > 0$. Then there exists $N = N(\epsilon) \in \mathbb{N}$ such that $n > N$ implies $|S - S_n| < \frac{\epsilon}{2}$. Further, since for $|x| < 1$ we have $(1 - x) \sum_{n=0}^{\infty} x^n = 1$, it follows for $0 < x < 1$ that

$$|g(x) - S| = \left| (1 - x) \sum_{n=0}^{\infty} (S_n - S)x^n \right|$$

$$\leq (1 - x) \sum_{n=0}^{N} |S_n - S| + \frac{\epsilon}{2}.$$

Now, for this $\epsilon > 0$ we can also find $\delta > 0$ such that $1 - \delta < x < 1$ yields

$$(1 - x) \sum_{n=0}^{N} |S_n - S| < \delta \sum_{n=0}^{N} |S_n - S| < \frac{\epsilon}{2},$$

implying that $|g(x) - S| < \epsilon$, or

$$\lim_{\substack{x \to 1 \\ x < 1}} \sum_{n=0}^{\infty} a_n x^n = \sum_{n=0}^{\infty} a_n.$$

11.    a) Since for $|x| < 1$ we have the Taylor expansion

$$\ln(1 + x) = \sum_{l=1}^{\infty} (-1)^{l+1} \frac{x^l}{l}$$

Abel's convergence theorem gives

$$\lim_{x \to 1} \ln(1 + x) = \ln 2 = \sum_{l=1}^{\infty} \frac{(-1)^{l+1}}{l}.$$

718

For the second equality we just have to note

$$\frac{(-1)^{2l}}{2l} + \frac{(-1)^{2l-1}}{2l-1} = \frac{2l+1-2l}{(2l)(2l-1)}.$$

b) We can use the Taylor series for arctan:

$$\arctan x = \sum_{k=0}^{\infty} (-1)^k \frac{x^k}{2k+1}$$

and Abel's theorem gives

$$\frac{\pi}{4} = \arctan 1 = \sum_{k=0}^{\infty} \frac{(-1)^k}{2k+1}.$$

12. In both cases we use the Taylor formula with the Lagrange remainder term.

a) With some $0 < \vartheta_1 < 1$ we have

$$\ln(1+x) = x - \frac{x^2}{2} + \frac{x^3}{3} - \frac{x^4}{4} + \frac{x^5}{5} \cdot \frac{1}{(1+\vartheta_1)^5}$$

$$> x - \frac{x^2}{2} + \frac{x^3}{3} - \frac{x^4}{4},$$

and for some $0 < \vartheta_2 < 1$ we find

$$\ln(1+x) = x - \frac{x^2}{2} + \frac{x^3}{3} - \frac{x^4}{4} \cdot \frac{1}{(1+\vartheta_2 x)^4}$$

$$< x - \frac{x^2}{2} + \frac{x^3}{3}.$$

b) For some $0 < \vartheta_1 < 1$ we find

$$\sqrt{1+x} = 1 + \frac{x}{2} - \frac{x^2}{8} + \frac{x^3}{16} - \frac{x^4}{128} \cdot \frac{1}{(1+\vartheta_1 x)^{\frac{7}{2}}}$$

$$< 1 + \frac{x}{2} - \frac{x^2}{8} + \frac{x^3}{16},$$

and with some $0 < \vartheta_2 < 1$ we get

$$\sqrt{1+x} = 1 + \frac{x}{2} - \frac{x^2}{8} + \frac{x^3}{16} \cdot \frac{1}{(1+\vartheta_2 x)^{\frac{5}{2}}}$$

$$> 1 + \frac{x}{2} - \frac{x^2}{8}.$$

719

13.     a) We know that for $|x| < 1$ we have

$$\sum_{n=0}^{\infty} x^n = \frac{1}{1-x},$$

therefore

$$\left(\frac{1}{1-x}\right)^2 = \left(\sum_{n=0}^{\infty} x^n\right)\left(\sum_{m=0}^{\infty} x^m\right)$$

$$= \sum_{k=0}^{\infty} c_k x^k$$

where

$$c_k = \sum_{l=0}^{k} a_l b_{k-l}$$

with $a_l = 1, b_l = 1$ for all $l$, hence

$$c_k = \sum_{l=0}^{k} 1 = k+1,$$

which implies

$$\left(\frac{1}{1-x}\right)^2 = \sum_{k=0}^{\infty}(k+1)x^k.$$

b) We note that

$$\frac{\cos x}{1-x} = (\cos x)\left(\frac{1}{1-x}\right) = \left(\sum_{k=0}^{\infty}(-1)^k \frac{x^{2k}}{(2k)!}\right)\left(\sum_{l=0}^{\infty} x^l\right) = \sum_{m=0}^{\infty} c_k x^k$$

where with $a_n = \begin{cases} \frac{(-1)^k}{(2k)!}, & n = 2k \\ 0, & n = 2k-1 \end{cases}$ and $b_m = 1$ it follows that

$$c_k = \sum_{j=0}^{k} a_j b_{k-j} = \sum_{j=0}^{k} a_j.$$

For $k = 2l$ we find

$$c_{2l} = \sum_{j=0}^{l} a_{2j} = \sum_{j=1}^{l}(-1)^j \frac{1}{(2j)!}$$

and for $k = 2l + 1$ we have

$$c_{2l+1} = \sum_{j=0}^{l} a_{2j} = \sum_{j=1}^{l}(-1)^j \frac{1}{(2j)!},$$

implying the result.

14. From our assumptions we deduce first

$$(f \cdot g)(x) = \sum_{k=0}^{\infty} \frac{(f \cdot g)^{(k)}(0)}{k!} x^k,$$

and therefore it remains to prove that

$$\frac{(f \cdot g)^{(k)}(0)}{k!} = \sum_{l=0}^{k} \frac{f^{(l)}(0)}{l!} \frac{g^{(k-l)}(0)}{(k-l)!}.$$

However, Leibniz's rule for higher order derivatives, see Corollary 21.12, gives

$$(f \cdot g)^{(k)}(0) = \sum_{l=0}^{k} \binom{k}{l} f^{(l)}(0) g^{(k-l)}(0)$$

and since

$$\frac{1}{k!} \binom{k}{l} = \frac{1}{l!} \frac{1}{(k-l)!}$$

the result follows.

**Chapter 30**

1. Since

$$\frac{k^3 - 1}{k^3 + 1} = \frac{(k-1)(k^2 + k + 1)}{(k+1)(k^2 - k + 1)}$$

$$= \frac{(k-1)((k+1)^2 - (k+1) + 1)}{(k+1)(k^2 - k + 1)}$$

we find for $N \in \mathbb{N}$

$$\prod_{k=2}^{N} \frac{k^3 - 1}{k^3 + 1} = \prod_{k=2}^{N} \frac{(k-1)((k+1)^2 - (k+1) + 1)}{(k+1)(k^2 - k + 1)}$$

$$= \prod_{k=2}^{N} \left( \frac{k-1}{k+1} \right) \prod_{k=2}^{N} \frac{(k+1)^2 - (k+1) + 1}{k^2 - k + 1}$$

$$= \frac{2}{((N-1)+1)(N+1)} \cdot \frac{(N+1)^2 - (N+1) + 1}{4 - 2 + 1}$$

$$= \frac{2(N^2 + N + 1)}{3N(N+1)}.$$

Thus we have

$$\prod_{k=2}^{\infty} \frac{k^3 - 1}{k^3 + 1} = \lim_{N \to \infty} \prod_{k=2}^{N} \frac{k^3 - 1}{k^3 + 1} = \lim_{N \to \infty} \frac{2(N^2 + N + 1)}{3N(N+1)} = \frac{2}{3}.$$

721

b) We note that

$$1 + \frac{1}{l(l+2)} = \frac{(l+1)^2}{l(l+2)}$$

and therefore

$$\prod_{l=1}^{N}\left(1 + \frac{1}{l(l+2)}\right) = \prod_{l=1}^{N}\frac{(l+1)^2}{l(l+2)}$$

$$= \prod_{l=1}^{N}\frac{l+1}{l}\prod_{l=1}^{N}\frac{l+1}{l+2}$$

$$= \frac{N+1}{1}\cdot\frac{2}{N+2} = \frac{2(N+1)}{N+2},$$

which yields

$$\prod_{l=1}^{\infty}\left(1 + \frac{1}{l(l+2)}\right) = \lim_{N\to\infty}\prod_{l=1}^{N}\left(1 + \frac{1}{l(l+2)}\right)$$

$$= \lim_{N\to\infty}\frac{2(N+1)}{N+2} = 2.$$

2. We first observe that for $b_k \geq 0$ we have

$$(*) \qquad (1-b_1)\cdot\ldots\cdot(1-b_n) \geq 1 - (b_1 + \cdots + b_n).$$

Indeed, for $n = 1$ we have equality and if $(*)$ holds for $n$, then

$$(1-b_1)\cdot\ldots\cdot(1-b_n)(1-b_{n+1}) \geq (1 - (b_1 + \cdots + b_n))(1 - b_{n+1})$$

$$= 1 - (b_1 + \cdots + b_n) - b_{n+1} + (b_1 + \cdots + b_n)b_{n+1}$$

$$\geq 1 - (b_1 + \cdots + b_{n+1}).$$

Now assume that $\sum_{k=1}^{\infty} a_k$ converges. Then there exists $N \in \mathbb{N}$ such that $\sum_{k=N}^{\infty} a_k < \frac{1}{2}$. For $n > N$ we find

$$P_n = \prod_{k=1}^{n}(1 - a_k) = P_{N-1}\prod_{k=N}^{n}(1 - a_k),$$

or

$$\frac{P_n}{P_{N-1}} = \prod_{k=N}^{n}(1 - a_k) \geq 1 - \sum_{k=N}^{n}a_k > \frac{1}{2},$$

implying that $\frac{P_n}{P_{N-1}}$ is bounded from below and since for $n > N$ we have $0 < 1 - a_n < 1$ it follows that $\left(\frac{P_n}{P_{N-1}}\right)_{n\in\mathbb{N}}$ is also monotone decreasing, hence it has a limit $p \in [\frac{1}{2}, 1]$. Therefore we find

$$\prod_{k=1}^{\infty}(1 - a_k) = \lim_{n\to\infty} \prod_{k=1}^{n}(1 - a_k)$$

$$= P_{N-1} \lim_{n\to\infty} \prod_{k=N}^{n}(1 - a_n)$$

$$= P_{N-1}P \neq 0.$$

Conversely, suppose that $\sum_{k=1}^{\infty} a_k$ diverges. In order to have convergence of $\prod_{k=1}^{\infty}(1 - a_k)$ it is necessary that $\lim_{k\to\infty}(1 - a_k) = 1$, i.e. $\lim_{k\to\infty} a_k = 0$. We assume now that $\lim_{k\to\infty} a_k = 0$, otherwise the divergence of $\prod_{k=1}^{\infty}(1 - a_k)$ would follow immediately. Since $a_k \geq 0$ we deduce $0 \leq a_k \leq 1$ for all $k \geq N$ with some $N \in \mathbb{N}$. For $0 \leq x \leq 1$ we have $1 - x \leq e^{-x}$ and therefore with $n \geq N$

$$0 \leq \prod_{k=N}^{n}(1 - a_k) \leq e^{-\sum_{k=N}^{n} a_k}$$

and the divergence of $\sum_{k=1}^{\infty} a_k$ implies now that $\lim_{n\to\infty} \prod_{k=N}^{n}(1 - a_k) = 0$, which yields that $\prod_{k=1}^{\infty}(1 - a_k)$ diverges to 0.

3. We want to use Lemma 30.5 and hence we need a control on $\ln(1 + a_k)$. Since $\sum_{k=1}^{\infty} a_k$ converges, hence $\lim_{k\to\infty} a_k = 0$, there exists $N \in \mathbb{N}$ such that for $k \geq N$ we have $|a_k| < \frac{1}{2}$. Now we apply the Taylor formula with Lagrange remainder, see Theorem 29.14, to $\ln(1 + x)$, $|x| < \frac{1}{2}$, to find

$$\ln(1 + x) = x - \frac{x^2}{2(1 + \xi)^2}, 0 < |\xi| < \frac{1}{2},$$

or

$$\frac{2}{9} < \frac{1}{2(1 + \xi)^2} < 2.$$

a) From the considerations made above it follows that for $k \geq N$

$$\ln(1 + a_k) = a_k - \vartheta_k a_k^2, \frac{2}{9} < \vartheta_k < 2.$$

If $\sum_{k=1}^{\infty} a_k^2$ converges, then $\sum_{k=1}^{\infty} \vartheta_k a_k^2 \leq 2 \sum_{k=1}^{\infty} a_k^2$ and it follows that $\sum_{k=1}^{\infty} \ln(1 + a_k)$ converges. If however $\prod_{k=1}^{\infty}(1 + a_k)$ converges then $\sum_{k=1}^{\infty} \ln(1 + a_k)$ converges, implying first the convergence of $\sum_{k=1}^{\infty} \vartheta_k a_k^2$ and since $\frac{2}{9} < \vartheta_k$ the convergence of $\sum_{k=1}^{\infty} a_k^2$ follows.

b) Now suppose that $\sum_{k=1}^{\infty} a_k^2$ diverges. From our previous considerations we deduce for $k \geq N$

$$a_k - \ln(1 + a_k) > \frac{2}{9} a_k^2,$$

723

and since $\lim_{k \to \infty} |a_k| = 0$ it follows that $\sum_{k=1}^{\infty} \ln(1 + a_k)$ must diverge to $-\infty$. Consequently $\prod_{k=1}^{\infty}(1 + a_k)$ diverges and conversely, the divergence of $\sum_{k=1}^{\infty}(1 + a_k)$, i.e. the divergence of $\prod_{k=1}^{\infty}(1 + a_k)$, implies the divergence of $\sum_{k=1}^{\infty} a_k^2$.

4. If $\prod_{k=1}^{\infty}(1+a_k)$ converges absolutely, then it converges and consequently $\sum_{k=1}^{\infty} \ln(1+ a_k)$ converges. Moreover, we must have $\lim_{k \to \infty} a_k = 0$ thus for some $N \in \mathbb{N}$ it follows that $a_k > -1$ if $k \geq N$. Since $\prod_{k=1}^{\infty}(1 + a_k) = \prod_{k=1}^{N-1}(1 + a_k)\prod_{k=N}^{\infty}(1 + a_k)$, and a finite rearrangement cannot change the value of the infinite product, we may assume that $a_k > -1$ for all $k \in \mathbb{N}$. In this case, with $P = \prod_{k=1}^{\infty}(1 + a_k)$ and $S = \sum_{k=1}^{\infty} \ln(1 + a_k)$ we have $P = \exp(S)$. If we can show that $\sum_{k=1}^{\infty} \ln(1 + a_k)$ converges absolutely, then we can rearrange the series without changing its value, see Theorem 18.27. But the equality $P = \exp(S)$ then implies that we can also rearrange the product $\prod_{k=1}^{\infty}(1 + a_k)$ without changing its value. Thus it remains to prove that the absolute convergence of the product $\prod_{k=1}^{\infty}(1 + a_k)$ implies the absolute convergence of the series $\sum_{k=1}^{\infty} \ln(1 + a_k)$. From Proposition 30.10 we deduce that $\sum_{k=1}^{\infty} a_k$ converges absolutely. Moreover, since $\lim_{k \to \infty} a_k = 0$ we find

$$\lim_{k \to \infty} \frac{|\ln(1 + a_k)|}{|a_k|} = 1,$$

or $\frac{1}{2} \leq \frac{|\ln(1+a_k)|}{|a_k|} \leq 2$ for $k$ sufficiently large implying the absolute convergence of $\sum_{k=1}^{\infty} \ln(1 + a_k)$.

5.　　a) For $|x| < 1$ we find

$$(1 + x^{2^k})(1 - x^{2^k}) = 1 - x^{2k+1}$$

which implies

$$\prod_{k=0}^{N}(1 + x^{2^k}) = \prod_{k=0}^{N} \frac{1 - x^{2^{k+1}}}{1 - x^{2^k}} = \frac{1 - x^{2^{N+1}}}{1 - x}$$

and therefore

$$\prod_{k=0}^{\infty}(1 + x^{2^k}) = \lim_{N \to \infty} \prod_{k=0}^{N}(1 + x^{2^k}) = \lim_{N \to \infty} \frac{1 - x^{2^{N+1}}}{1 - x} = \frac{1}{1 - x}.$$

b) First we observe that for $x = 0$ the product has the value 1 and the right hand side converges for $x \to 0$ to 1. Now, for $x \neq 2^k(\frac{\pi}{2} + l\pi)$ we have $\cos \frac{x}{2^k} \neq 0$ as well as $\sin \frac{x}{2^k} \neq 0$. Using $\sin(2\varphi) = 2 \sin \varphi \cos \varphi$ we find

$$\cos \frac{x}{2^j} = \frac{1}{2} \frac{\sin \frac{x}{2^{j-1}}}{\sin \frac{x}{2^j}},$$

and consequently

$$\prod_{j=1}^{N} \cos \frac{x}{2^j} = \prod_{j=1}^{N} \frac{1}{2} \frac{\sin \frac{x}{2^{j-1}}}{\sin \frac{x}{2^j}} = \frac{\sin x}{2^N \sin \frac{x}{2^N}}.$$

724

Since $2^N \sin \frac{x}{2^N} = x \left( \frac{\sin \frac{x}{2^N}}{\frac{x}{2^N}} \right)$ we eventually get

$$\prod_{j=1}^{\infty} \cos \frac{x}{2^j} = \lim_{N \to \infty} \prod_{j=1}^{N} \frac{1}{2} \frac{\sin \frac{x}{2^{j-1}}}{\sin \frac{x}{2^j}} = \lim_{N \to \infty} \frac{\sin x}{2^N \sin \frac{x}{2^N}}$$

$$= \frac{\sin x}{x} \lim_{N \to \infty} \left( \frac{\sin \frac{x}{2^N}}{\frac{x}{2^N}} \right) = \frac{\sin x}{x}.$$

Finally, for $x = \frac{\pi}{2}$ we derive

$$\prod_{j=1}^{\infty} \cos \frac{\pi}{2^{j+1}} = \frac{\sin \frac{\pi}{2}}{\frac{\pi}{2}} = \frac{2}{\pi}.$$

## Chapter 31

1. From Theorem 31.12, the Legrendre duplication formula, we find for $n \in \mathbb{N}$

$$\Gamma \left( n + \frac{1}{2} \right) = \frac{\sqrt{\pi} \Gamma(2n)}{2^{2n-1} \Gamma(n)}$$

$$= \frac{\sqrt{\pi}(2n-1)!}{4^n \cdot \frac{1}{2}(n-1)!}$$

$$= \frac{\sqrt{\pi}(2n)!}{4^n n!} \cdot \frac{n}{\frac{1}{2} \cdot 2n}$$

$$= \frac{\sqrt{\pi}(2n)!}{4^n n!}.$$

2. Using the substitution $r = st$ we find

$$\int_0^{\infty} t^{\alpha} e^{-st} dt = \int_0^{\infty} \left( \frac{r}{s} \right)^{\alpha} e^{-r} \frac{1}{s} dr$$

$$= \frac{1}{s^{\alpha+1}} \int_0^{\infty} r^{\alpha} e^{-r} dr = \frac{\Gamma(\alpha+1)}{s^{\alpha+1}}.$$

Note that we applied the change of variable formula to an improper integral. Meanwhile we have seen several times, in particular in the context of the $\Gamma$-function, how to derive a result as the above one by looking first at $\int_{\epsilon}^R t^{\alpha} e^{-st} dt$ and then passing to the limit. For a function $f : (0, \infty) \to \mathbb{R}$ such that $F(s) := \int_0^{\infty} f(t) e^{-st} dt$ exists we call $F$ the **Laplace transform** of $f$.

3. The substitution $s = -\ln t$, i.e. $t = e^{-s}$ yields

$$\int_0^1 \left( \ln \frac{1}{t} \right)^{x-1} dt = \int_0^1 (-\ln t)^{x-1} dt$$

$$= -\int_{\infty}^0 s^{x-1} \left( e^{-s} \right) ds = \int_0^{\infty} s^{x-1} e^{-s} ds$$

$$= \Gamma(x).$$

For $x = \frac{3}{2}$ we find using $\Gamma\left(\frac{1}{2}\right) = \sqrt{\pi}$

$$\int_0^1 \left(\ln\frac{1}{t}\right)^{\frac{1}{2}} dt = \Gamma\left(\frac{3}{2}\right) = \frac{1}{2}\Gamma\left(\frac{1}{2}\right) = \frac{\sqrt{\pi}}{2},$$

and for $x = \frac{1}{2}$ we find

$$\int_0^1 \left(\ln\frac{1}{t}\right)^{-\frac{1}{2}} dt = \Gamma\left(\frac{1}{2}\right) = \sqrt{\pi}.$$

4. We use formula (31.14) to find

$$\frac{\Gamma'(1)}{\Gamma(1)} = -\gamma - \frac{1}{1} + \sum_{k=1}^{\infty}\left(\frac{1}{k} - \frac{1}{k+1}\right).$$

Since $\Gamma(1) = 1$ and since

$$\sum_{k=1}^{N}\left(\frac{1}{k} - \frac{1}{k+1}\right) = 1 - \frac{1}{2} + \frac{1}{2} - \frac{1}{3} + \cdots - \frac{1}{N} + \frac{1}{N} - \frac{1}{N+1},$$

we have

$$\sum_{k=1}^{\infty}\left(\frac{1}{k} - \frac{1}{k+1}\right) = 1,$$

and we find

$$\Gamma'(1) = -\gamma.$$

Note, that if we can justify

$$\frac{d}{dx}\Gamma(x)\Big|_{x=1} = \int_0^{\infty} \frac{d}{dx}\left(t^{x-1}\right)\Big|_{x=1} e^{-t} dt$$

we would obtain

$$\int_0^{\infty} (\ln t)e^{-t} dt = -\gamma.$$

5.   a) We again use formula (31.14) to get with $\psi(1) = \frac{\Gamma'(1)}{\Gamma(1)} = -\gamma$ that

$$\psi(x) - \psi(1) = -\frac{1}{x} + \sum_{k=1}^{\infty}\left(\frac{1}{k} - \frac{1}{k+x}\right)$$

$$= -\sum_{k=0}^{\infty}\left(\frac{1}{k+x} - \frac{1}{k+1}\right).$$

b) Since

$$\Gamma(x+n) = (x+n-1)(x+n-2)\cdot\ldots\cdot x\Gamma(x),$$

726

we have

$$\ln \Gamma(x+n) = \ln(x+n-1) + \ln(x+n-2) + \cdots + \ln x + \ln \Gamma(x),$$

and therefore

$$\psi(n+x) = \frac{d}{dx}\ln\Gamma(x+n)$$

$$= \frac{1}{x+n-1} + \frac{1}{x+n-2} + \cdots + \frac{1}{x} + \frac{d}{dx}\ln\Gamma(x)$$

$$= \frac{1}{x} + \cdots + \frac{1}{x+n-1} + \psi(x).$$

6. Starting with

$$B(x,y) = \int_0^1 t^{x-1}(1-t)^{y-1}dt,$$

the substitution $t = \frac{s}{1+s}$ yields

$$\int_0^1 t^{x-1}(1-t)^{y-1}dt$$

$$= \int_0^\infty \frac{s^{x-1}}{(s+1)^{x-1}}\left(1 - \frac{s}{1+s}\right)^{y-1}\frac{1}{(1+s)^2}ds$$

$$= \int_0^\infty \frac{s^{x-1}}{(s+1)^{x-1}}\frac{1}{(s+1)^{y-1}}\frac{1}{(1+s)^2}ds$$

$$= \int_0^\infty \frac{s^{x-1}}{(s+1)^{x+y}}ds.$$

7. We apply the result of Problem 6:

$$\int_0^\infty \frac{x^5}{(1+x)^7}dx = B(6,1),$$

and now we use Theorem 31.11 which states

$$B(x,y) = \frac{\Gamma(x)\Gamma(y)}{\Gamma(x+y)}.$$

Thus

$$B(6,1) = \frac{\Gamma(6)\Gamma(1)}{\Gamma(7)} = \frac{5!0!}{6!} = \frac{1}{6}.$$

Therefore we have proved that

$$\int_0^\infty \frac{x^5}{(1+x)^7}dx = \frac{1}{6}.$$

727

8. We apply Theorem 31.2 and Corollary 31.3 in combination with the formula

$$B(x,y) = \frac{\Gamma(x)\Gamma(y)}{\Gamma(x+y)}$$

to find

$$B(x,y) = \frac{e^{-\gamma x}}{x} \prod_{k=1}^{\infty} \frac{e^{\frac{x}{k}}}{1+\frac{x}{k}} \frac{e^{-\gamma y}}{y} \prod_{k=1}^{\infty} \frac{e^{\frac{y}{k}}}{1+\frac{y}{k}} \cdot$$

$$\cdot (x+y)e^{\gamma(x+y)} \prod_{k=1}^{\infty} \left(1+\frac{x+y}{k}\right) e^{-\frac{(x+y)}{k}}$$

$$= \frac{x+y}{xy} \prod_{k=1}^{\infty} \frac{1+\frac{x+y}{k}}{\left(1+\frac{x}{k}\right)\left(1+\frac{y}{k}\right)}.$$

## Chapter 32

1. For the partition $Z$, where $x_0 = 0, x_j = \frac{1}{j}$, $j \in \mathbb{N}_{2k}$ and $x_{2k+1} = 1$ we find that when $j = 2l$ is even a typical term in the variation sum is

$$|f(x_j) - f(x_{j-1})| = \left| \frac{1}{j} \cos l\pi - \frac{1}{j-1} \cos\left(l-\frac{1}{2}\right)\pi \right| = \frac{1}{j} = \frac{1}{2l}$$

and if $j = 2l + 1$ is odd

$$|f(x_j) - f(x_{j-1})| = \left| \frac{1}{j} \cos\left(l+\frac{1}{2}\right)\pi - \frac{1}{j-1} \cos l\pi \right| = \frac{1}{j-1} = \frac{1}{2l}.$$

This yields

$$V_Z(f) = \frac{1}{2} \sum_{l=1}^{k} \frac{1}{l}$$

which diverges for $k \to \infty$.

2. First we note that $||g(x_j)| - |g(x_{j-1})|| \le |g(x_j) - g(x_{j-1})|$ which implies for every partition $Z$ of $[a,b]$ that

$$V_Z(|g|) \le V_Z(g),$$

hence if $g \in BV([a,b])$ then $|g| \in BV([a,b])$. Next we note that $0 \in BV([a,b])$ and since

$$\max(f,g) = \frac{1}{2}(f+g+|f-g|)$$

and

$$\min(f,g) = \frac{1}{2}(f+g-|f-g|)$$

we deduce from the fact that $BV([a,b])$ is a vector space and the first part of our solution that $g^+, g^-$ as well as $\max(f,g)$ and $\min(f,g)$ belong to $BV([a,b])$.

3. Let $A = \inf |g|$ and $Z(x_0, \ldots, x_n)$ a partition of $[a, b]$. It follows that

$$\left| \frac{1}{g(x_k)} - \frac{1}{g(x_{k-1})} \right| = \frac{|g(x_{k-1}) - g(x_k)|}{|g(x_k)||g(x_{k-1})|}$$

which implies

$$V_Z\left(\frac{1}{g}\right) = \sum_{k=1}^{n} \left| \frac{1}{g(x_k)} - \frac{1}{g(x_{k-1})} \right|$$

$$= \sum_{k=1}^{n} \frac{|g(x_{k-1}) - g(x_k)|}{|g(x_k)||g(x_{k-1})|}$$

$$\leq \frac{1}{A^2} \sum_{k=1}^{n} |g(x_k) - g(x_{k-1})| = \frac{1}{A^2} V_Z(g)$$

and taking the supremum over all partitions $Z$ we arrive at $V\left(\frac{1}{g}\right) \leq \frac{1}{A^2} V(g)$.

4. For a partition $Z(x_0, \ldots, x_n)$ of $[a, b]$ we find

$$|F(x_k) - F(x_{k-1})| = \left| \int_{x_{k-1}}^{x_k} f(t)dt \right| \leq \int_{x_{k-1}}^{x_k} |f(t)|dt$$

and therefore

$$V_Z(F) = \sum_{k=1}^{n} |F(x_k) - F(x_{k-1})| \leq \sum_{k=1}^{n} \int_{x_{k-1}}^{x_k} |f(t)|dt = \int_{a}^{b} |f(t)|dt,$$

i.e. $V_Z(F) \leq \int_a^b |f(t)|dt$ for all partitions $Z$ implying that $V(F) \leq \int_a^b |f(t)|dt$. Now we prove the converse inequality. Let $m_k := \min\{|f(t)||t \in [x_{k-1}, x_k]\}$. By the mean value theorem for the Riemann integral there exists $\xi_k \in [x_{k-1}, x_k]$ such that

$$F(x_k) - F(x_{k-1}) = f(\xi_k)(x_k - x_{k-1})$$

implying

$$|F(x_k) - F(x_{k-1})| = |f(\xi_k)|(x_k - x_{k-1}) \geq m_k(x_k - x_{k-1})$$

and consequently

$$V_Z(F) = \sum_{k=1}^{n} |F(x_k) - F(x_{k-1})| \geq \sum_{k=1}^{n} m_k(x_k - x_{k-1}).$$

Taking the supremum over all partitions $Z$ of $[a, b]$ we find

$$V(F) \geq \sup_Z \sum_{k=1}^{n} m_k(x_k - x_{k-1}) = \int_{a}^{b} |f(t)|dt,$$

where the last equality follows from Theorem 25.24 when observing that $m_k = f(\eta_k)$ for some $\eta_k \in [x_{k-1}, x_k]$.

5.    a) This is trivial: we only need to take $m = 1$.

b) If $f : [a, b] \to \mathbb{R}$ is Lipschitz continuous, i.e. $|f(x) - f(y)| \leq \kappa|x - y|$ for all $x, y \in [a, b]$ with some $\kappa > 0$, then we find for $\epsilon > 0$ with $\delta = \frac{\epsilon}{\kappa}$ that with $(a_j, b_j)$ as in the definition

$$\sum_{j=1}^{m}(b_j - a_j) < \delta = \frac{\epsilon}{\kappa} \text{ implies } \sum_{j=1}^{m}\kappa(b_j - a_j) < \epsilon$$

and therefore

$$\sum_{j=1}^{m}|f(b_j) - f(a_j)| \leq \sum_{j=1}^{m}\kappa(b_j - a_j) < \epsilon.$$

c) Let $f : [a, b] \to \mathbb{R}$ be absolutely continuous. For $\epsilon = 1$ there exists $\delta > 0$ such that $\sum_{j=1}^{m}(b_j - a_j) < \delta$ (where $(a_j, b_j)$ is as in the definition) implying $\sum_{j=1}^{m}|f(b_j) - f(a_j)| < 1$. In particular we have $V_\alpha^\beta(f) \leq 1$ for every interval $[\alpha, \beta] \subset [a, b]$ with $\beta - \alpha < \delta$. Given $\delta > 0$ sufficiently small there exists $k \in \mathbb{N}$ such that $k\delta < b - a$, and intervals $I_j \subset [a, b]$ such that $\lambda^{(1)}(I_j) < \delta, j = 1, \ldots, k$, and $[a, b] \subset \cup_{j=1}^{k}I_j$. It follows that $V_a^b(f) \leq \sum_{j=1}^{k}V_{I_j}(f) \leq k = \frac{b-a}{\delta}$, hence $f$ has bounded variation.

6. Since the constant functions are obviously absolutely continuous we need to prove that with $f, g : [a, b] \to \mathbb{R}$ absolutely continuous the functions $f + g$ and $f \cdot g$ are absolutely continuous too. The absolute continuity of $f + g$ follows from the triangle inequality: if we know that for every $\epsilon > 0$ there exists $\delta > 0$ such that for $(a_j, b_j) \subset [a, b], j = 1, \ldots, m$, it follows that $\sum_{j=1}^{m}(b_j - a_j) < \delta$ implies $\sum_{j=1}^{m}|f(b_j) - f(a_j)| < \frac{\epsilon}{2}$ and $\sum_{j=1}^{m}|g(b_j) - g(a_j)| < \frac{\epsilon}{2}$ then we have of course that

$$\sum_{j=1}^{m}|(f + g)(b_j) - (f + g)(a_j)| \leq \sum_{j=1}^{m}|f(b_j) - f(a_j)| + \sum_{j=1}^{m}|g(b_j) - g(a_j)|$$
$$< \frac{\epsilon}{2} + \frac{\epsilon}{2} = \epsilon.$$

In order to prove that $f \cdot g$ is absolutely continuous we first note that $f$ and $g$ must be bounded, i.e. $\|f\|_\infty < \infty$ and $\|g\|_\infty < \infty$. For an interval $(a_j, b_j) \subset [a, b]$ we find

$$|(f \cdot g)(b_j) - (f \cdot g)(a_j)| = |f(b_j)g(b_j) - f(a_j)g(a_j)|$$
$$\leq |f(b_j)g(b_j) - f(a_j)g(b_j)| + |f(a_j)g(b_j) - f(a_j)g(a_j)|$$
$$\leq \|g\|_\infty|f(b_j) - f(a_j)| + \|f\|_\infty|g(b_j) - g(a_j)|$$

Thus with $\|f\|_\infty \leq M, \|g\|_\infty \leq M$, given $\epsilon > 0$, choose $\delta > 0$ such that for $(a_j, b_j), j = 1, \ldots, m$ and $(a_j, b_j) \subset [a, b]$, from $\sum_{j=1}^{m}b_j - a_j < \delta$ it follows that $\sum_{j=1}^{m}|f(b_j) - f(a_j)| < \frac{\epsilon}{M}$ and $\sum_{j=1}^{m}|g(b_j) - g(a_j)| < \frac{\epsilon}{M}$. This implies

$$\sum_{j=1}^{m}|(f \cdot g)(b_j) - (f \cdot g)(a_j)| < \frac{\epsilon}{M} + \frac{\epsilon}{M} = \epsilon.$$

7. For $f \in C([a,b]) \cup BV([a,b])$ we define $F$ by $F(x) := \int_a^x f(t)dt$. Let $x, y \in [a,b]$, it follows that

$$|F(y) - F(x)| = \left| \int_x^y f(t)dt \right| \leq \int_x^y |f(t)|dt \leq M|x - y|$$

for $M = ||f||_\infty < \infty$, note that continuous functions on a compact interval are bounded as are functions of bounded variation. Thus $F : [a,b] \to \mathbb{R}$ is Lipschitz continuous and therefore absolutely continuous.

# References

[1] Beals, R., and Wong, R., *Special Functions. A Graduate Text.* Cambridge Studies in Advanced Mathematics, Vol. 126. Cambridge University Press, 2010.

[2] Dieudonné, J., *Grundzüge der modernen Analysis, 2.* Aufl. Logik und Grundlagen der Mathematik Bd. 8. Friedrich Vieweg & Sohn, Braunschweig 1972.

[3] Endl. K., und Luh, W., *Analysis I, 3.* Aufl. Akademische Verlagsgesellschaft, Wiesbaden 1975.

[4] Garling, D.J.H., *A Course in Mathematical Analysis, Vol. I. Foundations and Elementary Real Analysis.* Cambridge University Press, Cambridge 2013.

[5] Heuser, H., *Lehrbuch der Analysis. Teil 1.* B.G. Teubner Verlag, Stuttgart 1980.

[6] Kaczor, W.J., and Nowak, M.T., *Problems in Mathematical Analysis II.* Students Mathematical Library, Vol. 12. American Mathematical Society, Providence R.I., 2001.

[7] Landau, E., *Grundlagen der Analysis.* Akademische Verlagsgesellschaft, Leipzig 1930.

[8] Lin, M., *The AM-GM inequality and the CBS inequality are equivalent.* The Mathematical Intelligencer 34. 2 (2012), 6.

[9] Maligranda, L., *The AM-GM inequality is equivalent to the Bernoulli inequality.* The Mathematical Intelligencer 34. 1 (2012), 1-2.

[10] Markuschewitsch, A.I., *Rekursive Folgen.* Kleine Ergänzungsreihe zu den Hochschulbuecher für Mathematik Bd. 11. VEB Deutscher Verlag der Wissenschaften, Berlin 1955.

[11] Rudin, W., *Principles of Mathematical Analysis, 3rd ed.* McGraw-Hill International Editions, Mathematical Series. McGraw-Hill Book Company, Singapore 1976.

[12] Schilling, R.L., *Measures, Integration and Martingales.* Cambridge University Press, Cambridge 2005.

# Mathematicians Contributing to Analysis

**Abel,** Niels Henrik (1802-1829).

**Archimedes,** (ca. 287B.C.-212B.C.).

**Banach,** Stefan (1892-1945).

**Bernoulli,** Jakob I (1654-1705).

**Bernstein,** Sergej Natanowitsch (1880-1968).

**Bessel,** Friedrich Wilhelm (1784-1846).

**Bohr,** Harald (1887-1951).

**Bolzano,** Bernard (1781-1848).

**Borel,** Emile (1871-1956).

**Bunyakovsky,** Viktor Jakovlevitsh (1805-1859).

**Cantor,** Georg (1845-1918).

**Cauchy,** Augustin-Louis (1789-1857).

**Cohen,** Paul J. (1934-2007).

**de Morgan,** Auguste (1806-1871).

**Dedekind,** Richard (1831-1916).

**Descartes,** René (1596-1650).

**Dirichlet,** Joham Peter Gustav, Lejeume- (1805-1859).

**du Bois-Reymond,** Paul (1831-1889).

**Euclid,** (ca. 325B.C.-265B.C.).

**Euler,** Leonard (1707-1783).

**Faà di Bruno,** Francesco (1825-1888).

**Fermat,** Pierre (1601-1655).

**Fibonacci,** Leonardo of Pisa, called- (ca. 1170-ca.1250).

**Fourier,** Jean-Baptiste Joseph (1768-1830).

**Fraenkel,** Abraham (1891-1965).

**Gödel,** Kurt (1906-1978).

**Hölder,** Otto (1859-1937).
**Heine,** Heinrich Edward (1821-1881).
**Hospital,** Guillaume Francois Antoine de l' (1661-1704).

**Jensen,** Johan (1859-1925).

**Lagrange,** Joseph-Louis (1736-1813).
**Laplace,** Pierre Simon (1749-1827).
**Lebesgue,** Henri (1875-1941).
**Legrendre,** Adrian-Marie (1752-1833).
**Leibniz,** Gottfried Wilhelm (1646-1716).
**Lindelöf,** Ernst (1870-1946).
**Lipschitz,** Rudolf (1832-1903).

**Minkowski,** Hermann (1864-1909).
**Mollerup,** Peter Johannes (1872-1937).

**Newton,** Isaac (1613-1727).

**Pascal,** Blaise (1623-1662).
**Peano,** Guisepp (1858-1939).
**Picard,** Emile (1856-1941).
**Poincaré,** Henri (1854-1912).
**Pythagoras,** (ca.580B.C.-ca.500B.C.).

**Raabe,** Josef Ludwig (1801-1859).
**Riemann,** Bernhard (1826-1866).
**Rolle,** Michel (1652-1719).

**Schwarz,** Hermann Amandus (1843-1921).
**Stirling,** James (1692-1770).

**Taylor,** Brook (1685-1731).

**Wallis,** John(1616-1703).
**Weierstrass,** Karl Theodor Wilhelm (1815-1897).

**Zermelo,** Ernst Friedrich Ferdinand (1871-1953).

# Subject Index

Printed in the United States
By Bookmasters